Undergraduate Lecture Notes in Physics

Series editors

Neil Ashby, University of Colorado, Boulder, CO, USA

William Brantley, Department of Physics, Furman University, Greenville, SC, USA

Matthew Deady, Physics Program, Bard College, Annandale-on-Hudson, NY, USA

Michael Fowler, Department of Physics, University of Virginia, Charlottesville, VA, USA

Morten Hjorth-Jensen, Department of Physics, University of Oslo, Oslo, Norway

Undergraduate Lecture Notes in Physics (ULNP) publishes authoritative texts covering topics throughout pure and applied physics. Each title in the series is suitable as a basis for undergraduate instruction, typically containing practice problems, worked examples, chapter summaries, and suggestions for further reading.

ULNP titles must provide at least one of the following:

- An exceptionally clear and concise treatment of a standard undergraduate subject.
- A solid undergraduate-level introduction to a graduate, advanced, or non-standard subject.
- A novel perspective or an unusual approach to teaching a subject.

ULNP especially encourages new, original, and idiosyncratic approaches to physics teaching at the undergraduate level.

The purpose of ULNP is to provide intriguing, absorbing books that will continue to be the reader's preferred reference throughout their academic career.

More information about this series at http://www.springer.com/series/8917

Albrecht Lindner · Dieter Strauch

A Complete Course on Theoretical Physics

From Classical Mechanics to Advanced Quantum Statistics

 Springer

Albrecht Lindner
Pinneberg, Germany

Dieter Strauch
Theoretical Physics
University of Regensburg
Regensburg, Germany

ISSN 2192-4791   ISSN 2192-4805   (electronic)
Undergraduate Lecture Notes in Physics
ISBN 978-3-030-04359-9   ISBN 978-3-030-04360-5   (eBook)
https://doi.org/10.1007/978-3-030-04360-5

Library of Congress Control Number: 2018961698

This Springer imprint is published by the registered company Springer Nature Switzerland AG
The registered company address is: Gewerbestrasse 11, 6330 Cham, Switzerland

*In memory of Albrecht Lindner (1935–2005),
scientist, teacher, friend*

Preface

This textbook is a translation of the third German edition of *Grundkurs Theoretische Physik* (*A Basic Course on Theoretical Physics*), originally published by Teubner, Stuttgart, Germany. Actually, this edition is much more than a typical textbook since it offers a mixture of basic and advanced material of all of the fundamental disciplines of theoretical physics in one volume, whence it may well serve also as a reference book. The large number of cross-references will guide the reader from the basic experimental observations to the construction of a "unified" theory, and the present compactness should ensure that the reader does not get lost along the way.

A wide range of problems invite the reader to tackle further applications at various stages of sophistication, and a list of textbooks offers the way forward to possible open questions.

The material itself and the way it is presented is due to the late Albrecht Lindner. My contribution is restricted merely to the translation into the English language; in fact, my sincerest gratitude goes to Dr. Steven Lyle who corrected the translation in manly places; whatever remains of insufficient vocabulary or grammar is due to my limited mastery of the language. The only changes I have made are to adjust to the publisher's requirements, made some changes in the numerical tables as to be expected from May 2019 on, and adapt the list of textbooks to an English-speaking readership.

I am proud, nevertheless, to present this book to the English-speaking community.

Regensburg, Germany   Dieter Strauch

Preface to the First German Edition

Like the standard course in theoretical physics, the present book introduces the physics of particles under the heading *Classical Mechanics*, the physics of fields under *Electromagnetism*, quantum physics under *Quantum Mechanics I*, and statistical physics under *Thermodynamics and Statistics*. Besides these branches, which would form a curriculum for *all* students of physics, there is a complement entitled *Quantum Mechanics II*, for those who wish to obtain a deeper understanding of the theory, which discusses scattering problems, quantization of fields, and Dirac theory (as an example of relativistic quantum mechanics).

The goal here is to stress the interrelations between the individual subjects. In an introductory chapter, there is a summary of the most important parts mathematical tools repeatedly needed in the different branches of physics. These constitute the *mathematical foundation* for rationalizing our practical experience, since we wish to describe our observations as precisely as possible.

The selection of material was mainly inspired by our local physics diploma curriculum. Only in a few places did I go beyond those limits, e.g., in Sect. 4.6 (quantum theory and dissipation), Sect. 5.2 (three-body scattering), and Sect. 5.4 (quasi-particles, quantum optics), since I have the impression that the essentials can also be worked out rather easily in these areas.

Section 5.5 on the Dirac equation also differs from the standard presentation, because I prefer the Weyl representation over the standard representation—despite my intention to avoid any special representation as far as possible. In this respect, I am grateful to my colleagues Till Anders (Munich), Dietmar Kolb (Kassel), und Gernot Münster (Münster) for their valuable comments on my drafts.

Thanks go also to numerous students in Hamburg and especially to Dr. Heino Freese and Dr. Adolf Kitz for many questions and suggestions, and various forms of support. The general interest in my notes encourages me to present these now to a larger community.

(Notes on figure production are left out here—D.S.)

Hamburg, Germany   Albrecht Lindner
Fall 1993

Preface to the Second German Edition

The text has been improved at many places, in particular in Sects. 3.5 and 5.4, and all figures have been inserted with pstricks. In addition, three-dimensional objects now appear in central instead of of parallel perspetive.

Hamburg, Germany
Summer 1996

Albrecht Lindner

Preface to the Third German Edition

The *Basic Course* (*Grundkurs*) was discovered in a third, extensively revised edition, after Albrecht Lindner, a passionate teacher, unexpectedly passed away. As one of those rare textbooks which presents a complete curriculum of theoretical physics in a single volume—compact and simultaneously profound—it should be offered to the teacher and student community. In the present third edition the material has been revised in many places, and the number of figures has been approximately doubled. Also in this edition is an additional chapter containing numerous problems.

My contribution here is restricted to adjusting the material to the changed appearance required by the Teubner publishing company.

Regensburg, Germany
Spring 2011

Dieter Strauch

Contents

Chapter 1
Basics of Experience

1.1 Vector Analysis

1.1.1 Space and Time

Space and time are two basic concepts which, according to Kant, inherently or innately determine the form of all experience in an a priori manner, thereby making possible experience as such: only in space and time can we arrange our sensations. [According to the doctrines of evolutionary cognition, what is innate to us has developed phylogenetically by adaption to our environment. This is why we only notice the insufficiency of these "self-evident" concepts under extraordinary circumstances, e.g., for velocities close to that of light (c_0) or actions of the order of Planck's quantum h. We shall tackle such "weird" cases later—in electromagnetism and quantum mechanics. For the time being, we want to make sure we can handle our familiar environment.]

To do this, we introduce a continuous parameter t. Like every other *physical quantity* it is composed of number and unit (for example, a second 1 s = 1 min/60 = 1 h/3600). The larger the unit, the smaller the number. *Physical quantities do not depend on the unit—likewise equations between physical quantities.* Nevertheless, the opposite is sometimes seen, as in: "We choose units such that the velocity of light c assumes the value 1". In fact, the *concept* of velocity is thereby changed, so that instead of the velocity v, the *ratio* v/c is taken here as the velocity, and ct as time or x/c as length.

The zero time ($t = 0$) can be chosen arbitrarily, since basically only the time difference, i.e., the duration of a process, is important. A differentiation with respect to time ($\mathrm{d}/\mathrm{d}t$) is often marked by a dot over the differentiated quantity, i.e., $\mathrm{d}x/\mathrm{d}t \equiv \dot{x}$.

In empty *space* every direction is equivalent. Here, too, we may choose the zero point freely and, starting from this point, determine the position of other points in a coordinate-free notation by the *position vector* **r**, which fixes the distance and direction of the point under consideration. This coordinate-free type of notation is

© Springer Nature Switzerland AG 2018
A. Lindner and D. Strauch, *A Complete Course
on Theoretical Physics*, Undergraduate Lecture Notes in Physics,
https://doi.org/10.1007/978-3-030-04360-5_1

particularly advantageous when we want to exploit the assumed homogeneity of
space. However, conditions often arise (i.e., when there is axial or spherical symmetry) which are best taken care of in special coordinates. We are free to choose a
coordinate system. We only require that it determine all positions uniquely. This we
shall treat in the next section.

Besides the position vector $\mathbf{r}$, there are other quantities in physics with both
value and direction, e.g., the velocity $\mathbf{v} = \dot{\mathbf{r}}$, the acceleration $\mathbf{a} = \dot{\mathbf{v}}$, the momentum
$\mathbf{p} = m\mathbf{v}$, and the force $\mathbf{F} = \dot{\mathbf{p}}$. The appropriate means to handle such quantities is
vector algebra, with which we shall be extensively concerned in this section. This
method allows us to encompass both the value and the direction of the quantities
under consideration much better than using components, which, moreover, depend
on the coordinate system.

For the time being—namely for plane and three-dimensional problems—we
understand a *vector* as a quantity with value and direction, which can be represented as an arrow of corresponding length. (Generally, vectors are mathematical
entities, which can be added together or multiplied by a number, with the usual rules
of calculation being valid.) Sometimes they are denoted by a letter with an arrow
atop. The value (the length) of $\mathbf{a}$ is denoted by a or $|\mathbf{a}|$.

1.1.2 Vector Algebra

From two vectors $\mathbf{a}$ and $\mathbf{b}$, their sum $\mathbf{a} + \mathbf{b}$ may be formed according to the construction of parallelograms (as the diagonal), as shown in Fig. 1.1. From this follows
the commutative and associative law of vector addition:

$$\mathbf{a} + \mathbf{b} = \mathbf{b} + \mathbf{a} , \qquad (\mathbf{a} + \mathbf{b}) + \mathbf{c} = \mathbf{a} + (\mathbf{b} + \mathbf{c}) .$$

The product of the vectors $\mathbf{a}$ with a scalar (i.e., directionless) factor α is understood
as the vector $\alpha\,\mathbf{a} = \mathbf{a}\,\alpha$ with the same (for $\alpha < 0$ opposite) direction and with value
$|\alpha|\,a$. In particular, $\mathbf{a}$ and $-\mathbf{a}$ have the same value, but opposite directions. For $\alpha = 0$
the *zero vector* $\mathbf{0}$ results, with length 0 and undetermined direction.

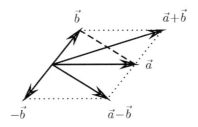

Fig. 1.1 Sum and difference of vectors $\mathbf{a}$ and $\mathbf{b}$. The vectors may be shifted in parallel, e.g., $\mathbf{a} - \mathbf{b}$
can also lie on the *dashed straight line*

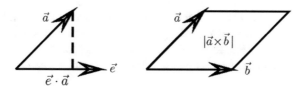

Fig. 1.2 Scalar and vector products: $\mathbf{e} \cdot \mathbf{a}$ is the component of $\mathbf{a}$ in the direction of the unit vector $\mathbf{e}$, and $| \mathbf{a} \times \mathbf{b} |$ is the area shown

The *scalar product* (*inner product*) $\mathbf{a} \cdot \mathbf{b}$ of the two vectors $\mathbf{a}$ and $\mathbf{b}$ is the product of their values times the cosine of the enclosed angle ϕ_{ab} (see Fig. 1.2 left):

$$\mathbf{a} \cdot \mathbf{b} \equiv a\, b \, \cos \phi_{ab} \ .$$

The dot between the two factors is important for the scalar product—if it is missing, then it is the *tensor product* of the two vectors, which will be explained in Sect. 1.2.4— with $\mathbf{a} \cdot \mathbf{b}\, \mathbf{c} \neq \mathbf{a}\, \mathbf{b} \cdot \mathbf{c}$, if $\mathbf{a}$ and $\mathbf{c}$ have different directions, i.e., if $\mathbf{a}$ is not a multiple of $\mathbf{c}$. Consequently, one has

$$\mathbf{a} \cdot \mathbf{b} = \mathbf{b} \cdot \mathbf{a}$$

and

$$\mathbf{a} \cdot \mathbf{b} = 0 \qquad \Longleftrightarrow \qquad \mathbf{a} \perp \mathbf{b} \quad \text{or} \quad a = 0 \quad \text{or} \quad b = 0 \ .$$

If the two vectors are oriented perpendicularly to each other ($\mathbf{a} \perp \mathbf{b}$), then they are also said to be *orthogonal*. Obviously, $\mathbf{a} \cdot \mathbf{a} = a^2$ holds. Vectors with value 1 are called *unit vectors*. Here they are denoted by $\mathbf{e}$. Given three Cartesian, i.e., pairwise perpendicular unit vectors $\mathbf{e}_x, \mathbf{e}_y, \mathbf{e}_z$, all vectors can be decomposed in terms of these:

$$\mathbf{a} = \mathbf{e}_x\, a_x + \mathbf{e}_y\, a_y + \mathbf{e}_z\, a_z \ ,$$

with the *Cartesian components*

$$a_x \equiv \mathbf{e}_x \cdot \mathbf{a} \ , \quad a_y \equiv \mathbf{e}_y \cdot \mathbf{a} \ , \quad a_z \equiv \mathbf{e}_z \cdot \mathbf{a} \ .$$

Here the components will usually be written after the unit vectors. This is particularly useful in quantum mechanics, but also meaningful otherwise, since the coefficients depend on the expansion basis. Since for a given basis $\mathbf{a}$ is fixed by its three components (a_x, a_y, a_z), $\mathbf{a}$ is thus often given as this *row vector*, or as a *column vector*, with the components written one below the other. However, the coordinate-free notation $\mathbf{a}$ is in most cases more appropriate to formal calculations, e.g., $\mathbf{a} + \mathbf{b}$ combines the three expressions $a_x + b_x$, $a_y + b_y$, and $a_z + b_z$. Because $\mathbf{e}_x \cdot \mathbf{e}_x = 1$, $\mathbf{e}_x \cdot \mathbf{e}_y = 0$ (and cyclic permutations $\mathbf{e}_y \cdot \mathbf{e}_y = 1$, $\mathbf{e}_y \cdot \mathbf{e}_z = 0$ and so on), one clearly has

$$\mathbf{a} \cdot \mathbf{b} = a_x\, b_x + a_y\, b_y + a_z\, b_z \ .$$

Hence it also follows that $\mathbf{a} \cdot (\mathbf{b} + \mathbf{c}) = \mathbf{a} \cdot \mathbf{b} + \mathbf{a} \cdot \mathbf{c}$.

The *vector product* (*outer product*) $\mathbf{a} \times \mathbf{b}$ of the two vectors $\mathbf{a}$ and $\mathbf{b}$ is another vector which is oriented perpendicularly to both and which forms with them a right-hand screw, like the thumb, forefinger, and middle finger of the right hand. Its value is equal to the area of the parallelogram spanned by $\mathbf{a}$ and $\mathbf{b}$ (see Fig. 1.2 right):

$$|\mathbf{a} \times \mathbf{b}| = a\,b\,\sin\phi_{ab} \;.$$

Hence it also follows that

$$\mathbf{a} \times \mathbf{b} = -\mathbf{b} \times \mathbf{a}\,, \qquad \mathbf{a} \times (\mathbf{b} + \mathbf{c}) = \mathbf{a} \times \mathbf{b} + \mathbf{a} \times \mathbf{c}\,,$$

and

$$\mathbf{a} \times \mathbf{b} = \mathbf{0} \qquad \Longleftrightarrow \qquad \mathbf{a} \parallel \mathbf{b} \;\; \text{or} \;\; a = 0 \;\; \text{or} \;\; b = 0\,.$$

Using a right-handed Cartesian coordinate system, we have

$$\mathbf{e}_x \times \mathbf{e}_y = \mathbf{e}_z \qquad \text{(and cyclic permutations } \mathbf{e}_y \times \mathbf{e}_z = \mathbf{e}_x,\; \ldots)\,,$$

and also $\mathbf{e}_x \times \mathbf{e}_x = \mathbf{0}$, etc., whence

$$\mathbf{a} \times \mathbf{b} = \mathbf{e}_x\,(a_y\,b_z - a_z\,b_y) + \mathbf{e}_y\,(a_z\,b_x - a_x\,b_z) + \mathbf{e}_z\,(a_x\,b_y - a_y\,b_x)\,.$$

This implies

$$\mathbf{a} \times (\mathbf{b} \times \mathbf{c}) = (\mathbf{c} \times \mathbf{b}) \times \mathbf{a} = \mathbf{b}\,\mathbf{c}\cdot\mathbf{a} - \mathbf{c}\,\mathbf{a}\cdot\mathbf{b}\,.$$

(This decomposition also follows without calculation because the product depends linearly upon its three factors, lies in the plane spanned by $\mathbf{b}$ and $\mathbf{c}$, vanishes for $\mathbf{b} \propto \mathbf{c}$, and points in the direction of $\mathbf{b}$ for $\mathbf{c} = \mathbf{a} \perp \mathbf{b}$.) According to the last equation, every vector $\mathbf{a}$ can be decomposed into its component along a unit vector $\mathbf{e}$ and its component perpendicular to it:

$$\mathbf{a} = \mathbf{e}\,\mathbf{e}\cdot\mathbf{a} - \mathbf{e} \times (\mathbf{e} \times \mathbf{a})\,.$$

In addition, it satisfies the *Jacobi identity* (note the cyclic permutation)

$$\mathbf{a} \times (\mathbf{b} \times \mathbf{c}) + \mathbf{b} \times (\mathbf{c} \times \mathbf{a}) + \mathbf{c} \times (\mathbf{a} \times \mathbf{b}) = \mathbf{0}\,.$$

The scalar product of a vector with a vector product, viz.,

$$\mathbf{a} \cdot (\mathbf{b} \times \mathbf{c}) = \mathbf{b} \cdot (\mathbf{c} \times \mathbf{a}) = \mathbf{c} \cdot (\mathbf{a} \times \mathbf{b})\,,$$

is called the *(scalar) triple product* of the three vectors. It is positive or negative, if $\mathbf{a}$, $\mathbf{b}$, and $\mathbf{c}$ form a right- or left-handed triad, respectively. Its value gives the volume of the parallelepiped with edges $\mathbf{a}$, $\mathbf{b}$, and $\mathbf{c}$. In particular, $\mathbf{e}_x \cdot (\mathbf{e}_y \times \mathbf{e}_z) = 1$.

In this context, the concept of a *matrix* is useful. An $M \times N$ matrix A is understood as an entity made of $M \times N$ "matrix elements", arranged in M rows and N columns: A_{ik} ($i \in \{1, \dots, M\}, k \in \{1, \dots, N\}$), e.g.,

$$A = \begin{pmatrix} A_{11} & A_{12} & A_{13} \\ A_{21} & A_{22} & A_{23} \\ A_{31} & A_{32} & A_{33} \end{pmatrix} \quad \Longleftrightarrow \quad \tilde{A} = \begin{pmatrix} A_{11} & A_{21} & A_{31} \\ A_{12} & A_{22} & A_{32} \\ A_{13} & A_{23} & A_{33} \end{pmatrix}.$$

The *transposed matrix* $\tilde{A}$ just introduced has elements $\tilde{A}_{ik} = A_{ki}$, hence N rows and M columns. We shall mainly be concerned with *square matrices*, which have equal numbers of rows and columns, i.e., $M = N$. The *matrix product* of A and B is

$$C = AB \quad \text{with} \quad C_{ik} = \sum_{j=1}^{N} A_{ij} B_{jk},$$

which is, of course, defined only if the number of columns of A is the same as the number of rows of B. We have $\widetilde{AB} = \tilde{B}\,\tilde{A}$.

If we now combine the 3×3 Cartesian components of the vectors $\mathbf{a}$, $\mathbf{b}$, and $\mathbf{c}$ in the form of a matrix, its *determinant*

$$\begin{vmatrix} a_x & a_y & a_z \\ b_x & b_y & b_z \\ c_x & c_y & c_z \end{vmatrix} \equiv a_x (b_y c_z - b_z c_y) + a_y (b_z c_x - b_x c_z) + a_z (b_x c_y - b_y c_x)$$

$$= a_x (b_y c_z - b_z c_y) + b_x (c_y a_z - c_z a_y) + c_x (a_y b_z - a_z b_y)$$

is equal to the triple product $\mathbf{a} \cdot (\mathbf{b} \times \mathbf{c})$. For determinants, we have

$$\det \tilde{A} = \det A \quad \text{and} \quad \det (AB) = \det A \times \det B.$$

Therefore, also

$$\mathbf{a} \cdot (\mathbf{b} \times \mathbf{c})\ \mathbf{f} \cdot (\mathbf{g} \times \mathbf{h}) = \begin{vmatrix} \mathbf{a} \cdot \mathbf{f} & \mathbf{a} \cdot \mathbf{g} & \mathbf{a} \cdot \mathbf{h} \\ \mathbf{b} \cdot \mathbf{f} & \mathbf{b} \cdot \mathbf{g} & \mathbf{b} \cdot \mathbf{h} \\ \mathbf{c} \cdot \mathbf{f} & \mathbf{c} \cdot \mathbf{g} & \mathbf{c} \cdot \mathbf{h} \end{vmatrix}.$$

Moreover, from $(\mathbf{a} \times \mathbf{b}) \cdot \mathbf{c} = \mathbf{a} \cdot (\mathbf{b} \times \mathbf{c})$ and replacing $\mathbf{c}$ by $\mathbf{c} \times \mathbf{d}$, it follows that

$$(\mathbf{a} \times \mathbf{b}) \cdot (\mathbf{c} \times \mathbf{d}) = (\mathbf{a} \cdot \mathbf{c})(\mathbf{b} \cdot \mathbf{d}) - (\mathbf{a} \cdot \mathbf{d})(\mathbf{b} \cdot \mathbf{c}) \equiv \begin{vmatrix} \mathbf{a} \cdot \mathbf{c} & \mathbf{a} \cdot \mathbf{d} \\ \mathbf{b} \cdot \mathbf{c} & \mathbf{b} \cdot \mathbf{d} \end{vmatrix},$$

the determinant of a 2×2 matrix, and in particular,

$$(\mathbf{a} \times \mathbf{b}) \cdot (\mathbf{a} \times \mathbf{b}) = a^2 b^2 - (\mathbf{a} \cdot \mathbf{b})^2,$$

which, of course, follows from $\sin^2 \phi_{ab} = 1 - \cos^2 \phi_{ab}$.

Table 1.1 Space-inversion behavior

Type	Original image	Mirror image
Polar vector	↑	↓
Axial vector	$-\uparrow-$	$-\uparrow-$

It is not allowed to divide by vectors—neither scalar products nor vector products can be decomposed uniquely in terms of their factors, as can be seen from the examples $\mathbf{a} \cdot \mathbf{b} = 0$ and $\mathbf{a} \times \mathbf{b} = \mathbf{0}$.

In the context of the vector product, we have to consider the fact that only in three-dimensional space can a third vector be assigned uniquely as a vector normal to two vectors. Otherwise a perpendicular direction cannot be fixed uniquely, and no direction can be given in the sense of the right-hand rule. In fact, in Sect. 3.4.3, in order to extend the three-dimensional space to the four-dimensional space-time continuum of the theory of special relativity, we change from the vector product to a skew-symmetric matrix (or a tensor of second rank) which, in three-dimensional space, has three independent elements, just like every vector.

Actually, we also have to distinguish between *polar vectors* (like the position vector $\mathbf{r}$ and the velocity $\mathbf{v} = \dot{\mathbf{r}}$) and *axial vectors* (e.g., the vector product of two polar vectors), because they behave differently under a space inversion (with respect to the origin): the direction of a polar vector is reversed, while the direction of an axial vector is preserved. Correspondingly the triple product of three polar vectors is a *pseudo-scalar*, because it changes its sign under space inversion. Axial vectors can actually be viewed as rotation axes with sense of rotation and not as arrows—they are *pseudo-vectors* (Table 1.1).

Inversion involves a special change of coordinates: it cannot be composed of infinitesimal transformations, like rotations and translations. General properties of coordinate transformations will be treated in the next section. Until then we will thus assume only right-handed Cartesian coordinate systems with $\mathbf{e}_x \times \mathbf{e}_y = \mathbf{e}_z$ (and cyclic permutations).

1.1.3   Trajectories

If a vector depends upon a parameter, then we speak of a vector function. The vector function $\mathbf{a}(t)$ is continuous at t_0, if it tends to $\mathbf{a}(t_0)$ for $t \to t_0$. With the same limit $t \to t_0$, the vector differential $d\mathbf{a}$ and the first derivative $d\mathbf{a}/dt$ is introduced. These quantities may be formed for every Cartesian component, and we have

$$d(\mathbf{a} + \mathbf{b}) = d\mathbf{a} + d\mathbf{b}, \qquad d(\alpha\mathbf{a}) = \alpha\, d\mathbf{a} + \mathbf{a}\, d\alpha,$$
$$d(\mathbf{a} \cdot \mathbf{b}) = \mathbf{a} \cdot d\mathbf{b} + \mathbf{b} \cdot d\mathbf{a}, \qquad d(\mathbf{a} \times \mathbf{b}) = \mathbf{a} \times d\mathbf{b} - \mathbf{b} \times d\mathbf{a}.$$

Obviously, $\mathbf{a} \cdot d\mathbf{a}/dt = \frac{1}{2} d(\mathbf{a} \cdot \mathbf{a})/dt = \frac{1}{2} da^2/dt = a\, da/dt$ holds. In particular the derivative of a unit vector is always perpendicular to the original vector—if it does not vanish.

As an example of a vector function, we investigate $\mathbf{r}(t)$, the path of a point as a function of the time t. Thus we want to consider also the velocity $\mathbf{v} = \dot{\mathbf{r}}$ and the acceleration $\mathbf{a} = \ddot{\mathbf{r}}$ rather generally. The time is not important for the trajectories as geometrical lines. Therefore, instead of the time t we introduce the path length s as a parameter and exploit $ds = |d\mathbf{r}| = v\,dt$.

We now take three mutually perpendicular unit vectors $\mathbf{e}_T$, $\mathbf{e}_N$, and $\mathbf{e}_B$, which are attached to every point on the trajectory. Here $\mathbf{e}_T$ has the direction of $\mathbf{v}$:

$$\text{tangent vector} \qquad \mathbf{e}_T \equiv \frac{d\mathbf{r}}{ds} = \frac{\mathbf{v}}{v}.$$

For a straight path, this vector is already sufficient for the description. But in general the

$$\text{path curvature} \qquad \kappa \equiv \left| \frac{d\mathbf{e}_T}{ds} \right| = \left| \frac{d^2\mathbf{r}}{ds^2} \right|$$

is different from zero. In order to get more insight into this parameter we consider a plane curve of constant curvature, namely, the circle with $s = R\,\varphi$. For $\mathbf{r}(\varphi) = \mathbf{r}_0 + R(\cos\varphi\,\mathbf{e}_x + \sin\varphi\,\mathbf{e}_y)$, we have $\kappa = |d^2\mathbf{r}/d(R\varphi)^2| = R^{-1}$. Instead of the curvature κ, its reciprocal, the

$$\text{curvature radius} \qquad R \equiv \frac{1}{\kappa},$$

can also be used to determine the curve. Hence as a further unit vector we have the

$$\text{normal vector} \qquad \mathbf{e}_N \equiv R\,\frac{d\mathbf{e}_T}{ds} = R\,\frac{d^2\mathbf{r}}{ds^2}.$$

Since it has the direction of the derivative of the unit vector $\mathbf{e}_T$, it is perpendicular to $\mathbf{e}_T$. Now we may express the velocity and the accelerations because $\dot{\mathbf{e}}_T = (d\mathbf{e}_T/ds)\,v = (v/R)\,\mathbf{e}_N$ as follows:

$$\mathbf{v} \equiv \dot{\mathbf{r}} = v\,\mathbf{e}_T, \qquad \mathbf{a} \equiv \ddot{\mathbf{r}} = \dot{v}\,\mathbf{e}_T + \frac{v^2}{R}\,\mathbf{e}_N.$$

Thus there is a *tangential acceleration* $\mathbf{a} \cdot \mathbf{e}_T \equiv a_T = \dot{v}$, if the value of the velocity changes, and a *normal acceleration* $\mathbf{a} \cdot \mathbf{e}_N \equiv a_N = v^2/R$, if the direction of the velocity changes. From this decomposition we can also see why motions are often investigated either along a straight line or along a uniformly traveled circle—then only a_T or only a_N appears.

If the curve leaves the plane spanned by $\mathbf{e}_T$ and $\mathbf{e}_N$, then the

$$\text{binormal vector} \qquad \mathbf{e}_B \equiv \mathbf{e}_T \times \mathbf{e}_N$$

also changes with s. Because $d\mathbf{e}_T/ds = \kappa\mathbf{e}_N$, its derivative with respect to s is equal to $\mathbf{e}_T \times d\mathbf{e}_N/ds$. This expression (perpendicular to $\mathbf{e}_T$) must be proportional to $\mathbf{e}_N$, because derivatives of unit vectors do not have components in their direction. Since $e_N = e_B \times e_T$, besides

$$\frac{de_T}{ds} = \kappa\, e_N, \quad \text{the derivatives} \quad \frac{de_B}{ds} = -\tau\, e_N \quad \text{and} \quad \frac{de_N}{ds} = \tau\, e_B - \kappa\, e_T$$

appear with the *torsion* τ, also called the *winding* or *second curvature*. For a right-hand thread, one has $\tau > 0$, and for a left-hand thread, $\tau < 0$. The relation

$$\tau = R^2 \left(\frac{dr}{ds} \times \frac{d^2r}{ds^2} \right) \cdot \frac{d^3r}{ds^3}$$

also holds, because of $\tau = e_B \cdot (de_N/ds)$ and $e_B = e_T \times e_N$. (Here it is unimportant for the winding whether the curvature depends upon s.)

With the *Darboux vector*

$$\delta = \kappa\, e_B + \tau\, e_T,$$

the expressions just obtained for the derivatives of the three unit vectors with respect to the curve length s (*Frenet–Serret formulas*) can be combined to yield

$$\frac{de_\bullet}{ds} = \delta \times e_\bullet \quad \text{with} \quad e_\bullet \in \{e_T, e_N, e_B\}.$$

As long as neither the first nor the second curvature changes along the curve, the Darboux vector is constant: $d\kappa/ds = 0 = d\tau/ds \implies d\delta/ds = 0$, because $\kappa\, de_B/ds = -\tau\, de_T/ds$. The curve winds around it. An example will follow in Sect. 2.2.5, namely the spiral curve of a charged particle in a homogeneous magnetic field: in this case the Darboux vector is $\delta = -q\mathbf{B}/(mv)$. The curves with constant δ thus depend upon the initial velocity v_0. Among these are also circular orbits (perpendicular to δ) and straight lines (along $\pm\delta$), where admittedly a straight line has vanishing curvature ($\kappa = 0$), and the concept of the second curvature (winding) thus has no meaning. The quantities δ and v_0 yield the winding $\tau = \delta \cdot v_0/v_0$ and curvature κ (≥ 0) because of $\delta^2 = \kappa^2 + \tau^2$. The radius h and the helix angle α (with $|\alpha| \leq \frac{1}{2}\pi$) of the associated thread follow from $h = \kappa/\delta^2$ and $\alpha = \arctan \tau/\kappa$. [With $\mathbf{r} = \mathbf{r}_0 + h\,(\cos\varphi\, e_x + \sin\varphi\, e_y + \tan\alpha\,\varphi\, e_z)$ and $s\cos\alpha = h\,\varphi$ and because of $\tan\alpha = \tau/\kappa$, the scalar triple product expression for τ yields the equation $\cos^2\alpha = h/R$.] The geometrical meaning of the curvature radius R and radius h is thus the reciprocal of the length of the Darboux vector (see Fig. 1.3).

If the curve traveled is given by the functions $y(x)$ and $z(x)$ in Cartesian coordinates, then we have

$$\frac{d^2r}{ds^2} = \frac{d}{ds}\left(\frac{dr}{ds}\right) = \frac{d}{dx}\left(\frac{dr}{dx}\frac{dx}{ds}\right)\frac{dx}{ds},$$

and because $ds^2 = dx^2 + dy^2 + dz^2$, we also have $dx/ds = 1/\sqrt{1 + y'^2 + z'^2}$ with $y' \equiv dy/dx$ and $z' \equiv dz/dx$. Hence, the square of the path curvature is given by

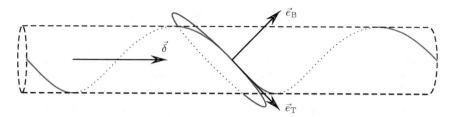

Fig. 1.3 Spiral curve around the constant Darboux vector $\boldsymbol{\delta}$ oriented to the right (constant curvature and winding, here with $\kappa = \tau$). Shown are also the tangent and binormal vectors of the moving frame and the tangential circle. Not shown is the normal vector $\mathbf{e}_N = \mathbf{e}_B \times \mathbf{e}_T$, which points toward the symmetry axis

$$\kappa^2 = \frac{(y'z'' - y''z')^2 + y''^2 + z''^2}{(1 + y'^2 + z'^2)^3}$$

and the torsion by

$$\tau = \frac{y''z''' - y'''z''}{(1 + y'^2 + z'^2)^3} \frac{1}{\kappa^2}.$$

For the curvature, we have $\kappa \geq 0$, while τ is negative for a left-hand thread.

1.1.4   Vector Fields

If a vector is associated with each position, we speak of a vector field. With scalar fields, a scalar is associated with each position. The vector field $\mathbf{a}(\mathbf{r})$ is only continuous at $\mathbf{r}_0$ if all paths approaching $\mathbf{r}_0$ have the same limit. For scalar fields, this is already an essentially stronger requirement than in one dimension.

Instead of drawing a vector field with arrows at many positions, it is often visualized by a *set of field lines*: at every point of a field line the tangent points in the direction of the vector field. Thus $\mathbf{a} \parallel d\mathbf{r}$ and $\mathbf{a} \times d\mathbf{r} = \mathbf{0}$.

For a given vector field many integrals can be formed. In particular, we often have to evaluate integrals over surfaces or volumes. In order to avoid double or triple integral symbols, the corresponding differential is often written immediately after the integral symbol: dV for the volume, $d\mathbf{f}$ for the surface integral, e.g., $\int d\mathbf{f} \times \mathbf{a}$ instead of $-\int \mathbf{a} \times d\mathbf{f}$ (in this way the unnecessary minus sign is avoided for the introduction of the curl density or rotation on p. 13). Here $d\mathbf{f}$ is perpendicular to the related *surface element*. However, the sign of $d\mathbf{f}$ still has to be fixed. In general, we consider the surface of a volume V, which will be denoted here by (V). Then $d\mathbf{f}$ points outwards. Corresponding to (V), the edge of an area A is denoted by (A).

An important example of a scalar integral is the *line integral* $\int d\mathbf{r} \cdot \mathbf{a}(\mathbf{r})$ along a given curve $\mathbf{r}(t)$. If the parameter t determines the points on the curve uniquely, then the line integral

$$\int d\mathbf{r} \cdot \mathbf{a}\,(\mathbf{r}) = \int dt \, \frac{d\mathbf{r}}{dt} \cdot \mathbf{a}\,(\mathbf{r}\,(t))$$

is an ordinary integral over the scalar product $\mathbf{a} \cdot d\mathbf{r}/dt$. Another example of a scalar integral is the surface integral $\int d\mathbf{f} \cdot \mathbf{a}\,(\mathbf{r})$ taken over a given area A or over the surface (V) of the volume V.

Besides the scalar integrals, vectorial integrals like $\int dV \, \mathbf{a}$, $\int d\mathbf{f} \times \mathbf{a}$, and $\int d\mathbf{r} \times \mathbf{a}$ can arise, e.g., the x-component of $\int dV \, \mathbf{a}$ is the simple integral $\int dV \, a_x$.

Different forms are also reasonable through differentiation: vector fields can be deduced from scalar fields, and scalar fields (but also vector fields and tensor fields) from vector fields. These will now be considered one by one. Then the operator ∇ will always turn up. The symbol ∇, an upside-down Δ, resembles an Ancient Greek harp and hence is called *nabla*, after W. R. Hamilton (see 122).

1.1.5 Gradient (Slope Density)

The gradient of a scalar function $\psi\,(\mathbf{r})$ is the vector field

$$\mathrm{grad}\,\psi \equiv \nabla \psi\,, \quad \text{with} \quad \nabla \psi \cdot d\mathbf{r} \equiv d\psi \equiv \psi\,(\mathbf{r} + d\mathbf{r}) - \psi\,(\mathbf{r})\,.$$

This is clearly perpendicular to the area $\psi = \mathrm{const.}$ at every point and points in the direction of $d\psi > 0$ (see Fig. 1.4). The value of the vector $\nabla \psi$ is equal to the derivative of the scalar function $\psi\,(\mathbf{r})$ with respect to the line element in this direction. In Cartesian coordinates, we thus have

$$\nabla \psi = \mathbf{e}_x \, \frac{\partial \psi}{\partial x} + \mathbf{e}_y \, \frac{\partial \psi}{\partial y} + \mathbf{e}_z \, \frac{\partial \psi}{\partial z} = \left(\mathbf{e}_x \, \frac{\partial}{\partial x} + \mathbf{e}_y \, \frac{\partial}{\partial y} + \mathbf{e}_z \, \frac{\partial}{\partial z} \right) \psi\,.$$

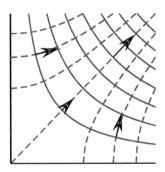

Fig. 1.4 Gradient $\nabla \psi$ of a scalar field $\psi\,(\mathbf{r})$ represented by *arrows*. Contour lines with constant ψ are drawn as *continuous red* and field lines (slope lines) of the *gradient field* as *dashed blue*. In the example considered here, both families of curves contain only hyperbolas (and their asymptotes)

Here $\partial\psi/\partial x$ is the *partial derivative* of $\psi(x, y, z)$ with respect to x for constant y and z. (If other quantities are kept fixed instead, then special rules have to be considered, something we shall deal with in Sect. 1.2.7.)

The gradient is also obtained as a limit of a vectorial integral:

$$\nabla\psi = \lim_{V \to 0} \frac{1}{V} \int_{(V)} d\mathbf{f}\,\psi(\mathbf{r})\ .$$

If we take a cube with infinitesimal edges dx, dy, and dz, we have on the right-hand side as x-component $(dx\,dy\,dz)^{-1}\{dy\,dz\,\psi(x + dx, y, z) - dy\,dz\,\psi(x, y, z)\} = \partial\psi/\partial x$, and similarly for the remaining components. Hence, also

$$\int_V dV\,\nabla\psi = \int_{(V)} d\mathbf{f}\,\psi\ ,$$

because a finite volume can be divided into infinitesimal volume elements, and for continuous ψ, contributions from adjacent planes cancel in pairs. With this surface integral the gradient can be determined even if ψ is not differentiable (singular) at individual points—the surface integral depends only upon points in the neighbourhood of the singular point, where everything is continuous. (In Sect. 1.1.12, we shall consider the example $\psi = 1/r$.)

Corresponding to $d\psi = (d\mathbf{r} \cdot \nabla)\,\psi$, we shall also write in the following

$$d\mathbf{a} = (d\mathbf{r} \cdot \nabla)\,\mathbf{a} = dx\,\frac{\partial\mathbf{a}}{\partial x} + dy\,\frac{\partial\mathbf{a}}{\partial y} + dz\,\frac{\partial\mathbf{a}}{\partial z}\ .$$

We also attribute a meaning to the operation $\nabla\,\mathbf{a}$, but notice that there is no scalar product between ∇ and $\mathbf{a}$ (rather it is the *dyadic* or *tensor product*, as shown in the next section), but there is a scalar product between $d\mathbf{r}$ and ∇. Then for a *Taylor series*, we may write

$$\psi(\mathbf{r} + d\mathbf{r}) = \psi(\mathbf{r}) + (d\mathbf{r} \cdot \nabla)\psi + \tfrac{1}{2}\,(d\mathbf{r} \cdot \nabla)^2\psi + \cdots\ ,$$

where all derivatives are to be taken at the position $\mathbf{r}$.

1.1.6 Divergence (Source Density)

While a vector field has been derived from a scalar field with the help of the gradient, the divergence associates a scalar field with a vector field:

$$\mathrm{div}\,\mathbf{a} \equiv \nabla \cdot \mathbf{a} \equiv \lim_{V \to 0} \frac{1}{V} \int_{(V)} d\mathbf{f} \cdot \mathbf{a}\ .$$

For the same cube as in the last section, the right-hand expression yields

$$
\begin{aligned}
\frac{1}{dx\,dy\,dz}\,[& dy\,dz\,\{a_x(x+dx,\,y,\,z)-a_x(x,\,y,\,z)\} \\
& +dz\,dx\,\{a_y(x,\,y+dy,\,z)-a_y(x,\,y,\,z)\} \\
& +dx\,dy\,\{a_z(x,\,y,\,z+dz)-a_z(x,\,y,\,z)\}]=\frac{\partial a_x}{\partial x}+\frac{\partial a_y}{\partial y}+\frac{\partial a_z}{\partial z}\,,
\end{aligned}
$$

as suggested by the notation $\nabla \cdot \mathbf{a}$, i.e., a scalar product between the vector operator ∇ and the vector $\mathbf{a}$. With this we have also proven *Gauss's theorem*

$$
\int_V dV\,\nabla\cdot\mathbf{a}=\int_{(V)}d\mathbf{f}\cdot\mathbf{a}\,,
$$

since for any partition of the finite volume V into infinitesimal ones and for a continuous vector field $\mathbf{a}$, the contributions of adjacent planes cancel in pairs. The integrals here may even enclose points at which $\mathbf{a}\,(\mathbf{r})$ is singular (see Fig. 1.5 left). We shall discuss this in more detail in Sect. 1.1.12.

The integral $\int d\mathbf{f}\cdot\mathbf{a}$ over an area is called the *flux* of the vector field $\mathbf{a}\,(\mathbf{r})$ through this area (even if $\mathbf{a}$ is not a current density). In this picture, the integral over the closed area (V) describes the source strength of the vector field, i.e., how much more flows into V than out. The divergence is therefore to be understood as a *source density*. A vector field is said to be source-free if its divergence vanishes everywhere. (If the source density is negative, then "drains" predominate.)

The concept of a *field-line tube* is also useful (we discussed field lines in Sect. 1.1.4). Its walls are everywhere parallel to $\mathbf{a}\,(\mathbf{r})$. Therefore, there is no flux through the walls, and the flux through the end faces is equal to the volume integral of $\nabla\cdot\mathbf{a}$. For a source-free vector field ($\nabla\cdot\mathbf{a}=0$), the flux flowing into the field-line tube through one end face emerges again from the other.

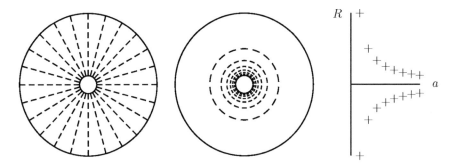

Fig. 1.5 Fields between coaxial walls. On the *left* and in the *center*, the walls are drawn as *continuous lines* and the field lines as *dashed lines*. On the *left*, the field is curl-free and has sources on the walls, while in the *center* it is source-free and has curls on the wall, if in both cases the field strength $|a|=a$ decays with increasing distance R from the axis as shown in the *right-hand graph*, i.e., in such a way that aR is constant

1.1.7 Curl (Vortex Density)

The curl (rotation) of the vector field $\mathbf{a}\,(\mathbf{r})$ is the vector field

$$\text{rot}\,\mathbf{a} \equiv \nabla \times \mathbf{a} \equiv \lim_{V \to 0} \frac{1}{V} \int_{(V)} d\mathbf{f} \times \mathbf{a}\;.$$

For the above-mentioned cube with the edges dx, dy, dz, the x-component of the right-hand expression is equal to

$$\frac{1}{dx\,dy\,dz}\,[+dz\,dx\,\{a_z(x, y+dy, z) - a_z(x, y, z)\}$$

$$-dx\,dy\,\{a_y(x, y, z+dz) - a_y(x, y, z)\}] = \frac{\partial a_z}{\partial y} - \frac{\partial a_y}{\partial z}\;.$$

With $\partial_i \equiv 1/\partial x_i$, we thus have

$$\nabla \times \mathbf{a} = \mathbf{e}_x \left(\frac{\partial a_z}{\partial y} - \frac{\partial a_y}{\partial z} \right) + \mathbf{e}_y \left(\frac{\partial a_x}{\partial z} - \frac{\partial a_z}{\partial x} \right) + \mathbf{e}_z \left(\frac{\partial a_y}{\partial x} - \frac{\partial a_x}{\partial y} \right) \equiv \begin{vmatrix} \mathbf{e}_x & \mathbf{e}_y & \mathbf{e}_z \\ \partial_x & \partial_y & \partial_z \\ a_x & a_y & a_z \end{vmatrix}\;,$$

which is the vector product of the operators ∇ and $\mathbf{a}$. This explains the notation $\nabla \times \mathbf{a}$. Moreover, we have

$$\int_V dV\,\nabla \times \mathbf{a} = \int_{(V)} d\mathbf{f} \times \mathbf{a}$$

for all continuous vector fields, although they may become singular point-wise, and even along lines, as will become apparent shortly.

An important result is *Stokes's theorem*

$$\int_A d\mathbf{f} \cdot (\nabla \times \mathbf{a}) = \int_{(A)} d\mathbf{r} \cdot \mathbf{a}\;,$$

where $d\mathbf{f}$ is taken in the rotational sense on the edge (A) and forms a right-hand screw. The right-hand side is the *rotation (curl)* of $\mathbf{a}$, that is, the line integral of $\mathbf{a}$ along the edge of A. In order to get an insight into the theorem, consider an infinitesimal rectangle in the yz-plane. On the left, we have

$$\int_A d\mathbf{f} \cdot (\nabla \times \mathbf{a}) = \int dy\,dz \left(\frac{\partial a_z}{\partial y} - \frac{\partial a_y}{\partial z} \right),$$

and on the right

$$\int_{(A)} d\mathbf{r} \cdot \mathbf{a} = \int dy\, a_y(x, y, z) - \int dy\, a_y(x, y, z + dz)$$

$$+ \int dz\, a_z(x, y + dy, z) - \int dz\, a_z(x, y, z) \ .$$

The first two integrals on the right-hand side together result in $-\int dy\, (\partial a_y/\partial z)\, dz$, the last two in $\int dz\, (\partial a_z/\partial y)\, dy$. This implies

$$\int_{(A)} d\mathbf{r} \cdot \mathbf{a} = \int dy\, dz \left(\frac{\partial a_z}{\partial y} - \frac{\partial a_y}{\partial z} \right) .$$

The theorem holds thus for an infinitesimal area. A finite area can be divided into sufficiently small ones, where adjacent lines do not contribute, since the integration paths from adjacent areas are opposite to each other.

According to Stokes's theorem we may also set

$$\mathbf{e}_A \cdot (\nabla \times \mathbf{a}) = \lim_{A \to 0} \frac{1}{A} \int_{(A)} d\mathbf{r} \cdot \mathbf{a} \ ,$$

where the unit vector $\mathbf{e}_A$ is perpendicular to the area A and $d\mathbf{r}$ forms a right-hand screw with $\mathbf{e}_A$. The curl density $\nabla \times \mathbf{a}$ can be introduced more pictorially with this equation than with the one mentioned first, and even for vector fields which are singular along a line (perpendicular to the area). Therefore, the inner "conductor" in Fig. 1.5 may even be an arbitrarily thin "wire".

For $\nabla \times \mathbf{a} \neq \mathbf{0}$, the vector field has a non-vanishing rotation, or *vortex*. If $\nabla \times \mathbf{a}$ vanishes everywhere, then the field is said to be curl-free (vortex-free).

1.1.8 Rewriting Products. Laplace Operator

Given various fields, the linear differential operators gradient, divergence, and rotation assign other fields to them. They have the following properties:

$$\nabla(\phi\,\psi) = \phi\,\nabla\psi + \psi\,\nabla\phi \ ,$$
$$\nabla \cdot (\psi\,\mathbf{a}) = \psi\,\nabla \cdot \mathbf{a} + \mathbf{a} \cdot \nabla\psi \ ,$$
$$\nabla \times (\psi\,\mathbf{a}) = \psi\,\nabla \times \mathbf{a} - \mathbf{a} \times \nabla\psi \ ,$$
$$\nabla \cdot (\mathbf{a} \times \mathbf{b}) = \mathbf{b} \cdot (\nabla \times \mathbf{a}) - \mathbf{a} \cdot (\nabla \times \mathbf{b}) \ ,$$
$$\nabla \times (\mathbf{a} \times \mathbf{b}) = (\mathbf{b} \cdot \nabla)\,\mathbf{a} - \mathbf{b}\,(\nabla \cdot \mathbf{a}) - (\mathbf{a} \cdot \nabla)\,\mathbf{b} + \mathbf{a}\,(\nabla \cdot \mathbf{b}) \ ,$$
$$\nabla\,(\mathbf{a} \cdot \mathbf{b}) = (\mathbf{b} \cdot \nabla)\,\mathbf{a} + \mathbf{b} \times (\nabla \times \mathbf{a}) + (\mathbf{a} \cdot \nabla)\,\mathbf{b} + \mathbf{a} \times (\nabla \times \mathbf{b}) \ .$$

All these equations can be proven by decomposing into Cartesian coordinates and using the product rule for derivatives. For the last three, however, it is better to refer to Sect. 1.1.2 (and the product rule) and place ∇ between the other two vectors, so that this operator then acts only on the last factor (see Problem 3.1). Since

$$\nabla \cdot \mathbf{r} = 3, \quad \nabla \times \mathbf{r} = \mathbf{0}, \quad (\mathbf{a} \cdot \nabla) \, \mathbf{r} = \mathbf{a}$$

(Problem 3.2), we find in particular

$$\nabla \cdot (\psi \, \mathbf{r}) = 3\psi + \mathbf{r} \cdot \nabla \psi \, ,$$
$$\nabla \times (\psi \, \mathbf{r}) = -\mathbf{r} \times \nabla \psi \, ,$$
$$(\mathbf{a} \cdot \nabla) \, \psi \mathbf{r} = \mathbf{a} \, \psi + \mathbf{r} \, (\mathbf{a} \cdot \nabla \psi) \, ,$$

and

$$\nabla \cdot (\mathbf{a} \times \mathbf{r}) = \mathbf{r} \cdot (\nabla \times \mathbf{a}) \, ,$$
$$\nabla \times (\mathbf{a} \times \mathbf{r}) = 2\mathbf{a} + (\mathbf{r} \cdot \nabla) \, \mathbf{a} - \mathbf{r} \, (\nabla \cdot \mathbf{a}) \, ,$$
$$\nabla \, (\mathbf{a} \cdot \mathbf{r}) = \mathbf{a} + (\mathbf{r} \cdot \nabla) \, \mathbf{a} + \mathbf{r} \times (\nabla \times \mathbf{a}) \, .$$

These equations are generally applicable and save us lengthy calculations—we shall use them often. Besides these, we also have

$$\nabla r^n = n \, r^{n-2} \, \mathbf{r} \, ,$$

not only for integer numbers n, but also for fractions. Furthermore, if ψ and $\mathbf{a}$ have continuous derivatives with respect to their coordinates, then the order of the derivatives may be interchanged, viz.,

$$\nabla \times \nabla \psi = \mathbf{0} \quad \text{and} \quad \nabla \cdot (\nabla \times \mathbf{a}) = 0 \, .$$

Hence, *gradient fields are curl-free (vortex-free), and curl fields are source-free.* Point-like singularities do not alter these results.

The operator Δ in the expression

$$\Delta \psi \equiv \nabla \cdot \nabla \psi$$

is called the *Laplace operator*. For a final reformulation, we make use once again of a result in Sect. 1.1.2, namely $\mathbf{b} \cdot \mathbf{c} \, \mathbf{a} = \mathbf{c} \, (\mathbf{b} \cdot \mathbf{a}) - \mathbf{b} \times (\mathbf{c} \times \mathbf{a})$, whence

$$\Delta \, \mathbf{a} \equiv \nabla \cdot \nabla \mathbf{a} = \nabla (\nabla \cdot \mathbf{a}) - \nabla \times (\nabla \times \mathbf{a}).$$

Therefore, this operator can act on scalars $\psi (\mathbf{r})$ and vectors $\mathbf{a} (\mathbf{r})$. In Cartesian coordinates it reads in both cases

$$\Delta = \frac{\partial^2}{\partial x^2} + \frac{\partial^2}{\partial y^2} + \frac{\partial^2}{\partial z^2} \, .$$

According to Gauss's theorem,

$$\int_{(V)} d\mathbf{f} \cdot \nabla \psi = \int_V dV \, \Delta \psi \, , \qquad \text{thus} \qquad \Delta \psi = \lim_{V \to 0} \frac{1}{V} \int_{(V)} d\mathbf{f} \cdot \nabla \psi \, .$$

The Laplace operator is thus to be understood as the limit of a surface integral. It is apparently only different from zero if $\nabla \psi$ changes on the surface (V). A further important relation is

$$\nabla \cdot (\psi \nabla \phi - \phi \nabla \psi) = \psi \, \Delta \phi - \phi \, \Delta \psi \, ,$$

which can be derived from the above equations.

   According to Gauss's theorem a source- and curl-free field has to vanish every-where, if it vanishes on the surface ("at infinity"). Every curl-free vector field can be represented as a gradient field $\nabla \psi$, where ψ obeys the *Laplace equation* $\Delta \psi = 0$ everywhere, because the field is also taken to be source-free. Hence, we have $\nabla \cdot \psi \nabla \psi = \nabla \psi \cdot \nabla \psi$, according to Gauss's theorem $\int_{(V)} d\mathbf{f} \cdot \psi \nabla \psi = \int_V dV \, \nabla \psi \cdot \nabla \psi$. The left-hand side has to be zero, and on the right the integrand is nowhere negative, whence it has to vanish everywhere.

1.1.9   Integral Theorems for Vector Expressions

The concepts gradient, divergence, and rotation follow from the equations

$$\int_V dV \, \nabla \psi = \int_{(V)} d\mathbf{f} \, \psi \, ,$$

$$\int_V dV \, \nabla \cdot \mathbf{a} = \int_{(V)} d\mathbf{f} \cdot \mathbf{a} \qquad \text{(Gauss's theorem)},$$

$$\int_V dV \, \nabla \times \mathbf{a} = \int_{(V)} d\mathbf{f} \times \mathbf{a} \, .$$

Dividing a finite volume into infinitesimal parts, the contributions of adjacent planes cancel in pairs. Corresponding to these, we found in Sect. 1.1.7 [the first expression is, of course, also equal to $\int_A (d\mathbf{f} \times \nabla) \cdot \mathbf{a}$]

$$\int_A d\mathbf{f} \cdot (\nabla \times \mathbf{a}) = \int_{(A)} d\mathbf{r} \cdot \mathbf{a} \qquad \text{(Stokes's theorem)},$$

$$\int_A d\mathbf{f} \times \nabla \psi = \int_{(A)} d\mathbf{r} \, \psi \, .$$

The last equation can be proven like Stokes's theorem. Likewise, we may also derive the following equation:

$$\int_A (d\mathbf{f} \times \nabla) \times \mathbf{a} = \int_{(A)} d\mathbf{r} \times \mathbf{a} \,.$$

If we take the area element $d\mathbf{f} = \mathbf{e}_x \, dy \, dz$ once again, then using the vector product expansion on p. 4, the integrand on the left-hand side is equal to $\nabla(\mathbf{e}_x \cdot \mathbf{a}) - \mathbf{e}_x \nabla \cdot \mathbf{a}$. On the right, one has the same, namely, $dz \, \mathbf{e}_z \times (\partial \mathbf{a}/\partial y) \, dy - dy \, \mathbf{e}_y \times (\partial \mathbf{a}/\partial z) \, dz$.

In addition, since $\nabla \cdot (\psi \mathbf{a}) = \psi \, \nabla \cdot \mathbf{a} + \mathbf{a} \cdot \nabla \psi$ Gauss's theorem implies

$$\int_{(V)} d\mathbf{f} \cdot \psi \mathbf{a} = \int_V dV \, (\psi \, \nabla \cdot \mathbf{a} + \mathbf{a} \cdot \nabla \psi) \,.$$

(Here the left- and right-hand sides should be interchanged, i.e., the triple integral should be simplified to a double integral.) Hence, we deduce the *first* and *second* *Green theorems*

$$\int_{(V)} d\mathbf{f} \cdot \psi \, \nabla \phi = \int_V dV \, (\psi \, \Delta \phi + \nabla \phi \cdot \nabla \psi) \,,$$

$$\int_{(V)} d\mathbf{f} \cdot (\psi \, \nabla \phi - \phi \, \nabla \psi) = \int_V dV \, (\psi \, \Delta \phi - \phi \, \Delta \psi) \,.$$

Taking ψ as the Cartesian component of a vector $\mathbf{b}$, we may also infer

$$\int_{(V)} (d\mathbf{f} \cdot \mathbf{a}) \, \mathbf{b} = \int_V dV \, \{\mathbf{b} \, (\nabla \cdot \mathbf{a}) + (\mathbf{a} \cdot \nabla) \, \mathbf{b}\} \,.$$

Since $\mathbf{b} = \mathbf{r}$ and $(\mathbf{a} \cdot \nabla) \, \mathbf{r} = \mathbf{a}$, it also follows that

$$\int_V dV \, \mathbf{a} = \int_{(V)} (d\mathbf{f} \cdot \mathbf{a}) \, \mathbf{r} - \int_V dV \, \mathbf{r} \, (\nabla \cdot \mathbf{a}) \,.$$

The volume integral over a source-free vector field $\mathbf{a}$ is thus always zero if $\mathbf{a}$ vanishes on the surface (V) .

Finally, we should mention the equation

$$\int_{(V)} d\mathbf{f} \times \psi \, \mathbf{a} = \int_V dV \, (\psi \, \nabla \times \mathbf{a} - \mathbf{a} \times \nabla \psi) \,,$$

where we have used $\nabla \times (\psi \mathbf{a}) = \psi \, \nabla \times \mathbf{a} - \mathbf{a} \times \nabla \psi$.

1.1.10 Delta Function

In the following, we shall often use the Dirac delta function. Therefore, its properties
are compiled here, even though it does not actually belong to vector analysis, but to
general analysis (and in particular to integral calculus).

We start with the *Kronecker symbol*

$$\delta_{ik} = \begin{cases} 0 & \text{for } i \neq k , \\ 1 & \text{for } i = k . \end{cases}$$

It is useful for many purposes. In particular we may use it to filter out the k th element
of a sequence $\{f_i\}$:

$$f_k = \sum_i f_i \, \delta_{ik} .$$

Here, of course, within the sum, one of the i has to take the value k. Now, if we
make the transition from the countable (discrete) variables i to a continuous quantity
x, then we must also generalize the Kronecker symbol. This yields Dirac's *delta
function* $\delta(x - x')$. It is defined by the equation

$$f(x') = \int_a^b f(x)\,\delta(x - x')\,\mathrm{d}x \qquad \text{for } a < x' < b , \text{ zero otherwise} ,$$

where $f(x)$ is an arbitrary continuous test function. If the variable x (and hence also
$\mathrm{d}x$) is a physical quantity with unit $[x]$, the delta function has the unit $[x]^{-1}$.

Obviously, the delta function $\delta(x - x')$ is not an ordinary function, because it has
to vanish for $x \neq x'$ and it has to be singular for $x = x'$, so that the integral becomes
$\int \delta(x - x')\,\mathrm{d}x = 1$. Consequently, we have to extend the concept of a function:
$\delta(x - x')$ is a *distribution*, or *generalized function*, which makes sense only as a
weight factor in an integrand, while an ordinary function $y = f(x)$ is a map $x \rightarrow y$.
*Every equation in which the delta function appears without an integral symbol is an
equation between integrands*: on both sides of the equation, the integral symbol and
the test function have been left out.

The delta function is the derivative of the Heaviside *step function*:

$$\varepsilon(x - x') = \begin{cases} 0 & \text{for } x < x' \\ 1 & \text{for } x > x' \end{cases} \qquad \Longrightarrow \qquad \delta(x) = \varepsilon'(x) .$$

At the discontinuity, the value of the step function is not usually fixed, although the
mean value $1/2$ is sometimes taken, whence it becomes point symmetric. The step
function is often called the *theta function* and noted by θ (or Θ) instead of ε (contrary to the IUPAP recommendation). The derivative of the step function vanishes for
$x \neq x'$, while $\int_a^b \varepsilon'(x - x')\,\mathrm{d}x \equiv \varepsilon(b - x') - \varepsilon(a - x')$ is equal to one for
$a < x' < b$ and zero for other values of x'.

Hence, using

$$\varepsilon(x) = \frac{1}{2} + \frac{1}{\pi} \lim_{\varepsilon \to +0} \arctan \frac{x}{\varepsilon} ,$$

we find the important equations

$$\delta(x) = \frac{1}{\pi} \lim_{\varepsilon \to +0} \frac{\varepsilon}{x^2 + \varepsilon^2} = \frac{1}{2\pi i} \lim_{\varepsilon \to +0} \left(\frac{1}{x - i\varepsilon} - \frac{1}{x + i\varepsilon} \right) \equiv \frac{1}{2\pi i} \left(\frac{1}{x - io} - \frac{1}{x + io} \right) .$$

We may thus represent the generalized function $\delta(x)$ as a limit of ordinary functions which are concentrated ever more sharply at only one position. According to the last equation it is practical here to decompose the delta function in the complex plane into two functions with the same pole for $\pm io$ with opposite residues, then to take the limit $o \to +0$.

Clearly, we also have

$$\frac{i}{x + io} = 2\pi\, \delta(x) + \frac{i}{x - io} = \pi\, \delta(x) + \frac{i}{2} \left(\frac{1}{x + io} + \frac{1}{x - io} \right) ,$$

if we make use of $\pi\, \delta(x) = \frac{1}{2}i\, \{(x + io)^{-1} - (x - io)^{-1}\}$ for the second reformulation. Here, the expression in the last bracket vanishes for $x^2 \ll o^2$, while it turns into $2x/(x^2 + o^2) \approx 2/x$ for $x^2 \gg o^2$. This can be exploited for the *principal-value integral* (the principal value) P..., a kind of opposite to the delta function, because it leaves out the singular position x' in the integration, with equally small paths on either side of it:

$$P \int_a^b \frac{f(x)\, dx}{x - x'} \equiv \lim_{\varepsilon \to +0} \left(\int_a^{x' - \varepsilon} + \int_{x' + \varepsilon}^b \right) \frac{f(x)\, dx}{x - x'} .$$

Like the delta function, the symbol P also makes sense only in the context of an integral. Hence we may also write the equation above as

$$\frac{1}{x \pm io} = \frac{P}{x} \mp i\pi\, \delta(x) .$$

This result is obtained rather crudely here, because the infinitesimal quantity o is supposed to be arbitrarily small, but nevertheless different from zero. It can be proven using the residue theorem from the theory of complex functions. To this end, we consider

$$\int_{-\infty}^{\infty} \frac{f(x)\, dx}{x - (x' - io)} \pm \int_{-\infty}^{\infty} \frac{f(x)\, dx}{x - (x' + io)} = \left(\int_{C_1} \pm \int_{C_2} \right) \frac{f(z)\, dz}{z - z'} ,$$

with the two integrations running from left to right because of C_1 (above) and C_2 (below the symmetry axis) in Fig. 1.6 for regular test functions $f(z)$. In the complex

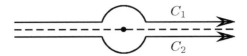

Fig. 1.6 Integration paths C_1 and C_2 (*continuous lines*) to determine the principal value and the residues. The (real) symmetry axis is shown by the *dashed line*

z-plane the integrand only has the pole at $z' = x' - io$ in the lower half-plane and at $x' + io$ in the upper half-plane, whence the indicated integrations can be performed.

The difference between the two integrals is equal to $-\oint f(x)(x - x')^{-1}\, dx$, according to the *residue theorem*, thus equal to $-2\pi i\, f(x')$. In the sum of the two integrals the contributions from the half circles cancel, since for $z = z' + \varepsilon \exp(i\phi)$, we have $dz = i\varepsilon \exp(i\phi)\, d\phi = i(z - z')\, d\phi$, and what remains is twice the principal value, which is what was to be shown. Hence, we have proven our claim that $(x \pm io)^{-1} = P x^{-1} \mp i\pi\, \delta(x)$.

Since $x\, \delta(x) = 0$, the integrand may even be divided by functions which have zeros:

$$A = B \qquad \Longleftrightarrow \qquad \frac{A}{x} = \frac{B}{x} + C\, \delta(x)\ .$$

The constant C in the integrals can be fixed, provided that we also fix the integration path across the singularity (e.g., as for the principal value integral).

An important property of the delta function is

$$\delta(a\,x) = \frac{1}{|a|}\, \delta(x)\ ,$$

because both sides are equal to $d\varepsilon(y)/dy$ for $y = ax$. In particular, the delta function is even, i.e., $\delta(-x) = \delta(x)$. Hence we can even infer $\int_0^\infty \delta(x)\, dx = \frac{1}{2}$. If instead of ax we take a function $a(x)$ as argument, and if $a(x)$ has only one-fold zeros x_n, then it follows that

$$\delta(a(x)) = \sum_n \frac{\delta(x - x_n)}{|a'(x_n)|}\ ,$$

and in particular also that $\delta(x^2 - x_0^2) = \{\delta(x - x_0) + \delta(x + x_0)\}/(2|x_0|)$. In addition, $\iint f(x)\, \delta(x - y)\, \delta(y - x')\, dx\, dy = \int f(y)\, \delta(y - x')\, dy = f(x') = \int f(x)\, \delta(x - x')\, dx$ delivers the equation

$$\int \delta(x - y)\, \delta(y - x')\, dy = \delta(x - x')\ .$$

This is similar to the defining equation of the delta function, in which we allowed only for ordinary, continuous functions as test functions.

For the *n th derivative of the delta function*, n partial integrations (for $a < x' < b$, zero otherwise) result in

$$\int_a^b f(x)\,\delta^{(n)}(x - x')\,\mathrm{d}x = (-)^n\, f^{(n)}(x')\,,$$

because the limits do not contribute. It thus follows that $x\,\delta'(x) = -\delta(x)$, which we shall need in quantum theory (Sect. 4.3.2) for the real-space representation of the momentum operator, viz., $\mathbf{P} \,\widehat{=}\, (\hbar/i)\,\nabla$.

If, in the interval $a \le x \le b$, we have a complete *orthonormal set* of functions $\{g_n(x)\}$, i.e., a series of functions with the properties

$$\int_a^b g_n{}^*(x)\, g_{n'}(x)\,\mathrm{d}x = \delta_{nn'}$$

as well as $f(x) = \sum_n g_n(x)\, f_n$ for all (square-integrable) functions $f(x)$, then after interchange of summation and integration, we have $f_n = \int_a^b g_n{}^*(x)\, f(x)\,\mathrm{d}x$ for the expansion coefficients, and hence $\sum_n \int_a^b g_n(x')\, g_n{}^*(x)\, f(x)\,\mathrm{d}x = f(x')$, which leads to

$$\delta(x - x') = \sum_n g_n{}^*(x)\, g_n(x')\,.$$

Each complete set of functions delivers a representation of the delta function, i.e., it can be expanded in terms of ordinary functions.

In particular, we can expand the delta function in the interval $-a \le x \le a$ in terms of a *Fourier series*: we have $g_n(x) = 1/\sqrt{2a}\, \exp(inx\pi/a)$ with $n \in \{0,\ \pm 1,\ \pm 2,\ \ldots\}$ and (the result is even in $x - x'$)

$$\delta(x - x') = \frac{1}{2a} \sum_n \exp\frac{in\pi(x - x')}{a} \qquad \text{for } -a \le x \le a\,.$$

For $a \to \infty$, we can even go over to a *Fourier integral*. For very large a, the sequence $k_n = n\pi/a$ becomes nearly continuous. Therefore, we replace the sum $\sum_n f(k_n)\,\Delta k$ with $\Delta k = \pi/a$ by its associated integral

$$\delta(x - x') = \frac{1}{2\pi} \int_{-\infty}^{\infty} \exp\{ik(x - x')\}\,\mathrm{d}k \qquad \text{for } -\infty < x < \infty\,.$$

For the Fourier expansion, we therefore take $g(k, x) = 1/\sqrt{2\pi}\, \exp(ikx)$. We now have the basics for the *Fourier transform*, which we shall discuss in the next section.

The integral from $-\infty$ to $+\infty$ can be decomposed into the one from $-\infty$ to 0 plus the one from 0 to $+\infty$. But with $k \to -k$, we have $\int_{-\infty}^0 \exp(ikx)\,\mathrm{d}k = \int_0^\infty \exp(-ikx)\,\mathrm{d}k$, so this part delivers the complex-conjugate of the other part. Therefore, we infer $\mathrm{Re}\int_0^\infty \exp(ikx)\,\mathrm{d}k = \pi\,\delta(x)$ or

$$\delta(x) = \frac{1}{\pi} \int_0^\infty \cos kx\,\mathrm{d}k \qquad \text{and} \qquad \varepsilon(x) = \frac{1}{2} + \frac{1}{\pi} \int_0^\infty \frac{\sin kx}{k}\,\mathrm{d}k\,.$$

On the other hand, the usual integration rules for $\int_0^\infty \exp(ikx)\,dk$ deliver the expression $(ix)^{-1}\exp(ikx)|_{k=0}^{k=\infty}$. For real x, this is undetermined for $k \to \infty$. But if x contains an (even very small) positive imaginary part, then it vanishes for $k \to \infty$. We include this small positive imaginary part of x as before through $x + io$ (with real x):

$$\int_0^\infty \exp(ikx)\,dk = \frac{i}{x + io} = \pi\,\delta(x) + i\,\frac{P}{x}\,.$$

We have already proven this for the real part of the integral, because the real part of the right-hand side has turned out to be equal to $\pi\,\delta(x)$. But then the equation holds also for the imaginary part, because the proof used only general properties of integrals.

1.1.11   Fourier Transform

If the region of definition is infinite on both sides, we use

$$f(x) = \int_{-\infty}^\infty g(k,\,x)\,f(k)\,dk\,, \qquad f(k) = \int_{-\infty}^\infty g^*(k,\,x)\,f(x)\,dx\,,$$

with $g(k,\,x) = 1/\sqrt{2\pi}\,\exp(ikx)$:

$$f(x) = \frac{1}{\sqrt{2\pi}}\int_{-\infty}^\infty \exp(+ikx)\,f(k)\,dk\,,$$

$$f(k) = \frac{1}{\sqrt{2\pi}}\int_{-\infty}^\infty \exp(-ikx)\,f(x)\,dx\,.$$

Generally, $f(x)$ and $f(k)$ are different functions of their arguments, but we would like to distinguish them only through their argument. [The less symmetric notation $f(x) = \int \exp(ikx)\,F(k)\,dk$ with $F(k) = f(k)/\sqrt{2\pi}$ is often used. This avoids the square root factor with the agreement that $(2\pi)^{-1}$ always appears with dx.] Instead of the pair of variables $x \leftrightarrow k$, the pair $t \leftrightarrow \omega$ is also often used.

Important properties of the Fourier transform are

$$f(x) = f^*(x) \qquad \Longleftrightarrow \qquad f(k) = f^*(-k)\,,$$
$$f(x) = g(x)\,h(x) \qquad \Longleftrightarrow \qquad f(k) = \frac{1}{\sqrt{2\pi}}\int_{-\infty}^\infty g(k-k')\,h(k')\,dk'\,,$$
$$f(x) = g(x-x') \qquad \Longleftrightarrow \qquad f(k) = \exp(-ikx')\,g(k)\,.$$

For a periodic function $f(x) = f(x - l)$ the last relation leads to the condition $k_n = 2\pi\,n/l$ with $n \in \{0, \pm 1, \pm 2, \ldots\}$, thus to a *Fourier series* instead of the integral. In addition, by Fourier transform, all *convolution integrals* $\int g(x - x')\,h(x')\,dx'$ can clearly be turned into products $\sqrt{2\pi}\,g(k)\,h(k)$ (Problem 3.9), which are much easier to handle.

If $f(x)$ vanishes for all $x < 0$, then $f(x) = \varepsilon(x)\,f(x)$ holds with the step function mentioned in the last section, e.g., for "causal functions" $f(t)$, which depend upon the time t. Then the Fourier transform yields the relation

$$f(x) = \varepsilon(x)\,f(x) \qquad \Longleftrightarrow \qquad f(k) = \frac{\mathrm{i}}{\pi}\,\mathrm{P}\int_{-\infty}^{\infty} \mathrm{d}k'\,\frac{f(k')}{k - k'}\ .$$

Here, due to the factors i in the Fourier transformed $f(k)$, the real and imaginary parts are related to each other in such a way that only the one or the other (for all k) needs to be measured. This relation is sometimes called the *Kramers–Kronig* or *dispersion relation*, even though it also actually exploits the fact that $f(x)$ is real, whence the integration has to be performed over just half the region, viz., 0 to ∞.

Another result that is often useful is *Parseval's equation*

$$\int_{-\infty}^{\infty} \mathrm{d}x\ g^*(x)\,h(x) = \int_{-\infty}^{\infty} \mathrm{d}k\ g^*(k)\,h(k)\ .$$

In order to prove it, we expand the left-hand side according to Fourier and obtain the integral $(2\pi)^{-1} \int \mathrm{d}x\,\mathrm{d}k\,\mathrm{d}k'\,\exp\{\mathrm{i}(k - k')x\}\,g^*(k')\,h(k)$. After integration over x, we encounter the delta function $2\pi\,\delta(k - k')$ and can then also integrate easily over k', which yields the right-hand side. In particular, $\int \mathrm{d}x\,|f(x)|^2 = \int \mathrm{d}k\,|f(k)|^2$.

Table 1.2 shows some of the Fourier transforms commonly encountered. To prove the last relation in the table, we have to use a square addition in the exponent and the integral $\int_{-\infty}^{\infty} \exp(-x^2)\,\mathrm{d}x = \sqrt{\pi}$, the latter following from

$$\iint_{-\infty}^{\infty} \exp(-x^2 - y^2)\,\mathrm{d}x\,\mathrm{d}y = 2\pi \int_0^{\infty} \mathrm{e}^{-s}\,\frac{1}{2}\mathrm{d}s = \pi\ , \quad \text{with}\ \ s = r^2 = x^2 + y^2\ .$$

Table 1.2 Some functions and their Fourier transforms

$f(x)$	$f(k)$
$\delta(x - x')$	$\dfrac{\exp(-\mathrm{i}kx')}{\sqrt{2\pi}}$
$\dfrac{1}{2a}\,\varepsilon(a^2 - x^2)$	$\dfrac{1}{\sqrt{2\pi}}\,\dfrac{\sin(ak)}{ak}$
$\varepsilon(x)\,\exp(-\lambda x)$	$\dfrac{1}{\sqrt{2\pi}}\,\dfrac{1}{\lambda + \mathrm{i}k}$ if $\mathrm{Re}\,\lambda > 0$
$\exp\dfrac{-(x - x')^2}{2\Delta^2}$	$\Delta\,\exp\dfrac{-\Delta^2 k^2}{2}\,\exp(-\mathrm{i}kx')$

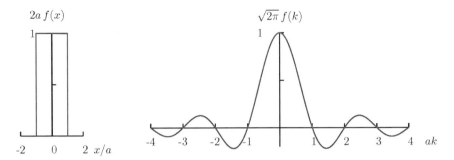

Fig. 1.7 Fourier transform (*left, red*) of the *box function* (*right, blue*). This is useful, e.g., for the refraction from a slit

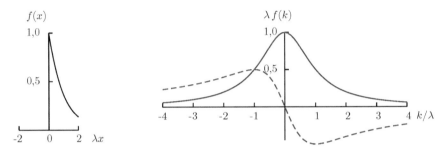

Fig. 1.8 Fourier transform (*right*) of the *truncated exponential function* $f(x) = \varepsilon(x) \exp(-\lambda x)$ (*left*). This is useful for decay processes, if x stands for the time and k for the angular frequency. Here the *dashed blue curve* shows the real part and the *continuous red curve* the imaginary part of $\lambda f(k)$. The Kramers–Kronig relation relates these real and imaginary parts

From the first example with $x' = 0$, the Fourier transform of a constant is a delta function, and from the fourth example with $x' = 0$, the Fourier transform of a Gaussian function is a Gaussian function again. The second relation is represented in Fig. 1.7 and the third in Fig. 1.8.

Correspondingly, in three dimensions with $\mathbf{k}$ as *wave vector* (more on p. 137), we have

$$\delta(\mathbf{k} - \mathbf{k}') = \frac{1}{(2\pi)^3} \int_{-\infty}^{\infty} d^3r \, \exp\{i\,(\mathbf{k} - \mathbf{k}') \cdot \mathbf{r}\} \,,$$

$$f(\mathbf{r}) = \frac{1}{\sqrt{2\pi}^3} \int_{-\infty}^{\infty} d^3k \, \exp(+i\,\mathbf{k} \cdot \mathbf{r}) \, f(\mathbf{k}) \,,$$

$$f(\mathbf{k}) = \frac{1}{\sqrt{2\pi}^3} \int_{-\infty}^{\infty} d^3r \, \exp(-i\,\mathbf{k} \cdot \mathbf{r}) \, f(\mathbf{r}) \,.$$

Here, d^3r is used for the volume element dV in real space and correspondingly d^3k for the volume element in reciprocal space. In Cartesian coordinates, we then have $\delta(\mathbf{r} - \mathbf{r}') = \delta(x - x')\,\delta(y - y')\,\delta(z - z')$.

From the expansion

$$\mathbf{a}\,(\mathbf{r}) = \frac{1}{\sqrt{2\pi}^{\,3}} \int d^3k \; \exp(i\mathbf{k} \cdot \mathbf{r}) \, \mathbf{a}(\mathbf{k})$$

of a vector field $\mathbf{a}\,(\mathbf{r})$, since Fourier expansions are unique, it follows that

$$\nabla \times \mathbf{a}\,(\mathbf{r}) = \mathbf{b}\,(\mathbf{r}) \qquad \Longleftrightarrow \qquad i\mathbf{k} \times \mathbf{a}\,(\mathbf{k}) = \mathbf{b}\,(\mathbf{k})$$

and

$$\nabla \cdot \mathbf{a}\,(\mathbf{r}) = b\,(\mathbf{r}) \qquad \Longleftrightarrow \qquad i\mathbf{k} \cdot \mathbf{a}\,(\mathbf{k}) = b\,(\mathbf{k}) \,.$$

If, for example, the curly bracket in $\int d^3k \; \exp(i\mathbf{k} \cdot \mathbf{r}) \, \{i\mathbf{k} \times \mathbf{a}\,(\mathbf{k}) - \mathbf{b}(\mathbf{k})\}$ vanishes for all $\mathbf{k}$, then of course the integral also does for all $\mathbf{r}$. Rotation-free fields thus have Fourier component $\mathbf{a}\,(\mathbf{k})$ in the direction of the wave vector (*longitudinal field* $\mathbf{a}_{\text{long}}$). In contrast, source-free fields have Fourier component $\mathbf{a}\,(\mathbf{k})$ perpendicular to the wave vector (*transverse field* $\mathbf{a}_{\text{trans}}$). According to p. 4, the decomposition

$$\mathbf{a}(\mathbf{k}) = \mathbf{e}_k\,(\mathbf{e}_k \cdot \mathbf{a}(\mathbf{k})) - \mathbf{e}_k \times (\mathbf{e}_k \times \mathbf{a}(\mathbf{k})) \,, \quad \text{with} \quad \mathbf{e}_k \equiv \frac{\mathbf{k}}{k} \,,$$

therefore splits up into a longitudinal and a transverse part, i.e., into the vortex-free and the source-free part.

Some important examples of Fourier transforms in the three-dimensional space are listed on p. 410.

1.1.12 Calculation of a Vector Field from Its Sources and Curls

Every vector field that is continuous everywhere and vanishes at infinity can be uniquely determined from its sources and curls (rotations, vortices):

$$\mathbf{a}\,(\mathbf{r}) = -\nabla \int dV' \, \frac{\nabla' \cdot \mathbf{a}\,(\mathbf{r}')}{4\pi\,|\mathbf{r} - \mathbf{r}'|} + \nabla \times \int dV' \, \frac{\nabla' \times \mathbf{a}\,(\mathbf{r}')}{4\pi\,|\mathbf{r} - \mathbf{r}'|} \,.$$

The first term here becomes fixed by the sources of $\mathbf{a}$ and, like every pure gradient field, is vortex-free, while the second, like every pure vortex field, is source-free and becomes fixed by the vortex of $\mathbf{a}$. The operator ∇' acts on the coordinate $\mathbf{r}'$, while ∇ acts on the coordinate $\mathbf{r}$ and therefore may be interchanged with the integration.

The decomposition is unique. If there were two different vector fields $\mathbf{a}_1$ and $\mathbf{a}_2$ with the same sources and curls, then $\mathbf{a}_1 - \mathbf{a}_2$ would have neither sources nor curls, and in addition would vanish at infinity. But according to p. 16, $\mathbf{a}_1 = \mathbf{a}_2$ has to hold.

To prove the claim, we evaluate $\nabla \cdot \mathbf{a}$ and $\nabla \times \mathbf{a}$:

$$\nabla \cdot \mathbf{a} = \frac{-1}{4\pi} \int dV' \, \nabla' \cdot \mathbf{a}\,(\mathbf{r}') \, \Delta \frac{1}{|\mathbf{r} - \mathbf{r}'|} \, ,$$

$$\nabla \times \mathbf{a} = \frac{-1}{4\pi} \int dV' \left\{ \nabla' \times \mathbf{a}\,(\mathbf{r}') \, \Delta \frac{1}{|\mathbf{r} - \mathbf{r}'|} - \nabla \left(\nabla \cdot \frac{\nabla' \times \mathbf{a}\,(\mathbf{r}')}{|\mathbf{r} - \mathbf{r}'|} \right) \right\} .$$

Still, $\mathbf{a}\,(\mathbf{r})$ could contain a constant term, which would affect neither $\nabla \cdot \mathbf{a}$ nor $\nabla \times \mathbf{a}$, but $\mathbf{a} = \mathbf{0}$ has to hold at infinity and this fixes this term uniquely. Now we show—and this is sufficient for the proof—that

$$\Delta \frac{1}{|\mathbf{r} - \mathbf{r}'|} = -4\pi \, \delta(\mathbf{r} - \mathbf{r}') \, ,$$

and that the last term in $\nabla \times \mathbf{a}$ does not contribute. With $\mathbf{r}' = \mathbf{0}$ and recalling from Sect. 1.1.8 that $\nabla r^n = n \, r^{n-2} \, \mathbf{r}$, we have

$$\Delta \frac{1}{r} = \nabla \cdot \nabla \frac{1}{r} = -\nabla \cdot \frac{\mathbf{r}}{r^3} = -\left(\frac{\nabla \cdot \mathbf{r}}{r^3} + \mathbf{r} \cdot \nabla \frac{1}{r^3} \right) = -\left(\frac{3}{r^3} + \mathbf{r} \cdot \frac{-3\mathbf{r}}{r^5} \right) .$$

This expression vanishes for $r \neq 0$. On the other hand, if we evaluate the source strength at the origin using Gauss's theorem with a sphere of radius $r > 0$ around it, we have

$$\int dV \, \nabla \cdot \nabla \frac{1}{r} = \int d\mathbf{f} \cdot \nabla \frac{1}{r} = -\frac{1}{r^2} \int d\mathbf{f} \cdot \mathbf{e}_r = -4\pi \, .$$

This shows the first part of the proof, since $\delta(\mathbf{r} - \mathbf{r}')$ vanishes for $\mathbf{r} \neq \mathbf{r}'$ and $\int dV \, \delta(\mathbf{r} - \mathbf{r}')$ is equal to 1. In addition, with $\mathbf{b} = \nabla' \times \mathbf{a}(\mathbf{r}')$, which depends only upon $\mathbf{r}'$, but not upon $\mathbf{r}$, we have

$$\nabla \left(\nabla \cdot \frac{\mathbf{b}}{|\mathbf{r} - \mathbf{r}'|} \right) = \nabla \left(\mathbf{b} \cdot \nabla \frac{1}{|\mathbf{r} - \mathbf{r}'|} \right) = (\mathbf{b} \cdot \nabla) \, \nabla \frac{1}{|\mathbf{r} - \mathbf{r}'|} \, .$$

Since $\nabla |\mathbf{r} - \mathbf{r}'|^{-1} = -\nabla' |\mathbf{r} - \mathbf{r}'|^{-1}$, this is equal to $(\mathbf{b} \cdot \nabla') \, \nabla' |\mathbf{r} - \mathbf{r}'|^{-1}$, and using $\int_{(V)} d\mathbf{f} \cdot \mathbf{b} \, \mathbf{a} = \int_V dV \, \{ \mathbf{a} \nabla \cdot \mathbf{b} + (\mathbf{b} \cdot \nabla) \, \mathbf{a} \}$ (see p. 17), it therefore delivers

$$\int_V dV' \, \nabla \left(\nabla \cdot \frac{\mathbf{b}}{|\mathbf{r} - \mathbf{r}'|} \right) = \int_V dV' \, (\mathbf{b} \cdot \nabla') \, \nabla' \frac{1}{|\mathbf{r} - \mathbf{r}'|}$$

$$= \int_{(V)} d\mathbf{f}' \cdot \mathbf{b} \, \nabla' \frac{1}{|\mathbf{r} - \mathbf{r}'|} - \int_V dV' \, \nabla' \frac{1}{|\mathbf{r} - \mathbf{r}'|} \, \nabla' \cdot \mathbf{b} \, .$$

Since $\nabla' \cdot \mathbf{b} = \nabla' \cdot (\nabla' \times \mathbf{a}\,(\mathbf{r}')) = 0$, the last integral does not contribute. For the surface integral, we take a sphere with sufficiently large radius r'. Its surface area is $4\pi r'^2$, while $\nabla |\mathbf{r} - \mathbf{r}'|^{-1}$ is equal to r'^{-2} there. Thus we only have to require that $\nabla \times \mathbf{a}$ vanishes at the surface with $r' \to \infty$ and everything is proven.

According to the relation $\Delta|\mathbf{r}-\mathbf{r}'|^{-1} = -4\pi\,\delta(\mathbf{r}-\mathbf{r}')$ just proven, the solution of the inhomogeneous differential equation $\Delta\Phi = \phi(\mathbf{r})$ (*Poisson equation*) can be represented as an integral over the inhomogeneity $\phi(\mathbf{r})$ with suitable weight factor. This is called the *Green function* $G(\mathbf{r},\mathbf{r}')$ of the Laplace operator:

$$\Delta G(\mathbf{r},\mathbf{r}') = \delta(\mathbf{r}-\mathbf{r}') \qquad \Longleftrightarrow \qquad G(\mathbf{r},\mathbf{r}') = \frac{-1}{4\pi}\,\frac{1}{|\mathbf{r}-\mathbf{r}'|}\;.$$

In particular, it yields the solutions of the differential equations

$$\Delta\Phi = \phi(\mathbf{r}) \qquad \text{and} \qquad \Delta\mathbf{A} = \mathbf{a}(\mathbf{r})\;,$$

i.e., of $\nabla\cdot\nabla\Phi = \phi$ and of $\nabla(\nabla\cdot\mathbf{A}) - \nabla\times(\nabla\times\mathbf{A}) = \mathbf{a}$ with $\Phi\sim 0$ and $\mathbf{A}\sim\mathbf{0}$ for $r\to\infty$. In electromagnetism, we shall meet them in the context of the scalar potential (Sect. 3.1.3) and the vector potential (Sect. 3.2.8). These solutions are

$$\Phi(\mathbf{r}) = \int dV'\,G(\mathbf{r},\mathbf{r}')\,\phi(\mathbf{r}') \qquad \text{and} \qquad \mathbf{A}(\mathbf{r}) = \int dV'\,G(\mathbf{r},\mathbf{r}')\,\mathbf{a}(\mathbf{r}')\;.$$

By partial integration, they have the properties

$$\nabla\,\Phi = \int dV'\,G(\mathbf{r},\mathbf{r}')\,\nabla'\phi(\mathbf{r}')\;,$$

$$\nabla\cdot\mathbf{A} = \int dV'\,G(\mathbf{r},\mathbf{r}')\,\nabla'\cdot\mathbf{a}(\mathbf{r}')\;,$$

$$\nabla\times\mathbf{A} = \int dV'\,G(\mathbf{r},\mathbf{r}')\,\nabla'\times\mathbf{a}(\mathbf{r}')\;.$$

Here, we used the fact that Φ and $\mathbf{A}$ vanish at infinity, whence the inhomogeneities ϕ and $\mathbf{a}$ vanish faster by two orders. Thus, if $\mathbf{a}$ is source- or curl-free, the solution $\mathbf{A}$ of the Poisson equation $\Delta\mathbf{A} = \mathbf{a}$ is likewise.

The theorem proven in this section is called the *principal theorem of vector analysis*. It assumes that the source and curl densities are known everywhere—these fix the vector fields.

1.1.13 Vector Fields at Interfaces

If $\nabla\cdot\mathbf{a}$ or $\nabla\times\mathbf{a}$ are different from zero only on a sheet, the volume integrals just mentioned simplify to surface integrals. Correspondingly, instead of $\nabla\cdot\mathbf{a}$ and $\nabla\times\mathbf{a}$, we now introduce the *surface divergence* and *surface rotation*. They have different units from $\nabla\cdot\mathbf{a}$ and $\nabla\times\mathbf{a}$, related to the area instead of the volume:

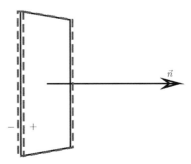

Fig. 1.9 View of a sheet of discontinuity of a vector field. *Dashed red lines* show the envelope

$$\text{Div } \mathbf{a} \equiv \nabla_A \cdot \mathbf{a} \equiv \lim_{V \to 0} \frac{1}{A} \int_{(V)} d\mathbf{f} \cdot \mathbf{a} \,,$$

$$\text{Rot } \mathbf{a} \equiv \nabla_A \times \mathbf{a} \equiv \lim_{V \to 0} \frac{1}{A} \int_{(V)} d\mathbf{f} \times \mathbf{a} \,.$$

Here, V is the volume of a thin layer, covering the latter surface A (see Fig. 1.9). Even though A is infinitesimally small, it nevertheless has dimensions that are large compared with the layer thickness, so only the faces contribute to the surface integrals of the layer. With **n** as unit normal vector to the face, pointing "from minus to plus", we may then write

$$\nabla_A \cdot \mathbf{a} = \mathbf{n} \cdot (\mathbf{a}_+ - \mathbf{a}_-) \,,$$

$$\nabla_A \times \mathbf{a} = \mathbf{n} \times (\mathbf{a}_+ - \mathbf{a}_-) \,.$$

Thus, if the vector field **a** changes in a step-like manner at a sheet (from $\mathbf{a}_-$ to $\mathbf{a}_+$), then for $\delta\mathbf{a} \parallel \mathbf{n}$, it has an area divergence (discontinuous normal component like, e.g., at the interface on the left in Fig. 1.5) and for $\delta\mathbf{a} \perp \mathbf{n}$, it has an area rotation (discontinuous tangential component like, e.g., at the interface on the right in Fig. 1.5).

1.2 Coordinates

1.2.1 Orthogonal Transformations and Euler Angles

In order to perform sums, we now prefer to write $\mathbf{e}_1$, $\mathbf{e}_2$, $\mathbf{e}_3$ instead of $\mathbf{e}_x$, $\mathbf{e}_y$, $\mathbf{e}_z$. In addition, the coordinate origin will be assumed fixed here for every coordinate transformation. Displacements would be easy to include.

For the transition from a Cartesian frame $\{\mathbf{e}_1, \mathbf{e}_2, \mathbf{e}_3\}$ to one rotated about the origin $\{\mathbf{e}_1', \mathbf{e}_2', \mathbf{e}_3'\}$, we have

$$\mathbf{e}_i' = \sum_k (\mathbf{e}_i' \cdot \mathbf{e}_k)\, \mathbf{e}_k \equiv \sum_k D_{ik}\, \mathbf{e}_k$$

and

$$\mathbf{e}_k = \sum_i (\mathbf{e}_k \cdot \mathbf{e}_i')\, \mathbf{e}_i' = \sum_i D_{ik}\, \mathbf{e}_i' \ .$$

Since $\mathbf{e}_k \cdot \mathbf{e}_l = \delta_{kl} = \mathbf{e}_k' \cdot \mathbf{e}_l'$,

$$\sum_i D_{ik}\, D_{il} = \delta_{kl} = \sum_i D_{ki}\, D_{li} \ , \quad \text{and in addition} \quad D_{ik} = D_{ik}{}^* \ .$$

These equations may be written as matrix equations, if we understand D_{ik} as the element of the matrix D in row i and column k. Then, if $\widetilde{D}$ is the transpose of D (with $\widetilde{D}_{ik} = D_{ki}$), we have

$$\widetilde{D}\, D = 1 = D\, \widetilde{D} \quad (\text{so } D^{-1} = \widetilde{D}) \ , \quad \text{and in addition} \quad D = D^* \ .$$

This is called an *orthogonal transformation*. If $D^{-1} = \widetilde{D}^* \equiv D^\dagger$, the transformation is *unitary*. Real unitary transformations are thus orthogonal transformations. Because $\det(D_2 D_1) = \det D_2 \cdot \det D_1$ and $\det \widetilde{D} = \det D$ (see p. 5), orthogonal transformations have $\det D = \pm 1$. Depending on the sign, we distinguish between *proper orthogonal transformations* with

$$\det D = +1$$

and *improper orthogonal transformations* with $\det D = -1$. Only the proper ones are connected continuously to the identity and therefore correspond to rotations. But if we go over from a right- to a left-handed frame, then this is an improper transformation, in particular, $D_{ik} = -\delta_{ik}$, i.e., $D = -1$, corresponds to a *space reflection* (*inversion or parity operation*).

Carrying out two rotations D_1 and D_2 one after the other amounts to doing a single rotation $D = D_2 D_1$, because $\widetilde{D}\, D = \widetilde{D_2 D_1}\, D_2 D_1 = \widetilde{D}_1 \widetilde{D}_2 D_2 D_1 = 1$ and $D\, \widetilde{D} = D_2 D_1 \widetilde{D}_1 \widetilde{D}_2 = 1$. However, the resulting rotation depends on the order, that is, in general $D_1 D_2 \neq D_2 D_1$, e.g., for finite rotations about different axes.

For the Cartesian components of a vector $\mathbf{a}$, we have

$$a_k \equiv \mathbf{e}_k \cdot \mathbf{a} \ , \qquad a_i' \equiv \mathbf{e}_i' \cdot \mathbf{a} = \sum_k D_{ik}\, a_k \ .$$

Instead of going over to a rotated coordinate system, we may also stick with the reference frame and rotate all objects. In both cases we change the Cartesian components of every vector $\mathbf{a}$. However, the rotation of an object through an angle α corresponds to the opposite rotation of the coordinate systems, through the angle $-\alpha$, and

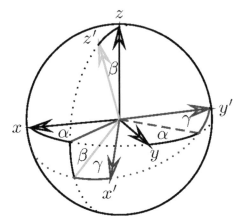

Fig. 1.10 The *Euler angles* α, β, γ, used to describe the transition from unprimed to primed coordinates. The *dashed line* is the *line of nodes* $\mathbf{e}_z \times \mathbf{e}_{z'}$. The sequence is black → blue → green → red. The initial equator is *black* and the last one *red*

vice versa. Therefore, with column matrices A' and A and with the *rotation matrix* D, we write

$$A' = D\,A\,, \quad \text{or} \quad \mathbf{a}' = D\,\mathbf{a}\,.$$

Here, the second equation refers to a rotation of the vectors, because $\mathbf{a}$ and $\mathbf{a}'$ should be fixed independently of the coordinate system. Correspondingly, we may also write the scalar product $\mathbf{a} \cdot \mathbf{b}$ as a matrix product $\widetilde{A}B$ of a row and of a column vector, for which their Cartesian components are necessary. Then we find $\widetilde{DA}DB = \widetilde{A}\widetilde{D}DB = \widetilde{A}B$, implying that $\mathbf{a}' \cdot \mathbf{b}' = \mathbf{a} \cdot \mathbf{b}$, as it should be for a scalar product. (In the next section, we will obtain the scalar product for other coordinate systems.)

Because of $\widetilde{1} = 1$ the requirement $\widetilde{D}D = 1$ constitutes six conditions in three dimensions, and $\frac{1}{2}N(N+1)$ conditions in N dimensions. Consequently, orthogonal transformations in three dimensions depend upon three real parameters. A rotation can be fixed uniquely by specifying these, e.g., by specifying the (axial) rotation vector in the direction of the rotation axis, with value equal to the rotation angle, or by specifying the three *Euler angles* α, β, γ, with which one goes over from the original frame $\{\mathbf{e}_x, \mathbf{e}_y, \mathbf{e}_z\}$ to the rotated one $\{\mathbf{e}_{x'}, \mathbf{e}_{y'}, \mathbf{e}_{z'}\}$ (see Fig. 1.10):

- The first Euler angle α fixes the *azimuth*, i.e., $\{\mathbf{e}_x, \mathbf{e}_y, \mathbf{e}_z\} \to \{\mathbf{e}_{\bar{x}}, \mathbf{e}_{\bar{y}}, \mathbf{e}_{\bar{z}}\}$ with $\mathbf{e}_{\bar{z}} = \mathbf{e}_z$, while the other axes move in a horizontal plane P_1.
- The second Euler angle β describes the *polar distance* (motion of the z-direction), i.e., $\{\mathbf{e}_{\bar{x}}, \mathbf{e}_{\bar{y}}, \mathbf{e}_{\bar{z}}\} \to \{\mathbf{e}_{\bar{x}'}, \mathbf{e}_{\bar{y}'}, \mathbf{e}_{\bar{z}'}\}$, with $\mathbf{e}_{\bar{y}'} = \mathbf{e}_{\bar{y}}$. The new $\mathbf{e}_{\bar{x}'}$ and $\mathbf{e}_{\bar{y}'}$ axes span a plane P_2 inclined at an angle β to the horizontal. The two planes P_1 and P_2 intersect along $\mathbf{e}_{\bar{y}'} = \mathbf{e}_{\bar{y}}$.
- The third Euler angle γ describes the rotation about the new $\bar{z}'$ direction, that is, $\{\mathbf{e}_{\bar{x}'}, \mathbf{e}_{\bar{y}'}, \mathbf{e}_{\bar{z}'}\} \to \{\mathbf{e}_{x'}, \mathbf{e}_{y'}, \mathbf{e}_{z'}\}$, with $\mathbf{e}_{\bar{z}'} = \mathbf{e}_{z'}$, and the other axes moving on the plane P_2. The common axis is along $\mathbf{e}_{\bar{z}'} = \mathbf{e}_{z'}$, the so-called line of nodes.

The first two Euler angles are called the *azimuth* and *polar distance* of the new **z**-axis in the old system, while the third Euler angle gives the angle between the new **y**-axis and the *line of nodes*. This line of nodes forms a right-handed system with the old and the new **z**-axes.

In some cases the Euler angles are defined differently, namely with a left-handed frame or the angles between the line of nodes and the **x**-axes instead of the **y**-axes, but the simple assignment of α to the azimuth of the new **z**-axis is then lost.

We now have

$$D = D_\alpha\, D_\beta\, D_\gamma$$

with

$$D_\alpha \triangleq \begin{pmatrix} \cos\alpha & -\sin\alpha & 0 \\ \sin\alpha & \cos\alpha & 0 \\ 0 & 0 & 1 \end{pmatrix}, \quad D_\beta \triangleq \begin{pmatrix} \cos\beta & 0 & \sin\beta \\ 0 & 1 & 0 \\ -\sin\beta & 0 & \cos\beta \end{pmatrix},$$

and D_γ like D_α, but γ instead of α, because D_α and D_γ describe rotations about the (old) **z**-axis, D_β a rotation about the **y**-axis. If it were the coordinate system that were rotated, then every sine would have the opposite sign, because of the opposite rotation. Of course, starting from the Euler angles, we can evaluate the rotation vector, and vice versa, but we shall not discuss that here. Further properties are derived in Problems 2.1–2.3.

1.2.2 General Coordinates and Their Base Vectors

So far all quantities have been written in a coordinate-free manner as far as possible— Cartesian coordinates and unit vectors have occasionally been useful only for conversions. Sometimes *curvilinear coordinates* are more appropriate, e.g., spherical coordinates (r, θ, φ) or cylindrical coordinates (r, φ, z), where circles also appear as coordinate lines. Still, for these two examples the coordinates are orthogonal to each other everywhere. We are thus dealing here with curvilinear rectangular coordinates. But we would like to allow also for *oblique coordinates*. These are convenient, e.g., for crystallography, and they also provide with a suitable framework for relativity theory. Curvilinear oblique coordinates are what restrict us the least.

Even though a three-dimensional space is assumed throughout the following, most of the discussion can be transferred easily to higher dimensions. We shall hint at the special features of three-dimensional space in the appropriate place, namely, for axial vectors.

As usual, from now on we will write $(x^1, x^2, x^3) = \{x^i\}$ for the coordinate triple of coordinates, despite the risk here of confusing i with a power. In addition, instead of the Cartesian unit vectors, we introduce two sorts of *base vectors*. In crystal physics, $\mathbf{g}_i$ is called a *lattice vector* and $\mathbf{g}^i$ (except for a factor of 2π) a *reciprocal lattice vector*, but restricted to linear coordinates with constant base vectors:

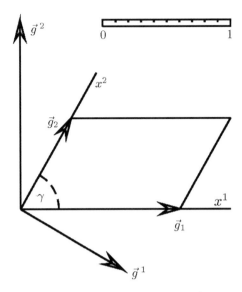

Fig. 1.11 *Oblique coordinates* are indicated here by lines with $\delta x^i = 1$. Shown are their covariant base vectors $\mathbf{g}_i$ and also their contravariant base vectors $\mathbf{g}^i$. If $\mathbf{g}_1$ and $\mathbf{g}_2$ form an angle γ and if these vectors have lengths g_1 and g_2, respectively, then the lengths of the contravariant base vectors are $g^i = 1/(g_i \sin \gamma)$ (from $\mathbf{g}^i \cdot \mathbf{g}_k = \delta^i_k$). Oblique coordinates appear, e.g., if for unequal masses two-body coordinates are transformed to center-of-mass and relative coordinates (see Fig. 2.7)

$$
\begin{aligned}
\text{covariant base vectors} \quad (g\ i\ \text{down}) \quad \mathbf{g}_i &\equiv \frac{\partial \mathbf{r}}{\partial x^i} \,, \\
\text{contravariant base vectors} \quad (g\ i\ \text{up}) \quad \mathbf{g}^i &\equiv \nabla x^i \,.
\end{aligned}
$$

In these equations the index i on the right-hand side is really a lower or upper index.

The *covariant base vector* $\mathbf{g}_i$ is tangent to the coordinate line x^i (all other coordinates remain fixed), and the *contravariant base vector* $\mathbf{g}^i$ is perpendicular to the surface $x^i = $ const. (all other coordinates may change) (see Fig. 1.11). For rectangular coordinates, $\mathbf{g}_i$ and $\mathbf{g}^i$ have the same direction, but for oblique ones, they do not. For rectangular coordinates the two base vectors generally have different lengths. Only for Cartesian coordinates are covariant and contravariant base vectors equal, viz., to the corresponding unit vectors (see Problems 3.10 to 3.12).

The two scalar products

$$
\begin{aligned}
g_{ik} &\equiv \mathbf{g}_i \cdot \mathbf{g}_k = \frac{\partial \mathbf{r}}{\partial x^i} \cdot \frac{\partial \mathbf{r}}{\partial x^k} = g_{ki} \,, \\
g^{ik} &\equiv \mathbf{g}^i \cdot \mathbf{g}^k = \nabla x^i \cdot \nabla x^k = g^{ki} \,,
\end{aligned}
$$

depend on the chosen coordinates (because all base vectors depend on them), but not the scalar products of covariant and contravariant base vectors,

$$\mathbf{g}_i \cdot \mathbf{g}^k = \mathbf{g}^k \cdot \mathbf{g}_i = \nabla x^k \cdot \frac{\partial \mathbf{r}}{\partial x^i} = \frac{\partial x^k}{\partial x^i} = \delta_i^k = \begin{cases} 0 & \text{for } i \neq k \text{ ,} \\ 1 & \text{for } i = k \text{ .} \end{cases}$$

Covariant and contravariant base vectors each form an expansion basis. Therefore, also

$$\mathbf{a} = \sum_i \mathbf{g}_i \, (\mathbf{g}^i \cdot \mathbf{a}) = \sum_i \mathbf{g}^i \, (\mathbf{g}_i \cdot \mathbf{a}) \text{ ,}$$

in particular, $\mathbf{g}^k = \sum_i \mathbf{g}_i \, g^{ik}$, $\mathbf{g}_k = \sum_i \mathbf{g}^i \, g_{ik}$, and

$$\sum_i g_{ik} \, g^{il} = \mathbf{g}_k \cdot \mathbf{g}^l = \delta_k^l \text{ .}$$

This very decisively generalizes the decomposition into Cartesian unit vectors, not only to curvilinear, but also to oblique coordinates. With the useful concepts

covariant component of $\mathbf{a}$: $\quad a_i \equiv \mathbf{g}_i \cdot \mathbf{a}$
and contravariant component of $\mathbf{a}$: $\quad a^i \equiv \mathbf{g}^i \cdot \mathbf{a}$

and with $\mathbf{a} = \sum_i \mathbf{g}_i \, a^i = \sum_i \mathbf{g}^i \, a_i$, we thus obtain

$$a_i = \sum_k g_{ik} \, a^k \text{ ,} \qquad a^i = \sum_k g^{ik} \, a_k \text{ ,} \qquad \text{and} \qquad \mathbf{a} \cdot \mathbf{b} = \sum_i a_i \, b^i \text{ .}$$

Covariant and contravariant components can be converted into each other, referred to as raising and lowering indices. With the scalar product, covariant and contravariant components always appear. We shall always meet sums of products where the index in the factors appears one up and one down. Therefore, we generally use *Einstein's summation convention*, according to which, for these index positions, the summation symbol is left out. This is indeed what we shall do below (from Sect. 3.4.3 on).

1.2.3 Coordinate Transformations

New and old quantities are usually denoted with and without a prime, respectively. In view of various indices being added, a bar will be used instead of the prime in this book.

With a change of coordinates, the behavior depends decisively on the position of the indices. Since $\partial/\partial \bar{x}^i = \sum_k (\partial x^k/\partial \bar{x}^i)\,(\partial/\partial x^k)$, on the one hand, and since we also have $\bar{\mathbf{g}}^i \cdot d\mathbf{r} = d\bar{x}^i = \sum_k (\partial \bar{x}^i/\partial x^k)\,dx^k$, with $dx^k = \mathbf{g}^k \cdot d\mathbf{r}$, on the other, the transition $x^i \to \bar{x}^i$ is connected to the following equations, the order of factors being irrelevant. Here the coefficients form a matrix, the row index being given by the numerator and the column index by the denominator:

$$\bar{\mathbf{g}}_i = \sum_k \mathbf{g}_k \frac{\partial x^k}{\partial \bar{x}^i}, \qquad \bar{\mathbf{g}}^i = \sum_k \frac{\partial \bar{x}^i}{\partial x^k} \mathbf{g}^k,$$

$$\bar{a}_i = \sum_k a_k \frac{\partial x^k}{\partial \bar{x}^i}, \qquad \bar{a}^i = \sum_k \frac{\partial \bar{x}^i}{\partial x^k} a^k.$$

Here, $\bar{a}_i \equiv \mathbf{a} \cdot \bar{\mathbf{g}}_i$ and $\bar{a}^i \equiv \mathbf{a} \cdot \bar{\mathbf{g}}^i$. With the change of coordinates, the base vectors change, but not the other vectors $\mathbf{a}$. Covariant and contravariant quantities have transformation matrices inverse to each other:

$$\sum_k \frac{\partial \bar{x}^i}{\partial x^k} \frac{\partial x^k}{\partial \bar{x}^j} = \frac{\partial \bar{x}^i}{\partial \bar{x}^j} = \delta^i_j.$$

The system of equations $\mathrm{d}\bar{x}^i = \sum_k (\partial \bar{x}^i / \partial x^k)\, \mathrm{d}x^k$ can be written as a matrix equation:

$$\begin{pmatrix} \mathrm{d}\bar{x}^1 \\ \mathrm{d}\bar{x}^2 \\ \mathrm{d}\bar{x}^3 \end{pmatrix} = \begin{pmatrix} \dfrac{\partial \bar{x}^1}{\partial x^1} & \dfrac{\partial \bar{x}^1}{\partial x^2} & \dfrac{\partial \bar{x}^1}{\partial x^3} \\[2mm] \dfrac{\partial \bar{x}^2}{\partial x^1} & \dfrac{\partial \bar{x}^2}{\partial x^2} & \dfrac{\partial \bar{x}^2}{\partial x^3} \\[2mm] \dfrac{\partial \bar{x}^3}{\partial x^1} & \dfrac{\partial \bar{x}^3}{\partial x^2} & \dfrac{\partial \bar{x}^3}{\partial x^3} \end{pmatrix} \begin{pmatrix} \mathrm{d}x^1 \\ \mathrm{d}x^2 \\ \mathrm{d}x^3 \end{pmatrix}.$$

The transformation matrix is called the *Jacobi matrix* or *functional matrix*. Naturally, it also exists for space dimensions other than three.

For two successive transformations, the two associated Jacobi matrices can be combined in a single product matrix. If the second transformation is the transformation back to the original coordinates, then the result is the unit matrix: the inverse transformation is described by the inverse matrix. This exists only if the *Jacobi determinant* (*functional determinant*), viz.,

$$\frac{\partial\,(\bar{x}^1, \bar{x}^2, \bar{x}^3)}{\partial\,(x^1, x^2, x^3)} \equiv \begin{vmatrix} \dfrac{\partial \bar{x}^1}{\partial x^1} & \dfrac{\partial \bar{x}^1}{\partial x^2} & \dfrac{\partial \bar{x}^1}{\partial x^3} \\[2mm] \dfrac{\partial \bar{x}^2}{\partial x^1} & \dfrac{\partial \bar{x}^2}{\partial x^2} & \dfrac{\partial \bar{x}^2}{\partial x^3} \\[2mm] \dfrac{\partial \bar{x}^3}{\partial x^1} & \dfrac{\partial \bar{x}^3}{\partial x^2} & \dfrac{\partial \bar{x}^3}{\partial x^3} \end{vmatrix},$$

does not vanish, and likewise the determinant of the inverse Jacobi matrix, because the two coordinate systems should be treated on an equal footing.

1.2.4 The Concept of a Tensor

We generalize the expressions derived so far for a vector field and denote as a *tensor* of rank $n + m$ (with n covariant and m contravariant indices) a quantity whose components transform under a change of coordinates according to

$$\bar{T}^{i_1...i_m}_{k_1...k_n} = \sum_{j_1...l_n} \frac{\partial \bar{x}^{i_1}}{\partial x^{j_1}} \cdots \frac{\partial \bar{x}^{i_m}}{\partial x^{j_m}} \frac{\partial x^{l_1}}{\partial \bar{x}^{k_1}} \cdots \frac{\partial x^{l_n}}{\partial \bar{x}^{k_n}} T^{j_1...j_m}_{l_1...l_n} .$$

Scalars are tensors of zeroth rank and *vectors* are tensors of first rank. If $T(x)$ is a scalar field, then the new function $\bar{T}(\bar{x})$ should have the same value for the coordinates $\bar{x}$ as the old function $T(x)$ for the old coordinates $x = f(\bar{x})$, whence we should have $\bar{T}(\bar{x}) = T(f(\bar{x}))$ without further transformation matrices. In contrast, for a gradient field with $\nabla T_i \equiv \nabla T \cdot \mathbf{g}_i$, because $\mathbf{g}_i = \partial \mathbf{r}/\partial x^i$ and $\nabla T_i = \partial T/\partial x^i$, we have

$$\nabla \bar{T}_k \equiv \frac{\partial \bar{T}(\bar{x})}{\partial \bar{x}^k} = \sum_i \frac{\partial T(x)}{\partial x^i} \frac{\partial x^i}{\partial \bar{x}^k} \equiv \sum_i \nabla T_i \frac{\partial x^i}{\partial \bar{x}^k} ,$$

showing that this is a vector field.

Tensors of the same type can be added, and the (tensor) product of a tensor of nth rank with a tensor of mth rank is a tensor of rank $n + m$:

$$T^{i_1...i_n} T^{k_1...k_m} = T^{i_1...i_n k_1...k_m} .$$

Of course, some covariant components may occur on the left- and right-hand sides. But one can also lower the tensorial rank by *contracting* the tensor:

$$\sum_i T^{i\, i_1...i_m}_{i\, k_1...k_n} = T^{i_1...i_m}_{k_1...k_n} ,$$

because covariant and contravariant components transform inversely to each other. (Here, too, the summation symbol is often left out, using the Einstein summation convention.) A special case of this is the scalar product of two vectors,

$$\sum_i a_i b^i = \mathbf{a} \cdot \mathbf{b} = \sum_i a^i b_i = \sum_i \bar{a}_i \bar{b}^i .$$

Generally, a tensor of nth rank can be contracted with n vectors to produce a scalar. This fixes tensors in a coordinate-free way. In Sect. 2.2.10, for example, we shall introduce the moment of inertia I, which is a tensor of second rank. The tensor product $I\omega$ delivers the vector $\mathbf{L}$ (angular momentum) and $\frac{1}{2}\omega \cdot \mathbf{L}$ a scalar (kinetic energy), where I is contracted twice with the vector ω.

The *trace* of a square matrix is the sum of its diagonal elements: $\sum_i I^i{}_i = \operatorname{tr} I$, which is the contraction of a tensor of second rank to a scalar. In fact, $\operatorname{tr} I$ remains unchanged under a change of coordinates.

The change of coordinates under a rotation on p. 30 led to the matrix equation $A' = DA$ for a column vector A. Correspondingly, $\mathbf{L} = I\omega$ reads $L = I\Omega$ as a matrix equation where L and Ω are column matrices and I is a square matrix. For a rotation we have $L' = DL$, $\Omega' = D\Omega$, and $\Omega = D^{-1}\Omega'$, respectively, and hence $L' = DID^{-1}\Omega'$, so $L' = I'\Omega'$ with $I' = DID^{-1}$. Here we now write $L^i = \sum_k I^i{}_k \omega^k$ and

$$\bar{I}^i{}_k = \sum_{jl} \frac{\partial \bar{x}^i}{\partial x^j} \frac{\partial x^l}{\partial \bar{x}^k} I^j{}_l \,, \quad \text{with} \quad \sum_j \frac{\partial \bar{x}^i}{\partial x^j} \frac{\partial x^j}{\partial \bar{x}^k} = \delta^i{}_k \,.$$

The last equation corresponds to $DD^{-1} = 1$.

The quantities g^{ik} and g_{ik} introduced above are tensors of second rank. Since

$$d\mathbf{r} \cdot d\mathbf{r} = \sum_{ik} \frac{\partial \mathbf{r}}{\partial x^i} \cdot \frac{\partial \mathbf{r}}{\partial x^k} dx^i\, dx^k = \sum_{ik} g_{ik}\, dx^i\, dx^k \,,$$

we call (g_{ik}) the *metric tensor*. The matrices (g_{ik}) and (g^{ik}) are diagonal for rectangular coordinates, but not for oblique coordinates. With Cartesian coordinates, they are unit matrices.

The indices of all tensors can be raised or lowered using the tensors g_{ik} and g^{ik}, as we have seen already in Sect. 1.2.2 for vectors. Similarly,

$$T^{ik} = \sum_j g^{ij} T_j{}^k = \sum_{jl} g^{ij} g^{kl} T_{jl} \,,$$

and similarly, $T_{ik} = \sum_{jl} g_{ij} g_{kl} T^{jl}$.

If an equation holds in Cartesian coordinates and if it holds as a tensor equation, then it holds also in general coordinates. If a tensor of second rank is symmetric or antisymmetric, $T^{ik} = \pm T^{ki}$, then it has this property in every coordinate system.

The (scalar) triple product of the three base vectors $\mathbf{g}_1, \mathbf{g}_2, \mathbf{g}_3$ is denoted by ε_{123}. Generally, we have

$$\varepsilon_{ijk} \equiv \mathbf{g}_i \cdot (\mathbf{g}_j \times \mathbf{g}_k) = \frac{\partial \mathbf{r}}{\partial x^i} \cdot \left(\frac{\partial \mathbf{r}}{\partial x^j} \times \frac{\partial \mathbf{r}}{\partial x^k} \right) = \frac{\partial(x, y, z)}{\partial(x^i, x^j, x^k)} \,.$$

This is the *totally anti-symmetric (Levi-Civita) tensor* of third rank. Under a change of coordinates, ε_{ijk} transforms like a tensor with three lower indices and changes sign for the interchange of two indices. Therefore, we only need to evaluate ε_{123}. This component can be traced back to the determinant of (g_{ik}) because, according to p. 5, we have

$$\{\mathbf{g}_i \cdot (\mathbf{g}_j \times \mathbf{g}_k)\}^2 = \begin{vmatrix} \mathbf{g}_i \cdot \mathbf{g}_i & \mathbf{g}_i \cdot \mathbf{g}_j & \mathbf{g}_i \cdot \mathbf{g}_k \\ \mathbf{g}_j \cdot \mathbf{g}_i & \mathbf{g}_j \cdot \mathbf{g}_j & \mathbf{g}_j \cdot \mathbf{g}_k \\ \mathbf{g}_k \cdot \mathbf{g}_i & \mathbf{g}_k \cdot \mathbf{g}_j & \mathbf{g}_k \cdot \mathbf{g}_k \end{vmatrix} .$$

The (scalar) triple product of three real vectors is always real, and only zero if they are coplanar (in which case the coordinates would be useless). Therefore, the determinant is positive. We thus have

$$\varepsilon_{123} = \pm\sqrt{g} , \quad \text{with} \quad g \equiv \det(g_{ik}) > 0 ,$$

where the plus sign corresponds to a right-handed coordinate system and the minus sign to a left-handed one. (In particular, for a "reflection at the origin", i.e., for $x^i \to -x^i$ for all i, the sign of ε_{123} switches.) In addition,

$$\varepsilon^{ijk} = \mathbf{g}^i \cdot (\mathbf{g}^j \times \mathbf{g}^k) = \frac{\partial(x^i, x^j, x^k)}{\partial(x, y, z)} ,$$

and hence, according to p. 5,

$$\varepsilon_{ijk}\, \varepsilon^{lmn} = \begin{vmatrix} \delta_i^l & \delta_i^m & \delta_i^n \\ \delta_j^l & \delta_j^m & \delta_j^n \\ \delta_k^l & \delta_k^m & \delta_k^n \end{vmatrix} .$$

We deduce that $\varepsilon_{123}\, \varepsilon^{123} = 1$, but also

$$\sum_i \varepsilon_{ijk}\, \varepsilon^{imn} = \begin{vmatrix} \delta_j^m & \delta_j^n \\ \delta_k^m & \delta_k^n \end{vmatrix}$$

and

$$\sum_{ij} \varepsilon_{ijk}\, \varepsilon^{ijn} = 2\, \delta_k^n .$$

This equation is often useful.

The last paragraph is true only in three-dimensional space. Only there is the vector product determined uniquely—otherwise the direction perpendicular to two given directions is not determined. (But a totally antisymmetric tensor can also be introduced for spaces of different dimensions via the functional determinant.)

Hence, in three dimensions we have

$$\mathbf{g}_k \times \mathbf{g}_l = \sum_i \mathbf{g}^i\, \varepsilon_{ikl} \quad \text{and} \quad \mathbf{a} \times \mathbf{b} = \sum_{ikl} \mathbf{g}^i\, a^k\, b^l\, \varepsilon_{ikl} .$$

The volume element is the parallelepiped spanned by the line elements $(\partial \mathbf{r}/\partial x^i)\,dx^i$,

$$dV = |\mathbf{g}_1 \cdot (\mathbf{g}_2 \times \mathbf{g}_3)\,dx^1\,dx^2\,dx^3| = |\varepsilon_{123}\,dx^1\,dx^2\,dx^3| = \sqrt{g}\,|dx^1\,dx^2\,dx^3|$$
$$= \left|\frac{\partial (x,y,z)}{\partial (x^1,x^2,x^3)}\,dx^1\,dx^2\,dx^3\right|.$$

In addition to $|dx^1\,dx^2\,dx^3|$, the functional determinant of the associated coordinates appears.

The area element $d\mathbf{f}(1)$ is related to the vector $\mathbf{g}^1$ which is perpendicular to the area $x^1 = \text{const.}$ of the parallelepiped. Its scalar product with the vector $\mathbf{g}_1\,dx^1$ results in $\varepsilon_{123}\,dx^1\,dx^2\,dx^3$. Hence, we infer that

$$d\mathbf{f}(1) = \mathbf{g}_2 \times \mathbf{g}_3\,dx^2\,dx^3 = \varepsilon_{123}\,\mathbf{g}^1\,dx^2\,dx^3 ,$$

with the value $df(1) = \sqrt{g\,g^{11}}\,|dx^2\,dx^3|$. As we shall soon see, these expressions are useful for vector analysis—and, by the way, also for relativity theory. (Of course, cyclic permutation of the three numbers 1, 2, 3 is allowed in this paragraph.)

1.2.5 Gradient, Divergence, and Rotation in General Coordinates

For general coordinates, we find the expressions

$$\nabla\psi = \sum_i \mathbf{g}^i\,(\mathbf{g}_i \cdot \nabla\psi) = \sum_i \mathbf{g}^i\,\frac{\partial\psi}{\partial x^i} ,$$
$$\nabla \cdot \mathbf{a} = \lim_{V\to 0}\frac{1}{V}\int_{(V)} d\mathbf{f}\cdot\mathbf{a} = \frac{1}{\sqrt{g}}\sum_i \frac{\partial}{\partial x^i}\left(\sqrt{g}\,a^i\right) ,$$
$$\Delta\psi = \frac{1}{\sqrt{g}}\sum_{ik}\frac{\partial}{\partial x^i}\left(\sqrt{g}\,g^{ik}\,\frac{\partial\psi}{\partial x^k}\right).$$

However, the corresponding surface integrals for gradient and rotation are not useful here, because $d\mathbf{f}(i)$ can change its direction. Therefore, for the still missing curl density, we start from Stokes's theorem, viz., $\int d\mathbf{f}\cdot(\nabla\times\mathbf{a}) = \oint d\mathbf{r}\cdot\mathbf{a} = \oint\sum_i a_i\,dx^i$, and hence infer the equation $\sqrt{g}\,\mathbf{g}^1\cdot(\nabla\times\mathbf{a}) = \partial a_3/\partial x^2 - \partial a_2/\partial x^3$. Since $\sqrt{g}\,\varepsilon^{123} = -\sqrt{g}\,\varepsilon^{132} = 1$, we may also write $\sqrt{g}\sum_{kl}\varepsilon^{1kl}\,\partial a_l/\partial x^k$ for the right-hand side:

$$\nabla\times\mathbf{a} = \sum_{ikl}\varepsilon^{ikl}\,\mathbf{g}_i\,\frac{\partial a_l}{\partial x^k} .$$

Now we have all the quantities mentioned in the title (and the Laplace operator) in general coordinates.

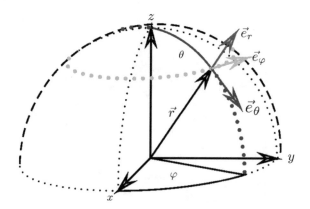

Fig. 1.12 Spherical coordinates r, θ, φ and their unit vectors, with $\mathbf{g}_1 = \mathbf{e}_r = \mathbf{r}/r$ (*red*), $\mathbf{g}_2 = r\,\mathbf{e}_\theta$ (*blue*), and $\mathbf{g}_3 = r\sin\theta\,\mathbf{e}_\varphi$ (*green*). Here, the angles φ and θ correspond to the "meridian" and "latitude", respectively, in geodesy. However, the *polar distance* θ is measured from the north pole (always positive), and the "latitude" from the equator

For rectangular coordinates, much is simplified here. In particular, (g_{ik}) and (g^{ik}) are diagonal, and $\mathbf{g}_i$ and $\mathbf{g}^i$ have the same direction $\mathbf{e}_i$. Only their lengths are different:

$$\mathbf{e}_i = \frac{\mathbf{g}_i}{g_i} = \mathbf{g}^i\,g_i\,, \quad \text{with}\quad g_i{}^2 = g_{ii} = \frac{1}{g^{ii}}\quad \text{and}\quad g_i > 0\,.$$

Hence, $\mathbf{dr} = \sum_i \mathbf{e}_i\,g_i\,dx^i$ and $\sqrt{g} = g_1\,g_2\,g_3$, together with $a_i = (\mathbf{a}\cdot\mathbf{e}_i)\,g_i$ and $a^i = (\mathbf{a}\cdot\mathbf{e}_i)\,/\,g_i$. We thus obtain

$$\nabla\psi = \sum_i \mathbf{e}_i\,\frac{1}{g_i}\,\frac{\partial\psi}{\partial x^i}\,,$$

$$\nabla\cdot\mathbf{a} = \frac{1}{g_1 g_2 g_3}\sum_i \frac{\partial\,g_1 g_2 g_3\,a^i}{\partial x^i}\,,$$

$$\Delta\psi = \frac{1}{g_1 g_2 g_3}\sum_i \frac{\partial}{\partial x^i}\frac{g_1 g_2 g_3}{g_i{}^2}\frac{\partial\psi}{\partial x^i}\,,$$

$$\mathbf{e}_1\cdot(\nabla\times\mathbf{a}) = \frac{1}{g_2 g_3}\left(\frac{\partial a_3}{\partial x^2} - \frac{\partial a_2}{\partial x^3}\right)\quad \text{(and cyclic permutations).}$$

The most important examples are, on the one hand, *spherical coordinates*, for which $\mathbf{dr} = \mathbf{e}_r\,dr + \mathbf{e}_\theta\,r\,d\theta + \mathbf{e}_\varphi\,r\sin\theta\,d\varphi$ (see Fig. 1.12):

$$\nabla\psi = \mathbf{e}_r\,\frac{\partial\psi}{\partial r} + \mathbf{e}_\theta\,\frac{1}{r}\,\frac{\partial\psi}{\partial\theta} + \mathbf{e}_\varphi\,\frac{1}{r\sin\theta}\,\frac{\partial\psi}{\partial\varphi}\,,$$

$$\nabla\cdot\mathbf{a} = \frac{1}{r^2}\,\frac{\partial r^2 a_r}{\partial r} + \frac{1}{r\sin\theta}\,\frac{\partial\sin\theta\,a_\theta}{\partial\theta} + \frac{1}{r\sin\theta}\,\frac{\partial a_\varphi}{\partial\varphi}\,,$$

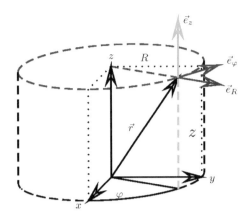

Fig. 1.13 Cylindrical coordinates R, φ, z and their unit vectors. Instead of the Cartesian coordinates x, y, the *polar coordinates* R and φ appear, so $\mathbf{g}_3 = \mathbf{e}_z$ (*green*), $\mathbf{g}_1 = \mathbf{e}_R = \mathbf{R}/R$ (*red*), and $\mathbf{g}_2 = R\,\mathbf{e}_\varphi = \mathbf{e}_z \times \mathbf{R}$ (*blue*)

$$\Delta \psi = \frac{1}{r^2} \frac{\partial}{\partial r} r^2 \frac{\partial \psi}{\partial r} + \frac{1}{r^2 \sin\theta} \frac{\partial}{\partial \theta} \sin\theta \frac{\partial \psi}{\partial \theta} + \frac{1}{r^2 \sin^2\theta} \frac{\partial^2 \psi}{\partial \varphi^2},$$

$$\nabla \times \mathbf{a} = \mathbf{e}_r \frac{1}{r \sin\theta} \left(\frac{\partial \sin\theta\, a_\varphi}{\partial \theta} - \frac{\partial a_\theta}{\partial \varphi} \right) + \mathbf{e}_\theta \frac{1}{r} \left(\frac{1}{\sin\theta} \frac{\partial a_r}{\partial \varphi} - \frac{\partial r a_\varphi}{\partial r} \right)$$

$$+ \mathbf{e}_\varphi \frac{1}{r} \left(\frac{\partial r a_\theta}{\partial r} - \frac{\partial a_r}{\partial \theta} \right),$$

and on the other hand, *cylindrical coordinates*, for which $d\mathbf{r} = \mathbf{e}_R\, dR + \mathbf{e}_\varphi\, R\, d\varphi + \mathbf{e}_z\, dz$ (see Fig. 1.13):

$$\nabla \psi = \mathbf{e}_R \frac{\partial \psi}{\partial R} + \mathbf{e}_\varphi \frac{1}{R} \frac{\partial \psi}{\partial \varphi} + \mathbf{e}_z \frac{\partial \psi}{\partial z},$$

$$\nabla \cdot \mathbf{a} = \frac{1}{R} \frac{\partial R\, a_R}{\partial R} + \frac{1}{R} \frac{\partial a_\varphi}{\partial \varphi} + \frac{\partial a_z}{\partial z},$$

$$\Delta \psi = \frac{1}{R} \frac{\partial}{\partial R} R \frac{\partial \psi}{\partial R} + \frac{1}{R^2} \frac{\partial^2 \psi}{\partial \varphi^2} + \frac{\partial^2 \psi}{\partial z^2},$$

$$\nabla \times \mathbf{a} = \mathbf{e}_R \left(\frac{1}{R} \frac{\partial a_z}{\partial \varphi} - \frac{\partial a_\varphi}{\partial z} \right) + \mathbf{e}_\varphi \left(\frac{\partial a_R}{\partial z} - \frac{\partial a_z}{\partial R} \right) + \mathbf{e}_z \frac{1}{R} \left(\frac{\partial R\, a_\varphi}{\partial R} - \frac{\partial a_R}{\partial \varphi} \right).$$

In many cases, the fields ψ or $\mathbf{a}$ depend *only* on r (*isotropy*) or R (*cylindrical symmetry*), respectively—then we need only ordinary derivatives in spherical or cylindrical coordinates, instead of partial derivatives.

For rectilinear coordinates there are also simplifications with constant base vectors, because then g remains the same everywhere:

$$(\nabla \psi)_i = \frac{\partial \psi}{\partial x^i} \ ,$$

$$\nabla \cdot \mathbf{a} = \sum_i \frac{\partial a^i}{\partial x^i} \ ,$$

$$\Delta \psi = \sum_{ik} g^{ik} \frac{\partial^2 \psi}{\partial x^i \, \partial x^k} \ ,$$

$$(\nabla \times \mathbf{a})^i = \sum_{kl} \varepsilon^{ikl} \frac{\partial a_l}{\partial x^k} \ , \qquad \text{e.g.,} \quad (\nabla \times \mathbf{a})^1 = \frac{1}{\sqrt{g}} \left(\frac{\partial a_3}{\partial x^2} - \frac{\partial a_2}{\partial x^3} \right).$$

The next section should only be read by those who want to enter into more detail—it is not needed to understand the following. Section 1.2.7 will be important only for thermodynamics.

1.2.6 Tensor Extension, Christoffel Symbols

In deriving a gradient field from a scalar field, the rank of a tensor increases by one. This *tensor extension* through differentiation also arises for tensors of higher rank, but in this case variable base vectors require additional terms. In particular, we have

$$\frac{\partial \mathbf{g}_k}{\partial x^l} = \sum_i \mathbf{g}^i \left(\mathbf{g}_i \cdot \frac{\partial \mathbf{g}_k}{\partial x^l} \right) = \sum_i \mathbf{g}^i \, \{kl, i\}$$

$$= \sum_i \mathbf{g}_i \left(\mathbf{g}^i \cdot \frac{\partial \mathbf{g}_k}{\partial x^l} \right) = \sum_i \mathbf{g}_i \left\{ \begin{matrix} i \\ kl \end{matrix} \right\} \ ,$$

with the *Christoffel symbols of the first kind*

$$\{kl, i\} \equiv \frac{\partial \mathbf{g}_k}{\partial x^l} \cdot \mathbf{g}_i = \frac{\partial^2 \mathbf{r}}{\partial x^k \, \partial x^l} \cdot \mathbf{g}_i = \frac{\partial g_{ik}}{\partial x^l} - \mathbf{g}_k \cdot \frac{\partial \mathbf{g}_i}{\partial x^l} = \{lk, i\}$$

$$= \frac{\partial g_{ik}}{\partial x^l} - \{il, k\} = \frac{1}{2} \left(\frac{\partial g_{ik}}{\partial x^l} + \frac{\partial g_{il}}{\partial x^k} - \frac{\partial g_{kl}}{\partial x^i} \right)$$

and the *Christoffel symbols of the second kind*

$$\left\{ \begin{matrix} i \\ kl \end{matrix} \right\} \equiv \frac{\partial \mathbf{g}_k}{\partial x^l} \cdot \mathbf{g}^i = \frac{\partial^2 \mathbf{r}}{\partial x^k \, \partial x^l} \cdot \mathbf{g}^i = \left\{ \begin{matrix} i \\ lk \end{matrix} \right\} = \sum_j g^{ij} \, \{kl, j\} \ .$$

Despite the last equation, the new symbols are generally not tensors of third rank, because they contain second derivatives. Therefore, we shall avoid the notations Γ_{ikl} for $\{kl, i\}$ and Γ^i_{kl} for $\{^i_{kl}\}$.

From these equations, it follows immediately that

$$\frac{1}{2}\frac{\partial g_{kk}}{\partial x^l} = \{lk,\,k\} = \{kl,\,k\}\,.$$

For rectangular coordinates, all g_{ik} with $i \neq k$ vanish. If $\frac{1}{2}\,\partial g_{kk}/\partial x^l = g_{kk}\,\{^k_{kl}\}$ holds, and in addition for $k \neq l$, this is equal to $-\{kk,\,l\} = -g_{ll}\,\{^l_{kk}\}$, because $g_{kl} = 0$. Since $\mathbf{g}_i \cdot \mathbf{g}^k$ is constant, we have finally

$$\frac{\partial \mathbf{g}^k}{\partial x^l} = \sum_i \mathbf{g}^i\left(\mathbf{g}_i \cdot \frac{\partial \mathbf{g}^k}{\partial x^l}\right) = -\sum_i \mathbf{g}^i\left(\frac{\partial \mathbf{g}_i}{\partial x^l}\cdot\mathbf{g}^k\right) = -\sum_i \mathbf{g}^i\,\{^k_{il}\}\,.$$

For the derivatives of the vector field, we have

$$\frac{\partial \mathbf{a}}{\partial x^l} = \sum_k \mathbf{g}^k\left(\mathbf{g}_k \cdot \frac{\partial \mathbf{a}}{\partial x^l}\right) = \sum_k \mathbf{g}_k\left(\mathbf{g}^k \cdot \frac{\partial \mathbf{a}}{\partial x^l}\right)\,.$$

These coefficients are referred to as *covariant derivatives*:

$$a_{k;l} \equiv \mathbf{g}_k \cdot \frac{\partial \mathbf{a}}{\partial x^l} = \frac{\partial a_k}{\partial x^l} - \mathbf{a}\cdot\frac{\partial \mathbf{g}_k}{\partial x^l} = \frac{\partial a_k}{\partial x^l} - \sum_i a_i\,\{^i_{kl}\}\,,$$

$$a^k_{\;;l} \equiv \mathbf{g}^k \cdot \frac{\partial \mathbf{a}}{\partial x^l} = \frac{\partial a^k}{\partial x^l} - \mathbf{a}\cdot\frac{\partial \mathbf{g}^k}{\partial x^l} = \frac{\partial a^k}{\partial x^l} + \sum_i a^i\,\{^k_{il}\}\,.$$

They are clearly tensors of second rank, obtained by differentiation from tensors of first rank.

These observations can be applied to the velocity and acceleration. Since

$$\frac{d\mathbf{g}_j}{dt} = \sum_i \frac{\partial \mathbf{g}_j}{\partial x^i}\frac{dx^i}{dt} = \sum_{ik} \mathbf{g}_k\,\{^k_{ij}\}\frac{dx^i}{dt}\,,$$

we obtain

$$\mathbf{v} \equiv \frac{d\mathbf{r}}{dt} = \sum_k \frac{\partial \mathbf{r}}{\partial x^k}\frac{dx^k}{dt} = \sum_k \mathbf{g}_k\frac{dx^k}{dt} \quad\Longrightarrow\quad v^k = \frac{dx^k}{dt}\,,$$

and since $\mathbf{a} = \dot{\mathbf{v}} = \sum_k(\mathbf{g}_k\ddot{x}^k + \dot{\mathbf{g}}_k\dot{x}^k)$, we find

$$a^k = \frac{d^2 x^k}{dt^2} + \sum_{ij}\{^k_{ij}\}\frac{dx^i}{dt}\frac{dx^j}{dt}\,.$$

For motion along the coordinate line x^k, the first term $\ddot{x}^k$ here describes the tangential accelerations and the rest the normal accelerations. This decomposition was already explained on p. 7.

1.2.7 Reformulation of Partial Differential Quotients

In the analysis of functions of multiple variables, partial derivatives appear. Here we restrict to two variables for reasons of simplicity, but generalization is straightforward. The main interest here is in the transformation to new variables (coordinates).

For a function f of the two variables x and y, we have

$$df(x, y) = \frac{\partial f(x, y)}{\partial x}\, dx + \frac{\partial f(x, y)}{\partial y}\, dy \ .$$

It is common to leave out the arguments of f and instead attach the fixed parameter to the differential quotient as a lower index. Hence the equation appears in the form

$$df = \left(\frac{\partial f}{\partial x}\right)_y dx + \left(\frac{\partial f}{\partial y}\right)_x dy \ , \quad \text{with} \quad \left(\frac{\partial f}{\partial x}\right)_f = 0 \ \text{and} \ \left(\frac{\partial f}{\partial f}\right)_x = 1 \ .$$

From here various relations can be derived.

If we divide by df and form the limit $df \to 0$ with constant y or x, respectively, i.e., for $dy = 0$ or $dx = 0$, respectively, then

$$1 = \left(\frac{\partial f}{\partial x}\right)_y \left(\frac{\partial x}{\partial f}\right)_y = \left(\frac{\partial f}{\partial y}\right)_x \left(\frac{\partial y}{\partial f}\right)_x \ .$$

The derivative of a function is thus equal to the reciprocal of the derivative of the inverse function, as suggested by the notation (due to Leibniz). On the other hand, if we divide by dy and form the limit $dy \to 0$ for fixed f, we have $0 = (\partial f/\partial x)_y (\partial x/\partial y)_f + (\partial f/\partial y)_x$, whence the noteworthy equation

$$\left(\frac{\partial f}{\partial y}\right)_x = -\left(\frac{\partial f}{\partial x}\right)_y \left(\frac{\partial x}{\partial y}\right)_f \ .$$

We thus see that the fixed and the changed variable can be exchanged. This equation may also be written in the form $(\partial f/\partial x)_y (\partial x/\partial y)_f (\partial y/\partial f)_x = -1$ if we consider the reciprocal of $(\partial f/\partial y)_x$.

If we replace a variable with a new one, e.g., y with $g(x, y)$, then from $df = (\partial f/\partial x)_g\, dx + (\partial f/\partial g)_x \{(\partial g/\partial x)_y dx + (\partial g/\partial y)_x dy\}$, we may deduce the two important equations

$$\left(\frac{\partial f}{\partial x}\right)_y = \left(\frac{\partial f}{\partial x}\right)_g + \left(\frac{\partial f}{\partial g}\right)_x \left(\frac{\partial g}{\partial x}\right)_y$$

and

$$\left(\frac{\partial f}{\partial y}\right)_x = \left(\frac{\partial f}{\partial g}\right)_x \left(\frac{\partial g}{\partial y}\right)_x \ .$$

According to the first equation, the fixed variable can be changed. The second corresponds to the chain rule for ordinary derivatives.

In the last product, if we swap the fixed and adjustable pair of variables and then apply the chain rule twice, it follows that

$$\left(\frac{\partial f}{\partial g}\right)_x \left(\frac{\partial g}{\partial f}\right)_y = \left(\frac{\partial x}{\partial y}\right)_f \left(\frac{\partial y}{\partial x}\right)_g .$$

Here, the pair (f, g) is exchanged with the pair (x, y). By the way,

$$\left(\frac{\partial f}{\partial g}\right)_x \left(\frac{\partial g}{\partial f}\right)_y + \left(\frac{\partial y}{\partial g}\right)_x \left(\frac{\partial g}{\partial y}\right)_f = 1 .$$

For the proof, we can trace $(\partial f/\partial g)_x$ back to $(\partial f/\partial g)_y$ and then exploit the equations above.

If in $(\partial f/\partial g)_x$ we use the chain rule with the variable y and then in $(\partial y/\partial g)_x$ exchange the fixed and the adjustable variable, we also have

$$\left(\frac{\partial f}{\partial g}\right)_x \left(\frac{\partial g}{\partial x}\right)_y = -\left(\frac{\partial f}{\partial y}\right)_x \left(\frac{\partial y}{\partial x}\right)_g .$$

This corresponds to the replacement $y \leftrightarrow g$. In addition,

$$\left(\frac{\partial f}{\partial g}\right)_x \left(\frac{\partial g}{\partial y}\right)_f = -\left(\frac{\partial f}{\partial x}\right)_g \left(\frac{\partial x}{\partial y}\right)_f .$$

This can be understood by replacing $x \leftrightarrow g$ in the first factor.

1.3   Measurements and Errors

1.3.1   Introduction

The search for laws prepares the ground on which the principles of nature are built. We generalize by relating comparable things. Of course, this has its limitations. When are two things equal to each other, and when are they only similar? The following is important for all measurements, but also for quantum theory and for thermodynamics and statistics.

We consider an arbitrary physical quantity which we assume does not change with time and can be measured repeatedly, e.g., the length of a rod or the oscillation period of a pendulum. Each measurement is carried out in terms of a "multiple of a scale unit". It may be that a tenth of the unit can be estimated, but certainly not essentially finer divisions. An uncertainty is therefore attached to each of our measured values, and this uncertainty can be estimated rather simply.

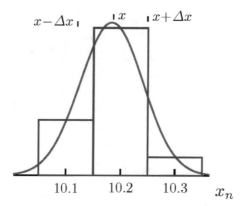

Fig. 1.14 Frequency distribution of the measurement series $\{x_n\}$ mentioned in the text. The more often the same value is measured, the higher the associated column (*blue*). The adjusted *red* bell-shaped curve is symmetrical with respect to the mean value ($x = 10.183$), and the half-width ($\Delta x = 0.058$) corresponds to the "measurement error"

It is more difficult to find a statement about how well an instrument is adjusted and whether there are further systematic errors. We will not deal with these questions here, but we do want to be able to estimate the bounds on the error from the statistical fluctuations of our measured data.

In particular, if we repeat our measurement in order to ensure against erroneous readings, then the values x_n ($n \in \{1, \ldots, N\}$) may not all be equal, e.g., we may find three times 10.1 scale units (that is, three values with $10.05 < x_n < 10.15$), eight times 10.2 (eight values with $10.15 < x_n < 10.25$), and one 10.3 (with $10.25 < x_n < 10.35$) in an arbitrary order. Apparently, there are always "measurement errors", the origin of which we do not know. (Systematic errors can be estimated separately.) Therefore, we have to assign a greater uncertainty than the assumed scale fineness to the results of our measurements.

Hence, from the N readings $\{x_n\}$ of our measurement, we would like to determine a *measurement result with error estimate* in the form $x \pm \Delta x$. For the example mentioned, the result is 10.183 ± 0.058, as will be shown shortly, often abbreviated to 10.183(58). This example is shown in Fig. 1.14. The error estimate here presents only a frame for the actual error: improved measurement readings may also lie outside the error limits given previously. If we compare, e.g., the error analysis for the fundamental constants of the year 1999 (see p. 623) with the ones from 1986, we obtain Table 1.3. Only the value for the Boltzmann constant k has remained within the old error limits. The Avogadro constant (N_A) and Planck constant (h) came to lie outside the old limits, as did the value for the elementary charge e. The error limits for the gravitational constant G even went up by more than two orders.

It is pointless to give the error to more than the two leading digits, and the mean value more exactly than the error. This is forgotten by many laypeople, if they communicate their computational result "exactly", with far too many digits.

Table 1.3 Improvement of precision with time

Quantity	Relative uncertainty 1986 in 10^{-7}	Relative change 1999/1986 in 10^{-7}	Relative uncertainty 1999 in 10^{-7}
e	± 3	-5.6	± 0.4
h	± 6	-10.3	± 0.8
N_A	± 6	$+8.6$	± 0.8
k	± 87	-56	± 17

1.3.2   *Mean Value and Average Error*

After N measurements of x, we have a sequence of measured values $\{x_1, \ldots, x_N\}$. These values are generally not all equal, but we want to assume that their fluctuations are purely random, and we shall only deal with such errors in the following.

Since none of the measurement readings should be preferred, the true value x_0 is assumed to be near the *mean value*

$$\overline{x} \equiv \frac{1}{N} \sum_{n=1}^{N} x_n \, ,$$

because deviations may occur equally often to higher or lower values: $x_0 \approx \overline{x}$. Our best estimate for the true value x_0 is the mean value $\overline{x}$.

Here, the less the values x_n deviate from $\overline{x}$, the more we trust the approximation $x_0 \approx \overline{x}$. From the fluctuations, we deduce a measure Δx for the uncertainty in our estimate. To do this, we take the squares $(x_n - \overline{x})^2$ of the deviations rather than their absolute values $|x_n - \overline{x}|$, because the squares are differentiable, while the absolute values are not, something we shall exploit in Sect. 1.3.7. However, we may take their mean value

$$\overline{(x - \overline{x})^2} = \overline{x^2 - 2\overline{x}\,x + \overline{x}^2} = \overline{x^2} - 2\overline{x}\,\overline{x} + \overline{x}^2 = \overline{x^2} - \overline{x}^2$$

as a measure for the uncertainty only in the limit of many measurements, not just a small number of measurements. So, for a single measurement nothing whatsoever can be said about the fluctuations. For a second measurement, we would have only a first clue about the fluctuations. In fact, we shall set

$$(\Delta x)^2 = \frac{1}{N-1} \sum_{n=1}^{N} (x_n - \overline{x})^2 = \frac{N}{N-1} \overline{(x - \overline{x})^2} \, ,$$

as will be justified in the following sections. Here we shall rely on a simple special case of the law of error propagation. But this law can also be proven rather easily in its general form and will be needed for other purposes. Therefore, we prove it generally now, whereupon the last equation can be derived easily. To this end, however, we have to consider general properties of error distributions.

1.3.3 Error Distribution

We presume that the errors are distributed in a purely random manner. Then the error probability can be derived from sufficiently many readings of the measurement ($N \gg 1$). From the relative occurrences of the single values, we can determine the probability $\rho(\varepsilon)\, d\varepsilon$ that the error lies between ε and $\varepsilon + d\varepsilon$. The probability density $\rho(\varepsilon)$ is characterized essentially by the *average error* σ, as the following considerations show.

Each probability distribution ρ has to be normalized to unity and may not take negative values: $\int \rho(\varepsilon)\, d\varepsilon = 1$ and $\rho(\varepsilon) \geq 0$ for all ε ($\in$ R). In addition, we expect $\rho(\varepsilon)$ to be essentially different from zero only for $\varepsilon \approx 0$ and to tend to zero monotonically with increasing $|\varepsilon|$. The distribution is also assumed to be an even function, at least in the important region around the zero point: $\rho(\varepsilon) = \rho(-\varepsilon)$. Hence, $\int \varepsilon\, \rho(\varepsilon)\, d\varepsilon = 0$. The next important feature is the width of the distribution. It can be measured with the second moment, the average of the *squared errors* σ (≥ 0), also called the *mean square fluctuation* or *variance*,

$$\sigma^2 \equiv \int \varepsilon^2\, \rho(\varepsilon)\, d\varepsilon \ .$$

Note, however, that the mean square error is not finite for all allowable error distributions, e.g., for the *Lorentz distribution* $\rho(\varepsilon) = \gamma/\{\pi(\varepsilon^2 + \gamma^2)\}$, which is instead characterized by half the *Lorentz half-width* γ—more on that in the discussion around Fig. 5.6.

From the probability distribution $\rho(\varepsilon)$, we can evaluate the *expectation value* $\langle f \rangle$ of any function $f(\varepsilon)$. Each value of the function is summed with its associated weight:

$$\langle f \rangle \equiv \int_{-\infty}^{\infty} f(\varepsilon)\, \rho(\varepsilon)\, d\varepsilon \ .$$

In particular, $\langle \varepsilon^n \rangle = \int \varepsilon^n \rho(\varepsilon)\, d\varepsilon$.

For the error distribution in the following we use only on the properties $\langle \varepsilon^0 \rangle = 1$, $\langle \varepsilon^1 \rangle = 0$, and $\langle \varepsilon^2 \rangle = \sigma^2$, among which only the middle one might be disputed—the first is obvious by normalization, the last fixes the average error σ.

If, however, we want to write down the probability $W(\lambda)$ for an error $|\varepsilon| \leq \lambda\sigma$, i.e.,

$$W(\lambda) = \int_{-\lambda\sigma}^{\lambda\sigma} \rho(\varepsilon)\, d\varepsilon \ ,$$

we have to know $\rho(\varepsilon)$ in more detail. Detailed statistical investigations suggest the *normal or Gaussian distribution*—this will become apparent in Sect. 6.1.4:

$$\rho(\varepsilon) = \frac{\exp\left(-\frac{1}{2}\varepsilon^2/\sigma^2\right)}{\sqrt{2\pi}\, \sigma} \ .$$

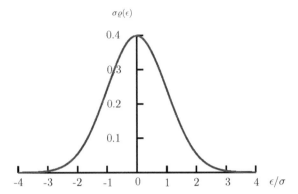

Fig. 1.15 *Normal distribution of the error*. Gauss function (bell-shaped curve). In order for all average errors σ (> 0) to result in the same curve, the probability ρ for the error ε times the average error as a function of the ratio ε/σ is shown here. The area is unity for all σ

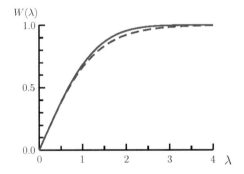

Fig. 1.16 *Error integral* $W(\lambda)$ (*blue*): the probability of errors with $|\varepsilon| \leq \lambda\sigma$ (σ is the average error). The *dashed red* curve is the function $\tanh(\sqrt{2/\pi}\,\lambda)$ for comparison

Figure 1.15 shows this function and Fig. 1.16 the associated *error integral* $W(\lambda)$. The error integral is related to the *error function*

$$\text{erf}\,x \equiv \frac{2}{\sqrt{\pi}} \int_0^x \exp\left(-y^2\right) \mathrm{d}y = W(\sqrt{2}\,x)\,,$$

for which the following expansions are useful:

$$\text{erf}\,x = \begin{cases} \dfrac{2}{\sqrt{\pi}} \displaystyle\sum_{n=0}^{\infty} \dfrac{(-)^n}{n!}\, \dfrac{x^{2n+1}}{2n+1}\,, \\[2ex] 1 - \dfrac{\exp\left(-x^2\right)}{\sqrt{\pi}\,x}\left(1 - \displaystyle\sum_{n=0} \dfrac{(-)^n\,(2n+1)!!}{(2x^2)^{n+1}}\right) & \text{for } x \gg 1. \end{cases}$$

The second series is *semi-convergent*, i.e., it does not converge for $n \to \infty$, but approximates the function sufficiently well for finite n $(< x)$.

From Fig. 1.16, we see that, for the normal distribution, slightly more than $\frac{2}{3}$ of all values have an error $|\varepsilon| \leq \sigma$ and barely 5% an error $|\varepsilon| > 2\sigma$.

1.3.4 Error Propagation

We now start from K physical quantities x_k with average errors σ_k and consider the derived quantity $y = f(x_1, \ldots, x_K)$. Here all the quantities x_k will be independent of each other. What is then the average error in y?

To begin with, the error ε in $f(x_1, \ldots, x_K)$ is to first order

$$\varepsilon = \sum_{k=1}^{K} \frac{\partial f}{\partial x_k} \varepsilon_k ,$$

and hence

$$\sigma^2 = \langle \varepsilon^2 \rangle$$

$$= \int \cdots \int_{-\infty}^{\infty} \left(\sum_{k=1}^{K} \frac{\partial f}{\partial x_k} \varepsilon_k \right)^2 \rho(\varepsilon_1, \ldots, \varepsilon_K) \, d\varepsilon_1 \cdots d\varepsilon_K$$

$$= \left\langle \sum_{k=1}^{K} \frac{\partial f}{\partial x_k} \varepsilon_k \cdot \sum_{l=1}^{K} \frac{\partial f}{\partial x_l} \varepsilon_l \right\rangle = \sum_{k,l=1}^{K} \frac{\partial f}{\partial x_k} \frac{\partial f}{\partial x_l} \langle \varepsilon_k \varepsilon_l \rangle .$$

Since the quantities x_k and x_l should not depend upon each other, they are not correlated to each other (the property x_l does not care how large x_k is—correlations will be investigated in more detail in Sect. 6.1.5). With $\langle \varepsilon_k^2 \rangle = \sigma_k^2$ this leads to

$$\langle \varepsilon_k \varepsilon_l \rangle = \begin{cases} \langle \varepsilon_k \rangle \langle \varepsilon_l \rangle & \text{for } k \neq l , \\ \sigma_k^2 & \text{for } k = l . \end{cases}$$

Here, $\langle \varepsilon_k \rangle = 0$ holds for all k (and l). Therefore, the *law of error propagation* follows:

$$\sigma^2 = \sum_{k=1}^{K} \left(\frac{\partial f}{\partial x_k} \right)^2 \sigma_k^2 .$$

In the proof, no normally distributed errors were necessary—thus other distributions with the properties $\langle \varepsilon_k^0 \rangle = 1$, $\langle \varepsilon_k^1 \rangle = 0$, and $\langle \varepsilon_k^2 \rangle = \sigma_k^2$ deliver the error propagation law and with it the basis for all further proofs in this section. In particular, we may invoke this law for repeated measurements of the same quantity, as we shall now do.

1.3.5 Finite Measurement Series and Their Average Error

If we consider the expression

$$\langle x \rangle \approx \overline{x} \equiv \frac{1}{N} \sum_{n=1}^{N} x_n$$

as $\overline{x} = f(x_1, \ldots, x_N)$, then we can use it in the law of error propagation and deduce that $\partial f / \partial x_n = N^{-1}$. Hence, all single measurements enter into the error estimate with the same weight—as already for the estimated value x_0.

In order to determine the error σ_n, we think of an average over several measurement series, each with N measurements. In this way, we can introduce the *average error of the single measurement* and find that all single measurements have the same average error Δx. Therefore, the law of error propagation for N equal terms $N^{-2}(\Delta x)^2$ delivers

$$(\Delta \overline{x})^2 = N \cdot N^{-2} (\Delta x)^2 = \frac{(\Delta x)^2}{N} .$$

The average error $\Delta \overline{x}$ in the mean value of the measurement series is thus the $\sqrt{N}$ th part of the average error in a single measurement: the more often measurements are made, the more accurate is the determination of the mean value. However, because of the square root factors, the accuracy can be increased only rather slowly.

Since we do not know the true value x_0 itself, but only its approximation $\overline{x}$, we still have to account for its uncertainty $\Delta \overline{x}$ in order to determine the average error of the single measurement:

$$(\Delta x)^2 = \overline{(x - x_0)^2} = \overline{(x - \overline{x} + \overline{x} - x_0)^2} = \overline{(x - \overline{x})^2} + 2 \overline{(x - \overline{x})} (\overline{x} - x_0) + (\overline{x} - x_0)^2 .$$

Here, $\overline{x - \overline{x}} = \overline{x} - \overline{x} = 0$ and thus $\overline{(x - \overline{x})^2} = \overline{x^2} - \overline{x}^2$ is rather easy to evaluate. For $(\overline{x} - x_0)^2$, we take $(\Delta \overline{x})^2 = (\Delta x)^2 / N$. Hence, because $1 - N^{-1} = N^{-1}(N - 1)$, the average error of the single measurement is

$$(\Delta x)^2 = \frac{N}{N - 1} \overline{(x - \overline{x})^2} ,$$

as claimed previously (see p. 46). And so we have the announced proof. For sufficiently large N, we may write $(\Delta x)^2 = \overline{x^2} - \overline{x}^2$. The expression Δx is referred to as the *uncertainty* of x in quantum theory (see p. 275).

1.3.6 Error Analysis

How should we modify the result obtained so far if the same quantity is measured in different ways: first as $x_1 \pm \Delta x_1$, then as $x_2 \pm \Delta x_2$, and so on? What is then the most probable value for x_0, and what average error does it have?

If the readings of the measurement were taken with the same instrument and equally carefully, the difference in the average errors stems from values x_n from measurement series of different lengths. According to the last section, the average error of every single measurement in such a measurement series should be equal to $\Delta x_n \sqrt{N_n}$, and this independently of n in each of the measurement series. Therefore, the mentioned values x_n should contribute with the weight

$$\rho_n = \frac{N_n}{\sum_k N_k} = \frac{1}{(\Delta x_n)^2} \bigg/ \sum_k \frac{1}{(\Delta x_k)^2} \ ,$$

whence $\bar{x} = \sum_n \rho_n x_n$ is the properly weighted mean value. The error propagation law delivers

$$\sigma^2 = \sum_n \rho_n^2 \, \sigma_n^2 = \frac{1}{(\sum_k (\Delta x_k)^{-2})^2} \sum_n \frac{1}{(\Delta x_n)^4} (\Delta x_n)^2 = \frac{\sum_n (\Delta x_n)^{-2}}{(\sum_k (\Delta x_k)^{-2})^2}$$

$$= \frac{1}{\sum_n (\Delta x_n)^{-2}} \qquad \Longrightarrow \qquad \frac{1}{(\Delta x)^2} = \sum_n \frac{1}{(\Delta x_n)^2} \ .$$

The more detailed the readings of the measurement, the more important they are for the mean value and for the (un)certainty of the results. These considerations are only then valid without restriction, if the values are compatible with each other within their error limits. If they lie further apart from each other, then we have to take

$$(\Delta x)^2 = \frac{1}{N-1} \frac{1}{\sum_n (\Delta x_n)^{-2}} \sum_n \frac{(x_n - \bar{x})^2}{(\Delta x_n)^2} \ .$$

Note that, if the values x_n do not lie within the error limits, then systematic errors may be involved.

Thus, these two equations answer the questions raised in the general case, where measurements are taken with different instruments and different levels of care: to each value x_n, we must attach the relative weight $1/(\Delta x_n)^2$.

1.3.7 Method of Least Squares

A further generalization is necessary if the readings of measurement happen to be along a straight line, but scatter about it due to random errors. What are then the best values a and b for $\{y_n = a\, x_n + b\}$? More generally, we can fit a power series, a Fourier series, or some series of known functions.

We always want to determine the readings of measurements as precisely as possible, in order to make the average error small as possible. This requirement is effective under general conditions. Thus the values a and b of the fitting line are to be determined from the conditions

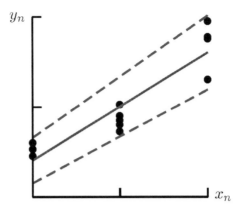

Fig. 1.17 Example of a fitting line through 12 pairs of measurement values (•). The *continuous line* shows $y = ax + b$, and the *dashed lines* show the upper and lower error limits $(a + \Delta a) x + (b + \Delta b)$ and $(a - \Delta a) x + (b - \Delta b)$, respectively. (In a beginners' lab course, we can thus establish *Hooke's law* by showing how the length y of a copper wire depends linearly upon the load x)

$$\sum_{n=1}^{N} (y_n - ax_n - b)^2 = \min(a, b) ,$$

i.e., $\partial \sum_{n=1}^{N} (y_n - ax_n - b)^2 / \partial a = 0 = \partial \sum_{n=1}^{N} (y_n - ax_n - b)^2 / \partial b$. From the last, we have

$$b = \frac{\sum_n (y_n - ax_n)}{N} = \bar{y} - a\bar{x} ,$$

and from the condition above,

$$a = \frac{\overline{xy} - \bar{x}\,\bar{y}}{\overline{x^2} - \bar{x}^2} = \frac{\overline{(x - \bar{x})(y - \bar{y})}}{\overline{(x - \bar{x})^2}} .$$

Here, the first fraction is easy to evaluate, the last less easy to interpret—hence the reformulation. We have thus answered our question as to which values for a and b are the best. For an example, see Fig. 1.17.

To calculate the average errors Δa and Δb, we have to consider the fact that pairs of values (x_n, y_n) are always associated and only the error in each pair counts, not the error in x_n and y_n separately. Therefore, for reasons of simplicity we take the error in x_n as an additional error in y_n and then take a and b in the law of error propagation as functions of y_n. From $b = \bar{y} - a\bar{x}$, it then follows that

$$(\Delta b)^2 = (\Delta \bar{y})^2 + \bar{x}^2 (\Delta a)^2 .$$

From $a = \{\sum_n (x_n - \bar{x}) y_n\} / \{N(\overline{x^2} - \bar{x}^2)\}$, we obtain $\partial a / \partial y_n = (x_n - \bar{x}) / \{N(\overline{x^2} - \bar{x}^2)\}$, and thus, finally,

$$(\Delta a)^2 = \sum_{n=1}^{N} \frac{(x_n - \overline{x})^2}{N^2 (\overline{x^2} - \overline{x}^2)^2} (\Delta y_n)^2 .$$

If all errors Δy_n in the pairs of values are equally large, then $(\Delta \overline{y})^2 = (\Delta y)^2/N$ and hence

$$(\Delta a)^2 = \frac{(\Delta y)^2}{N (\overline{x^2} - \overline{x}^2)} \quad \text{and} \quad (\Delta b)^2 = \overline{x^2} (\Delta a)^2 .$$

We still lack a prescription for calculating the average error Δy in a single measurement. From the original equation $y = ax + b$, where two pairs of values are now necessary in order to determine a and b, we have

$$(\Delta y)^2 = \frac{1}{N - 2} \sum_{n=1}^{N} (y_n - a x_n - b)^2 .$$

More generally, with K parameters we would have the denominator $N - K$, because the equation $(\Delta \overline{x})^2 = (\Delta x)^2/N$ in one dimension becomes $(\Delta \overline{y})^2 = K (\Delta y)^2/N$ in K dimensions, and this can then be used in $(\Delta y)^2 = \overline{(y - \overline{y})^2} + (\Delta \overline{y})^2$.

List of Symbols

We stick closely to the recommendations of the *International Union of Pure and Applied Physics* (IUPAP) and the *Deutsches Institut für Normung* (DIN). These are listed in *Symbole, Einheiten und Nomenklatur in der Physik* (Physik-Verlag, Weinheim 1980) and are marked here with an asterisk. However, one and the same symbol may represent different quantities in different branches of physics. Therefore, we have to divide the list of symbols into different parts (Table 1.4).

Table 1.4 Standard notation and symbols

	Symbol	Name	Page reference
*	t	Time	1
*	$\mathbf{r}$	Position vector	1
*	V	Volume	9
	(V)	Surface of a volume V	9
*	A	Area	9
	(A)	Boundary of an area A	9
1	$d\mathbf{r}$	Path element vector	7
1	$d\mathbf{f}$	Surface element vector	9
1	dV, d^3r	Volume element	9, 24
*	$\mathbf{a} \cdot \mathbf{b}$	Scalar product of $\mathbf{a}$ and $\mathbf{b}$	3
*	$\mathbf{a} \times \mathbf{b}$	Vector product of $\mathbf{a}$ and $\mathbf{b}$	4

(continued)

Table 1.4 (continued)

	Symbol	Name	Page reference
∗	**a b**	Dyadic product of **a** and **b**	11
∗	$\mathbf{e}_x$	Unit vector $\mathbf{x}/x$	3
∗[2]	∇	Nabla operator	10
∗[2]	$\nabla\phi$	Gradient of a scalar field ϕ	10
∗[2]	$\nabla\cdot\mathbf{a}$	Divergence of a vector field **a**	11
∗[2]	$\nabla\times\mathbf{a}$	Rotation (curl) of a vector field **a**	13
	$\nabla_A\cdot\mathbf{a}$	Area divergence of a vector field **a**	27
	$\nabla_A\times\mathbf{a}$	Area rotation of a vector field **a**	27
∗	Δ	Laplace operator	15
∗	δ_{ik}	Kronecker symbol	18
∗	$\delta(x)$	Dirac delta function	18
∗	δx	Variation of x	58, 139
∗	$\varepsilon(x)$	Discontinuity function (theta function, step function)	18
	P I	Principal value of I	19
∗	**k**	Wave vector	24
∗	$\widetilde{D}$	Transpose of the matrix D	5
∗	D^{-1}	Inverse of the matrix D	29
∗	D^*	Conjugate of the matrix D	29
∗	$D^\dagger$	Adjoint of the matrix D	29
∗	det D	Determinant of the matrix D	5
∗	tr D	Trace of the matrix D	36
	$\mathbf{g}_i$	Covariant base vector $\partial\mathbf{r}/\partial x^i$	31
	$\mathbf{g}^i$	Contravariant base vector ∇x^i	31
	a_i	Covariant component of **a**	33
	a^i	Contravariant component of **a**	33
∗	$\overline{x}$	Mean value of x	46
	Δx	Uncertainty in x	46

[1]Total differentials are written with an upright d rather than an italic d. We stick to this throughout.
[2]In the recommended notation there is no vector arrow above ∇, even though it is a vector operator.

Suggestions for Further Reading

1. J. Arfken, H.J. Weber, *Mathematical Methods for Physicists*, 6th edn. (Elsevier Academic, Burlington MA, 2005)
2. E. Ph Blanchard, *Brüning: Mathematical Methods in Physics: Distributions, Hilbert Space Operators, and Variational Methods* (Springer Science + Business, Media, 2003)
3. S. Hassani, *Mathematical Physics-A Modern Introduction to Its Foundations* (Springer, Berlin, 2013)
4. A. Sommerfeld: *Lectures on Theoretical Physics 6-Partial Differential Equations in Physics* (Academic, London, 1949/1953)
5. H. Triebel, *Analysis and Mathematical Physics* (Springer, Berlin, 1986)

Chapter 2
Classical Mechanics

2.1 Basic Concepts

2.1.1 Force and Counter-Force

The best known example of a force is the *gravitational force*. If we let go of our book, it falls downwards. The Earth attracts it. Only with a counter-force can we prevent it from falling, as we clearly sense when we are holding it. Instead of our hand, we can use something to fix it in place. We can even measure the counter-force with a spring balance, e.g., in the unit of force called the newton, denoted $N = kg \, m/s^2$.

Each force has a strength and a direction and can be represented by a vector—if several forces act on the same point mass, then the total force is found using the addition law for vectors. As long as our book is at rest, the gravitational and counter-force cancel each other and the total force vanishes. Therefore, the book remains in equilibrium.

Forces act between bodies. In the simplest case, we consider only two bodies. It is this case to which *Newton's third law* refers: *Two bodies act on each other with forces of equal strength, but with opposite direction.* This law is often phrased also as the equation "force = counter-force" or "action = reaction", even though they refer only to their moduli. If body j acts on body i with the force $\mathbf{F}_{ij}$, then

$$\mathbf{F}_{ij} = -\mathbf{F}_{ji} \, .$$

According to this, no body is preferred over any another. They are all on an equal footing.

We often have to deal with *central forces*. Then,

$$\mathbf{F}_{ij} = \mp F(r_{ij}) \, \mathbf{e}_{ij} \, , \quad \text{with} \quad \mathbf{e}_{ij} \equiv \frac{\mathbf{r}_{ij}}{r_{ij}} \quad \text{and} \quad \mathbf{r}_{ij} \equiv \mathbf{r}_i - \mathbf{r}_j = -\mathbf{r}_{ji} \, ,$$

© Springer Nature Switzerland AG 2018
A. Lindner and D. Strauch, *A Complete Course*
on Theoretical Physics, Undergraduate Lecture Notes in Physics,
https://doi.org/10.1007/978-3-030-04360-5_2

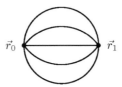

Fig. 2.1 The force can only be derived from a potential energy if the work needed to move against the force from the point $\mathbf{r}_0$ to point $\mathbf{r}_1$ does not depend upon the path between these points, i.e., only for $\oint \mathbf{F} \cdot d\mathbf{r} = 0$ *for all closed paths*

where we have a minus sign for an attractive force and a plus sign for a repulsive force (see Fig. 1.1). Clearly, they have the required symmetry.

The force between two magnetic dipoles $\mathbf{m}_i$ and $\mathbf{m}_j$ is not a central force, but a *tensor force*:

$$\mathbf{F}_{ij} = \frac{3\mu_0}{4\pi} \left(\frac{\mathbf{m}_i \cdot \mathbf{e}_{ij}}{r_{ij}{}^4} \, \mathbf{m}_j + \frac{\mathbf{m}_j \cdot \mathbf{e}_{ij}}{r_{ij}{}^4} \, \mathbf{m}_i + \frac{\mathbf{m}_i \cdot \mathbf{m}_j - 5\,\mathbf{m}_i \cdot \mathbf{e}_{ij}\,\mathbf{m}_j \cdot \mathbf{e}_{ij}}{r_{ij}{}^4} \, \mathbf{e}_{ij} \right) .$$

This expression is derived in Sect. 3.2.9 and presented in Fig. 3.12. It depends on the directions of the three vectors $\mathbf{m}_i$, $\mathbf{m}_j$, and $\mathbf{r}_{ij}$, but also has the required symmetry.

Newton's third law also holds for changing positions $\mathbf{r}_{ij}(t)$, but we shall deal with this in the next section. For the time being, we restrict ourselves to statics. The total force of the bodies j on a test body i is thus

$$\mathbf{F}_i = \sum_j \mathbf{F}_{ij} .$$

This force will generally change with the position $\mathbf{r}_i$ of the test body, if the other positions $\mathbf{r}_j$ are kept fixed. We now want to investigate this in more detail and write $\mathbf{r}$ instead of $\mathbf{r}_i$.

2.1.2   Work and Potential Energy

It may be easier to work with a scalar field than with a vector field. Therefore, we derive the force field $\mathbf{F}(\mathbf{r})$ from a scalar field, viz., the *potential energy* $V(\mathbf{r})$:

$$\mathbf{F} = -\nabla V .$$

But for this to work, since $\nabla \times \nabla V = \mathbf{0}$, $\mathbf{F}$ has to be curl-free, i.e., the integral $\oint \mathbf{F} \cdot d\mathbf{r}$ has to vanish along each closed path. We conclude that a potential energy can only be introduced if the *work*

$$A \equiv \int_{\mathbf{r}_0}^{\mathbf{r}_1} \mathbf{F} \cdot d\mathbf{r}$$

depends solely upon the initial and final points $\mathbf{r}_0$ and $\mathbf{r}_1$ of the path, but not on the actual path taken in-between (see Fig. 2.1). (Instead of the abbreviation A, the symbol W is often used, but we shall use W in Sect. 2.4.7 for the *action function*.) Generally, on a very short path $d\mathbf{r}$, an amount of work $\delta A = \mathbf{F} \cdot d\mathbf{r}$ is done. Here we write δA instead of dA, because δA is a very small (infinitesimally small) quantity, but not necessarily a differential one. It is only a differential quantity if there is a potential energy, hence if $\mathbf{F}$ is curl-free and can be obtained by differentiation:

$$dV \equiv \nabla V \cdot d\mathbf{r} = -\mathbf{F} \cdot d\mathbf{r} \equiv -\delta A .$$

For the example of the central and tensor forces mentioned in the last section, a potential energy can be given, but it cannot for velocity-dependent forces, i.e., neither for the frictional nor for the Lorentz force (acting on a moving charge in a magnetic field). We shall investigate these counter-examples in Sect. 2.3.4.

If there is a potential energy, then according to the equations above it is determined only up to an additive constant. The zero of V can still be chosen at will and the constant "adjusted" in some suitable way. The zero of V is set at the point of vanishing force. If it vanishes for $r \to \infty$, then it follows that

$$V(\mathbf{r}) = -\int_{\infty}^{\mathbf{r}} \mathbf{F}(\mathbf{r}') \cdot d\mathbf{r}' .$$

But it should be noted once again that this is unique only for $\nabla \times \mathbf{F} = \mathbf{0}$, that is, only then is there a potential energy.

For a *homogeneous force field*, the force $\mathbf{F}$ does not depend on the position. Then the expression $V = -\mathbf{F} \cdot \mathbf{r}$ fits. Likewise, for a central force with $F \propto r^n$, the potential energy is easily found:

$$\mathbf{F} = c\, r^n \frac{\mathbf{r}}{r} \quad \Longleftrightarrow \quad V = \frac{-c}{n+1} r^{n+1} ,$$

if $n \neq -1$, otherwise $V = -c \ln (r/r_0)$, with an arbitrary gauge constant r_0. Note that $V(\infty) = 0$ holds only for $n < -1$.

If we can derive the two-body force $\mathbf{F}_{ij}$ from a potential energy V_{ij}, we have (with $\mathbf{r}_j$ kept fixed)

$$\mathbf{F}_{ij} = -\nabla_i V_{ij} , \quad \text{with} \quad \nabla_i V_{ij} \cdot d\mathbf{r}_i = dV_{ij} .$$

Newton's third law now delivers $-\nabla_i V_{ij} = \nabla_j V_{ji}$ (on the right-hand side, $\mathbf{r}_i$ is kept fixed and $\mathbf{r}_j$ variable) with $\nabla_i = -\nabla_j$, since $\mathbf{r}_{ij} = -\mathbf{r}_{ji}$ here, so we have $d\mathbf{r}_i = -d\mathbf{r}_j$. Therefore, with a convenient gauge, we can obtain the symmetry

$$V_{ij} = V_{ji} .$$

Hence a many-body problem has the total potential energy

$$V = \sum_{i<j} V_{ij} = \tfrac{1}{2} \sum_{i \neq j} V_{ij} \ ,$$

because each pair (ij) is to be counted only once. (It is often taken for granted that V_{ii} vanishes and the summation is then simply over i and j, without indicating $i \neq j$.)

2.1.3 Constraints: Forces of Constraint, Virtual Displacements, and Principle of Virtual Work

We can often replace forces by geometric constraints. If the test body has to remain on a plane, we should decompose the force acting on it into its tangential and normal components—because it is only the tangential component that is decisive for the equilibrium (as long as there is no *static friction*, since this depends upon the normal component). The normal component describes only how strongly the body presses on the support, e.g., a sphere on a tabletop.

Geometrically conditioned forces are called *forces of constraint* $\mathbf{Z}$. In equilibrium, the external forces cancel, whence $\sum_i \mathbf{F}_i = \sum_i \mathbf{Z}_i$. We now consider *virtual* changes in the configuration of an experimental setup. In our minds, we alter the positions slightly, while respecting the constraints, in order to find out how much of it is rigid and how much is flexible. These alterations (*variations*) will be denoted by $\delta \mathbf{r}$. If there is no perturbation due to static friction, then the forces of constraint are perpendicular to the permitted alterations of position, and therefore the displacement $\delta \mathbf{r}$ does not contribute to the work. Since $\mathbf{Z}_i \cdot \delta \mathbf{r}_i = 0$, we find the extremely useful *principle of virtual work*:

$$\sum_i \mathbf{F}_i \cdot \delta \mathbf{r}_i = 0 \ .$$

In equilibrium, the virtual work of the externally applied forces vanishes. We do not need to calculate the forces of constraint here—instead, only the geometrical constraints must be obeyed. If only curl-free forces are involved, then the associated potential energy of the total system also suffices. Equilibrium prevails if it does not change under a virtual displacement: $\nabla V \cdot \delta \mathbf{r} \equiv \delta V = 0$.

For a lever, the virtual work can be evaluated in a particularly easy way with a virtual rotation, because the length R of the lever arm does not change, and therefore we may set $\delta \mathbf{r} = \delta \boldsymbol{\varphi} \times \mathbf{R}$, if $\delta \boldsymbol{\varphi}$ points in the direction of the axis of rotation (right-hand rule) and if $\mathbf{R}$ points from the axis of rotation to the point where the force acts (see Fig. 2.2). For the virtual work we obtain therefore

$$\delta A = \mathbf{F} \cdot \delta \mathbf{r} = \mathbf{F} \cdot (\delta \boldsymbol{\varphi} \times \mathbf{R}) = (\mathbf{R} \times \mathbf{F}) \cdot \delta \boldsymbol{\varphi} = \mathbf{M} \cdot \delta \boldsymbol{\varphi} \ ,$$

with the *torque* $\mathbf{M}$ of the force $\mathbf{F}$ on the lever arm $\mathbf{R}$ defined by

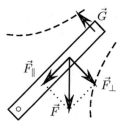

Fig. 2.2 Lever law. The rigid lever transmits all those forces to the axle bearing (*open circle*) which do not have an angular momentum with respect to the axis—they are canceled by forces of constraint, here by $\mathbf{F}_\parallel$. Equilibrium prevails, if the torque due to the force $\mathbf{F}$ and the counter-force $\mathbf{G}$ cancel each other, as indicated by the two hyperbola branches

$$\mathbf{M} \equiv \mathbf{R} \times \mathbf{F}.$$

Since $\delta\varphi_i$ for a rigid lever is the same everywhere, equilibrium prevails here if the sum of the torques vanishes. The principle of virtual work then implies the *lever law*

$$\sum_i \mathbf{M}_i = \mathbf{0}, \quad \text{with} \quad \mathbf{M}_i = \mathbf{R}_i \times \mathbf{F}_i.$$

The equilibrium of the lever depends upon the torques, i.e., the vector products of lever arm and force.

2.1.4 General Coordinates and Forces

This section and the next actually touch upon the subject of Lagrangian mechanics, wherein, by a clever choice of coordinates, problems can be made soluble which otherwise would be intractable. In the static case, many things are much simpler than for time-dependent phenomena. It is this that we want to exploit here, and then begin by solving several examples (Problems 2.7–2.10), in order to get used to these notions. The lever law introduces us to this way of thinking.

Very often the solubility of a problem depends on a choice of coordinates which can lead to mathematical as well as physical simplifications. For example, for the two-body problem we employ center-of-mass and relative coordinates, because the forces only depend upon the relative coordinate. At best, we choose the coordinates such that each constraint removes a variable and hence only the remaining ones survive as variables, e.g., in the case of the lever, we use cylindrical coordinates, because then the forces of constraint determine R and z, and only the angular coordinate φ can vary. Then for an N-body problem, we do not need the $3N$ coordinates of the real space, but only $f \leq 3N$ coordinates in the *configuration space*. Here f is called the number of *degrees of freedom* of the mechanical problem.

In most textbooks, generalized coordinates are denoted by q_k. But here we shall adopt the notation x^k used in relativity theory and lattice dynamics. This is explained in detail in Sects. 1.2.2–1.2.5.

The variables

$$x^k = x^k(t, \mathbf{r}_1, \ldots, \mathbf{r}_N), \quad \text{with} \quad k \in \{1, \ldots, f\} \quad \text{and} \quad f \le 3N,$$

can be Cartesian coordinates, but also curvilinear (e.g., spherical or cylindrical) coordinates, or even oblique ones—for which ∇x^i is not perpendicular to ∇x^k and therefore we have to distinguish between x^k and x_k.

The f *generalized coordinates* x^k (in addition to parameters like t) will describe the given problem completely:

$$\mathbf{r}_i = \mathbf{r}_i(t, x^1, \ldots, x^f), \quad \text{for all } i \in \{1, \ldots, N\}.$$

Correspondingly, we have for the virtual displacements, keeping time fixed,

$$\delta\mathbf{r}_i = \sum_{k=1}^{f} \frac{\partial \mathbf{r}_i}{\partial x^k} \, \delta x^k, \quad \text{at} \quad \delta t = 0,$$

and for the principle of virtual work,

$$0 = \sum_{i=1}^{N} \mathbf{F}_i \cdot \delta\mathbf{r}_i = \sum_{k=1}^{f} F_k \, \delta x^k,$$

with the *generalized forces*

$$F_k \equiv \sum_{i=1}^{N} \mathbf{F}_i \cdot \frac{\partial \mathbf{r}_i}{\partial x^k}, \quad \text{or} \quad F_k = -\frac{\partial V}{\partial x^k},$$

if the external forces can be associated with a potential energy. The notation F_k with lower index k corresponds to the convention of Sect. 1.2.2, while the indices i are used here to count the particles. The x^k do not need to be lengths and the F_k are not necessarily forces in the usual sense, but $F_k \, \delta x^k$ has to be an energy. Thus, according to the last section, the generalized force "torque" corresponds to the generalized coordinate "angle".

In static equilibrium, all the F_k are equal to zero if none of the x^k depends upon the others. However, the constraints do not always have such simple properties: not every constraint fixes a coordinate and leaves the remaining ones undetermined. Therefore, we now want to treat a more general case.

2.1.5 Lagrangian Multipliers and Lagrange Equations of the First Kind

If, for the moment, we do not consider the $3N - f$ constraints for N point masses, then in addition to the f generalized coordinates x^k introduced so far, $3N - f$ further coordinates x^κ (with $\kappa \in \{f + 1, \ldots, 3N\}$) are still required, and these are in fact determined by the constraints. We assume that these constraints are given in the form of equations:

$$\Phi_n(t, x^1, \ldots, x^{3N}) = 0, \quad \text{for all } n \in \{1, \ldots, 3N - f\}.$$

Here, equations for differential forms suffice, because only the following $3N - f$ equations have to be valid for arbitrary parameter variations at fixed time:

$$\sum_{k=1}^{f} \frac{\partial \Phi_n}{\partial x^k} \delta x^k + \sum_{\kappa=f+1}^{3N} \frac{\partial \Phi_n}{\partial x^\kappa} \delta x^\kappa = 0, \quad \text{with} \quad \delta t = 0,$$

where the coefficients do not need to be differential quotients—this becomes necessary, when we trace the forces back to a potential energy. Now we want to make use of the fact that only the f variations δx^k are free, but the remaining δx^κ depend upon the former and require the $3N - f$ *Lagrangian (undetermined) multipliers* λ_n (one Lagrangian multiplier for each constraint) to satisfy

$$\sum_{n=1}^{3N-f} \lambda_n \frac{\partial \Phi_n}{\partial x^\kappa} = -F_\kappa, \quad \text{with} \quad F_\kappa \equiv \sum_{i=1}^{N} \mathbf{F}_i \cdot \frac{\partial \mathbf{r}_i}{\partial x^\kappa}.$$

This is an inhomogeneous linear system of $3N - f$ equations with the same number κ of dependent variables. Once we have determined all Lagrangian multipliers λ_n from this, then the relation

$$\sum_\kappa F_\kappa \delta x^\kappa = -\sum_{\kappa n} \lambda_n \frac{\partial \Phi_n}{\partial x^\kappa} \delta x^\kappa = \sum_{nk} \lambda_n \frac{\partial \Phi_n}{\partial x^k} \delta x^k, \quad \text{at} \quad \delta t = 0,$$

implies the following expression for the principle of virtual work (with $\delta t = 0$):

$$\sum_{i=1}^{N} \mathbf{F}_i \cdot \delta \mathbf{r}_i = \sum_{k=1}^{f} F_k \delta x^k + \sum_{\kappa=f+1}^{3N} F_\kappa \delta x^\kappa = \sum_{k=1}^{f} \left(F_k + \sum_{n=1}^{3N-f} \lambda_n \frac{\partial \Phi_n}{\partial x^k} \right) \delta x^k.$$

This has to vanish for arbitrary δx^k. The bracket has to be zero for all f values k. Here, the Lagrangian multipliers have to be chosen such that the bracket vanishes also for the remaining $3N - f$ values κ. We thus have generally for all $l \in \{1, \ldots, 3N\}$,

$$F_l + \sum_{n=1}^{3N-f} \lambda_n \frac{\partial \Phi_n}{\partial x^l} = 0 , \quad \text{with} \quad F_l \equiv \sum_{i=1}^{N} \mathbf{F}_i \cdot \frac{\partial \mathbf{r}_i}{\partial x^l} \quad \text{at} \quad \delta t = 0 .$$

These are essentially the *Lagrange equations of the first kind*. For time-dependent problems, only the inertial forces are missing, and this will be treated later, in the context of d'Alembert's principle (Sect. 2.3.1).

We consider a plane problem as an example. Let $z = 0$ be given. Then we can leave out the z-coordinate right away or, using the position vector $\mathbf{r}$ in three-dimensional space, calculate with the constraint $\Phi = z = 0$. With the coordinates $(x^1, x^2, x^3) = (x, y, z)$, we have $0 = F_1 = F_2 = F_3 + \lambda$ in equilibrium. Here, the Lagrangian multiplier λ is equal to the force of constraint $-F_3$, while the forces in the plane have to vanish. (Further examples can be found in Problems 2.7–2.10.)

Since we could also have required $\lambda_n \Phi_n = 0$ instead of the constraint $\Phi_n = 0$, only the product $\lambda_n \Phi_n$ has a physical meaning, but not the Lagrangian multiplier λ_n itself.

If the external forces can be derived from a potential energy, then for all $l \in 1, \ldots, 3N$ in equilibrium, we also have

$$F_l = -\frac{\partial V}{\partial x^l} \quad \text{and} \quad \frac{\partial}{\partial x^l}\left(V - \sum_{n=1}^{3N-f} \lambda_n \Phi_n\right) = 0 .$$

Consequently, the forces of constraint are obsolete, if we subtract the constraints with suitable Lagrangian multipliers from the potential energy.

2.1.6 The Kepler Problem

The three laws due to Kepler[1] lead uniquely to the acceleration

$$\ddot{\mathbf{r}} = -C \frac{\mathbf{r}}{r^3} , \quad \text{with} \quad C = 1.33 \times 10^{20} \frac{\text{m}^3}{\text{s}^2} ,$$

as we will prove immediately. It is more usual to start with the gravitational law and deduce Kepler's laws, something we shall do only afterwards. It is customary to infer the possible types of motion from a given coupling. But to begin with, we solve here the so-called *inverse problem*, that is, we infer the coupling from the observed motion, just as one derives the interaction from scattering experiments. In this context, Lenz's vector results as a conserved quantity, something that is not easy to explain with the usual procedure.

---

[1] Johannes Kepler (1571–1630), among other things, imperial mathematician and astronomer in Prague and then Professor in Linz (Austria).

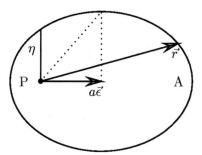

Fig. 2.3 An ellipse with eccentricity $\varepsilon = 2/3$, its left focus (*filled circle*), the distance η of the ellipse from the focus, perpendicular to the main axis, a ray $\mathbf{r}$, and the vector $a\boldsymbol{\varepsilon}$ to the center, where a is the semi-major axis. The *straight dotted lines* have length a and b, and $a^2 = b^2 + a^2\varepsilon^2$, according to Pythagoras. The apex P is called the perihelion and A the aphelion (from the Greek *helios* for the Sun)

According to Kepler's first law, each planet moves along an ellipse with the Sun at the focus.

Both celestial objects will be treated as point-like.

Consider an ellipse with semi-major axis a. For each point on the ellipse specified by the vector $\mathbf{r}$ with origin at one of the foci, the sum of the distances from the foci, viz., $2a = r + \sqrt{(\mathbf{r} - 2a\boldsymbol{\varepsilon}) \cdot (\mathbf{r} - 2a\boldsymbol{\varepsilon})}$ is fixed. Here, $a\boldsymbol{\varepsilon}$ is the vector from one of the foci to the center of the ellipse, as shown in Fig. 2.3. Hence it follows that $(2a - r)^2 = r^2 - 4a\,\boldsymbol{\varepsilon} \cdot \mathbf{r} + 4\,a^2\varepsilon^2$, and we have

$$r - \boldsymbol{\varepsilon} \cdot \mathbf{r} = a\,(1 - \varepsilon^2) \equiv \eta\,,$$

where η is the distance of the ellipse from the focus, measured perpendicular to the symmetry axis, i.e., at $\mathbf{r} \perp \boldsymbol{\varepsilon}$. This is the starting equation for what follows. The number ε is the *eccentricity* of the ellipse. The vector $\boldsymbol{\varepsilon}$ is the *Lenz vector* which will be important later on because it is a characteristic of the orbit, hence a constant of the motion. (The vector $\mathbf{A} = -m^2 C\,\boldsymbol{\varepsilon}$ is often taken as the Lenz vector.) The area of the ellipse is $A = \pi ab = \pi a^2 \sqrt{1 - \varepsilon^2}$, something we shall need for Kepler's third law.

Note that our starting equation has not yet fixed a plane orbit, if we take $\mathbf{r}$ as a vector in three dimensions. The plane orbit is required by Kepler's second law (in vector form). In addition, the equation for fixed $\eta > 0$ comprises further plane orbits:

$$\varepsilon = 0:\ \text{circle}\,, \qquad 0 < \varepsilon < 1:\ \text{ellipse}\,,$$
$$\varepsilon = 1:\ \text{parabola}\,, \qquad \varepsilon > 1:\ \text{hyperbola branch}\,.$$

If η is negative, then the branch is described from the other focus, but for $\varepsilon \leq 1$ there is no longer a real solution. Still, we would like to allow for the generalization to $\varepsilon \geq 1$. In this way, we include orbits of meteorites, but also the motion of electrical point charges in the Coulomb field of other point charges.

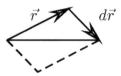

Fig. 2.4 The triangle spanned by **r** and d**r** has area $dA = \frac{1}{2}|\mathbf{r} \times d\mathbf{r}|$ (see Fig. 1.2). For the *area–velocity law*, we use $d\mathbf{r} = \dot{\mathbf{r}}\,dt$

Differentiating $r^2 = \mathbf{r} \cdot \mathbf{r}$ with respect to time, the starting equation yields

$$\dot{r} = \frac{\mathbf{r} \cdot \dot{\mathbf{r}}}{r} = \boldsymbol{\varepsilon} \cdot \dot{\mathbf{r}} \quad \Longrightarrow \quad \left(\frac{\mathbf{r}}{r} - \boldsymbol{\varepsilon}\right) \cdot \dot{\mathbf{r}} = 0 \; .$$

(As an aside, note that $\dot{r}^2$ is *not* equal to $\dot{\mathbf{r}} \cdot \dot{\mathbf{r}}$, as we can see immediately from a circular orbit with $\dot{r} = 0$, but $\dot{\mathbf{r}} \neq \mathbf{0}$.) Thus, $\dot{\mathbf{r}}$ is perpendicular to $\mathbf{r}/r - \boldsymbol{\varepsilon}$. Here we have

$$\frac{d}{dt}\frac{\mathbf{r}}{r} = \frac{\dot{\mathbf{r}}}{r} - \frac{\mathbf{r}}{r^2}\dot{r} = \frac{\dot{\mathbf{r}}\,(\mathbf{r} \cdot \mathbf{r}) - \mathbf{r}\,(\dot{\mathbf{r}} \cdot \mathbf{r})}{r^3} = \frac{(\mathbf{r} \times \dot{\mathbf{r}}) \times \mathbf{r}}{r^3} \; ,$$

and therefore a further differentiation with respect to time yields

$$\left(\frac{\mathbf{r}}{r} - \boldsymbol{\varepsilon}\right) \cdot \ddot{\mathbf{r}} = -\frac{d}{dt}\left(\frac{\mathbf{r}}{r} - \boldsymbol{\varepsilon}\right) \cdot \dot{\mathbf{r}} = -\frac{(\mathbf{r} \times \dot{\mathbf{r}}) \cdot (\mathbf{r} \times \dot{\mathbf{r}})}{r^3}$$

as a further consequence of Kepler's first law. This equation for $\ddot{\mathbf{r}}$ makes a statement about the normal acceleration, since $(\mathbf{r}/r - \boldsymbol{\varepsilon}\,)$ is perpendicular to $\dot{\mathbf{r}}$.

According to Kepler's second law, the ray **r** *traces equal areas* $dA = \frac{1}{2}|\mathbf{r} \times \dot{\mathbf{r}}\,dt|$ *in equal times* dt.

This is also called the area–velocity law (see Fig. 2.4). Here, **r** and $\dot{\mathbf{r}}$ always span the same plane. Consequently, the product $\mathbf{r} \times \dot{\mathbf{r}}$ is constant:

$$\mathbf{r} \times \dot{\mathbf{r}} = \mathbf{c} \quad \Longrightarrow \quad \mathbf{r} \times \ddot{\mathbf{r}} = \mathbf{0} \quad \Longrightarrow \quad \ddot{\mathbf{r}} = f(\mathbf{r})\,\mathbf{r} \; .$$

(Later on, we shall introduce the momentum $m\dot{\mathbf{r}}$ and the orbital angular momentum $\mathbf{L} = \mathbf{r} \times m\dot{\mathbf{r}}$, where m is the reduced mass, explained in more detail in Sect. 2.2.2. The angular momentum is a constant of the motion in the non-relativistic context: according to the area–velocity law, the orbital angular momentum is conserved.) Using the above-mentioned relation here, we obtain

$$-\frac{c^2}{r^3} = f(\mathbf{r})\,(r - \boldsymbol{\varepsilon} \cdot \mathbf{r}) = \eta\,f(\mathbf{r}) \quad \Longrightarrow \quad f(\mathbf{r}) = -\frac{c^2}{\eta\,r^3} \; .$$

The acceleration is always oriented towards the focus for $\eta > 0$ (away from the focus for $\eta < 0$) and decreases as r^{-2} with distance r.

Fig. 2.5 For the Kepler problem, the *velocity* $\dot{\mathbf{r}}$ traces a circle about the center $-\mathbf{c} \times \boldsymbol{\varepsilon}/\eta$ with radius c/η if $\eta > 0$ (for $\varepsilon > 1$, it is only a section of a circle, because then $\mathbf{r}$ traces only one hyperbola branch). At perihelion (P), the speed is greatest, and at aphelion it is smallest, with the ratio equal to $(1+\varepsilon) : (1-\varepsilon)$

The orbit runs perpendicular to $\mathbf{c} = \mathbf{r} \times \dot{\mathbf{r}}$. Therefore, the velocity $\dot{\mathbf{r}}$, which has to be perpendicular to both $\mathbf{c}$ and $\mathbf{r}/r - \boldsymbol{\varepsilon}$, also satisfies $\dot{\mathbf{r}} \propto \mathbf{c} \times (\mathbf{r}/r - \boldsymbol{\varepsilon})$. The missing factor follows from $\mathbf{c} = \mathbf{r} \times \dot{\mathbf{r}}$, because $(\mathbf{r} \times \dot{\mathbf{r}}) \times (\mathbf{r}/r - \boldsymbol{\varepsilon})$ is equal to

$$\dot{\mathbf{r}}\,\mathbf{r} \cdot (\mathbf{r}/r - \boldsymbol{\varepsilon}) - \mathbf{r}\,\dot{\mathbf{r}} \cdot (\mathbf{r}/r - \boldsymbol{\varepsilon}) = \dot{\mathbf{r}}\,(r - \mathbf{r} \cdot \boldsymbol{\varepsilon}) = \dot{\mathbf{r}}\,\eta \,,$$

so

$$\eta\,\dot{\mathbf{r}} = \mathbf{c} \times \left(\frac{\mathbf{r}}{r} - \boldsymbol{\varepsilon}\right) .$$

Since $\mathbf{c}$ is perpendicular to $\mathbf{r}$, all vectors $\mathbf{c} \times \mathbf{r}/r$ have the fixed length c. Therefore, $\dot{\mathbf{r}}$ describes a circle about the center $-\mathbf{c} \times \boldsymbol{\varepsilon}/\eta$ with radius c/η (see Fig. 2.5).

Since $\boldsymbol{\varepsilon}$ and $\mathbf{r}$ are perpendicular to $\mathbf{c}$, the last equation delivers $\mathbf{c} \times \dot{\mathbf{r}}\,\eta/c^2 = \boldsymbol{\varepsilon} - \mathbf{r}/r$ or

$$\frac{\mathbf{r}}{r} + \frac{\eta}{c^2}\,\mathbf{c} \times \dot{\mathbf{r}} = \boldsymbol{\varepsilon} \,.$$

Thus, the left vector is a constant of the motion (namely, Lenz's vector), as is $\mathbf{r} \times \dot{\mathbf{r}} = \mathbf{c}$.

The two Kepler laws discussed so far can be derived only for pure two-body problems. However, other planets (and moons) perturb, so those laws are valid only approximately, as we shall see in Sect. 2.2.6. There we shall also see that, for Kepler's third law, the mass of the planet has to be negligible compared to the mass of the Sun. With *Kepler's third law* the properties of different planets can be compared with each other:

The cubes of the semi-major axes a of all planets behave like the squares of the periods T.

Indeed,

$$a^3 = \frac{C}{(2\pi)^2}\,T^2 \,, \quad \text{with} \quad C = 1.33 \times 10^{20}\,\frac{\mathrm{m}^3}{\mathrm{s}^2} \,.$$

According to the second law, the area $A = \pi a^2 \sqrt{1 - \varepsilon^2}$ of the ellipse (see p. 63) is equal to $\frac{1}{2}cT$ and thus $T = 2\pi a^2 \sqrt{1 - \varepsilon^2}/c$, so we have

$$C = a^3/(T/2\pi)^2 = c^2/\{a\,(1 - \varepsilon^2)\} = c^2/\eta\ .$$

The abbreviation η introduced above may therefore be replaced by c^2/C:

$$r - \boldsymbol{\varepsilon} \cdot \mathbf{r} = \frac{c^2}{C}\,,\qquad \dot{\mathbf{r}} = \frac{C}{c^2}\,\mathbf{c} \times \left(\frac{\mathbf{r}}{r} - \boldsymbol{\varepsilon}\right),\qquad \ddot{\mathbf{r}} = -C\,\frac{\mathbf{r}}{r^3}\,,$$

$$\mathbf{c} = \mathbf{r} \times \dot{\mathbf{r}}\,,\qquad \boldsymbol{\varepsilon} = \frac{\mathbf{r}}{r} + \frac{\mathbf{c} \times \dot{\mathbf{r}}}{C}\,,\qquad \text{with}\ \ \mathbf{c} \cdot \boldsymbol{\varepsilon} = 0\ .$$

Thus, the Kepler problem is uniquely characterized by the two fixed vectors $\mathbf{c}$ and $\boldsymbol{\varepsilon}$ and the constant C (or a), where $\mathbf{c}$ and $\boldsymbol{\varepsilon}$ are perpendicular to each other. This gives 6 independent parameters: three Euler angles fix the orbital plane and the direction of the major axis, while two further parameters determine the lengths of the axes and the sixth the period.

If conversely we would like to infer the orbit from the acceleration $\ddot{\mathbf{r}} = -C\,\mathbf{r}/r^3$, then Kepler's second law follows immediately from $\ddot{\mathbf{r}} \parallel -\mathbf{r}$. Therefore, we may introduce the vector $\mathbf{c} = \mathbf{r} \times \dot{\mathbf{r}}$ as a constant of the motion. It is perpendicular to the orbit, i.e., to $\mathbf{r}$ and $\dot{\mathbf{r}}$. A further constant of the motion follows from

$$\frac{d}{dt}\left(\frac{\mathbf{r}}{r} + \frac{\mathbf{c} \times \dot{\mathbf{r}}}{C}\right) = \frac{(\mathbf{r} \times \dot{\mathbf{r}}) \times \mathbf{r}}{r^3} + \frac{\mathbf{c} \times \ddot{\mathbf{r}}}{C} = \frac{\mathbf{c} \times \mathbf{r}}{r^3} - \frac{\mathbf{c} \times \mathbf{r}}{r^3} = \mathbf{0}\,,$$

namely Lenz's vector

$$\boldsymbol{\varepsilon} = \frac{\mathbf{r}}{r} + \frac{\mathbf{c} \times \dot{\mathbf{r}}}{C}\ .$$

This can be solved for $\dot{\mathbf{r}}$, because $\mathbf{c} \cdot \dot{\mathbf{r}} = 0$, and we can also take the scalar product with $\mathbf{r}$:

$$\dot{\mathbf{r}} = \frac{C}{c^2}\,\mathbf{c} \times \left(\frac{\mathbf{r}}{r} - \boldsymbol{\varepsilon}\right)\qquad \text{and}\qquad \mathbf{r} \cdot \boldsymbol{\varepsilon} = r - \frac{c^2}{C}\ .$$

From this, we obtain (for $C > 0$) the elliptical orbit with the focus as the origin (Kepler's first law). We can thus derive all Kepler's laws from the single equation $\ddot{\mathbf{r}} = -C\,\mathbf{r}/r^3$, because the third follows from the other two if C is the same for all planets. Instead of c^2/C, we have used the geometric quantity η above.

Since $\boldsymbol{\varepsilon} = \mathbf{r}/r + \mathbf{c} \times \dot{\mathbf{r}}\,\eta/c^2$ and $\mathbf{c} \cdot \dot{\mathbf{r}} = 0$, we obtain the relation $\varepsilon^2 = \boldsymbol{\varepsilon} \cdot \boldsymbol{\varepsilon} = 1 - 2\eta/r + \dot{\mathbf{r}} \cdot \dot{\mathbf{r}}\,\eta^2/c^2$ for the square of the Lenz vector $\boldsymbol{\varepsilon}$, and since $c^2/\eta = C$, the square of the velocity is given by

$$v^2 \equiv \dot{\mathbf{r}} \cdot \dot{\mathbf{r}} = 2C/r - (1 - \varepsilon^2)\,C/\eta\ .$$

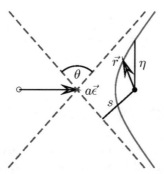

Fig. 2.6 The hyperbola branch $r - \boldsymbol{\varepsilon} \cdot \mathbf{r} = \eta$ for eccentricity $\varepsilon = 3/2$ (*red*) with the two foci (*full circle*, $\eta > 0$, attractive force and *open circle*, $\eta < 0$, repulsive force) and the asymptotes (*dashed blue lines*) in the directions to the initial and final points. In addition to the ray $\mathbf{r}$, the vector $a\boldsymbol{\varepsilon}$ to the center and the length η can be seen as in Fig. 2.3. The turning point is at a distance a from the center. In addition, the scattering angle θ and the collision parameter s are shown

This relation can also be derived from the conservation of energy (p. 78), because $\frac{1}{2}mv^2$ is the kinetic energy and $-mC/r$ the potential.

For a *circular orbit* about the zero point, the two foci coincide and $\mathbf{r} \cdot \dot{\mathbf{r}} = 0$. For constant orbital angular momentum, $\mathbf{r} \times \dot{\mathbf{r}} = \mathbf{c}$ is also conserved. Then, with $\boldsymbol{\omega} \equiv \mathbf{c}/r^2$,

$$\dot{\mathbf{r}} = \boldsymbol{\omega} \times \mathbf{r} .$$

We shall encounter this differential equation repeatedly. With $\mathbf{r}(0) \perp \boldsymbol{\omega}$, it is solved by $\mathbf{r}(t) = \mathbf{r}(0)\cos(\omega t) + \omega^{-1}\boldsymbol{\omega} \times \mathbf{r}(0)\sin(\omega t)$. Note that, if $\mathbf{r}(0)$ also has a component in the direction of $\boldsymbol{\omega}$, then this is conserved.

Let us thus look at the *hyperbolic orbit* (with $\varepsilon > 1$) (see Fig. 2.6). The directions of their asymptotes are determined by $r - \boldsymbol{\varepsilon} \cdot \mathbf{r} = 0$ and $\varepsilon \cos \varphi = 1$, where φ is half the opening angle. It is convenient here to define the *scattering angle* $\theta = \pi - 2\varphi$ and obtain $\sin \frac{1}{2}\theta = \varepsilon^{-1}$ and $\cot \frac{1}{2}\theta = \sqrt{\varepsilon^2 - 1}$. This can be expressed in terms of v_∞, because $v_\infty{}^2 = (\varepsilon^2 - 1)(C/c)^2$ and thus $\cot \frac{1}{2}\theta = v_\infty c/C$. If we then introduce the *collision parameter* s (distance of the asymptotes from the foci) with $c = s\,v_\infty$, we obtain $\cot \frac{1}{2}\theta = s\,v_\infty{}^2/C$.

This result is useful for the *Rutherford cross-section*, which describes the angular distribution for the elastic scattering of point charges q by a point charge q'— whatever enters the circular ring $2\pi s\,ds$ is scattered into the cone opening $2\pi \sin \theta\,d\theta$ (shown on the left of Fig. 5.5):

$$\frac{d\sigma}{d\Omega} = \left| \frac{2\pi\,s\,ds}{2\pi\,\sin\theta\,d\theta} \right| = \left| \frac{\frac{1}{2}\,ds^2}{d\cos\theta} \right| = \frac{C^2}{2\,v_\infty{}^4}\left| \frac{d\cot^2 \frac{1}{2}\theta}{d\cos\theta} \right| .$$

Since $\cos\theta = \cos^2 \frac{1}{2}\theta - \sin^2 \frac{1}{2}\theta = 2\cos^2 \frac{1}{2}\theta - 1$ and

$$\frac{d \cot^2 \frac{1}{2}\theta}{d \cos \theta} = \frac{1}{2} \frac{d}{d \cos^2 \frac{1}{2}\theta} \frac{\cos^2 \frac{1}{2}\theta}{1 - \cos^2 \frac{1}{2}\theta} = \frac{1}{2 \sin^4 \frac{1}{2}\theta} \ ,$$

we obtain

$$\frac{d\sigma}{d\Omega} = \frac{C^2}{4} \frac{1}{(v_\infty \sin \frac{1}{2}\theta)^4} = \frac{4\,C^2}{|\mathbf{v} - \mathbf{v}'|^4} \ ,$$

where $\mathbf{v}$ is the initial velocity and $\mathbf{v}'$ the final velocity. Here, $C = qq'/(4\pi \varepsilon_0\, m)$ with the reduced mass m, explained in more detail in Sect. 2.2.2, and the electric field constant ε_0, if the charges q and q' are given in coulomb (see p. 165 ff.). This is also obtained in (non-relativistic) quantum mechanics, as will be shown in Sect. 5.2.3. The scattering cross-section integrated over all directions Ω diverges, because the Coulomb force extends too far out. In reality, it will be screened by further charges.

2.1.7   Summary: Basic Concepts

We set the notion of *force* $\mathbf{F}$ as a basic ingredient of mechanics. The next section will be concerned with a different possibility, and we shall thus derive further quantities. In particular, a force can do *work* $\int \mathbf{F} \cdot d\mathbf{r}$ along a path. If this work depends only upon the initial and final point of the path and not upon the path in-between, then we may set $\mathbf{F} = -\nabla V$ and work with the simpler, scalar *potential energy* V.

According to Newton's third law, two bodies act on each other via equal but oppositely directed forces, which are not necessarily central forces.

A special kind of forces are the *forces of constraint*. They originate from geometric constraints and do no work. Therefore, they do not need to be accounted for as forces due to virtual displacements—instead, the geometric constraint has to be obeyed for all displacements $\delta\mathbf{r}$. If we can write the constraint as an equation $\Phi = 0$, then it can be accounted for by a Lagrangian parameter for the potential energy: $\delta\,(V - \lambda\,\Phi) = 0$ in equilibrium.

In addition to these notions, decisive for statics, we have also treated the Kepler problem as an example for kinematics. From Kepler's first and second laws (dating to 1609), we could infer $\ddot{\mathbf{r}} \propto -\mathbf{r}/r^3$. The missing factor here is the same for all planets, according to Kepler's third law (dating to 1619). With these laws, their motions can be described by a single differential equation—the different orbits follow from the corresponding initial conditions.

2.2 Newtonian Mechanics

2.2.1 Force-Free Motion

Newton[2] took the *inertial law* due to Galileo (1564–1642) as his *first axiom* in 1687:

If no force acts on it (also no frictional force), a body remains in its state of rest or of uniform rectilinear motion—it is inertial.

Here uniform rectilinear motion and the state of rest are equivalent. Different points of view are permitted, at rest and moving, as long as they are not accelerated relative to one other. Such allowable reference frames will be called *inertial frames*. In these frames, force-free bodies obey the inertial law. In contrast, bodies on curved orbits are always accelerated, according to Sect. 1.1.3. As a measure for uniform rectilinear motion, it is natural to think of the velocity. But Newton introduced instead the *momentum*

$$\mathbf{p} \equiv m\,\mathbf{v} = m\,\dot{\mathbf{r}}$$

as *motional quantity*. This is the velocity weighted by the *inertial mass m*. We shall encounter the notion of inertial mass in the context of the scattering laws in Sect. 2.2.3. For the moment it is sufficient for our purposes to note that each invariable body has a fixed mass, which depends neither upon time nor upon the position or the velocity, and is therefore a *conserved quantity*: a quantity is said to be conserved if it does not change with time. (Burning rockets and growing avalanches are "variable bodies", whose mass does not remain constant in time.) Therefore, the inertial law may also be called the *momentum conservation law* (often called *momentum conservation* for short):

$$\frac{\mathrm{d}\mathbf{p}}{\mathrm{d}t} \equiv \dot{\mathbf{p}} = \mathbf{0}\,, \quad \text{for force-free motion.}$$

If no force acts, the momentum is conserved (inertial law, law of persistence).

According to the theory of special relativity (Sect. 3.4), a body cannot move faster than the speed of light ($c = 299\,792\,458$ m/s), and therefore one actually has to set

$$\mathbf{v} = \gamma\,\frac{\mathrm{d}\mathbf{r}}{\mathrm{d}t}\,, \quad \text{with} \quad \gamma \equiv \frac{1}{\sqrt{1 - (\mathrm{d}\mathbf{r}/\mathrm{d}t)^2/c^2}}\,.$$

We then have $\mathbf{p} = m\,\gamma\,\mathrm{d}\mathbf{r}/\mathrm{d}t$. The factor γ is notably different from 1 only for $v \approx c$, as is clear from Fig. 3.23. Therefore, the simple non-relativistic calculation

---

[2]Isaac Newton (1643–1727) was professor in Cambridge from 1669–1701, Master of the Royal Mint in London in 1699, and President of the Royal Society of London in 1703.

fully suffices for many applications. To a similarly good approximation, "fixed" stars always remain at the same position and deliver a generalized reference frame.

As long as m does not depend on time and we consider force-free motion, then in addition to the polar vector $\mathbf{p}$, a scalar and an axial vector remain conserved, namely, the *kinetic energy T* and the *angular momentum* $\mathbf{L}$:

$$T \equiv \frac{1}{2m}\,\mathbf{p}\cdot\mathbf{p} = \frac{m}{2}\,\mathbf{v}\cdot\mathbf{v} \quad \text{and} \quad \mathbf{L} \equiv \mathbf{r}\times\mathbf{p}\,,$$

since for fixed $\mathbf{p}$ and m, the quantity T is also conserved—and for $\dot{\mathbf{p}} = \mathbf{0}$, we also have $\dot{\mathbf{L}} = \dot{\mathbf{r}}\times\mathbf{p} = \mathbf{p}\times\mathbf{p}/m = \mathbf{0}$. Altogether then, we have

$$\dot{T} = 0\,, \quad \dot{\mathbf{p}} = \mathbf{0}\,, \quad \text{and} \quad \dot{\mathbf{L}} = \mathbf{0}\,,$$

if no force acts.

In what follows, it will be useful to view the kinetic energy T as a scalar field of the variables $\mathbf{v}$. Hence, in the velocity space, we may also take the momentum $\mathbf{p}$ as the gradient of T (and use lower indices according to p. 35):

$$\mathbf{p} = \nabla_v T \quad \text{and} \quad p_k = \frac{\partial T}{\partial v^k}\,.$$

This will help us later in Lagrangian and Hamiltonian mechanics, but also for the separation into center-of-mass and relative motion.

2.2.2   Center-of-Mass Theorem

We have just introduced the mass m as a constant factor in $\mathbf{p} = m\,\mathbf{v}$. It has not yet been explained why we need the momentum at all in addition to the velocity. This changes only when there are several masses $m_1, m_2, \ldots$, for the above-mentioned laws are valid not only for a single body, but also for several bodies, which normally act on each other and thus exert forces—as long as there are no external forces acting on the bodies. According to Newton's third law (force equal to counter-force), the forces between the bodies cancel each other. Therefore, without external forces there is also no force on the system as a whole. This system we can treat as a single body. Its momentum is composed of the individual momenta and is conserved:

$$\mathbf{P} \equiv \sum_i \mathbf{p}_i\,, \quad \dot{\mathbf{P}} = \mathbf{0}\,, \quad \text{if no external forces act.}$$

The masses thus weight the individual velocities. Hence, for two bodies without external forces, we have $\dot{\mathbf{p}}_1 + \dot{\mathbf{p}}_2 = \mathbf{0}$, but $\dot{\mathbf{p}}_1 = -\dot{\mathbf{p}}_2 \neq \mathbf{0}$ if they act on each other.

If we introduce the total mass M and the position $\mathbf{R}$ of the center of mass,

$$M \equiv \sum_i m_i , \quad \mathbf{R} \equiv \frac{1}{M} \sum_i m_i \, \mathbf{r}_i ,$$

then for $\mathbf{F} = \mathbf{0}$, it moves with the constant velocity

$$\mathbf{V} \equiv \dot{\mathbf{R}} = \frac{1}{M} \sum_i m_i \, \dot{\mathbf{r}}_i = \frac{1}{M} \sum_i \mathbf{p}_i = \frac{1}{M} \mathbf{P} .$$

The total momentum is thus equal to the momentum of the center of mass. It is conserved if there are no external forces (*center-of-mass law*)—and hence according to the last section, the kinetic energy and the angular momentum of the center of mass remain conserved.

For many-body problems it is helpful to introduce center-of-mass and relative vectors instead of the position vectors $\mathbf{r}_i$. We shall show this for the case of two point masses with $M = m_1 + m_2$:

$$\mathbf{R} \equiv \frac{m_1 \, \mathbf{r}_1 + m_2 \, \mathbf{r}_2}{M} \quad \text{and} \quad \mathbf{r} \equiv \mathbf{r}_2 - \mathbf{r}_1 .$$

(For more point masses, we must proceed stepwise. After the two-body problem, the third is to be treated with respect to the center of mass of the first two, and so on. This leads to the *Jacobi coordinates*. In view of this, we thus take $\mathbf{r}_2 - \mathbf{r}_1$ and not $\mathbf{r}_1 - \mathbf{r}_2$ as the relative vector.) For this, it is convenient to write

$$\begin{pmatrix} \mathbf{R} \\ \mathbf{r} \end{pmatrix} = \begin{pmatrix} m_1/M & m_2/M \\ -1 & 1 \end{pmatrix} \begin{pmatrix} \mathbf{r}_1 \\ \mathbf{r}_2 \end{pmatrix} .$$

The determinant of the matrix here is equal to 1 and thus the map is area-preserving (see Fig. 2.7). (In more than two dimensions, the corresponding volume should remain conserved, whence the functional determinant has also to be equal to 1, as discussed further in Sect. 1.2.4.) Therefore, conversely, we have

$$\begin{pmatrix} \mathbf{r}_1 \\ \mathbf{r}_2 \end{pmatrix} = \begin{pmatrix} 1 & -m_2/M \\ 1 & m_1/M \end{pmatrix} \begin{pmatrix} \mathbf{R} \\ \mathbf{r} \end{pmatrix} ,$$

since the *inverse* of a 2×2 matrix is

$$\begin{pmatrix} a & b \\ c & d \end{pmatrix}^{-1} = \frac{1}{ad - bc} \begin{pmatrix} d & -b \\ -c & a \end{pmatrix} ,$$

something we shall use repeatedly.

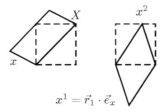

$$x^1 = \vec{r}_1 \cdot \vec{e}_x$$

Fig. 2.7 With the change from two-body to center-of-mass and relative coordinates (*left*) or vice versa (*right*), we make a transformation from rectangular to oblique coordinates which is not angle-preserving, shown here for the x-components and $m_1 = 2m_2$. The unit square (*dashed lines*) turns into a rhomboid of equal area (see Fig. 1.11). For $m_1 = m_2$, we have a rectangle on the left and a rhombus on the right

The same matrices appear for the transition $(\mathbf{v}_1, \mathbf{v}_2) \leftrightarrow (\mathbf{V}, \mathbf{v})$, because they remain conserved for the derivative with respect to time. Because $\mathbf{v}_1 = \mathbf{V} - \mathbf{v}\, m_2/M$ and $\mathbf{v}_2 = \mathbf{V} + \mathbf{v}\, m_1/M$, the *kinetic energy* is

$$T = \frac{m_1 v_1{}^2 + m_2 v_2{}^2}{2} = \frac{MV^2 + \mu v^2}{2} \;, \quad \text{with} \quad \mu \equiv \frac{m_1 m_2}{M} \;.$$

Only this *reduced mass* μ is important for relative motions. Hence, in T the mixed term $\mathbf{V} \cdot \mathbf{v}$ vanishes if we introduce a relative vector $\mathbf{r} \propto \mathbf{r}_2 - \mathbf{r}_1$ in addition to the center-of-mass vector $\mathbf{R}$. The center-of-mass and relative motion then already decouple. With $\mathbf{r} = \mathbf{r}_2 - \mathbf{r}_1$, we even obtain an area-preserving map (hence also $\mathbf{r}_1 \times \mathbf{r}_2 = \mathbf{R} \times \mathbf{r}$), but not an angle-preserving map, because the matrices are not orthogonal. Since we have already made a transition from $T(\mathbf{v}_1, \mathbf{v}_2)$ to $T(\mathbf{V}, \mathbf{v})$, we can also easily derive the momenta as gradients in velocity space:

$$\mathbf{P} = M\,\mathbf{V}\,, \quad \mathbf{p} = \mu\,\mathbf{v} \quad \Longrightarrow \quad T = \frac{p_1{}^2}{2m_1} + \frac{p_2{}^2}{2m_2} = \frac{P^2}{2M} + \frac{p^2}{2\mu}\;.$$

We already know the expression for $\mathbf{P}$. Clearly, the two momenta $\mathbf{P}$ and $\mathbf{p}$ can be expressed as linear combinations of $\mathbf{p}_1$ and $\mathbf{p}_2$, viz.:

$$\begin{pmatrix} \mathbf{P} \\ \mathbf{p} \end{pmatrix} = \begin{pmatrix} 1 & 1 \\ -m_2/M & m_1/M \end{pmatrix} \begin{pmatrix} \mathbf{p}_1 \\ \mathbf{p}_2 \end{pmatrix},$$

or

$$\begin{pmatrix} \mathbf{p}_1 \\ \mathbf{p}_2 \end{pmatrix} = \begin{pmatrix} rr m_1/M & -1 \\ m_2/M & 1 \end{pmatrix} \begin{pmatrix} c\mathbf{P} \\ \mathbf{p} \end{pmatrix},$$

noting that the momentum transformations are also area-preserving. In addition, we find for the *angular momentum*

$$\mathbf{L} = \mathbf{r}_1 \times \mathbf{p}_1 + \mathbf{r}_2 \times \mathbf{p}_2 = \mathbf{R} \times \mathbf{P} + \mathbf{r} \times \mathbf{p}\;.$$

If no external forces act, the forces depend only upon $\mathbf{r}$ (and possibly upon $\mathbf{v}$), and we only need to deal with the relative motion. With this, the two-body problem is reduced to a single-body problem and has become essentially easier.

The center-of-mass frame stands out here: if we choose the center of mass as origin, then $\mathbf{P} = \mathbf{0}$ and consequently, $\mathbf{p}_2 = -\mathbf{p}_1 = \mathbf{p}$.

2.2.3 Collision Laws

If two bodies collide without external forces acting, then the relative motion changes, but not the motion of the center of mass: $\mathbf{P}' = \mathbf{P}$. (Primed quantities will be used to describe the final state.) As far as the relative motion is concerned we need further information. In the following we consider only the motion before and after the collision, not during the collision—therefore we do not care about the forces between the collision partners. These are necessary, however, if we need to determine the scattering angle. In genuine scattering theory (see, e.g., Sects. 5.1 and 5.2), the interaction between the partners is indispensable.

In addition to elastic scattering, we need also to deal with inelastic processes, but without exchange of mass, i.e., the collision partners keep their masses, but during the collision, their relative motion could possibly lose energy which is converted into work of deformation, rotational energy, or heat. (With exchange of mass the equations become less clear, but in principle, the situation is no more difficult to treat.) Here, we introduce the *heat tone* $Q = (p'^2 - p^2)/2\mu$—for elastic scattering $p' = p$ and hence $Q = 0$. In contrast, for completely inelastic scattering, we have $p' = 0$, and thus $Q = -p^2/2\mu$. The ratio p'/p is abbreviated to

$$\xi \equiv \frac{p'}{p}, \quad \text{with} \quad p' = \sqrt{p^2 + 2\mu\,Q}\ .$$

For elastic scattering $\xi = 1$ and for completely inelastic scattering $\xi = 0$.

The relative momenta $\mathbf{p}'$ and $\mathbf{p}$ may have different moduli and also different directions. Therefore, we set $\mathbf{p}' = \xi\,D\,\mathbf{p}$ with the rotation operator D given in Sect. 1.2.1. Then with $\mathbf{P}' = \mathbf{P}$, according to the last section, we obtain

$$\begin{pmatrix} \mathbf{p}'_1 \\ \mathbf{p}'_2 \end{pmatrix} = \frac{1}{M} \begin{pmatrix} m_1 + \xi\,m_2\,D & m_1 - \xi\,m_1\,D \\ m_2 - \xi\,m_2\,D & m_2 + \xi\,m_1\,D \end{pmatrix} \begin{pmatrix} \mathbf{p}_1 \\ \mathbf{p}_2 \end{pmatrix}.$$

For a completely inelastic collision ($\xi = 0$), we thus have $\mathbf{v}'_1 = \mathbf{v}'_2 = \mathbf{V}$.

Rather simple situations also occur for collisions between two mass *points*, because they take place only for $\mathbf{r} = \mathbf{0}$. In this case the conservation of angular momentum leads to $D = -1$. If we consider here an elastic collision, then $\xi\,D = -1$ and hence,

$$\begin{pmatrix} \mathbf{p}_1' \\ \mathbf{p}_2' \end{pmatrix} = \frac{1}{M} \begin{pmatrix} m_1 - m_2 & 2m_1 \\ 2m_2 & m_2 - m_1 \end{pmatrix} \begin{pmatrix} \mathbf{p}_1 \\ \mathbf{p}_2 \end{pmatrix}.$$

In the special case where $m_2 = m_1$, it follows that $\mathbf{p}_1' = \mathbf{p}_2$ and $\mathbf{p}_2' = \mathbf{p}_1$: for equal masses the momenta (velocities) are exchanged. In contrast, for $m_2 \ll m_1$, it follows that $\mathbf{p}_1' \approx \mathbf{p}_1 + 2\mu\,\mathbf{v}_2$ and $\mathbf{p}_2' \approx 2\mu\mathbf{v}_1 - \mathbf{p}_2$ with $\mu \approx m_2$: only the small mass significantly changes its velocity.

Let us return to the collision of extended particles, but choose $\mathbf{p}_1 = \mathbf{0}$ and in this "laboratory frame" derive $\mathbf{p}_{1L}'$ and $\mathbf{p}_{2L}'$—here and in the following we indicate clearly whether the quantity refers to the laboratory frame (L) or the center-of-mass frame (S). However, this is unnecessary for $\mathbf{P}$ and $\mathbf{p}$: since $\mathbf{P}_S = \mathbf{0}$, the total momentum $\mathbf{P}$ should always refer to the laboratory frame, and the relative momentum $\mathbf{p}$ does not depend on the reference frame. According to the last section, $\mathbf{p} = \mathbf{p}_{2S} = -\mathbf{p}_{1S}$, and for $\mathbf{p}_{1L} = \mathbf{0}$, we have $\mathbf{P} = \mathbf{p}_{2L}$ and also $\mathbf{p} = (m_1/M)\,\mathbf{p}_{2L}$, as well as $p^2/2\mu = (m_1/M)\,T_{2L}$. This can be used to determine the parameter ξ:

$$\xi = \sqrt{1 + \frac{2\mu\,Q}{p^2}} = \sqrt{1 + \frac{M}{m_1}\frac{Q}{T_{2L}}}.$$

Since $\mathbf{p}_{1L} = \mathbf{0}$ and $\mathbf{p}_{2L} = (M/m_1)\,\mathbf{p}$, we now have

$$\mathbf{p}_{1L}' = (1 - \xi D)\,\mathbf{p} \qquad \text{and} \qquad \mathbf{p}_{2L}' = (\frac{m_2}{m_1} + \xi D)\,\mathbf{p}.$$

The scattering normal $\mathbf{n} = \mathbf{p} \times \mathbf{p}'/|\mathbf{p} \times \mathbf{p}'|$ in Fig. 2.8 points into the viewing direction, so the vector $\mathbf{n} \times \mathbf{p}$ points to the right. Therefore, in the center-of-mass frame, as a function of the scattering angle (θ_S), the rotated vector $D\mathbf{p}$ can be

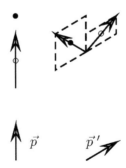

Fig. 2.8 A mass (*open circle*) collides with a mass twice as heavy (*closed circle*) at rest. The momenta before (*left*) and after (*right*) the collision are indicated in the laboratory frame (*top*) and the centre-of-mass frame (*bottom*). In the latter, only the relative momenta $\mathbf{p}$ and $\mathbf{p}'$ before and after are shown, as the total momentum $\mathbf{P}$ is conserved. Here an elastic collision was assumed, whence the *dashed lines* around the *full circle* form a rhombus and are in the ratio $m_2 : m_1$. The large rhombus angle is clearly equal to $\pi - \theta_S$ and twice as large as $\theta_{\bullet L}$. For equal masses the two objects fly off from each other at right angles

expanded in terms of $\mathbf{p}$ and $\mathbf{n} \times \mathbf{p}$: $D\mathbf{p} = \cos \theta_S \, \mathbf{p} + \sin \theta_S \, \mathbf{n} \times \mathbf{p}$. We then obtain for the recoil momentum $\mathbf{p}'_{1L}$ and for the momentum of the colliding particle $\mathbf{p}'_{2L}$,

$$\mathbf{p}_{1L}' = (\ 1 \ - \xi \cos \theta_S) \, \mathbf{p} - \xi \sin \theta_S \, \mathbf{n} \times \mathbf{p} \, ,$$

$$\mathbf{p}_{2L}' = (\frac{m_2}{m_1} + \xi \cos \theta_S) \, \mathbf{p} + \xi \sin \theta_S \, \mathbf{n} \times \mathbf{p} \, .$$

Hence, noting that we should always have $0 \leq \theta \leq \pi$, and in addition that $\tan \theta_{iL} = |\mathbf{p}_{iL}' \times \mathbf{p}|/(\mathbf{p}'_{iL} \cdot \mathbf{r})$ and $(\mathbf{n} \times \mathbf{r}) \times \mathbf{p} = -\mathbf{n}$, the scattering angle in the laboratory frame of the scattering particle (the one that is impinged upon) is given by

$$\tan \theta_{1L} = \frac{\sin \theta_S}{\xi^{-1} - \cos \theta_S} \, ,$$

while for the scattered (impinging) particle,

$$\tan \theta_{2L} = \frac{\sin \theta_S}{\zeta + \cos \theta_S} \, , \quad \text{with} \quad \zeta \equiv \frac{m_2}{\xi \, m_1} \, .$$

From this we can conclude that, for elastic collision ($\xi = 1$), as in Fig. 2.8, $\theta_{1L} = \frac{1}{2}(\pi - \theta_S)$ and for $\zeta = 1$, $\theta_{2L} = \frac{1}{2}\theta_S$, so that for equal masses (and elastic collision) $\theta_{1L} + \theta_{2L} = \frac{1}{2}\pi$.

For a given θ_S, there is a value θ_{1L} and a value of θ_{2L}, and for $\xi \leq 1$, $\theta_{1L} \leq \frac{1}{2}\pi$. In most cases, we do not consider the recoil, and instead of θ_{2L}, we simply write θ_L. Since $\sin \theta_S \cos \theta_L = (\zeta + \cos \theta_S) \sin \theta_L$, the relation between θ_L and θ_S can thus be written as

$$\zeta \sin \theta_L = \sin (\theta_S - \theta_L) \quad \Longleftrightarrow \quad \theta_S = \theta_L + \arcsin (\zeta \sin \theta_L) \, .$$

This relation is shown in Fig. 2.9 for different values of ζ. For $\zeta > 1$, only values $\theta_L \leq \arcsin \zeta^{-1}$ occur, and for each θ_L below this bound, there are two values of θ_S.

For the moduli of the momenta, we find

$$p_{1L}' = p \sqrt{1 - 2\xi \cos \theta_S + \xi^2}$$

and

$$p_{2L}' = p \xi \sqrt{1 + 2\zeta \cos \theta_S + \zeta^2} \, .$$

The recoil momentum p'_{1L} is equal to the momentum transfer in the laboratory or center-of-mass frame ($|\mathbf{p}_{2L}' - \mathbf{p}_{2L}| = |\mathbf{p}'_{2S} - \mathbf{p}_{2S}|$). For an elastic collision, it is equal to $2p \sin \frac{1}{2}\theta_S$. Since $\cos \theta_L = \mathbf{p}_{2L}' \cdot \mathbf{p}/(p'_{2L} \, p)$, we have

Fig. 2.9 Relation between
the scattering angles in
laboratory and
center-of-mass coordinates
for $\zeta = 1/2$ and $3/4$ (*dashed
red*), for $\zeta = 1$ (*full
magenta*), and for $\zeta = 4/3$
and 2 (*dotted blue*)

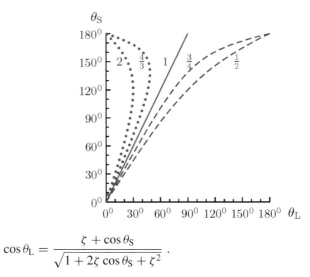

$$\cos\theta_L = \frac{\zeta + \cos\theta_S}{\sqrt{1 + 2\zeta\cos\theta_S + \zeta^2}} \, .$$

Hence one obtains the ratio $d\Omega_L/d\Omega_S = |d\cos\theta_L/d\cos\theta_S|$, namely

$$\frac{d\Omega_L}{d\Omega_S} = \frac{|1 + \zeta\ \cos\theta_S|}{\sqrt{1 + 2\zeta\cos\theta_S + \zeta^2}^{\,3}} \, ,$$

whence the scattering cross-sections can be converted from the laboratory to the center-of-mass frame (or vice versa).

In principle the mass ratio can be determined from the velocities before and after the collision, even if it is inelastic, whence $\xi = |\mathbf{v}_2' - \mathbf{v}_1'|/v_2 \neq 1$: for a central collision ($\theta_S = \pi$) and since $v_2' : v_2 = (m_2 - \xi m_1) : (m_2 + m_1)$, we have

$$\frac{m_2}{m_1} = \frac{\xi v_2 + v_2'}{v_2 - v_2'} \, .$$

For all other collisions, the momenta perpendicular to the original one cancel each other (momentum conservation): $m_2 : m_1 = v_{1\perp}' : v_{2\perp}'$. Further supplements can be found in Problems 2.15–2.18.

2.2.4 Newton's Second Law

Newton's law of motion is understood as his second axiom:

Each force $\mathbf{F}$ *on a freely mobile body changes its momentum according to*

$$\mathbf{F} = \dot{\mathbf{p}} \, .$$

The inertial law, referring to the case $\mathbf{F} = \mathbf{0}$, seems to be a special case. But this was taken as defining an inertial system, because only then can the mass and momentum be introduced as observables. Since $d\mathbf{p} = \mathbf{F}\,dt$, the force often appears in an integral over $\mathbf{F}\,dt$, which is referred to as the *impulse (impulsive force)*. In $\mathbf{p} = m\dot{\mathbf{r}}$, we can often take the mass as constant, whereupon

$$\mathbf{F} = m\,\ddot{\mathbf{r}}\,.$$

In relativistic dynamics, the factor $\gamma = 1/\sqrt{1 - v^2/c^2}$ also enters our considerations, because we must refer to the proper time, as will be shown in Sect. 3.4.10.

The equation $\mathbf{F} = \dot{\mathbf{p}}$ can be applied to rotational motion. Since $\dot{\mathbf{r}} \parallel \mathbf{p}$, it is clear that $d\,(\mathbf{r} \times \mathbf{p})/dt = \mathbf{r} \times \dot{\mathbf{p}} = \mathbf{r} \times \mathbf{F} = \mathbf{M}$, and since $\mathbf{r} \times \mathbf{p} = \mathbf{L}$, we conclude that

$$\mathbf{M} = \dot{\mathbf{L}}\,.$$

A torque on a mobile body changes its angular momentum.

For an invariable mass, the law of motion delivers a differential equation of second order:

$$\ddot{\mathbf{r}} = \frac{\mathbf{F}(t,\ \mathbf{r},\ \dot{\mathbf{r}})}{m}\,.$$

This differential equation has to be integrated, because we are interested in the orbit, and from $\mathbf{r}\,(t)$ we can derive the velocity. Then, for each integration an integration constant occurs (here, actually an integration vector). The law of motion leaves us with the choice of initial position and velocity, so the general solution $\mathbf{r}$ of the differential equation depends upon t, $\mathbf{r}_0$, and $\dot{\mathbf{r}}_0$. These values have to determine the solution uniquely, otherwise the force is unphysical.

If the force does not depend upon the velocity, but only on position $\mathbf{r}$ and possibly time t, we speak of a given *force field*. Since for a given force the acceleration $\ddot{\mathbf{r}}$ is inversely proportional to the mass, we consider the field $\mathbf{F}/m$, and for curl-free force fields, the *potential* $\Phi \equiv V/m$ instead of the potential energy V. Only then is the force field independent of the test body, and we have

$$\ddot{\mathbf{r}} = -\nabla\Phi\,, \quad \text{with} \quad \Phi \equiv \frac{V}{m}\,,$$

if $\nabla \times \mathbf{F} = \mathbf{0}$ and $\dot{m} = 0$.

2.2.5 Conserved Quantities and Time Averages

If a force acts, $\mathbf{F} \neq \mathbf{0}$, then the momentum is no longer a conserved quantity because it changes with time. But let us consider also the two conserved quantities encountered so far, the kinetic energy and the angular momentum: what are their derivatives with respect to time when a force acts? If we assume a constant mass, we obtain

$$\frac{\mathrm{d}T}{\mathrm{d}t} = \frac{1}{2m} \frac{\mathrm{d}}{\mathrm{d}t} \mathbf{p} \cdot \mathbf{p} = \frac{\mathbf{p}}{m} \cdot \frac{\mathrm{d}\mathbf{p}}{\mathrm{d}t} = \mathbf{v} \cdot \mathbf{F} = \mathbf{F} \cdot \frac{\mathrm{d}\mathbf{r}}{\mathrm{d}t} \ .$$

For a time-independent force, we thus find $\mathrm{d}T = \mathbf{F} \cdot \mathrm{d}\mathbf{r}$. If moreover the force field is curl-free, then it can be derived from a potential energy V, and because $\mathrm{d}V = \nabla V \cdot \mathrm{d}\mathbf{r} = -\mathbf{F} \cdot \mathrm{d}\mathbf{r}$, we clearly have $\mathrm{d}T = -\mathrm{d}V$. (If the force depends upon time, then neither $\mathrm{d}T = \mathbf{F} \cdot \mathrm{d}\mathbf{r}$ nor $\oint \mathbf{F} \cdot \mathrm{d}\mathbf{r} = 0$ can be inferred, and it then depends on the time span over which the work is done.) Thus, there is a *conservation law for the energy*, viz.,

$$E \equiv T + V \ ,$$

if V (or the associated force $\mathbf{F}$) does not depend on time.

In the following sections, we shall discuss several examples with curl-free forces, to which a potential can therefore be assigned. An important counter-example is provided by the *Lorentz force*

$$\mathbf{F} = q \ (\mathbf{E} + \mathbf{v} \times \mathbf{B}) \ ,$$

which acts on an electric charge q in an electromagnetic field specified by $\mathbf{E}$ and $\mathbf{B}$. The Maxwell equations $\nabla \times \mathbf{E} = -\partial \mathbf{B}/\partial t$ and $\nabla \cdot \mathbf{B} = 0$ imply that $\mathbf{F}$ has the curl density

$$\nabla \times \mathbf{F} = -q \left(\frac{\partial \mathbf{B}}{\partial t} + (\mathbf{v} \cdot \nabla) \mathbf{B} \right) = -q \dot{\mathbf{B}} \ ,$$

since $\mathbf{r}$ and $\mathbf{v}$ have to be treated as mutually independent variables, whereupon $\nabla \times (\mathbf{v} \times \mathbf{B}) = \mathbf{v} \nabla \cdot \mathbf{B} - (\mathbf{v} \cdot \nabla) \mathbf{B}$. Even if the magnetic field does not depend on position (only on time), the force field has curls. Then there is no potential energy, unless we introduce a generalized potential energy as in Sect. 2.3.4. In any case, here (with $\mathbf{E} = \mathbf{0}$ and constant mass) the equation of motion is

$$\dot{\mathbf{v}} = \boldsymbol{\omega} \times \mathbf{v} \ , \quad \text{with} \quad \boldsymbol{\omega} = \frac{-q\mathbf{B}}{m} \ .$$

The value of $\boldsymbol{\omega}$ is the *cyclotron frequency*. (Note that we encountered a similar differential equation for the circular orbit, but for $\mathbf{r}$ rather than $\mathbf{v}$, on p. 67.) Even though a force acts here, the kinetic energy is still conserved because the Lorentz force is always perpendicular to $\mathbf{v}$ and thus does not change v. Therefore, if we set $\mathbf{v} = v \mathbf{e}_\mathrm{T}$ with fixed v, it follows that $\dot{\mathbf{e}}_\mathrm{T} = \boldsymbol{\omega} \times \mathbf{e}_\mathrm{T}$, or $\mathrm{d}\mathbf{e}_\mathrm{T}/\mathrm{d}s = v^{-1} \boldsymbol{\omega} \times \mathbf{e}_\mathrm{T}$. We have already met this differential equation on p. 8. Quite generally, the charge moves on a helical orbit in the homogeneous magnetic field, with fixed Darboux vector $\tau \mathbf{e}_\mathrm{T} + \kappa \mathbf{e}_\mathrm{B} = \boldsymbol{\omega}/v$.

The other conserved quantity introduced so far, the angular momentum $\mathbf{L} = \mathbf{r} \times \mathbf{p}$, is only a conserved quantity if the torque $\mathbf{M}$ vanishes, i.e., if $\mathbf{F}$ is a central force, since $d\mathbf{L}/dt = \mathbf{M}$ according to the last section. This is the case, e.g., if the potential has spherical symmetry:

$$\Phi(\mathbf{r}) = \Phi(r) \quad \Longrightarrow \quad \nabla\Phi = \frac{d\Phi}{dr}\frac{\mathbf{r}}{r} \quad \Longrightarrow \quad \frac{d\mathbf{L}}{dt} = \mathbf{0} \, .$$

Note that here only the angular momentum with respect to the symmetry center is conserved. It is only if no force acts at all that it is conserved with respect to any point. For cylindrical symmetry, thus if Φ does not depend upon the angle coordinate φ, at least the angular momentum component along the symmetry axis is conserved.

Of course, mean values taken over time are also conserved. This is important for the *virial theorem*, which says that, if $\mathbf{r}$ and $\mathbf{p}$ always stay finite (and the mass always the same), then for the time-averaged value of the *virial* $\mathbf{r} \cdot \mathbf{F}$,

$$\overline{\mathbf{r} \cdot \mathbf{F}} = -2\,\overline{T} \, .$$

Hence, if $\mathbf{r}$ and $\mathbf{p}$ always stay finite, then so does the auxiliary quantity $G(t) = \mathbf{r} \cdot \mathbf{p}$. For sufficiently long times τ, the quantity $\tau^{-1}\{G(\tau) - G(0)\}$ thus vanishes, and this is the mean value of $\dot{G} = \mathbf{v} \cdot \mathbf{p} + \mathbf{r} \cdot \mathbf{F} = 2T + \mathbf{r} \cdot \mathbf{F}$ between 0 and τ. For example, for a central force $\mathbf{F} = cr^n\,\mathbf{r}/r$, the virial is equal to cr^{n+1}, and hence, according to p. 57, it is equal to $-(n+1)\,V$. This theorem leads here to $\overline{T} = \frac{1}{2}(n+1)\,\overline{V}$. In particular, for a harmonic oscillation, we have $n = 1$ and thus $\overline{T} = \overline{V}$, while for the gravitational and Coulomb forces $n = -2$, and thus $\overline{T} = -\frac{1}{2}\overline{V}$. The virial theorem must not be applied to the hyperbolic orbit, because $\mathbf{r} \cdot \mathbf{p}$ does not remain finite.

2.2.6 Planetary Motion as a Two-Body Problem, and Gravitational Force

If there are no external forces, the total momentum is conserved. Then we are concerned only with the relative motion. An important example is application to the Sun–Earth system, which may be viewed approximately as a two-body problem, although the Moon and the other planets should be accounted for in a more accurate solution.

Here *gravity* acts, that is, the force between *massive bodies*. So far, the term "mass" has always been understood as inertial mass. But in fact, the active gravitational mass m_1 exerts a force

$$\mathbf{F}_{21} = -G\,\frac{m_2 m_1}{|\mathbf{r}_2 - \mathbf{r}_1|^2}\,\frac{\mathbf{r}_2 - \mathbf{r}_1}{|\mathbf{r}_2 - \mathbf{r}_1|}$$

on the passive gravitational mass m_2, where G is the *gravitational constant* (see p. 623). But from experience, we may assume the active and passive gravitational masses and the inertial mass to be the same, at least to an accuracy of one part in 10^{11})—this is the basis of general relativity theory.

Exactly as the Sun (S) attracts the Earth (E) and hence exerts the force $\mathbf{F}_{\mathrm{ES}}$, the Earth attracts the Sun with the opposite force $\mathbf{F}_{\mathrm{SE}}$ according to Newton's third law. Hence we infer

$$\dot{\mathbf{p}}_{\mathrm{E}} = \mathbf{F}_{\mathrm{ES}} = -\mathbf{F}_{\mathrm{SE}} = -\dot{\mathbf{p}}_{\mathrm{S}} \, ,$$

and $\dot{\mathbf{P}} \equiv \dot{\mathbf{p}}_{\mathrm{S}} + \dot{\mathbf{p}}_{\mathrm{E}} = \mathbf{0}$ for the center of mass. Then, according to p. 72,

$$\dot{\mathbf{p}} \equiv \frac{m_{\mathrm{S}}\,\dot{\mathbf{p}}_{\mathrm{E}} - m_{\mathrm{E}}\,\dot{\mathbf{p}}_{\mathrm{S}}}{m_{\mathrm{S}} + m_{\mathrm{E}}} = \dot{\mathbf{p}}_{\mathrm{E}} = \mathbf{F}_{\mathrm{ES}} = -G\,\frac{m_{\mathrm{E}}\,m_{\mathrm{S}}}{r^2}\,\frac{\mathbf{r}}{r} \, ,$$

for the relative momentum.

Once again, only the relative coordinate is of interest—no external force acts on the center of mass, as long as the influence of other celestial objects remains negligible. Since $\mathbf{p} = \mu\dot{\mathbf{r}}$ with $\mu = m_{\mathrm{S}}\,m_{\mathrm{E}}/(m_{\mathrm{S}} + m_{\mathrm{E}})$,

$$\ddot{\mathbf{r}} = -G\,\frac{m_{\mathrm{S}} + m_{\mathrm{E}}}{r^2}\,\frac{\mathbf{r}}{r} \, .$$

Therefore, the first two Kepler laws are valid not only with the Sun at the coordinate origin, but also for the relative motion. With the third law, however, we have

$$\frac{a^3}{T^2} = G\,\frac{m_{\mathrm{S}} + m_{\mathrm{E}}}{4\pi^2} \, ,$$

i.e., for every planet there is another "constant", since $a^3/T^2 = C/4\pi^2$ holds with $\ddot{\mathbf{r}} = -C\mathbf{r}/r^3$, according to p. 65. However, the mass ratio of planet to Sun is less than 0.001 even for Jupiter. In addition, we have neglected the mutual attractions of the other planets and moons. This perturbation can be accounted for approximately. This is how, from the perturbed orbit of Uranus, Leverrier deduced the presence of the as yet unknown planet Neptune, a jewel of celestial mechanics. Incidentally, Kepler had already noticed that Jupiter and Saturn did not travel on purely elliptical orbits—these two neighboring planets are the heaviest in the Solar System and therefore perturb each other with a particularly strong force. Likewise, returning comets move on elliptical orbits about the Sun which are sensitive to perturbations (see Problem 2.11).

The gravitational force acts "not only in heaven, but also on Earth". All objects are pulled toward the Earth—they have *weight*. However, this notion is used with different meanings. In the international system (SI), the (gravitational) mass is understood, but in any everyday context, the associated gravitational force. If we buy 1 kg of flour, we actually want to have the associated mass, but when we weigh it, we use the force with which the Earth attracts this mass. Physicists should stick to the international system and also take "weight" as mass.

2.2.7 *Gravitational Acceleration*

According to the gravitational law, at its surface, the Earth exerts the gravitational force

$$\mathbf{F} = m\,\mathbf{g} \quad \text{with} \quad \mathbf{g} = -\frac{G\,m_E}{R^2}\frac{\mathbf{R}}{R},$$

on a body of mass m, if we assume a spherically symmetric Earth. Here the gravitational acceleration $\mathbf{g}$ is assumed constant, as long as the distance R from the center of the Earth changes only negligibly (since the Earth rotates about its axis, we should also take into account the position-dependent centrifugal force). The vector $-\mathbf{R}/R$ is a unit vector which, at the surface of the Earth, points "vertically downwards". The gravitational acceleration $\mathbf{g}$ thus follows from the mass m_E and radius R of the Earth and the gravitational constant G.

According to this equation, the total mass m_E can be considered as concentrated at the center of the Earth when evaluating the gravitational force on a test body near the surface of the Earth. For the proof, since a scalar field is easier to work with than the associated force field, we consider the gravitational potential

$$\Phi(\mathbf{r}) = -\frac{G\,m_E}{r} \quad \Longleftrightarrow \quad \mathbf{F}(\mathbf{r}) = -G\,\frac{m\,m_E}{r^2}\frac{\mathbf{r}}{r},$$

which we derive for $r \geq R$ from

$$\Phi(\mathbf{r}) = -G\int \frac{\rho(\mathbf{r}')\,d^3r'}{|\mathbf{r}-\mathbf{r}'|}.$$

Here we assume the density distribution to be spherically symmetric, i.e., $\rho(\mathbf{r}') = \rho(r')$, although it does not need to be homogeneous (actually, the Earth's mantle has a lower density than the core). Thus let

$$m_E = \int \rho(\mathbf{r}')\,d^3r' = 4\pi \int \rho(r')\,r'^2 dr'.$$

In order to evaluate the potential, we expand $|\mathbf{r}-\mathbf{r}'|^{-1}$ in powers of $s = r'/r < 1$ and introduce the angle θ between $\mathbf{r}'$ and $\mathbf{r}$:

$$\frac{1}{|\mathbf{r}-\mathbf{r}'|} = \frac{1}{r\sqrt{1-2s\cos\theta+s^2}} = \frac{1}{r}\sum_{n=0}^{\infty} P_n(\cos\theta)\,s^n.$$

The expansion coefficients $P_n(\cos\theta)$ are called *Legendre polynomials*. We shall meet them occasionally, e.g., in electrostatics (see p. 181) and in quantum theory with the spherical functions (see p. 334 ff). The first of these are (see Fig. 2.10)

Fig. 2.10 Legendre
polynomials $P_n(z)$ with n
from 0 to 5. *Continuous
curves*: Even n. *Dashed
curves*: Odd n. It can be
proven recursively that
$P_n(1) = 1$ and
$P_n(-z) = (-)^n P_n(z)$ for all
$n \in \{0, 1, 2, \ldots\}$ and that n
gives the number of zeros of
the given function

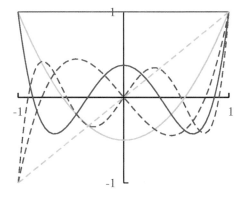

$$P_0(z) = 1 , \qquad P_1(z) = z , \qquad P_2(z) = \tfrac{1}{2} (3z^2 - 1) , \qquad \ldots \ .$$

The remaining ones can be obtained via the *recursion relation*

$$(n+1) P_{n+1}(z) - (2n+1) z P_n(z) + n P_{n-1}(z) = 0 ,$$

which follows from the *generating function* (see the power series above)

$$\frac{1}{\sqrt{1 - 2sz + s^2}} = \sum_{n=0}^{\infty} P_n(z) s^n , \quad \text{for } |s| < 1 .$$

This means that, if we differentiate this equation with respect to s and then multiply it
by $1 - 2sz + s^2$, we obtain $(z - s) \sum_n P_n(z) s^n = (1 - 2sz + s^2) \sum_n n P_n(z) s^{n-1}$,
and hence by comparing coefficients, the recursion relation is proven.

In addition, the Legendre polynomials have the property (important for us)

$$\int_{-1}^{1} P_n(z) P_{n'}(z) \, dz = \int_0^{\pi} P_n(\cos\theta) P_{n'}(\cos\theta) \sin\theta \, d\theta = \frac{2}{2n + 1} \delta_{nn'} ,$$

whence they form a (complete) *orthogonal system* for $-1 \leq z \leq 1$. This can be
proven using the generating function of the Legendre polynomials. For $|s| < 1$ and
$|t| < 1$, this delivers

$$\frac{1}{\sqrt{1 - 2sz + s^2} \sqrt{1 - 2tz + t^2}} = \sum_{mn} P_m(z) P_n(z) s^m t^n .$$

But now, if we cancel a factor of $\sqrt{2t} + \sqrt{2s}$,

$$\int_{-1}^{1} \frac{dz}{\sqrt{1 - 2sz + s^2}\sqrt{1 - 2tz + t^2}}$$

$$= -\frac{1}{\sqrt{st}} \ln\left(\sqrt{2t\,(1 - 2sz + s^2)} + \sqrt{2s\,(1 - 2tz + t^2)}\,\right)\Bigg|_{-1}^{+1}$$

$$= \frac{1}{\sqrt{st}} \ln \frac{\sqrt{2t}\,(1 + s) + \sqrt{2s}\,(1 + t)}{\sqrt{2t}\,(1 - s) + \sqrt{2s}\,(1 - t)} = \frac{1}{\sqrt{st}} \ln \frac{1 + \sqrt{st}}{1 - \sqrt{st}}\,.$$

Hence, since $\ln(1 + x) = \sum_{n=0}^{\infty} x^n / n$ for $|x| < 1$, it follows that

$$\int_{-1}^{1} \frac{dz}{\sqrt{1 - 2sz + s^2}\,\sqrt{1 - 2tz + t^2}} = \sum_{n=0}^{\infty} \frac{2}{2n + 1}\,(st)^n, \quad \text{for } |st| < 1\,.$$

Comparing coefficients proves the claim.

Further properties of the Legendre polynomials are given on p. 334, and more can be found in, e.g., [1].

Since we started from a spherically symmetric density distribution $\rho(\mathbf{r}') = \rho(r')\,P_0$, after integrating over all directions, only the term with $n = 0$ remains:

$$\int \frac{\rho(\mathbf{r}')\,d^3 r'}{|\mathbf{r} - \mathbf{r}'|} = \frac{2 \times 2\pi}{r} \int \rho(r')\,r'^2\,dr' = \frac{m_E}{r}\,.$$

This means that we can perform calculations as though the mass of the Earth were concentrated at its center. (Problem 2.20 is also instructive here.)

2.2.8 Free-Fall, Thrust, and Atmospheric Drag

If we calculate with the same gravitational acceleration $\mathbf{g}$ everywhere on the surface of the Earth, then, according to Newton's law of motion, we obtain

$$\ddot{\mathbf{r}} = \mathbf{g}\,, \qquad \dot{\mathbf{r}} = \mathbf{v}_0 + \mathbf{g}\,t\,, \qquad \mathbf{r} = \mathbf{r}_0 + \mathbf{v}_0\,t + \tfrac{1}{2}\mathbf{g}\,t^2\,.$$

According to pp. 57 and 77, a gravitational potential

$$\Phi(\mathbf{r}) = -\,\mathbf{g} \cdot \mathbf{r}$$

is associated with the constant acceleration. The gauge here is such that the potential vanishes at the surface of the Earth, where the coordinate origin is taken. If a body loses height h, its potential energy decreases by mgh. For free fall, the kinetic energy increases by this amount, so its velocity goes from zero to $v = \sqrt{2gh}$.

If the body is thrown through air instead of empty space, then it loses momentum to the air molecules it collides with. The number of collisions per unit time increases linearly with its velocity, and in each collision, it loses on average a fraction of its momentum determined by the mass ratio. Hence we have to set the frictional force proportional to $-v\,\mathbf{v}$ (*Newtonian friction*, not Stokes friction, which would be proportional to $\mathbf{v}$, as, e.g., later on p. 99) and write with $\beta > 0$,

$$\dot{\mathbf{v}} = \mathbf{g} - \beta^2\, g\, v\, \mathbf{v}\ .$$

For objects surrounded by fluids one normally writes $c_{\mathrm{w}}\frac{1}{2}\rho A\, v^2$ for the frictional force, where c_{w} is the *drag coefficient*, ρ the density of the medium (here the air), and A the cross-section of the body perpendicular to the air stream. Streamlined bodies have the smallest drag coefficient, namely 0.055. As far as the author is aware, the above non-linear differential equation can be solved in closed form only in one dimension. Therefore we assume that $\mathbf{v}_0$ is parallel to the vertical and measure v in the direction of $\mathbf{g}$. Then we have

$$\frac{dv}{dt} = g\,(1 - \beta^2 v^2) \qquad \Longrightarrow \qquad \frac{dv}{1 - \beta^2 v^2} = g\,dt\ .$$

After separation of variables, we can integrate and obtain

$$v = \frac{1}{\beta}\,\frac{\beta v_0 + \tanh(\beta g t)}{1 + \beta v_0\,\tanh(\beta g t)}\ .$$

Consequently, the velocity changes at first linearly with time, $v \approx v_0 + (1 - \beta^2 v_0{}^2)\,gt$, and finally becomes constant (incidentally faster than the horizontal component of $\mathbf{v}$ which tends to zero):

$$v \approx \frac{1}{\beta}\left(1 - 2\,\frac{1 - \beta v_0}{1 + \beta v_0}\,\exp\left(-2\beta g t\right) + \cdots\right)\ .$$

For $x \gg 1$, i.e., $\tanh x \approx -2\exp\left(-2x\right)$, and with $b = \beta v_0$ and $\mathrm{e} = 2\exp(-2x)$, we have

$$\frac{b + 1 - \mathrm{e}}{1 + b\,(1 - \mathrm{e})} = \frac{1 - \mathrm{e}/(1 + b)}{1 - b\mathrm{e}/(1 + b)}\ ,$$

and because $|\mathrm{e}| \ll 1$, we may replace $\{1 - b\mathrm{e}/(1 + b)\}^{-1}$ approximately by $1 + b\mathrm{e}/(1 + b)$ and likewise $\{1 - \mathrm{e}/(1 + b)\}\{1 + b\mathrm{e}/(1 + b)\}$ by $1 - \mathrm{e}\,(1 - b)/(1 + b)$. The body is accelerated until its gravitational force and frictional force cancel each other. It then permanently loses potential energy without increasing its kinetic energy—the energy has now completely turned into frictional energy (heat).

Note that a solution initially changing linearly in time and ending up exponentially approaching a constant velocity also occurs for free fall with Stokes's friction, viz., $\dot{\mathbf{v}} = \mathbf{g} - \alpha\mathbf{v}$. Then we have $\mathbf{v} = \alpha^{-1}\mathbf{g} + (\mathbf{v}_0 - \alpha^{-1}\mathbf{g})\,\exp\left(-\alpha t\right)$, where $\mathbf{v}_0$ and $\mathbf{g}$ may span a plane. As can be seen from Fig. 2.11, it is better in any case to calculate with this approximation than to neglect friction completely.

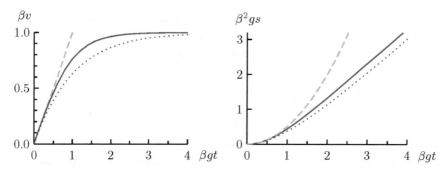

Fig. 2.11 Free fall from rest with friction with the air. *Continuous red curves*: Newtonian friction. *Dotted blue curves*: Stokes's friction (here, $\alpha = \beta g$). *Dashed green curves*: Without friction (with appropriate scaling $v = gt$ and $s = \frac{1}{2} gt^2$)

2.2.9 Rigid Bodies

The parts of a rigid body keep always their relative distances. Therefore, we shall refer the position vectors of the mass elements dm to a fixed point in the body, usually the center of mass $\mathbf{R} = \int \mathbf{r}\, dm / M$:

$$\mathbf{r}' = \mathbf{r} - \mathbf{R} \quad \Longrightarrow \quad \int \mathbf{r}'\, dm = \mathbf{0}\ .$$

The vectors $\mathbf{r}'$ have constant lengths, so we can infer the equations

$$\frac{d}{dt}\,(\mathbf{r}' \cdot \mathbf{r}') = 2\,\mathbf{r}' \cdot \frac{d\mathbf{r}'}{dt} = 0 \quad \Longrightarrow \quad \frac{d\mathbf{r}'}{dt} = \boldsymbol{\omega} \times \mathbf{r}'\ .$$

Here $\boldsymbol{\omega}$ is an axial vector in the direction of the axis of rotation (right-hand rule) and with the value of the angular velocity, as already introduced on p. 67 for circular motion. It describes the rotation of the rigid body and does not depend on the position $\mathbf{r}'$. For all i and k, $\mathbf{r}'_i \cdot \mathbf{r}'_k$ will not depend on time, whence $\dot{\mathbf{r}}'_i \cdot \mathbf{r}'_k + \mathbf{r}'_i \cdot \dot{\mathbf{r}}'_k = (\boldsymbol{\omega}_i - \boldsymbol{\omega}_k) \cdot (\mathbf{r}'_i \times \mathbf{r}'_k)$ must always vanish, and $\boldsymbol{\omega}_i = \boldsymbol{\omega}_k$ must hold.

From these considerations, we may generally decompose the motion of each point of the body into that of the reference point and a rotational motion:

$$\dot{\mathbf{r}} = \mathbf{V} + \boldsymbol{\omega} \times \mathbf{r}'\ .$$

For the total momentum,

$$\mathbf{P} = \int \dot{\mathbf{r}}\, dm = M\,\mathbf{V} + \boldsymbol{\omega} \times \int \mathbf{r}'\, dm\ ,$$

where we may write $\rho(\mathbf{r}')\,\mathrm{d}^3r'$ instead of $\mathrm{d}m$. The last term vanishes because $\int \mathbf{r}'\,\mathrm{d}m = \mathbf{0}$. The expressions for the *angular momentum* and the *kinetic energy* (see p. 72) then simplify to (otherwise there would be further terms):

$$\mathbf{L} \equiv \int \mathbf{r} \times \dot{\mathbf{r}}\,\mathrm{d}m = \mathbf{R} \times \mathbf{P} + \int \mathbf{r}' \times (\boldsymbol{\omega} \times \mathbf{r}')\,\mathrm{d}m \ ,$$

$$T \equiv \tfrac{1}{2} \int \dot{\mathbf{r}} \cdot \dot{\mathbf{r}}\,\mathrm{d}m = \tfrac{1}{2}\,M\,V^2 + \tfrac{1}{2}\,I_\omega\,\omega^2 \ .$$

Here I_ω is the moment of inertia of the body with respect to the axis $\mathbf{e}_\omega = \boldsymbol{\omega}/\omega$, which must go through the center of mass:

$$I_\omega \equiv \int (\mathbf{e}_\omega \times \mathbf{r}')^2\,\mathrm{d}m = \int \{r'^{\,2} - (\mathbf{e}_\omega \cdot \mathbf{r}')^2\}\,\rho(\mathbf{r}')\,\mathrm{d}^3r' \ .$$

More precisely, we should write $I_{\omega\mathrm{CM}}$, because the axis of rotation goes through the center of mass. For a rotation about the origin, $\mathbf{e}_\omega \times \mathbf{r}'$ is to be replaced by $\mathbf{e}_\omega \times (\mathbf{R} + \mathbf{r}')$, and therefore both moments of inertia differ by the non-negative quantity

$$I_\omega - I_{\omega\mathrm{CM}} = M\,(\mathbf{e}_\omega \times \mathbf{R})^2 \ ,$$

i.e., by the mass multiplied by the square of the distances of the center of mass from the axis of rotation. This is *Steiner's theorem*. It is very helpful, because we may then choose the origin of our coordinate systems in a more convenient place.

2.2.10 Moment of Inertia

In general, the moment of inertia I_ω also depends upon the rotational orientation $\mathbf{e}_\omega$. This is what we shall investigate now. Here we let the center of mass remain at rest and thus take it as the reference point of the fixed body system. We shall write $\mathbf{r}$ instead of $\mathbf{r}'$ as we have done so far. Then, because $\dot{\mathbf{r}} = \boldsymbol{\omega} \times \mathbf{r}$, we obtain the expression

$$\mathbf{L} = \int \mathbf{r} \times (\boldsymbol{\omega} \times \mathbf{r})\,\mathrm{d}m$$

for the angular momentum of the rigid body, which is also important for the kinetic energy of the rotation (the *rotational energy*), since $(\boldsymbol{\omega} \times \mathbf{r})^2 = (\boldsymbol{\omega} \times \mathbf{r}) \cdot (\boldsymbol{\omega} \times \mathbf{r}) = \boldsymbol{\omega} \cdot \{\mathbf{r} \times (\boldsymbol{\omega} \times \mathbf{r})\}$ delivers

$$T = \tfrac{1}{2}\,I_\omega\,\omega^2 = \tfrac{1}{2} \int (\boldsymbol{\omega} \times \mathbf{r})^2\,\mathrm{d}m = \tfrac{1}{2}\,\boldsymbol{\omega} \cdot \mathbf{L} \ ,$$

corresponding to $T = \frac{1}{2}\mathbf{v} \cdot \mathbf{p}$ for rectilinear motion. Clearly, $\mathbf{L}$ and $\boldsymbol{\omega}$ depend on each other linearly, but may have different directions. If we write

$$\mathbf{L} = I\,\boldsymbol{\omega}\,,$$

then I is a linear operator—more precisely a tensor of second rank, because it assigns a vector linearly to another vector. If we decompose

$$\mathbf{L} = \int \mathbf{r} \times (\boldsymbol{\omega} \times \mathbf{r})\,\mathrm{d}m = \int \{\boldsymbol{\omega}\,r^2 - \mathbf{r}\,(\mathbf{r} \cdot \boldsymbol{\omega})\}\,\mathrm{d}m$$

in terms of Cartesian components, e.g., $L_x = \int \omega_x r^2 - x\,(x\omega_x + y\omega_y + z\omega_z)\,\mathrm{d}m$, we arrive at the system of linear equations

$$\begin{pmatrix} L_x \\ L_y \\ L_z \end{pmatrix} = \begin{pmatrix} I_{xx} & I_{xy} & I_{xz} \\ I_{yx} & I_{yy} & I_{yz} \\ I_{zx} & I_{zy} & I_{zz} \end{pmatrix} \begin{pmatrix} \omega_x \\ \omega_y \\ \omega_z \end{pmatrix}$$

with

$$\begin{aligned} I_{xx} &= \int\ (r^2 - x^2)\ \mathrm{d}m = \int\ (y^2 + z^2)\ \rho(\mathbf{r})\ \mathrm{d}^3r\,, \\ I_{xy} &= \int\ (-xy)\ \ \mathrm{d}m = \int\ (-xy)\ \ \rho(\mathbf{r})\ \mathrm{d}^3r\ \ = I_{yx}\,, \end{aligned}$$

and cyclic permutations thereof. The 3×3 matrix is symmetric and has thus only six (real) independent elements. The three on the diagonal are called the *moments of inertia*, the remaining ones (without minus signs) the *deviation moments*, i.e., deviation of the direction of $\mathbf{L}$ from the direction of $\boldsymbol{\omega}$.

In the next section it will turn out that, for a suitable choice of axes, all the deviation moments vanish. In addition to the three *principal moments of inertia* on the diagonal, three further parameters are then required to fix the orientation of the axes, e.g., the Euler angles. This transition to diagonal form is called the *principal axis transformation*.

2.2.11 Principal Axis Transformation

If I is diagonal, there are three *eigenvectors* $\mathbf{u}_i$, for which $I\,\mathbf{u}_i$ is in the direction of $\mathbf{u}_i$, namely the three column vectors with two components equal to zero. Since I is a linear operator, the value of $\mathbf{u}_i$ is of no interest here. We take unit vectors. The factors I_i in the equation $I\,\mathbf{u}_i = I_i\,\mathbf{u}_i$ are called *eigenvalues*. If I is not diagonal, then we still have to rotate. Only DID^{-1} can then be diagonal and correspondingly $D\mathbf{u}_i$ is an eigenvector with two vanishing components. We therefore consider

$$(I - I_i\,\mathbb{1})\,\mathbf{u}_i = 0\,,$$

and determine I_i and $\mathbf{u}_i$ from this homogeneous linear system of equations. As is well known, it is only soluble if its determinant vanishes:

$$\det(I - I_i \, 1) = 0 \; .$$

This *characteristic equation*, involving 3×3-matrices, leads to an equation of third order with three solutions I_1, I_2, I_3, actually, to three such equations, viz., $0 = \det(I - I_1 \, 1) = \det(I - I_2 \, 1) = \det(I - I_3 \, 1)$. These solutions are all real, because I is real and symmetric. They would still be real if I were only Hermitian, i.e., if $I = I^\dagger \; \Leftrightarrow \; \widetilde{I} = I^*$. As for the orthogonal transformations, we may write the eigenvectors $\mathbf{u}_i$ as a column matrix U_i, and for $I_i \, U_i$ its three elements multiplied by the number I_i. Therefore,

$$I_i \, U_j{}^\dagger \, U_i = U_j{}^\dagger \, (I \, U_i) = U_j{}^\dagger \, I^\dagger \, U_i = (I \, U_j)^\dagger \, U_i = I_j{}^* \, U_j{}^\dagger \, U_i \; .$$

It follows that the eigenvalues are real, because we may set $j = i$ and have $U_i{}^\dagger U_i = 1$, and since $(I_i - I_j{}^*) \, U_j{}^\dagger U_i = 0$, it also follows that the eigenvectors corresponding to different eigenvalues $I_i \neq I_j$ are orthogonal to each other, since in the given case, we have $U_j = U_j{}^* \Leftrightarrow U_j{}^\dagger = \widetilde{U}_j$. If, however, two eigenvalues are equal, the two eigenvectors need not be perpendicular to each other, but then any vector from the subspace spanned by the two eigenvectors is an eigenvector, so any pair of mutually orthogonal unit vectors may be chosen from this set: then all eigenvectors are pairwise orthogonal to each other. Since the diagonal elements of I are sums of squares, the eigenvalues here are not only real, but positive-definite, i.e., non-negative (positive or zero).

When determining the principal moments of inertia, symmetry considerations are often helpful—then we can avoid the diagonalization of I. Here, axial symmetry is not necessary. Reflection symmetry with respect to a plane suffices. Then from the distribution with $\rho \, (x, y, z) = \rho \, (-x, y, z)$, symmetric in the yz-plane, it follows that $I_{xy} = - \int xy \, \rho \, \mathrm{d}^3r = I_{yx}$ as well as $I_{xz} = I_{zx}$ vanish, whence I_{xx} is a principal moment of inertia. The normal to the mirror plane is a principal axis of the moment of inertia.

For a *plane* mass distribution, the moment of inertia with respect to the normal to the plane is composed of the moments of inertia of the two mutually perpendicular axes in the plane. Hence, if we choose the x and y axes in the plane ($z = 0$), we find $I_{xx} = \int y^2 \, \mathrm{d}m$, $I_{yy} = \int x^2 \, \mathrm{d}m$, and $I_{zz} = I_{xx} + I_{yy}$. See also Problems 2.24–2.26.

Of course, we may order the eigenvectors so that they form a right-handed frame. Then with a rotation D we arrive at these new unit vectors and, as on p. 36, we may also set $I' = D I D^{-1}$. The sum and product of the eigenvalues can thus be determined even without a principal axis transformation. Because $\mathrm{tr}(AB) = \mathrm{tr}(BA)$ and $\det(AB) = \det(BA)$ (and $D^{-1}D = 1$), the trace and the determinant are conserved under the principal axis transformation.

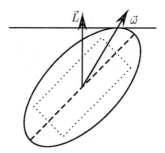

Fig. 2.12 Poinsot's construction. The inertial ellipsoid rolls on the invariant plane, i.e., the plane tangential to the inertial ellipsoid at the contact point of the angular velocity $\boldsymbol{\omega}$. The angular momentum $\mathbf{L}$ is perpendicular to this plane. As an example, of a body with the inertial ellipsoid shown here (*continuous curve*) we take an appropriate cylinder (*dotted line*). The principal moment of inertia about the symmetry axis (*dashed curve*) is $\frac{1}{2} m R^2$, while the one perpendicular to it is $\frac{1}{4} m \left(R^2 + \frac{1}{3} l^2\right)$

For an arbitrary axis of rotation $\boldsymbol{\omega}$, since $\boldsymbol{\omega} = \sum_i \mathbf{u}_i \, (\mathbf{u}_i \cdot \boldsymbol{\omega})$, the moment of inertia is

$$I_\omega = \mathbf{e}_\omega \cdot \int \mathbf{r} \times (\mathbf{e}_\omega \times \mathbf{r}) \, dm = \mathbf{e}_\omega \cdot I \, \mathbf{e}_\omega = \sum_i I_i \, (\mathbf{u}_i \cdot \mathbf{e}_\omega)^2 \; .$$

It can thus be evaluated rather easily from the principal moments of inertia. Hence, they only have to be weighted with the squares of the directional cosines of $\boldsymbol{\omega}$ along the principal axes of inertia fixed in the body.

When the principal moments of inertia are known, the equation

$$T(\boldsymbol{\omega}) = \tfrac{1}{2} \, I_\omega \, \omega^2 = \tfrac{1}{2} \left(I_1 \, \omega_1{}^2 + I_2 \, \omega_2{}^2 + I_3 \, \omega_3{}^2\right)$$

represents an ellipsoid in the variables $\boldsymbol{\omega}$, with semi-axes $\sqrt{2T/I_i}$. This is the *inertial ellipsoid*. Clearly, $\partial T / \partial \omega_i = I_i \, \omega_i = L_i$, or in vector notation, just as we had $\nabla_v T = \mathbf{p}$,

$$\nabla_\omega T = \mathbf{L} \; .$$

For a given $\boldsymbol{\omega}$, the angular momentum $\mathbf{L}$ is perpendicular to the tangential plane of the inertial ellipsoid at the point of contact of $\boldsymbol{\omega}$ (*Poinsot's construction*) (see Fig. 2.12). Conversely, for a given angular momentum, the rotation vector can be found at each time using the inertial ellipsoid.

If no torque acts, then $T = \frac{1}{2} \, \boldsymbol{\omega} \cdot \mathbf{L}$ and $\mathbf{L}$ are constant, and so also is the projection of $\boldsymbol{\omega}$ onto the the spatially fixed angular momentum. The point of contact of $\boldsymbol{\omega}$ then moves on an *invariant plane* perpendicular to the angular momentum. The inertial ellipsoid rolls on this plane and the center of mass is at a constant distance from this plane. This motion is also called *nutation* (see Fig. 2.13). Instead of "nutation",

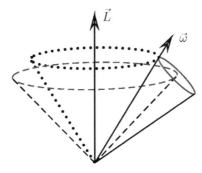

Fig. 2.13 Nutation of the figure axis (*dashed line*) for an axially symmetric moment of inertia. Here the *polhode cone* (*continuous curve*) rolls on the *herpolhode cone* (*dotted curve*). As in Fig. 2.12, an elongated top is assumed here—otherwise the polhode cone does not roll outside the herpolhode cone, but rather inside it. Quantitatively, this rolling is described by the Euler equations (without torque) in Sect. 2.2.12

regular precession is occasionally used, since for a precession the angular momentum changes because a torque acts.

For an axially symmetric moment of inertia, $\boldsymbol{\omega}(t)$ generates the spatially fixed *herpolhode cone* on which the the body-fixed *polhode cone* rolls about the figure axis. For an axially symmetric moment of inertia, the rotation of the figure axis about the angular momentum axis degenerates to a *nutation cone*.

2.2.12 Accelerated Reference Frames and Fictitious Forces

So far the laws have been valid in arbitrary inertial systems. But in accelerated reference frames, "fictitious forces" also appear. We shall deal with those here.

In a rectilinear accelerated (body-fixed) system with $\mathbf{r}_K = \mathbf{r}_T - \mathbf{r}_N$, the acceleration $\ddot{\mathbf{r}}_K$ differs from that in the inertial system ($\ddot{\mathbf{r}}_T$) by the acceleration of the origin, $\ddot{\mathbf{r}}_N$. In particular, from $m\,\ddot{\mathbf{r}} = \mathbf{F}$, we have

$$m\,\ddot{\mathbf{r}}_K = \mathbf{F} - m\,\ddot{\mathbf{r}}_N \ .$$

The last term is the additional *inertial force* in the accelerated system.

But in a rotating reference frame, e.g., fixed in the Earth, according to our considerations about rigid bodies and for arbitrary vectors $\mathbf{x}$ (the origin of all position vectors being fixed), we have

$$\left(\frac{d\mathbf{x}}{dt}\right)_T = \left(\frac{d\mathbf{x}}{dt}\right)_K + \boldsymbol{\omega} \times \mathbf{x}_K \quad \Longleftrightarrow \quad \left(\frac{d\mathbf{x}}{dt}\right)_K = \left(\frac{d\mathbf{x}}{dt}\right)_T - \boldsymbol{\omega} \times \mathbf{x}_K \ .$$

In particular, $\mathbf{v}_K = \mathbf{v}_T - \boldsymbol{\omega} \times \mathbf{r}_K$. Taking this as an operator equation

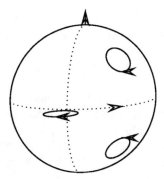

Fig. 2.14 Coriolis force on the Earth. Our "laboratory" rotates eastwards (indicated by the *arrow* at the equator and the *rotation vector* at the north pole). Rectilinear motions are thus deflected: motions restricted to the horizontal are thus deflected to the right in the northern hemisphere and to the left in the southern hemisphere (see also Problem 2.29)

$$\left(\frac{\mathrm{d}}{\mathrm{d}t} \bullet \right)_{\mathrm{T}} = \left(\frac{\mathrm{d}}{\mathrm{d}t} \bullet + \boldsymbol{\omega} \times \bullet \right)_{\mathrm{K}} ,$$

we can easily obtain the second derivative with respect to time:

$$\left(\frac{\mathrm{d}^2}{\mathrm{d}t^2} \bullet \right)_{\mathrm{T}} = \left(\frac{\mathrm{d}}{\mathrm{d}t} \bullet + \boldsymbol{\omega} \times \bullet \right)_{\mathrm{K}}^2$$
$$= \left(\frac{\mathrm{d}^2}{\mathrm{d}t^2} \bullet + \dot{\boldsymbol{\omega}} \times \bullet + 2\,\boldsymbol{\omega} \times \frac{\mathrm{d}}{\mathrm{d}t} \bullet + \boldsymbol{\omega} \times (\boldsymbol{\omega} \times \bullet) \right)_{\mathrm{K}} .$$

Hence, $\mathbf{a}_{\mathrm{T}} = \mathbf{a}_{\mathrm{K}} + \dot{\boldsymbol{\omega}} \times \mathbf{r}_{\mathrm{K}} + 2\,\boldsymbol{\omega} \times \mathbf{v}_{\mathrm{K}} + \boldsymbol{\omega} \times (\boldsymbol{\omega} \times \mathbf{r}_{\mathrm{K}})$, and the force equation is

$$m\,\mathbf{a}_{\mathrm{K}} = \mathbf{F} - m\,\dot{\boldsymbol{\omega}} \times \mathbf{r}_{\mathrm{K}} - 2\,m\,\boldsymbol{\omega} \times \mathbf{v}_{\mathrm{K}} - m\,\boldsymbol{\omega} \times (\boldsymbol{\omega} \times \mathbf{r}_{\mathrm{K}}) .$$

The last term is the well-known *centrifugal force*. It points away from the axis of rotation. If $\mathbf{r}_\perp$ is the part of $\mathbf{r}_{\mathrm{K}}$ perpendicular to $\boldsymbol{\omega}$ (measured from the axis of rotation), the centrifugal force is equal to $m\omega^2 \mathbf{r}_\perp$.

The term $-2m\,\boldsymbol{\omega} \times \mathbf{v}_{\mathrm{K}}$ is the *Coriolis force*, named after G.-G. Coriolis,[3] which occurs only for moving bodies and is formally similar to the Lorentz force $-q\mathbf{B} \times \mathbf{v}$. On the Earth, it is weak compared to the attraction of the Earth. Therefore, we express the rotational vector $\boldsymbol{\omega}$ in terms of the local unit vectors of the spherical coordinates (θ, φ) (see Fig. 1.12): $\boldsymbol{\omega} = \omega\,(\cos\theta\ \mathbf{e}_r - \sin\theta\ \mathbf{e}_\theta)$. The part $2\omega\cos\theta\ \mathbf{v} \times \mathbf{e}_r$ deflects horizontal motions in the northern hemisphere ($0 \le \theta < \frac{1}{2}\pi$) to the right, and in the southern hemisphere ($\frac{1}{2}\pi < \theta \le \pi$) to the left (see Fig. 2.14). Among other things, it rotates the oscillation plane of *Foucault's pendulum*. The remainder $2\omega\sin\theta\ \mathbf{e}_\theta \times \mathbf{v}$ is strongest at the equator and deflects uprising masses to the west, i.e., against the rotational orientation the Earth.

---

[3] Gustave-Gaspard Coriolis (1792–1843).

The equation $\dot{\mathbf{L}} = \mathbf{M}$, which is valid in the inertial system, is more complicated in the rotating system because $d\mathbf{L}_T/dt = d\mathbf{L}_K/dt + \boldsymbol{\omega} \times \mathbf{L}_K$, i.e.,

$$\dot{\mathbf{L}} = \mathbf{M} - \boldsymbol{\omega} \times \mathbf{L} \,,$$

where we now leave out the index K for $\mathbf{L}$ (the torque refers further on to the inertial system). On the other hand, the angular momentum and the rotational vector are related to each other in a simpler way, because in the inertial system the moment of inertia (of a rigid body) does not change with time. In particular, if we introduce Cartesian coordinates along the principal axes of the moment of inertia, such that $L_i = I_i \, \omega_i$, then it follows that

$$I_1 \, \dot{\omega}_1 = M_1 + \omega_2 \omega_3 \, (I_2 - I_3) \,, \quad \text{and cyclic permutations.}$$

These are the *Euler equations for the rigid body*. We shall investigate these now for $\mathbf{M} = \mathbf{0}$, namely for the *free top*, and deal with the heavy top ($\mathbf{M} \neq \mathbf{0}$) in Sect. 2.4.10.

Since $\dot{\boldsymbol{\omega}} = \mathbf{0}$, the spherically symmetric top (with $I_1 = I_2 = I_3$) always rotates about a fixed axis. With the axially symmetric top ($I_1 = I_2 \neq I_3$), only the component along the symmetry axis is conserved ($\dot{\omega}_3 = 0 \Rightarrow \omega_3$ and L_3 constant). With the fixed vector

$$\boldsymbol{\Omega} \equiv \frac{I_3 - I_1}{I_1} \, \omega_3 \, \mathbf{e}_3 \,,$$

and because $\dot{\omega}_1 = -\Omega \, \omega_2 \;\; \dot{\omega}_2 = \Omega \, \omega_1 \;\; \dot{\omega}_3 = 0$, the Euler equations (for $I_1 = I_2$) can be taken together as

$$\dot{\boldsymbol{\omega}} = \boldsymbol{\Omega} \times \boldsymbol{\omega} \,.$$

Thus the vector $\boldsymbol{\omega}$ moves with angular frequency Ω on a polhode cone about the body-fixed figure axis. The opening angle of the cone is determined by the integration constants, e.g., energy and value of the angular momentum.

For a three-axis inertial ellipsoid ($I_1 \neq I_2 \neq I_3$), all three components of $\boldsymbol{\omega}$ change in the course of time. Then Poinsot's construction can lead us to the result. In any case, $T = \frac{1}{2} \boldsymbol{\omega} \cdot \mathbf{L}$ is a constant of the motion (if no torque acts) and therefore $\mathbf{L} = \boldsymbol{\nabla}_\omega T$. Problem 2.28 will also be instructive here.

2.2.13 *Summary of Newtonian Mechanics*

Newton identified three basic laws for non-relativistic mechanics: the inertial law which says that force-free bodies move in a uniform rectilinear way or are at rest (this allows us to draw conclusions about mass ratios in collision processes), the equation $\dot{\mathbf{p}} = \mathbf{F}$ (where $\mathbf{p}$ is an abbreviation for $m\mathbf{v}$), and the law of "action and reaction". Without the action of a force, the momentum $\mathbf{p}$ is conserved—we only need to investigate those motions that are affected by forces. We have explained this in some detail for collisions and the motion of planets. Here the bodies were treated as point masses. We then also treated extended rigid bodies, describing their motion about the center of mass with the Euler equations. In accelerating reference frames, fictitious forces must also be accounted for, e.g., the centrifugal and Coriolis forces.

2.3 Lagrangian Mechanics

2.3.1 D'Alembert's Principle

We could have considered many more applications of Newtonian mechanics. Basically, there will be no new physical effects in the next few sections. These will only appear in electrodynamics (relativity theory), quantum mechanics, and statistical mechanics. But with new notions and better mathematical methods, we can often simplify the workload and even obtain a complete mastery of it. In particular, we shall deal more easily with "geometric constraints" (forces of constraint)—this is accomplished by Lagrangian mechanics.[4]

Here we generalize the notion of momentum and, in addition to the *mechanical momentum* $m\mathbf{v}$ considered so far, introduce also the *canonical momentum* $\mathbf{p}$. Therefore, instead of the usual letter $\mathbf{p}$, we shall always write $m\mathbf{v}$ for the mechanical momentum from now on.

To begin with, we generalize the principle of virtual work (p. 58) of statics to time-dependent processes, i.e., to *d'Alembert's principle*. Here the inertial force $-\mathrm{d}(m\mathbf{v})/\mathrm{d}t$ appears as a new force:

$$\sum_i \left(\mathbf{F}_i - \frac{\mathrm{d}(m_i \mathbf{v}_i)}{\mathrm{d}t} \right) \cdot \delta\mathbf{r}_i = 0 , \quad \text{for} \quad \delta t = 0 .$$

As long as we neglect frictional forces, forces of constraint do not contribute, i.e., $\mathbf{Z}_i \cdot \delta\mathbf{r}_i = 0$. Then we only need to account for the remaining forces. For the determination of the force of constraint for accelerated bodies, we have to use the expression $\mathbf{Z} = m\dot{\mathbf{v}} - \mathbf{F}$, and the body presses against the geometrically formulated boundaries with the opposite force.

If, for example, we enforce a curved orbit with the curvature radius R for a given velocity $\mathbf{v}$, then according to p. 7, the normal acceleration is $v^2/R\,\mathbf{e}_\mathrm{N}$. A force of constraint equal to $m\,(v^2/R) = m\,\omega^2\,R$ will thus be necessary, if no further force acts—only then will the centrifugal force be canceled. Since inertial forces occur only for accelerations, they can be taken as fictitious forces, and can be "transformed away" in an accelerated reference frame. To do this we generally require curvilinear

---

[4]Joseph Louis de Lagrange (1736–1813) became professor in Turin in 1755, was Euler's successor in Berlin in 1766, and became professor in Paris in 1787.

coordinates—this idea leads to general relativity theory, where we use the fact that the gravitational and inertial masses are always equal.

As long as no forces of constraint occur, we do not need d'Alembert's principle, as we have seen in the last section. But otherwise this principle is very useful—in statics the principle of virtual work may be employed repeatedly. And now we even know the generalization to changes in time.

Correspondingly, we can generalize the *Lagrangian equations of the first kind* from statics (see p. 61) to time-dependent processes:

$$\mathbf{F} + \sum_n \lambda_n \, \nabla \Phi_n = \frac{\mathrm{d}(m\mathbf{v})}{\mathrm{d}t} \ .$$

This equation refers to one particle—as in statics it can be generalized to more particles. Then further coordinates and masses are involved.

2.3.2 Constraints

We already know an example of constraints from the case of the rigid body: instead of introducing $3N$ independent coordinates (degrees of freedom) for N point masses, six are sufficient, because for a rigid body the remaining ones can be chosen as fixed—clearly an example of "geometrical" constraints. Something like this has already been encountered in statics: for a displacement along a line, there is only one degree of freedom, for the displacement on a plane there are only two. A constraint is said to be *holonomic* or *integrable* if it can be brought into the form $\Phi\,(t, \mathbf{r}_1, \ldots, \mathbf{r}_N) = 0$. (The Greek *holos* means *whole* or *perfect*, implying that it can be integrated.) If the constraint refers to velocities or if it can be expressed only differentially or as an inequality, e.g., confinement within a volume, then we are dealing with a "non-holonomic" (non-integrable) condition. (Sometimes constraints given as inequalities are referred to as *unilateral* or *bilateral*, because the forces of constraint act only in one direction or two.) If a constraint does not depend explicitly upon time then it is said to be *scleronomous* (*skleros* means fixed or rigid), otherwise *rheonomous* (*rheos* means flowing). In statics, we always assumed holonomic and scleronomous constraints. They are barely simpler than the differentials—they occur, e.g., when a wheel rolls on a plane. Then its rotation is related to the motion of the contact point (Problem 2.7).

Instead of constraints, we can also introduce forces of constraint which ensure that the constraints are respected: constraints and forces of constraint are two pictures for the same situation, because both allow us to deal with the motion of the body. However, geometrical constraints are intuitively descriptive, while forces of constraint often have to be computed, something that is necessary, however, when designing machines in order to determine forces and loads.

In general, constraints couple the equations of motion. But for holonomic constraints, the number of independent variables can often be reduced by a clever choice of coordinates. Then positions can no longer be described by three-vectors, and the coordinates are often different physical quantities, appearing, e.g., as angles or amplitudes of a Fourier decomposition. In Hamiltonian mechanics, we may also take (angular) momentum components and energy as new variables.

In the following, we shall neglect kinetic friction. Then the forces of constraint do not lead to tangential acceleration, but just a normal acceleration, whence they cannot perform work, being perpendicular to the allowed displacements—as long as no kinetic friction perturbs the system, we do not need to account for forces of constraint in the energy conservation law.

If the constraints lead to a single degree of freedom and are scleronomous, then the energy (conservation) law helps—so instead of one differential equation of second order, only one of first order remains to be solved (with the energy as integration constant):

$$E = \frac{m}{2} v^2 + V(x) \qquad \Longrightarrow \qquad \frac{dx}{dt} = \sqrt{\frac{E - V(x)}{m/2}} \ .$$

Of course, there can only be curl-free forces here, otherwise there is no potential energy.

2.3.3 Lagrange Equations of the Second Kind

For time-dependent problems, we start from d'Alembert's principle, i.e., from the equation $\sum_i \{\mathbf{F}_i - d(m_i \mathbf{v}_i)/dt\} \cdot \delta \mathbf{r}_i = 0$ for $\delta t = 0$. Since

$$\delta \mathbf{r}_i = \sum_{k=1}^{f} \frac{\partial \mathbf{r}_i}{\partial x^k} \delta x^k \ ,$$

where the δx^k do not depend upon each other (otherwise Lagrangian parameters are still necessary)—or in particular if there is only one $\delta x^k \neq 0$—and since

$$F_k \equiv \sum_{i=1}^{N} \mathbf{F}_i \cdot \frac{\partial \mathbf{r}_i}{\partial x^k} \ ,$$

we find the equations

$$F_k = \sum_i \frac{d(m_i \mathbf{v}_i)}{dt} \cdot \frac{\partial \mathbf{r}_i}{\partial x^k} \qquad \text{for } k \in \{1, \ldots, f\} \ .$$

The right-hand side can be simplified:

$$\frac{\mathrm{d}(m\mathbf{v})}{\mathrm{d}t} \cdot \frac{\partial \mathbf{r}}{\partial x^k} = \frac{\mathrm{d}}{\mathrm{d}t}\left(m\mathbf{v} \cdot \frac{\partial \mathbf{r}}{\partial x^k}\right) - m\mathbf{v} \cdot \frac{\partial \mathbf{v}}{\partial x^k} \, ,$$

and since $\mathbf{v} = \dot{\mathbf{r}}$, we also have $\partial \mathbf{r}/\partial x^k = \partial \mathbf{v}/\partial \dot{x}^k$, because t is treated as the orbital parameter, whence

$$\frac{\mathrm{d}(m\mathbf{v})}{\mathrm{d}t} \cdot \frac{\partial \mathbf{r}}{\partial x^k} = \frac{\mathrm{d}}{\mathrm{d}t}\left(m\mathbf{v} \cdot \frac{\partial \mathbf{v}}{\partial \dot{x}^k}\right) - m\mathbf{v} \cdot \frac{\partial \mathbf{v}}{\partial x^k} \, .$$

But now we have $\mathbf{v} \cdot \mathrm{d}\mathbf{v} = \frac{1}{2}\,\mathrm{d}(\mathbf{v} \cdot \mathbf{v}) = \mathrm{d}T/m$ and therefore, with $T = \sum_i T_i$,

$$\sum_i \frac{\mathrm{d}(m_i v_i)}{\mathrm{d}t} \cdot \frac{\partial \mathbf{r}_i}{\partial x^k} = \frac{\mathrm{d}}{\mathrm{d}t}\left(\frac{\partial T}{\partial \dot{x}^k}\right) - \frac{\partial T}{\partial x^k} \, .$$

Finally, the f equilibrium conditions $F_k = 0$ can be generalized to

$$F_k \equiv \sum_i \mathbf{F}_i \cdot \frac{\partial \mathbf{r}_i}{\partial x^k} = \frac{\mathrm{d}}{\mathrm{d}t}\left(\frac{\partial T}{\partial \dot{x}^k}\right) - \frac{\partial T}{\partial x^k} \, , \quad \text{for } k \in \{1, \dots, f\} \, .$$

These are the *generalized Lagrange equations of the second kind*. In general, however, we also assume that the external forces can be derived from a potential energy:

$$\mathbf{F}_i = -\nabla_i V(\mathbf{r}_1, \dots, \mathbf{r}_N) \quad \Longrightarrow \quad F_k = -\sum_i \nabla_i V \cdot \frac{\partial \mathbf{r}_i}{\partial x^k} \equiv -\frac{\partial V}{\partial x^k} \, .$$

Since $\partial V/\partial \dot{x}^k = 0$, we introduce the *Lagrange function*

$$L = T - V \, ,$$

and we obtain the *Lagrange equations of second kind (Euler–Lagrange equations)*

$$\frac{\mathrm{d}}{\mathrm{d}t}\left(\frac{\partial L}{\partial \dot{x}^k}\right) - \frac{\partial L}{\partial x^k} = 0 \, , \quad \text{for } k \in \{1, \dots, f\} \, .$$

Many problems of mechanics can be solved with these. We need only the scalar
Lagrange function L and a convenient choice of coordinates.

Let us consider as an example the plane motion of a particle of mass m under
arbitrary (but not necessarily curl-free) forces. In Cartesian coordinates x, y, we have

$$T = \tfrac{1}{2}m\,(\dot{x}^2 + \dot{y}^2) \, ,$$

and consequently,

$$\frac{\partial T}{\partial \dot{x}} = m\,\dot{x}\;, \quad \frac{\partial T}{\partial \dot{y}} = m\,\dot{y}\;, \quad \frac{\partial T}{\partial x} = \frac{\partial T}{\partial y} = 0\;.$$

Therefore, for constant mass, the Lagrange equations lead to Newton's relation $\mathbf{F} = m\,\ddot{\mathbf{r}}$. In contrast, in (curvilinear) polar coordinates r, φ, we have

$$T = \tfrac{1}{2}m\,(\dot{r}^2 + r^2\dot{\varphi}^2)\;,$$

and consequently,

$$\frac{\partial T}{\partial \dot{r}} = m\,\dot{r}\;, \quad \frac{\partial T}{\partial \dot{\varphi}} = m\,r^2\,\dot{\varphi}\;, \quad \frac{\partial T}{\partial r} = m\,r\,\dot{\varphi}^2\;, \quad \frac{\partial T}{\partial \varphi} = 0\;.$$

With $F_r \equiv \mathbf{F}\cdot\partial\mathbf{r}/\partial r = \mathbf{F}\cdot\mathbf{r}/r$ and $F_\varphi \equiv \mathbf{F}\cdot\partial\mathbf{r}/\partial\varphi = \mathbf{F}\cdot r\,\mathbf{e}_\varphi$ (according to p. 40), whence $F_\varphi = \mathbf{F}\cdot(\mathbf{n}\times\mathbf{r}) = (\mathbf{r}\times\mathbf{F})\cdot\mathbf{n} = \mathbf{M}\cdot\mathbf{n}$, we have

$$F_r = m\,\ddot{r} - m\,r\,\dot{\varphi}^2 \quad \text{and} \quad F_\varphi = \frac{\mathrm{d}}{\mathrm{d}t}(m\,r^2\,\dot{\varphi})\;.$$

According to the first equation for the radial motion, in addition to F_r, the centrifugal force $m\,r\dot{\varphi}^2$ is accounted for, and for $\dot{\varphi}$ we have so far set ω, e.g., on p. 91. The second equation has been written so far as $\mathbf{M} = \mathrm{d}\mathbf{L}/\mathrm{d}t$, because $\mathbf{L}\cdot\mathbf{n} = mr^2\dot{\varphi}$. From our new viewpoint, it is the same equation as $\mathbf{F} = \mathrm{d}(m\mathbf{v})/\mathrm{d}t$, only expressed in other coordinates.

2.3.4 Velocity-Dependent Forces and Friction

For time- and velocity-dependent forces, there is no potential energy and thus also no Lagrange function as yet. But in fact, a *generalized potential energy* U with the property

$$F_k = -\frac{\partial U}{\partial x^k} + \frac{\mathrm{d}}{\mathrm{d}t}\left(\frac{\partial U}{\partial \dot{x}^k}\right)\;, \quad \text{for } k \in \{1,\dots,f\}\;,$$

also suffices, because then the *generalized Lagrange function*

$$L = T - U$$

obeys the Lagrange equations of the second kind.

The most important example is the Lorentz force on a charge in an electromagnetic field:

$$\mathbf{F} = q\,(\mathbf{E} + \mathbf{v}\times\mathbf{B})\;.$$

In order to derive this from a generalized potential energy U, we employ the two Maxwell equations

$$\nabla \times \mathbf{E} = -\frac{\partial \mathbf{B}}{\partial t} , \qquad \nabla \cdot \mathbf{B} = 0 .$$

According to this the two vector fields $\mathbf{E}$ and $\mathbf{B}$ are related to each other and can be derived from a scalar potential Φ and a *vector potential* $\mathbf{A}$, taken at the coordinates of the test body:

$$\mathbf{E} = -\nabla \Phi - \frac{\partial \mathbf{A}}{\partial t} , \qquad \mathbf{B} = \nabla \times \mathbf{A} .$$

The two potentials Φ and $\mathbf{A}$ are functions of t and $\mathbf{r}$ (but not $\mathbf{v}$). Hence the position of the test body depends upon the time, and therefore total and partial derivatives are to be distinguished from each other: $d\mathbf{A}/dt - \partial \mathbf{A}/\partial t = (\mathbf{v} \cdot \nabla) \mathbf{A}$. But since $\mathbf{r}$ and $\mathbf{v}$ are to be treated as independent variables, we may set $\mathbf{v} \times (\nabla \times \mathbf{A}) = \nabla (\mathbf{v} \cdot \mathbf{A}) - (\mathbf{v} \cdot \nabla) \mathbf{A}$, because the terms to be expected formally $-\mathbf{A} \times (\nabla \times \mathbf{v}) - (\mathbf{A} \cdot \nabla) \mathbf{v}$ do not contribute. This leads to

$$\mathbf{F} = q\left(-\nabla (\Phi - \mathbf{v} \cdot \mathbf{A}) - \frac{d\mathbf{A}}{dt}\right) .$$

Therefore, the generalized potential energy for the Lorentz force on a charge q in the electromagnetic field is

$$U = q (\Phi - \mathbf{v} \cdot \mathbf{A}) .$$

However, the potentials are not yet uniquely determined. In particular, we may still have *gauge transformations*: $\Phi' = \Phi + \partial\Psi/\partial t$ and $\mathbf{A}' = \mathbf{A} - \nabla\Psi$ deliver the same fields $\mathbf{E}$ and $\mathbf{B}$ as Φ and $\mathbf{A}$. This gauge invariance of the fields leads to the fact that $U' = q (\Phi' - \mathbf{v} \cdot \mathbf{A}') = U + q\, d\Psi/dt$ can be taken as the generalized potential energy, corresponding to the undetermined Lagrange function (an example is given in Problem 2.31)

$$L' = L - \frac{dG}{dt} .$$

We will come back to the gauge dependence of the Lagrange function in Sect. 2.4.5. There it will also be understood why we write G here instead of $q\Psi$, because there G is a *generating function* (*generator*) of a canonical transformation. (However, here G depends only upon t and x^k, while there it may also depend on further variables.)

For friction there is no generalized potential energy U. Then we have to take

$$\frac{d}{dt}\left(\frac{\partial L}{\partial \dot{x}^k}\right) - \frac{\partial L}{\partial x^k} = f_k ,$$

where f_k contains all forces which cannot be derived from a generalized potential energy U.

There are many examples where the frictional force is proportional to the velocity, i.e., $\mathbf{f} = -\alpha \mathbf{v}$, which is called *Stokes friction*, contrasted with Newtonian friction, where $\mathbf{f} \propto -v\mathbf{v}$ (e.g., for the case of free fall), e.g., laminar flow (only turbulent flow leads to a squared term) or electrical loop currents with Ohm resistance. Stokes-type friction also occurs in the Langevin equation (Sect. 6.2.7). Then we may set

$$\mathbf{f} = -\nabla_v \mathscr{F}, \quad \text{with} \quad \mathscr{F} = \frac{\alpha}{2} \mathbf{v} \cdot \mathbf{v} \quad \text{and} \quad \alpha > 0,$$

where $\mathscr{F}$ is Rayleigh's *dissipation function*. It supplies half the power which the system has to give off because of the friction: $dA = -\mathbf{f} \cdot d\mathbf{r} = -\mathbf{f} \cdot \mathbf{v} \, dt = \alpha v^2 \, dt = 2 \mathscr{F} \, dt$. In this case we also need two scalar functions, L and $\mathscr{F}$, to derive the equation of motion (and to describe the thermal expansion).

But for this friction and $\partial L / \partial \dot{x} = m\dot{x}$, we can also take the Lagrange function $L \exp(\alpha t / m)$ (now time-dependent). The unknown term $\alpha \dot{x}$ then appears in addition to $d(\partial L / \partial \dot{x})/dt - \partial L / \partial x$.

2.3.5 Conserved Quantities. Canonical and Mechanical Momentum

The Lagrange function $L(t, x, \dot{x})$ contains the velocity in a non-linear way. Therefore, the Lagrange equation (without friction!),

$$\frac{d}{dt}\left(\frac{\partial L}{\partial \dot{x}^k}\right) = \frac{\partial L}{\partial x^k},$$

is a differential equation of second order, because $\ddot{x}$ occurs. We search for "solutions" $C(t, x, \dot{x}) = 0$, which are differential equations of only first order.

This is straightforward if L does not depend on x^k, but on $\dot{x}^k$:

$$\frac{\partial L}{\partial x^k} = 0 \quad \Longrightarrow \quad \frac{d}{dt}\frac{\partial L}{\partial \dot{x}^k} = 0 \quad \Longrightarrow \quad \frac{\partial L}{\partial \dot{x}^k} = \text{const.}$$

The assumption $\partial L / \partial x^k = 0$ is justified if we can move the origin of x^k with impunity, i.e., if we can add an arbitrary constant to x^k. For example, the dynamics of a rotating wheel does not depend on the angle coordinate φ, but only on the angular velocity $\dot{\varphi}$. Therefore, all coordinates which do not appear in L are said to be *cyclic* —a further important example of cyclic coordinates is given in Problem 2.32.

Generally, $\partial L / \partial \dot{x}^k$ is called the *canonical momentum* conjugate to x^k:

$$p_k \equiv \frac{\partial L}{\partial \dot{x}^k} \quad \Longrightarrow \quad \dot{p}_k = \frac{\partial L}{\partial x^k}.$$

(Here we have the decisive quantity for Hamiltonian mechanics, as we shall see in the next section.) A free point mass has $L = \frac{m}{2}\mathbf{v} \cdot \mathbf{v}$ and $\mathbf{p} = m\mathbf{v}$, whence $\mathbf{p} = \nabla_v L$. For rotations, we have $L = \frac{1}{2}I\dot{\varphi}^2$ and we obtain $p_\varphi = I\dot{\varphi}$ as the canonical momentum, i.e., the *angular momentum*, or more precisely, the angular momentum component along the corresponding axis of rotation. This holds even if a potential energy V also appears. If, however, a point mass with charge q moves in an electromagnetic field, then $L = \frac{m}{2}\mathbf{v} \cdot \mathbf{v} - q\,(\Phi - \mathbf{v} \cdot \mathbf{A}\,)$, and hence the canonical momentum is

$$\mathbf{p} = m\mathbf{v} + q\mathbf{A}\,.$$

It differs from the *mechanical momentum* $m\mathbf{v}$ by the additional term $q\mathbf{A}$ and depends on the gauge, whence $\mathbf{p}' = \mathbf{p} - \nabla G$ is a canonical momentum.

In the following, $\mathbf{p}$ will always denote the canonical momentum and $m\mathbf{v}$ the mechanical one. Therefore, we may no longer call $\dot{\mathbf{p}}$ a force $\mathbf{F}$, because we have

$$\frac{\mathrm{d}(m\mathbf{v}\,)}{\mathrm{d}t} = \dot{\mathbf{p}} - q\dot{\mathbf{A}}\,,$$

and according to the last section,

$$\mathbf{F} = -\nabla U - q\dot{\mathbf{A}}\,.$$

Consequently,

$$\frac{\mathrm{d}(m\mathbf{v}\,)}{\mathrm{d}t} = \mathbf{F} \quad\Longrightarrow\quad \dot{\mathbf{p}} = -\nabla U$$

delivers a noteworthy result.

A homogeneous magnetic field $\mathbf{B}$ can be obtained from the vector potential $\mathbf{A} = \frac{1}{2}\mathbf{B} \times \mathbf{r}$ (among others), and since $\Phi \equiv 0$, this leads to $-\nabla U = \frac{1}{2}q\,\mathbf{v} \times \mathbf{B}$. Here $\dot{\mathbf{p}}$ is thus equal to *half* the Lorentz force.

In a constant and homogeneous magnetic field, which thus does not depend upon either t or $\mathbf{r}$, neither the mechanical nor the canonical momentum is conserved. However, since $m\dot{\mathbf{v}} = q\dot{\mathbf{r}} \times \mathbf{B}$, only the *pseudo-momentum*

$$\mathbf{K} \equiv m\mathbf{v} + q\,\mathbf{B} \times \mathbf{r}$$

is conserved. In fact, on the helical orbit, only the mechanical momentum in the field direction is conserved ($\mathbf{K}_\parallel = m\mathbf{v}_\parallel$). Perpendicular to it, there is a circular orbit, and we use $\boldsymbol{\omega} = -q\mathbf{B}/m$ from p. 78. Using

$$m\mathbf{v}_\perp = m\boldsymbol{\omega} \times \mathbf{r} + \mathbf{K}_\perp = m\boldsymbol{\omega} \times (\mathbf{r} - \mathbf{r}_\mathrm{A})\,,$$

which implies

$$\mathbf{K}_\perp = m\mathbf{r}_\mathrm{A} \times \boldsymbol{\omega}\,,$$

we infer on the helical axis

$$\mathbf{r}_A = \frac{\boldsymbol{\omega}}{\omega} \times \frac{\mathbf{K}}{m\omega} \, ,$$

and the radius $v_\perp / \omega$.

The canonical momentum conjugate to a cyclic variable is conserved according to what was said above, i.e., $\dot{p} = \partial L / \partial x = 0$. Therefore, for (infinitesimal) translational invariance, the momentum is conserved, and for isotropy (rotational invariance), the angular momentum is conserved.

If L does not depend explicitly on time, then, according to the Lagrange equation, we have

$$\frac{dL}{dt} = \sum_k \left(\frac{\partial L}{\partial x^k} \dot{x}^k + \frac{\partial L}{\partial \dot{x}^k} \frac{d\dot{x}^k}{dt} \right) = \sum_k \left(\dot{p}_k \, \dot{x}^k + p_k \, \ddot{x}^k \right) = \frac{d}{dt} \sum_k p_k \, \dot{x}^k \, .$$

Thus, for $\partial L / \partial t = 0$, the sum $\sum_k p_k \, \dot{x}^k - L$ is also a conserved quantity (*constant of the motion*). Here $\sum_k p_k \, \dot{x}^k$ is equal to $2T$ if the kinetic energy T is a homogeneous function of second order in the velocity, thus if $T(k\mathbf{v}) = k^2 \, T(\mathbf{v})$ holds for all real k, which according to *Euler's identity* for continuously differentiable T is equivalent to $\mathbf{v} \cdot \nabla_v T(\mathbf{v}) = 2 \, T(\mathbf{v})$. (For time-independent constraints, but not for time-dependent ones, T is homogeneous of second order.) Thus for $\partial L / \partial t = 0$ and $2T = \mathbf{v} \cdot \nabla_v T$, the quantity $2T - L$ is conserved. If there is in addition a potential energy V, then $L = T - V$ and the energy $T + V$ is conserved.

2.3.6 Physical Pendulum

Here we discuss a rigid body of mass m with moment of inertia I with respect to a (horizontal) axis of rotation: a *plane pendulum*. (A *rotational pendulum* would move freely about a point, as discussed in Sect. 2.4.10. A *mathematical pendulum* is a point mass which moves. It has $I = m \, s^2$, but is otherwise not easy to treat. On the other hand, friction is neglected for the time being. It will be accounted for in the next section.) The angle θ gives the displacement from the equilibrium position (see Fig. 2.15).

In this situation, the kinetic and potential energies are

$$T = \tfrac{1}{2} I \dot{\theta}^2 \quad \text{and} \quad V = 2 \, I \, \omega^2 \sin^2 \tfrac{1}{2} \theta \, , \quad \text{with } \omega^2 \equiv \frac{m \, g \, s}{I} \, .$$

As in the case of free fall (Sect. 2.2.8), we have assumed here that the gravitational field of the Earth is homogeneous. As an aside, formally the same expression holds for an electric dipole moment $\mathbf{p}$ in a homogeneous electric field $\mathbf{E}$, because there the potential energy is $V = -\mathbf{p} \cdot \mathbf{E}$ (see Sect. 3.1.4), and for a magnetic moment $\mathbf{m}$

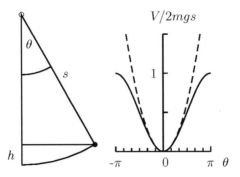

Fig. 2.15 Plane pendulum. The center of mass (*full circle*) is a distance s from the axis of rotation (*open circle*) and a height $h = s\,(1-\cos\theta) = 2s\sin^2\frac{1}{2}\theta$ above the equilibrium position. *Right*: Potential energy V (relative to $2mgs$) as a function of θ. *Dashed curve*: Approximation for harmonic oscillation

in a homogeneous magnetic field **B**, since $V = -\mathbf{m}\cdot\mathbf{B}$, according to Sect. 3.2.9. The following considerations can also be transferred to the pendulum motion of an undamped *compass needle*, because for V, the origin is not important and $I\omega^2$ is then the product of the dipole moment and the field strength.

As stressed in Sect. 2.3.2, for such problems with a single unknown and time-independent energy $T + V$, conservation of energy is useful:

$$E = 2\,I\,\{(\tfrac{1}{2}\dot\theta)^2 + \omega^2\sin^2\tfrac{1}{2}\theta\}\;.$$

According to this, $(\tfrac{1}{2}\dot\theta)^2 = E/2I - \omega^2\sin^2\tfrac{1}{2}\theta$, which is a differential equation of first order for the unknown function $\theta(t)$.

Small pendulum amplitudes are generally considered, and we may set $\sin\frac{1}{2}\theta \approx \frac{1}{2}\theta$. This leads to the differential equation $E = \frac{1}{2}I\,(\dot\theta^2 + \omega^2\theta^2)$ for *harmonic oscillation*, viz.,

$$\theta(t) = \theta_0\cos(\omega t) + (\dot\theta_0/\omega)\sin(\omega t) = \widehat\theta\cos(\omega t - \phi)\;,$$

with the initial values $\theta(0) \equiv \theta_0 = \widehat\theta\cos\phi$ and $\dot\theta(0) \equiv \dot\theta_0 = \omega\widehat\theta\sin\phi$. The amplitude $\widehat\theta$ then follows from $\widehat\theta^2 = 2E/I\omega^2 = \theta_0{}^2 + (\dot\theta_0/\omega)^2$ and the *phase shift* (at zero time) ϕ from $\tan\frac{1}{2}\phi = (\widehat\theta - \theta_0)\,\omega/\dot\theta_0$. Note that we use the equation $\tan\frac{1}{2}\phi = (1 - \cos\phi)/\sin\phi$, not the more suggestive $\tan\phi = \sin\phi/\cos\phi$, because this gives ϕ uniquely only up to an even multiple of π. As the integration constant we thus have either the energy E (or, respectively, $\widehat\theta$) and the phase shift ϕ or the initial values θ_0 and $\dot\theta_0$.

However, we would like also to allow for larger pendulum amplitudes, and for that we use the abbreviation (with $k \geq 0$)

$$k^2 \equiv \frac{E}{2I\omega^2} = \frac{E}{2mgs} \quad\text{and}\quad x \equiv \tfrac{1}{2}\theta\;.$$

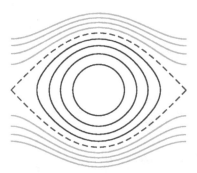

Fig. 2.16 Pendulum trajectories in phase space ($y \propto p$). These are solutions of the equation $y^2 + \sin^2 x = k^2$, here for k^2 from 0.2 to 1.8 in steps of 0.2 and a periodicity interval $-\frac{1}{2}\pi \leq x \leq \frac{1}{2}\pi$. Thus $x \stackrel{\wedge}{=} \frac{1}{2}\theta$ and $y \stackrel{\wedge}{=} \dot{x}/\omega$. The *dashed red curve* ($k^2 = 1$) is the *separatrix*—it separates the rotating solutions (*green*) from the librations (*blue*). The curves are always plotted clockwise ↻: for $x > 0$, the velocity decreases ($\ddot{x} < 0$), for $x < 0$, it increases. This happens also for the damped oscillation (see Fig. 2.21)

We then have the non-linear differential equation $k^2 = \omega^{-2}\dot{x}^2 + \sin^2 x$. So far we have restricted ourselves to $k \ll 1$ and so have been able to use $\sin x \approx x$. In this way, in the $(x, \dot{x})$-plane, we obtained an ellipse with semi-axes k and $k\omega$. With increasing k (<1), the ellipse increases in size and changes shape—in fact, it no longer remains an ellipse. For $k = 1$, we have $\dot{x} = \pm\omega \cos x$. The requirement $|\sin x| \leq k$ limits the x values for $k < 1$ (then $k = \sin \frac{1}{2}\widehat{\theta}$), but no longer for $k > 1$. Hence, the pendulum rotates (see Fig. 2.16). In all cases, the highest angular velocity is $\dot{\theta}_{max} = 2\omega k = \sqrt{2E/I}$. For $k \gg 1$, the term $\sin^2 x$ is negligible compared to $\omega^{-2}\dot{x}^2$ and the pendulum then rotates with constant angular velocity $\dot{\theta}$.

In the differential equation $\dot{x}^2 = \omega^2 (k^2 - \sin^2 x)$, the variables can be separated:

$$\omega \, dt = \frac{dx}{\sqrt{k^2 - \sin^2 x}} \, .$$

We first consider the oscillations (the case $k < 1$) and then the rotating solutions ($k > 1$). In both cases, we choose the zero time (i.e., the second fitting parameter in addition to k or E) at $\theta(0) = 0$ (with $\dot{\theta} > 0$).

For $k < 1$, we transform $\sin x = k \sin z$, thus $\cos x \, dx = k \cos z \, dz$: the denominator becomes $k \cos z$ and $dx/\sqrt{k^2 - \sin^2 x}$ becomes $dz/\sqrt{1 - k^2 \sin^2 z}$. Then we arrive at the *incomplete elliptic integral of the first kind* (in the Legendre normal form)

$$F(\varphi \mid k^2) \equiv \int_0^\varphi \frac{dz}{\sqrt{1 - k^2 \sin^2 z}} \, ,$$

and hence,

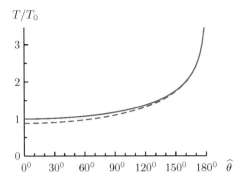

Fig. 2.17 Dependence of the oscillation period T on the pendulum amplitude, here in relation to the oscillation period $T_0 = 2\pi/\omega$ for small amplitude. *Dashed blue curve*: Limiting curve $(2/\pi) \ln(4/\cos\frac{1}{2}\widehat{\theta})$ for large amplitude. *Continuous red curve*: Complete elliptic integral of the first kind $K(\sin^2\frac{1}{2}\widehat{\theta})$ up to a factor $\frac{1}{2}\pi$ (see Fig. 2.33)

$$\omega t = F(\arcsin \frac{\sin(\frac{1}{2}\theta)}{k} \mid k^2) \ .$$

This equation yields the *oscillation period* T (see Fig. 2.17), because for $\frac{1}{4}T$, we have $\sin\frac{1}{2}\theta = k$ or $\varphi = \frac{1}{2}\pi$:

$$\tfrac{1}{4}\omega T = F(\tfrac{1}{2}\pi \mid k^2) \equiv K(k^2) \ .$$

Here $K(k^2)$ is a *complete elliptic integral of the first kind*. (More details on the special functions mentioned here can be found, e.g., in [1], or in particular [2].)

The Legendre normal form of the the elliptic integrals mentioned here depends on a circular function. If we take $\sin z$ as integration variable t, then the incomplete elliptic integral reads

$$F(\varphi \mid k^2) = \int_0^{\sin\varphi} \frac{dt}{\sqrt{(1-t^2)(1-k^2t^2)}} \ ,$$

and the complete elliptic integral

$$K(k^2) = \int_0^1 \frac{dt}{\sqrt{(1-t^2)(1-k^2t^2)}} \ .$$

Thus we only need a purely algebraic integrand.

If the pendulum oscillates with small angle amplitudes, then $k^2 \approx 0$. If we expand the integrand for $k^2 < 1$ in terms of powers of k^2 and integrate term by term, this yields

Fig. 2.18 The amplitude of the elliptic functions, $\varphi =$ am F, during a quarter period for $k^2 = 0$ (*black*), 0.5 (*red*), 0.9 (*blue*), and 0.99 (*green*). This is needed for the Jacobi functions, e.g., sine amplitudes (see Fig. 2.31). The dependence of the inverse functions F(φ) can also be read off

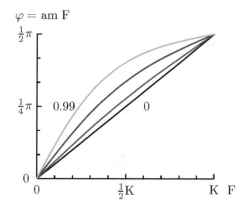

$$K(k^2) = \frac{\pi}{2} \sum_{n=0}^{\infty} \frac{(2n)!^2}{2^{4n} n!^4} k^{2n} , \quad \text{for } k^2 < 1 ,$$

and thus $T = 2\pi\omega^{-1} (1 + \frac{1}{4}k^2 + \frac{9}{64}k^4 + \cdots)$. Only for amplitudes larger than $23°$ does the bracket deviate by more than 1% from unity. In the special case $k^2 = 1$, the oscillation period T increases beyond all limits, because for $k' = \sqrt{1 - k^2} \ll 1$ it is

$$K(k^2) = \sum_{n=0}^{\infty} \binom{-\frac{1}{2}}{n} \left(\ln \frac{4}{k'} - 2 \sum_{j=1}^{2n} \frac{(-)^{j-1}}{j} \right) k'^{2n} = \ln \frac{4}{\sqrt{1 - k^2}} + \cdots ,$$

as proven in Fig. 3.14. We shall use these relations in electrodynamics.

In order to obtain the amplitudes as a function of time, however, we also need the inverse functions of the incomplete elliptic integrals of the first kind, namely the (*angle*) *amplitude* of F (see Fig. 2.18):

$$\tau = F(\varphi \,|\, k^2) \quad \Longleftrightarrow \quad \varphi = \text{am}(\tau | k^2) \equiv \text{am}\, \tau .$$

Then our result with $\tau = \omega t$ can be brought into the form (see Fig. 2.19)

$$\sin(\tfrac{1}{2}\theta) = k \sin(\text{am}\, \tau) \equiv k \,\text{sn}(\tau \,|\, k^2) \equiv k \,\text{sn}\, \tau .$$

The Jacobi elliptic function *sinus amplitudinis* sn τ arises. It is odd in τ and, like all elliptic functions, it is doubly periodic, if we allow for a complex arguments:

$$\text{sn}\, \tau = \text{sn}\{\tau + 4\,K(k^2)\} = \text{sn}\{\tau + 2i\,K(1-k^2)\} .$$

For $k^2 = 0$, it is $\sin \tau$, and for $k^2 = 1$ (with $K \to \infty$), it is $\tanh \tau$.

For the rotating solutions (with $k > 1$), the calculation is easy, because here, even without the above-mentioned transformation $x \to z$, the differential equation

Fig. 2.19 Pendulum
amplitude θ for one period
when $k^2 = 0.5$ (*red*), 0.9
(*blue*), and 0.99 (*green*)

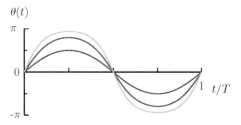

$\omega\,dt = dx/\sqrt{k^2 - \sin^2 x} = k^{-1}\,dx/\sqrt{1 - k^{-2}\sin^2 x}$ leads to an incomplete elliptic
integral of the first kind:

$$\omega t = \frac{F(\tfrac{1}{2}\theta \mid k^{-2})}{k} \quad\text{and}\quad \tfrac{1}{2}\theta = \mathrm{am}(k\omega t \mid k^{-2}) .$$

For $\theta = \pi$, we have half a rotation and the time $K(k^{-2})/k\omega$.

2.3.7 Damped Oscillation

If we had restricted ourselves to small displacements above, then we would still have
had the simple differential equation for the *harmonic oscillation*:

$$\ddot{x} + \omega_0{}^2 x = 0 .$$

Multiplying by $\dot{x}$ and integrating over t, we deduce the "conservation of energy",

$$\dot{x}^2 + \omega_0{}^2 x^2 = \text{const.}$$

But the harmonic oscillation can also be perturbed by other additional terms—in
particular, it normally decays, i.e., it is *damped*. We now write ω_0 for the ω used so
far, because the angular frequency of the oscillation depends upon the damping, as
we shall see shortly.

We assume Stokes's friction because only comparably small velocities occur and
therefore a term linear in $\dot{x}$ will contribute more than a squared one. We thus consider
the differential equation

$$\ddot{x} + 2\gamma\,\dot{x} + \omega_0{}^2 x = 0 , \quad\text{with}\quad \gamma > 0 .$$

In the solutions, γ can be viewed as the *decay coefficient* and γ^{-1} as the *decay or
relaxation time*. Because of the damping, the conservation of energy does not help, but
because the linear differential equation is homogeneous, the ansatz $x = c\,\exp(-i\,\omega t)$
leads to (see Fig. 2.20)

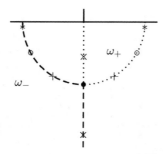

Fig. 2.20 Dependence of the pair $\omega_\pm$ in the complex ω-plane with increasing damping. For $\gamma \ll \omega_0$, they start from $\pm\omega_0$ ($*$) and move symmetrically towards each other on a semi-circle (from $*$ to $\circ$ to $+$) until, for $\gamma = \omega_0$, they coincide at $-i\omega_0$ ($\bullet$). Because $|\omega_+\omega_-| = \omega_0{}^2$ for $\gamma > \omega_0$, they move apart again as mirror points ($\times$) of the circle on the imaginary axis. Damped oscillations occur only for negative imaginary part

Fig. 2.21 Damped oscillations for $\gamma = \omega_0/10$. As in Fig. 2.16, $\dot{x}/\omega_0$ is represented as a function of x, with equal time intervals between neighboring points ($\bullet$). For other initial values, the figure is rotated about the origin ($\circ$), where all orbits end. This is the *attractor* of all orbits

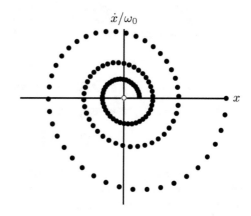

$$\omega^2 + 2i\gamma\omega = \omega_0{}^2 \implies \omega_\pm = \pm\sqrt{\omega_0{}^2 - \gamma^2} - i\gamma .$$

In the following, the angular frequency

$$\Omega \equiv \sqrt{|\omega_0{}^2 - \gamma^2|}$$

will be useful, because $\omega_\pm = \pm\Omega - i\gamma$ for $\gamma < \omega_0$ and $\omega_\pm = -i\,(\gamma \mp \Omega)$ with $\gamma > \Omega$ for $\gamma > \omega_0$.

Hence we have two linearly independent solutions $\exp(-i\,\omega_\pm t)$. Note that, for $\gamma = \omega_0$, the two solutions $x_\pm$ coincide, but their difference at the transition $\gamma \to \omega_0$ is to a first approximation proportional to $t\,\exp(-\gamma t)$, which then delivers a linearly independent solution. Therefore, we can adjust $x(t)$ to the initial values x_0 and $\dot{x}_0$ (see Fig. 2.21):

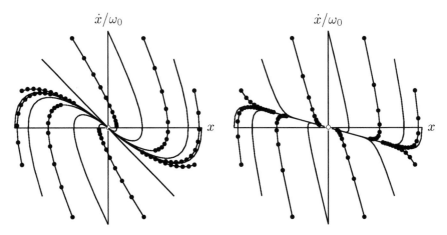

Fig. 2.22 *Left*: Critically damped oscillation ($\gamma = \omega_0$). *Right*: Supercritically damped oscillation (with $\gamma = 2\omega_0$). The representation is the same as in the last figure, except that here the trajectories depend upon the initial conditions, but all finish at the origin

$$\gamma < \omega_0 : \quad x = \exp(-\gamma t) \left(x_0 \cos \Omega t + \frac{\dot{x}_0 + \gamma x_0}{\Omega} \sin \Omega t \right),$$

$$\gamma = \omega_0 : \quad x = \exp(-\gamma t) \qquad \left(x_0 + (\dot{x}_0 + \gamma x_0) t \right),$$

$$\gamma > \omega_0 : \quad x = \exp(-\gamma t) \left(x_0 \cosh \Omega t + \frac{\dot{x}_0 + \gamma x_0}{\Omega} \sinh \Omega t \right).$$

Except for the exponential factor in front of the brackets, the last two brackets no longer describe periodic motion. What we have here is in fact *aperiodic damping*: for $\gamma = \omega_0$, *critical damping*, for $\gamma > \omega_0$, *supercritical damping* (see Fig. 2.22).

2.3.8 Forced Oscillation

For the time being, we assume a force acting periodically with angular frequency ω and consider the inhomogeneous linear differential equation

$$\ddot{x} + 2\gamma \dot{x} + \omega_0{}^2 x = c \cos(\omega t) .$$

On the right-hand side, we could could have assumed a Fourier integral, and then we would have to superpose the corresponding solutions. The general solution is composed of the general solution of the homogeneous equation treated above and a special solution of the inhomogeneous equation. The special solution describes here the long-time behavior (with $\gamma > 0$), because the solutions of the homogeneous equation decay exponentially in time—they are important only for the initial process and are needed to satisfy the initial conditions.

For a special solution, we make the ansatz

$$x = C \cos(\omega t - \phi) = C\,[\cos\phi\,\cos(\omega t) + \sin\phi\,\sin(\omega t)]\;.$$

Hence, for $\phi \neq 0$, the solution is delayed with respect to the exciting oscillation. Therefore, we set $\omega t - \phi$ and expect $\phi \geq 0$. In order to ensure that ϕ is unique (mod 2π), we require that C should have the same sign as c. With this ansatz and after comparing coefficients of $\cos(\omega t)$ and $\sin(\omega t)$, the differential equation leads to the conditions

$$(\omega_0{}^2 - \omega^2)\cos\phi + 2\gamma\omega\sin\phi = c/C > 0\;,$$
$$(\omega_0{}^2 - \omega^2)\sin\phi - 2\gamma\omega\cos\phi = 0\;,$$

which we can solve for the unknown C and ϕ. For unique determination of ϕ, we first consider $\omega = \omega_0$ and find here $\phi = \frac{1}{2}\pi \bmod 2\pi$. Hence, $0 \leq \phi \leq \pi$ has to hold. Therefore, we derive ϕ from $\tan(\frac{1}{2}\pi - \phi) = \cot\phi = (\omega_0{}^2 - \omega^2)/2\gamma\omega$ and use $\sin\phi = 1/\sqrt{1 + \cot^2\phi}$ for c/C (see Fig. 2.23):

$$C = \frac{c}{\sqrt{(\omega_0{}^2 - \omega^2)^2 + 4\gamma^2\omega^2}} \quad \text{and} \quad \phi = \frac{\pi}{2} - \arctan\frac{\omega_0{}^2 - \omega^2}{2\gamma\omega}\;.$$

For $\omega \approx \omega_0$, the ratio C/c is very large. For $\gamma \neq 0$, the maximum lies at somewhat lower frequencies than ω_0. However, for larger amplitudes, the starting equation is no longer valid, because then the free oscillation becomes anharmonic. Note also that the *phase shift* ϕ increases with ω. For $\omega \ll \omega_0$, it is negligible, for $\omega = \omega_0$, it takes the value $\frac{1}{2}\pi$, and for $\omega \gg \omega_0$, it tends to π. The higher the driving frequency, the more the forced oscillation is delayed, until it finally oscillates in opposite phase. This transition from in-phase to opposite-phase becomes ever more sudden with decreasing damping γ.

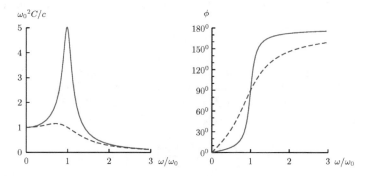

Fig. 2.23 Forced oscillation. *Left*: Amplitude of the ratio $\omega_0{}^2 C/c$ as a function of ω/ω_0. *Right*: Phase shift ϕ for $\gamma = 0.1\,\omega_0$ (*continuous red curve*) and for $\gamma = 05\,\omega_0$ (*dashed blue curve*)

Somewhat more concise is the treatment using complex variables. The ansatz

$$x = \mathrm{Re}\,\{\mathscr{C}\exp(-\mathrm{i}\,\omega t)\}\,,\quad \text{with}\quad \mathscr{C} = C\exp(\mathrm{i}\,\phi)$$

leads via the differential equation to $(\omega_0{}^2 - \omega^2 - 2\mathrm{i}\,\gamma\omega)\,\mathscr{C} = c$. With

$$\omega_\pm = \pm\sqrt{\omega_0{}^2 - \gamma^2} - \mathrm{i}\,\gamma$$

from the last section, or $\omega^2 + 2\mathrm{i}\,\gamma\omega - \omega_0{}^2 = (\omega - \omega_+)(\omega - \omega_-)$, we then arrive at

$$\mathscr{C} = \frac{c}{(\omega - \omega_-)(\omega_+ - \omega)} = \frac{c}{\omega_+ - \omega_-}\left(\frac{1}{\omega - \omega_-} - \frac{1}{\omega - \omega_+}\right).$$

For $\gamma \neq \omega_0$, the amplitude $\mathscr{C}$ thus has two simple poles below the real ω-axis (see Fig. 2.20). This representation is particularly suitable when the driving force is not purely harmonic and therefore has to be integrated (according to Fourier)—this is then straightforward using the theorem of residues.

In addition, in many cases it is not only the long-time behavior that is of interest. Therefore, we still wish to generalize the previous considerations. To this end, we shall transform the inhomogeneous linear differential equation

$$\ddot{x}(t) + 2\gamma\dot{x}(t) + \omega_0{}^2 x(t) = f(t)\,,$$

with a *Laplace transform*, viz.,

$$x \longrightarrow \mathscr{L}\{x\} \equiv \int_0^\infty \exp(-st)\,x(t)\,\mathrm{d}t\,,$$

into an algebraic equation [3], where $\mathrm{Re}\,s > 0$ has to hold. Naturally, the solution here still has to undergo the inverse Laplace transform. Note that the great advantage of the Laplace transform over the similar Fourier transform is the fact that only *one* integration limit is unrestricted. The Laplace-transformed derivative $\dot{x}$ is equal to

$$\mathscr{L}\{\dot{x}\} = s\,\mathscr{L}\{x\} - x(+0)\,,$$

since partial integration delivers $\int_0^\infty e^{-st}\dot{x}\,\mathrm{d}t = e^{-st}x\big|_{t=0}^{t=\infty} + s\int_0^\infty e^{-st}x\,\mathrm{d}t$. The region $t < 0$ is of no interest. Hence for $t = 0$, x may even jump from $x(-0)$ to finite $x(+0)$. Since $\mathscr{L}\{\ddot{x}\} = s(s\mathscr{L}\{x\}-x(0))-\dot{x}(0)$ and $s^2 + 2\gamma s + \omega_0{}^2 = (s + \mathrm{i}\,\omega_+)(s + \mathrm{i}\,\omega_-)$, the original differential equation leads to

$$\mathscr{L}\{x\} = \frac{\mathscr{L}\{f\} + (s+2\gamma)\,x(0) + \dot{x}(0)}{(s + \mathrm{i}\,\omega_+)(s + \mathrm{i}\,\omega_-)}\,.$$

The result may also be written as

$$\mathscr{L}\{x\} = \mathscr{L}\{x_0\} + \mathscr{L}\{g\} \cdot \mathscr{L}\{f\} ,$$

with

$$\mathscr{L}\{g\} \equiv \frac{1}{(s + i\omega_+)(s + i\omega_-)} = \frac{i}{2\sqrt{\omega_0^2 - \gamma^2}} \left(\frac{1}{s + i\omega_+} - \frac{1}{s + i\omega_-} \right),$$

if $x_0(t)$ solves the associated homogeneous differential equation under the given initial conditions: $\ddot{x}_0 + 2\gamma \dot{x}_0 + \omega_0^2 x_0 = 0$, along with $x_0(0) = x(0)$ and $\dot{x}_0(0) = \dot{x}(0)$. According to the last section, we can determine this auxiliary quantity.

The product of the Laplace-transformed $\mathscr{L}\{g\} \cdot \mathscr{L}\{f\}$ comes from a convolution integral:

$$x(t) = x_0(t) + \int_0^t g(t - t') f(t') \, dt' .$$

Since we are only interested here in $0 \le t' \le t$, we may then amend both functions g and f so that they vanish for negative arguments. Then we may also integrate from $-\infty$ to $+\infty$. This leads to the convolution theorem, as for the Fourier transform on p. 22, because the Laplace transform

$$\mathscr{L}\{F\} = \iint_{-\infty}^{\infty} \exp(-st)\, g(t - t')\, f(t')\, dt\, dt'$$

arises for the function $F(t) = \int_{-\infty}^{\infty} g(t - t') f(t') \, dt'$. And because we have $\exp(-st) = \exp\{-s(t - t')\} \exp(-st')$ with the new integration variables $\tau = t - t'$ (and equal integration limits for τ and t), this double integral can be split into the product of the Laplace-transformed functions of g and f, as required.

In order to determine g, we compare the expression $\{(s + i\omega_+)(s + i\omega_-)\}^{-1}$ for $\mathscr{L}\{g\}$ with that for $\mathscr{L}\{x\}$. The two Laplace-transformed functions are apparently equal, if $x(0) = 0$, $\dot{x}(0) = 1$ holds and f vanishes—the oscillation is not forced. If we set $\tau \equiv t - t'$, then for $g(\tau)$, the constraints are

$$g(0) = 0 , \quad \dot{g}(0) = 1 , \quad \text{and} \quad \ddot{g} + 2\gamma \dot{g} + \omega_0^2 g = 0 .$$

Consequently, according to the last section, we already know $g(\tau)$. In particular, we have $g(\tau) = \exp(-\gamma \tau) \Omega^{-1} \sin(\Omega \tau)$ with $\Omega = \sqrt{\omega_0^2 - \gamma^2}$ for $\gamma < \omega_0$ (see Fig. 2.24).

Note that the integral often extends to ∞, where $g(\tau)$ then has to vanish for $\tau < 0$. This function remains continuous, but its first derivative at $\tau = 0$ has to jump from zero to one. This leads to the differential equation $\ddot{g}(\tau) + 2\gamma \dot{g}(\tau) + \omega_0^2 g(\tau) = \delta(\tau)$, thus the starting equation with $f(\tau) = \delta(\tau)$ as inhomogeneity. Generally, solutions of linear differential equations with the delta function as inhomogeneity are called *Green functions*. Using these, the solutions for other inhomogeneities can be represented

Fig. 2.24 Green function $g(\tau)$ (for the damped oscillator). For $\tau \neq 0$, it satisfies the homogeneous differential equation $\ddot{g} + 2\gamma\dot{g} + \omega_0{}^2 g = 0$. For $\tau = 0$, its first derivative jumps by one. Hence, the second derivative is given there by the delta function $\delta(\tau)$

as convolution integrals (Problem 2.38). We encountered the Green function for the Laplace operator on p. 26, and the one here will be generalized in Sect. 2.3.10.

If for finite damping only the long-time behavior is of interest, then we may leave out $x_0(t)$ and take $-\infty$ as the lower integration limit. Then we arrive at a convolution integral from $-\infty$ to $+\infty$.

2.3.9   Coupled Oscillations and Normal Coordinates

So far we have restricted ourselves to oscillations of just one coordinate. Now we consider several coordinates ($f > 1$), e.g., a double pendulum (one hanging from the other) or several point masses coupled to each other by springs (atoms in a molecule or in a crystal). Here we start from a conservative system with the potential energy $V(x^1, \ldots, x^f)$ and choose the origin of all f coordinates x^k in their equilibrium position. Then all forces vanish:

$$F_k = -\left.\frac{\partial V}{\partial x^k}\right|_0 = 0, \quad \text{for}\ \ k \in \{1, \ldots, f\}\,.$$

We assume a stable equilibrium, i.e., small displacements from the equilibrium cost energy. Then the extremum of V has to be a local minimum, and for the corresponding gauge, according to Taylor, we have

$$V = \tfrac{1}{2}\sum_{kl}\left.\frac{\partial^2 V}{\partial x^k\,\partial x^l}\right|_0 x^k\,x^l \equiv \tfrac{1}{2}\sum_{kl} A_{kl}\,x^k\,x^l\,, \quad \text{with}\ \ A_{kl} = A_{lk} = A_{kl}{}^*\,,$$

if we neglect higher-order terms—the pendulum is just barely displaced, and no anharmonic forces act between the atoms. Here the coefficients do not depend upon the time t.

In addition, we need the kinetic energy, for which we make an ansatz of the form

$$T = \tfrac{1}{2}\sum_{kl} B_{kl}\,\dot{x}^k\,\dot{x}^l\,, \quad \text{with}\ \ B_{kl} = B_{lk} = B_{kl}{}^*\,,$$

where it is assumed that these coefficients do not depend on time (which is approximately true only occasionally). In any case, no linear terms in $\dot{x}^k$ should appear, because otherwise they would change sign for $t \to -t$.

For $k \in \{1, \ldots, f\}$ and since $A_{kl} = A_{lk}$ and $B_{kl} = B_{lk}$, the Lagrange equations now deliver

$$0 = \frac{d}{dt} \frac{\partial L}{\partial \dot{x}^k} - \frac{\partial L}{\partial x^k} = \sum_l B_{kl} \ddot{x}^l + \sum_l A_{kl} x^l .$$

If we take A and B as square matrices and $(x^1, \ldots, x^f)$ as a row vector $\tilde{x}$, we then have

$$V = \tfrac{1}{2} \tilde{x} A x \quad \text{and} \quad T = \tfrac{1}{2} \tilde{\dot{x}} B \dot{x} ,$$

and also

$$B \ddot{x} + A x = 0 ,$$

or $\ddot{x} = -B^{-1} A x$. For one degree of freedom ($f = 1$), we could have written simply ω^2 instead of the matrix product $B^{-1}A$.

Now we would like to make a transition to new coordinates, called *normal coordinates* x', relative to which the matrices A and B become diagonal, the oscillations thus become decoupled, and the solutions are already known. The total energy is then the sum of the energies of the individual decoupled oscillators.

If we set

$$x = C x' \quad \Longrightarrow \quad V = \tfrac{1}{2} \tilde{x}' \tilde{C} A C x' \text{ and } T = \tfrac{1}{2} \tilde{\dot{x}}' \tilde{C} B C \dot{x}' ,$$

then we search for a matrix C, which diagonalizes $\tilde{C} A C$ as well as $\tilde{C} B C$. Here we choose the free factor—only the product $C x'$ is fixed—such that $\tilde{C} B C = 1$ holds. Then the diagonal elements λ of $\tilde{C} A C$ are the squares of the angular frequencies. These are the frequencies with which the normal coordinates oscillate. The amplitudes and phases are adjusted to the initial values.

In this case, $A x$ becomes $A C x'$, and with $\ddot{x}' = -\lambda x'$, $B\ddot{x}$ becomes $-\lambda B C x'$. The vector $C x'$ will be denoted by c and we shall seek f such column vectors and combine them to form the matrix C. Finally, from $B \ddot{x} + A x = 0$, we have

$$(A - \lambda B) c = 0 , \quad \text{with} \quad A = A^* = \tilde{A} , \quad B = B^* = \tilde{B} , \quad c = c^* .$$

The homogeneous linear system of equations $(A - \lambda B) c = 0$ is an *eigenvalue problem*, because it is soluble only for suitable *eigenvalues* λ_k. With these, we determine the *eigenvectors* c_k. Despite the fact that in general the number of degrees of freedom is $f \neq 3$, this eigenvalue problem differs from that of the principal axis transformation for the moment of inertia in Sect. 2.2.11 in that B was a unit matrix there.

The eigenvalues can be determined from the *characteristic equation*

$$\det (A - \lambda B) = 0 .$$

Since we are working with Hermitian matrices with f rows, there are f real eigenvalues λ_k and associated eigenvectors c_k, which then follow from

$$(A - \lambda_k B)\, c_k = 0 .$$

These eigenvectors are determined only up to a factor, which we shall soon choose in a convenient way. So if we combine the total set of f column vectors $\{c_k\}$, each with f components, to form an *eigenvector matrix $C = (c_1, \ldots, c_f)$*, we arrive at

$$\widetilde{C} B C = 1 .$$

With the help of the kth diagonal element of this matrix and an appropriate "normalization factor" (a scale transformation), we can choose the kth (row and) column vector and make all non-diagonal elements—in different row and column vectors—equal to zero. This can be seen immediately for different eigenvalues $\lambda_k \neq \lambda_{k'}$, because $(\lambda_k - \lambda_{k'})\, \widetilde{c}_k B c_{k'}$ is the same as $\widetilde{c}_k (A - A) c_{k'}$. But for equal eigenvalues $\lambda_k = \lambda_{k'}$ (*degeneracy*) all linear combinations of these eigenvectors are still eigenvectors, and this freedom can be exploited for $\widetilde{C} B C = 1$. The matrix $\widetilde{C} A C \equiv \Lambda$ is then also diagonal. Thus we have

$$2\,T = \widetilde{\dot{x}}\, B\, \dot{x} = \widetilde{\dot{x}'}\, \widetilde{C} B C\, \dot{x}' = \widetilde{\dot{x}'}\, 1\, \dot{x}' ,$$
$$2\,V = \widetilde{x}\, A\, x = \widetilde{x'}\, \widetilde{C} A C\, x' = \widetilde{x'}\, \Lambda\, x' .$$

In the new coordinates, the kinetic and potential energy no longer contain mixed terms. The f harmonic oscillations are *decoupled* in the normal coordinates.

The eigenvalues λ of Λ are the squares of the desired angular frequencies because they represent the harmonic oscillation in the expression $\frac{1}{2} m\, (\dot{x}^2 + \omega^2 x^2)$ for the energy.

For example, for two coupled oscillations ($f = 2$), we thus arrive at the eigenfrequencies

$$\omega_\pm{}^2 = \frac{K \pm \sqrt{K^2 - 4\, \det A\, \det B}}{2\, \det B} , \quad \text{with} \quad K = A_{11} B_{22} + A_{22} B_{11} - 2 A_{12} B_{12} .$$

To these belong the two eigenvectors $c_\pm$, each with two components, whose ratio is

$$\frac{c_{2\pm}}{c_{1\pm}} = -\frac{A_{11} - \omega_\pm{}^2 B_{11}}{A_{12} - \omega_\pm{}^2 B_{12}} ,$$

and which are normalized via $c_{1\pm}^{-2} = B_{11} + 2B_{12}\,(c_{2\pm}/c_{1\pm}) + B_{22}\,(c_{2\pm}/c_{1\pm})^2$. With this,

$$C = \begin{pmatrix} c_{1+} & c_{1-} \\ c_{2+} & c_{2-} \end{pmatrix},$$

and its inverse matrix (see p. 71) can be calculated. In normal coordinates, the solutions read

$$x_\pm' = x_{0\pm}'\cos(\omega_\pm t) + \frac{\dot{x}_{0\pm}'}{\omega_\pm}\sin(\omega_\pm t)\,,$$

where the coefficients $x_{0\pm}'$ and $\dot{x}_{0\pm}'$ follow from the initial conditions:

$$x_0' = C^{-1}x_0 \quad\text{and}\quad \dot{x}_0' = C^{-1}\dot{x}_0\,.$$

Note that, according to p. 102, we may thus write also $x_\pm' = \widehat{x}_\pm'\cos(\omega_\pm t - \phi_\pm)$. Since all unknown quantities have then been determined from the matrix elements of A and B and from the initial values, the solution $x = Cx'$ can finally be calculated (Problems 2.39–2.42).

If the two eigenfrequencies are nearly equal ($\omega_+ \approx \omega_-$), then *beats* are formed, i.e., the oscillation amplitudes change periodically, and this all the more clearly as the amplitudes are close to one another. From

$$x_i = c_{i+}\widehat{x}_+'\cos(\omega_+ t - \phi_+) + c_{i-}\widehat{x}_-'\cos(\omega_- t - \phi_-)\,,$$

with $\omega_+ > \omega_-$ and positive amplitude $c_{i\pm}\widehat{x}_\pm'$ abbreviated to $C_{i\pm}$, together with the notation $\omega_\pm = \Omega \pm \omega$ and $\phi_\pm = \phi \pm \varphi$, it follows that

$$x_i = +(C_{i+} + C_{i-})\cos(\Omega t - \phi)\cos(\omega t - \varphi)$$
$$-(C_{i+} - C_{i-})\sin(\Omega t - \phi)\sin(\omega t - \varphi)\,.$$

Since $\omega_+ \approx \omega_-$, we have $\omega \ll \Omega$, whence the amplitude of the oscillation changes periodically with the angular frequency Ω according to

$$\sqrt{(C_{i+} - C_{i-})^2 + 4C_{i+}C_{i-}\cos^2(\omega t - \varphi)}\,.$$

Examples are shown in Fig. 2.25.

If one of the eigenvalues is zero, then this is not an oscillation, but free motion. If no external forces act, in a first step we separate out the center-of-mass motion and also the rotation of a rigid body. The following considerations are necessary only for the relative motion.

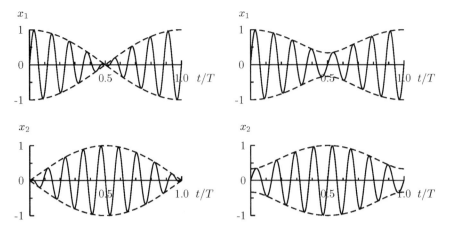

Fig. 2.25 Examples of the displacements of two coupled oscillations as a function of time t during a period T. *Left*: Equal amplitudes of $c\widehat{x}'$. *Right*: For the amplitude ratio 1:2. In both cases, $\phi_\pm = \frac{1}{2}\pi$ and $8\omega_+ = 9\omega_-$. The oscillation amplitudes are shown by *dashed lines*

2.3.10 Time-Dependent Oscillator. Parametric Resonance

If the parameters kept fixed so far are assumed now to change rhythmically with time, then this affects the stability of the system. This is observed for a child's swing, where the moment of inertia fluctuates in the course of time, and for a pendulum if its support oscillates vertically up and down. In both cases we encounter *Hill's differential equation*

$$\ddot{x} + f(t)\, x = 0 , \quad \text{with} \quad f(t+T) = f(t) = f^*(t) ,$$

which we shall discuss now. We shall often write ω^2 instead of f, even though f may also become negative. In the end, generalizations of the functions $\cos(\omega t)$ and $\sin(\omega t)$ are obtained, which belong to constant $f > 0$. Incidentally, Hill's differential equation also arises in the quantum theory of crystals and in the theory of charged particles in a synchrotron with alternating gradients, although t is then a position coordinate. The *Bloch function* is encountered in the context of a periodic potential.

We take the two (presently unknown) *fundamental solutions* x_1 and x_2 with the properties $x_1(0) = 1 = \dot{x}_2(0)$ and $\dot{x}_1(0) = 0 = x_2(0)$. Their *Wronski determinant* $x_1\dot{x}_2 - \dot{x}_1 x_2$ has the value 1 for all t, this being the value for $t = 0$ and a constant, because its derivative vanishes. All remaining solutions of the differential equation can be expanded in terms of this basis. We clearly have $x(t) = x(0)\, x_1(t) + \dot{x}(0)\, x_2(t)$ since this expression satisfies the differential equation and the initial conditions. We may thus write

$$\begin{pmatrix} cx(t) \\ \dot{x}(t) \end{pmatrix} = \begin{pmatrix} ccx_1(t) & x_2(t) \\ \dot{x}_1(t) & \dot{x}_2(t) \end{pmatrix} \begin{pmatrix} cx(0) \\ \dot{x}(0) \end{pmatrix} \equiv U(t) \begin{pmatrix} cx(0) \\ \dot{x}(0) \end{pmatrix} ,$$

and with this obtain an area-preserving *time-shift matrix* $U(t)$ (since det $U = 1$).

We would now like to exploit the periodicity—so far we have not used it for the time shift and, of course, we could also have introduced the matrix for other factors f.

The *Floquet operator* $U(T)$ will be important for us: for given initial conditions it delivers $x(T)$ and $\dot{x}(T)$, and we have

$$x(t + T) = x(T) \, x_1(t) + \dot{x}(T) \, x_2(t) \, ,$$

because this expression satisfies the initial conditions and, since $f(t) = f(t + T)$, it also satisfies the differential equation.

Therefore, we look for the eigenvalues $\sigma_\pm$ of $U(T)$. For a 2×2 matrix U, they follow from $\sigma^2 - \sigma \, \mathrm{tr}U + \det U = 0$ and, because of $\det U = 1$, satisfy the equations $\sigma_+ \sigma_- = 1$ and $\sigma_+ + \sigma_- = \mathrm{tr}U$. We thus set $\sigma_\pm = \exp(\pm i \phi)$ and determine ϕ from $\mathrm{tr}U = 2 \cos \phi$, which is uniquely possible only up to an integer multiple of π. However, we require in addition that ϕ should depend continuously on f, and set $\phi = 0$ for $f \equiv 0$. (For $f \equiv 0$, we have $\mathrm{tr}U = 2$ because $x_1 = 1$ and $x_2 = t$.) Since x_1 and x_2 are real initially and remain so for all times, $\mathrm{tr}U$ will also be real. Since $\cos(\alpha + i\beta) = \cos \alpha \cosh \beta - i \sin \alpha \sinh \beta$, this means that either ϕ has to be real ($\beta = 0$ for $|\mathrm{tr}U| \leq 2$) or its real part has to be an integer multiple of π ($\alpha = n\pi$ for $|\mathrm{tr}U| \geq 2$). For $|\mathrm{tr}U| < 2$, we thus have $|\sigma_\pm| = 1$, and for $|\mathrm{tr}U| > 2$, it is clear that $|\sigma_\pm| \neq 1$. (We will return to the degeneracy for $|\mathrm{tr}U| = 2$.)

For the two eigensolutions (*Floquet solutions*), we have $x_\pm(t + T) = \sigma_\pm x_\pm(t)$. Then for $|\mathrm{tr}U| > 2$, their moduli change by the factor $|\sigma_\pm| \neq 1$ for each additional T. For $t \to \infty$, one of them exceeds all limits, while for $t \to -\infty$, it is the other that does so. Therefore, they are said to be (Lyapunov) *unstable*. For $|\mathrm{tr}U| > 2$ all solutions of the differential equation are unstable, because they are linear compositions of both of these eigensolutions. In contrast, for $|\mathrm{tr}U| < 2$, the eigensolutions change only by a complex factor of absolute value one with the time increment T—here all solutions are *stable*, and we may choose $x_- = x_+{}^*$.

Except for the factor $\sigma_\pm{}^{t/T} = \exp(\pm i \, \phi t / T)$, the Floquet solutions have period $T = 2\pi / \Omega$ and can therefore be represented by a Fourier series or a Laurent series. These solutions are linearly independent if there is no degeneracy. For degeneracy ($|\mathrm{tr}U| = 2$ or $\sigma_\pm{}^2 = 1$), there are stable as well as unstable solutions: $x(t) = Q(t) + t P(t)$ with periodic P and Q (for $\sigma_\pm = +1$ with period T, for $\sigma_\pm = -1$ with period $2T$). Here the differential equation for x can be satisfied if $\ddot{P} + f P = 0$ and $\ddot{Q} + f Q = -2\dot{P}$. The expansion coefficients in the Fourier series depend on the function $f(t)$.

The special case $f(t) = \frac{1}{4}\Omega^2(a - 2q \cos \Omega t)$, *Mathieu's differential equation*, has been thoroughly investigated (see, e.g., [1, 3]). It also arises in the separation of the wave equation $(\triangle + k^2) \, u = 0$ in elliptic coordinates, where only periodic solutions make sense—and they then acquire the special eigenvalues $a(q)$. The curves $a(q)$ (see Fig. 2.26) separate the regions of stable and unstable Mathieu functions—thus also allowed and non-allowed energy bands in crystal fields with the potential energy $V(x) = V_0 \cos(kx)$, because there we have $a \stackrel{\frown}{=} 8mE/(\hbar k)^2$ and

Fig. 2.26 Stability chart of the functions solving the special Hill differential equation $\ddot{x} + (\frac{1}{2}\Omega)^2(a - 2q\cos\Omega t)\,x = 0$, here for $0 \le q \le 8$ and $-5 \le a \le 15$. *Curves* indicate the stability limits. For $q = 0$, we must have $a \ge 0$, while for $q > 0$, the region splits into *bands* which become ever narrower, but also allow for $a < 0$

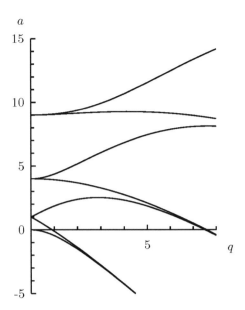

$q \mathrel{\hat=} 4mV_0/(\hbar k)^2$ (see Fig. 2.27). The computation of the Mathieu functions and their stability chart is explained in more detail in Sect. 2.4.11.

Simplifications are generally available if f is an even function, thus if $f(-t) = f(t)$ holds. In particular, x_1 is then even and x_2 odd, whence $x(T - t) = x(T)\,x_1(t) - \dot{x}(T)\,x_2(t)$. If this is used for $t = T$ for the two fundamental solutions x_1 and x_2, we obtain $\dot{x}_2(T) = x_1(T)$. For even f, we thus have $\cos\phi = x_1(T)$. Therefore, the solutions for $|x_1(T)| < 1$ are then stable and otherwise unstable. In addition, not only $x(t)$ but also $x(-t)$ now solves the given differential equation. Therefore, we may now also set $x_-(t) = x_+(-t)$ and $P_-(t) = P_+(-t)$.

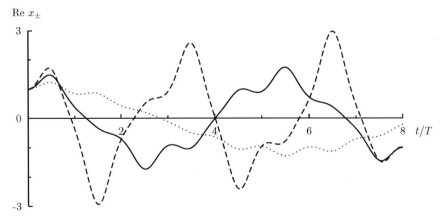

Fig. 2.27 Real part of the Mathieu functions $x_\pm$ for $0 \le t \le 8T$ and $a = 0$, for $q = 1/4$ (*dotted curve*), $q = 2/4$ (*continuous curve*), and $q = 3/4$ (*dashed curve*)

Finally, we consider the weakly time-dependent oscillator:

$$f(t) = \omega_0^2 \left\{ 1 + \varepsilon\, a(t) \right\}, \quad \text{with} \quad \varepsilon \ll 1 \quad \text{and} \quad a(t+T) = a(t)\,.$$

Here $\mathrm{tr}U \approx 2\cos(\omega_0 T)$ holds, whence $\phi \approx \omega_0 T$. Therefore, the stability is only at risk if $\omega_0 T$ is an integer multiple of π, thus if the period T of f or a is a half or integer multiple of the period $T_0 = 2\pi/\omega_0$ of the basic oscillation. Even for very small fluctuations in the moments of inertia, an (undamped) swinging effect comes about. This instability is called *parametric resonance*. It is particularly pronounced for $T = \frac{1}{2}T_0$ because, according to Fig. 2.26, the first unstable band for $a \approx 1$ is particularly close to the axis $q = 0$ and ever smaller for the higher ones ($a \approx 2^2, 3^2, \ldots$): when swinging on a child's swing, we must move on the way back and forth, and anyone who does that too rarely will not get into motion, whatever the effort.

Our starting equation also holds for a linear frictional force. Hence, if we start from $\ddot{y} + 2\gamma\dot{y} + h(t)\, y = 0$ and set $y = \exp(-\gamma t)\, x$, then with $f = h - \gamma^2$, we arrive at the starting equation. Naturally, the factor $\exp(-\gamma t)$ strengthens the stability, because γ is positive and only $t > 0$ is of interest. Now the solutions with $|\mathrm{Im}\phi| \leq \gamma T$ are still stable.

For a forced oscillation $\ddot{y} + 2\gamma\dot{y} + h(t)\, y = f(t)$, we may make the ansatz

$$y(t) = y_0(t) + \int_0^t g(t, t')\, f(t')\, \mathrm{d}t'$$

for the solution. If h did not depend on t, we might simplify the *Green function* $g(t, t')$ to $g(t - t')$, as was shown in Sect. 2.3.8. Correspondingly, we now have to require $g(t, t) = 0$, $\dot{g}(t, t) = 1$, and $\ddot{g} + 2\gamma\dot{g} + h\, g = 0$, for $0 \leq t' \leq t$. If we replace (as there) the upper integration limit by ∞, then $g(t, t') = 0$ has to hold for $t' > t$, and therefore $\ddot{g} + 2\gamma\dot{g} + h\, g = \delta(t - t')$ must be valid. If x_1 and x_2 are linearly independent solutions of the homogeneous differential equation $\ddot{x} + (h - \gamma^2)\, x = 0$, then all these requirements can be satisfied with

$$g(t, t') = \exp\{-\gamma\, (t - t')\}\, \frac{x_1(t')\, x_2(t) - x_1(t)\, x_2(t')}{x_1(t')\, \dot{x}_2(t') - \dot{x}_1(t')\, x_2(t')}\,, \quad \text{for } t \geq t' \text{ (zero otherwise)}\,.$$

In particular, for $t \neq t'$, this expression satisfies the differential equation, it vanishes for $t = t'$, and its first derivative with respect to t makes a jump there from 0 to 1. The above-mentioned Wronski determinant $x_1\dot{x}_2 - \dot{x}_1 x_2$ appears in the denominator.

Incidentally, $g(t, t')$ does not need to vanish for $t < t'$, if we account for the contribution to the initial values $y(0)$ and $\dot{y}(0)$ (thus modify y_0). The Green function only has to satisfy the differential equation $\ddot{g} + 2\gamma\dot{g} + hg = \delta(t - t')$. This can be done with

$$g(t, t') = \exp\{-\gamma\, (t - t')\}\, \frac{x_1(t_<)\, x_2(t_>)}{x_1\dot{x}_2 - \dot{x}_1 x_2}\,,$$

where $t_<$ is the smaller and $t_>$ the larger of the two values t and t', and again the Wronski determinant appears in the denominator. Here, $\dot{g}$ jumps by 1 for $t = t'$, but is not zero at lower t.

2.3.11 Summary: Lagrangian Mechanics

In Sect. 2.1, we already anticipated some important aspects of Lagrangian mechanics, although we restricted ourselves there to time-independent phenomena. Geometric constraints can often be incorporated through the use of appropriate coordinates in a simpler way than by the associated forces of constraint. In particular, it is often the case that fewer variables (generalized coordinates) depend on the time—otherwise the constraints have to be accounted for by Lagrangian parameters in the Lagrange equations of the first kind. To this end, we generalized the principle of virtual work to d'Alembert's principle by taking into account inertial forces.

With a convenient choice of coordinates, we have an N-body problem in the "configuration space" with $f (\leq 3N)$ dimensions and Lagrange equations of the second kind

$$F_k = \frac{\mathrm{d}}{\mathrm{d}t} \frac{\partial T}{\partial \dot{x}^k} - \frac{\partial T}{\partial x^k} \ .$$

Here, the generalized forces $F_k = \sum_i \mathbf{F}_i \cdot \partial \mathbf{r}_i / \partial x^k$ are often derived from a potential energy. The forces may even depend upon the velocity, since there may also be a generalized potential energy U with the property

$$F_k = \frac{\mathrm{d}}{\mathrm{d}t} \frac{\partial U}{\partial \dot{x}^k} - \frac{\partial U}{\partial x^k} \ .$$

Then we can use the Lagrange function

$$L = T - U$$

for calculations with the equations

$$\frac{\mathrm{d}}{\mathrm{d}t} \frac{\partial L}{\partial \dot{x}^k} - \frac{\partial L}{\partial x^k} = 0 \ .$$

Several applications have been discussed and exemplified for these methods. With the canonical momentum

$$p_k = \frac{\partial L}{\partial \dot{x}^k}$$

which is conjugate to x^k, we may also write $\dot{p}_k = \partial L/\partial x^k$, or $\dot{\mathbf{p}} = \nabla L$ with $\mathbf{p} = \nabla_v L$. This canonical momentum is to be distinguished from the mechanical momentum $m\mathbf{v}$, e.g., $\mathbf{p} = m\mathbf{v} + q\mathbf{A}$ holds if the vector potential $\mathbf{A}$ acts on the electric charge q (the curl of $\mathbf{A}$ is the magnetic field $\mathbf{B}$). If L does not explicitly depend on time, then $\sum_k p_k \dot{x}^k - L$ is a constant of the motion. Furthermore, the conjugate momenta are conserved for all cyclic variables, i.e., for those x^k that do not appear in L, p_k does not depend on time.

We have investigated examples of various oscillations (harmonic, anharmonic, damped, forced, and coupled). Note that, while the solutions of linear differential equations change continuously with the initial conditions, this is different for non-linear ones, as illustrated by the example of the (anharmonic) pendulum near the separatrix.

2.4 Hamiltonian Mechanics

2.4.1 Hamilton Function and Hamiltonian Equations

According to the last section, when a (generalized) potential is given, we may always start from the Euler–Lagrange equation

$$\frac{\mathrm{d}}{\mathrm{d}t}\frac{\partial L}{\partial \dot{x}} = \frac{\partial L}{\partial x} \ .$$

Here x stands for an arbitrary generalized position coordinate x^k and $\dot{x}$ for its velocity $\dot{x}^k$. But in Sect. 2.3.5, it already turned out that, instead of the velocity $\dot{x}$, it is often better to consider the canonical momentum

$$p \equiv \frac{\partial L}{\partial \dot{x}} \qquad \Longrightarrow \qquad \dot{p} = \frac{\partial L}{\partial x}$$

conjugate to x. From now on, instead of the velocity $\dot{x}$, we shall always take this momentum p as an independent variable and investigate everything in the *phase space* (x, p), as we have already done for the pendulum orbits in Fig. 2.16 (see Fig. 2.28). Here we may still gauge arbitrarily—only then does the canonical momentum depend uniquely on the velocity. This greater freedom is occasionally of use and often also provides a deeper understanding of the interrelations.

The new variable p is the derivative of L with respect to the variable $\dot{x}$ (hereafter $\dot{x}$ will be replaced by p). Therefore, a *Legendre transformation* is necessary. Instead of the Lagrange function $L(t, x, \dot{x})$ with

$$\mathrm{d}L = \frac{\partial L}{\partial t}\,\mathrm{d}t + \frac{\partial L}{\partial x}\,\mathrm{d}x + \frac{\partial L}{\partial \dot{x}}\,\mathrm{d}\dot{x} \quad \text{and} \quad \frac{\partial L}{\partial \dot{x}} = p \ ,$$

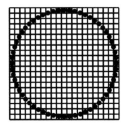

Fig. 2.28 Representation of a harmonic oscillation in the (two-dimensional) phase space (with convenient scales for the x- and the p-coordinate—otherwise we obtain an ellipse). The points (•) are traversed clockwise. The phase-space units may not be arbitrarily small, according to quantum physics—otherwise there would be a contradiction with Heisenberg's uncertainty relation

we have to take the *Hamilton function*[5] $H(t, x, p)$ with

$$\mathrm{d}H = \frac{\partial H}{\partial t}\,\mathrm{d}t + \frac{\partial H}{\partial x}\,\mathrm{d}x + \frac{\partial H}{\partial p}\,\mathrm{d}p \quad \text{and} \quad H = p\,\dot{x} - L \ .$$

In particular, the last equation implies $\mathrm{d}H = \dot{x}\,\mathrm{d}p + p\,\mathrm{d}\dot{x} - \mathrm{d}L$ or $\mathrm{d}H = \dot{x}\,\mathrm{d}p - (\partial L/\partial t)\,\mathrm{d}t - (\partial L/\partial x)\,\mathrm{d}x$. Comparing this expression with the one before, we then find

$$\frac{\partial H}{\partial t} = -\frac{\partial L}{\partial t}\ , \quad \frac{\partial H}{\partial x} = -\frac{\partial L}{\partial x}\ , \quad \frac{\partial H}{\partial p} = \dot{x}\ .$$

We reformulate the middle relation with the Lagrange equation and find that, for the *conjugate variables* x^k and p_k, and with the *Hamilton function*

$$H \equiv \sum_k p_k\,\dot{x}^k - L\ ,$$

we obtain the *Hamilton equations*

$$\dot{x}^k = \frac{\partial H}{\partial p_k}\ , \qquad \dot{p}_k = -\frac{\partial H}{\partial x^k}\ .$$

These are very general and we shall thus refer to them as the *canonical equations*. In Lagrangian mechanics, there is one differential equation of second order for each degree of freedom, whereas in Hamiltonian mechanics, there are always two differential equations of first order. In addition, one has

$$\frac{\mathrm{d}H}{\mathrm{d}t} = \frac{\partial H}{\partial t} + \sum_k \Big(\frac{\partial H}{\partial x^k}\frac{\partial H}{\partial p_k} - \frac{\partial H}{\partial p_k}\frac{\partial H}{\partial x^k}\Big) = \frac{\partial H}{\partial t}\ .$$

---

[5]William Rowan Hamilton (1805–1865).

If further the Hamilton function H does not depend explicitly on time, then it remains a conserved quantity along all orbits.

If there is a potential energy V (and hence also $L = T - V$), and if T is a homogeneous function of second order in the velocities (so that $p\dot{x} = 2T$, according to p. 101), we have $H = T + V$, so H is an energy. But we shall find shortly that H and E may also be different.

For a non-relativistic particle of mass m and charge q in an electromagnetic field, we infer the Hamilton function from the Lagrange function

$$L = \frac{m}{2}\, \mathbf{v} \cdot \mathbf{v} - q\, (\Phi - \mathbf{v} \cdot \mathbf{A})$$

in Sect. 2.3.5 (p. 100) as $H = \mathbf{p} \cdot \mathbf{v} - L$. To this end, we only have to express the velocity $\mathbf{v}$ in terms of the canonical momentum $\mathbf{p} = m\,\mathbf{v} + q\,\mathbf{A}$ (see p. 100). Since $(m\mathbf{v} + q\mathbf{A}) \cdot \mathbf{v} - L = \frac{m}{2}\mathbf{v} \cdot \mathbf{v} + q\,\Phi$ and $\mathbf{v} = (\mathbf{p} - q\,\mathbf{A})/m$ this leads to

$$H(t,\, \mathbf{r},\, \mathbf{r}) = \frac{(\mathbf{p} - q\,\mathbf{A}) \cdot (\mathbf{p} - q\,\mathbf{A})}{2m} + q\,\Phi\,.$$

If the magnetic field $\mathbf{B}$ depends neither on time nor on position, then according to p. 100, we may use the vector potential $\mathbf{A} = \frac{1}{2}\mathbf{B} \times \mathbf{r}$ with $q\mathbf{B} = -m\boldsymbol{\omega}$ (see p. 78), where ω is the associated cyclotron frequency. It then follows from $m\dot{\mathbf{r}} = \mathbf{p} - q\mathbf{A} = \mathbf{p} + \frac{1}{2}m\,\boldsymbol{\omega} \times \mathbf{r}$ or from $\dot{\mathbf{r}} = \boldsymbol{\nabla}_p\, H$ that

$$\dot{\mathbf{r}} = \frac{\mathbf{p}}{m} + \frac{\boldsymbol{\omega} \times \mathbf{r}}{2}\,.$$

In addition, for $\Phi = 0$, $\dot{\mathbf{p}} = -\boldsymbol{\nabla} H$ (in agreement with $\dot{\mathbf{p}} = \frac{1}{2}\mathbf{F} = \frac{1}{2}q\,\mathbf{v} \times \mathbf{B}$ on p. 100) delivers

$$\dot{\mathbf{p}} = \tfrac{1}{2}\,\boldsymbol{\omega} \times (\mathbf{p} + \tfrac{1}{2}m\,\boldsymbol{\omega} \times \mathbf{r})\,,$$

and thus,

$$\dot{\mathbf{p}} = \tfrac{1}{2}m\,\boldsymbol{\omega} \times \dot{\mathbf{r}}\,.$$

We have already integrated these differential equations on p. 100.

According to p. 98, for a *gauge transformation* $\Phi' = \Phi + \partial\Psi/\partial t$, $\mathbf{A}' = \mathbf{A} - \boldsymbol{\nabla}\Psi$, the Lagrange function is transformed into $L' = L - dG/dt$ with $G = q\,\Psi$, and the canonical momentum into $\mathbf{p}' = \mathbf{p} - \boldsymbol{\nabla}G$ (see p. 100). Since $dG/dt = \partial G/\partial t + \boldsymbol{\nabla}G \cdot \mathbf{v}$, the Hamilton function is

$$H' = \sum_k p_k'\,\dot{x}^{k\prime} - L' = H + \frac{\partial G}{\partial t}\,.$$

The term $\partial G/\partial t$ may depend on position and time—this is more than the arbitrariness in the choice of the zero energy. Therefore, the Hamilton function agrees with the

energy only for an appropriate gauge. A more detailed investigation is available in
[4]. The scalar potential Φ may not depend upon time! So only then can $q\Phi(\mathbf{r})$ be
a potential energy $V(\mathbf{r})$ and $H - V$ a homogeneous function of second order in the
velocity—consequently, H is an energy. If the electric field $\mathbf{E}$ depends on time, then
this has to be included in the vector potential $\mathbf{A}$, or more precisely, in its sources,
because its curl determines the magnetic field $\mathbf{B}$. For a time-dependent force, its path
integral depends upon the amount of time needed to traverse this path. The force field
is then not always curl-free and therefore cannot be derived from a potential energy.

In the Lagrangian formalism, we find that p_k is a constant of the motion if L does
not depend on x^k, i.e., if x^k is a cyclic coordinate. This leads to $0 = \partial L/\partial x^k = \dot{p}_k = -\partial H/\partial x^k$ in the Hamiltonian formalism: then x^k does not appear in H. Hence,
the conservation of momentum and angular momentum follows immediately for
each system with only internal forces, for which H does not involve center-of-mass
coordinates.

2.4.2   Poisson Brackets

The Poisson brackets for functions $u(t, x, p)$ and $v(t, x, p)$ are defined by

$$[u, v] \equiv \sum_k \left(\frac{\partial u}{\partial x^k} \frac{\partial v}{\partial p_k} - \frac{\partial u}{\partial p_k} \frac{\partial v}{\partial x^k} \right)$$

and have the properties (with constant α and β)

$$[u, v] = -[v, u] \, ,$$
$$[uv, w] = u\,[v, w] + [u, w]\,v \, ,$$
$$[\alpha u + \beta v, w] = \alpha\,[u, w] + \beta\,[v, w] \, .$$

In addition, the *Jacobi identity* holds:

$$[u, [v, w]] + [v, [w, u]] + [w, [u, v]] = 0 \, ,$$

as for the vector product on p. 4. This is proved using $\partial u/\partial x = u_x$, $\partial u/\partial p = u_p$
and similarly for v and w instead of u in Problem 2.43. The Hamilton equations
lead to

$$[u, H] = \sum_k \left(\frac{\partial u}{\partial x^k} \dot{x}^k + \frac{\partial u}{\partial p_k} \dot{p}_k \right) = \frac{du}{dt} - \frac{\partial u}{\partial t} \, ,$$

and for arbitrary u, we deduce

$$\frac{du}{dt} = \frac{\partial u}{\partial t} + [u, H] \, .$$

If u does not depend explicitly on time, then $\dot{u}$ is equal to the Poisson bracket of u with the Hamilton function H. In particular, we obtain

$$\dot{x}^k = [x^k, H], \quad \dot{p}_k = [p_k, H],$$

instead of the Hamilton equations.

Since position and momentum coordinates do not depend on each other, we also have

$$[x^i, x^j] = 0 = [p_i, p_j], \quad [x^i, p_j] = \delta^i_j = \begin{cases} 1 & \text{for } i = j, \\ 0 & \text{for } i \neq j. \end{cases}$$

These equations will play an important role for the transition to quantum mechanics, where the quantities will be replaced by (Hermitian) operators and the Poisson brackets by commutators (divided by $i\hbar$). Connections can also be found with these results in thermodynamics (statistical mechanics), namely, with the *Liouville equation*. The latter gives the time dependence of the probability density ρ in phase space and states that $d\rho/dt = 0$:

$$\frac{d\rho}{dt} = 0 \quad \Longrightarrow \quad \frac{\partial \rho}{\partial t} + [\rho, H] = 0.$$

Whatever is altered in a volume element of the phase space happens because of the equations of motion. This equation is proven in Sect. 2.4.4. With the probability density ρ, the mean values $\overline{A}$ of functions $A(t, x, p)$ can be evaluated from $\overline{A} = \int \rho A \, dx \, dp$.

2.4.3 Canonical Transformations

We would now like to choose new coordinates in phase space (still for fixed time), and possibly also a new Hamilton function, such that the canonical equations are still valid. In the Lagrangian formalism, we only considered transformations in the configuration space, which has only half as many coordinates. For the moment we restrict ourselves to just one degree of freedom and leave out the index k. Then the Poisson bracket $[u, v]$ is the same as the functional determinant $\partial(u, v)/\partial(x, p)$. Since

$$\frac{\partial (u, v)}{\partial (x, p)} = \frac{\partial (u, v)}{\partial (x', p')} \frac{\partial (x', p')}{\partial (x, p)},$$

it only remains the same for transformations of the phase-space coordinates when the functional determinant of the new phase-space coordinates is equal to 1, viz.,

$$\frac{\partial (x', p')}{\partial (x, p)} \equiv [x', p'] = 1 ,$$

i.e., if the map is area-preserving. (If we no longer require the restriction $f = 1$, then this constraint is necessary, but not sufficient for canonical transformations. We shall deal with this later.) If we write

$$\begin{pmatrix} dx' \\ dp' \end{pmatrix} = K \begin{pmatrix} dx \\ dp \end{pmatrix} , \quad \text{with} \quad K = \begin{pmatrix} \dfrac{\partial x'}{\partial x} & \dfrac{\partial x'}{\partial p} \\ \dfrac{\partial p'}{\partial x} & \dfrac{\partial p'}{\partial p} \end{pmatrix} ,$$

then, because $[x', p'] = 1$, for the inverse K^{-1} of this 2×2 matrix given by the formula on p. 71, we have

$$K^{-1} = \begin{pmatrix} \dfrac{\partial x}{\partial x'} & \dfrac{\partial x}{\partial p'} \\ \dfrac{\partial p}{\partial x'} & \dfrac{\partial p}{\partial p'} \end{pmatrix} = \begin{pmatrix} \dfrac{\partial p'}{\partial p} & -\dfrac{\partial x'}{\partial p} \\ -\dfrac{\partial p'}{\partial x} & \dfrac{\partial x'}{\partial x} \end{pmatrix} .$$

The two matrices must have equal elements. This results in the four equations

$$\frac{\partial x'}{\partial x} = \frac{\partial p}{\partial p'} , \quad \frac{\partial x'}{\partial p} = -\frac{\partial x}{\partial p'} , \quad \frac{\partial p'}{\partial x} = -\frac{\partial p}{\partial x'} , \quad \frac{\partial p'}{\partial p} = \frac{\partial x}{\partial x'} .$$

Here one alone actually suffices (e.g., the first), because the remaining ones follow from this one according to p. 44, in particular the second from

$$\left(\frac{\partial x'}{\partial p}\right)_x \left(\frac{\partial p}{\partial p'}\right)_{x'} = -\left(\frac{\partial x'}{\partial x}\right)_p \left(\frac{\partial x}{\partial p'}\right)_{x'} ,$$

the third from

$$\left(\frac{\partial p}{\partial p'}\right)_{x'} \left(\frac{\partial p'}{\partial x}\right)_p = -\left(\frac{\partial p}{\partial x'}\right)_{p'} \left(\frac{\partial x'}{\partial x}\right)_p ,$$

and the fourth from

$$\left(\frac{\partial x'}{\partial x}\right)_p \left(\frac{\partial x}{\partial x'}\right)_{p'} = \left(\frac{\partial p}{\partial p'}\right)_{x'} \left(\frac{\partial p'}{\partial p}\right)_x .$$

This we generalize now to $f > 1$ for time-independent canonical transformations. With $i, k \in \{1, \ldots, f\}$, we obtain the following constraints:

$$\frac{\partial x^{i\prime}}{\partial x^k} = \frac{\partial p_k}{\partial p_i{}'} , \quad \frac{\partial x^{i\prime}}{\partial p_k} = -\frac{\partial x^k}{\partial p_i{}'} , \quad \text{and} \quad \frac{\partial p_i{}'}{\partial x^k} = -\frac{\partial p_k}{\partial x^{i\prime}} , \quad \frac{\partial p_i{}'}{\partial p_k} = \frac{\partial x^k}{\partial x^{i\prime}} .$$

Here for the first (and last) equation, the notation with upper and lower indices from Sect. 1.2.2 turns out to be quite successful, and the remaining equations follow therefrom. In fact, these equations ensure that

$$\dot{x}^{k\prime} = \sum_l \left(\frac{\partial x^{k\prime}}{\partial x^l} \dot{x}^l + \frac{\partial x^{k\prime}}{\partial p_l} \dot{p}_l \right) = + \sum_l \left(\frac{\partial p_l}{\partial p_k'} \frac{\partial H}{\partial p_l} + \frac{\partial x^l}{\partial p_k'} \frac{\partial H}{\partial x^l} \right) = + \frac{\partial H}{\partial p_k'} \;,$$

$$\dot{p}_k' = \sum_l \left(\frac{\partial p_k'}{\partial x^l} \dot{x}^l + \frac{\partial p_k'}{\partial p_l} \dot{p}_l \right) = - \sum_l \left(\frac{\partial p_l}{\partial x^{k\prime}} \frac{\partial H}{\partial p_l} + \frac{\partial x^l}{\partial x^{k\prime}} \frac{\partial H}{\partial x^l} \right) = - \frac{\partial H}{\partial x^{k\prime}} \;.$$

If, for a time-independent transformation, we have $H' = H$, then the canonical equations remain untouched. Therefore, the name canonical transformation makes sense.

The linear transformation (with functional determinant 1)

$$\begin{pmatrix} x' \\ p' \end{pmatrix} = \begin{pmatrix} a_{xx} & a_{xp} \\ a_{px} & a_{pp} \end{pmatrix} \begin{pmatrix} x \\ p \end{pmatrix}, \quad \text{with} \quad \det a = 1 \;,$$

is clearly also canonical. In particular, we may choose $a_{xx} = a_{pp} = \cos \alpha$ and $a_{px} = -a_{xp} = \sin \alpha$, i.e., rotate in phase space. Therefore, the identity (with $\alpha = 0$) is canonical, but so also is the transformation $x' = p$, $p' = -x$ (with $\alpha = -\frac{1}{2}\pi$). This shows clearly that the meaning of position and momentum coordinates becomes blurred for the canonical equations—therefore q is often written preferentially for the generalized position coordinates rather than x. Moreover, the canonical transformations are essentially more general than the point transformations which are the only ones allowed in the Lagrangian formalism, i.e., in the latter, only the coordinates could be chosen, but not the velocities.

Let us consider the example of a linear harmonic oscillation with

$$H(x, p) = \frac{p^2 + m^2 \omega^2 x^2}{2m} \;.$$

Here only the squares of x and p appear. Therefore, by a non-linear canonical transformation, a cyclic coordinate x' can be introduced. We make a transition to polar coordinates, which are suggested according to Fig. 2.28:

$$x = \frac{f(p') \sin x'}{m\omega} \;, \quad p = f(p') \cos x' \quad \Longrightarrow \quad H' = \frac{f^2(p')}{2m} \;.$$

The transformation is only canonical if $f(p')$ obeys the constraint $f \, df/dp' = m\omega$ (since $m\omega \det K^{-1} = f \, df/dp'$). The associated differential form $f \, df = m\omega \, dp'$ is easily integrated:

$$\tfrac{1}{2} f^2(p') = m\omega p' \quad \Longrightarrow \quad H' = \omega p' \;.$$

No integration constant is added here because it would only move the zero energy. Now the Hamilton equations are very simple and are easily integrated:

$$\dot{x}' = +\frac{\partial H'}{\partial p'} = \omega \qquad \Longrightarrow \qquad x' = \omega(t - t_0) \,,$$

$$\dot{p}' = -\frac{\partial H'}{\partial x'} = 0 \qquad \Longrightarrow \qquad p' = \text{const} = \frac{H'}{\omega} \,.$$

If we write E_0 instead of H' for the total energy, then because $f^2(p') = 2m E_0$ and with the abbreviations $\hat{p} = \sqrt{2m E_0}$ and $\hat{x} = \hat{p}/(m\omega)$ for the original variables, we obtain

$$x = \hat{x} \, \sin(\omega(t - t_0)) \,, \qquad p = \hat{p} \, \cos(\omega(t - t_0)) \,.$$

As expected, we have had to integrate two differential equations of first order instead of one of second order. The integration constants E_0 and t_0 can be adjusted to the initial values.

For a charged point mass in a homogeneous magnetic field, we only search for the motion perpendicular to this field and, according to p. 123, the Hamilton function is

$$H(x, y, p_x, p_y) = \frac{(p_x - \frac{1}{2}m\omega y)^2 + (p_y + \frac{1}{2}m\omega x)^2}{2m} \,.$$

We carry out the canonical transformation

$$x' = \frac{x}{2} + \frac{p_y}{m\omega} \,, \qquad p_x{}' = p_x - \frac{m\omega}{2} y \,,$$

$$y' = \frac{x}{2} - \frac{p_y}{m\omega} \,, \qquad p_y{}' = p_x + \frac{m\omega}{2} y \,.$$

The proof that it is truly canonical is rather cumbersome at the present stage, because here there are four derivatives of the primed quantities with respect to the unprimed ones to be determined, and likewise many derivatives of the inverse functions, but at the end of Sect. 2.4.5, there is a *generating function* of this transformation, which simplifies the proof (see Problems 2.47–2.48). The Hamilton function now reads $(p_x{}'^2 + m^2\omega^2 x'^2)/2m$. The coordinates y' and $p_y{}'$ are cyclic, and we recognize the Hamilton function of a linear harmonic oscillation with the cyclotron frequency as angular frequency. The two cyclic coordinates are related to the pseudo-momentum $\mathbf{K}$ (treated on p. 100):

$$\mathbf{K} = \mathbf{p} + \tfrac{1}{2}q \, \mathbf{B} \times \mathbf{r} = \mathbf{p} - \tfrac{1}{2}m \, \boldsymbol{\omega} \times \mathbf{r} \,,$$

whence $K_x = p_y{}'$ and $K_y = -m\omega y'$. It was introduced earlier as a conserved quantity and delivered the center of the circular orbit. Here it is also clear that $\mathbf{K} \cdot \mathbf{K}/2m$ belongs to a linear oscillation with the cyclotron frequency as the angular frequency.

The angular momentum is given by

$$L_z = xp_y - yp_x = \frac{1}{\omega}\left(H' - \frac{\mathbf{K} \cdot \mathbf{K}}{2m}\right) .$$

We can thus split H' into ωL_z and $\mathbf{K} \cdot \mathbf{K}/2m$.

2.4.4 Infinitesimal Canonical Transformations. Liouville Equation

An infinitesimal canonical transformation is defined by

$$x' = x + \frac{\partial g(x, p)}{\partial p}\,\varepsilon , \qquad p' = p - \frac{\partial g(x, p)}{\partial x}\,\varepsilon ,$$

if ε is small enough to be able to neglect terms of the order of ε^2 compared to 1 in the functional determinant, and thus use the fact that $\partial^2 g/\partial p\,\partial x = \partial^2 g/\partial x\,\partial p$ (for which g has to be twice continuously differentiable). In particular, also

$$x' = x + \dot{x}\,\mathrm{d}t = x + \frac{\partial H}{\partial p}\,\mathrm{d}t , \qquad p' = p + \dot{p}\,\mathrm{d}t = p - \frac{\partial H}{\partial x}\,\mathrm{d}t ,$$

is a canonical transformation:

We can interpret the time evolution of the system as a canonical transformation.

This yields *Liouville's theorem*, regarding the time dependence of the probability density in phase space, thus of the weight with which each volume element of the phase space contributes to a statistical ensemble (e.g., for the molecules of an ideal gas—more on that in Sect. 6.2.3). In particular, the density has to have the property $\rho'(t, x', p')\,\mathrm{d}x'\,\mathrm{d}p' = \rho(t, x, p)\,\mathrm{d}x\,\mathrm{d}p$ because, despite its motion, each phase-space element keeps its probability content. Since each canonical transformation is area-preserving, it follows that

$$\rho'(t, x', p') = \rho(t, x, p) \qquad \Longrightarrow \qquad \frac{\mathrm{d}\rho}{\mathrm{d}t} = 0 ,$$

and hence the *Liouville (continuity) equation*

$$\frac{\partial \rho}{\partial t} + [\rho, H] = 0 .$$

In equilibrium, ρ does not depend explicitly on time. Then $[\rho, H] = 0$.

Table 2.1 Generators and infinitesimal transformations

Generating function g	Change	Infinitesimal transformation
H	$\mathrm{d}t$	$x' = x + \dot{x}\,\mathrm{d}t\,,\quad p' = p + \dot{p}\,\mathrm{d}t$
p	$\mathrm{d}x$	$x' = x + \mathrm{d}x\,,\quad p' = p$
p_φ	$\mathrm{d}\varphi$	$\varphi' = \varphi + \mathrm{d}\varphi\,,\quad p'_\varphi = p_\varphi$

The above function $g(x, p)$ is usually called the *generating function (generator)* of the infinitesimal canonical transformation. In particular, the Hamilton function H generates a time shift, the momentum p a change in position, and the angular momentum p_φ a rotation, as listed in Table 2.1.

For Cartesian coordinates in the last row, the generating function $L_z = x\,p_y - y\,p_x$ is to be taken. This delivers

$$x' = x - y\,\mathrm{d}\varphi\,,\quad p'_x = p_x - p_y\,\mathrm{d}\varphi\,,$$
$$y' = y + x\,\mathrm{d}\varphi\,,\quad p'_y = p_y + p_x\,\mathrm{d}\varphi\,,$$

as required for a rotation through the angle $\mathrm{d}\varphi$ about the z-axis. Generally, we require as generating function the quantity canonically conjugate to the differential variable, so we also view the time t and the Hamilton function (energy) H as canonically conjugate to each other.

2.4.5 Generating Functions

Finite and time-dependent canonical transformations can also be derived from generating functions. To this end, we start preferably from the gauge dependence of the Lagrange function (see p. 98), and $L = p\dot{x} - H$. Since $L' = L - \dot{G}$, we have

$$\mathrm{d}G = (L - L')\,\mathrm{d}t = (H' - H)\,\mathrm{d}t + p\,\mathrm{d}x - p'\,\mathrm{d}x'\,.$$

If we now make the ansatz that G and x' are functions of t, x, and p, we obtain

$$\mathrm{d}G = \frac{\partial G}{\partial t}\,\mathrm{d}t + \frac{\partial G}{\partial x}\,\mathrm{d}x + \frac{\partial G}{\partial p}\,\mathrm{d}p\,,$$
$$\mathrm{d}x' = \frac{\partial x'}{\partial t}\,\mathrm{d}t + \frac{\partial x'}{\partial x}\,\mathrm{d}x + \frac{\partial x'}{\partial p}\,\mathrm{d}p\,.$$

Therefore, we infer

$$\frac{\partial G}{\partial t} = H' - H - p'\,\frac{\partial x'}{\partial t}\,,$$

$$\frac{\partial G}{\partial x} = p - p' \, \frac{\partial x'}{\partial x} \, ,$$

$$\frac{\partial G}{\partial p} = -p' \, \frac{\partial x'}{\partial p} \, .$$

The transformation is canonical if the two mixed derivatives

$$\frac{\partial^2 G}{\partial p \, \partial x} = 1 - \frac{\partial p'}{\partial p} \frac{\partial x'}{\partial x} - p' \, \frac{\partial^2 x'}{\partial p \, \partial x} \, ,$$

$$\frac{\partial^2 G}{\partial x \, \partial p} = - \frac{\partial p'}{\partial x} \frac{\partial x'}{\partial p} - p' \, \frac{\partial^2 x'}{\partial x \, \partial p} \, ,$$

agree with each other, and likewise those of $x'(t, x, p)$. Then, in particular, we have

$$\frac{\partial x'}{\partial x} \frac{\partial p'}{\partial p} - \frac{\partial x'}{\partial p} \frac{\partial p'}{\partial x} \equiv \frac{\partial (x', p')}{\partial (x, p)} \equiv [x', p'] = 1 \, .$$

Thus x', p', and H' have to obey the partial differential equations for G above (derivatives of G with respect to t, x, and p). In particular $H' = H$ holds if G and x' do not depend explicitly on time.

In the last section, we introduced generating functions $g(x, p)$ for the infinitesimal transformations. We now ask how they are connected with $G(x, p)$. Since

$$x' = x + \varepsilon \, \frac{\partial g}{\partial p} \quad \text{and} \quad p' = p - \varepsilon \, \frac{\partial g}{\partial x} \, ,$$

then up to terms of order ε^2, we have

$$\frac{\partial G}{\partial x} = p - \left(p - \varepsilon \, \frac{\partial g}{\partial x} \right) \left(1 + \varepsilon \, \frac{\partial^2 g}{\partial x \, \partial p} \right) \approx \varepsilon \, \frac{\partial}{\partial x} \left(g - p \, \frac{\partial g}{\partial p} \right) \, ,$$

$$\frac{\partial G}{\partial p} = - \left(p - \varepsilon \, \frac{\partial g}{\partial x} \right) \varepsilon \, \frac{\partial^2 g}{\partial p^2} \approx -\varepsilon p \, \frac{\partial^2 g}{\partial p^2} = \varepsilon \, \frac{\partial}{\partial p} \left(g - p \, \frac{\partial g}{\partial p} \right) \, .$$

Therefore, we may take $G(x, p) \approx \varepsilon \, (g - p \, \partial g / \partial p)$ and obtain a unique connection between G and g, whereupon both shall be referred to as generating functions.

Likewise, we may also take x and p as functions of x' and p' or any other pair of old and new phase-space coordinates as functions of the other pair. However, different generating functions appear then. Later we will denote them by G and include the associated variables. So, with $G(t, x', p')$, $x(t, x', p')$, and $p(t, x', p')$, for example, we have

$$\frac{\partial G}{\partial t} = H' - H + p \, \frac{\partial x}{\partial t} \, , \qquad \frac{\partial G}{\partial x'} = -p' + p \, \frac{\partial x}{\partial x'} \, , \qquad \text{and} \qquad \frac{\partial G}{\partial p'} = p \, \frac{\partial x}{\partial p'} \, .$$

Here, too, x and p result from partial differential equations.

But if the generating function depends upon a primed and an unprimed variable (except for the time, which is not transformed, i.e., $t = t'$), then even simpler algebraic equations follow instead of the (partial) differential equations. So we require

$$dG(t, x, x') = \frac{\partial G}{\partial t}\, dt + \frac{\partial G}{\partial x}\, dx + \frac{\partial G}{\partial x'}\, dx' \,,$$

because of the starting equation

$$H' = H + \frac{\partial G}{\partial t} \,, \qquad p = \frac{\partial G}{\partial x} \,, \qquad \text{and} \qquad p' = -\frac{\partial G}{\partial x'} \,.$$

If the mixed derivatives $\partial^2 G/\partial x \partial x'$ and $\partial^2 G/\partial x' \partial x$ are equal, then it follows that $\partial p/\partial x' = -\partial p'/\partial x$, whence the transformation is canonical if in addition $p = \partial G/\partial x$ can be solved for x'. Further generating functions follow from the Legendre transformations:

$$\begin{aligned} G(t, x, x') &= G(t, x, p') - p' x' \\ &= G(t, p, x') + p x \\ &= G(t, p, p') + p x - p' x' \,. \end{aligned}$$

Actually, here we should use four different notations instead of just G, and these are often written G_1, G_2, G_3, and G_4, i.e., *generating functions of type 1, type 2, type 3, and type 4*. However, only their variables are important. Each of these generating functions depends on one primed and one unprimed variable, except for the time. Thus we obtain the list in Table 2.2—in all these cases we also have

$$H' = H + \frac{\partial G}{\partial t} \,,$$

with the other variables held fixed in each case. The remaining constraints for the canonical transformation are then also fulfilled, because one constraint already takes care of $\det K = 1$. However, there are not always all four. Thus, the identity can be generated by $G(x, p') = xp'$, for example, while this is not satisfied by the transformed function $G(x, x') = (x - x') p'$.

Table 2.2 Different generating functions

Generating function	Fixed variables		Reason	
$G(t, x, x')$	$p = +\dfrac{\partial G}{\partial x}$,	$p' = -\dfrac{\partial G}{\partial x'}$	$\dfrac{\partial p}{\partial x'} = -\dfrac{\partial p'}{\partial x}$	
$G(t, x, p')$	$p = +\dfrac{\partial G}{\partial x}$,	$x' = +\dfrac{\partial G}{\partial p'}$	$\dfrac{\partial p}{\partial p'} = +\dfrac{\partial x'}{\partial x}$	
$G(t, p, x')$	$x = -\dfrac{\partial G}{\partial p}$,	$p' = -\dfrac{\partial G}{\partial x'}$	$\dfrac{\partial x}{\partial x'} = +\dfrac{\partial p'}{\partial p}$	
$G(t, p, p')$	$x = -\dfrac{\partial G}{\partial p}$,	$x' = +\dfrac{\partial G}{\partial p'}$	$\dfrac{\partial x}{\partial p'} = -\dfrac{\partial x'}{\partial p}$	

For functions with several pairs of parameters, mixing is also allowed. Thus the generating function $x_1 p_1' + x_2 x_2'$ leads to $x_1' = x_1$, $p_1' = p_1$ and $x_2' = p_2$, $p_2' = -x_2$. With the first pair, nothing is changed here, while for the second pair, position and momentum swap names.

The canonical transformation $x = \sqrt{2p'/(m\omega)}\sin x'$, $p = \sqrt{2m\omega p'}\cos x'$ (see p. 127) with a harmonic oscillation can be generated by the function

$$G(x, x') = \frac{m\omega}{2}\, x^2 \cot x' \,,$$

because it leads to $p = m\omega x \cot x'$ and $p' = \frac{1}{2}m\omega x^2 \sin^{-2} x'$.

The following canonical transformation for a point charge (with mass m) in the homogeneous magnetic field can be derived from the generating function (Problem 2.47)

$$G(x, p_x', p_y, p_y') = x\,\frac{p_x' + p_y'}{2} + p_y\,\frac{p_x' - p_y'}{m\omega}\,,$$

whence it can be proven easily that the transformation mentioned on p. 128 is truly canonical.

2.4.6 Transformations to Moving Reference Frames. Perturbation Theory

An important application is transformations to moving reference frames. We investigate in particular

$$H = H_0(p) + H_1(x, p)\,,$$

in which x is cyclic with respect to H_0, but not with respect to the total Hamilton function. For $H_1 = 0$, the condition $\partial H_0/\partial x = 0$ leads to constant $p = p_0$ and

$$\dot{x} = \left.\frac{\partial H_0}{\partial p}\right|_{p=p_0} \equiv v_0 \qquad \Longrightarrow \qquad x = v_0 t + x_0\,.$$

With the generalized case $H_1 \neq 0$, we now take the canonical transformation

$$x' = x - v_0 t - x_0\,, \qquad p' = p - p_0\,,$$

which can be derived from the generating function

$$G(t, x, p') = (x - v_0 t - x_0)(p_0 + p')\,,$$

with $p = \partial G/\partial x$ and $x' = \partial G/\partial p'$. Since $H' = H + \partial G/\partial t$, we have

$$H' = H_0(p_0+p') + H_1(v_0\,t+x_0+x',\ p_0+p') - v_0\,(p_0+p')\ .$$

These equations have been derived without approximations.

But these are often useful also for *perturbation theory*, if one has the solution for H_0, but not for H. If we have $|H_1| \ll |H_0|$, then for not too long times, x' and p' will also be small compared to x and p, because they even vanish for $H_1 = 0$. Here we may still choose x_0 such that, for $t = 0$, $|H_1|$ is as small as possible compared to $|H_0|$. The perturbation theory then works as follows. In

$$\dot{x}'(t, x', p') = +\frac{\partial H'}{\partial p'} = \frac{\partial H_0}{\partial p'} + \frac{\partial H_1}{\partial p'} - v_0\ ,$$

$$\dot{p}'(t, x', p') = -\frac{\partial H'}{\partial x'} = -\frac{\partial H_1}{\partial x'}\ ,$$

we first set x' and p' equal to 0 on the right, and thus find solutions to $\dot{x}'(t, 0, 0)$ and $\dot{p}'(t, 0, 0)$. Here the integration constant has to be fixed in such a way that x' and p' vanish for $t = 0$. With these approximations we can improve the expressions on the right of the differential equations and evaluate the next approximation, i.e., the next order in the Taylor expansion. Where possible, we may even be able to identify the complete solutions.

If we consider as an example a harmonic oscillation and the free motion as unperturbed (a coarse approximation, where here actually $\overline{V} = \overline{T}$ holds),

$$H_0 = \frac{p^2}{2m}\ ,\quad H_1 = \frac{m\omega^2 x^2}{2}\ ,$$

delivers

$$H' = \frac{(p_0 + p')^2}{2m} + \frac{m\omega^2}{2}\,(v_0\,t + x')^2 - v_0\,(p_0 + p')\ .$$

With this and because of $v_0 = \partial H_0/\partial p|_{p_0} = p_0/m$, we have

$$\dot{x}' = \frac{p_0 + p'}{m} - v_0 = \frac{p'}{m}\ ,\quad \dot{p}' = -m\omega^2\,(v_0\,t + x')\ ,$$

and consequently $p' \approx -\frac{1}{2!}\,p_0\omega^2 t^2$ and $x' \approx -\frac{1}{3!}\,v_0\omega^2 t^3$. The next order delivers the additional terms $\frac{1}{4!}\,p_0\omega^4 t^4$ for p' and $\frac{1}{5!}\,v_0\omega^4 t^5$ for x'. In fact, the correct solution is

$$p = p_0\,\cos(\omega t) = p_0 + p'\ ,\quad x = (v_0/\omega)\,\sin(\omega t) = v_0 t + x'\ ,$$

with $x(0) \equiv x_0 = 0$.

2.4.7 Hamilton–Jacobi Theory

The Hamilton–Jacobi theory is a further application of time-dependent canonical transformations and will be explained briefly here. Note that, in his book (see the suggestions for textbooks on p. 162), H. Goldstein devotes a whole chapter to this subject. Unfortunately, he, along with many others, does not comply with the IUPAP recommendations: the quantities W (action function) and S (characteristic function) are used by him in the opposite notation S and W, respectively.

In this theory the Hamilton function is transformed canonically to zero. Then all new variables x' and p' are conserved quantities, fixed by the initial values. Here, the generating function is the associated *Hamilton action function* $W(t, x, p')$. Because $H'(t, x', p') = H(t, x, p) + \partial W(t, x, p')/\partial t$ for $H' = 0$ and because $p = \partial W/\partial x$, W has to satisfy the *Hamilton–Jacobi differential equation*

$$\frac{\partial W}{\partial t} + H\left(t, x, \frac{\partial W}{\partial x}\right) = 0 \,.$$

Since here p' does not depend on time, we have

$$\frac{dW}{dt} = \frac{\partial W}{\partial t} + \frac{\partial W}{\partial x}\, \dot{x} = p\dot{x} - H = L \quad \Longrightarrow \quad W = \int L\, dt \,.$$

The integration constant is left out here, because we may still find a suitable gauge. The single partial differential equation of Hamilton and Jacobi replaces all f pairs of ordinary differential equations in the Hamilton theory! However, it is difficult to solve, because the momenta in the Hamilton function and hence the required functions mostly appear squared. But the theory is useful for formal considerations. Using this we shall be able to discover in particular a connection with *geometrical optics (ray optics)*. Note that we have so far expressed all laws as differential equations and taken, e.g., the Lagrange function L as the quantity to start from. Now L is the derivative of the "anti-derivative" W, so the action has to be viewed as the original quantity.

The choice of the new momenta p' is not unique. Functions of it are also allowable, and we shall choose their structure to be as simple as possible. Of course, the associated coordinates $x' = \partial W/\partial p'$ depend upon this choice. In any case, x' and p' are constants of the motion, which have to be adjusted to the initial values. After that, $x(t, x', p')$ and $p(t, x', p')$ can be obtained.

If the Hamilton function does not depend on time, the ansatz

$$W(t, x, p') = S(x, p') - E\,t$$

suffices, since it leads from $p = \partial W/\partial x$ to $p = \partial S/\partial x$ and from the Hamilton–Jacobi equation to

$$H\left(x, \frac{\partial S}{\partial x}\right) = E \ ,$$

and H should also be taken as energy. Since S depends only on x and p', it is sometimes called the *reduced action*, but usually the *characteristic function*. This can be concluded from $\partial S/\partial x = p$ and leads to a sheet in phase space:

$$S = \int p \, dx \ , \quad \text{or} \quad S = \int \sum_{k=1}^{f} p_k \, dx^k \ , \quad \text{with } f > 1 \ .$$

Again, the integration constant vanishes here for a suitable gauge. For periodic motions (oscillations or rotations), we also introduce the *phase integral* (sometimes called the *action variable*), taken along the closed path in phase space, viz.,

$$J = \oint p \, dx \ ,$$

or several action variables J_k for more periodic degrees of freedom. According to quantum theory (*Bohr–Sommerfeld quantization rule*), this quantity cannot change continuously, but only in steps of the action quantum h (see also p. 367).

We may take one of the new momenta p'_k as energy. Then the associated coordinate $x^{k'}$ is connected to the choice of the zero time, as we show now for a simple example. If a coordinate oscillates harmonically, then $H = (P^2 + m^2\omega^2 x^2)/2m$ leads to

$$\frac{1}{2m} \left(\frac{\partial S}{\partial x}\right)^2 + \frac{m}{2} \omega^2 x^2 = E \ .$$

From this we could immediately conclude $S = \int (\partial S/\partial x) \, dx$ by integration, with the result $S = \frac{1}{2}m\omega \, x\sqrt{2E/m\omega^2 - x^2} + E/\omega \ \arcsin(\sqrt{m\omega^2/2E} \, x)$. But this is unnecessary, since with $x' = \partial W/\partial p'$ and $p' = E$, we can also immediately obtain

$$x' = \frac{\partial W}{\partial E} = \frac{\partial S}{\partial E} - t = \frac{\partial}{\partial E} \int \frac{\partial S}{\partial x} \, dx - t \ ,$$

and hence then, with $x' = -t_0$,

$$t - t_0 = \frac{1}{\omega} \int \frac{dx}{\sqrt{2E/m\omega^2 - x^2}} = \frac{1}{\omega} \arcsin \sqrt{\frac{m\omega^2}{2E}} x \ .$$

This is the solution $x = \hat{x} \ \sin[\omega(t - t_0)]$ with amplitude $\hat{x} = \sqrt{2p'/m\omega^2}$ and the second adjusted parameter $t_0 = -x'$ mentioned on p. 128. Note that, inserting the solution $x(t)$ into the expression for S, we can also obtain $W(t) = E/(2\omega) \ \sin[2\omega(t - t_0)]$ and $dW/dt = L$, implying $J = ET = 2\pi \ E/\omega$ for the phase integral.

In order to understand the properties of W and S, we have to start from a time-independent Hamilton function. At time zero, W and S agree. If, in configuration space (i.e., the space of coordinates $\mathbf{x}$), we investigate the areas of constant W values or S values as functions of time, then the sheets of the S values stay constant, while the sheets of constant W values move like a wave front. The latter follows in particular from $dW/dt = 0$, thus $\nabla W \cdot \dot{\mathbf{x}} - E = 0$ or $\mathbf{p} \cdot \dot{\mathbf{x}} = E$. The larger the momentum p, the smaller the velocity of the wave for given energy.

In order to understand what kind of wave this is, we consider the wave equation

$$\triangle \psi - \frac{1}{c^2} \frac{\partial^2 \psi}{\partial t^2} = 0 \,,$$

where c is the *phase velocity* of the wave, as can be seen from the ansatz for the solution $\psi \propto \exp\{i(\mathbf{k} \cdot \mathbf{r} - \omega t)\}$, which contains the *wave vector* $\mathbf{k}$ with $k \equiv 2\pi/\lambda$ and the *angular frequency* $\omega \equiv 2\pi/T$, where λ is the wavelength and T the oscillation period of the wave. For the differential equation to be satisfied, $ck = \omega$ or $c = \lambda/T$ has to hold. In an inhomogeneous medium, the wavelength depends on the position, and so also does the phase velocity. For this notion of a wave to make sense at all, we would like to assume that both vary only slowly on their paths. We thus restrict ourselves to waves of very short wavelength or very high wave number k and call the smallest of the occurring wave numbers k_0. Then we can make an ansatz

$$\psi = \exp\{A(\mathbf{r}) + ik_0 \left(S(\mathbf{r}) - c_0 t\right)\}$$

for the solution of the wave equation in the inhomogeneous medium with $c_0 = \omega/k_0$, real amplitude $\exp[A(\mathbf{r})]$, and real path *eikonal* $S(\mathbf{r})$. (The word *eikonal* is reminiscent of the Greek $\varepsilon\iota\kappa\omega\nu$, meaning picture or icon. With the mapping of an object point $\mathbf{r}_0$ on the image point $\mathbf{r}_1$, both points are singular points of the wave areas, and the optical paths for all connecting rays are equal to $S(\mathbf{r}_1) - S(\mathbf{r}_0)$. The eikonal is related to the characteristic function, as we shall see soon.) In particular, this ansatz leads to $\nabla \psi = \psi \, \nabla(A + ik_0 S)$ and

$$\triangle \psi = \psi \left\{\triangle(A + ik_0 S) + \nabla(A + ik_0 S) \cdot \nabla(A + ik_0 S)\right\} \,,$$

which, according to the wave equation, should agree with $-(c_0 k_0/c)^2 \, \psi$. Then, after separation into real and imaginary parts, we infer

$$\triangle A + \nabla A \cdot \nabla A + k_0^2 \left(n^2 - \nabla S \cdot \nabla S\right) = 0 \,,$$

with the position-dependent *refractive index* $n \equiv c_0/c$ and

$$\triangle S + 2\nabla A \cdot \nabla S = 0 \,.$$

The refractive index should barely vary, according to the assumption about the wavelength: k_0 should be sufficiently large. With this we obtain the *eikonal equation* of geometrical optics, viz.,

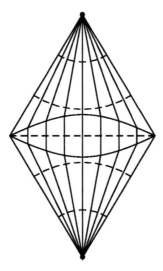

Fig. 2.29 Geometrical optics and classical mechanics (beam path and particle path) have much in common, here shown for a lens with refractive index $n = 2$. But note that the refractive index for a wave corresponds to the ratio c_0/c, in contrast to the ratio v/v_0 for a particles. Actually, we have to distinguish between phase and particle velocity. *Dashed lines* are the wave fronts. Those of W move in the course of time, but not those of S. The wave fronts are singular at the object and image points ($\bullet$)

$$\nabla S \cdot \nabla S = n^2 \; ,$$

an inhomogeneous differential equation of first order and second degree. (It holds only in the limit of short wavelengths, because otherwise we would also have to take into account $\triangle A + \nabla A \cdot \nabla A = 0$: ∇A would have to have only drains, because its source density would be $\nabla \cdot \nabla A = \triangle A = -\nabla A \cdot \nabla A \leq 0$.) If we integrate to find the eikonal $S(\mathbf{r})$, then from the second differential equation $\triangle S + 2\nabla A \cdot \nabla S = 0$, we obtain the gradient of the amplitude function A in the direction of the gradient of S. Perpendicular to it, the gradient of A remains undetermined. In this plane it may even vary in steps, whence, according to geometrical optics, *rays* are possible. The wave propagates along ∇S, (see Fig. 1.4) perpendicular to the wave fronts $S = $ const. (see Fig. 2.29).

With the Hamilton–Jacobi equation for $H = \frac{1}{2m}\,\mathbf{p} \cdot \mathbf{p} + V(\mathbf{r})$, we arrive at

$$\nabla S \cdot \nabla S = 2m \left\{E - V(\mathbf{r})\right\} \; ,$$

and hence also at the eikonal equation with $n^2 = 2m \left\{E - V(\mathbf{r})\right\}$, which however is not a pure number, and where the "characteristic function" appears instead of the eikonal. Classical mechanics can describe the motion of particles of mass m with the same differential equation as geometrical optics. This holds for waves of negligible wavelength. Conversely, the propagation of light can be viewed as the motion of particles (photons), as long as the wavelength is sufficiently small.

2.4.8 Integral Principles

So far we have derived the basic laws from differential equations, e.g., from the Lagrange equations of the second kind

$$\frac{d}{dt}\left(\frac{\partial L}{\partial \dot{x}^k}\right) = \frac{\partial L}{\partial x^k} \, .$$

However, for the problem under consideration, there has to be a potential energy, or at least a generalized U. But these differential equations can also be related to integral expressions via the variational calculus. Then there is no need for a potential energy, and the basic laws can be interpreted a different way. This is also important for our general understanding.

In the variational calculus, we seek functions $x(t)$ that make an integral

$$I = \int_{t_0}^{t_1} f(t, x, \dot{x}) \, dt$$

extremal under constraints. Here the boundaries t_0 and t_1 are given as fixed, or at least connected to constraints that deliver fixed boundaries after a transformation $t \to t'$. The values of the function are also given at those boundaries, viz., $\delta x(t_0) = 0 = \delta x(t_1)$, but not their derivatives $\dot{x}(t_0)$ and $\dot{x}(t_1)$.

If we search for the "extremal" $x(t)$ for the regime between t_0 and t_1, then initially, in addition to x, we also have to allow for $x + \delta x$ and hence, in addition to $\dot{x}$, also for $\dot{x} + \delta \dot{x}$. Here, to begin with, the variations always refer to the same time: $\delta t = 0$ (see Fig. 2.30). Consequently, we have $\delta \dot{x} = \delta \, dx/dt = d \, \delta x/dt$, and therefore (with partial integration for the second equation)

$$\delta I = \int_{t_0}^{t_1} \left(\frac{\partial f}{\partial x} \delta x + \frac{\partial f}{\partial \dot{x}} \frac{d \, \delta x}{dt}\right) dt = \frac{\partial f}{\partial \dot{x}} \delta x \Big|_{t_0}^{t_1} + \int_{t_0}^{t_1} \left\{\frac{\partial f}{\partial x} - \frac{d}{dt}\left(\frac{\partial f}{\partial \dot{x}}\right)\right\} \delta x \, dt \, .$$

Now, for δI to vanish for arbitrary δx,

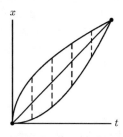

Fig. 2.30 Path variation with $\delta t = 0$ but $\delta x \neq 0$ along *dashed lines*. Since $\delta x(t_0) = 0 = \delta x(t_1)$, each permitted orbit ends at the points shown by the dots (●). Since t_1 may follow arbitrarily quickly after t_0, $\dot{x}(t_0)$ and $\dot{x}(t_1)$ effectively vary

$$\delta I = 0 \quad \text{at} \quad \delta t = 0 \qquad \Longleftrightarrow \qquad \int_{t_0}^{t_1} \delta f \, \mathrm{d}t = 0 \quad \text{at} \quad \delta t = 0 \,,$$

whence (uniquely) we must satisfy *Euler's differential equation*

$$\frac{\mathrm{d}}{\mathrm{d}t}\left(\frac{\partial f}{\partial \dot{x}}\right) - \frac{\partial f}{\partial x} = 0 \,.$$

Correspondingly, for $f(t, x^1, \ldots, x^f, \dot{x}^1, \ldots, \dot{x}^f)$, one of the extremal conditions delivers a total of f such differential equations of second order.

From the Lagrange equations of the second kind, it follows that the *action function* W introduced on p. 135 takes an extremum, yielding *Hamilton's principle*:

$$\delta W \equiv \delta \int_{t_0}^{t_1} L \, \mathrm{d}t = 0 \,, \quad \text{at} \quad \delta t = 0 \,.$$

Among all possible paths the one with extremal W is realized. We usually replace L by $T - V$. But Hamilton's principle holds even if there is no potential energy at all. This can be understood with d'Alembert's principle $(m\dot{\mathbf{v}} - \mathbf{F}) \cdot \delta\mathbf{r} = 0$, implying that $\mathbf{F} \cdot \delta\mathbf{r} = \delta A$ and $\dot{\mathbf{v}} \cdot \delta\mathbf{r} = \mathrm{d}(\mathbf{v} \cdot \delta\mathbf{r})/\mathrm{d}t - \mathbf{v} \cdot \delta\dot{\mathbf{r}}$ hold with $m\mathbf{v} \cdot \delta\dot{\mathbf{r}} = \delta T$. Since $\mathbf{v} \cdot \delta\mathbf{r}$ vanishes at the integration limits, we thus obtain $\int(\delta T + \delta A) \, \mathrm{d}t = 0$. This we may also write as (*general Hamilton principle*)

$$\delta \int_{t_0}^{t_1} T \, \mathrm{d}t + \int_{t_0}^{t_1} \delta A \, \mathrm{d}t = 0 \,, \quad \text{at} \quad \delta t = 0 \,.$$

Note that the virtual work δA makes sense, but the work A does not generally as such. Only if a potential energy V produces the (external) forces do we have

$$\delta T + \delta A = \delta(T - V) = \delta L \,,$$

and then the variation can be moved in front of the integral.

Hamilton's principle does not depend on the choice of coordinates. Arbitrary (unique) transformations of t and of the generalized coordinates x^k are permitted. We only need to be able to give T and V or, respectively, δA. With this, we have a general basis for the problems of mechanics, and even for friction. If there is a potential energy and hence also a Lagrange function, then from the same principle we can immediately conclude that

$$L' = L - \frac{\mathrm{d}G}{\mathrm{d}t}$$

is also an allowable Lagrange function (gauge invariance, see p. 98).

Another integral principle is the *action principle* (due to Maupertuis, Leibniz, Euler, Lagrange), for which, however, it is not the action W that is varied, but the *characteristic function* (*reduced action*) S, and where the energy is held fixed instead of the time (and likewise the integration limits $\mathbf{r}_0$ and $\mathbf{r}_1$). In addition, the Hamilton function need not depend on time for S to be formed. In particular, $S = Et + W$ with $E = -\partial W/\partial t$, and therefore $\delta S = t\,\delta E + E\,\delta t + \delta W$. Then from $\delta W = 0$ for $\delta t = 0$, we have $\delta S = 0$ for $\delta E = 0$:

$$\delta S = \delta \int_{\mathbf{r}_0}^{\mathbf{r}_1} \mathbf{p} \cdot \mathrm{d}\mathbf{r} = 0 \,, \quad \text{for} \quad \delta E = 0 \,.$$

The action principle is often written in the form

$$\delta \int_{t_0}^{t_1} 2T\,\mathrm{d}t = 0 \,, \quad \text{for} \quad \delta E = 0 \,.$$

In fact, $\mathrm{d}S$ is not only equal to $p\,\mathrm{d}x$, but also to $2T\,\mathrm{d}t$, because for $\mathrm{d}E = 0$, we can derive $\mathrm{d}S = 2T\,\mathrm{d}t$ from $\mathrm{d}S = \mathrm{d}W + E\,\mathrm{d}t$ and $\mathrm{d}W = L\,\mathrm{d}t$ with $L = T - V = 2T - E$. However, we must remember here that the integration limits will now also be varied, because times of different lengths are necessary for the different paths between $\mathbf{r}_0$ and $\mathbf{r}_1$, if the kinetic energy is determined by a potential energy.

For a force-free motion neither T nor V is altered, and thus

$$\delta \int_{t_0}^{t_1} \mathrm{d}t = \delta\{t_1 - t_0\} = 0 \,, \quad \text{with constant } T \text{ and } V \,.$$

This *principle of least time* due to Fermat had already been applied by Hero of Alexandria to the refraction of light. (It could also be a principle of latest arrival, because we only search for an extremum with the variational calculus. Therefore, I have also avoided the name *principle of least action* for the action principle.) Here the position coordinates are missing in the Hamilton function, e.g., $S = \mathbf{p} \cdot \mathbf{x}$. With the characteristic function and for the action function, each cyclic coordinate x leads to a term px, which comprises the whole x-dependence!

So far, for all transformations, the time t has not been altered, but treated as an invariant parameter. If we had altered it in addition to the position and momentum coordinates, then we would have had to keep fixed another parameter τ in the variation—some parameter has to mark the progress along the path. Then since

$$L = \sum_{k=1}^{f} p_k \frac{\mathrm{d}x^k}{\mathrm{d}t} - H \,,$$

a generalized Hamilton principle has the form

$$\delta \int_{\tau_0}^{\tau_1} \left(\sum_{k=1}^{f} p_k \frac{\mathrm{d}x^k}{\mathrm{d}\tau} - H \frac{\mathrm{d}t}{\mathrm{d}\tau} \right) \mathrm{d}\tau = 0 \,, \quad \text{with } \delta\tau = 0 \,.$$

This suggests taking t as a further coordinate x^0 and $-H$ as its conjugate momentum:

$$\delta \int_{\tau_0}^{\tau_1} \sum_{k=0}^{f} p_k \frac{\mathrm{d}x^k}{\mathrm{d}\tau} \, \mathrm{d}\tau = 0 \,, \quad \text{with } \delta\tau = 0 \,.$$

After a canonical transformation here, p_k' and x'^k would appear, with $-p_0'$ as the new Hamilton function and x'^0 the new time. With a generating function $G(x^k, p_k')$, we obtain $f+1$ pairs of equations

$$p_k = \frac{\partial G}{\partial x^k} \,, \quad x'^k = \frac{\partial G}{\partial p_k'} \,, \quad \text{for } k \in \{0, \ldots, f\} \,.$$

These more general equations are only necessary for time-dependent Hamilton functions. As an example of this, we consider the time-dependent oscillator, in Sect. 2.4.11.

2.4.9 Motion in a Central Field

For a central field the angular momentum is conserved. We may restrict ourselves to a plane orbit with polar coordinates r and φ. According to p. 97, since $p_r = \partial L/\partial \dot{r} = m\dot{r}$ and $p_\varphi = \partial L/\partial \dot{\varphi} = mr^2\dot{\varphi}$, we obtain for the kinetic energy

$$T = \frac{m}{2} \, (\dot{r}^2 + r^2 \, \dot{\varphi}^2) = \frac{1}{2m} \left(p_r^2 + \frac{p_\varphi^2}{r^2} \right) \,.$$

Since φ does not appear in $L = T - V(r)$, the component of the angular momentum perpendicular to the plane of motion, p_φ, is a constant of the motion. Since the energy E is also conserved, conservation of energy can be used:

$$\dot{r}^2 = \frac{2}{m} \left\{ E - V(r) - \frac{p_\varphi^2}{2mr^2} \right\} \,, \quad \dot{\varphi} = \frac{p_\varphi}{mr^2} \,.$$

The last term inside the curly brackets comes from the centrifugal force. Part of the energy appears because of the *centrifugal potential* as rotational energy. In the ordinary differential equation $\dot{r} = f(r)$, the variables can be separated and then integrated:

$$t - t_0 = \int \frac{m \, \mathrm{d}r}{\sqrt{2m \, \{E - V(r)\} - (p_\varphi/r)^2}} \,.$$

Hence $t(r)$ or $r(t)$ can be obtained. Then the last expression for $\dot{\varphi}$ no longer contains any unknown term. This equation supplies the *area–velocity law*: $r^2\dot{\varphi} = (\mathbf{r} \times \mathbf{v}) \cdot \mathbf{n} = p_\varphi/m$. The integration constants are E, p_φ, r_0, and φ_0.

In many cases, we desire only the equation $r(\varphi)$ of the orbit. Then we use

$$\frac{dr}{d\varphi} = \frac{\dot{r}}{\dot{\varphi}} = \sqrt{2m\{E - V(r)\} - (p_\varphi/r)^2} \; \frac{r^2}{p_\varphi} \;,$$

and separate again in terms of variables. If the radicand vanishes, we have to expect a circular orbit, since then $\dot{r} = 0$ and thus $r = r_0$ and $\varphi = p_\varphi t/(mr_0{}^2) + \varphi_0$ (if $p_\varphi \neq 0$ and $r_0 > 0$).

The Hamilton–Jacobi equation for this problem reads

$$\frac{\partial W}{\partial t} + \frac{1}{2m}\left\{\left(\frac{\partial W}{\partial r}\right)^2 + \frac{1}{r^2}\left(\frac{\partial W}{\partial \varphi}\right)^2\right\} + V(r) = 0 \;.$$

Since t and φ do not occur in H, we may set $W = S(r) + p_\varphi \varphi - Et$, and from the last differential equation, we obtain

$$S = \int \sqrt{2m\{E - V(r)\} - (p_\varphi/r)^2} \; dr \;.$$

This expression also delivers the orbit equation, because it yields $\varphi' = \partial W/\partial p_\varphi = \partial S/\partial p_\varphi + \varphi$. According to this, r and φ are then related, as we have found before from $dr/d\varphi$:

$$\varphi - \varphi' = \int \frac{p_\varphi \, dr}{r^2 \sqrt{2m\{E - V(r)\} - (p_\varphi/r)^2}} \;.$$

Likewise, we could also have arrived immediately at $-t_0 = \partial W/\partial E = \partial S/\partial E - t$.

From the beginning we have only considered plane orbits. If this plane is still unknown, then spherical coordinates are suggested. Then we have

$$T = \frac{m}{2}(\dot{r}^2 + r^2\,\dot{\theta}^2 + r^2\sin^2\theta\,\dot{\varphi}^2) = \frac{1}{2m}\left(p_r{}^2 + \frac{p_\theta{}^2}{r^2} + \frac{p_\varphi{}^2}{r^2\sin^2\theta}\right) \;,$$

with $p_\theta = mr^2\,\dot{\theta}$ and (the new) $p_\varphi = mr^2\sin^2\theta\,\dot{\varphi}$. With $W = S - Et$, this leads to the the Hamilton–Jacobi equation

$$\frac{1}{2m}\left\{\left(\frac{\partial S}{\partial r}\right)^2 + \frac{1}{r^2}\left(\frac{\partial S}{\partial \theta}\right)^2 + \frac{1}{r^2\sin^2\theta}\left(\frac{\partial S}{\partial \varphi}\right)^2\right\} + V(r) = E \;.$$

Since φ does not appear here, we have a conserved quantity

$$\frac{\partial S}{\partial \varphi} = p_\varphi \;,$$

in addition to the energy E. For a central force, each component of the angular momentum is conserved, thus also the square of the angular momentum, which we denote here as

$$p_{\theta,\varphi}{}^2 = \left(\frac{\partial S}{\partial \theta}\right)^2 + \frac{1}{\sin^2\theta}\left(\frac{\partial S}{\partial \varphi}\right)^2 .$$

From this we conclude

$$\frac{1}{2m}\left\{\left(\frac{\partial S}{\partial r}\right)^2 + \frac{p_{\theta,\varphi}{}^2}{r^2}\right\} + V(r) = E .$$

Here, p_φ is no longer of interest, but only the conserved quantities $p_{\theta,\varphi}$ and E. *For central forces there is a degeneracy*, because different p_φ lead to the same $p_{\theta,\varphi}{}^2$. The last equation once again delivers the above-mentioned expression for $\dot r$, since

$$p_r = \frac{\partial W}{\partial r} = \frac{\partial S}{\partial r} = \sqrt{2m\left\{E - V(r)\right\} - \frac{p_\varphi{}^2}{r^2}}$$

is equal to $m\dot r$.

2.4.10   Heavy Symmetrical Top and Spherical Pendulum

If the center of mass of a pendulum moves on a spherical surface, we have a spherical pendulum—or even a heavy top, if the body rotates about the axis connecting the hinge and the center of mass. The spherical pendulum is not much simpler to treat than the heavy top, and clearly a special case of the top, which we would like to deal with anyway.

If the center of mass does not lie on the vertical through the rotational point, the gravitational force exerts a torque and changes the angular momentum along the horizontal direction. Hence, consideration of the "free" top in Sect. 2.2.11 is no longer adequate. The kinetic energy of the top reads most simply in Cartesian coordinates along the principal axes of the moment of inertia fixed in the body:

$$T = \tfrac{1}{2}\left(I_1\,\omega_1{}^2 + I_2\,\omega_2{}^2 + I_3\,\omega_3{}^2\right) .$$

On the other hand, the Euler angles are suitable coordinates to describe the motion in space. Therefore we express $\boldsymbol\omega$ using the Euler angles and their derivatives with respect to time.

In the body-fixed system, the space-fixed **z**-axis has polar angles β and $\pi - \gamma$ (see Fig. 1.10). Therefore, for a rotational vector proportional to $\dot\alpha$, it follows that (Problem 2.4)

$$\boldsymbol{\omega}_\alpha = \dot{\alpha}\ \{\sin\beta\ (-\cos\gamma\ \mathbf{e}_1 + \sin\gamma\ \mathbf{e}_2) + \cos\beta\ \mathbf{e}_3\}\ .$$

Correspondingly, $\boldsymbol{\omega}_\beta = \dot{\beta}\ (\sin\gamma\ \mathbf{e}_1 + \cos\gamma\ \mathbf{e}_2)$ and $\boldsymbol{\omega}_\gamma = \dot{\gamma}\ \mathbf{e}_3$, whence

$$
\begin{aligned}
\omega_1 &= -\dot{\alpha}\ \sin\beta\cos\gamma + \dot{\beta}\ \sin\gamma\ , \\
\omega_2 &= \ \ \dot{\alpha}\ \sin\beta\sin\gamma + \dot{\beta}\ \cos\gamma\ , \\
\omega_3 &= \ \ \dot{\alpha}\ \cos\beta \qquad\qquad\quad +\dot{\gamma}\ .
\end{aligned}
$$

Hence we have $\omega_1^2 + \omega_2^2 = \dot{\alpha}^2\ \sin^2\beta + \dot{\beta}^2$. Since with s as the distance of the center of mass from the rotational point, the potential energy is

$$V = mgs\cos\beta\ ,$$

we shall restrict in the following to a symmetrical top ($I_1 = I_2$) or a symmetric pendulum. Then, since

$$T = \tfrac{1}{2}\ I_1\ (\dot{\alpha}^2\ \sin^2\beta + \dot{\beta}^2) + \tfrac{1}{2}\ I_3\ (\dot{\alpha}\ \cos\beta + \dot{\gamma})^2\ ,$$

α and γ are cyclic coordinates, $\partial H/\partial\alpha = 0 = \partial H/\partial\gamma$, and thus the associated generalized momenta—the angular-momentum components along the lab-fixed and the body-fixed $\mathbf{z}$-axes—are constants of the motion:

$$p_\gamma = \frac{\partial L}{\partial\dot{\gamma}} = I_3\ (\dot{\alpha}\ \cos\beta + \dot{\gamma}) = \text{const.},$$

$$p_\alpha = \frac{\partial L}{\partial\dot{\alpha}} = I_1\ \dot{\alpha}\ \sin^2\beta + p_\gamma\ \cos\beta = \text{const.}$$

(If $p_\gamma = 0$, then we have a *spherical pendulum* instead of the top—for the plane pendulum, $p_\alpha = 0$ also holds.) Only $p_\beta = \partial L/\partial\dot{\beta} = I_1\ \dot{\beta}$ still depends on time. But this is therefore a one-dimensional problem, which we simply solve using the conservation of energy—then we avoid a differential equation of second order:

$$H = \frac{1}{2I_1}\left\{p_\beta^2 + \left(\frac{p_\alpha - p_\gamma\cos\beta}{\sin\beta}\right)^2\right\} + \frac{p_\gamma^2}{2I_3} + mgs\cos\beta$$

is a constant of the motion. Hence we now have to determine $\beta(t)$. The expression for p_α leads to a linear differential equation of first order for $\alpha(t)$, and the expression for p_γ to a similar equation for $\gamma(t)$.

In order to avoid the transcendent circular functions, we set

$$\cos\beta = z \quad\Longrightarrow\quad \dot{\beta} = \frac{-\dot{z}}{\sqrt{1 - z^2}}\ ,$$

and then obtain

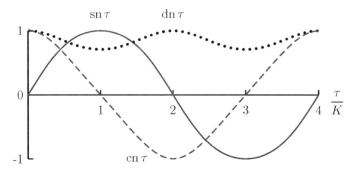

Fig. 2.31 The three Jacobi elliptic functions $\mathrm{sn}(\tau|k^2)$ (*continuous red*), $\mathrm{cn}(\tau|k^2)$ (*dashed blue*), and $\mathrm{dn}(\tau|k^2)$ (*dotted black*) for the parameter $k^2 = \frac{1}{2}$. Compare also with Fig. 2.18

$$\frac{I_1}{2}\,\dot{z}^2 = (1-z^2)\left(H - \frac{p_\gamma{}^2}{2\,I_3} - mgs\,z\right) - \frac{(p_\alpha - p_\gamma\,z)^2}{2\,I_1} \equiv mgs\,f(z)\ .$$

Here, $f(z)$ is a polynomial of third order in z, which is important for us only in the regime $-1 \le z \le 1$, and there also only for $f(z) \ge 0$. Now $f(z)$ is positive for $z \gg 1$ and negative for $z = \pm 1$ (or zero in the special case of a top with perpendicular axis of rotation and therefore without torque). Thus only the two lower zeros of $f(z)$ are relevant here. The differential equation can be solved with the Jacobi function $\mathrm{sn}(\tau|k^2)$ mentioned on p. 105. For this as for the other elliptic functions, it is customary (see, e.g., [1]) to number the zeros z_i of the polynomials in order of decreasing value, viz., $z_1 > z_2 > z_3$. The zero time can be chosen as the integration constant:

$$z(t) = z_3 + (z_2 - z_3)\,\mathrm{sn}^2\!\left(\sqrt{\frac{mgs}{2I_1}\,(z_1 - z_3)}\,(t - t_0)\ \Big|\ \frac{z_2 - z_3}{z_1 - z_3}\right)\ .$$

The derivative of $\mathrm{sn}\,\tau$ is equal to the product of the Jacobi elliptic functions *cosinus amplitudinis* $\mathrm{cn}\,\tau$ and *delta amplitudinis* $\mathrm{dn}\,\tau$ (see Fig. 2.31):

$$\mathrm{cn}(\tau|k^2) \equiv \cos(\mathrm{am}(\tau|k^2))\ ,$$
$$\mathrm{dn}(\tau|k^2) \equiv \sqrt{1 - k^2\,\mathrm{sn}^2(\tau|k^2)}\ .$$

Consequently, in addition to $\mathrm{sn}(\tau|k^2) = \sin(\mathrm{am}(\tau|k^2))$ and $\mathrm{sn}'(\tau|k^2) = \mathrm{cn}(\tau|k^2) \cdot \mathrm{dn}(\tau|k^2)$, we have

$$\mathrm{sn}^2(\tau|k^2) = 1 - \mathrm{cn}^2(\tau|k^2) = \frac{1 - \mathrm{dn}^2(\tau|k^2)}{k^2}\ .$$

The above-mentioned expression $z(t)$ therefore satisfies the original differential equation $\dot{z}^2 = (z - z_1)(z - z_2)(z - z_3)\,2mgs/I_1$ for $z_3 \le z \le z_2 < z_2 z_1$. The figure axis of the heavy top thus tumbles back and forth between two circles of latitude

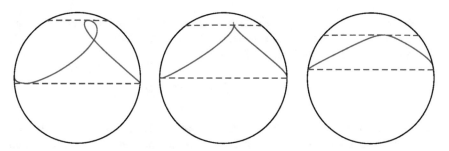

Fig. 2.32 Orbits of the body axis of a heavy symmetric top (*red line*). *Left*: With loops. *Centre*: With peaks. *Right*: With simple passes. *Dashed blue lines* are the limiting circles of latitude of the intersections of the figure axis on the sphere

$\beta_{2,3} = \arccos z_{2,3}$ (with $\beta_2 \leq \beta_3$). For the first return to the old circle of latitude, half an "oscillation" is performed. Thus the oscillation period is

$$T = 4\sqrt{\frac{I_1}{2mgs}} \int_{z_3}^{z_2} \frac{dz}{\sqrt{f(z)}} = 2\sqrt{\frac{2I_1}{mgs}} \frac{2}{\sqrt{z_1-z_3}} \, K\left(\frac{z_2-z_3}{z_1-z_3}\right).$$

As with the plane pendulum (see p. 104), we thus arrive at a complete elliptic integral K, however, we still have to determine the three solutions z_i (see Fig. 2.32).

For the tumbling motion, there are simple passes, but also loops or peaks. This can be read off from the zeros of $I_1\dot{\alpha} = (p_\alpha - p_\gamma \cos\beta)/\sin^2\beta$, which are determined by $p_\alpha - p_\gamma z$: for $z_3 < p_\alpha/p_\gamma < z_2$, there are loops, for p_α/p_γ equal to z_3 or z_2, there are peaks, and otherwise (with $p_\alpha/p_\gamma < z_3$ or $p_\alpha/p_\gamma > z_2$), neither loops nor peaks. This clearly holds also for the *force-free top* (with $mgs = 0$), which was already dealt with in Sect. 2.2.12.

Peaks occur, e.g., for the frequent initial condition $\dot{\alpha}(0) = \dot{\beta}(0) = 0$, for motions with an energy as small as possible, because $\dot{\alpha}(0) = 0$ delivers $z(0) = p_\alpha/p_\gamma$, and since $\dot{\beta}(0) = 0$, $\dot{z}$ also vanishes initially and hence so does $f(z)$. We thus start from one of the limiting circles of latitude with a peak. In fact, the nutation starts from the upper circle of latitude (z_2), because there the potential energy is highest, whence the kinetic energy is lowest. For these initial conditions, we already know the zero z_2 of $f(z)$, viz.,

$$z_2 = \frac{p_\alpha}{p_\gamma} = \frac{1}{mgs}\left(H - \frac{p_\gamma^2}{2I_3}\right),$$

and can determine the other zero z_3 more easily from a second-order equation, because

$$mgs \, f(z) = (z_2 - z)\left(mgs\,(1 - z^2) - \frac{p_\gamma^2}{2I_1}(z_2 - z)\right)$$

delivers $mgs\,(1-z_3{}^2) = [p_\gamma{}^2/(2I_1)]\,(z_2-z_3)$. For a fast top, $p_\gamma{}^2/(2I_3) \gg mgs$ holds. If now I_1 is not very much greater than I_3, then because $0 \le z_3{}^2 \le 1$, it follows that $z_3 \approx z_2$. Therefore, we obtain

$$z_2 - z_3 \approx \frac{mgs}{p_\gamma{}^2/(2I_1)}\,\sin^2\beta(0)\,,$$

i.e., the faster the top rotates, the less its nutation. It can also happen that the two circles of latitude coincide—then z and hence β are constant, as are $\dot\alpha$ and $\dot\gamma$, and we have *regular precession*. For very small nutation compared to the precession, we speak of *pseudo-regular precession*.

The differential equation $I_1\dot\alpha = (p_\alpha - p_\gamma z)/(1 - z^2)$ for the Euler angles α can be reformulated in the following way using $\dot\alpha = (d\alpha/dz)\,\dot z = \sqrt{2mgs/I_1}\sqrt{f(z)}\,d\alpha/dz$:

$$\frac{d\alpha}{dz} = \sqrt{\frac{p_\gamma{}^2/(2I_1)}{mgs}}\,\frac{p_\alpha/p_\gamma - z}{2\sqrt{f(z)}}\left(\frac{1}{1+z} + \frac{1}{1-z}\right),$$

with $f(z) = (z-z_1)(z-z_2)(z-z_3)$ and $z_1 > z_2 \ge z > z_3$. The solution of this differential equation can be given with the help of the *incomplete elliptic integral of the third kind*

$$\Pi(n;\varphi\,|k^2) \equiv \int_0^\varphi \frac{d\psi}{(1-n\sin^2\psi)\sqrt{1-k^2\sin^2\psi}}$$

$$= \int_0^{\sin\varphi} \frac{dt}{(1-nt^2)\sqrt{(1-t^2)(1-k^2t^2)}}\,,$$

and with the integral of the first kind $F(\varphi\,|k^2)$ from Sect. 2.3.6. With the abbreviations

$$g(z) \equiv \sqrt{\frac{z-z_3}{z_2-z_3}} \quad\text{and}\quad k^2 = \frac{z_2-z_3}{z_1-z_3}\,,$$

both with values between 0 and 1, we have in particular

$$\int_{z_3}^z \frac{q-t}{p-t}\,\frac{dt}{\sqrt{f(t)}} = \frac{2}{\sqrt{z_1-z_3}}\left\{\frac{q-p}{p-z_3}\,\Pi\left(\frac{z_2-z_3}{p-z_3}\,\arcsin g(z)\,\bigg|\,k^2\right)\right.$$

$$\left. + F\left(\arcsin g(z)\,\bigg|\,k^2\right)\right\}.$$

Therefore, after an oscillation period T, the body axis does not return to the initial point, in contrast to what happens with the plane pendulum, but precesses about the angle

Fig. 2.33 Complete elliptic integrals of the first kind $K(k^2)$ (*continuous green*), of the second kind $E(k^2)$ (*continuous red*), and of the third kind $\Pi(n \mid k^2)$ (*dashed black*), where n changes in steps of $1/4$ (top $3/4$, bottom -1). We also have $\Pi(0 \mid k^2) = K(k^2)$

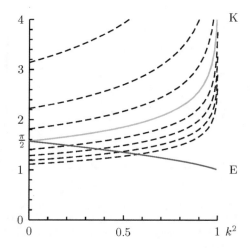

$$\Delta\alpha = 4\sqrt{\frac{p_\gamma^2/(2I_1)}{mgs\,(z_1 - z_3)}} \left\{ \frac{1 + p_\alpha/p_\gamma}{1 + z_3}\,\Pi\left(-\frac{z_2 - z_3}{1 + z_3}\;;\;\frac{\pi}{2}\,\middle|\,k^2\right) \right.$$
$$\left. -\frac{1 - p_\alpha/p_\gamma}{1 - z_3}\,\Pi\left(+\frac{z_2 - z_3}{1 - z_3}\;;\;\frac{\pi}{2}\,\middle|\,k^2\right)\right\}.$$

Due to the argument $\frac{1}{2}\pi$ here, *complete elliptic integrals of the third kind* occur, written for short $\Pi(n \mid k^2)$ (see Fig. 2.33).

2.4.11 Canonical Transformation of Time-Dependent Oscillators

The time-dependent oscillator investigated in Sect. 2.3.10 offers an instructive example of how a canonical transformation can transform a time-dependent Hamilton function into a time-independent one.

According to Floquet, Hill's differential equation $\ddot{x} + f(t)\,x = 0$ with $f(t + T) = f(t)$ also has quasi-periodic solutions $x_F(t) = y(t)\,\exp(i\phi t/T)$ with $y(t + T) = y(t)$. Here, ϕ is real for stable solutions, to which we would like to restrict ourselves here, even if then not all periodic functions $f(t)$ are allowed. We now take x_F and x_F^* as the fundamental system and set $w = (\dot{x}_F\, x_F^* - x_F\, \dot{x}_F^*)/(2i) = w^* > 0$. (It will turn out that w corresponds to an angular frequency. The similarity with ω is intended. For $w < 0$, we have to swap $x_F \leftrightarrow x_F^*$.) The value w does not depend on t, because it is the Wronski determinant of the two solutions, except for the factor $2i$ in the denominator. Two real fundamental solutions are often taken, which behave for $t \approx 0$ like the circular functions $\cos(wt)$ and $\sin(wt)$. Here we prefer $\exp(\pm iwt)$ for $t \approx 0$.

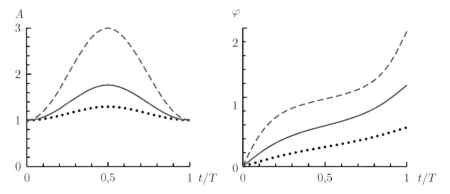

Fig. 2.34 Solutions of the Mathieu differential equation $\ddot{x} = \frac{1}{2}q\Omega^2 \cos \Omega t \, x$ for $q = 1/4$ (*dotted black*), $q = 2/4$ (*continuous red*), and $q = 3/4$ (*dashed blue*). Amplitude A (*left*) and phase φ (*right*) of the Floquet solutions as a function of t/T. The amplitude has period T, while the phase increases by ϕ during this time

In the following, it will be useful to set $x = A \exp(i\varphi)$ with real functions $A(t)$ and $\varphi(t)$. From Hill's differential equation, we then have the two equations

$$\ddot{A} + f\,A = \frac{w^2}{A^3} \quad \text{and} \quad \dot{\varphi} = \frac{w}{A^2}\,.$$

Here the quasi-periodicity of the Floquet solution x_F also delivers

$$A_F(t + T) = A_F(t) \quad \text{and} \quad \varphi_F(t + T) = \varphi_F(t) + \phi\,.$$

The amplitude A_F is thus strictly periodic, while the phase φ_F increases by ϕ with each period T. Note that $\phi > 0$ holds because $\dot{\varphi}_F = w/A^2 > 0$.

In the following, we leave out the index F and choose as initial conditions $A(0) = 1$, $\dot{A}(0) = 0$, and $\varphi(0) = 0$. Then w is also uniquely determined.

As an important example we consider the Mathieu differential equation. As in Sect. 2.3.10, $\ddot{x} + f(t)\,x = 0$ with $f(t) = \frac{1}{4}\Omega^2 (a - 2q \cos \Omega t)$. Figure 2.34 shows the amplitude and phase of the Floquet solutions, and Fig. 2.27 its real part. Since the amplitude is periodic, it can be expanded in a Fourier series. We consider now

$$A^2(t) = \sum_{n=0}^{\infty} b_n \cos(n\Omega t) \quad \Longrightarrow \quad \varphi(t) = \int_0^t \frac{w\,dt'}{\sum_{n=0}^{\infty} b_n \cos n\Omega t'}\,,$$

since its Fourier coefficients converge quickly to 0 as $q^n/(n!)^2$ and can be determined from a recursion relation. (This is shown in [5].) The Wronski determinant w becomes imaginary at the stability limits. Note that, in the unstable region, the same recursion relation holds for an expansion $A^2 = \sum_n b_n \cosh(\Omega t)$. The phase φ follows numerically from the above-mentioned integral expression using the Simpson method.

If we now take the generating function $G(t, p, x') = -A\,p\,x' + \frac{1}{2}m\,A\,\dot{A}\,x'^2$ (thus with $x = -\partial G/\partial p = A\,x'$ and $p' = -\partial G/\partial x' = A\,p - m\,A\,\dot{A}\,x'$), then from $H = \frac{1}{2m}\,p^2 + \frac{m}{2}\,f\,x^2$, we have

$$H' = H + \frac{\partial G}{\partial t} = \frac{p'^{\,2}}{2m\,A^2} + \frac{m}{2}\left(\ddot{A} + f\,A\right)A\,x'^2 \,,$$

since $\partial G/\partial t = -\dot{A}\,p\,x' + \frac{1}{2}m(\dot{A}^2 + A\dot{A})x'^2$. For $t = 0$, we should have $x' = x$ and $p' = p$, thus $A(0) = 1$ and $\dot{A}(0) = 0$. Because $\ddot{A} + f\,A = w^2/A^3$, we arrive at

$$H' = \frac{I\,w}{A^2}\,, \quad \text{with} \quad I\,w \equiv \frac{p'^{\,2}}{2m} + \frac{m\,w^2}{2}\,x'^2 \,,$$

and because $\dot{I} = [I, H'] = [I, I]\,w/A^2 = 0$, I does not depend upon t and is thus an invariant. Since $w/A^2 = d\varphi/dt$, it is clearly appropriate here to use the phase instead of the time. For each observable B not explicitly depending on time, we then have

$$\frac{dB}{dt} = [B, I]\,\frac{d\varphi}{dt} \quad \Longrightarrow \quad \frac{dB}{d\varphi} = [B, I]\,.$$

In order to determine the function $B(\varphi)$, we therefore only need to know the invariant I. In particular, the position and momentum can then be determined. (Neither φ nor I nor H' depend on the choice of scale for w: for $A \to c\,A'$, we have in particular $w \to c^2 w$, $x' \to c^{-1}x'$, and $p' \to c\,p'$.)

The invariant I does indeed help for the computation of the time dependence (of, e.g., position and momentum), because $H' = I\,w/A^2$ is a Hamilton function, but H' is not an energy. For this, the gauge is chosen such that the Hamilton function is composed of a *potential* and a *kinetic energy* according to p. 124. This works with

$$E = \frac{(p - m\,F\,x)^2}{2m} + \frac{m}{2}\,\overline{f}\,x^2\,, \quad \text{if } \dot{F} = f - \overline{f} \text{ (and } \overline{F} = 0)\,.$$

Once again, the bar indicates the time average ($\overline{F}$ need not be zero, but this choice makes $\overline{F^2}$ as small as possible, which has advantages), and thus $\frac{m}{2}\,\overline{f}\,x^2$ is a potential energy. The given expression for E via the generating function

$$G(t, p, x') = A\left\{\tfrac{1}{2}m\,(\dot{A} + A\,F)\,x'^2 - p\,x'\right\}$$

leads to the above-mentioned form $H' = I\,w/A^2$, thus also allowed as a Hamilton function. Because $\dot{x} = \partial E/\partial p = (p - m\,F\,x)/m$, the part $(p - m\,F\,x)^2/(2m)$ can be viewed as a kinetic energy $\frac{m}{2}\,\dot{x}^2$. Since $\dot{p} = -\partial E/\partial x = (p - m\,F\,x)\,F - m\,\overline{f}\,x = m\,(\dot{x}\,F - \overline{f}\,x)$, it turns out that $\ddot{x} = -f\,x$.

2.4.12 Summary: Hamiltonian Mechanics

When searching for the time dependence, we tend to rely on conserved quantities. Therefore, momenta are often better to use than velocities. In the Hamiltonian formalism, *canonical* transformations between position and momentum coordinates are permitted. Here, the difference between the two kinds of variables is blurred: we only talk about *canonical variables in phase space*. Because of the greater freedom in the choice of the phase space coordinates, even more suitable coordinates for a problem can be found than in Lagrangian mechanics.

Moreover, formally, Hamiltonian mechanics is to be preferred because the Hamilton function H is the generating function of infinitesimal variations in time. The Liouville equation can be derived from this (important for statistical mechanics), and the Poisson brackets are also useful in quantum mechanics.

According to the Hamilton–Jacobi theory, the Hamilton equations

$$\dot{x}^k = \frac{\partial H}{\partial p_k} \, , \qquad \dot{p}_k = -\frac{\partial H}{\partial x^k} \, ,$$

can be combined into a single partial differential equation which is useful also in light-ray optics, viz.,

$$\frac{\partial W}{\partial t} + H\left(t, x, \frac{\partial W}{\partial x}\right) = 0 \, ,$$

where W is the action

$$W = \int L \, \mathrm{d}t \, .$$

Conversely, $\mathrm{d}W/\mathrm{d}t$ delivers the Lagrange function and everything that follows likewise from derivatives.

The goal, namely to treat problems with many degrees of freedom with a single equation, is therefore achieved by Hamilton's principle

$$\delta W = 0 \, , \quad \text{at} \quad \delta t = 0 \, .$$

Since $\delta W = \int_{t_0}^{t_1} \delta(T + A) \, \mathrm{d}t$, it may even be applied to cases for which no potential energy exists, and hence there is neither a Lagrange function nor a Hamilton function.

Problems

Problem 2.1 Determine the 3×3 matrix of the rotation operator D for a body as a function of the Euler angles α, β, γ, which are introduced in Fig. 1.10 on p. 30. (2 P)

Problem 2.2 Verify the result for the following 7 special cases: no rotation, $180°$ rotation about the x-, y-, z-axis, and $90°$ rotation about the x-, y-, z-axis. Which Euler angles belong to these 7 cases?
Hint: Here, occasionally only $\alpha + \gamma$ or $\alpha - \gamma$ are determined. (7 P)

Problem 2.3 Which Euler angles $\{\bar{\alpha}, \bar{\beta}, \bar{\gamma}\}$ belong to the inverse rotations? (Note that $0 \leq \bar{\alpha} < 2\pi, 0 \leq \bar{\beta} \leq \pi$, and $0 \leq \bar{\gamma} < 2\pi$.) (3 P)

Problem 2.4 The rotation operator and Euler angles are needed to describe a top. For this application, the original coordinate system is the laboratory system, the new system is the body-fixed system. Let the unit vectors be $\mathbf{l}_x$, $\mathbf{l}_y$, $\mathbf{l}_z$ or $\mathbf{k}_x$, $\mathbf{k}_y$, $\mathbf{k}_z$. Let $\omega_\gamma = \dot{\gamma}\mathbf{k}_z$ and determine the corresponding decomposition into $\omega_\beta = \dot{\beta}\mathbf{e}_{\text{line of nodes}}$ and $\omega_\alpha = \dot{\alpha}\mathbf{l}_z$ in the body-fixed and lab-fixed systems.
  Let a rotation be $D = D_\alpha D_\beta$. How do the vector A and the matrix M given by

$$
A = \begin{pmatrix} a_x \\ a_y \\ a_z \end{pmatrix} \quad \text{and} \quad M = \begin{pmatrix} 0 & a_z & -a_y \\ -a_z & 0 & a_x \\ a_y & -a_x & \end{pmatrix}
$$

transform under the rotation D? Note that the fact that $M' = DMD^{-1} = DM\tilde{D}$ shows the same behavior under a rotation as $A' = DA$ is connected to the notion of axial vector. (8 P)

Problem 2.5 Is the tensor force $\mathbf{F} = \frac{3\mu_0}{4\pi} \mathbf{T}$ with (see p. 56)

$$
\mathbf{T}(\mathbf{r}) = \frac{(\mathbf{m} \cdot \mathbf{r})\,\mathbf{m}' + (\mathbf{m}' \cdot \mathbf{r})\,\mathbf{m} + (\mathbf{m} \cdot \mathbf{m}')\,\mathbf{r}}{r^5} - 5\,\frac{(\mathbf{m} \cdot \mathbf{r})(\mathbf{m}' \cdot \mathbf{r})\,\mathbf{r}}{r^7}
$$

curl-free?
Hint: To investigate the singularity for $\mathbf{r} = \mathbf{0}$, we may encircle the origin and apply Stokes's theorem $\nabla \times \mathbf{F} = \lim_{A \to 0} \frac{1}{A} \int_{(A)} d\mathbf{r} \cdot \mathbf{F}$.) (8 P)

Problem 2.6 Determine the potential energy V for Problem 2.5, and check that $\mathbf{F} = -\nabla V$. (4 P)

Problem 2.7 A circular disk of radius R rolls, without sliding, on the x, y-plane. In addition to the two coordinates (x, y) of the point of contact, the three Euler angles α, β, γ arise, because the normal to the circular disc has spherical coordinates (β, α), and γ describes the rotation of the disc. The problem requires five coordinates *with finite ranges*, having five degrees of freedom (see Fig. 2.35). However, the static friction also delivers two differential conditions between the coordinates *on the infinitely small scale*:

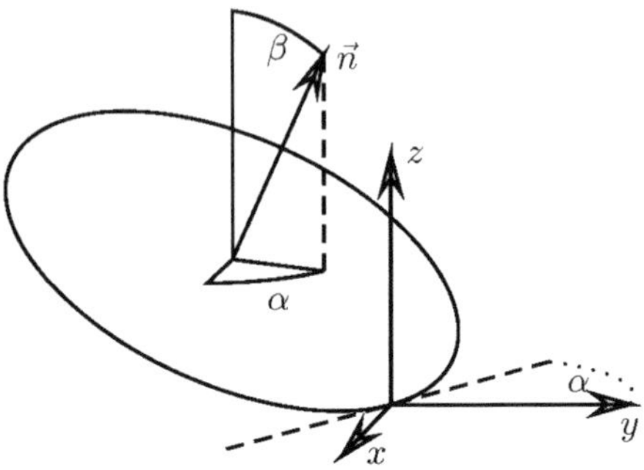

Fig. 2.35 Rolling circular disc with normal **n**. The Euler angles α and β are shown, but not the rolling angle γ. The orbit arises from the rotation about the normal in the positive sense

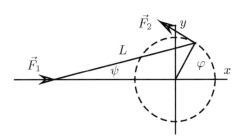

Fig. 2.36 Crank motion. The pinion runs on a circle of radius R and moves a connecting rod of length L

- How do the constraints for the virtual displacements read?
- How many degrees of freedom does the disc have on the infinitely small scale?
- Why do the equations $\Phi(\alpha, \beta, \gamma, x, y) = 0$ here lead to inner contradictions? (Why are the constraints non-holonomous)?

(8 P)

Problem 2.8 How do the Lagrange equations of the first kind read in statics if the constraints are given only in differential form (as in the last problem), namely through $\sum_{m=1}^{3N} \phi_{nm} \, \delta x^m = 0$? Here n counts the $3N - f$ constraints.

Use this for Problem 2.7 to determine the Lagrangian parameters, and interpret the connection found between the generalized forces. Show in particular that, in the contact plane, a tangential force acts on the disc, that F_γ cancels its torque, and that both F_α and F_β are equal to zero.   (8 P)

Problem 2.9 How strong does the force $F_2(F_1, \varphi)$ at the crank in Fig. 2.36 have to be for equilibrium? Determine this using the principle of virtual work.   (4 P)

Problem 2.10 What does one obtain for this crank from the Lagrange equations (Cartesian coordinates with origin at the center of rotation)? Do the results agree?   (8 P)

Problem 2.11 How much does the eccentricity ε differ from 1 for a given axis ratio $b : a \leq 1$ of an ellipse? Relate the difference between the distances at *aphelion* and *perihelion* for this ellipse to the mean value of these distances, and compare this with the axis ratio b/a. What follows then for small ε if we account only for linear, but no squared terms in ε? (For the orbits of planets around the Sun, $\varepsilon < 0.1$.) For Comet Halley, $\varepsilon = 0.9672760$. How are the two axes related to each other, and what is the ratio of the lowest to the highest velocity? (2 P)

Problem 2.12 Let the polar angle $\varphi = 0$ be associated with the aphelion of the orbit of the Earth (astronomers associate $\varphi = 0$ with the perihelion) and the polar angle φ_F with the beginning of spring. Then φ increases by $\pi/2$ at the beginning of the summer, autumn, and winter, respectively. In Hamburg, Germany, the lengths of the seasons are $T_{sp} = 92\,d\ 20.5\,h$, $T_{su} = 93\,d\ 14.5\,h$, $T_{fa} = 89\,d\ 18.5\,h$, $T_{wi} = 89\,d\ 0.5\,h$. Determine φ_{sp} and ε, neglecting squared terms in ε compared to the linear ones. (6 P)

Problem 2.13 By how much is the sidereal day shorter than the solar day (the time between two highest altitudes of the Sun)? By how much does the length of the solar day change in a year? (The result should be determined at least to a linear approximation as a function of ε. Then for $\varepsilon = 1/60$, the difference between the longest and shortest solar day follows absolutely.) (4 P)

Problem 2.14 Why are the following three theorems valid for the acceleration $\ddot{\mathbf{r}} = -k\mathbf{r}$ (and constant $k > 0$)?

- The orbit is an ellipse with the center $\mathbf{r} = \mathbf{0}$.
- The ray $\mathbf{r}$ moves over equal areas in equal time spans.
- The period T does not depend on the form of the orbital ellipse, but only on k.

Hint: Show that at certain times $\mathbf{r}$ and $\dot{\mathbf{r}}$ are perpendicular to each other. With such a time as the zero time, the problem simplifies enormously. (8 P)

Problem 2.15 What is the kinetic energy T_{2L}' of a mass m_2 in the laboratory system *after* the collision with another mass m_1 initially at rest, taken relative to its kinetic energy T_{2L} *before* the collision as a function of the scattering angle θ_S (in the center-of-mass system) and of the heat tone Q or the parameter $\xi = \sqrt{1 + (m_1+m_2)/m_1 \cdot Q/T_{2L}}$? How does this ratio read for equal masses and elastic scattering as a function of the scattering angle in the laboratory system? (4 P)

Problem 2.16 What is the angle between the directions of motion of two particles in the laboratory system after the collision? Consider the special case of elastic scattering and in particular of equal masses. (4 P)

Problem 2.17 Two smooth spheres with radii R_1 and R_2 collide with each other with the collision parameter s (see Fig. 2.37). How large is the scattering angle θ_S? (2 P)

Problem 2.18 How high is the mass m_1 of a body initially at rest, which has collided elastically with another body of mass m_2 and momentum $\mathbf{p}_2$, if it is scattered by $\theta_{2L} = 90°$ and keeps only the fraction q of its kinetic energy in the laboratory system? (3 P)

Fig. 2.37 For the collision of two smooth spheres in the center-of-mass system, the component of **p** in the direction of the line connecting the sphere centers becomes reversed, the component perpendicular to it is conserved (see Problem 2.17)

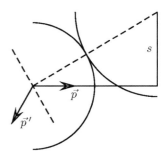

Problem 2.19   A spherical rain drop falls in a homogeneous gravitational field without friction through a saturated cloud. Its mass increases in proportion to its surface area with time. Which inhomogeneous linear differential equation follows for the velocity v, if instead of the time, we take the radius as independent variable? What is the solution of this differential equation? (In practice, we consider only the momentum $\propto r^3 v$ as an unknown function.) Compare with the free fall of a constant mass. (7 P)

Problem 2.20   How can we show using Legendre polynomials that the gravitational potential is constant within an inhomogeneous, but spherically symmetric hollow sphere, and therefore that it does not exert there a gravitational force on a test body. How does the potential read if a sphere with radius r_1 and homogeneous density ρ_1 is covered by a hollow sphere of homogeneous density ρ_2 and external radius r_2? (The Earth has a core, mainly of iron, and a mantle of SiO_2, MgO, FeO, and others, approximately 2900 km thick.) (6 P)

Problem 2.21   What height is reached by a ball thrown vertically upwards with velocity v_0? Consider the friction with the air (Newtonian friction) and determine the frictional work done, by integration as well as by comparing heights with and without friction. (8 P)

Problem 2.22   A horizontal plate oscillates harmonically up and down with amplitude A and oscillation period T. What inequality is obeyed by A and T if a loosely attached body on the plate does not lift off? (2 P)

Problem 2.23   A car at a speed of 20 km/h runs into a wall and is then evenly decelerated, until it stops, at which point it has been deformed by 30 cm. What is the deceleration during the collision? Can a weightlifter who can lift twice the weight of his body protect himself from hitting the steering wheel? If two such cars with relative velocity 40 km/h hit each other head-on, are the same processes valid for the single drivers as above, or do double or fourfold forces arise? (4 P)

Problem 2.24   Prove the following theorem: *For each plane mass distribution, the moment of inertia with respect to the normal of the plane is equal to the sum of the moments of inertia with respect to two mutually perpendicular axes in the plane.* (1 P)

Problem 2.25 Derive from that the main moments of inertia of a homogeneous cuboid with edge lengths a, b, and c. (1 P)

Problem 2.26 Determine the moment of inertia of the cuboid with respect to the edge c using three methods:

- As in Problem 2.25.
- Using Problem 2.25 but dividing up a correspondingly larger cuboid.
- Using Steiner's theorem.

(2 P)

Problem 2.27 Decide whether the following claim is correct: *The moment of inertia of a rod of mass M and length l perpendicular to the axis does not depend on the cross-section A, and with respect to an axis of rotation on the face is four times as large as with respect to an axis of rotation through the center of mass.* (2 P)

Problem 2.28 Prove the following: *Rotations about the axes of the highest and lowest moments of inertia are stable motions, while rotations about the axis of the middle moment of inertia are unstable.*
Hint: Use the Euler equations for the rigid body, and make an ansatz for the angular velocity $\boldsymbol{\omega} = \boldsymbol{\omega}_1 + \delta\boldsymbol{\omega}$ with constant $\boldsymbol{\omega}_1$ along a principal axis of the moment of inertia under small perturbations $\delta\boldsymbol{\omega} = \boldsymbol{\delta}\exp(\lambda t)$ perpendicular to it. This implies a constraint for $\lambda(I_1, I_2, I_3, \omega_1)$. (4 P)

Problem 2.29 How high is the Coriolis acceleration of a sphere shot horizontally with velocity v_0 at the north pole? Through which angle φ is it deflected during the time t? Through which angle does the Earth rotate during the same time? (2 P)

Problem 2.30 A uniform heavy rope of length l and mass μl hangs on a pulley of radius R and moment of inertia I, with the two rope ends initially at the same height. Then the pulley gets pushed with $\dot{\theta}(0) = \omega_0$. Neglect the friction of the pulley about its horizontal axis. As long as the rope presses on the pulley with the total force $F \geq F_0$, the static friction leads to the same (angular) velocity of rope and pulley— after that the rope slides down faster. How does the (angular) velocity depend on the time, up until the rope starts sliding? What is the difference in height of the ends of the rope at this time? (8 P)

Problem 2.31 Show that the homogeneous magnetic field $\mathbf{B} = B\mathbf{e}_z$ may be associated with the two vector potentials $\mathbf{A}_1 = \frac{1}{2}(\mathbf{B} \times \mathbf{r})$ and $\mathbf{A}_2 = Bx\mathbf{e}_y$ (gauge invariance). What scalar field ψ leads to $\nabla\psi = \mathbf{A}_1 - \mathbf{A}_2$? What is the difference between the associated Lagrange functions L_1 and L_2? Why is it that this difference does not affect the motion of a particle of charge q and mass m in the magnetic field $\mathbf{B}$? (6 P)

Problem 2.32 Two point masses interact with $V(|\mathbf{r}_1 - \mathbf{r}_2|)$ and are not subject to any external forces. How do the Lagrange equations (of the second kind) read in the center-of-mass and relative coordinates? (4 P)

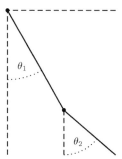

Fig. 2.38 Double pendulum made from rods of mass m_1 and m_2 and with moments of inertia I_1 and I_2 with respect to the hinges (•), which are separated by a distance l. The distances of the centers of mass of the rods from the hinges are s_1 and s_2, respectively

Problem 2.33 For the double pendulum in Fig. 2.38, determine T as well as V as a function of θ_1, θ_2, $\dot\theta_1$, and $\dot\theta_2$. How are these expressions simplified for small amplitudes?   (6 P)

Problem 2.34 In the last problem, let $\theta_1 = \theta_2 = 0$ for $t < 0$. At time $t = 0$, the upper pendulum obtains an impulse, in fact with angular momentum **L** with respect to its hinge. What initial values follow for $\dot\theta_1$ and $\dot\theta_2$, in particular for the mathematical pendulum?   (4 P)

Problem 2.35 A homogeneous sphere of mass M and radius r rolls on the inclined plane shown in Fig. 2.39 (with $\mathbf{g} \cdot \mathbf{e}_x = 0$). Its moment of inertia is $I = \frac{2}{5} Mr^2$. Determine its Lagrange function and the equations of motion for the coordinates (x, z) at the point of contact. (Here we use z instead of y, in anticipation of the next problem.)   (3 P)

Problem 2.36 Treat the corresponding problem if the plane is deformed into a cylindrical groove with radius R and axis parallel to $\mathbf{e}_z$ (see Fig. 2.40). (Instead of x, it is better to adopt the cylindrical coordinate φ with $\varphi = 0$ at the lowest position.) How large may $\dot\varphi(0)$ be at most, if we always have $|\varphi| \leq \frac{1}{2}\pi$ and $\varphi(0) = 0$?   (4 P)

Fig. 2.39 Oblique plane with inclination angle α (the angle between the downwards oriented normal and the vertical), whence $\mathbf{g} \cdot \mathbf{e}_z = g \sin\alpha$

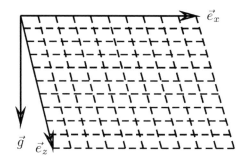

Fig. 2.40 Sphere in a groove. A sphere with radius r rolls on a circle of radius R, then with the angles ψ and φ shown, the relation $r(\psi + \varphi) = R\varphi$ holds

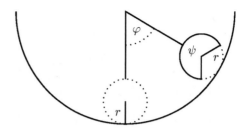

Problem 2.37 Determine the resonance angular frequency ω_R of a forced damped oscillation and show that the frequency ω_0 of the undamped oscillation is higher than ω_R. What is the ratio of the oscillation amplitude for ω_R to that for ω_0? What is the approximate result for $\gamma \ll \omega_0$? (4 P)

Problem 2.38 What differential equation and initial values are valid for the *Green function* $G(\tau)$ for the differential equation $\ddot{x} + 2\gamma\dot{x} + \omega_0^2 x = f(t)$ of the forced damped oscillation with solution written as $x(t) = \int_{-\infty}^{t} f(t')\, G(t - t')\, dt'$? Which $G(\tau)$ is the most general solution, independent of $f(t)$? With this, the solution of the differential equation may be traced back to a simple integration—check this for the example $f(t) = c\,\cos(\omega t)$ in the special case $\gamma = \omega_0\ (> 0)$. (9 P)

Problem 2.39 What equation of motion is supplied by the Lagrange formalism for the double pendulum investigated in Problem 2.33 in the angle coordinates θ_1 and θ_2, exactly on the one hand, and for restriction to small oscillations on the other (i.e., taking θ_1 and θ_2 and their derivatives to be small quantities)? (4 P)

Problem 2.40 Which normal frequencies $\omega_\pm$ result for this double pendulum? Determine the matrices A and B. Investigate also the special case of the mathematical pendulum with $s_1 = l$, and use the abbreviation $\sigma = s_2/s_1$ and $\mu = m_2/m_1$, where the normal frequencies are given here at best as multiples of the eigenfrequency ω_1 of the upper pendulum. (6 P)

Problem 2.41 Determine the normal frequencies and the matrices A, B, and C for the mathematical double pendulum with $s_2 = s_1 = l$ and $m_2 \ll m_1$. (Here one should use the fact that $\mu \ll 1$ holds in C—why does one have to calculate $\omega_\pm$ "more precisely by one order"?) (6 P)

Problem 2.42 What functions $\theta_1(t)$ and $\theta_2(t)$ belong to the just investigated mathematical double pendulum (with $\mu \ll 1$) for the following initial values: $\theta_1(0) = \theta_2(0) = 0$, $\dot{\theta}_1(0) = -\dot{\theta}_2(0) = \Omega$, which according to Problem 2.34 correspond to a collision against the upper pendulum for the double pendulum initially at rest? (Why do we only have to consider here the behavior of the normal coordinates?) Which angular frequencies do the beats have, and how does the amplitude of θ_1 behave in comparison to that of θ_2? (4 P)

Problem 2.43 Prove the Jacobi identity $[u, [v, w]] + [v, [w, u]] + [w, [u, v]] = 0$ for the Poisson brackets, with $[u, v] = u_x v_p - u_p v_x$ and $u_x = \partial u/\partial x$, etc.   (3 P)

Problem 2.44 Determine the Poisson brackets of the angular momentum component L_x with x, y, z, p_x, p_y, p_z, and L_y. Note that, by cyclic commutation, $[\mathbf{L}, \mathbf{r}]$, $[\mathbf{L}, \mathbf{p}]$, and $[L_i, L_k]$ are then also proven.   (5 P)

Problem 2.45 Under which constraints is the transformation $x' = \arctan(\alpha x/p)$, $p' = \beta x^2 + \gamma p^2$ a canonical one?   (3 P)

Problem 2.46 Is the transformation $x' = x^\alpha \cos \beta p$, $p' = x^\alpha \sin \beta p$ canonical?   (2 P)

Problem 2.47 Using the generating function

$$G(x, p_x', p_y, p_y') = x \frac{1}{2} \{p_x' + p_y'\} - p_y \{p_x' - p_y'\}/(qB) ,$$

show that the Hamilton function

$$H = \frac{1}{2m} \{(p_x + \frac{1}{2} q By)^2 + (p_y - \frac{1}{2} q Bx)^2\}$$

for a charged point mass in the plane perpendicular to a homogeneous magnetic field $B\mathbf{e}_z$ can be written as the Hamilton function of a linear harmonic oscillation. (3 P)

Problem 2.48 From this derive the transformation on p. 128. Show also, without using the generating function, that this transformation is canonical. Why does it not suffice here to compare the four derivatives $\partial x'/\partial x$, $\partial x'/\partial y$, $\partial y'/\partial x$, and $\partial y'/\partial y$ with $\partial p_x/\partial p_x'$, $\partial p_x/\partial p_y'$, $\partial p_y/\partial p_x'$, and $\partial p_y/\partial p_y'$, as seems to suffice according to p. 126? (Whence an additional comment is missing here.)   (4 P)

List of Symbols

We stick closely to the recommendations of the *International Union of Pure and Applied Physics* (IUPAP) and the *Deutsches Institut für Normung* (DIN). These are listed in *Symbole, Einheiten und Nomenklatur in der Physik* (Physik-Verlag, Weinheim 1980) and are marked here with an asterisk. However, one and the same symbol may represent different quantities in different branches of physics. Therefore, we have to divide the list of symbols into different parts (Table 2.3).

Table 2.3 Symbols used in mechanics

	Symbol	Name	Page reference
*	$\mathbf{v}$, $\dot{\mathbf{r}}$	Velocity	2
*	$\mathbf{a}$, $\dot{\mathbf{v}}$, $\ddot{\mathbf{r}}$	Acceleration	2
*	$\mathbf{F}$	Force	55
*	$\mathbf{M}$	Torque	70
	M	Total mass	71
*	m	Mass	69
*	μ	Reduced mass	72
*	A	Work	56
*	E	Energy	78
*	V	Potential energy	56
*	T	Kinetic energy	70
*	T	Oscillation period	104
*	ρ	Density (massdensity)	81
	ρ	Probability density	125
*	$\mathbf{p}$	Motional quantity, momentum	69, 93, 99
*	$\mathbf{L}$	Angular momentum	70
*	G	Gravitational constant	623, 79
	G	Generating function	130
*	$\mathbf{g}$	Free-fall acceleration	81
*	I	Moment of inertia	87
*	ω	Angular frequency	67
	$\boldsymbol{\omega}$	Angular velocity	67
	x^k	Generalized coordinate	60
*	p_k	Momentum canonical conjugate to x^k	93, 99
	F_k	Generalized force	60
*	L	Lagrange function	96
*	H	Hamilton function	122
*	W	Action function	135
*	S	Characteristic function	136
	$[u, v]$	Poisson bracket	124

References

1. M. Abramowitz, I.A. Stegun, *Handbook of Mathematical Functions* (Dover, New York, 1970)
2. P.F. Byrd, M.D. Friedman, *Handbook of Elliptic Integrals for Engineers and Physicists* (Springer, Berlin, 1954)
3. J. Meixner, F.W. Schäfke, G. Wolf, *Mathieu Functions and Spheroidal Functions and Their Mathematical Foundations* (Springer, Berlin, 1980)
4. D.H. Kobe, K.H. Yang, Eur. J. Phys. **80**, 236 (1987)
5. A. Lindner, H. Freese, J. Phys. A **27**, 5565 (1994)

Suggestions for Textbooks and Further Reading

6. W. Greiner, *Classical Mechanics—System of Particles and Hamiltonian Dynamics* (Springer, New York, 2010)

7. L.D. Landau, E.M. Lifshitz, *Course of Theoretical Physics. Volume 1—Mechanics*, 3rd edn. (Butterworth-Heinemann, Oxford, 1976)

8. W. Nolting, *Theoretical Physics 1—Classical Mechanics* (Springer, Berlin, 2016)

9. W. Nolting, *Theoretical Physics 2—Analytical Mechanics* (Springer, Berlin, 2016)

10. F. Scheck, *Mechanics—From Newton's Laws to Deterministic Chaos* (Springer, Berlin, 2010)

11. A. Sommerfeld, *Lectures on Theoretical Physics 1—Mechanics* (Academic, London, 1964)

12. D. Strauch, *Classical Mechanics* (Springer, Berlin, 2009)

13. W. Thirring, *Classical Mathematical Physics: Dynamical Systems and Field Theories*, 3rd edn. (Springer, New York, 2013)

14. G. Ludwig, *Einführung in die Grundlagen der Theoretischen Physik 1–4* (Vieweg, Braunschweig, 1974) (in German)

15. M. Mizushima, *Theoretical Physics: From Classical Mechanics to Group Theory of Microparticles* (Wiley, New York, 1972)

Chapter 3
Electromagnetism

3.1 Electrostatics

3.1.1 Overview of Electromagnetism

The basic equations of electromagnetism were found by Maxwell in 1862. They comprise not only electricity and magnetism, but also (wave) optics (as electromagnetic radiation)—and thus a very diverse range of phenomena. Actually, most of this was known before Maxwell, but he discovered the displacement current and thus also correctly connected the time-dependent electric and magnetic fields for non-conductors. Since then the concept of fields has been accepted.

We start from Coulomb's law giving the force between two charges, and from this derive the electric field. Then we consider its action on polarizable media and discriminate between microscopic and macroscopically averaged quantities. The essential basic concepts are electric charge and polarization.

We then consider moving charges and the Lorentz force. This will lead us to the concept of the magnetic field (the Biot–Savart law). Ampère's molecular currents in microscopic conductor loops produce magnetic moments, but otherwise cannot be verified (as currents). The magnetic moments of elementary particles with spin 1/2 (e.g., electrons) cannot even be attributed to currents in such microscopic conductor loops: like charges we have to accept them as non-derivable properties of these particles. Thus the coupling between two magnetic moments is likewise discarded as "basic", in contrast to the force between electric charges and Coulomb's law as the sole basis of electromagnetism—even if the scalar interaction between charges can be described in a simpler way than the tensor coupling between dipole moments.

The conservation law of charges and Faraday's induction law then result from Maxwell's equations:

© Springer Nature Switzerland AG 2018
A. Lindner and D. Strauch, *A Complete Course*
on Theoretical Physics, Undergraduate Lecture Notes in Physics,
https://doi.org/10.1007/978-3-030-04360-5_3

$$\nabla \times \mathbf{E} = -\frac{\partial \mathbf{B}}{\partial t}\,, \qquad\qquad \nabla \cdot \mathbf{B} = 0\,,$$

$$\nabla \cdot \mathbf{D} = \rho\,, \qquad\qquad \nabla \times \mathbf{H} = \mathbf{j} + \frac{\partial \mathbf{D}}{\partial t}\,.$$

The various quantities have the following names:

E electric field strength,   **B** magnetic displacement field (induction),
D electric displacement field,   **H** magnetic field strength,
ρ charge density,   **j** current density.

The term $\partial \mathbf{D}/\partial t$ is the density of the above-mentioned *displacement current.*

Maxwell's equations connect on the one hand **E** with **B** and on the other hand **D** with **H**. Therefore, **E** and **B** are also sometimes called field strengths, while **D** and **H** are referred to as excitations. The last two equations in particular contain further fields, viz., the charge and current densities. However, the two quantities **E** and **B** supply the force on a test charge. Here we have to know how **D** and **E** as well as **H** and **B** are connected—only then are the source and curl densities of the fields given, whereupon the basic theorem of vector analysis on p. 25 becomes applicable.

The wave equations for the fields result from Maxwell's equations with $\mathbf{D} \propto \mathbf{E}$ and $\mathbf{H} \propto \mathbf{B}$. Then waves can propagate in empty space with the velocity of light

$$c_0 = 299\,792\,458 \text{ m/s}\,.$$

This is the same in all inertial frames, which leads to Lorentz invariance, something we shall discuss after dealing with Maxwell's equations. Then the four equations for the three-vectors appearing above will be derived from two equations for four-vectors.

After that we shall consider the electromagnetic radiation field, which is produced by an accelerated charge, similar to the electric field of a charge at rest and the magnetic field of a uniformly moving charge.

Here we shall comply with the *international system of units* (SI). In addition to length, time, and mass with the units m, s, and kg, a basic electromagnetic quantity is introduced, namely the current strength with the unit A (ampere). Then further units are related to these, e.g.,

volt $V \equiv W/A$,   ohm $\Omega \equiv V/A \equiv S^{-1}$ (siemens),
coulomb $C \equiv A\,s$,   farad $F \equiv C/V = S\,s$,
weber $Wb \equiv V\,s$,   henry $H \equiv Wb/A = \Omega\,s$,
tesla $T \equiv Wb/m^2$.

In the international system of units, a *magnetic field constant* is necessary, viz.,

$$\mu_0 \equiv 4\pi \times 10^{-7} \text{ H/m} = 4\pi \times 10^{-7} \text{ N/A}^2\,,$$

and an *electric field constant*, viz.,

$$\varepsilon_0 \equiv \frac{1}{c_0{}^2\,\mu_0} = 8.854187817622\ldots \times 10^{-12}\ \text{F/m} .$$

Here, $\mu_0/4\pi$ appears in many equations for point charges and dipole moments, as does $1/4\pi\varepsilon_0 = c_0{}^2\,\mu_0/4\pi$, and $c_0\mu_0 = (c_0\varepsilon_0)^{-1} = 376.7303134618\ldots$ Ω is the so-called *wave resistance* in empty space, mentioned on p. 222.

However, in theoretical and atomic physics, the *Gauss system of units* is also often used. There, the electromagnetic quantities are introduced differently (despite the warning above: Coulomb's law is taken as the starting point from which Maxwell's equations have to be derived, while the international system starts from Maxwell's equations and deduces Coulomb's law), but irritatingly the same names and letters are used. If we denote the quantities in the Gauss system with an asterisk, we have

$$E^* = \sqrt{4\pi\,\varepsilon_0}\ E , \qquad\qquad B^* = \sqrt{4\pi/\mu_0}\ B ,$$
$$D^* = \sqrt{4\pi/\varepsilon_0}\ D , \qquad\qquad H^* = \sqrt{4\pi\,\mu_0}\ H ,$$
$$\rho^* = \rho/\sqrt{4\pi\,\varepsilon_0} , \qquad\qquad j^* = j/\sqrt{4\pi\,\varepsilon_0} .$$

Then Maxwell's equations appear in the form

$$\nabla \times E^* = -\frac{1}{c_0}\,\frac{\partial B^*}{\partial t} , \qquad\qquad \nabla \cdot B^* = 0 ,$$
$$\nabla \cdot D^* = 4\pi\,\rho^* , \qquad\qquad \nabla \times H^* = \frac{4\pi}{c_0}\,j^* + \frac{1}{c_0}\,\frac{\partial D^*}{\partial t} .$$

Here, further factors occur in Maxwell's equations. Particularly bothersome are the factors 4π. They occur in the Gauss system in plane problems and are missing in spherically symmetric ones. The difference between the two systems is dismissed as a problem of units, even though the equations deal with quantities that do not depend at all upon the chosen units (see Sect. 1.1.1). However, different notions generally have different units. Thus, in the Gauss system for B^*, the *gauss* (G) is used and for H^*, the *oersted* (Oe). They are both equal to $\sqrt{\text{g/cm s}^2}$, whence B^* and H^* are also easily confused. For the transition between the two unit systems, we have $10\ \text{kG} \triangleq 1\ \text{T}$ and $4\pi\ \text{mOe} \triangleq 1\ \text{A/m}$.

Particularly elaborate are the textbooks by Jackson and by Panofsky and Phillips (see the recommended textbooks on p. 274). The first employs the Gauss system in earlier editions, but since then both have used the international system.

3.1.2 Coulomb's Law—Far or Near Action?

In classical mechanics, mass is associated with all bodies. Some of them also carry *electric charge* Q, as becomes apparent from new forces—for point charges we usually write q. An electron, for example, has the charge

$$q_{\rm e} = -e = -1.602176462(63) \times 10^{-19}\ {\rm C},$$

and the proton the opposite charge. There are charges of both signs (in contrast to the mass, which is always positive) and the excess of positive or negative charge results in the charge Q of the body. We thus introduce the *charge density* $\rho(\mathbf{r})$, whereupon $Q = \int dV\, \rho(\mathbf{r})$.

According to Coulomb (1785), there is a force acting between two point charges q and q' (at rest) at the positions $\mathbf{r}$ and $\mathbf{r}'$ in empty space, which depends upon the distance as $|\mathbf{r} - \mathbf{r}'|^{-2}$ and which is proportional to the product qq' of the charges. Here the force is repulsive or attractive, depending on whether the charges have equal or opposite sign:

$$\mathbf{F} = \frac{1}{4\pi\varepsilon_0}\, \frac{qq'}{|\mathbf{r} - \mathbf{r}'|^2}\, \frac{\mathbf{r} - \mathbf{r}'}{|\mathbf{r} - \mathbf{r}'|}\,.$$

This is the force on the charge q. The one on q' (at $\mathbf{r}'$) is oriented oppositely, as required by Newton's third law (action = reaction, see p. 55). The factor $(4\pi\varepsilon_0)^{-1}$ is connected with the concept of charge in the international system—it is missing in the Gauss system. Here ε_0 is the *electric field constant*, and according to the last section,

$$\frac{1}{4\pi\varepsilon_0} \equiv \frac{c_0^2}{10^7}\, \frac{\rm H}{\rm m} = 8.987551787368\ldots \times 10^9\ \frac{\rm N\ m^2}{\rm C^2}\,.$$

Hence for electron and proton pairs, we have

$$\frac{e^2}{4\pi\varepsilon_0} = 2.307\,077\,06(19) \times 10^{-28}\ {\rm J\ m} = 1.439\,964\,392(57)\ {\rm eV\ nm},$$

where the last expression is suitable for atomic scales, and because eV nm = MeV fm, for nuclear physics.

Coulomb's law describes an action at a distance. But we may also introduce a field $\mathbf{E}(\mathbf{r})$ which surrounds the charge q' and acts on the test charge $q(\mathbf{r})$:

$$\mathbf{F} = q(\mathbf{r})\,\mathbf{E}(\mathbf{r})\,, \quad \text{with} \quad \mathbf{E}(\mathbf{r}) = \frac{1}{4\pi\varepsilon_0}\, \frac{q'}{|\mathbf{r} - \mathbf{r}'|^2}\, \frac{\mathbf{r} - \mathbf{r}'}{|\mathbf{r} - \mathbf{r}'|}\,.$$

This *electric field strength* $\mathbf{E}$ is conveniently given in N/C = V/m.

The concept of a field will be proven to be correct in the context of time-dependent phenomena, because actions propagate only with finite velocity, which contradicts the law of action at a distance. Therefore, we shall already use the field concept in electrostatics.

A point-like charge q' is thus associated with the electric field

$$\mathbf{E}(\mathbf{r}) = \frac{q'}{4\pi\varepsilon_0} \frac{\mathbf{r} - \mathbf{r}'}{|\mathbf{r} - \mathbf{r}'|^3} = -\frac{q'}{4\pi\varepsilon_0} \nabla \frac{1}{|\mathbf{r} - \mathbf{r}'|} ,$$

the source of which is the charge q' at the position $\mathbf{r}'$, according to p. 25, and which is irrotational (curl-free):

$$\nabla \cdot \mathbf{E} = \frac{q'}{\varepsilon_0} \delta(\mathbf{r} - \mathbf{r}') \quad \text{and} \quad \nabla \times \mathbf{E} = \mathbf{0} .$$

From the point-like charge, we extend the notion to an extended charge with charge density ρ'. So far we have been dealing with the special case of $\rho' = q' \, \delta(\mathbf{r} - \mathbf{r}')$ and now generalize this to

$$\nabla \cdot \mathbf{E} = \frac{\rho}{\varepsilon_0} \quad \text{and} \quad \nabla \times \mathbf{E} = \mathbf{0} .$$

(Here, and in the next few equations, we should have ρ' instead of ρ and Q' instead of Q, but temporarily there will only be field-creating charges and no test charges, so we prefer to simplify the notation.) However, this is allowed only if the fields of the various point charges superimpose linearly—and if these charges remain at their positions when we move the test charge around as a field sensor. (Because of induction, this is not justified for conductors, as will become apparent on p. 181.)

For charges distributed over a sheet, the normal component of the field strength thus has a discontinuity (see p. 28), while the tangential component is continuous:

$$\mathbf{n} \cdot (\mathbf{E}_+ - \mathbf{E}_-) = \frac{\rho_A}{\varepsilon_0} \quad \text{and} \quad \mathbf{n} \times (\mathbf{E}_+ - \mathbf{E}_-) = \mathbf{0} .$$

The two basic differential equations for the electrostatic field can be converted into integral equations using the theorems of Gauss and Stokes. Instead of the charge density ρ, only the charge $Q = \int dV \, \rho(\mathbf{r})$ enclosed in V is important:

$$\int_{(V)} d\mathbf{f} \cdot \mathbf{E} = \frac{Q}{\varepsilon_0} \quad \text{and} \quad \int_{(A)} d\mathbf{r} \cdot \mathbf{E} = 0 .$$

According to the last equation, we also have $\oint d\mathbf{r} \cdot \mathbf{F} = q \oint d\mathbf{r} \cdot \mathbf{E} = 0$: no work is needed to move a test charge on a closed path in an electrostatic field, since the field is irrotational. The charge-free space is also source-free. Therefore, the field lines, with tangents in the direction of the field, can be taken as the lines forming the walls of the flux tubes (see Fig. 3.1 and also p. 12). Figures 3.2 and 3.3 present examples. For two source points, we take a series of cones around the symmetry axis with increasing units of flux and then connect appropriate intersections (*Maxwell's construction*).

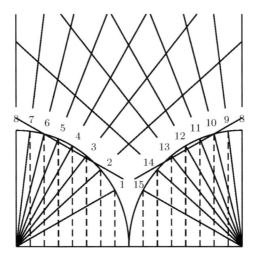

Fig. 3.1 Construction of field lines around point charges q. The same displacement field passes through the surface of spheres with radii $r \propto \sqrt{|q|}$ around q (here $q = q'$ is assumed, and thus equal spheres). Disks of equal thickness are shown with *dashed lines* and hence with walls of equal area $dA = 2\pi R \sin \alpha \, R|d\alpha| = 2\pi R \, |dz|$, and also equal flux. In the next two pictures, the intersections of the straight lines with equal parameter sum or difference are to be connected, because what flows into a quadrangle ◊ (solenoidal) (e.g., from below as ↗ and ↘), must also emerge again (in the example, diffracted at the wall of the field-line tubes |). See also Problem 3.13

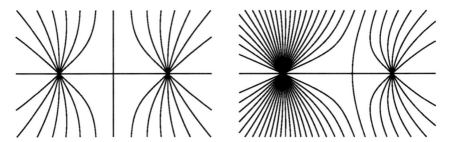

Fig. 3.2 Field lines of two like charges—the ratio of the charges *on the left* is 1:1 and *on the right* 3:1—with their saddle point between the two charges

3.1.3 Electrostatic Potential

The electrostatic force field is irrotational. Therefore, according to p. 25, we would like to attribute it to a scalar field Φ, which will be much easier to calculate with than the vector field:

$$\mathbf{E} = -\nabla \Phi \,, \quad \text{with} \quad \Phi(\mathbf{r}) \equiv \frac{1}{4\pi\varepsilon_0} \int dV' \, \frac{\rho(\mathbf{r}')}{|\mathbf{r} - \mathbf{r}'|} \,.$$

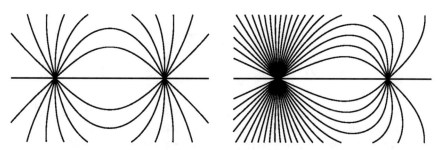

Fig. 3.3 Field lines of two unlike charges—ratio of the charges again 1:1 and 3:1 on the *left* and *right*, respectively

Φ is called the *electrostatic potential*, because it is connected with the potential energy E_{pot}. (Note that here, and in thermodynamics, we use V to denote the volume, so we cannot use this letter for the potential energy, as is possible in classical mechanics and quantum mechanics.) As is well known (see p. 56), we have $\mathbf{F} = -\nabla E_{\mathrm{pot}}$, so here $\mathbf{F} = q\,\mathbf{E} = -q\nabla\Phi$. Therefore,

$$E_{\mathrm{pot}} = q\,\Phi\,,$$

and in classical mechanics (see p. 77), $E_{\mathrm{pot}} = m\Phi$ with the mechanical potential Φ.

Between two points $\mathbf{r}_1$ and $\mathbf{r}_0$ of different potential, there is a *voltage*:

$$U \equiv \Phi(\mathbf{r}_1) - \Phi(\mathbf{r}_0) = \int_{\mathbf{r}_0}^{\mathbf{r}_1} d\mathbf{r}\cdot\nabla\Phi = -\int_{\mathbf{r}_0}^{\mathbf{r}_1} d\mathbf{r}\cdot\mathbf{E} = \int_{\mathbf{r}_1}^{\mathbf{r}_0} d\mathbf{r}\cdot\mathbf{E}\,.$$

It can be positive or negative, but we are often concerned only with its absolute value.

Since $\rho/\varepsilon_0 = \nabla\cdot(-\nabla\Phi)$, the potential follows from a linear differential equation with the charge density as inhomogeneous term, viz., the *Poisson equation*

$$\Delta\Phi = -\frac{\rho}{\varepsilon_0}\,.$$

To obtain unique solutions, we have to set boundary conditions (to *gauge*) the solution. The potential and its first derivatives must vanish at infinity, like the charge density.

This boundary condition can also be introduced via Green's second theorem (p. 17). Then one obtains the equation

$$\int_{(V')} d\mathbf{f}' \cdot \left(\Phi(\mathbf{r}') \, \nabla' \frac{1}{|\mathbf{r} - \mathbf{r}'|} - \frac{1}{|\mathbf{r} - \mathbf{r}'|} \nabla' \Phi(\mathbf{r}') \right)$$
$$= \int_{V'} dV' \left(\Phi(\mathbf{r}') \, \Delta' \frac{1}{|\mathbf{r} - \mathbf{r}'|} - \frac{1}{|\mathbf{r} - \mathbf{r}'|} \Delta' \Phi(\mathbf{r}') \right) .$$

Here the Poisson equation and $\Delta' |\mathbf{r} - \mathbf{r}'|^{-1} = -4\pi \, \delta(\mathbf{r} - \mathbf{r}')$ holds, according to p. 26. Hence we obtain the "Green function solution" (see, e.g., Fig. 1.5 for the cylindrical capacitor, with field lines on the left field and equipotential lines on the right):'

$$4\pi \; \Phi(\mathbf{r}) = \frac{1}{\varepsilon_0} \int_{V'} dV' \frac{\rho(\mathbf{r}')}{|\mathbf{r} - \mathbf{r}'|}$$
$$+ \int_{(V')} \frac{d\mathbf{f}' \cdot \nabla' \Phi(\mathbf{r}')}{|\mathbf{r} - \mathbf{r}'|} - \int_{(V')} d\mathbf{f}' \cdot \; \Phi(\mathbf{r}') \, \nabla' \frac{1}{|\mathbf{r} - \mathbf{r}'|} .$$

The first integral is no longer taken over the whole space. The two surface integrals account for all charges outside of it and occur as new boundary conditions. In particular, V' can also be a charge-free space, such that the first integral vanishes. Then the potential and field strength are uniquely fixed by Φ and $\nabla\Phi$ on the surface. In charge-free space, these two vary monotonically, as follows from the Poisson equation, so the field has no extremum there.

Incidentally, for a charge-free space, it is sufficient that either only Φ or only (the normal component of) $\nabla\Phi$ is given on its surface. In particular, according to Gauss's theorem, for $\Delta\Phi = 0$, we have

$$\int d\mathbf{f} \cdot \Phi\nabla\Phi = \int dV \, \nabla \cdot \Phi\nabla\Phi = \int dV \, \nabla\Phi \cdot \nabla\Phi .$$

If two solutions Φ_1 and Φ_2 of $\Delta\Phi = 0$ now satisfy the boundary conditions, then the surface integral of $\Phi \equiv \Phi_1 - \Phi_2$ vanishes because of $\Phi_1 = \Phi_2$ or $\mathbf{n} \cdot \nabla\Phi_1 = \mathbf{n} \cdot \nabla\Phi_2$. On the right, the integrand is nowhere negative. Consequently, everywhere in the considered volume, we have $\nabla\Phi_1 = \nabla\Phi_2$, so Φ_1 and Φ_2 differ at most by a constant, and this can eventually be fixed by the gauge.

In a finite regime, the same electric field can be generated by different charge distributions. The continuation across the boundaries is not unique. This should be considered if models for the charge distribution in inaccessible regions are presented.

3.1.4 Dipoles

So far we have allowed charges of both signs, but the test body should carry only charge of one sign, and as small as possible.

Totally new phenomena arise if the test body carries two point charges $\pm q$ of opposite sign. For simplicity, we assume that its total charge $Q = \int dV \rho(\mathbf{r})$ vanishes, otherwise we would also have to consider the properties of a *mono*pole, which have already been treated. An ideal *dipole* consists of two point charges $\pm q$ at the positions $\mathbf{r}_\pm = \pm\frac{1}{2}\,\mathbf{a}$, where a is as small as possible, but the product qa is nevertheless finite. We thus introduce the *dipole moment*

$$\mathbf{p} \equiv \int dV\,\mathbf{r}\,\rho(\mathbf{r})\,.$$

In the example considered, we would have $\mathbf{p} = q\mathbf{a}$. For finite a, higher *multipole moments* appear, i.e., integrals over ρ with weight factors other than $\mathbf{r}$ or 1, which we shall only discuss at the end of Sect. 3.1.7. If the total charge vanishes, the dipole moment does not depend upon the choice of the origin of $\mathbf{r}$.

However, in the following it will be advantageous, as in Sect. 2.2.2, to introduce center-of-charge and relative coordinates. Here we restrict ourselves to $Q = 0$ and choose $\mathbf{R} = \int dV\,\mathbf{r}\,|\rho|/\int dV\,|\rho|$ as "center of charge". We derive the *potential energy of the dipole* $\mathbf{p}$ in the electric field $\mathbf{E}$ from a series expansion of the potential around the center of the dipole:

$$\Phi(\mathbf{R} + \mathbf{r}) = \Phi(\mathbf{R}) - \mathbf{r} \cdot \mathbf{E}(\mathbf{R}) + \cdots \quad \text{because } \nabla\Phi = -\mathbf{E}\,.$$

For $Q = 0$ and with $E_{\text{pot}} = \int dV\rho(\mathbf{r})\,\Phi(\mathbf{r})$, this supplies the potential energy

$$E_{\text{pot}} = -\mathbf{p} \cdot \mathbf{E}\,.$$

Here the field strength is to be taken at the position of the dipole. For a homogeneous field, it does not depend on the position. Then there is no force $\mathbf{F} = -\nabla E_{\text{pot}}$ acting on the dipole—the forces on the different charges cancel each other in the homogeneous field.

In an inhomogeneous field, the forces acting on the two poles have different strengths. Then there remains an excess field

$$\mathbf{F} = -\nabla E_{\text{pot}} = \nabla(\mathbf{p} \cdot \mathbf{E}) = (\mathbf{p} \cdot \nabla)\mathbf{E}$$

acting on the dipole—its "center-of-charge coordinate". For the last equation, we have used $\nabla \times \mathbf{E} = \mathbf{0}$ and constant $\mathbf{p}$.

In addition, there is a *torque*

$$\mathbf{N} = \mathbf{p} \times \mathbf{E}\,,$$

if $\mathbf{p}$ and $\mathbf{E}$ do not have the same direction—then the potential energy is minimal (stable equilibrium)—or opposite directions (unstable equilibrium). (Note that the letter $\mathbf{M}$ common in classical mechanics is reserved for the magnetization in electromagnetism, and $\mathbf{r} \times \rho\mathbf{E} = \rho\,\mathbf{r} \times \mathbf{E}$.) The expression $\mathbf{p} \times \mathbf{E}$ supplies only the part expressible in relative coordinates. In addition, there is a part connected to $\mathbf{r}$, the "center-of-charge coordinate" (the position of the dipole, so far called $\mathbf{R}$), namely $\mathbf{r} \times (\mathbf{p} \cdot \nabla)\mathbf{E}$. Because $\mathbf{p} = (\mathbf{p} \cdot \nabla)\,\mathbf{r}$, the sum can also be combined to

$$\mathbf{N} = (\mathbf{p} \cdot \nabla)\,(\mathbf{r} \times \mathbf{E})\,.$$

However, in many cases, only the torque $\mathbf{p} \times \mathbf{E}$ with respect to the center of the dipole is of interest.

What field is generated by a dipole? To answer this question, we consider to begin with the potential of two point charges $\pm q'$ at the positions $\mathbf{r}'_{\pm} = \mathbf{r}' \pm \frac{1}{2}\mathbf{a}$ and investigate the limit $a \ll |\mathbf{r} - \mathbf{r}'_{\pm}|$. Since

$$\left|\mathbf{r} - \mathbf{r}' - \frac{1}{2}\mathbf{a}\,\right|^{-1} - \left|\mathbf{r} - \mathbf{r}' + \frac{1}{2}\mathbf{a}\,\right|^{-1} \approx -\mathbf{a} \cdot \nabla |\mathbf{r} - \mathbf{r}'|^{-1}\,,$$

we end up with

$$4\pi\varepsilon_0\,\Phi(\mathbf{r}) = -\mathbf{p}' \cdot \nabla \frac{1}{|\mathbf{r} - \mathbf{r}'|} = \frac{\mathbf{p}' \cdot (\mathbf{r} - \mathbf{r}')}{|\mathbf{r} - \mathbf{r}'|^3} = -\nabla \cdot \frac{\mathbf{p}'}{|\mathbf{r} - \mathbf{r}'|}\,.$$

Thus the scalar product of $\mathbf{p}'$ with the unit vector $\mathbf{e} \equiv (\mathbf{r} - \mathbf{r}')/|\mathbf{r} - \mathbf{r}'|$ from the source $\mathbf{r}'$ to the point $\mathbf{r}$ is important. The potential decays in inverse proportion to the square of the distance. The field strength $\mathbf{E} = -\nabla\Phi$ decays more strongly by one power of the distance:

$$4\pi\varepsilon_0\,\mathbf{E}(\mathbf{r}) = \nabla\left(\mathbf{p}' \cdot \nabla\frac{1}{|\mathbf{r} - \mathbf{r}'|}\right) = \frac{3\,\mathbf{p}'\cdot\mathbf{e}\,\mathbf{e} - \mathbf{p}'}{|\mathbf{r} - \mathbf{r}'|^3} - \frac{4\pi}{3}\,\mathbf{p}'\,\delta(\mathbf{r} - \mathbf{r}')\,.$$

An example is shown in Fig. 3.4. The last term appears because $|\mathbf{r} - \mathbf{r}'|^{-1}$ is discontinuous at $\mathbf{r} = \mathbf{r}'$. Thus, the volume integral around this point must still be considered (see Problem 3.8). For a point charge, a delta function appears at $\nabla \cdot \mathbf{E}$, thus ultimately with the derivative of the field strength—for the dipole this derivative is already included by taking the limit $a \to 0$. We usually only require the field outside the source, so this addition is not needed, but it does contribute to the average field, in particular, for N dipoles $\mathbf{p}'$ in a volume V with $\Delta\mathbf{E} = -\frac{1}{3}\varepsilon_0^{-1}\,N\mathbf{p}'/V$. We will take advantage of this in the next section, in the context of polarization.

But first we consider also dipole moments, which will be distributed evenly over a sheet df and lead to a dipole density $\mathbf{P}_A$. We then set $df\,\mathbf{P}_A = d\mathbf{f}\,P_A$. Note that

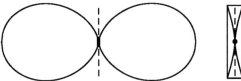

Fig. 3.4 Field lines of a dipole pointing along the *dashed symmetry axis*. *Right*: The field in the middle is magnified eight times. All other field lines are similar to the ones shown here, because point-like sources do not provide a length scale (see Problem 3.14)

P_A can also be negative, because we have already selected the direction of **df**, if the surface of a finite body is intended (see p. 9). In particular, **df** should then point outwards. We obtain the associated potential

$$\Phi(\mathbf{r}) = \frac{1}{4\pi\varepsilon_0} \int \frac{\mathrm{df}' \cdot (\mathbf{r} - \mathbf{r}')}{|\mathbf{r} - \mathbf{r}'|^3} P_A(\mathbf{r}') .$$

The fraction in the integrand gives the solid angle $d\Omega'$ subtended by the surface element df' at the point **r**, where the sign changes on crossing the surface. Therefore, upon crossing the dipole layer, the potential jumps by P_A/ε_0:

$$\Phi_+ - \Phi_- = \frac{P_A}{\varepsilon_0} ,$$

while, according to Sect. 3.1.2, upon penetrating a monopole layer, it is the field **E** that jumps, and hence the first derivative of Φ:

$$\mathbf{n} \cdot (\mathbf{E}_+ - \mathbf{E}_-) = -\mathbf{n} \cdot (\nabla\Phi_+ - \nabla\Phi_-) = \frac{\rho_A}{\varepsilon_0} .$$

We may therefore replace the boundary values on p. 170 by suitable mono and dipole densities on the surface of the considered volume. Then we have

$$\mathrm{df} \cdot \nabla\Phi = -\mathrm{d}f\, \mathbf{n} \cdot \mathbf{E} = \frac{\mathrm{d}f\, \rho_A}{\varepsilon_0} \quad \text{and} \quad -\mathrm{df}\, \Phi = \frac{\mathrm{d}f\, P_A}{\varepsilon_0} ,$$

if the potential (Φ_+) and the field ($\mathbf{E}_+$) vanish outside (see Fig. 3.5).

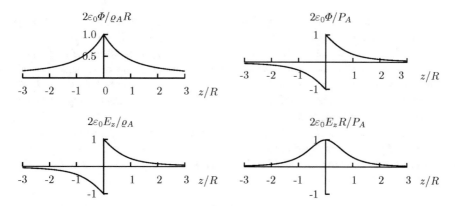

Fig. 3.5 Potential (*upper*) and field strength (*lower*) along the axes of circular disks of radius R. The disk is loaded with monopole charge density ρ_A (*left*) or with dipole charge density P_A (*right*), where the potential discontinuity at the disk also leads to a delta function. The curve *lower right* diverges at $z = 0$ (see Problems 3.18–3.19)

3.1.5 Polarization and Displacement Field

So far it has been presumed that the charge distribution is also known in the atomic interior. But in most cases such *microscopic* quantities are irrelevant. In *macroscopic* physics, knowledge of the average charge density is sufficient, if in addition the average density **P** of the dipole moments is also used. The average is taken over many atoms, as long as the volume ΔV of the averaging process is sufficiently small. For N molecules (ions) in ΔV with charges q_i and dipole moments $\mathbf{p}_i$, we have

$$\overline{\rho} \equiv \frac{1}{\Delta V} \sum_{i=1}^{N} q_i \quad \text{and} \quad \mathbf{P} \equiv \frac{1}{\Delta V} \sum_{i=1}^{N} \mathbf{p}_i \,.$$

P is called the (*electric*) *polarization*. According to the last section, it is associated with the potential

$$\Phi(\mathbf{r}) = -\frac{1}{4\pi\varepsilon_0} \nabla \cdot \int dV' \, \frac{\mathbf{P}(\mathbf{r}')}{|\mathbf{r} - \mathbf{r}'|} = \frac{1}{4\pi\varepsilon_0} \int dV' \, \mathbf{P}(\mathbf{r}') \cdot \nabla' \frac{1}{|\mathbf{r} - \mathbf{r}'|} \,.$$

The last expression can be rewritten. Since

$$\mathbf{P}(\mathbf{r}') \cdot \nabla' \frac{1}{|\mathbf{r} - \mathbf{r}'|} = \nabla' \cdot \frac{\mathbf{P}(\mathbf{r}')}{|\mathbf{r} - \mathbf{r}'|} - \frac{\nabla' \cdot \mathbf{P}}{|\mathbf{r} - \mathbf{r}'|} \,,$$

Gauss's theorem yields

$$\Phi(\mathbf{r}) = \frac{1}{4\pi\varepsilon_0} \int d\mathbf{f}' \cdot \frac{\mathbf{P}(\mathbf{r}')}{|\mathbf{r} - \mathbf{r}'|} - \frac{1}{4\pi\varepsilon_0} \int dV' \frac{\nabla' \cdot \mathbf{P}}{|\mathbf{r} - \mathbf{r}'|} \,.$$

A polarized medium then has the same field as the one due to the surface charge density $\rho_A' = \mathbf{n} \cdot \mathbf{P}$ and the space charge density $\rho' = -\nabla \cdot \mathbf{P}$. The minus sign is easy to understand. If we assume a rod of homogeneous polarization, then a positive charge results just there on its surface where the polarization has a sink. For ρ', we sometimes speak of *apparent charges*, because they actually belong to dipole moments and are not freely mobile. This concept is somewhat misleading, however, since the apparent charges do exist microscopically.

   If we integrate over the total space, then the surface integral does not contribute, since there is no matter at infinity. Clearly, we may replace the microscopic ("true") charge density ρ by the average $\overline{\rho}$ and the charge density $\rho' = -\nabla \cdot \mathbf{P}$ of the polarization:

$$\rho = \overline{\rho} - \nabla \cdot \mathbf{P} \,.$$

From the basic equation $\rho = \nabla \cdot \varepsilon_0 \mathbf{E}$ of microscopic electrostatics, we can in this way infer $\overline{\rho} = \nabla \cdot (\varepsilon_0 \mathbf{E} + \mathbf{P})$. Therefore, we introduce the *electric displacement field* (*displacement*) **D**, defined by

$$\mathbf{D} \equiv \varepsilon_0 \mathbf{E} + \mathbf{P} \, ,$$

and obtain as basic equations of macroscopic electrostatics

$$\nabla \cdot \mathbf{D} = \overline{\rho} \, , \qquad \nabla \times \mathbf{E} = \mathbf{0} \, .$$

The electric field remains irrotational because, according to our derivation, we may later calculate with a scalar potential. These basic equations also yield

$$\mathbf{n} \cdot (\mathbf{D}_+ - \mathbf{D}_-) = \overline{\rho}_A \, , \qquad \mathbf{n} \times (\mathbf{E}_+ - \mathbf{E}_-) = \mathbf{0} \, ,$$

and

$$\int_{(V)} d\mathbf{f} \cdot \mathbf{D} = Q \, , \qquad \int_{(A)} d\mathbf{r} \cdot \mathbf{E} = 0 \, .$$

Like the polarization $\mathbf{P}$, the electric displacement field $\mathbf{D}$ has the unit C/m^2.

So far we have viewed the dipole moments as being given. There are indeed molecules with permanent electric dipole moments; by allusion to paramagnetism, they are said to be *paraelectric*. But if no external field is applied and if the temperature is high, then the polarization averages to zero because of the disorder in the directions. The orientation increases with increasing field strength—and decreasing temperature. In addition, an electric field also shifts the charges in originally non-polar atoms and *induces* an electric dipole moment. In both cases, to a first approximation, $\mathbf{P}$ depends linearly upon $\mathbf{E}$:

$$\mathbf{P} = \chi_e \, \varepsilon_0 \mathbf{E} \, .$$

The *electric susceptibility* χ_e is a mere number. It is related to the *polarizability* α of the various molecules. To this end, we assume N equal molecules in the volume V, whence $\mathbf{P} = n\mathbf{p}$ with $n \equiv N/V$. Each individual molecule becomes polarized by the electric field $\mathbf{E}_0$ at its location, $\mathbf{p} = \alpha \varepsilon_0 \mathbf{E}_0$. In doing this, according to the last section, we assume that $\mathbf{E}_0$ differs from the average field strength $\mathbf{E}$ by $\frac{1}{3}\varepsilon_0^{-1} n\mathbf{p} = \frac{1}{3}\varepsilon_0^{-1}\mathbf{P}$:

$$\mathbf{P} = n\alpha \, (\varepsilon_0 \mathbf{E} + \tfrac{1}{3}\mathbf{P}) \qquad \Longrightarrow \qquad \mathbf{P} = \frac{n\alpha}{1 - n\alpha/3} \, \varepsilon_0 \mathbf{E} \, .$$

We deduce the *formula due to Clausius and Mosotti*, viz.,

$$\chi_e = \frac{n\alpha}{1 - n\alpha/3} \qquad \Longleftrightarrow \qquad \alpha = \frac{3}{n} \frac{\chi_e}{\chi_e + 3} \, ,$$

which has been derived here following [1].

However, in crystals there may be preferential directions, such that $\mathbf{P}$ is not then parallel to $\mathbf{E}$. In this case, χ_e is a symmetric tensor of second rank (an anti-symmetric part would supply an additional term $\mathbf{P}_a = \boldsymbol{\chi}_e \times \varepsilon_0 \mathbf{E}$ with a suitable vector $\boldsymbol{\chi}_e$, which contradicts the above-mentioned explanation for the polarization: even if there were microscopic screw axes, the polarization would nevertheless average out) with three *principal dielectric axes*, along which $\mathbf{P}$ is parallel to $\mathbf{E}$, but P/E is still different. There are also *ferroelectric* materials—in these a permanent polarization appears even when the field is switched off, and the dipole moments do not average out. In addition, χ_e does not remain constant at high fields because there are non-linear saturation effects. We will not go into all these special cases.

For the electric displacement field, it thus follows that

$$\mathbf{D} = (1 + \chi_e)\, \varepsilon_0 \mathbf{E} \equiv \varepsilon \mathbf{E}\,,$$

with the *permittivity (dielectric constant)* ε and the *relative dielectric constant* $\varepsilon_r \equiv \varepsilon/\varepsilon_0 = 1 + \chi_e$. This depends upon the temperature and for water is unusually high, namely equal to 80 at 20 °C and 55 at 100 °C. In crystals, ε is generally a (symmetric) tensor of second rank.

We will now always consider the two fields simultaneously, i.e., the electric field strength determined by the force $\mathbf{F} = q\,\mathbf{E}$ acting on a test charge q, and the electric field $\mathbf{D}$ given by the average charge density. When we do this, the relation between $\mathbf{D}$ and $\mathbf{E}$ is important ($\mathbf{D} = \varepsilon \mathbf{E}$), and we will restrict ourselves to scalar permittivities.

3.1.6 Field Equations in Electrostatics

In the following we will restrict ourselves to macroscopically measurable quantities and, following the usual practice, omit the bar over the charge density. Thus we start from the basic equations

$$\nabla \cdot \mathbf{D} = \rho\,, \qquad \nabla \times \mathbf{E} = \mathbf{0}\,, \qquad \text{and} \qquad \mathbf{D} = \varepsilon \mathbf{E}\,,$$

and consider now different cases.

Insulators do not contain mobile charges, and therefore in their interior we have

$$\nabla \cdot \mathbf{D} = 0\,, \qquad \nabla \times \mathbf{E} = \mathbf{0}\,.$$

Since they can be polarized, we have to distinguish carefully between $\mathbf{D}$ and $\mathbf{E}$. According to the second equation, we may replace the field strength $\mathbf{E}$ by $-\nabla\Phi$ and, using $\mathbf{D} = \varepsilon \mathbf{E}$ from the first equation, we obtain

$$\nabla \cdot \varepsilon \nabla \Phi = 0\,.$$

In particular, for constant permittivity, we obtain the *Laplace equation*

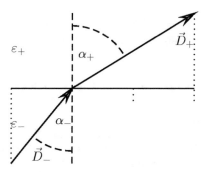

Fig. 3.6 Diffraction of the electric field entering into an insulator of higher permittivity (here $\varepsilon_+ = 2\varepsilon_-$). The force lines become diffracted away from the normal. In contrast, according to the optical *diffraction law* (see p. 220) the rays become diffracted towards the normal for $n_+ > n_-$ and, instead of $\tan\alpha$, we have in that case $\sin\alpha$

$$\Delta\Phi = 0 \,.$$

The boundary values then become physically decisive.

For two-dimensional problems, analytical functions in the complex plane are useful. A function $f(z) = \Phi(x,y) + i\Psi(x,y)$ is only differentiable if, regardless of the direction of approach,

$$\frac{\partial f}{\partial x} = \frac{\partial f}{i\partial y} \qquad \Longrightarrow \qquad \frac{\partial\Phi}{\partial x} + \frac{i\partial\Psi}{\partial x} = \frac{\partial\Phi}{i\partial y} + \frac{i\partial\Psi}{i\partial y} \,,$$

i.e., if the *Cauchy–Riemann equations* are satisfied:

$$\frac{\partial\Phi}{\partial x} = \frac{\partial\Psi}{\partial y} \qquad \text{and} \qquad \frac{\partial\Phi}{\partial y} = -\frac{\partial\Psi}{\partial x} \,.$$

These lead to the Laplace equations $\Delta\Phi = 0$ and $\Delta\Psi = 0$, and thus to $\nabla\Phi \cdot \nabla\Psi = 0$. If the entity $\{\Phi = \text{const.}\}$ represents equipotential lines, then the other entity $\{\Psi = \text{const.}\}$ represents field lines. For example, Fig. 1.4 corresponds to $f = z^2$.

At the *interface between insulators*, the normal components of **D** and the tangential components of **E** are continuous because

$$\mathbf{n} \cdot (\mathbf{D}_+ - \mathbf{D}_-) = 0 \,, \qquad \mathbf{n} \times (\mathbf{E}_+ - \mathbf{E}_-) = \mathbf{0} \,.$$

Hence, for scalar permittivity, it follows from $|\mathbf{n} \times \mathbf{E}_+|/|\mathbf{n} \cdot \mathbf{D}_+| = |\mathbf{n} \times \mathbf{E}_-|/|\mathbf{n} \cdot \mathbf{D}_-|$ that $\sin\alpha_+/(\varepsilon_+ \cos\alpha_+) = \sin\alpha_-/(\varepsilon_- \cos\alpha_-)$, where α is the angle between the field vector and the normal vector. Hence (see Fig. 3.6),

$$\frac{\tan\alpha_+}{\tan\alpha_-} = \frac{\varepsilon_+}{\varepsilon_-} \,.$$

In *homogeneous conductors*, charges can move freely. Therefore, for static equilibrium the field strength in the interior of the conductor must vanish, and with it also the polarization:

$$\mathbf{E} = \mathbf{D} = \mathbf{0} \,, \quad \text{and thus} \quad \Phi = \text{const.}$$

in the interior of homogeneous conductors.

At the *interface between insulator and conductor*, there may be surface charges, but no fields within the conductor. Therefore, the electric field lines in the insulator end perpendicularly at the interface:

$$\mathbf{n} \cdot \mathbf{D}_I = \rho_A \,, \quad \mathbf{n} \times \mathbf{E}_I = \mathbf{0} \,.$$

The subscript on $\mathbf{E}_I$ reminds us that we are considering the insulator, but it is not actually needed, since the fields vanish within the conductor.

At the *interface between two conductors*, the potential has a discontinuity, since their conduction electrons generally have different work functions. Upon contact between the two metals, charges move into the more strongly binding regime, until a corresponding counter-field has built up. Only then is there a static situation. Thus we find a *contact voltage*. The situation for the immersion of a metal in an electrolyte is similar, e.g., immersion of a copper rod in sulfuric acid, where some Cu^{++} ions become dissolved and hence a current flows until the negative loading of the rod has built up an electric counter-field.

All these fields caused by inhomogeneities are said to be *induced*, since they do not originate from an external charge, but from the structure of the material. We denote the induced field strength (as do Panofsky & Phillips) by $\mathbf{E}'$; another common notation is $\mathbf{E}^{(e)}$. In electrostatics, we have

$$\mathbf{E} + \mathbf{E}' = \mathbf{0}$$

in inhomogeneous conductors—in static equilibrium the induced field strength is canceled by the counter-field.

3.1.7 Problems in Electrostatics

In most cases, the field $\mathbf{E}(\mathbf{r})$ in the insulator is to be determined for a given form and position of the conductors and with a further requirement: we are given either the voltage

$$U = \Phi_1 - \Phi_0 = -\int_{\mathbf{r}_0}^{\mathbf{r}_1} d\mathbf{r} \cdot \mathbf{E}$$

between the conductors 0 and 1—where arbitrary initial and final points on the conductors may be taken (because the potential on each conductor is constant) and any path in-between, because the field is irrotational—or the charges

$$Q_i = \int_{A_i} df \, \rho_A = \int_{A_i} d\mathbf{f} \cdot \varepsilon \mathbf{E} = - \int_{A_i} d\mathbf{f} \cdot \varepsilon \nabla \Phi$$

on the conductor surfaces A_i. For two conductors with charges $Q > 0$ and $-Q$ and the voltage $U > 0$, Q and U are related to each other via a geometrical quantity, namely the *capacity*

$$C \equiv \frac{Q}{U} = \left| \frac{\int d\mathbf{f} \cdot \varepsilon \mathbf{E}}{\int d\mathbf{r} \cdot \mathbf{E}} \right|.$$

The best approach here is to solve the problem using Gauss's theorem or using the Laplace equation, and to adapt the coordinates to the boundary geometry. In the following, we consider some examples whose solution can be easily anticipated.

The *spherical capacitor* is a conducting sphere with charge Q and radius r_K in a comparably large (non-conducting) dielectric with scalar permittivity ε. This has a spherically symmetric field, which jumps from 0 to its maximum value at the charged surface—viewed from outside the sphere, it could also originate from a point charge at the center of the sphere:

$$\Phi(r) = \begin{cases} U, \\ U\, r_K/r, \end{cases} \qquad \mathbf{E} = \begin{cases} \mathbf{0} & \text{for } r < r_K, \\ U\, r_K\, \mathbf{r}/r^3 & \text{for } r > r_K, \end{cases}$$

with $U = Q/C$ and $Q = \int d\mathbf{f} \cdot \mathbf{D} = 4\pi r_K^2\, \varepsilon E(r_{K+}) = 4\pi \varepsilon\, U\, r_K$, whence the capacity is $C = 4\pi \varepsilon\, r_K$. (The potential has a kink at the charged surface.)

As a *cylindrical capacitor*, we take two coaxial conducting cylinders of length l, separated by a dielectric with scalar permittivity ε. If the inner cylinder (with radius R_i) carries the charge Q and the outer cylinder (with radius R_a) the charge $-Q$, then for $l \gg R_a$, the contribution of the cylinder ends may be neglected. Then in the dielectric there is a field strength decaying as R^{-1} which is the solution of Gauss's theorem $Q = \int d\mathbf{f} \cdot \varepsilon \mathbf{E}$, since the area of the inner cylinder walls is $A = 2\pi R_i\, l$, and $Q = 2\pi R\, l\, \varepsilon E(R)$ and $\Phi \propto \ln(R/R_a)$ in the capacitor:

$$\Phi(R) = \begin{cases} U, \\ U\, \dfrac{\ln(R/R_a)}{\ln(R_i/R_a)}, \\ 0, \end{cases} \qquad \mathbf{E} = \begin{cases} \mathbf{0} & \text{for} \quad R < R_i, \\ \dfrac{U}{\ln(R_a/R_i)} \dfrac{\mathbf{R}}{R^2} & \text{for } R_i < R < R_a, \\ \mathbf{0} & \text{for } R_a < R, \end{cases}$$

noting that $-\ln(R_i/R_a) = \ln(R_a/R_i)$. Hence we find $Q = 2\pi R_i\, l\, \varepsilon\, U / \{R_i \ln(R_a/R_i)\}$ and then $C = 2\pi \varepsilon\, l / \ln(R_a/R_i)$. For conductors (with the very small distance $d \equiv R_a - R_i \ll R_i$ and area $A = 2\pi R_i l$ for the inner conductor), and since $\ln(R_a/R_i) = \ln(1 + d/R_i) \approx d/R_i$, we may replace this by

$$C \approx \varepsilon \, \frac{A}{d} \,, \qquad E \approx \frac{|U|}{\mathrm{d}} \approx \frac{|Q|}{\varepsilon \, A} \,.$$

These equations are also valid for the *plate capacitor*, if boundary effects may be neglected.

When capacitors with capacities C_k are connected in parallel, the total capacity $C = Q/U = \sum C_k$, because $U = U_k$ and $Q = \sum_k Q_k$. For capacitors connected in series, we have $1/C = \sum_k 1/C_k$, because now $Q = Q_k$ and $U = \sum_k U_k = \sum_k Q/C_k$.

For a *point charge q at a distance a in front of a conducting plane* the field lines must end perpendicularly on the plane and must be irrotational in front of it. In order to find the field distribution, we imagine an *image charge* $-q$ at the same distance a behind the conductor surface—the field of the two point charges is shown on the left in Fig. 3.3. The total field of the two point charges satisfies the conditions in front of the plane. Hence, if we choose the center of this configuration as the origin, so that q is at $\mathbf{a}$ and $-q$ at $-\mathbf{a}$, we have been

$$\mathbf{E} = \frac{q}{4\pi\varepsilon} \left(\frac{\mathbf{r} - \mathbf{a}}{|\mathbf{r} - \mathbf{a}|^3} - \frac{\mathbf{r} + \mathbf{a}}{|\mathbf{r} + \mathbf{a}|^3} \right), \quad \text{for } \mathbf{r} \cdot \mathbf{a} > 0 \,, \quad \text{otherwise } \mathbf{0} \,.$$

This field is irrotational and has a source in front of the interface only at the position of the point charge. On the interface, $\mathbf{r} \cdot \mathbf{a} = 0$ holds, so $|\mathbf{r} \pm \mathbf{a}|^3 = (r^2 + a^2)^{3/2}$, and hence,

$$\mathbf{E} = -\frac{q}{2\pi\varepsilon} \, \frac{\mathbf{a}}{(r^2 + a^2)^{3/2}} \,.$$

Therefore, $\mathbf{E}$ is perpendicular to the plane as required. Behind the mirror there is no field. Therefore, we replace the imagined image charge now by a surface charge $\rho_A = \mathbf{n} \cdot \mathbf{D}$ on the plane, precisely in the sense of the last paragraph of Sect. 3.1.3. The image charge is replaced by an *induced charge* on the conductor surface. The total induced charge is, according to the last two equations, equal to the image charge:

$$\int \mathrm{d}f \, \rho_A = \int \mathrm{d}\mathbf{f} \cdot \varepsilon \mathbf{E} = -\frac{qa}{2\pi} \int_0^\infty \frac{2\pi \, R \, \mathrm{d}R}{(R^2 + a^2)^{3/2}} = \frac{qa}{\sqrt{R^2 + a^2}} \bigg|_0^\infty = -q \,.$$

Of course, the total charge of the conductor must be conserved. We have to imagine a charge $+q$ at infinity. For conductors of finite extension, it is important to know whether they are isolated or grounded—if necessary the image charge has to be neutralized by a further charge, e.g., for an ungrounded sphere, the additional charge has to be spread evenly over the surface.

With the help of image charges, the fields of other charged interfaces can be represented, e.g., for a conducting sphere (Problem 3.20) or for a separating plane to a non-conductor with a different permittivity. But then we have to calculate with different charges $q \neq q'$—the field-line pictures in the half-space inside the conductor are similar to those on the right in Figs. 3.2 or 3.3—and in the half-space of the other non-conductor, the field of a new source appears at the original position.

Each test charge leads to induced charges on the surrounding conductors and thus changes the field to be determined. Since this induction should remain negligible, the test charge must therefore be very small in comparison with the other charges. However, this is not possible for small distances because the induced charge is then very highly concentrated. Therefore, we may apply our concepts only to macroscopic objects.

If the microscopic charge density $\rho(\mathbf{r})$ is given, the potential and field strength follow from the Poisson equation $\Delta\Phi = -\rho/\varepsilon_0$ or the integral $\int dV' \, \rho(\mathbf{r}')/|\mathbf{r} - \mathbf{r}'|$. Here, we would like to separate the variables $\mathbf{r}$ and $\mathbf{r}'$. This is managed by *expanding in terms of Legendre polynomials* (see p. 81):

$$\frac{1}{|\mathbf{r} - \mathbf{r}'|} = \frac{1}{r} \sum_{n=0}^{\infty} P_n(\cos\theta) \left(\frac{r'}{r}\right)^n , \quad \text{for } r' < r \text{ and } \cos\theta \equiv \frac{\mathbf{r} \cdot \mathbf{r}'}{r \, r'} .$$

According to this, and in particular, for positions outside the field-creating charges, we may set

$$\Phi(\mathbf{r}) = \frac{1}{4\pi\varepsilon_0} \sum_{n=0}^{\infty} \frac{1}{r^{n+1}} \int dV' \, \rho(\mathbf{r}') \, r'^n \, P_n(\cos\theta') .$$

Upon integration, the angle between $\mathbf{r}$ and $\mathbf{r}'$ changes, so we have written here $\cos\theta'$. For $n = 0$, the integral supplies the charge Q' because $P_0 = 1$. (In Sect. 2.2.7 we integrated over the mass density and hence obtained the mass.) The next integral leads to $\mathbf{p}' \cdot \mathbf{r}/r$, because $r' P_1(\cos\theta') = \mathbf{r}' \cdot \mathbf{r}/r$, and the dipole moment is thus important. Generally, the integrals appearing here are called *multipole moments*. (According to Sect. 2.2.7, we have $(n + 1) P_{n+1}(z) - (2n + 1) z \, P_n(z) + n \, P_{n-1}(z) = 0$. However, numerical factors are often added to the multipole moments.) For a dipole of finite extension $(a \neq 0)$, there is, e.g., an additional octupole moment, but its influence decreases faster with the distance than that of the dipole moment (Problem 3.15).

Apart from the *spherical multipoles* of order 2^n just mentioned, there are (e.g., in ion optics) *axial multipoles* of order $2n$. In suitable cylindrical coordinates, their potentials are proportional to $R^n \cos(n\varphi)$.

3.1.8 Energy of the Electrostatic Field

The electric field carries energy, because according to p. 169, work is required to load a capacitor, i.e., the work $dW = U \, dQ$ to move the charge $dQ > 0$ from the cathode to the anode. Because $Q = C \, U$, if we let the voltage—or charge—increase from zero to its final value, we obtain

$$W = \tfrac{1}{2} C U^2 = \frac{Q^2}{2C}$$

for the energy stored in the capacitor.

Since $\rho = \nabla \cdot \mathbf{D}$ and therefore $\rho \Phi = \nabla \cdot (\Phi \mathbf{D}) - \mathbf{D} \cdot \nabla \Phi$, the expression $W = \tfrac{1}{2} Q U = \tfrac{1}{2} Q \Delta \Phi = \tfrac{1}{2} \int dV \, \rho \Phi$ can be rewritten. According to Gauss, the first term supplies a surface integral which vanishes at infinity, since $\Phi \mathbf{D}$ approaches zero as r^{-3}, and we obtain

$$W = \tfrac{1}{2} \int dV \, \rho \Phi = \tfrac{1}{2} \int dV \, \mathbf{D} \cdot \mathbf{E} \, .$$

The contribution to the last integral comes from all space containing fields, but the contribution to $\int dV \, \rho \Phi$, only from space containing charge. However, the *energy density* is only

$$w = \tfrac{1}{2} \mathbf{D} \cdot \mathbf{E} \, ,$$

because, subdividing the space, the interfaces should not contribute—and $\rho \Phi$ depends upon the gauge. In the sense of thermodynamics (see p. 575), we are dealing with the density of the *free* (fully usable) energy F. Temperature and volume here are the natural variables. The permittivity ε generally depends upon the temperature and the distances between the molecules. However, we follow the general custom and write w and not f. (The symbol u is often also used, but this is misleading, since U means the inner energy and not the free—fully exploitable—energy.)

Since $\mathbf{D} = \varepsilon_0 \mathbf{E} + \mathbf{P}$, the energy density is composed of two parts. Firstly, the field energy $\tfrac{1}{2} \varepsilon_0 \mathbf{E} \cdot \mathbf{E}$ "in vacuum", and secondly, the contribution $\tfrac{1}{2} \mathbf{P} \cdot \mathbf{E}$ from the dielectric—because according to p. 175 the dipole moment of polarizable molecules increases linearly with the applied field strength and requires the work $\int_0^{\mathbf{P}} \mathbf{E} \cdot d\mathbf{p} = \tfrac{1}{2} \mathbf{p} \cdot \mathbf{E}$. (This derivation succeeds only for $\mathbf{P} \propto \mathbf{E}$.)

3.1.9 Maxwell Stress Tensor in Electrostatics

Forces are transmitted from one space element to the next, in which case we can speak of near-action forces. Here this must also be true for empty space, because electric forces permeate even empty space. We expect space filled with fields to behave as an elastic medium, and this property is described by the Maxwell stress tensor.

In a continuous medium, the force $\mathbf{F}$ can be derived from a force density f, which will be denoted in this section by $\mathbf{f}$. For surface elements, we write $\mathbf{n} \, df$ or $\mathbf{g}^k \, df_k$. Then,

$$\mathbf{F} = \int dV \, \mathbf{f}(\mathbf{r}) \, .$$

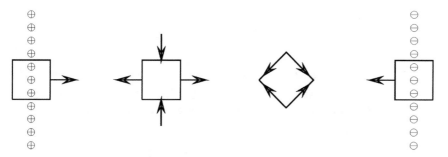

Fig. 3.7 Visualization of the Maxwell stress tensor for a homogeneous electric field (the field strength points from the ⊕ to the ⊖ charges). Shown are the surface tensions on four cubes—depending on their charge, the forces on opposite faces either cancel or supply the expected force (see Problems 3.21–3.22)

We decompose the force $\mathbf{f}\, dV$ acting on an infinitesimal cube $dV = dx\, dy\, dz$ into surface element times *surface tension* σ. For tensile and compressive forces, there are *normal stresses* perpendicular to the surface, while for shear forces, there are *shear stresses* on the surface (see Fig. 3.7). The mechanical stress (Latin *tensio*) is described by a *tensor* σ, viz.,

$$
dV\, f_x = \begin{pmatrix} dy\, dz\ \{\sigma_{xx}(x + dx, y, z) - \sigma_{xx}(x, y, z)\} \\ + dz\, dx\ \{\sigma_{xy}(x, y + dy, z) - \sigma_{xy}(x, y, z)\} \\ + dx\, dy\ \{\sigma_{xz}(x, y, z + dz) - \sigma_{xz}(x, y, z)\} \end{pmatrix}
$$
$$
= dV \left(\frac{\partial \sigma_{xx}}{\partial x} + \frac{\partial \sigma_{xy}}{\partial y} + \frac{\partial \sigma_{xz}}{\partial z} \right).
$$

The force density $\mathbf{f}$ is thus equal to the source density of the *stress tensor* σ—so far we have considered only divergences of vectors and have obtained scalars. According to Gauss's theorem, the volume integral of $\mathbf{f}$ can be converted into a surface integral. Adjacent interfaces do not contribute, provided that σ is continuous.

The stress tensor is useful in continuum mechanics, where near-action forces are assumed. Therefore, we use it now in electromagnetism. However, here we shall restrict ourselves to homogeneous matter with constant permittivity, since otherwise the problem is much more involved (see [2]). We thus start from

$$
\mathbf{F} = \int dV\, \rho\, \mathbf{E} = \int dV\, \mathbf{E}\, \nabla \cdot \mathbf{D}.
$$

In order to convert the integrand into the source density of a tensor, we use Sect. 1.1.8, adding $-\mathbf{D} \times (\nabla \times \mathbf{E})$ to the integrand. It does not contribute in electrostatics, since the field is irrotational—and the same procedure holds in magnetostatics, but where the field is solenoidal (source-free) and curls (vortices) appear instead.

For *rectilinear* (possibly oblique) coordinates, according to pp. 33–40 with fixed vectors $\mathbf{g}^i$, we find

$$\mathbf{E}\,\nabla\cdot\mathbf{D}-\mathbf{D}\times(\nabla\times\mathbf{E})=\sum_{ik}\mathbf{g}^{\,i}\left(\frac{\partial E_i\,D^k}{\partial x^k}-D^k\,\frac{\partial E_k}{\partial x^i}\right).$$

using $\mathbf{a}\times\mathbf{b}=\sum_{ikl}\mathbf{g}^i a^k b^l \varepsilon_{ikl}$ and the two equations $\nabla\cdot\mathbf{a}=\sum_k\frac{\partial a^k}{\partial x^k}$ and $(\nabla\times\mathbf{a})^i=\sum_{kl}\varepsilon^{ikl}\frac{\partial a_l}{\partial x^k}$, as well as the identity $\sum_l \varepsilon_{ikl}\varepsilon^{lmn}=\delta_i^m\delta_k^n-\delta_k^m\delta_i^n$. In addition, for homogeneous matter, i.e., an invariant permittivity tensor, we have

$$\sum_k D^k\,\frac{\partial E_k}{\partial x^i}=\sum_{kl}\varepsilon^{kl} E_l\,\frac{\partial E_k}{\partial x^i}=\sum_l E_l\,\frac{\partial D^l}{\partial x^i}=\sum_k\frac{\partial\frac{1}{2}E_k D^k}{\partial x^i}=\frac{\partial w}{\partial x^i}\,,$$

and thus $\sum_i \mathbf{g}^{\,i}\frac{\partial w}{\partial x^i}=\sum_k \mathbf{g}^k\frac{\partial w}{\partial x^k}=\sum_{ik}\mathbf{g}_i\,g^{ik}\frac{\partial w}{\partial x^k}$. Together these imply

$$\mathbf{E}\,\nabla\cdot\mathbf{D}-\mathbf{D}\times(\nabla\times\mathbf{E})=\sum_{ik}\mathbf{g}_i\,\frac{\partial}{\partial x^k}\left(E^i D^k-\tfrac{1}{2}g^{ik}\,\mathbf{E}\cdot\mathbf{D}\right).$$

Therefore, we introduce the *Maxwell stress tensor* (in rectilinear coordinates):

$$T^{ik}\equiv w\,g^{ik}-E^i\,D^k\,.$$

(In Cartesian coordinates, it has the trace $\mathrm{tr}T=3w-\mathbf{E}\cdot\mathbf{D}=w\geq 0$, and the trace does not depend upon the choice of coordinates. Some authors use it with opposite sign, but we shall see in Sect. 3.4.11 that this has disadvantages. In the form mentioned here, it is symmetric only for scalar permittivity. See also other forms discussed by Brevik.) With this, we find

$$\mathbf{f}+\sum_{ik}\mathbf{g}_i\,\frac{\partial T^{ik}}{\partial x^k}=\mathbf{0}\,.$$

Consequently, the force density vector $\mathbf{f}$ is related to the divergence of the tensor T.

Since we work with position-independent basic vectors $\mathbf{g}_i$ and may employ Gauss's theorem ($\mathrm{d}V=\mathrm{d}f_k\,\mathrm{d}x^k$ according to p. 38), we also find

$$\mathbf{F}+\sum_{ik}\mathbf{g}_i\int_{(V)}\mathrm{d}f_k\,T^{ik}=\mathbf{0}\,.$$

We can thus form Maxwell's stress tensor T^{ik} from the field strength, which expresses the force on a volume due to surface forces. Its diagonal elements supply the compressive or tensile stress on the surface pair with equal index, and its off-diagonal elements supply the shear stress on the remaining surface pairs.

3.1.10 Summary: Electrostatics

In electrostatics, we investigate the effects of charges Q and charge densities ρ at rest. All phenomena can be derived from Coulomb's law. It supplies the force between two point charges q and q' in vacuum:

$$\mathbf{F} = \frac{qq'}{4\pi\varepsilon_0} \frac{\mathbf{r} - \mathbf{r}'}{|\mathbf{r} - \mathbf{r}'|^3} \ .$$

From this action-at-a-distance law, we derived a field theory. We conceived of a test charge q and introduced a field strength $\mathbf{E}$:

$$\mathbf{F} = q(\mathbf{r})\, \mathbf{E}(\mathbf{r}) \ , \quad \text{with} \quad \boldsymbol{\nabla} \times \mathbf{E} = \mathbf{0} \quad \text{and} \quad \boldsymbol{\nabla} \cdot \mathbf{E} = \frac{\rho}{\varepsilon_0} \ .$$

However, the last equation is true only with the microscopic measurable charge density ρ, not with the macroscopic charge density $\overline{\rho}$, which accounts for freely moving charges.

We get round this difficulty by introducing dipole moments $\mathbf{p}$ and their density. This leads to the macroscopic concept of polarization $\mathbf{P}$. Its action on a test charge can be described by a charge density $-\boldsymbol{\nabla} \cdot \mathbf{P}$. With the electric displacement field

$$\mathbf{D} \equiv \varepsilon_0 \mathbf{E} + \mathbf{P} \ ,$$

we thereby obtain the source equation

$$\boldsymbol{\nabla} \cdot \mathbf{D} = \overline{\rho} \ .$$

If the connection between the field strength $\mathbf{E}$ and displacement field $\mathbf{D}$ is known, the field can be determined. Maxwell's equations of electrostatics read

$$\boldsymbol{\nabla} \times \mathbf{E} = \mathbf{0} \ , \quad \boldsymbol{\nabla} \cdot \mathbf{D} = \overline{\rho} \ , \quad \text{and} \quad \mathbf{D} = \varepsilon \mathbf{E} \ .$$

It is common to denote the measurable charge density by ρ. We will therefore omit the bar in the following. The first row of these equations yields

$$\int_{(A)} d\mathbf{r} \cdot \mathbf{E} = 0 \quad \text{and} \quad \int_{(V)} d\mathbf{f} \cdot \mathbf{D} = Q \ ,$$

and also

$$\mathbf{n} \times (\mathbf{E}_+ - \mathbf{E}_-) = \mathbf{0} \quad \text{and} \quad \mathbf{n} \cdot (\mathbf{D}_+ - \mathbf{D}_-) = \rho_A \ .$$

Since $\mathbf{E}$ is irrotational, this vector field can be attributed to a scalar potential Φ, with which calculations are greatly simplified:

$$\mathbf{E} = -\nabla \Phi , \qquad \nabla \cdot (\varepsilon \nabla \Phi) = -\rho .$$

Then for constant and scalar permittivity, the Poisson equation follows:

$$\Delta \Phi = -\frac{\rho}{\varepsilon} .$$

Here, a boundary condition is appended, namely that the potential should vanish at infinity—and that there should be no charge there. Instead of that, conditions may be introduced at the surface of the considered volumes.

In electrostatics, there are no fields in homogeneous conductors. Only at their interface with insulators are charges possible, and this supplies the boundary conditions:

$$\rho_A = \mathbf{n} \cdot \mathbf{D}_\mathrm{I} = -\mathbf{n} \cdot \varepsilon_\mathrm{I} \nabla \Phi , \qquad \mathbf{0} = \mathbf{n} \times \mathbf{E}_\mathrm{I} = -\mathbf{n} \times \nabla \Phi .$$

Here the index I refers to the adjacent insulator.

3.2   Stationary Currents and Magnetostatics

3.2.1   Electric Current

So far we have restricted ourselves to charges and dipole moments at rest. We shall now discard this restriction. We let the charges move and use the concept of *current density*:

$$\mathbf{j} \equiv \rho \mathbf{v} .$$

We call the current flux through the cross-section A of a conductor the *current strength*:

$$I \equiv \int_A \mathrm{d}\mathbf{f} \cdot \mathbf{j} .$$

For a cross-section that is small compared to the other dimensions of the conductor, we often replace

$$\mathbf{j}\, \mathrm{d}V \to I\, \mathrm{d}\mathbf{r} ,$$

where $\mathrm{d}\mathbf{r}$ is in the direction of $\mathbf{j}$, since $\mathrm{d}V\, \mathbf{j} \to \mathrm{d}\mathbf{r} \cdot \mathrm{d}\mathbf{f}\, \mathbf{j} = \mathrm{d}\mathbf{r}\, \mathrm{d}\mathbf{f} \cdot \mathbf{j}$.

There is a *conservation law for electric charges*: if the charge Q in a time-independent volume V changes, then it must flow through the surface of V. (We can also state that the volume associated with Q has changed—but this serves no purpose here.) Therefore, $dQ/dt \equiv \int_V dV\, \partial\rho/\partial t = -\int_{(V)} d\mathbf{f} \cdot \rho\mathbf{v}$. Note that the vector normal to the cross-section points to the outside. If this is true for $\rho\mathbf{v}$, then positive charge flows out, hence the minus sign. With Gauss's theorem, viz., $\int_{(V)} d\mathbf{f} \cdot \mathbf{j} = \int_V dV\, \nabla \cdot \mathbf{j}$, we have the *continuity equation*

$$\frac{\partial\rho}{\partial t} + \nabla \cdot \mathbf{j} = 0 ,$$

and for surface charges ρ_A,

$$\frac{\partial\rho_A}{\partial t} + \mathbf{n} \cdot (\mathbf{j}_+ - \mathbf{j}_-) = 0 ,$$

where $\mathbf{n}$ is again the unit vector perpendicular to the element of the cross-section, from front to back (from j_- to j_+). The continuity equation thus follows from charge conservation, and conversely, charge conservation from the continuity equation.

In this section, we shall deal with *stationary currents*—then the charge density does not change anywhere as time goes by ($\partial\rho/\partial t = 0$), and the current density is solenoidal ($\nabla \cdot \mathbf{j} = 0$ and $\mathbf{n} \cdot \mathbf{j} = 0$ at conductor surfaces). Only in the next section will we relax this restriction.

3.2.2 Ohm's Law

Electric currents are generated in conductors by electric fields. The fields exert a force on the charged particles and accelerate them. If we apply a voltage $U\,(> 0)$ at the ends of a conductor, then a current of strength $I\,(> 0)$ will flow. The ratio U/I is the *resistance R* of the conductor:

$$U = R\,I .$$

According to Ohm's law, the resistance depends on the properties of the conductor, but not on the applied voltage or the current. For a homogeneous conductor of length l and cross-section A, apart from its dimensions, it thus depends on the *conductivity* σ:

$$R = \frac{l}{A\,\sigma} .$$

Since $U = E\,l$ and $I = j\,A$, for a homogeneous conductor, we find the differential form of *Ohm's law*, viz.,

$$\mathbf{j} = \sigma\,\mathbf{E} .$$

In fact, the current density often depends linearly on the field strength. (The conductivity σ in some crystals is a tensor, because there are preferential directions—but we do not wish to deal with that here.) However, there are also counterexamples, as is to be expected, if we try to explain Ohm's law.

Actually, the field strength should accelerate the charges, since the field supplies a force, while the current density is proportional only to the velocity of the charged particles. This apparent contradiction in Ohm's law is resolved as for free fall by invoking friction (see p. 85, and in particular Fig. 2.11). In a metallic conductor, the electrons always lose the energy they acquire by collisions with the lattice, and hence move with a constant drift velocity. The associated *power* appears as *Joule heat*,

$$\mathbf{F} \cdot \mathbf{v} = \int dV \, \rho \mathbf{E} \cdot \mathbf{v} = \int dV \, \mathbf{j} \cdot \mathbf{E} = I \int d\mathbf{r} \cdot \mathbf{E} = I \, U \; ,$$

which heats the conductor. Furthermore, the conductivity often depends on the temperature, which limits Ohm's law.

Ohm's law cannot be applied as such to superconductors, which conduct currents loss-free, and then only at the surface of the conductor or in special tubes (superconductor of first or second kind, respectively).

In the given differential form, Ohm's law also holds only for homogeneous conductors (and insulators, which have $\sigma = 0$). For inhomogeneous conductors, we must also consider the induced field strength:

$$\mathbf{j} = \sigma \, (\mathbf{E} + \mathbf{E}') = \sigma \mathbf{E} + \mathbf{j}' \; ,$$

where the conductivity σ now also depends on position. The term $\mathbf{j}'$ refers to the additional current density at the sources.

Electric currents are immersed in magnetic fields, which in turn act on the currents—we shall now consider this. If we neglect this back-action, then stationary currents can be calculated easily, because for $\nabla \cdot \mathbf{j} = 0$, $\nabla \times \mathbf{E} = \mathbf{0}$, and $\mathbf{j} = \sigma \mathbf{E} + \mathbf{j}'$, we have

$$\nabla \cdot \sigma \mathbf{E} = -\nabla \cdot \mathbf{j}' \qquad \text{and} \qquad \nabla \times \mathbf{E} = \mathbf{0} \; ,$$
$$\mathbf{n} \cdot (\sigma_+ \mathbf{E}_+ - \sigma_- \mathbf{E}_-) = -\mathbf{n} \cdot \mathbf{j}'_A \qquad \text{and} \qquad \mathbf{n} \times (\mathbf{E}_+ - \mathbf{E}_-) = \mathbf{0} \; ,$$

where the potential may be introduced everywhere instead of the field strength ($\mathbf{E} = -\nabla \Phi$). The current density $\mathbf{j}'$ is thereby to be viewed as given. Hence for stationary currents, we have the same mathematical problem as for $\mathbf{E}$ or Φ in electrostatics, but with the conductivity σ instead of the permittivity ε and with $-\nabla \cdot \mathbf{j}'$ instead of the charge density ρ. If we can determine the capacity between two electrodes with a given form, then according to Ohm, the resistance between the same electrodes for a conductor satisfies

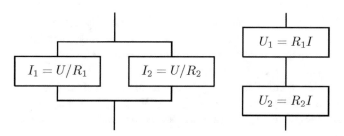

Fig. 3.8 For the proof Kirchhoff's laws. *Left*: Parallel connection. *Right*: Series connection

$$R\,C = \frac{\varepsilon}{\sigma}\,,$$

because from $I = \int_A d\mathbf{f} \cdot \sigma \mathbf{E} = (\sigma/\varepsilon) \int_A d\mathbf{f} \cdot \mathbf{D}$ with $Q = CU$ and $U = RI$. In particular, *Kirchhoff's laws* are obtained. The total resistance $R = U/I$ of the various individual resistors R_n depends on the type of connection:

$$\text{parallel connection (with } I = \textstyle\sum_n I_n) \qquad \frac{1}{R} = \sum_n \frac{1}{R_n}\,,$$
$$\text{series connection (with } U = \textstyle\sum_n U_n) \qquad R = \sum_n R_n\,.$$

This is illustrated in Fig. 3.8.

3.2.3 Lorentz Force

Moving charges (currents) are deflected by magnetic fields. There is a force acting on a point charge q moving with velocity $\mathbf{v}$ in a magnetic field $\mathbf{B}$, namely, the *Lorentz force*

$$\mathbf{F} = q\,\mathbf{v} \times \mathbf{B} \qquad \Longleftrightarrow \qquad \mathbf{F} = \int dV\,\mathbf{j} \times \mathbf{B}\,.$$

Note that, since $\mathbf{F}$ and $\mathbf{v}$ are polar vectors, $\mathbf{B}$ must be an axial vector. This velocity-dependent force was already mentioned on p. 78 and was generalized to the concept of potential energy (p. 98) and momentum (p. 100). Here then the acceleration is perpendicular to the velocity and the kinetic energy is therefore conserved. If we write $\dot{\mathbf{v}} = \boldsymbol{\omega} \times \mathbf{v}$, we have $\boldsymbol{\omega} = -(q/m)\,\mathbf{B}$ for the *cyclotron frequency*. For fixed $\mathbf{B}$, we find a helical orbit with the Darboux vector $\boldsymbol{\omega}/v$, and in particular with $\mathbf{v} \perp \mathbf{B}$, a circular orbit of radius $R = v/\omega$, because only then do the Lorentz force $m\omega v$ and the centrifugal force mv^2/R cancel each other.

However, there is no force acting on stationary currents in the homogeneous magnetic field, because for $\nabla \cdot \mathbf{j} = 0$, according to p. 17 or Problem 3.4, we also have $\int dV\,\mathbf{j} = 0$. Nevertheless we can measure this magnetic field, if we use the torque on a current loop. This will now be shown for very small conductor loops, since then a homogeneous magnetic field may be assumed.

For the torque, we require the volume integral of $\mathbf{r} \times (\mathbf{j} \times \mathbf{B}) = \mathbf{B} \cdot \mathbf{r}\,\mathbf{j} - \mathbf{r} \cdot \mathbf{j}\,\mathbf{B}$. Here we consider a little box around the conductor loop without current at its surface. Then, since $2\,\mathbf{r} \cdot \mathbf{j} = \mathbf{j} \cdot \nabla r^2 = \nabla \cdot (r^2 \mathbf{j}) - r^2 \nabla \cdot \mathbf{j}$ for a solenoidal current density, the volume integral of $\mathbf{r} \cdot \mathbf{j}$ vanishes according to Gauss's theorem, and for

$$2\,\mathbf{B} \cdot \mathbf{r}\,\mathbf{j} = \{\mathbf{B} \cdot \mathbf{r}\,\mathbf{j} + \mathbf{B} \cdot \mathbf{j}\,\mathbf{r}\} + (\mathbf{r} \times \mathbf{j}) \times \mathbf{B}\,,$$

the volume integral of the curly bracket vanishes as well, because we have

$$r_i\,j_k + j_i\,r_k = \mathbf{j} \cdot \nabla(r_k r_i) = \nabla \cdot (r_k r_i \mathbf{j}) - r_k r_i\,\nabla \cdot \mathbf{j}\,,$$

and therefore the same procedure as above is applicable, with $r_k r_i$ instead of r^2. Therefore, the homogeneous magnetic field $\mathbf{B}$ exerts a torque

$$\mathbf{N} = \tfrac{1}{2}\int dV\,(\mathbf{r} \times \mathbf{j}) \times \mathbf{B}$$

on the conductor loop. Hence the magnetic field can be determined and the concept of magnetic moment introduced.

3.2.4 Magnetic Moments

The last equation suggests introducing the *magnetic moment* of the conductor loop (or more precisely, its dipole moment)

$$\mathbf{m} \equiv \tfrac{1}{2}\int dV\,\mathbf{r} \times \mathbf{j}\,.$$

This is an axial vector. For a current of strength I around a plane sheet A, it has magnitude (see Fig. 2.4)

$$m = \tfrac{1}{2}I\left|\int \mathbf{r} \times d\mathbf{r}\right| = IA\,,$$

and points in the direction of the normal to the loop, in such a way that the current direction forms a right-hand screw around this axis.

Such a magnetic moment in a homogeneous magnetic field $\mathbf{B}$ experiences a torque

$$\mathbf{N} = \mathbf{m} \times \mathbf{B}\,,$$

as was shown before. (Instead of this, $\mathbf{N} = \mu_0 \mathbf{m} \times \mathbf{B}/\mu_0$ is often used and $\mu_0 \mathbf{m}$ is called the magnetic moment, whence for $\mathbf{m}$, the factor μ_0 is included—but this idea goes contrary to the IUPAP recommendation.)

If the current originates from a charge Q of mass M (both evenly distributed, so that $\rho/Q = \rho_M/M$) distributed along the closed orbit, the magnetic moment is related to the orbital angular momentum $\mathbf{L}$ by $\mathbf{m} = \frac{1}{2} \int dV \mathbf{r} \times \rho \mathbf{v} = \frac{1}{2}(Q/M)\,\mathbf{L}$. In atomic physics, the action quantum $\hbar$ is taken as a unit for the orbital angular momentum, along with the charge and mass of an electron ($Q = -e$ and $M = m_e$). Therefore, in that context, the magnetic moment is related to the *Bohr magneton* (see p. 623)

$$\mu_B = \frac{e\hbar}{2m_e} \,.$$

In atomic physics, magnetic moments are usually denoted by μ, but in macroscopic electromagnetism, this is already reserved for the permeability.

While electric dipole moments can be formed from monopoles, magnetic monopoles have not yet been observed. Such a thing would have to be a pseudo-scalar, because $\mathbf{m}$ is an axial vector. Since Dirac, it has not been excluded that there may be magnetic monopoles in elementary particle physics—it may just be that they have not yet been separated. In any case, all our macroscopic considerations work without magnetic monopoles.

3.2.5 Magnetization

As Fig. 3.9 shows, we can replace macroscopic current loops by many microscopic ones, and these then by magnetic moments, if we deal with the action of a magnetic field. It is therefore useful to introduce the density of magnetic moments on the surface or in the volume. If there are N magnetic moments $\mathbf{m}_i$ in the volume ΔV, then

$$\mathbf{M} = \frac{1}{\Delta V} \sum_{i=1}^{N} \mathbf{m}_i$$

is the associated *magnetization*, an axial vector like $\mathbf{m}$ or $\mathbf{r} \times \mathbf{v}$. (Here, many solid-state physicists include the factor μ_0 for the magnetic moment. This goes against the IUPAP recommendation.) It is sometimes also called the (*magnetic*) *polarization*.

As shown in the right-hand part of Fig. 3.9, $\mathbf{M}$ has curls, where the current density does not vanish. In fact, we find $\nabla \times \mathbf{M} = \mathbf{j}$, because if d is the distance between different current-loop planes, then in addition to $m = I \cdot A$, we also have $m = M \cdot d \cdot A$. The magnetization clearly has a discontinuity of I/d, i.e., by the surface current density $\mathbf{j}_A$, at the current-carrying surface, whence we arrive at $\nabla \times \mathbf{M} = \mathbf{j}$.

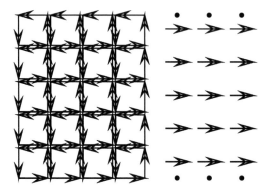

Fig. 3.9 *Left*: A macroscopic current loop (in the y, z-plane) is divided into 4×5 "microscopic" ones, each of which represents a magnetic moment. *Right*: A cuboid with three such planes with magnetic moments (*arrows*) is cut open (in the x, z-plane) and a current loop is associated with each one. The intersection points of the currents are indicated by the *black dots*

In atoms there are magnetic moments but no electric conduction currents, as can be verified in a magnetic field. Therefore, for the behavior in magnetic fields, we have to account for the magnetization in addition to macroscopic electric currents, and in microscopic electromagnetism, we have to introduce a "microscopic current density"

$$\mathbf{j} = \bar{\mathbf{j}} + \nabla \times \mathbf{M} .$$

(For the magnetic moments of elementary particles this is not justified, however, because their moment is connected to the spin and cannot be derived from a molecular current.) Since $\mathbf{j}$ differs from $\bar{\mathbf{j}}$ only by a rotational field, $\mathbf{j}$ is solenoidal like $\bar{\mathbf{j}}$.

Later we shall stick to macroscopic electromagnetism and always take only the macroscopic current density $\bar{\mathbf{j}}$ (leaving out the bar), but for the time being, $\mathbf{j}$ will be the microscopic current density.

3.2.6 Magnetic Fields

Even if Sect. 3.2.3 has already shown that a field $\mathbf{B}$ can be measured (by forces acting on magnetic moments or moving charges), we still have to deal with its generation: magnetic fields occur for magnetic moments as well as for electric currents.

Since there are no magnetic monopoles, the magnetic field $\mathbf{B}$ is solenoidal. In addition, we find from experiment that each microscopic current density is related to the circulation density of a magnetic field:

$$\nabla \cdot \mathbf{B} = 0 \quad \text{and} \quad \nabla \times \mathbf{B} = \mu_0 \, \mathbf{j} = \mu_0 \, (\bar{\mathbf{j}} + \nabla \times \mathbf{M}) ,$$

since we have the *Biot–Savart law*

$$\mathbf{B}(\mathbf{r}) = \frac{\mu_0}{4\pi}\, \nabla \times \int dV' \, \frac{\mathbf{j}\,(\mathbf{r}')}{|\mathbf{r}-\mathbf{r}'|} = -\frac{\mu_0}{4\pi} \int dV' \, \mathbf{j}\,(\mathbf{r}') \times \nabla \frac{1}{|\mathbf{r}-\mathbf{r}'|} \,.$$

∇ acts only on $\mathbf{r}$ and not on $\mathbf{r}'$, and hence we have $\nabla \times G\mathbf{j}' = -\mathbf{j}' \times \nabla G$. For sufficiently thin conductors, it follows that

$$\mathbf{B}(\mathbf{r}) = \frac{\mu_0\, I'}{4\pi}\, \nabla \times \int \frac{d\mathbf{r}'}{|\mathbf{r}-\mathbf{r}'|} = -\frac{\mu_0\, I'}{4\pi} \int d\mathbf{r}' \times \nabla \frac{1}{|\mathbf{r}-\mathbf{r}'|} \,.$$

For a given magnetization, $\nabla \times \mathbf{M}$ may of course appear instead of $\mathbf{j}$.

Since we have $\nabla \times \mathbf{B} = \mu_0\, \mathbf{j} = \mu_0\, (\bar{\mathbf{j}} + \nabla \times \mathbf{M})$ in the macroscopic theory, we set

$$\mathbf{B} = \mu_0\, (\mathbf{H} + \mathbf{M}) \,, \quad \nabla \times \mathbf{H} = \bar{\mathbf{j}} \,,$$

where $\mathbf{H}$ is called the *excitation* or *magnetic field strength* and $\mathbf{B}$ is referred to as the *displacement field of the magnetic field* or *magnetic induction*. Since $\mathbf{B}$ is a measure of the force on moving charges, it should actually be called the magnetic field strength, but if we compare electrostatics and magnetostatics, the choice of names is understandable, as we shall now show.

In *magnetostatics*, we deal with magnetized matter without electric currents, whence $\bar{\mathbf{j}} = \mathbf{0}$. Because $\nabla \cdot \mathbf{B} = 0$ and $\mathbf{B} = \mu_0\, (\mathbf{H} + \mathbf{M})$, we clearly then have

$$\nabla \times \mathbf{H} = \mathbf{0} \quad \text{and} \quad \nabla \cdot \mathbf{H} = -\nabla \cdot \mathbf{M} \,.$$

This is reminiscent of $\nabla \times \mathbf{E} = \mathbf{0}$ and $\nabla \cdot \mathbf{E} = -\varepsilon_0^{-1}\nabla \cdot \mathbf{P}$ for uncharged polarized matter in electrostatics. Since the excitation $\mathbf{H}$ is irrotational here, we may likewise introduce a scalar magnetic potential Φ_m by

$$\mathbf{H} = -\nabla \Phi_m \,,$$

where (see p. 174)

$$\Phi_m(\mathbf{r}) = -\frac{1}{4\pi} \int dV' \, \frac{\nabla' \cdot \mathbf{M}(\mathbf{r}')}{|\mathbf{r}-\mathbf{r}'|} = -\frac{1}{4\pi} \, \nabla \cdot \int dV' \, \frac{\mathbf{M}(\mathbf{r}')}{|\mathbf{r}-\mathbf{r}'|} \,.$$

The magnetic potential cannot be connected to a potential energy though (there are no magnetic monopoles), and it is a pseudo-scalar.

A tiny rod magnet with moment $\mathbf{m}'$ at the position $\mathbf{r}'$ thus produces the magnetic field (see p. 172 and Problem 3.8)

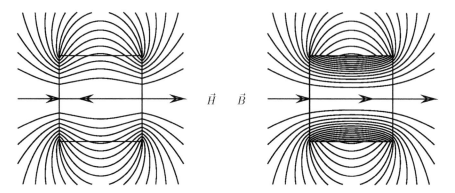

Fig. 3.10 Field lines of a permanent homogeneous magnetized cylinder. *Left*: **H** field. *Right*: **B** field. Except for the edges, the flux through the surface increases by one unit from line to line. The *right-hand figure* applies also to the **H** and **B** field of a current-carrying coil

$$\mathbf{H}(\mathbf{r}) = -\nabla \Phi_{\mathrm{m}} = \frac{1}{4\pi}\,\nabla\!\left(\nabla \cdot \frac{\mathbf{m}'}{|\mathbf{r}-\mathbf{r}'|}\right)$$

$$= \frac{1}{4\pi}\,\frac{3\,\mathbf{m}'\!\cdot\mathbf{e}\,\mathbf{e}-\mathbf{m}'}{|\mathbf{r}-\mathbf{r}'|^{3}} - \frac{\mathbf{m}'}{3}\,\delta(\mathbf{r}-\mathbf{r}')\,,\quad\text{with}\quad \mathbf{e} = \frac{\mathbf{r}-\mathbf{r}'}{|\mathbf{r}-\mathbf{r}'|}\,.$$

This is related to the magnetic induction field

$$\mathbf{B}(\mathbf{r}) = \mu_0 \mathbf{H} + \mu_0\,\mathbf{m}'\,\delta(\mathbf{r}-\mathbf{r}')\,.$$

Incidentally, this can also be written as $\mu_0/(4\pi)\,\nabla \times (\nabla \times \mathbf{m}'/|\mathbf{r}-\mathbf{r}'|)$, because $\nabla \times (\nabla \times \mathbf{a}) = \nabla(\nabla \cdot \mathbf{a}) - \Delta\mathbf{a}$, and according to p. 26, $\Delta|\mathbf{r}-\mathbf{r}'|^{-1} = -4\pi\delta(\mathbf{r}-\mathbf{r}')$.

For a homogeneous magnetized cylinder (in air) with the curved surface along **M**, the magnetization has sources only on the faces and curls only on the curved surface. They jump there from **M** to zero. Therefore, on the faces $\nabla_A \cdot \mathbf{M} = -\mathbf{n} \cdot \mathbf{M}$, and on the curved surface $\nabla_A \times \mathbf{M} = -\mathbf{n} \times \mathbf{M}$. The potential Φ_{m} of a circular face can be expressed with the help of a complete elliptic integral of the first kind (see Problem 3.17), and that of a homogeneous circular disk as an integral of it. If we have calculated Φ_{m} on the edge and on the faces, then the remaining values follow faster numerically via the Laplace equation (see [3]). Outside the cylinder, the two fields $\mu_0\mathbf{H}$ and **B** are equal, because $\mathbf{M} = \mathbf{0}$, while on the axis inside they are directed oppositely (see Fig. 3.10).

3.2.7 Basic Equations of Macroscopic Magnetostatics
with Stationary Currents

Once again we allow for electric currents $\bar{\mathbf{j}}$ and consider the *basic equations of macroscopic magnetostatics with stationary currents* derived at the beginning of the last section:

$$\nabla \cdot \mathbf{B} = 0\,, \quad \nabla \times \mathbf{H} = \bar{\mathbf{j}}\,, \quad \text{and} \quad \mathbf{B} = \mu \mathbf{H}\,.$$

These differential equations supply the boundary conditions

$$\mathbf{n} \cdot (\mathbf{B}_+ - \mathbf{B}_-) = 0 \quad \text{and} \quad \mathbf{n} \times (\mathbf{H}_+ - \mathbf{H}_-) = \bar{\mathbf{j}}_A\,,$$

and read in integral form

$$\int_{(V)} d\mathbf{f} \cdot \mathbf{B} = 0 \quad \text{and} \quad \int_{(A)} d\mathbf{r} \cdot \mathbf{H} = I\,.$$

Here $\bar{\mathbf{j}}_A$ denotes the macroscopic current density at the surface. It vanishes normally. It occurs only for superconductors of the first kind: then there is no magnetic field in the interior (*Meissner–Ochsenfeld effect*), and only the surface carries a current. The last equation is called *Ampère's circuital law*, and also in earlier years the *Ørsted law*. It relates the magnetic field and current strength in a particularly simple way and also contains the *right-hand rule*: the magnetic field encircles the current $\mathbf{I} = A\bar{\mathbf{j}}$ anticlockwise.

For example, for a straight normal conductor wire of circular cross-section with radius R_0 and constant current density $\bar{\mathbf{j}} = \mathbf{I}/(\pi R_0{}^2)$, the magnetic field in cylindrical coordinates R, φ, z about the wire axis is given by

$$\mathbf{H} = \frac{1}{2\pi} \begin{cases} \dfrac{\mathbf{I} \times \mathbf{R}}{R_0{}^2} & \text{for } R \le R_0\,, \\[2mm] \dfrac{\mathbf{I} \times \mathbf{R}}{R^2} & \text{for } R \ge R_0\,. \end{cases}$$

The right-hand rule requires $\mathbf{H}$ to be proportional to $\mathbf{I} \times \mathbf{R}$ (up to a positive factor), and Ampère's law fixes the absolute value. We have $2\pi R\, H(R)$ equal to $I\,(R/R_0)^2$ for $R \le R_0$ and equal to I for $R \ge R_0$. Of course, there is no arbitrarily long straight wire—therefore the realistic magnetic fields of stationary currents also decay at large distances more quickly than R^{-1}, in fact, like a dipole field as $(R^2 + z^2)^{-3/2}$.

In general, an applied magnetic field magnetizes a (magnetic) medium because it polarizes irregularly oriented moments. We therefore write

$$\mathbf{M} = \chi_{\mathrm{m}}\, \mathbf{H} \quad \text{and} \quad \mathbf{B} = \mu_0(1 + \chi_{\mathrm{m}})\, \mathbf{H} \equiv \mu \mathbf{H}\,,$$

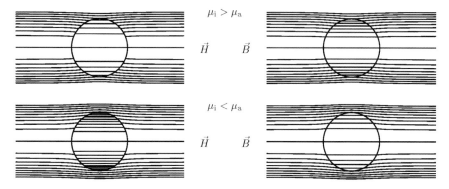

Fig. 3.11 Magnetic spheres. *Upper*: Paramagnetic. *Lower*: Diamagnetic. The sphere is brought into a homogeneous magnetic field. *Left*: **H** field. *Right*: **B** field. Both are axially symmetric, and we find $\nabla \cdot \mathbf{B} = 0$ and $\mathbf{B} = \mu \mathbf{H}$ in addition to $\nabla \times \mathbf{H} = \mathbf{0}$. The lower figure is always useful if the permeability inside is lower than outside, even if there is no diamagnet (in air). The figures are also valid for electric field lines for different permittivities and for current lines of stationary currents for different conductivities, as explained in the text: $\mathbf{H} \to \mathbf{E}$ and either $\mu \to \varepsilon$ and $\mathbf{B} \to \mathbf{D}$ or $\mu \to \sigma$ and $\mathbf{B} \to \bar{\mathbf{j}}$

with *permeability* μ (sometimes expressed as relative permeability $\mu_r = \mu/\mu_0$) and *magnetic susceptibility* $\chi_m = \mu_r - 1$. These are tensors, if there are preferential directions: **B** and **H** may have different directions. For a *ferromagnet*, **B** and **H** are not related to each other linearly. This is often represented in a *hysteresis curve* $M(H)$. (In a weak field typical values for these are $\mu_r \approx 500$.) For materials with smaller scalar permeability, we also distinguish between *paramagnets* with $\chi_m > 0$ or $\mu_r > 1$ and *diamagnets* with $\chi_m < 0$ or $0 < \mu_r < 1$. The dielectric susceptibility χ_e is always positive: paramagnetism can be explained as orientation of dipole moments, diamagnetism as a consequence of Lenz's law, which will be dealt with only in the next section.

If we consider a magnetic sphere (radius r_0) in a homogeneous magnetic field $\mathbf{H}_0$ (at long range), in addition to $\nabla \times \mathbf{H} = \mathbf{0}$ and because $\nabla \cdot \mathbf{B} = 0 = \nabla \cdot \mu \mathbf{H}$, we have $\mu_i \mathbf{n} \cdot \mathbf{H}_i = \mu_a \mathbf{n} \cdot \mathbf{H}_a$. The magnetic field is irrotational and only has sources on the surface. The associated discontinuity is related to the field of a dipole $\mathbf{m} = (\mu_i - \mu_a)/(\mu_i + 2\mu_a)\, r_0{}^3\, \mathbf{H}_0$ (except for a factor of 4π), because with

$$\mathbf{H}_i = \mathbf{H}_0 - \frac{\mathbf{m}}{r_0{}^3} \quad \text{and} \quad \mathbf{H}_a = \mathbf{H}_0 + \frac{3\,\mathbf{m} \cdot \mathbf{e}\,\mathbf{e} - \mathbf{m}}{r^3} \, , \quad \text{for} \quad \mathbf{e} = \frac{\mathbf{r}}{r} \, ,$$

all the above-mentioned conditions are satisfied. This result is illustrated in Fig. 3.11, where the pictures are also valid for electric field lines for different permittivities (because $\nabla \cdot \mathbf{D} = 0, \mathbf{D} = \varepsilon \mathbf{E}, \nabla \times \mathbf{E} = \mathbf{0}$) and for current lines of stationary currents for different conductivities (because $\nabla \cdot \bar{\mathbf{j}} = 0, \bar{\mathbf{j}} = \sigma \mathbf{E}, \nabla \times \mathbf{E} = \mathbf{0}$). To this end, we replace $\mathbf{H} \to \mathbf{E}$ and either $\mu \to \varepsilon$ and $\mathbf{B} \to \mathbf{D}$ or $\mu \to \sigma$ and $\mathbf{B} \to \bar{\mathbf{j}}$ (see also Problem 3.23).

3.2.8 Vector Potential

The displacement field **B** is always solenoidal and therefore a rotational field:

$$\mathbf{B} = \nabla \times \mathbf{A} .$$

A is called the *vector potential* and is a polar vector field because the induction is an axial field. Here the induction field **B** can be measured via the Lorentz force or by its action on magnetic moments, while the vector potential **A** represents only a computational tool and is not unique—only its curl is physically fixed, not its sources (and an additive constant). Therefore, a gradient field may also be added: $\mathbf{A}' = \mathbf{A} - \nabla\Psi$ would supply the same magnetic field as **A**. The vector potential must therefore be *gauged*, and in this case $\nabla \cdot \mathbf{A}$ is fixed along with a constant additive term (in most cases we require it to vanish for $r \to \infty$). For the *Coulomb gauge*, the vector potential is chosen solenoidal.

The equation $\nabla \times \mathbf{B} = \mu_0 \mathbf{j}$ does not depend on the gauge, but

$$\Delta\mathbf{A} = -\mu_0\,\mathbf{j}$$

does, since we only have $\Delta\mathbf{A} = -\nabla \times \mathbf{B}$ for a solenoidal vector potential given that $\Delta\mathbf{A} = \nabla (\nabla \cdot \mathbf{A}) - \nabla \times (\nabla \times \mathbf{A})$. On the other hand, $\Delta\mathbf{A} = -\mu_0\mathbf{j}$ (if $\mathbf{A} \to \mathbf{0}$ for $r \to \infty$ holds), and according to p. 27,

$$\mathbf{A}(\mathbf{r}) = \frac{\mu_0}{4\pi} \int dV' \, \frac{\mathbf{j}\,(\mathbf{r}')}{|\mathbf{r} - \mathbf{r}'|} \triangleq \frac{\mu_0\,I'}{4\pi} \int \frac{d\mathbf{r}'}{|\mathbf{r} - \mathbf{r}'|} \, ,$$

which yields the Biot–Savart law (see p. 193). For stationary currents, this vector potential is solenoidal, because for its source density we require $\mathbf{j}' \cdot \nabla G$, which can be rephrased as $G \, \nabla' \cdot \mathbf{j}' - \nabla' \cdot G\mathbf{j}'$ since $\nabla G = -\nabla' G$. According to Gauss's theorem, we then only require a surface where there is no current to prove the statement.

Here **j** is still the microscopic current density, and can also appear as the circulation density of a magnetization. In this case, we have $G \, \mathbf{j}' = G \, \nabla' \times \mathbf{M}' = \nabla' \times G\mathbf{M}' + \mathbf{M}' \times \nabla'G$. Therefore, the vector potential of a magnetic moment **m**' results in

$$\mathbf{A}(\mathbf{r}) = -\frac{\mu_0}{4\pi} \mathbf{m}' \times \nabla \frac{1}{|\mathbf{r} - \mathbf{r}'|} = \frac{\mu_0}{4\pi} \nabla \times \frac{\mathbf{m}'}{|\mathbf{r} - \mathbf{r}'|} \, ,$$

because the surface integral of $G\mathbf{M}'$ does not contribute.

For a homogeneous displacement field **B**, we may set $\mathbf{A}(\mathbf{r}) = \frac{1}{2}\mathbf{B} \times \mathbf{r}$, because then $\nabla \times \mathbf{A} = \mathbf{B}$, and the Coulomb gauge holds everywhere. Then the origin of **r** may be chosen arbitrarily—a constant is irrelevant. For other fields it is fixed by the condition $\mathbf{A} = \mathbf{0}$ for $r \to \infty$, which is not suitable for a homogeneous field.

The integral mentioned at the beginning does not need to be taken over the whole space, if we also take into account surface integrals (as for the scalar potential on

p. 169). We use the equation $\Delta \mathbf{A} = -\mu_0 \mathbf{j}$ and Green's second theorem, i.e., in $\int_V dV\,(\psi \Delta \phi - \phi \Delta \psi) = \int_{(V)} d\mathbf{f} \cdot (\psi \nabla \phi - \phi \nabla \psi)$, we replace the function ψ by $|\mathbf{r} - \mathbf{r}'|^{-1}$ and the function ϕ by the three components of the vector potential. It then follows that

$$4\pi\,\mathbf{A}(\mathbf{r}) = \int_{V'} dV' \frac{\mu_0\,\mathbf{j}\,(\mathbf{r}')}{|\mathbf{r} - \mathbf{r}'|}$$
$$+ \int_{(V')} \frac{d\mathbf{f}' \cdot \nabla'\,\mathbf{A}(\mathbf{r}')}{|\mathbf{r} - \mathbf{r}'|} - \int_{(V')} \left(d\mathbf{f}' \cdot \nabla' \frac{1}{|\mathbf{r} - \mathbf{r}'|} \right) \mathbf{A}(\mathbf{r}')\ .$$

In particular, we may choose V' such that there is no current: then the vector potential is fixed by its values and its first derivatives on the surface (V'). As in electrostatics (see the end of Sect. 3.1.3), then also in magnetostatics in a finite region, the same physical field can be generated in various ways (by distributions in space or on sheets). The continuation across the boundaries is not unique and allows various models. This has also been clearly demonstrated in the context of Fig. 3.9.

3.2.9   Magnetic Interaction

An inhomogeneous magnetic field exerts a force on a magnetic moment. If we use the equation $\int_{(A)} d\mathbf{r} \times \mathbf{B} = \int_A (d\mathbf{f} \times \nabla) \times \mathbf{B}$ of p. 17 and $m = IA$ of p. 190, then for a sufficiently small conductor loop, it follows that

$$\mathbf{F} = I \oint d\mathbf{r} \times \mathbf{B} = (\mathbf{m} \times \nabla) \times \mathbf{B}\ .$$

As for the electric dipole moment, we require likewise small extensions for the magnetic moment in order for the higher moments to become negligible. Here we may also write $\nabla(\mathbf{m} \cdot \mathbf{B}) - \mathbf{m}\,\nabla \cdot \mathbf{B}$, since the differential operator changes only $\mathbf{B}$. Given that $\mathbf{B}$ is always solenoidal, we find

$$\mathbf{F} = \nabla(\mathbf{m} \cdot \mathbf{B})\ .$$

Therefore, we may also introduce a *potential energy*

$$E_{\text{pot}} = -\mathbf{m} \cdot \mathbf{B}\ ,$$

and again, $\mathbf{F} = -\nabla E_{\text{pot}}$ holds. This corresponds to the expression $E_{\text{pot}} = -\mathbf{p} \cdot \mathbf{E}$ in electrostatics (see p. 171). There, because of $\nabla \times \mathbf{E} = 0$, we could also write $(\mathbf{p} \cdot \nabla)\mathbf{E}$ instead of $\nabla(\mathbf{p} \cdot \mathbf{E})$. In contrast, we have $\nabla \cdot \mathbf{B} = 0$ here, and therefore $\nabla(\mathbf{m} \cdot \mathbf{B})$ is also equal to $(\mathbf{m} \times \nabla) \times \mathbf{B}$. Furthermore, $\mathbf{p}$ and $\mathbf{E}$ are polar vectors, while $\mathbf{m}$ and $\mathbf{B}$ are axial.

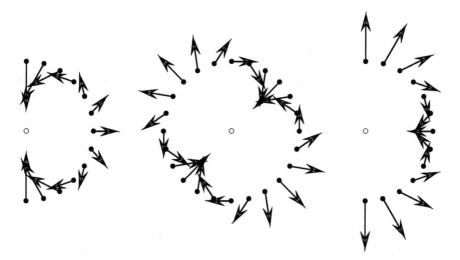

Fig. 3.12 Tensor force of a moment ↑ at the position o on a moment at the position •. Equal moments (↑), opposite moments (↓), in-between the perpendicular moment →

Hence the potential energy of two dipole moments $\mathbf{m}$ and $\mathbf{m}'$ at positions $\mathbf{r} \neq \mathbf{r}'$ is obtained as

$$E_{\mathrm{pot}} = \frac{\mu_0}{4\pi}\, \mathbf{m} \cdot \nabla\, \mathbf{m}' \cdot \nabla'\, \frac{1}{|\mathbf{r}-\mathbf{r}'|} = \frac{\mu_0}{4\pi}\, \frac{\mathbf{m}\cdot\mathbf{m}' - 3\,\mathbf{m}\cdot\mathbf{e}\,\mathbf{m}'\cdot\mathbf{e}}{|\mathbf{r}-\mathbf{r}'|^3}\,.$$

Here $\mathbf{e} \equiv (\mathbf{r}-\mathbf{r}')/|\mathbf{r}-\mathbf{r}'|$. (For the last equation, compare p. 172.) With $\mathbf{r} \neq \mathbf{r}'$, this yields

$$\begin{aligned}
\mathbf{F} = -\nabla E_{\mathrm{pot}} &= -\frac{\mu_0}{4\pi}\, \mathbf{m} \cdot \nabla\, \mathbf{m}' \cdot \nabla'\, \nabla\, \frac{1}{|\mathbf{r}-\mathbf{r}'|} \\
&= \frac{3\mu_0}{4\pi}\, \frac{\mathbf{m}\cdot\mathbf{e}\,\mathbf{m}' + \mathbf{m}'\cdot\mathbf{e}\,\mathbf{m} + (\mathbf{m}\cdot\mathbf{m}' - 5\,\mathbf{m}\cdot\mathbf{e}\,\mathbf{m}'\cdot\mathbf{e})\,\mathbf{e}}{|\mathbf{r}-\mathbf{r}'|^4}
\end{aligned}$$

for the force acting on $\mathbf{m}$. This force depends upon the directions of the three vectors $\mathbf{m}$, $\mathbf{m}'$, and $\mathbf{e}$, and does not always lie in the direction of $(\pm)\,\mathbf{e}$: it is not a central, but a *tensor force* (see Fig. 3.12).

We generalize the expression for E_{pot} to an extended magnetization:

$$E_{\mathrm{pot}} = -\int \mathrm{d}V\, \mathbf{M}\cdot\mathbf{B}\,.$$

The integrand can be rewritten: $\mathbf{M}\cdot(\nabla\times\mathbf{A}) = \nabla\cdot(\mathbf{A}\times\mathbf{M}) + \mathbf{A}\cdot(\nabla\times\mathbf{M})$. With Gauss's theorem (and no magnetization on the surface of V), and because $\nabla\times\mathbf{M} = \mathbf{j}$, it follows that

$$E_{\text{pot}} = -\int dV\, \mathbf{j}\,(\mathbf{r}) \cdot \mathbf{A}(\mathbf{r})\,.$$

(In Sect. 2.3.4, and in particular p. 98, we mentioned that the generalized potential energy $-q\,\mathbf{v}\cdot\mathbf{A}$ belongs to the velocity-dependent Lorentz force acting on point charges. This is in accord with $E_{\text{pot}} = -\int dV\,\mathbf{j}\cdot\mathbf{A}$.) Even though the vector potential can be re-gauged, the difference $\int dV\,\mathbf{j}\cdot\nabla\Psi = \int dV\,\{\nabla\cdot(\Psi\mathbf{j}) - \Psi\,\nabla\cdot\mathbf{j}\}$ does not contribute in the case of stationary currents because of Gauss's theorem (in finite current loops).

For the interaction energy of two conductors, we thus have

$$E_{\text{pot}} = -\frac{\mu_0}{4\pi} \iint dV\,dV'\,\frac{\mathbf{j}\,(\mathbf{r}) \cdot \mathbf{j}\,(\mathbf{r}')}{|\mathbf{r} - \mathbf{r}'|}\,.$$

In order to derive the associated force, we have to consider the position dependence of this potential energy. The two current loops change only their relative positions, but neither their current densities nor their form. The potential energy originates from the fact that two current loops are brought together from a very great distance and that forces then appear. We should therefore introduce the average separation $\mathbf{R}$ of the two conductors and consider the double integral

$$\iint d\mathbf{r}'' \cdot d\mathbf{r}'\,|\mathbf{R} + \mathbf{r}'' - \mathbf{r}'|^{-1}\,.$$

The force between the two current loops then follows from $\mathbf{F} = -\nabla_R E_{\text{pot}}$ as (*Ampère's force law*)

$$\mathbf{F} = \frac{\mu_0}{4\pi} \iint dV\,dV'\,\mathbf{j}\,(\mathbf{r}) \cdot \mathbf{j}\,(\mathbf{r}')\,\nabla\frac{1}{|\mathbf{r} - \mathbf{r}'|}\,.$$

According to this, parallel wires attract each other if electric currents flow in the same direction, and repel each other for currents that flow in opposite directions. In other words, currents of like sign are attracted, while like charges are repelled, because Coulomb's law contains $-c_0^2\,qq'$ instead of $\iint II'\,d\mathbf{r}\cdot d\mathbf{r}'$, something we shall be concerned with in the next section. Here $\mathbf{F}$ is the total force which the conductor with primed quantities exerts on the other (unprimed) one. Since $\nabla'G(\mathbf{r} - \mathbf{r}') = -\nabla G(\mathbf{r} - \mathbf{r}')$, it follows that $\mathbf{F}' = -\mathbf{F}$, as is required also by Newton's third law. (Current-carrying conductors do not exert a force on themselves. In this case primed and unprimed quantities must be interchangeable.) The factor $\mu_0/4\pi$ is connected with the chosen concept of current strength:

> If two parallel (straight) conductors of negligible cross-section a distance 1 m apart in vacuum each carry a current of 1 A, then they exert a force of 2×10^{-7} N per meter length on each other.

The double integral of $d\mathbf{r} \cdot d\mathbf{r}' \,\hat{=}\, dz\,dz'$ is important. We may restrict ourselves to a conductor element dz around $z = 0$. If the two conductors are separated by a distance

R, then since $\nabla |\mathbf{r} - \mathbf{r}'|^{-1} \widehat{=} |\partial (R^2 + z'^2)^{-1/2}/\partial R| = R (R^2 + z'^2)^{-3/2}$, the integral $\int dz' \, R \, (R^2 + z'^2)^{-3/2} = z' R^{-1} \, (R^2 + z'^2)^{-1/2}$ is to be taken from $-\infty$ to $+\infty$. We thus deduce the force per unit length to be

$$\frac{F}{l} = \frac{\mu_0 I I'}{2\pi R} \, .$$

Given the *magnetic field constant* $\mu_0 = 4\pi \times 10^{-7}$ N/A^2, we do indeed find the above-mentioned force from Ampère's law.

3.2.10 Inductance

For the interaction energy of two thin conductor loops, we find

$$E_{\text{pot}} = -I \oint d\mathbf{r} \cdot \mathbf{A} = -I \, I' \, L$$

with the *mutual inductance*

$$L \equiv \frac{\mu_0}{4\pi} \iint \frac{d\mathbf{r} \cdot d\mathbf{r}'}{|\mathbf{r} - \mathbf{r}'|} \, .$$

According to this, known as the *Neumann formula*, L is positive for currents in the same direction in coaxial loops. Figure 3.13 shows an example whose inductance we shall now calculate for radii R and R', and distance a.

Because $|\mathbf{r} - \mathbf{r}'|^2 = a^2 + (\mathbf{R} - \mathbf{R}') \cdot (\mathbf{R} - \mathbf{R}')$, we can find L from

$$L = \frac{\mu_0}{4\pi} \, R R' \int_0^{2\pi} d\varphi \int_0^{2\pi} d\varphi' \, \frac{\cos (\varphi - \varphi')}{\sqrt{a^2 + R^2 + R'^2 - 2RR' \cos (\varphi - \varphi')}} \, .$$

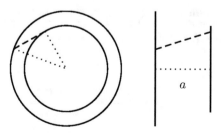

Fig. 3.13 Top and side view of two coaxial current loops (*continuous lines*) at the distance a. The line $|\mathbf{r} - \mathbf{r}'|$ connecting two points is shown by a *dashed line*. It can be calculated with the help of the *dotted lines* (radii R and R')

Fig. 3.14 Mutual inductance $L(k^2)$ of two coaxial current loops (radii R and R' at distance a) with $k^2 = 4RR'/\{a^2 + (R + R')^2\}$

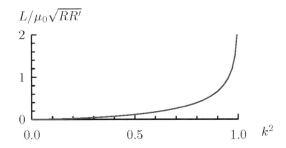

The double integral is equal to $2\pi \int_0^{2\pi} \{a^2 + R^2 + R'^2 - 2RR' \cos\psi\}^{-1/2} \cos\psi \, d\psi$. If we integrate only from 0 to π, we obtain half the value. With $z = \frac{1}{2}(\pi - \psi)$, it follows that $\cos\psi = -\cos(2z) = 2\sin^2 z - 1$ and $d\psi = -2dz$:

$$L = \mu_0 \sqrt{RR'}\, k \int_0^{\pi/2} \frac{2\sin^2 z - 1}{\sqrt{1 - k^2 \sin^2 z}} \, dz \,, \quad \text{with} \;\; k^2 \equiv \frac{4RR'}{a^2 + (R + R')^2} \,.$$

Note that here $k^2 < 1$, because we consider only separate conductor loops and we have $4RR' = (R + R')^2 - (R - R')^2$. We thus encounter the *complete elliptic integrals of first and second kind* (see p. 104 and Fig. 2.33):

$$K(k^2) \equiv \int_0^{\pi/2} \frac{dz}{\sqrt{1 - k^2 \sin^2 z}}$$

and

$$E(k^2) \equiv \int_0^{\pi/2} \sqrt{1 - k^2 \sin^2 z}\, dz \,,$$

Since $\sin^2 z = \{1 - (1 - k^2 \sin^2 z)\}/k^2$, this implies that

$$\int_0^{\pi/2} \frac{2\sin^2 z - 1}{\sqrt{1 - k^2 \sin^2 z}} \, dz = 2\frac{K(k^2) - E(k^2)}{k^2} - K(k^2) \,.$$

Finally,

$$L = \mu_0 \sqrt{RR'}\, \frac{2\,(K - E) - k^2\,K}{k} \,.$$

The mutual inductance of two coaxial circles can thus be reduced to elliptic integrals (see Fig. 3.14).

Particularly important is the special case $R \approx R' \gg a$, i.e., $k \approx 1$, of two close current loops. Then the integrand of E is approximately equal to $\cos z$, so $E \approx 1$ and $L \approx \mu_0 \sqrt{RR'}\,(K - 2)$. To calculate K for $k \approx 1$, a series expansion cannot be employed, since the indefinite integral for $k = 1$ diverges as $\ln \cot(\frac{1}{4}\pi - \frac{1}{2}x)$ at the upper boundary. But for the *incomplete elliptic integral of the first kind* (see p. 103)

$$F(\varphi \,|\, k^2) \equiv \int_0^\varphi \frac{dz}{\sqrt{1 - k^2 \sin^2 z}} \,, \qquad \text{thus} \qquad F(\tfrac{1}{2}\pi \,|\, k^2) = K(k^2) \,,$$

there exists the *ascending Landen transformation* (in k^2) $2z_1 = z + \arcsin(k \sin z)$ (see Problem 3.29), viz.,

$$F(\varphi \,|\, k^2) = \frac{2}{1 + k} F(\varphi_1 \,|\, k_1{}^2) \,,$$

with

$$k_1{}^2 = \frac{4k}{(1 + k)^2} \qquad \text{and} \qquad \varphi_1 = \frac{\varphi + \arcsin(k \sin \varphi)}{2} \,.$$

For $k^2 = 1 - \varepsilon$ and $\varphi = \frac{1}{2}\pi$, we have $k_1{}^2 \approx 1 - \frac{1}{16}\varepsilon^2$ and $\varphi_1 = \frac{1}{2}\pi - \delta\varphi$ with $\delta\varphi = \frac{1}{2} \arccos \sqrt{1 - \varepsilon} \approx \frac{1}{2}\sqrt{\varepsilon}$. Consequently, for the ascending transformation (in k^2), the upper boundary φ decreases, and now we may set

$$k_1{}^2 \approx 1 : \quad F(\frac{1}{2}\pi - \delta\varphi \,|\, 1) = \ln(\cot \frac{1}{2}\delta\varphi) \approx \ln(4/\sqrt{\varepsilon}) \,.$$

Hence for $k \approx 1$, we arrive at $K \approx \ln(4/\sqrt{1 - k^2})$ and obtain

$$L = \mu_0 \sqrt{RR'} \left(\ln \frac{4\,(R + R')}{\sqrt{a^2 + (R - R')^2}} - 2 \right), \quad \text{for} \quad R \approx R' \gg a \,,$$

i.e., for two nearby loops with like axis.

3.2.11 Summary: Stationary Currents and Magnetostatics

For electric currents, we use the current density $\mathbf{j} = \rho \mathbf{v}$ and the current strength $I = \int d\mathbf{f} \cdot \mathbf{j}$. Stationary currents are solenoidal. In the following, according to common practice, we write the averaged current density without the bar, since we would like to use only macroscopically measurable quantities anyway.

In many cases, we have Ohm's law in differential form

$$\mathbf{j} = \sigma \, \mathbf{E} + \mathbf{j}' \,.$$

Here σ is the conductivity and $\mathbf{j}'$ the current density at the current sources.

All electric currents are accompanied by a magnetic field. Hence we can also identify currents in atoms, which do not contribute to macroscopic electric currents. They can be understood via the magnetization $\mathbf{M}$ or via the magnetic moment $\mathbf{m} = \frac{1}{2} \int dV \mathbf{r} \times \mathbf{j}$. Hence, macroscopically,

$$\nabla \cdot \mathbf{B} = 0 \quad \text{and} \quad \nabla \times \mathbf{H} = \mathbf{j}, \quad \text{with} \quad \mathbf{B} = \mu_0 (\mathbf{H} + \mathbf{M}) = \mu \mathbf{H}.$$

Since the induction field is solenoidal, it derives from a vector potential $\mathbf{A}$ with the property $\mathbf{B} = \nabla \times \mathbf{A}$. For the Coulomb gauge ($\nabla \cdot \mathbf{A} = 0$) and using $\nabla \times \mathbf{B} = \mu_0 (\mathbf{j} + \nabla \times \mathbf{M})$, we have

$$\mathbf{A}(\mathbf{r}) = \frac{\mu_0}{4\pi} \int dV' \, \frac{\mathbf{j}(\mathbf{r}') + \nabla' \times \mathbf{M}}{|\mathbf{r} - \mathbf{r}'|}.$$

The magnetic field acts on a moving charge via the Lorentz force $\mathbf{F} = \int dV \mathbf{j} \times \mathbf{B}$. The force between two conductors with stationary currents is then given by the Ampère law

$$\mathbf{F} = \frac{\mu_0}{4\pi} \iint dV \, dV' \mathbf{j}(\mathbf{r}) \cdot \mathbf{j}(\mathbf{r}') \, \nabla \frac{1}{|\mathbf{r} - \mathbf{r}'|}.$$

Currents with like orientation in parallel conductors attract each other.

3.3   Electromagnetic Field

3.3.1   Charge Conservation and Maxwell's Displacement Current

The charge conservation law was expressed on p. 187 in the form of a continuity equation, viz.,

$$\partial \rho / \partial t + \nabla \cdot \mathbf{j} = 0.$$

Conversely, the continuity equation ensures charge conservation. Since $\rho = \nabla \cdot \mathbf{D}$, we thus also have

$$\nabla \cdot \left(\mathbf{j} + \frac{\partial \mathbf{D}}{\partial t} \right) = 0,$$

or according to Gauss's theorem,

$$0 = \int_{(V)} d\mathbf{f} \cdot \left(\mathbf{j} + \frac{\partial \mathbf{D}}{\partial t} \right) = I + \frac{dQ}{dt}.$$

As long as, e.g., the charge on the anode of a capacitor increases, a current will also flow, with a sink for the current density $\mathbf{j}$. If we connect the current loop with the capacitor in a Gedanken experiment, an electric current will flow in the conductor, while *Maxwell's displacement current* will flow in a non-conductor, with current density $\partial \mathbf{D}/\partial t$. If an electric field changes with time, then this is the corresponding current.

The sum of the conduction and displacement current densities is solenoidal, and hence is a rotational field. For stationary currents, it is the curl of the magnetic field $\mathbf{H}$—but this is in fact generally true:

$$\nabla \times \mathbf{H} = \mathbf{j} + \frac{\partial \mathbf{D}}{\partial t} \quad \Longleftrightarrow \quad \int_{(A)} \mathrm{d}\mathbf{r} \cdot \mathbf{H} = I + \frac{\mathrm{d}}{\mathrm{d}t} \int_A \mathrm{d}\mathbf{f} \cdot \mathbf{D} .$$

While a capacitor is being charged, there is thus a magnetic field around it, not only around the connecting wires. For the path integral $\oint \mathrm{d}\mathbf{r} \cdot \mathbf{H}$, only the boundary of the area A is of interest. If we choose two different sheets with the same boundary (A) for $\int \mathrm{d}\mathbf{f} \cdot \mathbf{D}$, then the values of the surface integrals differ by the charge Q enclosed by these two sheets. In fact, $I + \dot{Q}$ then no longer depends on the chosen area.

In insulators there is no conduction current, but at most a displacement current, while in conductors the displacement current is in most cases negligible compared to the electric current. If we have a periodic process with angular frequency ω, then for $j/\dot{D}$ this clearly depends on the ratio $\sigma/\varepsilon\omega$. Here most conductors have $\sigma/\varepsilon > 100$ THz. Therefore, the order of magnitude of the ratio $\sigma/\varepsilon\omega$ is only unity for frequencies common in optics.

As long as the displacement is negligible compared to the electric current, the currents are said to be *quasi-static*—for stationary currents all derivatives with respect to time vanish.

3.3.2 Faraday Induction Law and Lenz's Rule

As was just shown, the two equations $\nabla \cdot \mathbf{D} = \rho$ and $\nabla \times \mathbf{H} = \mathbf{j} + \partial \mathbf{D}/\partial t$ ensure charge conservation. If there were no free charges but only electric dipoles, we would have instead $\nabla \cdot \mathbf{D} = 0$ and $\partial \mathbf{D}/\partial t = \nabla \times \mathbf{H}$. This is noteworthy insofar as we would not find magnetic charges, but only magnetic dipoles—whence we already set up the equation $\nabla \cdot \mathbf{B} = 0$ in magnetostatics. Hence we can ask the question whether $\partial \mathbf{B}/\partial t$ is equal to the circulation density of a (polar) vector field, in particular, a vector field which is irrotational for time-independent phenomena.

In fact, we have the *Faraday induction law*,

$$\nabla \times \mathbf{E} = -\frac{\partial \mathbf{B}}{\partial t} .$$

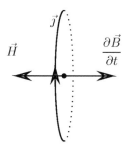

Fig. 3.15 Lenz's rule. The time-dependence of the magnetic field $\partial \mathbf{B}/\partial t = -\nabla \times \mathbf{E}$ induces a current density $\mathbf{j} = \sigma \mathbf{E}$ in the conductor loop. This current is accompanied by a magnetic field curl density $\nabla \times \mathbf{H} = \mathbf{j}$ such that, on the plane of the loop, $\mathbf{H}$ and $\partial \mathbf{B}/\partial t$ are oriented in opposite directions

A time-dependent magnetic field and the curl of the electric field are related: the magnetic field induces an electric current in a conductor loop. Every dynamo makes use of this. The sign in the induction law supplies the important *Lenz rule* (see Fig. 3.15): *the induced current works against its cause.*

In integral form, the induction law reads

$$\int_{(A)} d\mathbf{r} \cdot \mathbf{E} = -\frac{d}{dt} \int_A d\mathbf{f} \cdot \mathbf{B} \ .$$

Since $\nabla \cdot \mathbf{B} = 0$, the last expression depends only on the boundary of the area A. The left contour integral is called the *circulation voltage* or *induction voltage*. We note that the concept of electric voltage between two points introduced previously (p. 169) can now yield different values depending on the path in-between.

3.3.3 Maxwell's Equations

Now we have prepared sufficiently for the famous Maxwell equations, with which we can describe many phenomena of electricity and optics—including also $\mathbf{D} = \varepsilon \mathbf{E}$ and $\mathbf{B} = \mu \mathbf{H}$:

$$\nabla \times \mathbf{E} = -\frac{\partial \mathbf{B}}{\partial t} \ , \qquad\qquad \nabla \cdot \mathbf{B} = 0 \ ,$$
$$\nabla \cdot \mathbf{D} = \rho \ , \qquad\qquad \nabla \times \mathbf{H} = \mathbf{j} + \frac{\partial \mathbf{D}}{\partial t} \ .$$

These couple the electric and magnetic fields. It is thus better to speak of the total *electromagnetic field*. As integral equations, they read

$$\int_{(A)} d\mathbf{r} \cdot \mathbf{E} = -\frac{d}{dt} \int_A d\mathbf{f} \cdot \mathbf{B} \,, \qquad \int_{(V)} d\mathbf{f} \cdot \mathbf{B} = 0 \,,$$

$$\int_{(V)} d\mathbf{f} \cdot \mathbf{D} = Q \,, \qquad \int_{(A)} d\mathbf{r} \cdot \mathbf{H} = I + \frac{d}{dt} \int_A d\mathbf{f} \cdot \mathbf{D} \,.$$

because $\int_V dV \, \rho = Q$ and $\int_A d\mathbf{f} \cdot \mathbf{j} = I$. The boundary conditions for the transition at an interface are similar to those in the static case:

$$\mathbf{n} \times (\mathbf{E}_+ - \mathbf{E}_-) = 0 \,, \qquad \mathbf{n} \cdot (\mathbf{B}_+ - \mathbf{B}_-) = 0 \,,$$
$$\mathbf{n} \cdot (\mathbf{D}_+ - \mathbf{D}_-) = \rho_A \,, \qquad \mathbf{n} \times (\mathbf{H}_+ - \mathbf{H}_-) = \mathbf{j}_A \,.$$

In particular, there is no field $\mathbf{B}$ or $\mathbf{D}$ whose derivative with respect to time on the interface is singular like a delta function. There is at most a discontinuity like a step function. Its source density or circulation density may be singular like a delta function, but because $\delta(x) = \varepsilon'(x)$, there is only a finite discontinuity in the field, not an infinite one as for the delta function. Therefore, the derivatives of $\mathbf{B}$ and $\mathbf{D}$ with respect to time do not contribute to the surface curl density.

Clearly, the curl of the electric and the magnetic field are connected with time-dependent changes, while their sources are already known from statics. Therefore, in statics $\mathbf{E}$ and $\mathbf{H}$, or $\mathbf{D}$ and $\mathbf{B}$, are similar. But for time-dependent phenomena on the one hand $\mathbf{E}$ and $\mathbf{B}$ are connected, and on the other hand $\mathbf{D}$ and $\mathbf{H}$ are connected.

All Maxwell's equations were already known prior to Maxwell, except for the one involving the displacement current, but it is only by virtue of the latter that certain key phenomena such as charge conservation and electromagnetic waves can exist.

According to the Fourier transform $\mathbf{r} \to \mathbf{k}$ (see p. 22),

$$\mathbf{E}(t, \mathbf{r}) = \frac{1}{\sqrt{2\pi}^3} \int d^3k \, \exp(+i\mathbf{k} \cdot \mathbf{r}) \, \mathbf{E}(t, \mathbf{k}) \,,$$

$$\mathbf{E}(t, \mathbf{k}) = \frac{1}{\sqrt{2\pi}^3} \int d^3r \, \exp(-i\mathbf{k} \cdot \mathbf{r}) \, \mathbf{E}(t, \mathbf{r}) \,,$$

and correspondingly for $\mathbf{D}, \mathbf{B}, \mathbf{H}, \mathbf{j}$, and ρ, Maxwell's equations read

$$i\mathbf{k} \times \mathbf{E}(t, \mathbf{k}) = -\frac{\partial \mathbf{B}(t, \mathbf{k})}{\partial t} \,, \qquad i\mathbf{k} \cdot \mathbf{B}(t, \mathbf{k}) = 0 \,,$$

$$i\mathbf{k} \cdot \mathbf{D}(t, \mathbf{k}) = \rho(t, \mathbf{k}) \,, \qquad i\mathbf{k} \times \mathbf{H}(t, \mathbf{k}) = \mathbf{j}(t, \mathbf{k}) + \frac{\partial \mathbf{D}(t, \mathbf{k})}{\partial t} \,,$$

and the continuity equation

$$\frac{\partial \rho(t, \mathbf{k})}{\partial t} + i\mathbf{k} \cdot \mathbf{j}(t, \mathbf{k}) = 0 \,.$$

The real differential expressions in real space thus become complex in $\mathbf{k}$-space, but local expressions for the transverse and longitudinal parts of the field. In particular, the induction field is purely transverse:

$$\nabla \times \mathbf{E}_{\text{trans}} = -\frac{\partial \mathbf{B}_{\text{trans}}}{\partial t} \ , \qquad \mathbf{B}_{\text{long}} = \mathbf{0} \ .$$

In addition, $\nabla \cdot \mathbf{D}_{\text{long}} = \rho$ holds, and we can split up the fourth of Maxwell's equations:

$$\nabla \times \mathbf{H}_{\text{trans}} = \mathbf{j}_{\text{trans}} + \frac{\partial \mathbf{D}_{\text{trans}}}{\partial t} \quad \text{and} \quad \mathbf{0} = \mathbf{j}_{\text{long}} + \frac{\partial \mathbf{D}_{\text{long}}}{\partial t} \ .$$

With $\nabla \cdot \mathbf{D}_{\text{long}} = \rho$, the last equation leads to the continuity equation.

The fields are real in real space and, according to p. 22, have the symmetry $\mathbf{E}(t, \mathbf{k}) = \mathbf{E}^*(t, -\mathbf{k})$, and likewise for $\mathbf{D}$, $\mathbf{B}$, $\mathbf{H}$, $\mathbf{j}$, and ρ. In particular, for a point charge $\rho(t, \mathbf{r}) = q\, \delta(\mathbf{r} - \mathbf{r}')$ has (complex) Fourier transform $\rho(t, \mathbf{k}) = (2\pi)^{-3/2} q \exp(-i\mathbf{k} \cdot \mathbf{r}')$.

We derived the microscopic Maxwell equations from the "facts of observation". There are electric, but no magnetic charges; charges remain conserved; we find the force law due to Coulomb, the one due to Ampère (Lorentz), and also Faraday's induction law. The "macroscopic" Maxwell equations start from

$$\mathbf{D} = \varepsilon_0 \mathbf{E} + \mathbf{P} = \varepsilon \mathbf{E} \quad \text{and} \quad \mathbf{B} = \mu_0\, (\mathbf{H} + \mathbf{M}) = \mu \mathbf{H} \ ,$$

with averaged charge and current densities, the polarization $\mathbf{P}$, and the magnetization $\mathbf{M}$. Actually, we should have written $\mathbf{H} = \mathbf{B}/\mu_0 - \mathbf{M} = \mathbf{B}/\mu$ for the magnetic excitation, since $\mathbf{E}$ and $\mathbf{B}$ are related, and likewise $\mathbf{D}$ and $\mathbf{H}$.

In the following we shall always assume linear relations between $\mathbf{D}$ and $\mathbf{E}$ and/or $\mathbf{H}$ and $\mathbf{B}$, even though there are also "nonlinear effects", e.g., for hysteresis and for strong fields of the kind occurring in laser light. In addition, we calculate only with scalar relations—this is generally not allowed in crystal physics, where ε and μ are tensors. But even there, many phenomena can already be treated, and the calculations are then simple.

In addition, we have to observe Ohm's law:

$$\mathbf{j} = \sigma \mathbf{E} \quad \text{or} \quad U = R\, I \ .$$

To a first approximation, the conductivity σ and the resistance R do not depend on the applied field. (Here σ is actually a tensor.)

3.3.4 Time-Dependent Potentials

As long as the fields do not depend on time, they can be derived from the scalar potential Φ and the vector potential $\mathbf{A}$, as was shown in Sects. 3.1.3 and 3.2.8. This works even for time-dependent fields. The induction field in particular remains solenoidal, and therefore can still be derived from the curl of a vector potential:

$$\nabla \cdot \mathbf{B} = 0 \qquad \Longleftrightarrow \qquad \mathbf{B} = \nabla \times \mathbf{A} \,.$$

However, for time-dependent magnetic fields the electric field $\mathbf{E}$ has curls, and a gradient field $(-\nabla\Phi)$ is no longer sufficient, but since according to the last equation we have $\partial \mathbf{B}/\partial t = \nabla \times \partial \mathbf{A}/\partial t$, the induction law $\nabla \times \mathbf{E} = -\partial \mathbf{B}/\partial t$ now implies

$$\mathbf{E} = -\nabla\Phi - \frac{\partial \mathbf{A}}{\partial t} \,.$$

With the two quantities Φ and $\mathbf{A}$ (which have four components in total), we can thus determine the two vector fields $\mathbf{E}$ and $\mathbf{B}$ (with six components in total). It remains only to comply only with the two remaining Maxwell equations (where we assume $\mathbf{D} = \varepsilon\mathbf{E}$ and $\mathbf{B} = \mu\mathbf{H}$ with constant factors ε and μ, i.e., homogeneous matter). Since $\Delta\Phi = \nabla \cdot \nabla\Phi$, it follows that

$$\Delta\Phi = -\frac{\rho}{\varepsilon} - \frac{\partial}{\partial t} \nabla \cdot \mathbf{A} \,,$$

and since $\Delta\mathbf{A} = \nabla(\nabla \cdot \mathbf{A}) - \nabla \times (\nabla \times \mathbf{A})$,

$$\left(\Delta - \varepsilon\mu \frac{\partial^2}{\partial t^2}\right)\mathbf{A} = -\mu\mathbf{j} + \nabla\left(\nabla \cdot \mathbf{A} + \varepsilon\mu \frac{\partial\Phi}{\partial t}\right).$$

We do not use $\mathbf{j} = \sigma\mathbf{E}$, since here ρ and $\mathbf{j}$ are viewed as given. The potentials are not unique though, since the source of the vector potential has not yet been given. The magnetic field does not depend on it, and its influence on the electric field can be counteracted by a change in the scalar potential. Therefore, despite the *gauge transformation*

$$\Phi' = \Phi + \frac{\partial\Psi}{\partial t} \qquad \text{and} \qquad \mathbf{A}' = \mathbf{A} - \nabla\Psi \,,$$

with continuously differentiable Ψ, the same fields $\mathbf{E}$ and $\mathbf{B}$ result. Physical quantities do not depend on the gauge. The curl of the vector potential determines the magnetic field, and the sources determine $\Delta\Psi$. In the static case we were allowed to choose these sources arbitrarily, but now their time dependence shows up for the scalar potential. Every gauge transformation changes the longitudinal component of the vector potential and the scalar potential. Then it is clear that

$$\mathbf{E}_{\text{long}} = -\nabla\Phi - \frac{\partial \mathbf{A}_{\text{long}}}{\partial t} \,, \qquad\qquad \mathbf{B}_{\text{long}} = \mathbf{0} \,,$$
$$\mathbf{E}_{\text{trans}} = -\frac{\partial \mathbf{A}_{\text{trans}}}{\partial t} \,, \qquad\qquad \mathbf{B}_{\text{trans}} = \nabla \times \mathbf{A}_{\text{trans}} \,.$$

Longitudinal fields are irrotational, transverse ones solenoidal.

There are two possibilities for the gauge such that the equations for the scalar and the vector potential decouple. This can be seen immediately for the *Lorentz gauge*

$$\nabla \cdot \mathbf{A} + \varepsilon \mu \frac{\partial \Phi}{\partial t} = 0 \, ,$$

and in particular,

$$\left(\varepsilon \mu \frac{\partial^2}{\partial t^2} - \Delta \right) \Phi = \frac{\rho}{\varepsilon} \, , \qquad \left(\varepsilon \mu \frac{\partial^2}{\partial t^2} - \Delta \right) \mathbf{A} = \mu \mathbf{j} \, .$$

These formally similar equations will be preferred in the next section on Lorentz invariance. There is a retardation effect here: ρ and $\mathbf{j}$ are important at time $t' = t - |\mathbf{r} - \mathbf{r}'|/c$, showing that actions propagate with finite velocity. This will be explained in more detail in Sect. 3.5.1.

But for the moment we prefer to take the *Coulomb gauge* (*radiation gauge*, *transverse gauge*)

$$\nabla \cdot \mathbf{A} = 0 \, .$$

Even though initially this yields

$$\Delta \Phi = -\frac{\rho}{\varepsilon} \, , \qquad \left(\Delta - \varepsilon \mu \frac{\partial^2}{\partial t^2} \right) \mathbf{A} = -\mu \left(\mathbf{j} - \varepsilon \nabla \frac{\partial \Phi}{\partial t} \right) ,$$

according to p. 27, the Poisson equation $\Delta \Phi = -\rho/\varepsilon$ is solved by

$$\Phi(t, \, \mathbf{r}) = \frac{1}{4\pi\varepsilon} \int dV' \, \frac{\rho(t, \, \mathbf{r}')}{|\mathbf{r} - \mathbf{r}'|} \, ,$$

and with the continuity equation $\partial \rho / \partial t = -\nabla \cdot \mathbf{j}$ therefore leads to

$$\frac{\partial \Phi}{\partial t} = -\frac{1}{4\pi\varepsilon} \int dV' \, \frac{\nabla' \cdot \mathbf{j}(t, \, \mathbf{r}')}{|\mathbf{r} - \mathbf{r}'|} \, .$$

Thus according to p. 25, $\varepsilon \nabla \partial \Phi / \partial t$ comprises the part of the current density that originates in the sources, and therefore $\mathbf{j} - \varepsilon \nabla \partial \Phi / \partial t$ is the solenoidal (transverse) current density

$$\mathbf{j}_{\mathrm{trans}}(t, \, \mathbf{r}) \equiv \nabla \times \int dV' \, \frac{\nabla' \times \mathbf{j}(t, \, \mathbf{r}')}{4\pi \, |\mathbf{r} - \mathbf{r}'|} \, .$$

Consequently, the system of equations is also decoupled in the Coulomb gauge:

$$\left(\varepsilon \mu \frac{\partial^2}{\partial t^2} - \Delta \right) \mathbf{A} = \mu \, \mathbf{j}_{\mathrm{trans}} \, .$$

For this gauge, only a solenoidal current density is therefore of interest. This occurs, in particular, if there are no macroscopic charges (then even $\Phi \equiv 0$ holds), e.g., for the radiation field of single atoms. Therefore, it is sometimes called the radiation gauge. However, it does have a disadvantage: for each Lorentz transformation, a new gauge must be derived, because it is not Lorentz invariant.

3.3.5 Poynting's Theorem

The Maxwell equations imply in particular

$$\mathbf{E} \cdot \frac{\partial \mathbf{D}}{\partial t} + \mathbf{H} \cdot \frac{\partial \mathbf{B}}{\partial t} = \mathbf{E} \cdot (\nabla \times \mathbf{H} - \mathbf{j}) - \mathbf{H} \cdot (\nabla \times \mathbf{E}) = -\mathbf{j} \cdot \mathbf{E} - \nabla \cdot (\mathbf{E} \times \mathbf{H}) \ .$$

We recognize the expression $\mathbf{j} \cdot \mathbf{E}$ from p. 188 as the power density for the Joule heat, which does not arise in insulators. The power densities of the electric and magnetic fields are given on the left. If $\mathbf{D}$ and $\mathbf{E}$ are related to each other linearly, then the first term is the time-derivative of the known energy density $\frac{1}{2}\mathbf{E} \cdot \mathbf{D}$ of the electric field. If we also assume a linear relation between $\mathbf{H}$ and $\mathbf{B}$ (which is not allowed for ferromagnets because of hysteresis), then we may take the expression $\frac{1}{2}\mathbf{H} \cdot \mathbf{B}$ as the *energy density of the magnetic field*. It is positive-definite and is suggested in view of the similarity between the electric and magnetic field quantities. Thus we take

$$w = \tfrac{1}{2} \left(\mathbf{E} \cdot \mathbf{D} + \mathbf{H} \cdot \mathbf{B} \right)$$

as the energy density of a electromagnetic field and obtain *Poynting's theorem*:

$$\frac{\partial w}{\partial t} + \nabla \cdot (\mathbf{E} \times \mathbf{H}) = -\mathbf{j} \cdot \mathbf{E} \ .$$

If the Joule heat is missing, then this equation is similar to the continuity equation: $\mathbf{E} \times \mathbf{H}$ is the *energy flux density*, which is also called the *Poynting vector*:

$$\mathbf{S} \equiv \mathbf{E} \times \mathbf{H} \ .$$

In order to understand what it means for the stationary situation (with $\partial w/\partial t = 0$), we consider a finite piece of a conductor in Fig. 3.16. Here we have $\nabla \cdot \mathbf{S} = -\mathbf{j} \cdot \mathbf{E} = -\sigma E^2$.

Because $\mathbf{B} = \nabla \times \mathbf{A}$ and $\mathbf{H} \cdot (\nabla \times \mathbf{A}) = \nabla \cdot (\mathbf{A} \times \mathbf{H}) + \mathbf{A} \cdot (\nabla \times \mathbf{H})$ for quasi-stationary currents (i.e., for $\nabla \times \mathbf{H} = \mathbf{j}$ and no contribution from the surface integrals of $\mathbf{A} \times \mathbf{H}$), the now-justified ansatz $\frac{1}{2}\mathbf{H} \cdot \mathbf{B}$ for the energy density of the magnetic field leads to

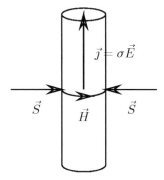

Fig. 3.16 Interpretation of the Poynting vector **S** for a stationary current along a wire of length l with radius R. Here **S** flows from the outside through the curved surface A and, because $E = U/l$ and $H = I/(2\pi R)$, it has the absolute value $S = UI/A$ there. The heat power UI generated inside then flows out through the curved surface, while the current flows through the faces

$$\tfrac{1}{2} \int dV \, \mathbf{H} \cdot \mathbf{B} = \tfrac{1}{2} \int dV \, \mathbf{j} \cdot \mathbf{A} \hat{=} \tfrac{1}{2} I \int d\mathbf{r} \cdot \mathbf{A} = \tfrac{1}{2} L \, I^2 \, ,$$

where L is now the *self-inductance* of the conductor. According to the Neumann formula

$$L = \frac{\mu_0}{4\pi} \iint \frac{d\mathbf{r} \cdot d\mathbf{r}'}{|\mathbf{r} - \mathbf{r}'|} \, ,$$

it can be determined, but no arbitrarily thin conductors can be taken, otherwise L diverges according to p. 203. We would then have to integrate over the mutual inductances of the various current lines (Problem 3.30).

For the energy of two stationary currents, we derived the expression $E_{\text{pot}} = -\int dV \, \mathbf{j} \cdot \mathbf{A}$ on p. 199. Despite the other sign, this does not contradict the value just found for the self-energy. In the previous case, the current distributions were given and the mutual position and orientation of the loops were changed for fixed current density, while now it is the geometrical situation that is kept fixed and the current strength increases from zero to the final value.

The energy of the electromagnetic field in thermodynamics is a "free energy". It can be fully used for work—more on that in Sect. 6.4.8. There, too, all energies will be split into products of intensive and extensive quantities, which disproves the microscopically suggested expression $\tfrac{1}{2}(\varepsilon_0 E^2 + B^2/\mu_0)$. Thermodynamically, D and H must appear in addition to E and B.

For static problems we left out integrals of the form

$$\int_V dV \, \boldsymbol{\nabla} \cdot \mathbf{S} = \int_{(V)} d\mathbf{f} \cdot \mathbf{S} \, ,$$

if integrations with boundaries at infinity were to be performed, since we assumed that the integrand would decrease more strongly at infinity than r^{-2}: in fact, E at least as r^{-2} and H at least as r^{-3} (monopole or dipole field). But for time-dependent situations, $\mathbf{E}$ and $\mathbf{H}$ then decrease rather slowly with the distance from the radiation source, whence the surface integral $\int d\mathbf{f} \cdot \mathbf{S}$ does not vanish even for very large volumes—we must still account for the radiation power, which we will only consider in Sect. 3.3.7.

3.3.6 Oscillating Circuits

If we connect a resistance R, an inductance L, and a capacity C in series to an AC voltage U, then the energy appears in three forms: in the resistance according to p. 188 as Joule heat $\int R I^2 \, dt$, in the inductance as magnetic energy $\frac{1}{2} L I^2$, and in the capacity as electric energy $\frac{1}{2} Q^2/C$. All three together must be supplied to the setup. We neglect the radiation power, which increases according to p. 264 as the fourth power of the frequency and barely contributes for quasi-stationary situations. Since $\dot{Q} = -I$, the total power is then $I \, (RI + L\dot{I} - Q/C)$. The expression in brackets must be equal to the applied voltage. The derivative with respect to time yields

$$ L \frac{d^2 I}{dt^2} + R \frac{dI}{dt} + \frac{1}{C} I = \frac{dU}{dt} , $$

which is the differential equation of a forced damped oscillation, as in Sect. 2.3.8. There the decay coefficient $\gamma \cong \frac{1}{2} R/L$ and the angular frequency $\omega_0 \cong 1/\sqrt{LC}$ were introduced, and it was shown that the initial eigenoscillation decays with time and that the solution then oscillates with the angular frequency ω of the source of the voltage. Therefore, we calculate in the final state, with

$$ U = \text{Re} \{\mathcal{U} \exp(-i\omega t)\} \quad \text{and} \quad I = \text{Re} \{\mathcal{I} \exp(-i\omega t)\} . $$

The ansatz $\exp(+i\omega t)$ is often made, and this leads to the opposite sign of i in the following equations. For our choice, which is also common in quantum theory, its value moves clockwise in the complex plane. $\mathcal{U}$ and $\mathcal{I}$ do not depend on time. In the course of time, their products with $\exp(-i\omega t)$ become purely real as well as also purely imaginary. Hence the differential equation leads to $(-\omega^2 L - i\omega R + C^{-1}) \, \mathcal{I} = -i\omega \, \mathcal{U}$ and then *Ohm's law for AC currents*, viz.,

$$ \mathcal{U} = \mathcal{Z} \, \mathcal{I} , \quad \text{with impedance} \quad \mathcal{Z} \equiv R + i \left(\frac{1}{\omega C} - \omega L \right) = R + i X . $$

It is composed of the *active resistance* R and the *reactance* X. The imaginary part shifts the phase between the voltage and current by $\phi = \arctan X/R$. The build-up of the electromagnetic field takes time—in the capacitor the voltage follows the current, while it precedes the current in the coil (see Fig. 3.17). Therefore, $|\phi| \leq \frac{1}{2}\pi$

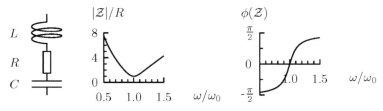

Fig. 3.17 Absorption circuit. Resonance for $\omega_0 = 1/\sqrt{LC}$. Here, $\omega_0 L = 5R$

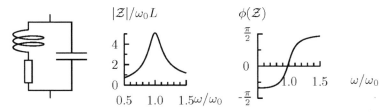

Fig. 3.18 Trap circuit. Resonance occurs for $\sqrt{\omega_0{}^2 - (R/L)^2}$. Note that here $\omega_0 L = 5R$

holds here, in contrast to the forced oscillation in Sect. 2.3.8 (see Fig. 2.23). For low frequencies ($\omega < 1/\sqrt{LC}$), it is determined mainly by the capacity, and for high frequencies by the inductance. (R does not depend on the frequency, as long as the conductivity does not depend on it, and it determines the power loss.) For $\omega = \omega_0 \equiv 1/\sqrt{LC}$, the reactance vanishes, and therefore the absolute value of the impedance, the *fictitious resistance* $Z = |\mathscr{Z}|$, is particularly small.

Corresponding to Kirchhoff's laws, we have added here the individual contributions of the three parts of the conductor. For parallel connection of a capacitor (capacity C) and a coil (inductance L and resistance R), we have in contrast $\mathscr{Z}^{-1} = (R - i\omega L)^{-1} - i\omega C$ (see Fig. 3.18):

$$\mathscr{Z} = \omega_0 L \, \frac{(R/\omega_0 L) + \mathrm{i}\,(\omega/\omega_0)\,\{(\omega/\omega_0)^2 - 1 + (R/\omega_0 L)^2\}}{(R/\omega_0 L)^2(\omega/\omega_0)^2 + \{(\omega/\omega_0)^2 - 1\}^2} \,.$$

The fictitious resistance is now highest for $\omega = \sqrt{\omega_0{}^2 - (R/L)^2}$, where it is equal to $(\omega_0 L)^2/R = L/(RC)$. Therefore, we also refer to such connections as *trap circuits* (and, if connected in series, as *absorption circuits*).

3.3.7 Momentum of the Radiation Field

With the force density $\rho\mathbf{E} + \mathbf{j} \times \mathbf{B}$, Maxwell's equations read

$$\rho\,\mathbf{E} + \mathbf{j} \times \mathbf{B} = \nabla \cdot \mathbf{D}\,\mathbf{E} + \left(\nabla \times \mathbf{H} - \frac{\partial \mathbf{D}}{\partial t}\right) \times \mathbf{B} \,.$$

Here, the last vector product can be rewritten

$$\mathbf{B} \times \frac{\partial \mathbf{D}}{\partial t} = -\frac{\partial (\mathbf{D} \times \mathbf{B})}{\partial t} + \mathbf{D} \times \frac{\partial \mathbf{B}}{\partial t} = -\frac{\partial (\mathbf{D} \times \mathbf{B})}{\partial t} - \mathbf{D} \times (\nabla \times \mathbf{E}) \ .$$

Because $\nabla \cdot \mathbf{B} = 0$, we therefore have

$$\mathbf{F} + \frac{d}{dt} \int dV \, \mathbf{D} \times \mathbf{B} = \int dV \, \{\mathbf{E} \, \nabla \cdot \mathbf{D} - \mathbf{D} \times (\nabla \times \mathbf{E})$$
$$+ \mathbf{H} \, \nabla \cdot \mathbf{B} - \mathbf{B} \times (\nabla \times \mathbf{H})\} \ .$$

We restrict ourselves to homogeneous matter, but allow also for anisotropic, preferential directions—then the permittivity and the permeability are tensors, and oblique coordinates can be useful, although at least rectilinear ones. According to p. 184, for homogeneous matter we have

$$\mathbf{E} \, \nabla \cdot \mathbf{D} - \mathbf{D} \times (\nabla \times \mathbf{E}) = \sum_{ik} \mathbf{g}_i \, \partial(E^i D^k - \frac{1}{2} g^{ik} \mathbf{E} \cdot \mathbf{D}) / \partial x^k \ ,$$

and likewise with H, B instead of E, D. Therefore, we now generalize Maxwell's stress tensor from p. 184 to include magnetic field contributions (it is symmetric only for isotropic media):

$$T^{ik} \equiv w \, g^{ik} - E^i D^k - H^i B^k \ ,$$

and according to Gauss's theorem and Sects. 1.2.4 and 1.2.5, obtain for time-dependent fields

$$\mathbf{F} + \frac{d}{dt} \int dV \, \mathbf{D} \times \mathbf{B} + \sum_{ik} \mathbf{g}_i \int_{(V)} df_k \, T^{ik} = 0 \ .$$

According to this, we have to view $\mathbf{D} \times \mathbf{B}$ as a *momentum density*. For isotropic media, it is equal to $\varepsilon \mu \mathbf{S}$ and then has the same direction as the energy flux density $\mathbf{S}$, but a different one for anisotropic media.

3.3.8 Propagation of Waves in Insulators

In insulators, i.e., if ρ and $\mathbf{j}$ vanish, and for constant ε and μ, we have

$$\left(\varepsilon \mu \frac{\partial^2}{\partial t^2} - \Delta \right) \mathbf{A}(t, \mathbf{r}) = 0 \ ,$$

according to Sect. 3.3.4 (see in particular p. 210), and this for both the Lorentz and the Coulomb gauge. This (homogeneous) *wave equation for a vector field* is also encountered for the electric and magnetic fields. In particular, in the insulator,

$$\nabla \times \mathbf{E} = -\frac{\partial \mathbf{B}}{\partial t} \,, \qquad \nabla \cdot \mathbf{B} = 0 \,, \qquad \nabla \cdot \mathbf{D} = 0 \,, \qquad \text{and} \quad \nabla \times \mathbf{H} = \frac{\partial \mathbf{D}}{\partial t} \,.$$

Hence, since $\Delta \mathbf{a} = \nabla(\nabla \cdot \mathbf{a}) - \nabla \times (\nabla \times \mathbf{a})$ for $\mathbf{D} = \varepsilon \mathbf{E}$ and $\mathbf{B} = \mu \mathbf{H}$, we have

$$\Delta \mathbf{E} = \quad \nabla \times \frac{\partial \mathbf{B}}{\partial t} = \quad \mu \frac{\partial}{\partial t} \, \nabla \times \mathbf{H} = \varepsilon \mu \, \frac{\partial^2 \mathbf{E}}{\partial t^2} \,,$$

$$\Delta \mathbf{B} = -\mu \nabla \times \frac{\partial \mathbf{D}}{\partial t} = -\mu \frac{\partial}{\partial t} \, \nabla \times \mathbf{D} = \varepsilon \mu \, \frac{\partial^2 \mathbf{B}}{\partial t^2} \,.$$

According to these wave equations, we find the *phase velocity c* from the permittivity ε and permeability μ :

$$\varepsilon \mu = c^{-2} \,, \quad \text{in particular in vacuum} \quad \varepsilon_0 \mu_0 = c_0^{-2} \,.$$

This is *Weber's equation*. In electromagnetism, in contrast to (non-relativistic) mechanics where all velocities are on an equal footing, a particular velocity is singled out. This is connected with the question of *Lorentz invariance*, discussed in the next section. If it is taken as an observational fact (Michelson experiment), charge conservation and Coulomb's law from the microscopic Maxwell equations can be derived from it, even without knowing anything about the magnetic field. However, the charge and magnetic moment of elementary particles are not properties on an equal footing.

The wave equation is a homogeneous partial differential equation of second order. In order to solve it, we take the *Fourier transform* (see Sect. 1.1.11) $\mathbf{A}(t, \mathbf{r}) \to \mathbf{A}(t, \mathbf{k})$. Hence with $\omega \equiv ck$, the partial differential equation can be simplified to

$$\left(\frac{\partial^2}{\partial t^2} - c^2 \Delta \right) \mathbf{A}(t, \mathbf{r}) = \mathbf{0} \qquad \Longrightarrow \qquad \left(\frac{\partial^2}{\partial t^2} + \omega^2 \right) \mathbf{A}(t, \mathbf{k}) = \mathbf{0} \,.$$

Since $\mathbf{A}(t, \mathbf{r})$ must be real, $\mathbf{A}^*(t, \mathbf{k}) = \mathbf{A}(t, -\mathbf{k})$. Therefore, the solution of the differential equation reads

$$\mathbf{A}(t, \mathbf{k}) = \frac{\mathbf{A}(\mathbf{k}) \, \exp(-i\omega t) + \mathbf{A}^*(-\mathbf{k}) \, \exp(+i\omega t)}{2} \,.$$

Here, the factor $1/2$ is arbitrary (it only has to be real), but it is nevertheless useful for what follows, because we then have $\mathbf{A}(\mathbf{k})$ from the initial values $\mathbf{A}(0, \mathbf{k}) = \frac{1}{2} \{\mathbf{A}(\mathbf{k}) + \mathbf{A}^*(-\mathbf{k})\}$ and $\partial \mathbf{A}(t, \mathbf{k})/\partial t|_{t=0} = -\frac{1}{2} i\omega \{\mathbf{A}(\mathbf{k}) - \mathbf{A}^*(-\mathbf{k})\}$ as

Fig. 3.19 Linearly polarized electromagnetic wave. The polarization plane (thus **E**) (*red curve*) lies in the plane of the page and **B** (*blue curve*) is perpendicular to it

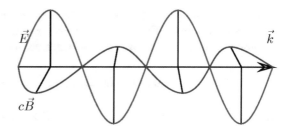

$$\mathbf{A}(\mathbf{k}) = \mathbf{A}(0,\ \mathbf{k}) + \frac{\mathrm{i}}{\omega} \left. \frac{\partial \mathbf{A}(t,\ \mathbf{k})}{\partial t} \right|_{t=0}.$$

Finally, because $\exp\{\mathrm{i}(\mathbf{k} \cdot \mathbf{r} + \omega t)\} = (\exp\{\mathrm{i}(-\mathbf{k} \cdot \mathbf{r} - \omega t)\})^*$ (and rewriting $\mathbf{k} \to -\mathbf{k}$), it follows that

$$\mathbf{A}(t,\ \mathbf{r}) = \frac{1}{\sqrt{2\pi}^{\,3}} \int \mathrm{d}^3 k \ \mathrm{Re}\left(\mathbf{A}(\mathbf{k}) \ \exp\{\mathrm{i}(\mathbf{k} \cdot \mathbf{r} - \omega t)\}\right),$$

with $\omega = ck$ and

$$\mathbf{A}(\mathbf{k}) = \frac{1}{\sqrt{2\pi}^{\,3}} \int \mathrm{d}^3 r \ \exp(-\mathrm{i}\mathbf{k} \cdot \mathbf{r}) \left(\mathbf{A}(0,\ \mathbf{r}) + \frac{\mathrm{i}}{\omega} \left. \frac{\partial \mathbf{A}(t,\ \mathbf{r})}{\partial t} \right|_{t=0}\right).$$

If we restrict ourselves to one value **k**, then this gives the propagation direction of the wave in which it travels through the homogeneous (and isotropic) medium with velocity $c = 1/\sqrt{\varepsilon\mu}$ and wavelength $\lambda = 2\pi/k$.

In a non-conductor, the fields **E** and **B** are solenoidal, thus transverse:

$$\mathbf{k} \cdot \mathbf{E}(t,\ \mathbf{k}) = 0 \quad \text{and} \quad \mathbf{k} \cdot \mathbf{B}(t,\ \mathbf{k}) = 0.$$

The vector potential is only solenoidal for the "transverse gauge" (Coulomb gauge)

$$\nabla \cdot \mathbf{A} = 0 \quad \Longrightarrow \quad \mathbf{k} \cdot \mathbf{A}(t,\ \mathbf{k}) = 0.$$

For the position and time dependence $\exp\{\mathrm{i}(\mathbf{k} \cdot \mathbf{r} - \omega t)\}$ of the fields, the equation $-\mathrm{i}\omega\,\mathbf{B}(\mathbf{k}) = -\mathrm{i}\mathbf{k} \times \mathbf{E}(\mathbf{k})$ follows from the induction law $\partial \mathbf{B}/\partial t = -\nabla \times \mathbf{E}$:

$$c\mathbf{B}(\mathbf{k}) = \mathbf{e}_k \times \mathbf{E}(\mathbf{k}), \quad \text{with} \quad \mathbf{e}_k = \frac{\mathbf{k}}{k}.$$

For $\omega \neq 0$, the three vectors **k**, **E**(**k**), and **B**(**k**) thus form a right-handed rectangular frame, and in homogeneous insulators we need only **E**(**k**) or **B**(**k**) (see Fig. 3.19).

However, this is not yet useful for the energy density $\frac{1}{2}(\mathbf{E} \cdot \mathbf{D} + \mathbf{H} \cdot \mathbf{B})$ and the energy flux density $\mathbf{E} \times \mathbf{H}$, since for a bilinear expression, a double integral over **k** and **k**′ would have to be performed. If we average over time, then we arrive at least at $\delta(\omega + \omega')$ or at $\delta(k + k')$, respectively, and if we average over space, also at $\delta(\mathbf{k} + \mathbf{k}')$. Here the Fourier components corresponding to $\mathbf{k}$ and $-\mathbf{k}$ are related, because the fields are real. We consider therefore the special case with fixed $\mathbf{k}$:

$$\mathbf{E}(t, \mathbf{r}) = \mathrm{Re}\Big(\mathbf{E}(\mathbf{k}) \exp\{i(\mathbf{k} \cdot \mathbf{r} - \omega t)\}\Big) .$$

The Maxwell equations require $\omega = ck$, $\mathbf{k} \cdot \mathbf{E}(\mathbf{k}) = 0$, and

$$c\mathbf{B}(t, \mathbf{r}) = \mathrm{Re}\Big(\mathbf{e}_k \times \mathbf{E}(\mathbf{k}) \exp\{i(\mathbf{k} \cdot \mathbf{r} - \omega t)\}\Big) .$$

Because $\mathrm{Re}z = \frac{1}{2}(z + z^*)$, the expression $\frac{1}{2}\mathbf{E}^*(\mathbf{k}) \cdot \mathbf{D}(\mathbf{k})$ follows for the time-averaged value of $\mathbf{E} \cdot \mathbf{D}$. For the mean value of $\mathbf{H} \cdot \mathbf{B}$, we find the same, because the fields are transverse. The average energy density is

$$\overline{w(t, \mathbf{r})} = \tfrac{1}{2}\mathbf{E}^*(\mathbf{k}) \cdot \mathbf{D}(\mathbf{k}) = \tfrac{1}{2}\mathbf{H}^*(\mathbf{k}) \cdot \mathbf{B}(\mathbf{k}) .$$

Therefore, from the average energy density $\overline{w}$, we can also determine the amplitude $\widehat{E}$ of the field strength:

$$\overline{w} = \varepsilon\,\overline{E^2} = \tfrac{1}{2}\varepsilon\,\widehat{E}^2 \quad\Longrightarrow\quad \widehat{E} = \sqrt{\frac{2\,\overline{w}}{\varepsilon}} .$$

This expression is needed, e.g., for the energy of interaction between a wave with energy $\hbar\omega = \overline{w}\,V$ and the dipole moment $\mathbf{p}$ of an atom, yielding

$$W = p\sqrt{2\hbar\omega/\varepsilon V}\,\cos(\omega t) .$$

For the mean value of the Poynting vector, we obtain

$$\overline{\mathbf{S}(t, \mathbf{r})} = c\,\overline{w(t, \mathbf{r})}\,\mathbf{e}_k .$$

Note that the bars are often left out, but the equations are valid only for the average. For the velocity $(\overline{\mathbf{S}}/\overline{w})$ of the energy flux, we thus obtain $c\,\mathbf{e}_k$, a vector of absolute value c in the propagation direction $\mathbf{k}$ of the wave. The momentum density $\varepsilon\mu\mathbf{S}$ has the same direction, and its absolute value is equal to $\overline{w}/c$, from Weber's equation $\varepsilon\mu = c^{-2}$. In Sect. 3.4.9, we shall also arrive at this ratio between energy and momentum for massless free particles.

A further feature of electromagnetic radiation is its *polarization direction*. Here we mean the oscillation direction of the electric field—the magnetic field oscillates perpendicular to it, since $\omega\mathbf{B}(\mathbf{k}) = \mathbf{k} \times \mathbf{E}(\mathbf{k})$. Therefore, one of the two unit vectors $\mathbf{e}_\parallel$ and $\mathbf{e}_\perp$ with $\mathbf{e}_\parallel \cdot \mathbf{e}_\perp = 0$ and $\mathbf{e}_\parallel \times \mathbf{e}_\perp = \mathbf{e}_k$ suffices for expansion of the field vectors. Then we have, e.g.,

$$\mathbf{E}(\mathbf{k}) = \mathbf{e}_\parallel \, E_\parallel + \mathbf{e}_\perp \, E_\perp \,.$$

The direction of the two unit vectors is thus not yet uniquely fixed. We are free to choose a preferred direction. For the example of diffraction, we take the plane of incidence as the preferred direction: $\mathbf{e}_\parallel$ lies in the plane, $\mathbf{e}_\perp$ is perpendicular to it.

The amplitudes $\mathbf{E}(\mathbf{k})$ are Fourier components of the real quantities $\mathbf{E}(t, \mathbf{r})$ and so have complex components $E_\parallel$ and $E_\perp$. Therefore, if we set $E = |E| \exp(i\beta)$, then in the plane $\mathbf{k} \cdot \mathbf{r} = 0$, it follows that

$$\mathbf{E}(t, \mathbf{r}) = \mathrm{Re}\{\mathbf{E}(\mathbf{k})\, \exp(-i\omega t)\} = \mathbf{e}_\parallel \, |E_\parallel| \cos(\omega t - \beta_\parallel) + \mathbf{e}_\perp \, |E_\perp| \cos(\omega t - \beta_\perp) \,.$$

Instead of the the two phases $\beta_\parallel$ and $\beta_\perp$, we use their difference $\delta\beta \equiv \beta_\perp - \beta_\parallel$ and their mean value $\overline{\beta} \equiv \frac{1}{2}(\beta_\parallel + \beta_\perp)$:

$$\mathbf{E}(t, \mathbf{r}) = \quad \{\mathbf{e}_\parallel \, |E_\parallel| + \mathbf{e}_\perp \, |E_\perp|\} \; \cos(\tfrac{1}{2}\delta\beta) \; \cos(\omega t - \overline{\beta})$$
$$- \{\mathbf{e}_\parallel \, |E_\parallel| - \mathbf{e}_\perp \, |E_\perp|\} \; \sin(\tfrac{1}{2}\delta\beta) \; \sin(\omega t - \overline{\beta}) \,.$$

In general, this is an *elliptically polarized wave*, because $\mathbf{a} \cos(\omega t - \overline{\beta}) +$ $\mathbf{b} \sin(\omega t - \overline{\beta})$ traces out an ellipse. For $\mathbf{a} \propto \mathbf{b}$, we obtain a piece of a straight line (*linearly polarized wave*) and for $a = b$ with $\mathbf{a} \perp \mathbf{b}$, a circle. Therefore, for $|E_\parallel| = |E_\perp|$ with $\delta\beta = \frac{1}{2}\pi \pmod \pi$, the wave is *circularly polarized*. For $\delta\beta = \pm\frac{1}{2}\pi$, the field rotates within a quarter period from the direction $\mathbf{e}_\parallel$ to $\pm\mathbf{e}_\perp$. In optics, we speak of left- or right-circularly polarized light, depending on how the field vector rotates when we view *against* the ray direction—anticlockwise or clockwise: $\delta\beta = +\frac{1}{2}\pi$ corresponds to left-circular polarization. In contrast, in particle physics, we view *along* the ray direction and for $\delta\beta = +\frac{1}{2}\pi$, we speak of *positive helicity* (*right-handedness*) and for $\delta\beta = -\frac{1}{2}\pi$, we speak of *negative helicity* (*left-handedness*).

Instead of linear polarization, we may of course expand in terms of circularly polarized light:

$$\mathbf{E}(\mathbf{k}) = \mathbf{e}_+ \, E_+ + \mathbf{e}_- \, E_- \,.$$

Because $\mathrm{Re}\{\mathbf{E}(\mathbf{k})\exp(-i\omega t)\} = \mathrm{Re}\mathbf{E}(\mathbf{k}) \cos(\omega t) + \mathrm{Im}\mathbf{E}(\mathbf{k}) \sin(\omega t)$, for circularly polarized light, $\mathrm{Re}\mathbf{E}(\mathbf{k})$ must be perpendicular to $\mathrm{Im}\mathbf{E}(\mathbf{k})$. This property must be satisfied by the vectors $\mathbf{e}_\pm$. We take *complex unit vectors* and set

$$\mathbf{e}_\pm \equiv \frac{\mathbf{e}_\parallel \pm i\mathbf{e}_\perp}{\sqrt{2}} \, \exp(i\varphi_\pm) \,,$$

where $\mathbf{e}_+$ is appropriate for positive helicity and $\mathbf{e}_-$ for negative. The phases $\varphi_\pm$ may be chosen arbitrarily, e.g., such that the coefficients $E_\pm$ are real. (Note that, in Sect. 5.5.1, we shall take the factor $\mp$ instead of $\exp(i\varphi_\pm)$.) In any case, we always have

$$\mathbf{e}_{\pm}{}^* \cdot \mathbf{e}_{\pm} = 1 \quad \text{and} \quad \mathbf{e}_{\pm}{}^* \cdot \mathbf{e}_{\mp} = 0 \,,$$

and hence $E_{\pm} = \mathbf{e}_{\pm}{}^* \cdot \mathbf{E}(\mathbf{k})$. In addition,

$$\mathbf{e}_{\pm}{}^* \times \mathbf{e}_{\pm} = \pm i\, e_k$$

is independent of the phase factor.

3.3.9   Reflection and Diffraction at a Plane

We consider the boundary plane between two insulators and let a plane wave with wave vector $\mathbf{k}_e$ fall onto the interface. Then there is a diffracted (transmitted) wave with wave vector $\mathbf{k}_d$, and a reflected wave with wave vector $\mathbf{k}_r$ (Problem 3.40) (see Fig. 3.20).

According to Maxwell's equations, we have the boundary conditions (see p. 207)

$$\mathbf{n} \times (\mathbf{E}_e + \mathbf{E}_r - \mathbf{E}_d) = 0 \,, \qquad \mathbf{n} \cdot (\mathbf{B}_e + \mathbf{B}_r - \mathbf{B}_d) = 0 \,,$$
$$\mathbf{n} \cdot (\mathbf{D}_e + \mathbf{D}_r - \mathbf{D}_d) = 0 \,, \qquad \mathbf{n} \times (\mathbf{H}_e + \mathbf{H}_r - \mathbf{H}_d) = 0 \,.$$

Since these always have to hold, all three waves must have the same angular frequency ω, because only then will their exponential functions $\exp(-i\omega t)$ always agree with each other. Likewise, for all positions $\mathbf{r}$ on the interface, we must require

$$\mathbf{k}_e \cdot \mathbf{r} = \mathbf{k}_r \cdot \mathbf{r} = \mathbf{k}_d \cdot \mathbf{r} \,,$$

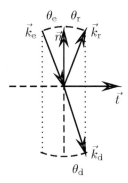

Fig. 3.20 Wave vectors $\mathbf{k}_e$, $\mathbf{k}_r$, and $\mathbf{k}_d$ at a beam splitter, an interface with the normal vector $\mathbf{n}$, the unit vector $\mathbf{t}$ in the plane of incidence, and the angles θ_e, θ_r, and θ_d. The three wave vectors have—as proven in the text—equal tangential components and $\mathbf{k}_e$ and $\mathbf{k}_r$ opposite normal components. In addition, $k_d/k_e = c_e/c_d$ holds, and the ratio of the indicated circular radii is thus equal to the refractive index n

since only then can the exponential functions $\exp(i\mathbf{k} \cdot \mathbf{r})$ be the same everywhere at the interface. If $\mathbf{r}$ is perpendicular to $\mathbf{k}_e$, then it is clearly also perpendicular to $\mathbf{k}_r$ and $\mathbf{k}_d$: all three vectors $\mathbf{k}_e$, $\mathbf{k}_r$, and $\mathbf{k}_d$ lie in the plane spanned by $\mathbf{k}_e$ and $\mathbf{n}$, the *plane of incidence*. If on the other hand we take a vector $\mathbf{r}$ along the intersecting line of the interface and the plane of incidence, namely the vector $\mathbf{t}$, then the three wave vectors must have equal tangential components:

$$k_e \sin \theta_e = k_r \sin \theta_r = k_d \sin \theta_d .$$

Now because $\omega = ck$, we also have $k_r = k_e$ and $c_d k_d = c_e k_e$, and therefore,

$$\sin \theta_e = \sin \theta_r \quad \text{and} \quad \frac{\sin \theta_e}{\sin \theta_d} = \frac{c_e}{c_d} = \sqrt{\frac{\varepsilon_d \mu_d}{\varepsilon_e \mu_e}} \equiv n ,$$

which is the *Snellius diffraction law* (see Fig. 3.20). The ratio c_e/c_d of the velocities is the *refractive index n*. One should not take the static values—the material constants depend upon the frequency (*dispersion*).

After the relations between the wave vectors, we now investigate those between the field amplitudes. To this end, it is useful to express all fields in terms of $\mathbf{E}(\mathbf{k})$ because, for linearly polarized light, the oscillation direction of the electric field is defined as the polarization direction:

$$\mathbf{D} = \varepsilon \mathbf{E} , \quad \mathbf{B} = \mathbf{e}_k \times \mathbf{E}/c , \quad \mathbf{H} = \mathbf{e}_k \times \mathbf{E}/\mu c .$$

The set of boundary conditions provides a system of equations. In order to solve these, we introduce the two unit vectors $\mathbf{t}$ and $\mathbf{b} = \mathbf{t} \times \mathbf{n}$ in addition to the normal vector $\mathbf{n}$ ($\mathbf{b}$ in Fig. 3.20 points toward the observer). With $\mathbf{k} = \mathbf{t} \; \mathbf{t} \cdot \mathbf{k} + \mathbf{n} \; \mathbf{n} \cdot \mathbf{k}$ and using the Snellius diffraction law, we find

$$\begin{aligned}
\mathbf{t} \cdot \mathbf{k}_e &= +k_e \sin \theta_e = + \mathbf{t} \cdot \mathbf{k}_r , & \mathbf{t} \cdot \mathbf{k}_d &= +k_d \sin \theta_d , \\
\mathbf{n} \cdot \mathbf{k}_e &= -k_e \cos \theta_e = - \mathbf{n} \cdot \mathbf{k}_r , & \mathbf{n} \cdot \mathbf{k}_d &= -k_d \cos \theta_d .
\end{aligned}$$

If we decompose these three $\mathbf{E}$ vectors into their *perpendicularly polarized components* $E_\perp \equiv \mathbf{E} \cdot \mathbf{b}$ (perpendicular to the plane of incidence) and their *parallel polarized components* $E_\parallel \equiv \mathbf{E} \cdot (\mathbf{b} \times \mathbf{e}_k)$ (in the plane of incidence),

$$\mathbf{E} = \mathbf{b} \, E_\perp + \mathbf{b} \times \mathbf{e}_k \, E_\parallel ,$$

then, because $\mathbf{k} \times \mathbf{E} = \mathbf{k} \times \mathbf{b} \, E_\perp + \mathbf{b} \, k \, E_\parallel$ and $\mathbf{k} \times \mathbf{b} = -\mathbf{n} \; \mathbf{t} \cdot \mathbf{k} + \mathbf{t} \; \mathbf{n} \cdot \mathbf{k}$, we have

$$\begin{aligned}
\mathbf{n} \cdot \quad \mathbf{E} &= \quad \mathbf{t} \cdot \mathbf{e}_k \, E_\parallel & , \\
\mathbf{n} \cdot (\mathbf{k} \times \mathbf{E}) &= \quad\quad\quad\quad -\mathbf{t} \cdot \mathbf{k} \, E_\perp , \\
\mathbf{n} \times (\mathbf{k} \times \mathbf{E}) &= \quad \mathbf{t} \, k \, E_\parallel - \mathbf{b} \; \mathbf{n} \cdot \mathbf{k} \, E_\perp , \\
\mathbf{n} \times \quad \mathbf{E} &= \mathbf{b} \; \mathbf{n} \cdot \mathbf{e}_k \, E_\parallel \quad\quad +\mathbf{t} \, E_\perp .
\end{aligned}$$

Hence the boundary conditions for the normal components yield

$$\varepsilon_e \sin\theta_e \; (E_{e\parallel} + E_{r\parallel}) = \varepsilon_d \sin\theta_d \; E_{d\parallel} \,, \qquad \frac{\sin\theta_e}{c_e} (E_{e\perp} + E_{r\perp}) = \frac{\sin\theta_d}{c_d} E_{d\perp} \,,$$

which are already contained in the requirements for the tangential components, if we take into account the Snellius diffraction law $\sin\theta_e : \sin\theta_d = c_e : c_d$ and Weber's equation:

$$\cos\theta_e \; (E_{e\parallel} - E_{r\parallel}) = \cos\theta_d \; E_{d\parallel} \,, \qquad\qquad E_{e\perp} + E_{r\perp} = E_{d\perp} \,,$$
$$\frac{1}{\mu_e c_e} (E_{e\parallel} + E_{r\parallel}) = \frac{1}{\mu_d c_d} E_{d\parallel} \,, \qquad \frac{\cos\theta_e}{\mu_e c_e} (E_{e\perp} - E_{r\perp}) = \frac{\cos\theta_d}{\mu_d c_d} E_{d\perp} \,.$$

Therefore, with

$$n' \equiv n \, \frac{\mu_e}{\mu_d} = \frac{c_e \, \mu_e}{c_d \, \mu_d}$$

—in insulators, in particular, $\mu \approx \mu_0$ and hence $n' \approx n$ (thus $n \approx \sqrt{\varepsilon_d/\varepsilon_e}$)—we obtain

$$\frac{E_{r\parallel}}{E_{e\parallel}} = \frac{n' \cos\theta_e - \cos\theta_d}{n' \cos\theta_e + \cos\theta_d} \,, \qquad\qquad \frac{E_{r\perp}}{E_{e\perp}} = \frac{\cos\theta_e - n' \cos\theta_d}{\cos\theta_e + n' \cos\theta_d} \,,$$
$$\frac{E_{d\parallel}}{E_{e\parallel}} = \frac{1}{n'} \left(1 + \frac{E_{r\parallel}}{E_{e\parallel}} \right) \,, \qquad\qquad \frac{E_{d\perp}}{E_{e\perp}} = 1 + \frac{E_{r\perp}}{E_{e\perp}} \,.$$

For the corresponding equations for the magnetic field strength **B**, the factor n is included in the lower row, because E and B differ by the velocity c. Note in addition that **B** oscillates in a direction perpendicular to **E**. Clearly, for perpendicular incidence and $n' = 1$, nothing is reflected, hence if the *wave resistance* $c\mu = \sqrt{\mu/\varepsilon}$ remains the same (the value for the vacuum is approximately 377Ω, according to p. 165).

If, after the approximation $n' \approx n$, we use the diffraction law $\sin\theta_e = n \sin\theta_d$, *Fresnel's equations* follow (see Fig. 3.21):

$$\frac{E_{r\parallel}}{E_{e\parallel}} = \frac{\tan(\theta_e - \theta_d)}{\tan(\theta_e + \theta_d)} \,, \qquad\qquad \frac{E_{r\perp}}{E_{e\perp}} = -\frac{\sin(\theta_e - \theta_d)}{\sin(\theta_e + \theta_d)} \,,$$
$$\frac{E_{d\parallel}}{E_{e\parallel}} = \frac{1}{\cos(\theta_e - \theta_d)} \frac{E_{d\perp}}{E_{e\perp}} \,, \qquad\qquad \frac{E_{d\perp}}{E_{e\perp}} = 1 + \frac{E_{r\perp}}{E_{e\perp}} \,.$$

Because

$$\tan(\alpha \pm \beta) = \frac{\sin\alpha\cos\beta \pm \cos\alpha\sin\beta}{\cos\alpha\cos\beta \mp \sin\alpha\sin\beta}$$

and $\cos^2\alpha + \sin^2\alpha = 1$, we have

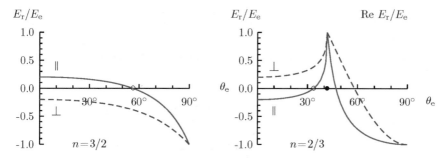

Fig. 3.21 Fresnel's equations for the transition from air to glass ($n = 3/2$) and back. Brewster angle ○. Limiting angle for total reflection ● (for larger angles, only $\mathrm{Re}\, E_r/E_e$ is shown)

$$\frac{\tan(\alpha - \beta)}{\tan(\alpha + \beta)} = \frac{\sin \alpha \cos \alpha - \sin \beta \cos \beta}{\sin \alpha \cos \alpha + \sin \beta \cos \beta} \, .$$

Part of the result could have been obtained without the calculation above. If the transmitted field strength oscillates in the direction of $\mathbf{k}_r$, then the reflected component $E_{r\parallel}$ is missing, i.e., $E_{r\parallel} = 0$ for $\mathbf{k}_r \perp \mathbf{k}_d$ or $\theta_d = 90° - \theta_r = 90° - \theta_e$. Since $n = \sin \theta_e / \sin \theta_d$, the *Brewster angle* is found to be

$$\theta_e = \arctan n \, ,$$

the reflected wave is linearly polarized, so $\mathbf{E}$ oscillates only perpendicularly to the plane of incidence. Note that, without the approximation $n' \approx n$, the Brewster angle is found to be $\arctan (n\sqrt{(n'^2 - 1)/(n^2 - 1)})$.

As a function of the angle of incidence and the refractive index in the approximation $n' \approx n$, it follows that

$$\frac{E_{r\parallel}}{E_{e\parallel}} = \frac{n^2 \cos \theta_e - \sqrt{n^2 - \sin^2 \theta_e}}{n^2 \cos \theta_e + \sqrt{n^2 - \sin^2 \theta_e}} \, , \qquad \frac{E_{r\perp}}{E_{e\perp}} = \frac{\cos \theta_e - \sqrt{n^2 - \sin^2 \theta_e}}{\cos \theta_e + \sqrt{n^2 - \sin^2 \theta_e}} \, ,$$

$$\frac{E_{d\parallel}}{E_{e\parallel}} = \frac{1}{n}\left(1 + \frac{E_{r\parallel}}{E_{e\parallel}}\right) , \qquad\qquad \frac{E_{d\perp}}{E_{e\perp}} = 1 + \frac{E_{r\perp}}{E_{e\perp}} \, ,$$

where we have used $\cos \theta_d = \sqrt{1 - \sin^2 \theta_d} = \sqrt{1 - n^{-2} \sin^2 \theta_e}$. For $n < 1$, there is a *limiting angle for total reflection*, viz., $\theta_e = \arcsin n$. For higher angles of incidence, the amplitude ratio E_r/E_e is complex (of absolute value 1) and the refractive index likewise. Linearly polarized radiation then becomes elliptically polarized, and the transmitted solution is damped. We shall not discuss this here, because we shall deal with damped solutions (in space) in the next section anyway. We sometimes speak of *evanescent* waves.

3.3.10 Propagation of Waves in Conductors

In contrast to the last two sections, we shall no longer restrict ourselves to $\sigma = 0$. Then,

$$\nabla \times \mathbf{E} = -\frac{\partial \mathbf{B}}{\partial t} \, , \qquad \nabla \cdot \mathbf{B} = 0 \, , \qquad \nabla \cdot \mathbf{D} = 0 \, , \qquad \nabla \times \mathbf{H} = \sigma \mathbf{E} + \frac{\partial \mathbf{D}}{\partial t} \, .$$

Here, electromagnetic energy is converted into heat and hence, for a homogeneous medium, the wave equations gain a damping term

$$\left\{ \Delta - \mu \left(\sigma + \varepsilon \frac{\partial}{\partial t} \right) \frac{\partial}{\partial t} \right\} \mathbf{E} = \mathbf{0} \, , \qquad \nabla \cdot \mathbf{E} = 0 \, ,$$

and likewise with $\mathbf{B}$ instead of $\mathbf{E}$. These are the *telegraph equations*.

If an external wave impinges on a conductor surface, then the fields depend periodically on time. We have to investigate the position dependence in the conductor. According to the telegraph equation, the ansatz

$$\mathbf{E}(t, \mathbf{r}) = \mathrm{Re}\Big(\mathbf{E}(\mathbf{k}') \, \exp\{i(\mathbf{k}' \cdot \mathbf{r} - \omega t)\} \Big)$$

for all positions in the conductor leads to the condition

$$k'^2 = \varepsilon \mu \, \omega^2 \left(1 + i \frac{\sigma}{\varepsilon \omega} \right) .$$

This can be satisfied for real ω only with a complex wave vector. A complex permittivity $\varepsilon \, (1 + i\sigma/\varepsilon\omega)$ is also often introduced. Here, for a scalar material with constant σ, ε, and μ, the real and imaginary parts of the wave vector have the same direction. The new feature in comparison with non-conductors is longitudinal damping. Therefore, we set

$$\mathbf{k}' = (\alpha + i\beta) \, \mathbf{k} \, ,$$

where as before $ck = \omega$ with $c = 1/\sqrt{\varepsilon\mu}$. Then we have

$$\exp\{i(\mathbf{k}' \cdot \mathbf{r} - \omega t)\} = \exp(-\beta \, \mathbf{k} \cdot \mathbf{r}) \, \exp\{i(\alpha \, \mathbf{k} \cdot \mathbf{r} - \omega t)\}$$

and $(\alpha + i\beta)^2 = 1 + i\sigma/\varepsilon\omega$, whence

$$\alpha = \sqrt{\tfrac{1}{2}\sqrt{1 + (\sigma/\varepsilon\omega)^2} + \tfrac{1}{2}} \qquad \text{and} \qquad \beta = \sqrt{\tfrac{1}{2}\sqrt{1 + (\sigma/\varepsilon\omega)^2} - \tfrac{1}{2}} \, .$$

Now, with increasing $\mathbf{k} \cdot \mathbf{r}$, the amplitude decreases. The wave is damped spatially. Since conductors usually have $\sigma/\varepsilon\omega \gg 1$, whereupon the electric current is large compared to the displacement current, we obtain the *decay length*

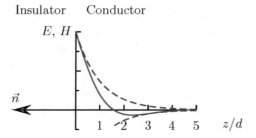

Fig. 3.22 Repulsion of the current. Decay of the alternating fields in the interior of a conductor—
dashed lines show their amplitude—here for $\sigma \gg \varepsilon\omega$ and hence $\alpha \approx \beta$. (When $\sigma/\varepsilon\omega < \infty$, there
is also a normal component of the magnetic field and a tangential component of the electric field)

$$d \equiv \frac{1}{\beta k} \approx \frac{1}{k}\sqrt{\frac{2\varepsilon\omega}{\sigma}} = \sqrt{\frac{2}{\sigma\mu\omega}} \,,$$

where the amplitude for perpendicular incidence is smaller than the factor $1/e$ at the
surface. High-frequency alternating currents are thus repelled from the interior of the
conductor, flowing only at the surface. This is referred to as *repulsion of the current*
or the *skin effect* (see Fig. 3.22). The higher the conductivity, the shorter the decay
length. For the *phase velocity*, we have $c' = \omega/\alpha k = c/\alpha$, and for $\sigma/\varepsilon\omega \gg 1$, we
thus have $\alpha \approx \beta \gg 1$, whence also $c' \approx c/\beta = \omega d$ and therefore $c' \ll c$.
    Since

$$\mathbf{k}' \cdot \mathbf{E}(\mathbf{k}') = 0 \,, \quad \omega\mathbf{B}(\mathbf{k}') = \mathbf{k}' \times \mathbf{E}(\mathbf{k}') \,, \quad \text{and} \quad \mathbf{k}' \cdot \mathbf{B}(\mathbf{k}') = 0 \,,$$

the three (complex) vectors $\mathbf{k}' = (\alpha + \mathrm{i}\beta)\,\mathbf{k}$, $\mathbf{E}(\mathbf{k}')$ and $\mathbf{B}(\mathbf{k}')$ are once again perpendicular to each other and still form a right-handed frame, but $\mathbf{E}$ and $\mathbf{B}$ differ in
phase and therefore no longer have the same nodes. If, as in Sect. 3.3.8, we average
over the time, we obtain

$$\overline{\mathbf{H}(t, \mathbf{r}) \cdot \mathbf{B}(t, \mathbf{r})} = \tfrac{1}{2}\,\mathbf{H}^*(\mathbf{k}') \cdot \mathbf{B}(\mathbf{k}')\,\exp(-2\beta\,\mathbf{k} \cdot \mathbf{r})$$
$$= \sqrt{1 + (\sigma/\varepsilon\omega)^2}\,\,\overline{\mathbf{E}(t, \mathbf{r}) \cdot \mathbf{D}(t, \mathbf{r})} \,,$$

where the square-root factor originates from $\mathbf{k}'^* \cdot \mathbf{k}'/k^2$. For most conductors, there
is much more energy in the magnetic field than in the electric field. Here now the
energy density decreases with increasing distance from the surface, in proportion to
$\exp(-2\beta\,\mathbf{k} \cdot \mathbf{r})$ (Problem 3.41).
    If a conductor is adjacent to an insulator, and if $\mathbf{n}$ points from the conductor to
the insulator, then we have the *boundary conditions*

$$\mathbf{n} \times (\mathbf{E}_{\mathrm{I}} - \mathbf{E}_{\mathrm{C}}) = \mathbf{0} \,, \qquad\qquad \mathbf{n} \cdot (\mathbf{B}_{\mathrm{I}} - \mathbf{B}_{\mathrm{C}}) = 0 \,,$$
$$\mathbf{n} \cdot (\mathbf{D}_{\mathrm{I}} - \mathbf{D}_{\mathrm{C}}) = \rho_A \,, \qquad\quad \mathbf{n} \times (\mathbf{H}_{\mathrm{I}} - \mathbf{H}_{\mathrm{C}}) = \mathbf{j}_A \,.$$

The fields do not enter an ideal conductor at all—it is fully screened by charges and currents on the surface ($E_C = \ldots = 0$). Therefore, the electric field lines end up perpendicular to the surface of an ideal conductor (without tangential component, i.e., $E_T = 0$), and the magnetic fields adapt to the surface (without normal component, i.e., $H_N = 0$). But if the conductivity is finite (normal conductor), a current is accompanied by a finite field in the current direction ($E_T \neq 0$), and there is no surface current density. Therefore the tangential component of $\mathbf{H}_C$ turns continuously into that of $\mathbf{H}_I$ and decays exponentially in the conductor (for $\omega \neq 0$) with increasing distance from the surface.

3.3.11 Summary: Maxwell's Equations

Two new quantities lead from statics to time-dependent phenomena: charge conservation (continuity equation) supplies Maxwell's displacement current $\partial \mathbf{D}/\partial t$, and Faraday's induction law connects $\partial \mathbf{B}/\partial t$ with $\nabla \times \mathbf{E}$, where the sign results in Lenz's rule. The induction field counteracts the change in the magnetic field. Hence we have the basic Maxwell equations:

$$\nabla \times \mathbf{E} = -\frac{\partial \mathbf{B}}{\partial t} , \qquad \nabla \cdot \mathbf{B} = 0 ,$$
$$\nabla \cdot \mathbf{D} = \rho , \qquad \nabla \times \mathbf{H} = \mathbf{j} + \frac{\partial \mathbf{D}}{\partial t} .$$

These differential equations correspond to integral equations,

$$\int_{(A)} d\mathbf{r} \cdot \mathbf{E} = -\frac{d}{dt} \int_A d\mathbf{f} \cdot \mathbf{B} , \qquad \int_{(V)} d\mathbf{f} \cdot \mathbf{B} = 0 ,$$
$$\int_{(V)} d\mathbf{f} \cdot \mathbf{D} = Q , \qquad \int_{(A)} d\mathbf{r} \cdot \mathbf{H} = I + \frac{d}{dt} \int_A d\mathbf{f} \cdot \mathbf{D} ,$$

and boundary conditions,

$$\mathbf{n} \times (\mathbf{E}_+ - \mathbf{E}_-) = \mathbf{0} , \qquad \mathbf{n} \cdot (\mathbf{B}_+ - \mathbf{B}_-) = 0 ,$$
$$\mathbf{n} \cdot (\mathbf{D}_+ - \mathbf{D}_-) = \rho_A , \qquad \mathbf{n} \times (\mathbf{H}_+ - \mathbf{H}_-) = \mathbf{j}_A .$$

Taking Fourier transforms with $\exp\{i(\mathbf{k} \cdot \mathbf{r} - \omega t)\}$, the four Maxwell equations read

$$\mathbf{k} \times \mathbf{E}(\omega, \mathbf{k}) = \omega \mathbf{B}(\omega, \mathbf{k}) , \qquad \mathbf{k} \cdot \mathbf{B}(\omega, \mathbf{k}) = 0 ,$$
$$\mathbf{k} \cdot \mathbf{D}(\omega, \mathbf{k}) = -i \rho(\omega, \mathbf{k}) , \qquad \mathbf{k} \times \mathbf{H}(\omega, \mathbf{k}) = -i\mathbf{j}(\omega, \mathbf{k}) - \omega \mathbf{D}(\omega, \mathbf{k}) .$$

In charge-free, homogeneous space, they lead to transverse waves, and they obey the telegraph equation, which is the same for $\mathbf{E}$ and $\mathbf{B}$. Here the three vectors $\mathbf{k}$, $\mathbf{E}$, and $\mathbf{B}$ are pairwise perpendicular to each other.

The time-dependent potentials $\Phi(t, \mathbf{r})$ and $\mathbf{A}(t, \mathbf{r})$ are useful:

$$\mathbf{E} = -\nabla\Phi - \frac{\partial \mathbf{A}}{\partial t} \quad \text{and} \quad \mathbf{B} = \nabla \times \mathbf{A}.$$

Then the first two Maxwell equations are automatically satisfied. However, the scalar potential Φ is determined only up to an additive term $\partial\Psi/\partial t$, and the vector potential $\mathbf{A}$ only up to its sources—it would have to be changed by $-\nabla\Psi$. The potentials may still be gauged to our advantage. Here Ψ or $\nabla \cdot \mathbf{A}$ is fixed. For the Coulomb gauge, we choose $\nabla \cdot \mathbf{A} = 0$, and for the Lorentz gauge, $\nabla \cdot \mathbf{A} = -\varepsilon\mu \, \partial\Phi/\partial t$. In both cases, the resulting system of equations is decoupled.

3.4 Lorentz Invariance

3.4.1 Velocity of Light in Vacuum

In contrast to the situation in mechanics, in electromagnetism a specific velocity is picked out, even if there is no matter in space which could supply a reference frame. This velocity is the velocity of light in vacuum, viz.,

$$c_0 = 299\,792\,458 \, \frac{\text{m}}{\text{s}}.$$

But in electromagnetism, no inertial system is special, because the four Maxwell equations are valid in all uniformly moving reference frames. In particular, the velocity of light in vacuum is the same in all inertial frames.

Due to this astonishing fact, we have to completely rethink the notion of velocity, and thus also the measurement of lengths and times. In particular, we need a signal velocity c_0 in order to fix equal times everywhere in space (coordinate system). In order to synchronize clocks at two points with constant separation $|\mathbf{r} - \mathbf{r}'|$, we send a signal from one point and expect it to arrive at the other point at the time $\Delta t = |\mathbf{r} - \mathbf{r}'|/c_0$. Without a signal velocity, we cannot synchronize clocks at different positions, and without clocks we cannot measure a velocity. The fastest velocity is that of light, a million times faster than sound in air. Therefore, we synchronize our clocks with light signals. (If there were some kind of action at a distance, with infinite propagation velocity, then of course we would use that to synchronize our clocks.)

Since c_0 is the same in all inertial frames, we may not start from a generally fixed (absolute) time, as we would in classical mechanics. There it is assumed that, for two inertial frames moving relative to one another, only the position coordinates transform, but not the time. That implies the validity of the

Galilean transformation: $t' = t$, $\mathbf{r}' = \mathbf{r} - \mathbf{v}t$.

But this can be valid only for $v \ll c_0$, because it does not contain the velocity of light in empty space.

3.4.2 Lorentz Transformation

We consider an inertial frame with unprimed coordinates $(t, \mathbf{r})$ and one with primed coordinates $(t', \mathbf{r}')$, moving uniformly with velocity $\mathbf{v}$ relative to the first, where the position vectors are given in Cartesian coordinates. We restrict ourselves to *homogeneous Lorentz transformations*: the origins $(0, \mathbf{0})$ of the two systems agree with each other. (*Inhomogeneous* Lorentz transformations contain four further parameters, since for them the zero point is also moved, and they form the *Poincaré group*.) Since otherwise no *event* is preferred, the two coordinate systems depend linearly on each other (via a real transformation matrix). The transition is reversible, and therefore their determinant must be either positive (a *proper Lorentz transformation*, continuously connected to the identity) or negative (*improper Lorentz transformation*, e.g., *space reflection*, also called the *parity operation*, $t' = t$, $\mathbf{r}' = -\mathbf{r}$, or *time reversal*, $t' = -t$, $\mathbf{r}' = \mathbf{r}$). If we include these two improper Lorentz transformations with the proper ones, then we obtain the *extended Lorentz group*. If the past remains behind and the future ahead, then the Lorentz transformation is *orthochronous* ($\partial t'/\partial t > 0$).

For *infinitesimal* Lorentz transformations, the matrix is barely different from the unit matrix, so no squared terms in this difference for $(c_0 t)^2 - r^2 = (c_0 t')^2 - r'^2$ need be accounted for. The additional terms form a skew-symmetric matrix with six (real) independent elements and lead for finite Lorentz transformations to six free parameters: three Euler angles and three parameters for the boost.

For the time being, we choose the axes such that $\mathbf{v}$ has only an x-component (> 0). Then $y = y'$ and $z = z'$, and only (t, x) and (t', x') depend on each other in a more involved way. At least in the two coordinate systems, the relative velocity will be denoted by $v = -v'$. Therefore, we require

$$x' = \gamma\,(x - vt) \quad \text{and} \quad x = \gamma\,(x' + vt')\,,$$

because the point $x' = 0$ moves away with velocity $v = x/t$ and the point $x = 0$ with the opposite velocity $-v = x'/t'$. The factor γ must be the same in the two equations, otherwise the two reference frames would differ fundamentally from one another. We determine γ from the requirement that, in the two systems, the same velocity of light c_0 must result. Then we have

$$c_0 \Delta t = \Delta x = \gamma\,(\Delta x' + v \Delta t')\,, \quad \text{as well as} \quad c_0 \Delta t' = \Delta x' = \gamma\,(\Delta x - v \Delta t)\,.$$

We must therefore have $c_0 \Delta t = \gamma\,(c_0 + v)\,\Delta t'$ and $c_0 \Delta t' = \gamma\,(c_0 - v)\,\Delta t$, and hence,

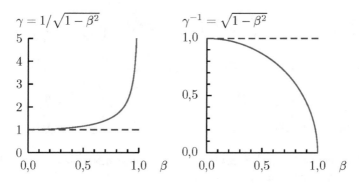

Fig. 3.23 Relations between the parameters β, γ, and γ^{-1}. The *dashed line* is the relation for the Galilean transformation

$$\frac{\Delta t}{\Delta t'} = \frac{\gamma\,(c_0 + v)}{c_0} = \frac{c_0}{\gamma\,(c_0 - v)} \quad\Longrightarrow\quad \gamma^2 = \frac{1}{1 - (v/c_0)^2} \;.$$

We therefore use the abbreviation

$$\boldsymbol{\beta} \equiv \frac{\mathbf{v}}{c_0} \quad\Longrightarrow\quad \gamma = \frac{1}{\sqrt{1 - \beta^2}} \;.$$

Since the coordinates remain real, $\beta \leq 1$ must hold, so $v \leq c_0$ and $\gamma \geq 1$ (see Fig. 3.23).

From $t' = (x/\gamma - x')/v$ and $x'/v = \gamma\,(x/v - t)$, it follows that $t' = (\gamma^{-1} - \gamma)\,x/v + \gamma t$. Here, $1 - \gamma^{-2} = \beta^2$, so $t' = \gamma\,(t - \beta x/c_0)$. If we combine $x' = \gamma\,(x - vt)$ $y' = y$ and $z' = z$ as a vector equation, we obtain finally the *Lorentz transformation*

$$t' = \gamma\left(t - \frac{\boldsymbol{\beta}\cdot\mathbf{r}}{c_0}\right) \quad\text{and}\quad \mathbf{r}' = \mathbf{r} + \frac{\gamma - 1}{\beta^2}\,\boldsymbol{\beta}\,\boldsymbol{\beta}\cdot\mathbf{r} - \gamma\boldsymbol{\beta}\,c_0 t \;.$$

Conversely, because $\boldsymbol{\beta}' = -\boldsymbol{\beta}$, we have

$$t = \gamma\left(t' + \frac{\boldsymbol{\beta}\cdot\mathbf{r}'}{c_0}\right) \quad\text{and}\quad \mathbf{r} = \mathbf{r}' + \frac{\gamma - 1}{\beta^2}\,\boldsymbol{\beta}\,\boldsymbol{\beta}\cdot\mathbf{r}' + \gamma\boldsymbol{\beta}\,c_0 t' \;.$$

In the limit of small velocities $v \ll c_0$, whence $\beta \ll 1$ and $\gamma \approx 1$, we arrive at the above-mentioned *Galilean transformation*

$$t' = t, \quad \mathbf{r}' = \mathbf{r} - \mathbf{v}t, \quad \text{or} \quad t = t', \quad \mathbf{r} = \mathbf{r}' + \mathbf{v}t \;.$$

But this holds only approximately because of the finite signal velocity c_0. Therefore, from now on, we shall only deal with the Lorentz transformation. In particular, for $\mathbf{v} \propto \mathbf{e}_x$, we have

$$\begin{pmatrix} c_0 t' \\ x' \end{pmatrix} = \begin{pmatrix} \gamma & -\beta\gamma \\ -\beta\gamma & \gamma \end{pmatrix} \begin{pmatrix} c_0 t \\ x \end{pmatrix}, \quad \text{or} \quad \begin{pmatrix} c_0 t \\ x \end{pmatrix} = \begin{pmatrix} \gamma & \beta\gamma \\ \beta\gamma & \gamma \end{pmatrix} \begin{pmatrix} c_0 t' \\ x' \end{pmatrix},$$

along with $y' = y$ and $z' = z$. With the consequences

$$\begin{aligned}
\Delta t' &= \gamma \left(\Delta t - \beta \frac{\Delta x}{c_0} \right), \\
\Delta x' &= \gamma \left(\Delta x - v \, \Delta t \right),
\end{aligned} \quad \text{or} \quad \begin{aligned}
\Delta t &= \gamma \left(\Delta t' + \beta \frac{\Delta x'}{c_0} \right), \\
\Delta x &= \gamma \left(\Delta x' + v \, \Delta t' \right),
\end{aligned}$$

we can compare rulers and clocks in reference frames moving relative to each other and derive two noteworthy phenomena.

The first is *Lorentz contraction*: the ends of a ruler of length Δx in its rest system must be measured simultaneously in the moving system and are found to be closer together:

$$\Delta t' = 0 : \quad \Delta x' = \frac{\Delta x}{\gamma} < \Delta x \ .$$

Conversely, thanks to the requirement $\Delta t = 0$, the length $\Delta x'$ in the oppositely moving (unprimed) system is also shorter, i.e., $\Delta x = \Delta x'/\gamma$. Moving lengths are shorter than the *proper length* in the rest system by the factor $1/\gamma = \sqrt{1 - \beta^2} < 1$. In addition to Lorentz contraction, owing to the finite light propagation time, a dilation by the factor $1/(1 - \beta)$ also occurs when frames approach one another and a compression by $1/(1 + \beta)$ when they move apart. The total factor $\sqrt{1 \pm \beta}/\sqrt{1 \mp \beta}$ is also shown in the middle of Fig. 3.26 (and see also Table 3.1, although reversed there, since frequencies are inversely proportional to wavelengths).

The second striking phenomenon is *relativistic time dilation*: times must be compared at the position of the clock in the rest system, and result in times in the moving system being dilated by the factor $\gamma > 1$ compared with the *proper time* (in the rest system):

$$\Delta x = 0 : \quad \Delta t' = \gamma \, \Delta t > \Delta t \ , \quad \text{or} \quad \Delta x' = 0 : \quad \Delta t = \gamma \, \Delta t' \ .$$

This effect must be included when determining the lifetimes of fast-moving particles: for $v \approx c_0$, the factor γ is significantly greater than 1.

The two phenomena can also be read off from the Minkowski diagram (Fig. 3.24). But it is worth making a few comments. The quantity $(c_0 t)^2 - x^2$ is a Lorentz invariant: $(c_0 t')^2 = \gamma^2 (c_0 t - \beta x)^2$ and $x'^2 = \gamma^2 (x - \beta c_0 t)^2$ imply $(c_0 t')^2 - x'^2 = \gamma^2 \{(c_0 t)^2 - x^2\} (1 - \beta^2)$ with $1 - \beta^2 = \gamma^{-2}$. Therefore, for a Lorentz transformation the *world points* $(c_0 t, x)$ in the Minkowski diagram move on a hyperbola, and for $c_0 t = x$, on the associated asymptote. We distinguish here the *time-like* region with $|c_0 t| > |x|$ and the *space-like* region with $|c_0 t| < |x|$ (actually, we should write $|\mathbf{r}|$ instead of $|x|$). The surface $|c_0 t| = |\mathbf{r}|$ is called the *light cone*. Time-like world points on the hyperbola $(c_0 t)^2 - x^2 = C^2 > 0$ then obey the parameter representation $c_0 t = C \cosh \phi$, $x = C \sinh \phi$. (For a space-like world point, $c_0 t$ is exchanged

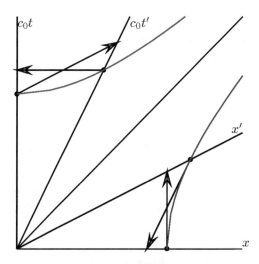

Fig. 3.24 Minkowski diagram. It has a spatial coordinate and the (reduced) time $c_0 t$ as axes and the light cone $|c_0 t| = |\mathbf{r}|$ as diagonal. A moving coordinate system is also shown. Its axes have slopes of β or β^{-1}. The scale transformation is indicated by the hyperbolic curves—they connect world points at equal positions (*blue curves*) or times (*red curves*). The two *arrows at bottom right* indicate the length contraction, those *top left* the time dilation: they point from the unit coordinate value in the rest systems to the axes of the moving systems, each parallel to the axes

with x.) With $\alpha = \text{arctanh}\,\beta$ and $\phi' = \phi - \alpha$, the above-mentioned Lorentz transformation, i.e., the transition to oblique space-time coordinates, is then simply

$$\left.\begin{array}{c} c_0 t = C \cosh \phi\,, \\ x = C \sinh \phi\,, \end{array}\right\} \quad \Longrightarrow \quad \left\{\begin{array}{c} c_0 t' = C \cosh \phi'\,, \\ x' = C \sinh \phi'\,, \end{array}\right.$$

if we employ the addition theorems for hyperbolic functions, namely, the relation $\cosh(\phi-\alpha) = \cosh \phi \cosh \alpha - \sinh \phi \sinh \alpha$ (with the special case $1 = \cosh^2 \alpha - \sinh^2 \alpha$) and $\sinh(\phi-\alpha) = \sinh \phi \cosh \alpha - \cosh \phi \sinh \alpha$.

3.4.3 Four-Vectors

The Lorentz transformation connects space and time and mixes their coordinates. Therefore, instead of the usual three-vectors in the normal space, we now take four-vectors in space and time. In order to have all four components as lengths, we use the path length $c_0 t$ of the light instead of the time, and take it as zeroth component:

$$(x^\mu) \mathrel{\widehat{=}} (x^0, x^k) \mathrel{\widehat{=}} (x^0, x^1, x^2, x^3) \equiv (c_0 t, x, y, z) \mathrel{\widehat{=}} (c_0 t, \mathbf{r})\,.$$

We let Greek indices (e.g., μ) run from 0 to 3, Latin indices (e.g., k) from 1 to 3.

As in Sect. 1.2.2, we also distinguish in four dimensions between covariant and contravariant vector components with different transformation behavior. Then for a Lorentz transformation, we have

$$dx'^{\mu} = \sum_{\nu=0}^{3} \frac{\partial x'^{\mu}}{\partial x^{\nu}} \, dx^{\nu} \, .$$

In the following we would like always to sum over doubly appearing indices from 0 to 3 if in an expression each occurs once as a subscript and once as a superscript (*Einstein summation convention*). In this way we avoid the bothersome notation of the summation sign—we often have to contract tensors and we have already used this idea to abbreviate the scalar product in vector algebra. With this and according to p. 33, we find

$$A'^{\mu} = \frac{\partial x'^{\mu}}{\partial x^{\nu}} A^{\nu} \qquad \text{for contravariant vector components,}$$

$$A'_{\mu} = A_{\nu} \frac{\partial x^{\nu}}{\partial x'^{\mu}} \qquad \text{for} \quad \text{covariant vector components.}$$

We may also read these two equations as matrix equations—they are linear transformations with symmetric transformation matrices which are inverse to each other. For the example considered in the last section, they read

$$\left(\frac{\partial x'^{\mu}}{\partial x^{\nu}} \right) \cong \begin{pmatrix} \gamma & -\beta\gamma & 0 & 0 \\ -\beta\gamma & \gamma & 0 & 0 \\ 0 & 0 & 1 & 0 \\ 0 & 0 & 0 & 1 \end{pmatrix} \quad \text{and} \quad \left(\frac{\partial x^{\nu}}{\partial x'^{\mu}} \right) \cong \begin{pmatrix} \gamma & \beta\gamma & 0 & 0 \\ \beta\gamma & \gamma & 0 & 0 \\ 0 & 0 & 1 & 0 \\ 0 & 0 & 0 & 1 \end{pmatrix} .$$

If the coordinate axes are not as well adjusted to the relative velocity as here, then of course not so many matrix elements $a^{\mu}{}_{\nu}$ will vanish. Generally, with $x'^{\mu} x'_{\mu} = x^{\nu} x_{\nu}$, we always have

$$a^{\mu}{}_{\nu} a_{\mu}{}^{\lambda} = g_{\nu}{}^{\lambda} \qquad \text{and} \qquad a^{\mu}{}_{\nu}{}^{*} = a^{\mu}{}_{\nu} \, .$$

The special case here suffices for the general principle.

The transition between covariant and contravariant components is, however, not as simple as for Cartesian coordinates. In particular, the velocity of light must have the same value in all coordinate systems, so the Lorentz invariant $(c_0 dt)^2 - \mathbf{dr} \cdot \mathbf{dr}$ must be a scalar:

$$dx_{\mu} \, dx^{\mu} = c_0^2 \, dt^2 - \mathbf{dr} \cdot \mathbf{dr} \, .$$

This is achieved by introducing the *Minkowski metric*

$$\left(g_{\mu\nu}\right) \widehat{=} \begin{pmatrix} 1 & 0 & 0 & 0 \\ 0 & -1 & 0 & 0 \\ 0 & 0 & -1 & 0 \\ 0 & 0 & 0 & -1 \end{pmatrix} \widehat{=} \left(g^{\mu\nu}\right).$$

This matrix is sometimes written also as diag $(1, -1, -1, -1)$. On p. 32, we have $g_{ii} = \mathbf{g}_i \cdot \mathbf{g}_i > 0$ and also $g^{ii} > 0$. This suggests introducing imaginary base vectors, but we shall not consider these further here, since we only need the metric to interchange upper and lower indices. To this end, we always take the above-mentioned fundamental tensor. Thus, since $x_\mu = g_{\mu\nu}x^\nu$, we have

$$(x_\mu) \widehat{=} (x_0, \ x_k) \widehat{=} (x_0, \ x_1, \ x_2, \ x_3) = (c_0 t, \ -x, \ -y, \ -z).$$

This suggests choosing the fundamental tensor with opposite sign, since then the space components remain unchanged for the transition from three to four dimensions. (It is also common to set $x^4 = i c_0 t$ and drop x^0.) But then physically sensible scalar products like $p_\mu p^\mu$ become negative. For this reason, we prefer, like many other authors, to stick to the choice just made.

For the transition from three to four dimensions, however, we encounter some difficulties with the concept of the vector product. In particular, $\mathbf{a} \times \mathbf{b}$ should be perpendicular to $\mathbf{a}$ and $\mathbf{b}$, but this is unique to three dimensions. In four dimensions, we may no longer refer to "axial vectors" as vectors. But if we take over the usual components of a vector product, then it transforms according to (with Latin indices running from 1 to 3)

$$a^{\prime i} b^{\prime j} - a^{\prime j} b^{\prime i} = \frac{\partial x^{\prime i}}{\partial x^k} \frac{\partial x^{\prime j}}{\partial x^l} \left(a^k b^l - a^l b^k\right),$$

thus as a tensor of second rank (see p. 35 and Problem 2.4). It is skew-symmetric:

$$T^{ij} = -T^{ji} = -T_i{}^j = +T^j{}_i = +T_j{}^i = -T^i{}_j = +T_{ij} = -T_{ji} .$$

In three dimensions, such a tensor has three independent components $T^{12} = -T^{21}$, $T^{23} = -T^{32}$ and $T^{31} = -T^{13}$, while $T^{11} = T^{22} = T^{33} = 0$. In the following, we shall also consider four-dimensional skew-symmetric tensors of second rank, with six independent components. They have the properties

$$T^{i0} = -T^{0i} = -T_i{}^0 = +T^0{}_i = -T_0{}^i = +T^i{}_0 = -T_{i0} = +T_{0i} ,$$

as follows immediately from the metric.

In addition, the circulation density of a vector field is a skew-symmetric tensor of second rank:

$$\frac{\partial a_j}{\partial x^i} - \frac{\partial a_i}{\partial x^j} = \frac{\partial \mathbf{a}}{\partial x^i} \cdot \mathbf{g}_j - \frac{\partial \mathbf{a}}{\partial x^j} \cdot \mathbf{g}_i .$$

Note that, because of the derivatives, all indices are taken to be covariant. For variable base vectors, the derivative of a vector component a_j with respect to x^i arises, and then also $\mathbf{a} \cdot \partial \mathbf{g}_j / \partial x^i$. We thus have to introduce the Christoffel symbols of Sect. 1.2.6, although for rotations, these contributions cancel: $\partial \mathbf{g}_j / \partial x^i = \partial^2 \mathbf{r} / (\partial x^i \partial x^j) = \partial \mathbf{g}_i / \partial x^j$. However, for space-time considerations, we now restrict ourselves to fixed base vectors anyway.

3.4.4 Examples of Four-Vectors

As a first example, we have already met the four-vector

$$(x^\mu) \mathrel{\widehat=} (c_0 t, \mathbf{r}) \qquad \Longleftrightarrow \qquad (x_\mu) \mathrel{\widehat=} (c_0 t, -\mathbf{r}) \,.$$

If we want to build the *velocity vector* (v^μ), we cannot simply differentiate with respect to time, since that would not be Lorentz invariant—we must differentiate with respect to the proper time τ (see p. 230). We have $\mathrm{d}t = \gamma \, \mathrm{d}\tau$, or $\mathrm{d}/\mathrm{d}\tau = \gamma \, \mathrm{d}/\mathrm{d}t$, and hence

$$(v^\mu) \mathrel{\widehat=} \gamma \, (c_0, \mathbf{v}) \qquad \Longleftrightarrow \qquad (v_\mu) \mathrel{\widehat=} \gamma \, (c_0, -\mathbf{v}) \qquad \text{and} \qquad v_\mu v^\mu = c_0^2 \,.$$

Thus only in the non-relativistic limit $v \ll c_0$ do we arrive at the usual notion of velocity, for then $\gamma \approx 1$. We can also derive this in a different way. Corresponding to velocity 0, we have the four-vector $(v^\mu) = (c_0, 0, 0, 0)$. If it undergoes a Lorentz transformation with the velocity $-v$ in the $\mathbf{x}$-direction, or

$$\Lambda^{-1} = \begin{pmatrix} \gamma & \beta\gamma & 0 & 0 \\ \beta\gamma & \gamma & 0 & 0 \\ 0 & 0 & 1 & 0 \\ 0 & 0 & 0 & 1 \end{pmatrix} \,,$$

then since the matrix is symmetric, we may multiply it by a row vector from the left or a column vector from the right to obtain the four-vector $(v'^\mu) = \gamma \, (c_0, v, 0, 0)$, and thus the same result as before.

This second idea allows us to derive the *addition law for velocities*. If the above-mentioned matrix acts on the four-vector with parallel velocity vectors, $\gamma_0 \, (c_0, v_0, 0, 0)$, it follows that

$$(v_\parallel'^\mu) \mathrel{\widehat=} \gamma \gamma_0 \, (1 + \beta\beta_0) \left(c_0, \frac{v + v_0}{1 + \beta\beta_0}, 0, 0 \right) \,,$$

and if it acts on the perpendicular velocity vector $\gamma_0 \, (c_0, 0, v_0, 0)$, we find

Fig. 3.25 With the velocity parameter $w = \mathrm{arctanh}\beta$, known as the *rapidity*, the *addition law for parallel velocities* reads $\beta = \dfrac{\beta_0 + \beta_1}{1 + \beta_0\beta_1}$ then simply $w = w_0 + w_1$. For $|\beta| \ll 1$, we have $w \approx \beta$

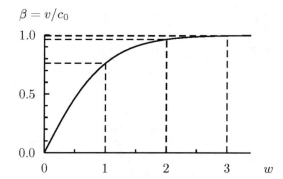

$$(v_\perp{}'^\mu) \,\hat{=}\, \gamma\gamma_0 \left(c_0,\, v,\, \frac{v_0}{\gamma},\, 0\right).$$

Here, v and v_0 are thus not equivalent: Lorentz transformations do not in general commute. The factors $\gamma\gamma_0 (1 + \beta\beta_0)$ or $\gamma\gamma_0$ are indeed the same as the quantity $\gamma' = (1 - \beta'^2)^{-1/2}$, as we shall now prove by showing that $\beta'^2 = 1 - \gamma'^{-2}$:

$$\left(\frac{\beta + \beta_0}{1 + \beta\beta_0}\right)^2 = 1 - \frac{(1 - \beta^2)(1 - \beta_0^2)}{(1 + \beta\beta_0)^2} = 1 - \frac{1}{\gamma^2\gamma_0^2 \,(1 + \beta\beta_0)^2}\,,$$

and

$$\beta^2 + \frac{\beta_0^2}{\gamma^2} = 1 - (1 - \beta^2)(1 - \beta_0^2)\,.$$

Incidentally, this is also equal to $\beta_0^2 + \beta^2\gamma_0^{-2}$. The addition law can be summarized by

$$\mathbf{v}' = \frac{1}{1 + \mathbf{v}\cdot\mathbf{v}_0/c_0^2}\left(\mathbf{v} + \frac{\mathbf{v}_0}{\gamma} + \mathbf{v}\,\frac{\gamma - 1}{\gamma}\,\frac{\mathbf{v}\cdot\mathbf{v}_0}{v^2}\right).$$

This equation also follows from $d\mathbf{r}'/dt' = d\mathbf{r}'/dt \cdot dt/dt'$ with the formulae for the Lorentz transformation on p. 229, if $d\mathbf{r}/dt = \mathbf{v}_0$ is used (see Fig. 3.25).

Only if all velocities are small compared to c_0 do we have $\mathbf{v}' = \mathbf{v} + \mathbf{v}_0$. Otherwise the velocity of light in vacuum could also be exceeded, but in fact, $v' = c_0$ if v or v_0 is equal to c_0. For parallel velocities, this follows immediately from

$$v' = (v + v_0)/(1 + \beta\beta_0)\,,$$

and for perpendicular velocities,

$$v'^2 = v^2 + v_0^2/\gamma^2 = c_0^2\,(\beta^2 + \beta_0^2 - \beta_0^2\beta^2)\,.$$

Table 3.1 Longitudinal and transverse Doppler effect

θ	0	$\frac{1}{2}\pi$	π
ω'/ω	$\sqrt{1-\beta}/\sqrt{1+\beta}$	$1/\sqrt{1-\beta^2}$	$\sqrt{1+\beta}/\sqrt{1-\beta}$
θ'	0	$\frac{1}{2}\pi + \arcsin\beta$	π

When $\beta = 1$ or $\beta_0 = 1$, the bracket is equal to 1.

If a medium with refractive index $n = c_0/c$ moves with velocity v and there is light travelling in it in the same direction, the velocity of this light will depend on this reference system (*Fizeau experiment* on the drag of light in moving bodies):

$$c' = \frac{v+c}{1+\beta/n} = c_0 \frac{\beta+1/n}{1+\beta/n} = c + \left(1 - \frac{1}{n^2}\right)v + \cdots .$$

The expression in brackets is called the (Fresnel) *drag coefficient*. However, for dispersion, in addition to $-n^{-2}$, it also contains the term $(\omega/n)\,\mathrm{d}n/\mathrm{d}\omega$.

The zero of a wave is determined by the phase $\omega t - \mathbf{k} \cdot \mathbf{r}$ and must not depend upon the choice of coordinates. The expression must be a Lorentz invariant and must therefore be written in the form of a scalar product $k^\mu x_\mu$. Consequently, we have

$$(k^\mu) \mathrel{\hat{=}} \left(\frac{\omega}{c_0},\ \mathbf{k}\right), \quad \text{with} \quad k^\mu k_\mu = 0 \quad (\text{because } \omega = c_0 k) .$$

With $t = \gamma\,(t' + \mathbf{v}\cdot\mathbf{r}'/c_0{}^2)$ and $\mathbf{r} = \mathbf{r}' + \{(\gamma-1)v^{-2}\mathbf{v}\cdot\mathbf{r}' + \gamma\,t'\}\mathbf{v}$ (see p. 229), and comparing coefficients in $\omega t - \mathbf{k}\cdot\mathbf{r} = \omega't' - \mathbf{k}'\cdot\mathbf{r}'$, we deduce that

$$\omega' = \gamma\,(\omega - \mathbf{v}\cdot\mathbf{k}) \quad \text{and} \quad \mathbf{k}' = \mathbf{k} + \left(\frac{\gamma-1}{v^2}\mathbf{v}\cdot\mathbf{k} - \frac{\gamma\,k}{c_0}\right)\mathbf{v} .$$

With $\omega = c_0 k$, this implies the *Doppler effect* for the frequency, viz.,

$$\omega' = \omega\gamma\,(1 - \beta\cos\theta) ,$$

where θ is the angle between $\mathbf{v}$ and $\mathbf{k}$. Thus the Doppler effect with $v\lambda = c_0$ yields the wavelength $\lambda' = \lambda/\{\gamma(1-\beta\cos\theta)\}$. Some example applications are given in Table 3.1 and Figs. 3.26 and 3.27.

With the factor γ, a *transverse* and a *quadratic Doppler effect* occur (this does not of course hold for the propagation of sound, as the velocity of sound is so much smaller than the velocity of light). In addition, the propagation direction is described differently (*aberration*). With the vectors

$$\mathbf{e} \equiv \frac{\mathbf{k}}{k} \quad \text{and} \quad \mathbf{e}' \equiv \frac{\mathbf{k}'}{k'} ,$$

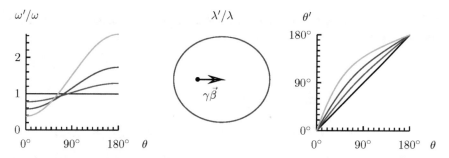

Fig. 3.26 Angular dependence of the frequency, wavelength, and deviation. In the *left and right figures*, *straight lines* refer to $\beta = 0$ (*black*), the curves to $\beta = \frac{1}{4}$ (*red*), $\frac{1}{2}$ (*blue*), and $\frac{3}{4}$ (*green*). The *middle picture* shows the ratio of the wavelengths λ'/λ in a polar diagram, namely the focal representation of an ellipse with semi-axes γ and 1, and hence eccentricity β (here 1/2). See also Fig. 3.31

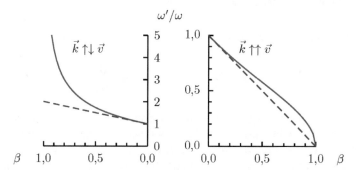

Fig. 3.27 Doppler effect. A frequency depends on how fast the detector moves relative to the emitter. *Left*: Decreasing distance. *Right*: Increasing distance. The *linear Doppler effect* is indicated by the *dashed line*

and using $k' = \omega'/c_0 = \gamma (k - \boldsymbol{\beta} \cdot \mathbf{k}) = \gamma k (1 - \boldsymbol{\beta} \cdot \mathbf{e})$, we deduce that

$$\mathbf{e}' = \frac{1}{\gamma (1 - \boldsymbol{\beta} \cdot \mathbf{e})} \left\{ \mathbf{e} + \left(\frac{\gamma - 1}{\beta^2} \, \boldsymbol{\beta} \cdot \mathbf{e} - \gamma \right) \boldsymbol{\beta} \right\},$$

and thus also $\boldsymbol{\beta} \cdot \mathbf{e}' = (\boldsymbol{\beta} \cdot \mathbf{e} - \beta^2)/(1 - \boldsymbol{\beta} \cdot \mathbf{e})$. Here $\boldsymbol{\beta} \cdot \mathbf{e}' = \beta \cos \theta'$, so

$$\cos \theta' = \frac{\cos \theta - \beta}{1 - \beta \cos \theta} \quad \Longrightarrow \quad \tan \theta' = \frac{\sin \theta}{\gamma (\cos \theta - \beta)} \, .$$

With increasing $|\beta|$, the difference between θ and θ' increases, although not for $\theta = 0$ and π (see Table 3.1 and Fig. 3.26). The motion of the Earth about the Sun produces an aberration of starlight $\leq 20.5''$.

For the concept of a density, it is important to note that the three-dimensional volume element $dV = dx \, dy \, dz$ is not invariant, because of the Lorentz contraction, while the rest volume $dV_0 = \gamma\, dV$ is. The charge does not change. With this, we then have $\rho\, dV = \rho_0\, dV_0$ and

$$\rho = \gamma\, \rho_0 \; .$$

From the charge and current density, we build a four-vector

$$(j^\mu) \;\equiv\; \rho_0\, (v^\mu) \,\widehat{=}\, \rho_0\, \gamma\, (c_0, \mathbf{v}) = (c_0\rho, \mathbf{j}) \; , \quad \text{with} \quad j_\mu j^\mu = (c_0\rho_0)^2 \; .$$

In particular, $j^{\,0} = c_0\rho$ and $\mathbf{j} = \rho\mathbf{v}$ as before, but ρ depends on the velocity through γ, i.e., through the Lorentz contraction.

3.4.5   Conservation Laws

In the following, we use the usual abbreviation

$$\partial_\mu \;\equiv\; \frac{\partial}{\partial x^\mu} \quad \text{and} \quad \partial^\mu \;\equiv\; \frac{\partial}{\partial x_\mu} \; .$$

Clearly, the components $\partial_\mu\psi$ transform covariantly and $\partial^\mu\psi$ contravariantly. We now prove the following theorem: If the four-dimensional source density $\partial_\mu j^\mu = 0$ of a four-vector vanishes everywhere and if this vector differs from zero only in a finite region of the three-dimensional space, then $\int dV\, j^0$ is constant for all times.

For the proof we extend Gauss's theorem to four dimensions:

$$\int d^4x\; \partial_\mu j^\mu = \int dS_\mu\; j^\mu \; ,$$

where $d^4x = c_0 dt\, dx\, dy\, dz = c_0 dt\, dV$ and dS_μ denotes a three-dimensional surface element for constant x^μ. Its sign (direction) is fixed in such a way that it is positive if its x^μ value is greater than in the considered volume (negative otherwise). Now we choose the surfaces $S_1{}^{()}$, $S_2{}^{()}$, and $S_3{}^{()}$ for large $|x^1|$, $|x^2|$, and $|x^3|$ such that $j = 0$ holds there. Figure 3.28 supplies the rest of the proof.

An important application is the continuity equation:

$$\frac{\partial\rho}{\partial t} + \nabla\cdot\mathbf{j} = 0 \qquad \Longleftrightarrow \qquad \partial_\mu j^\mu = 0 \; .$$

The theorem supplies

$$\int dV\, j^0 = \int dV\, c_0\rho = c_0\, Q$$

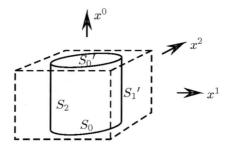

Fig. 3.28 j is restricted in finite space (the cylinder). $\partial_\mu j^\mu = 0$ then yields $\int \mathrm{d}S_0\, j^0 + \int \mathrm{d}S_0'\, j^{0'} = 0$ (the circular face S_0 is covered here and therefore not indicated). Due to the directional sense of the surface elements, the conservation law $\int \mathrm{d}V\, j^0 = \int \mathrm{d}V'\, j^{0'}$ follows

as a conserved quantity. The law of charge conservation follows from the continuity equation, and conversely, the continuity equation follows from charge conservation, something we already obtained in Sect. 3.2.1.

3.4.6 Covariance of the Microscopic Maxwell Equations

On p. 210, we found the microscopic Maxwell equations, viz.,

$$\frac{1}{c_0^2}\frac{\partial \Phi}{\partial t} + \nabla \cdot \mathbf{A} = 0\,,$$

$$\left(\frac{1}{c_0^2}\frac{\partial^2}{\partial t^2} - \Delta\right)\Phi = \frac{\rho}{\varepsilon_0}\,, \quad \text{and} \quad \left(\frac{1}{c_0^2}\frac{\partial^2}{\partial t^2} - \Delta\right)\mathbf{A} = \mu_0 \mathbf{j}\,,$$

with the help of the potentials Φ and $\mathbf{A}$ in the Lorentz gauge (the Coulomb gauge $\nabla \cdot \mathbf{A} = 0$ is not Lorentz invariant). With the first equation, we combine the scalar and vector potentials to yield the following *four-potential*:

$$\left(A^\mu\right) \cong \left(\frac{\Phi}{c_0},\, \mathbf{A}\right) \quad \Longrightarrow \quad \frac{\partial A^\mu}{\partial x^\mu} \equiv \partial_\mu A^\mu = 0\,.$$

Note that the equation $\partial_\mu A^\mu = 0$ does not result in a conservation law, since A^μ does not vanish sufficiently fast for large distances. In addition, using the other two equations, we generalize the Laplace operator to the *d'Alembert operator* (*quabla*)

$$\Box \equiv \frac{1}{c_0^2}\frac{\partial^2}{\partial t^2} - \Delta = \frac{\partial^2}{\partial x^\mu\, \partial x_\mu} = \partial_\mu \partial^\mu\,.$$

This is a Lorentz invariant, often taken with the opposite sign, in particular, if the other metric is used, with $g_{ik} = +\delta_{ik}$. If in addition to $\Phi = c_0 A^0$, we also take into account

$\rho = c_0^{-1} j^0$ and (by Weber's equation) $c_0^{-2} \varepsilon_0^{-1} = \mu_0$, then the above-mentioned inhomogeneous wave equations can be brought into the covariant form

$$\Box A^\mu = \mu_0 \, j^\mu , \quad \text{with} \quad \partial_\mu A^\mu = 0 .$$

In four-notation, the gauge transformation $\Phi' = \Phi + \partial\Psi/\partial t$, $\mathbf{A}' = \mathbf{A} - \nabla\Psi$ reads

$$A'^\mu = A^\mu + \partial^\mu \Psi ,$$

because $\partial_0 = \partial^0$ and $\partial_k = -\partial^k$, in addition to $A^0 = \Phi/c_0$.

With $\mathbf{B} = \nabla \times \mathbf{A}$ and $\mathbf{E} = -\nabla\Phi - \partial\mathbf{A}/\partial t$, noting that $A_k = -A^k$ and $A_0 = A^0$, we clearly have

$$B_x = \frac{\partial A_z}{\partial y} - \frac{\partial A_y}{\partial z} = -\partial^2 A^3 + \partial^3 A^2 = -\partial_2 A_3 + \partial_3 A_2 ,$$

$$E_x = -\frac{\partial \Phi}{\partial x} - \frac{\partial A_x}{\partial t} = c_0 (\partial^1 A^0 - \partial^0 A^1) = c_0 (\partial_0 A_1 - \partial_1 A_0) ,$$

and correspondingly for the other two components of $\mathbf{B}$ and $\mathbf{E}$. According to the last two columns, $\mathbf{E}/c_0$ and $\mathbf{B}$ can be combined in the form of a four-dimensional skew-symmetric tensor of second rank, the *electromagnetic field tensor*

$$F^{\mu\nu} \equiv \partial^\mu A^\nu - \partial^\nu A^\mu = -F^{\nu\mu} ,$$

or equivalently, $F_{\mu\nu} = \partial_\mu A_\nu - \partial_\nu A_\mu = -F_{\nu\mu}$:

$$\left(F^{\mu\nu}\right) = \begin{pmatrix} 0 & -E_x/c_0 & -E_y/c_0 & -E_z/c_0 \\ +E_x/c_0 & 0 & -B_z & B_y \\ +E_y/c_0 & B_z & 0 & -B_x \\ +E_z/c_0 & -B_y & B_x & 0 \end{pmatrix}$$

and

$$\left(F_{\mu\nu}\right) = \begin{pmatrix} 0 & +E_x/c_0 & +E_y/c_0 & +E_z/c_0 \\ -E_x/c_0 & 0 & -B_z & B_y \\ -E_y/c_0 & B_z & 0 & -B_x \\ -E_z/c_0 & -B_y & B_x & 0 \end{pmatrix} .$$

Unfortunately, the field tensor is not commonly denoted by B, rather than F, even though $\mathbf{B}$ is extended into four dimensions. For the extension of $\mathbf{j}$ to j^μ and $\mathbf{A}$ to A^μ, we are led by the space-like components, and likewise in the next section for the extension of $\mathbf{M}$ to $M^{\mu\nu}$. However, the field tensor is usually also amended with the factor c_0. Then it has the components of $\mathbf{E}$ and $c_0\mathbf{B}$ as elements.

Mixed derivatives commute with each other, if they are continuous. The Jacobi identity $\partial^\lambda(\partial^\mu A^\nu - \partial^\nu A^\mu) + \partial^\mu(\partial^\nu A^\lambda - \partial^\lambda A^\nu) + \partial^\nu(\partial^\lambda A^\mu - \partial^\mu A^\lambda) = 0$ yields

$$\partial^\lambda F^{\mu\nu} + \partial^\mu F^{\nu\lambda} + \partial^\nu F^{\lambda\mu} = 0 \ .$$

So far we have used two Maxwell equations, namely $\nabla \cdot \mathbf{B} = 0$ and $\partial \mathbf{B}/\partial t + \nabla \times \mathbf{E} = \mathbf{0}$, that is, precisely the two for which we have been able to introduce potentials. The other two microscopic Maxwell equations $\nabla \cdot \mathbf{E} = \rho/\varepsilon_0$ and $\nabla \times \mathbf{B} = \mu_0 (\mathbf{j} + \varepsilon_0 \partial \mathbf{E}/\partial t)$ can be combined if $\Box A^\nu = \mu_0 j^\nu$ and $\partial_\mu A^\mu = 0$ hold, to give

$$\partial_\mu F^{\mu\nu} = \partial_\mu (\partial^\mu A^\nu - \partial^\nu A^\mu) = \Box A^\nu - \partial^\nu \partial_\mu A^\mu = \mu_0 j^\nu \ .$$

Hence, we have $\mu_0 \partial_\nu j^\nu = \partial_\nu \partial_\mu F^{\mu\nu}$. The continuity equation $\partial_\nu j^\nu = 0$ now follows immediately from the antisymmetry of the field tensor, because $F^{\mu\nu} = -F^{\nu\mu}$, but $\partial_\nu \partial_\mu = +\partial_\mu \partial_\nu$, thus $\partial_\nu \partial_\mu F^{\mu\nu} = -\partial_\mu \partial_\nu F^{\nu\mu} = -\partial_\nu \partial_\mu F^{\mu\nu}$.

According to p. 100, the interaction density is equal to $\rho \, \Phi - \mathbf{j} \cdot \mathbf{A}$, which is the Lorentz invariant $j^\mu A_\mu$ in four-dimensional notation. Hence, we may also write $\partial_\mu F^{\mu\nu} = \mu_0 j^\nu$ as a *generalized Euler–Lagrange equation,* if we introduce the *Lagrange density*

$$\mathscr{L} = -\frac{F^{\mu\nu} F_{\mu\nu}}{4\mu_0} - j^\mu A_\mu \ ,$$

as a function of the A_μ and their derivatives $\partial_\mu A_\nu$. Using

$$\frac{\partial \mathscr{L}}{\partial(\partial_\mu A_\nu)} = \frac{\partial \mathscr{L}}{\partial F_{\kappa\lambda}} \frac{\partial F_{\kappa\lambda}}{\partial(\partial_\mu A_\nu)} = \frac{-F^{\kappa\lambda}}{2\mu_0} (\delta_\kappa^\mu \delta_\lambda^\nu - \delta_\lambda^\mu \delta_\kappa^\nu) = -\frac{F^{\mu\nu}}{\mu_0}$$

and $\partial_\mu F^{\mu\nu} = \mu_0 j^\nu = -\mu_0 \, \partial \mathscr{L}/\partial A_\nu$, we obtain the differential equation

$$\frac{\partial}{\partial x^\mu} \frac{\partial \mathscr{L}}{\partial(\partial_\mu A_\nu)} = \frac{\partial \mathscr{L}}{\partial A_\nu}$$

for the Lagrange density $\mathscr{L}$. This equation apparently generalizes the Euler–Lagrange equation in Sect. 2.3.3, viz.,

$$\frac{\mathrm{d}}{\mathrm{d}t}\left(\frac{\partial L}{\partial \dot{x}^k}\right) = \frac{\partial L}{\partial x^k} \ ,$$

where the time is no longer preferred over the space coordinates. Note that $\frac{1}{2} F^{\mu\nu} F_{\mu\nu} = \mathbf{B} \cdot \mathbf{B} - \mathbf{E} \cdot \mathbf{E}/c_0^2$.

3.4.7 Covariance of the Macroscopic Maxwell Equations

If we wish to use only macroscopically measurable notions, then instead of $\nabla \cdot \mathbf{E} = \rho/\varepsilon_0$ and $\nabla \times \mathbf{B} = \mu_0 (\mathbf{j} + \varepsilon_0 \partial \mathbf{E}/\partial t)$, or indeed $\partial_\mu F^{\mu\nu} = \mu_0 j^\nu$ we now have to take the Maxwell equations $\nabla \cdot \mathbf{D} = \overline{\rho}$ and $\nabla \times \mathbf{H} = \overline{\mathbf{j}} + \partial \mathbf{D}/\partial t$, i.e., in four-notation

$$\partial_\mu G^{\mu\nu} = \bar{j}^{\,\nu} \,,$$

with the skew-symmetric tensor

$$(G^{\mu\nu}) = \begin{pmatrix} 0 & -c_0 D_x & -c_0 D_y & -c_0 D_z \\ c_0 D_x & 0 & -H_z & H_y \\ c_0 D_y & H_z & 0 & -H_x \\ c_0 D_z & -H_y & H_x & 0 \end{pmatrix} \,,$$

which is the four-dimensional extension of the vectors $\mathbf{H}$, just as $F^{\mu\nu}$ is that of $\mathbf{B}$, and the four-vector of the average current density

$$(\bar{j}^{\,\mu}) \,\hat{=}\, (c_0 \bar{\rho},\ \bar{\mathbf{j}}) \,.$$

In doing this, we also generalize $\mathbf{D} = \varepsilon_0 \mathbf{E} + \mathbf{P}$, thus $\mathbf{E}/c_0 = \mu_0 c_0 \,(\mathbf{D} - \mathbf{P})$ and $\mathbf{B} = \mu_0 \,(\mathbf{H} + \mathbf{M})$ to

$$F^{\mu\nu} = \mu_0 \,(G^{\mu\nu} + M^{\mu\nu}) \,,$$

with the (skew-symmetric) magnetization tensor

$$(M^{\mu\nu}) = \begin{pmatrix} 0 & c_0 P_x & c_0 P_y & c_0 P_z \\ -c_0 P_x & 0 & -M_z & M_y \\ -c_0 P_y & M_z & 0 & -M_x \\ -c_0 P_z & -M_y & M_x & 0 \end{pmatrix} \,,$$

which extends the magnetization $\mathbf{M}$ to four dimensions. From this, we can easily establish the *magnetization current density* $\mathbf{j}_m$. The decomposition

$$j^\nu = \bar{j}^{\,\nu} + j_m^{\ \nu} \,,$$

with $j_m^{\ \nu} = j^\nu - \bar{j}^{\,\nu} = \mu_0^{-1} \partial_\mu F^{\mu\nu} - \partial_\mu G^{\mu\nu}$ leads to

$$j_m^{\ \nu} = \partial_\mu M^{\mu\nu} \,.$$

Note that there is therefore a continuity equation for the magnetization current, viz., $\partial_\nu j_m^{\ \nu} = 0$. Then,

$$(j_m^{\ \mu}) \,\hat{=}\, \left(-c_0 \nabla \cdot \mathbf{P},\ \frac{\partial \mathbf{P}}{\partial t} + \nabla \times \mathbf{M} \right) \,.$$

In electrostatics we have already encountered $\rho = \bar{\rho} - \nabla \cdot \mathbf{P}$ and in magnetostatics $\mathbf{j} = \bar{\mathbf{j}} + \nabla \times \mathbf{M}$. But the displacement current also contributes and results in the additional term $\partial \mathbf{P}/\partial t$.

The matrices $G_{\mu\nu}$ and $M_{\mu\nu}$ can be derived easily from $G^{\mu\nu}$ and $M^{\mu\nu}$ according to p. 233: $G_{0k} = -G^{0k} = -G_{k0}$ and $G_{ik} = G^{ik} = -G_{ki}$, and likewise for $M_{\mu\nu}$.

For given $\bar{j}^{\,\nu}$ and $M^{\mu\nu}$, the skew-symmetric tensors F and G are thus determined. Out of two Maxwell equations, just one equation has emerged in four-dimensional space.

3.4.8 Transformation Behavior of Electromagnetic Fields

Under a Lorentz transformation, the fields **E** and **B** (**D** and **H**) do not behave like vector fields A^{μ}, but the electromagnetic field tensors F and G are indeed tensors of second rank:

$$F'^{\mu\nu} = \partial_\kappa x'^{\mu}\, \partial_\lambda x'^{\nu}\, F^{\kappa\lambda}\ .$$

This system of equations corresponds to a matrix equation $F' = \tilde{\Lambda} F \Lambda$. The antisymmetry of $\tilde{F} = -F$ is transferred to $\tilde{F}' = \tilde{\Lambda}\tilde{F}\Lambda = -F'$, so only the six components with $\mu < \nu$ have to be determined. Since F is uniquely related to **E** and **B** according to p. 240, and likewise F' to **E**′ and **B**′, this means that the transformation properties of the fields can be derived using the matrices $\partial_\nu x'^{\mu}$ mentioned on p. 232. Then for a system moving with velocity **v**, we have the fields

$$\mathbf{E}_\parallel' = \mathbf{E}_\parallel\ , \qquad \mathbf{E}_\perp' = \gamma\left(\mathbf{E}_\perp + \mathbf{v}\times\mathbf{B}\right),$$
$$\mathbf{B}_\parallel' = \mathbf{B}_\parallel\ , \qquad \mathbf{B}_\perp' = \gamma\left(\mathbf{B}_\perp - \frac{\mathbf{v}\times\mathbf{E}}{c_0^2}\right).$$

These can be combined to give

$$\mathbf{E}' = \gamma\left(\mathbf{E} - \frac{\gamma-1}{\gamma}\frac{\mathbf{v}\cdot\mathbf{E}\,\mathbf{v}}{v^2} + \mathbf{v}\times\mathbf{B}\right),$$
$$\mathbf{B}' = \gamma\left(\mathbf{B} - \frac{\gamma-1}{\gamma}\frac{\mathbf{v}\cdot\mathbf{B}\,\mathbf{v}}{v^2} - \frac{\mathbf{v}\times\mathbf{E}}{c_0^2}\right).$$

Thus, the components of the electromagnetic field parallel to the velocity **v** remain unmodified, but not the perpendicular ones. In particular, in the non-relativistic limit $\gamma \approx 1$, it follows that

$$\mathbf{E}' \approx \mathbf{E} + \mathbf{v}\times\mathbf{B}\ , \qquad \mathbf{B}' \approx \mathbf{B} - \frac{\mathbf{v}\times\mathbf{E}}{c_0^2}\ .$$

(Note that the term $\mathbf{v}\times\mathbf{B}$ is well known, but not $\mathbf{v}\times\mathbf{E}/c_0^2$ due to technical limitations: we can produce strong magnetic fields, but strong electric fields are found only in the interior of atoms—because it is actually $c_0\mathbf{B}$ that should be compared with $\mathbf{E}$ in order to have equal units, i.e., $1\,\mathrm{T} \cong 3\,\mathrm{MV/cm}$, we should have considered $\mathbf{E} + \boldsymbol{\beta} \times c_0\mathbf{B}$ rather than $c_0\mathbf{B} - \boldsymbol{\beta} \times \mathbf{E}$.) Therefore, on a slowly moving electric point charge q an electromagnetic field acts with the *Lorentz force*

$$\mathbf{F} = q\,(\mathbf{E} + \mathbf{v} \times \mathbf{B})\,,$$

and on a moving magnetic moment $\mathbf{m}$, an electric field acts, because $\mathbf{F} = \nabla\,\mathbf{m}\cdot\mathbf{B}$ leads to

$$\mathbf{F} = \nabla\,\mathbf{m}\cdot\left(\mathbf{B} - \frac{\mathbf{v}\times\mathbf{E}}{c_0{}^2}\right).$$

In particular for a radially symmetric central field, we have

$$\mathbf{E} = -\nabla\Phi = -\frac{d\Phi}{dr}\frac{\mathbf{r}}{r}\,,$$

and hence $\mathbf{v}\times\mathbf{E} = r^{-1}\,(d\Phi/dr)\,(\mathbf{r}\times\mathbf{v})$. We thus arrive at the *spin–orbit coupling*, because there is a magnetic moment associated with a spin and $\mathbf{r}\times\mathbf{v}$ with an orbital angular momentum. According to this derivation, this is *not* a relativistic effect, despite what is often claimed.

Correspondingly, we can now establish the transformation properties of $\mathbf{D}$ and $\mathbf{H}$ from the behavior of the tensor G, which is the same as that of the tensor F. We only need to replace $\mathbf{E}$ by $c_0{}^2\,\mathbf{D}$ and $\mathbf{B}$ by $\mathbf{H}$, which yields

$$\mathbf{D}' = \gamma\left(\mathbf{D} - \frac{\gamma-1}{\gamma}\frac{\mathbf{v}\cdot\mathbf{D}\,\mathbf{v}}{v^2} + \frac{\mathbf{v}\times\mathbf{H}}{c_0{}^2}\right),$$

$$\mathbf{H}' = \gamma\left(\mathbf{H} - \frac{\gamma-1}{\gamma}\frac{\mathbf{v}\cdot\mathbf{H}\,\mathbf{v}}{v^2} - \mathbf{v}\times\mathbf{D}\right).$$

For the reverse transformation from the primed to the unprimed system, $\mathbf{v}$ is simply replaced by $-\mathbf{v}$, giving

$$\mathbf{E} = \gamma\left(\mathbf{E}' - \frac{\gamma-1}{\gamma}\frac{\mathbf{v}\cdot\mathbf{E}'\,\mathbf{v}}{v^2} - \mathbf{v}\times\mathbf{B}'\right).$$

Here the components of $\mathbf{E}$, $\mathbf{B}$, $\mathbf{D}$, and $\mathbf{H}$ along $\mathbf{v}$ remain unchanged.

3.4.9 Relativistic Dynamics of Free Particles

From the velocity we derive the (mechanical) momentum:

$$(p^\mu) \equiv m\,(v^\mu) \cong m\gamma\,(c_0,\,\mathbf{v})\,,\quad \text{with}\quad p_\mu p^\mu = (mc_0)^2\,.$$

Here, m stands for the mass (a relativistic invariant), often called the *rest mass*, while $m\gamma = m/\sqrt{1 - \beta^2}$ is called the *relativistic mass*, even though the factor γ belongs solely to the velocity—without it the velocity of light would not be the same in all inertial frames. It is thus a kinematic factor and has nothing to do with the mass. The zeroth component p^0 is connected to the energy:

$$p^0 = \frac{E}{c_0} \quad \Longrightarrow \quad E = m\gamma c_0^2 .$$

Note that the concept of the position–momentum pair corresponds to the time–energy pair (we neglect the potential energy and consider only free particles). The total energy E is composed of the *rest energy* mc_0^2 and the kinetic energy

$$T = E - mc_0^2 = m(\gamma - 1)c_0^2 = \tfrac{1}{2}mv^2 + \cdots .$$

By specifying the rest energy, we set the zero-point of the energy, so that is no longer arbitrary. We have thus set:

$$(p^\mu) \mathrel{\hat{=}} \left(\frac{E}{c_0}, \mathbf{p}\right), \quad \text{with} \quad E = m\gamma c_0^2 \quad \text{and} \quad \mathbf{p} = m\gamma \mathbf{v} .$$

With $p_\mu p^\mu = (mc_0)^2$, we conclude that $(E/c_0)^2 - \mathbf{p} \cdot \mathbf{p} = (mc_0)^2$, or again, restricting ourselves to the positive square root,

$$E = c_0 \sqrt{(mc_0)^2 + \mathbf{p} \cdot \mathbf{p}} .$$

From the previous pair of equations we conclude

$$\mathbf{p} = \frac{E}{c_0^2} \mathbf{v} .$$

(This holds for all $m \neq 0$, and hence we assume it also for $m = 0$.) For $m \neq 0$ and with $v \to c_0$ so that $\gamma \to \infty$, E and p also increase beyond all limits. In contrast, for $m = 0$, the relation $p_\mu p^\mu = (mc_0)^2$ yields $E = c_0 p$. This leads us to the unit vector $\mathbf{p}/p = \mathbf{v}/c_0$: in every inertial frame, *massless particles move with the velocity of light*.

In order to derive the Lagrange function for free particles, we use the integral principles (see Sect. 2.4.8) and take into account the fact that the proper time τ (but not the coordinate time t) is Lorentz invariant. According to p. 230, we have $dt = \gamma\, d\tau$. Now Hamilton's principle states that the *action function*

$$W = \int_{t_0}^{t_1} L\, dt = \int_{\tau_0}^{\tau_1} \gamma L\, d\tau$$

takes an extreme value. This must be valid for all reference frames. Consequently, γL must be Lorentz invariant. (However, we shall not introduce an abbreviation for γL). As explained on p. 250, L is connected to the Lagrange density $\mathscr{L}$ as used on p. 241. For free particles this function depends on the four-velocity, but not on the space-time coordinates. Then we only have to find out how γL depends on $v_\nu v^\nu$. Hence we investigate the ansatz $\gamma L = m f(v_\nu v^\nu)$, bearing in mind the requirement

$$mv^\mu = p^\mu = -\frac{\partial \gamma L}{\partial v_\mu} \ .$$

We already had the first equation at the beginning of this section. The second connects two contravariant quantities and generalizes $\mathbf{p} = \nabla_v L$ (see p. 99) with $v_k = -v^k$ to four dimensions. With $\partial v_\nu v^\nu / \partial v_\mu = 2 v^\mu$, this requirement can be satisfied by any function f with $\mathrm{d}f/\mathrm{d}(v_\nu v^\nu) = -1/2$. However, because $v_\nu v^\nu = c_0{}^2$, this does not seem to be unique. Hence the Lagrange function is often derived from *Fermat's principle*, valid for free particles according to p. 141, or the *geodesic principle*,

$$\delta \int_{t_0}^{t_1} \mathrm{d}t = 0 \ , \quad \text{or} \quad \delta \int_{s_0}^{s_1} \mathrm{d}s = 0 \ .$$

(For free particles, the velocity is constant, so the two expressions yield the same orbit.) If now σ increases monotonically with the proper time τ, but otherwise is an arbitrary parameter, we have

$$\mathrm{d}s = \sqrt{g_{\mu\nu} \frac{\mathrm{d}x^\mu}{\mathrm{d}\sigma} \frac{\mathrm{d}x^\nu}{\mathrm{d}\sigma}} \, \mathrm{d}\sigma \ .$$

Here the coordinates x^μ and their derivatives can be varied. Since the parameter σ does not need to be equal to the proper time, the inconvenient condition $v_\nu v^\nu = c_0{}^2$ does not apply for the variation. On the other hand, it may be equal to the proper time, and then the expression under the square root is equal to $v_\nu v^\nu$. Consequently, γL is equal to the square root of $v_\nu v^\nu$, up to a fixed factor, and this factor we derive from the requirement that, in the non-relativistic limit, we should have $L \approx T + \text{const.}$ with $T = \frac{1}{2} m \mathbf{v} \cdot \mathbf{v}$:

$$L = -\frac{mc_0}{\gamma} \sqrt{v_\nu v^\nu} = -mc_0 \sqrt{c_0{}^2 - \mathbf{v} \cdot \mathbf{v}} \approx -mc_0{}^2 + \frac{1}{2} m v^2 \ .$$

Since here (for free particles) the Lagrange function does not depend on the space-time coordinates, the Euler–Lagrange differential equation (p. 96) yields

$$\frac{\mathrm{d}p^\mu}{\mathrm{d}\tau} = 0 \ ,$$

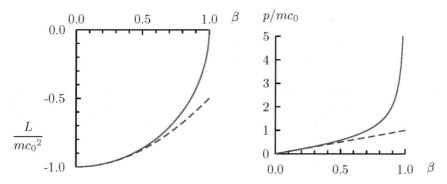

Fig. 3.29 Lagrange function and momentum of free particles as a function of $\beta = v/c_0$: non-relativistic (*dashed blue*) and relativistic (*continuous red*)

and hence also the energy and momentum conservation law for free particles (see Fig. 3.29).

3.4.10 Relativistic Dynamics with External Forces

In classical mechanics (see Sect. 2.3.4), we have already derived the generalized potential U for the interaction of a particle of charge q with an electromagnetic field, namely, $U = q\,(\Phi - \mathbf{v} \cdot \mathbf{A})$. After multiplying by γ, this expression is Lorentz invariant:

$$\gamma\,q\,(\Phi - \mathbf{v} \cdot \mathbf{A}) = q\,v_\mu A^\mu\;.$$

Here A^μ depends only on the space-time coordinates x^μ, but not on v^μ. Hence we obtain the *Lagrange function*

$$L = -\frac{mc_0\sqrt{v_\mu v^\mu} + q\,v_\mu A^\mu}{\gamma}\;.$$

This yields the *canonical conjugate momentum*

$$p^\mu = -\frac{\partial\,\gamma L}{\partial v_\mu} = mv^\mu + qA^\mu\;.$$

We have already considered its three-space components on p. 99, though not yet relativistically, and distinguished between the mechanical momentum and the canonical conjugate momentum. Its time component p^0 is related to the energy $E = c_0 p^0$, which now (with suitable gauge, see p. 124) also contains the potential energy $q\Phi$, with $A^0 = \Phi/c_0$, according to p. 239.

Important for the Lagrange equations is

$$\frac{\mathrm{d}p^\mu}{\mathrm{d}\tau} = m\,\frac{\mathrm{d}v^\mu}{\mathrm{d}\tau} + q\,\frac{\mathrm{d}A^\mu}{\mathrm{d}\tau}\,, \quad \text{with} \quad \frac{\mathrm{d}A^\mu}{\mathrm{d}\tau} = v_\nu\,\partial^\nu A^\mu\,,$$

because with $\dot{\mathbf{p}} = \nabla L$, this must be equal to $-\partial^\mu\,\gamma L$. With the above expression for the Lagrange function, it follows that $-\partial^\mu\,\gamma L = q\,v_\nu\,\partial^\mu A^\nu$. Then we arrive at the electromagnetic field tensor (see p. 240)

$$\frac{\mathrm{d}p^\mu}{\mathrm{d}\tau} = -\frac{\partial\,\gamma L}{\partial x_\mu} \qquad \Longrightarrow \qquad m\,\frac{\mathrm{d}v^\mu}{\mathrm{d}\tau} = q\,v_\nu\,F^{\mu\nu}\,.$$

Here, for $\mu = 0$, $v_\nu\,F^{\mu\nu}$ is equal to $(-\gamma\,\mathbf{v})\cdot(-\mathbf{E}/c_0) = \gamma\,\mathbf{v}\cdot\mathbf{E}/c_0$, and the space components can be combined into the three-vector $\gamma\,(\mathbf{E} + \mathbf{v}\times\mathbf{B})$. $F^\mu = q\,v_\nu\,F^{\mu\nu}$ is referred to as the *Minkowski force*:

$$F^\mu \equiv m\,\frac{\mathrm{d}v^\mu}{\mathrm{d}\tau}\,.$$

Its space components are greater by the factor γ than those of the Newtonian force. Its time component is related to the power $\gamma\,\mathbf{j}\cdot\mathbf{E}$.

The last equation also holds for forces other than electromagnetic ones.

3.4.11   Energy–Momentum Stress Tensor

We would like now to extend Maxwell's stress tensor to four dimensions. To this end, we go from the Minkowski force $F^\mu = q\,v_\nu\,F^{\mu\nu}$ over to a force density:

$$f^\mu = j_\nu\,F^{\mu\nu}\,.$$

With $\mu_0 j_\nu = \partial^\kappa F_{\kappa\nu}$, we have $\mu_0 f^\mu = (\partial^\kappa F_{\kappa\nu})\,F^{\mu\nu} = \partial^\kappa(F_{\kappa\nu}F^{\mu\nu}) - F_{\kappa\nu}\,\partial^\kappa F^{\mu\nu}$. The last term can be rewritten, because F is antisymmetric and the summation indices κ and ν may be renamed:

$$F_{\kappa\nu}\,\partial^\kappa F^{\mu\nu} = -\tfrac{1}{2}F_{\kappa\nu}\,(\partial^\nu F^{\mu\kappa} + \partial^\kappa F^{\nu\mu})\,.$$

It is then simplified using the Maxwell equations:

$$F_{\kappa\nu}\,\partial^\kappa F^{\mu\nu} = \tfrac{1}{2}F_{\kappa\nu}\,\partial^\mu F^{\kappa\nu} = \tfrac{1}{4}\,\partial^\mu F^{\lambda\nu}F_{\lambda\nu}\,.$$

Consequently, with $\partial^\mu = g^\mu{}_\kappa\partial^\kappa$, we find $\mu_0\,f^\mu = \partial^\kappa(F_{\kappa\nu}F^{\mu\nu} - \tfrac{1}{4}g^\mu{}_\kappa F^{\lambda\nu}F_{\lambda\nu})$. Therefore, the force density f^μ is the (four-dimensional) source density of a symmetric tensor:

$$f^\mu = -\partial_\kappa T^{\kappa\mu} \ , \quad \text{with} \quad T^{\kappa\mu} \equiv \frac{\frac{1}{4} g^{\kappa\mu} F^{\lambda\nu} F_{\lambda\nu} - F^\kappa{}_\nu F^{\mu\nu}}{\mu_0} = T^{\mu\kappa} \ .$$

If we restrict ourselves to $\mathbf{D} = \varepsilon_0 \mathbf{E}$ and $\mathbf{B} = \mu_0 \mathbf{H}$, then we can extend Maxwell's stress tensor, introduced on p. 215, with elements $T_{xx} = w - \varepsilon_0 E_x E_x - \mu_0 H_x H_x$ and $T_{xy} = -\varepsilon_0 E_x E_y - \mu_0 H_x H_y$ (and cyclic permutations), with the energy density $w = \frac{1}{2}(\mathbf{E} \cdot \mathbf{D} + \mathbf{H} \cdot \mathbf{B})$ and the Poynting vector $\mathbf{S} = \mathbf{E} \times \mathbf{H}$, into four dimensions:

$$\left(T^{\mu\nu} \right) = \begin{pmatrix} w & S_x/c_0 & S_y/c_0 & S_z/c_0 \\ S_x/c_0 & T_{xx} & T_{xy} & T_{xz} \\ S_y/c_0 & T_{yx} & T_{yy} & T_{yz} \\ S_z/c_0 & T_{zx} & T_{zy} & T_{zz} \end{pmatrix}, \quad \text{with} \quad \mathrm{tr}\,T = \sum_\mu T^\mu{}_\mu = 0 \ .$$

The stress tensor known from the static case is now completed with the Poynting vector and the energy density. According to p. 215, $\mathbf{S}/c_0{}^2$ is a momentum density, whence T is referred to as the *energy–momentum stress tensor*. Its space components are

$$f^i + \frac{1}{c_0{}^2} \frac{\partial S^i}{\partial t} + \frac{\partial T^{ik}}{\partial x^k} = 0 \ .$$

In addition, $f^0 = j_\nu F^{0\nu} = \mathbf{j} \cdot \mathbf{E}/c_0$, so $-\mathbf{j} \cdot \mathbf{E} = c_0 \partial_\kappa T^{\kappa 0} = \partial_t w + \nabla \cdot \mathbf{S}$. We already know this equation (p. 211) as Poynting's theorem.

3.4.12 Summary: Lorentz Invariance

Maxwell's equations ensure the same vacuum velocity of light in all inertial frames: the laws of electromagnetism are Lorentz invariant. The space-time description must be adjusted to this fact, something that leads to unusual consequences for high velocities. Just as time and space have to be combined to give $x^\mu \cong (c_0 t, \mathbf{r})$, so also do charge and current density to give $j^\mu \cong (c_0 \rho, \mathbf{j})$, energy and momentum to give $p^\mu \cong (E/c_0, \mathbf{p})$, scalar and vector potential to give $A^\mu \cong (\Phi/c_0, \mathbf{A})$, and angular frequency and wave vector to give $k^\mu \cong (\omega/c_0, \mathbf{k})$. By building skew-symmetric tensors $F^{\mu\nu}$ and $G^{\mu\nu}$ from $\mathbf{E}/c_0$ and $\mathbf{B}$ and from $c_0 \mathbf{D}$ and $\mathbf{H}$, respectively, the pairs of Maxwell equations for microscopic electromagnetism can each be combined into one equation, viz.,

$$\partial^\lambda F^{\mu\nu} + \partial^\mu F^{\nu\lambda} + \partial^\nu F^{\lambda\mu} = 0 \quad \text{and} \quad \partial_\mu F^{\mu\nu} = \mu_0 j^\nu \ ,$$

and those for macroscopic electromagnetism into

$$\partial^\lambda F^{\mu\nu} + \partial^\mu F^{\nu\lambda} + \partial^\nu F^{\lambda\mu} = 0 \quad \text{and} \quad \partial_\mu G^{\mu\nu} = \overline{j}^\nu \ .$$

In addition, the equation $f^\nu = -\partial_\mu T^{\mu\nu}$ with the (symmetric) energy–momentum stress-tensor

$$T^{\mu\nu} = \frac{\frac{1}{4} g^{\mu\nu} F_{\kappa\lambda} F^{\kappa\lambda} - F^\mu{}_\kappa F^{\nu\kappa}}{\mu_0}$$

combines Poynting's theorem and the relation between force density and Maxwell's stress tensor.

Lorentz invariance leads to the fact that, in classical mechanics, derivatives with respect to time must be replaced by derivatives with respect to the proper time, thereby introducing the factor γ. In particular, for free particles of mass m, we have the momentum $(p^\mu) \equiv m(v^\mu) \,\hat{=}\, m\gamma\,(c_0, \mathbf{v})$ with $p^0 = E/c_0$, or $E = m\gamma c_0^2$ and $\mathbf{p} = c_0^{-2} E\,\mathbf{v}$, and otherwise for particles with the charge q,

$$p^\mu = -\frac{\partial\gamma L}{\partial v_\mu} = m\,v^\mu + q\,A^\mu \ .$$

In the expression $(p^\mu) \,\hat{=}\, m\gamma\,(c_0, \mathbf{v})$, the factor γ belongs to the velocity, not to the mass—this is a Lorentz invariant, as is $v_\mu v^\mu$, but only because of the factor γ. There is no "velocity-dependent mass" (see L.B. Okun: Phys. Today 42, 6 (1989) 31–36.)

3.4.13 Supplement: Hamiltonian Formalism for Fields

On p. 241 the Lagrange function known from the mechanics of particles was extended to the Lagrange density $\mathscr{L}$ for the electromagnetic field. Here we present the transition to the Hamiltonian formulation, which is often applied to field quantization, even though there are other ways to derive the latter, as we shall see in Sect. 5.5.2.

After introducing the Lagrange density $\mathscr{L}$, Hamilton's principle reads

$$\delta \int d^4x^\mu\ \mathscr{L}(x^\mu, \eta, \frac{\partial\eta}{\partial x^\mu}) = 0\ , \quad \text{with}\quad \int d^3x^k\ \mathscr{L} = L\ ,$$

where the coordinates x^μ are given and the parameter (or parameters) η of the system are to be varied. For the electromagnetic field, η is equal to the four-potential A. Therefore, with the Einstein summation convention, we may set

$$\delta\mathscr{L} = \frac{\partial\mathscr{L}}{\partial\eta}\,\delta\eta + \frac{\partial\mathscr{L}}{\partial(\partial_\mu\eta)}\,\delta(\partial_\mu\eta)\ ,$$

using the abbreviation $\partial_\mu\eta \equiv \partial\eta/\partial x^\mu$. We may change the order of the derivative with respect to x^μ and the variation, i.e., $\delta(\partial_\mu\eta) = \partial\delta\eta/\partial x^\mu$, and integrate by parts. However, here η depends not only on the single coordinate x^μ, but also on the three remaining ones, and therefore the implicit dependence of the field quantity η on the

x^μ must also be accounted for, although in many textbooks there is only the partial derivative instead of the total derivative in the next equation:

$$\int dx^\mu \, \frac{\partial\mathscr{L}}{\partial(\partial_\mu\eta)} \frac{\partial\delta\eta}{\partial x^\mu} = \frac{\partial\mathscr{L}}{\partial(\partial_\mu\eta)} \, \delta\eta - \int dx^\mu \, \frac{d}{dx^\mu} \left(\frac{\partial\mathscr{L}}{\partial(\partial_\mu\eta)} \right) \delta\eta \,,$$

where there is of course no summation over μ. Since η is to be kept fixed at the integration limits during the variation, the first term on the right-hand side vanishes. Hence Hamilton's principle appears in the form

$$\int d^4x^\mu \left(\frac{\partial\mathscr{L}}{\partial\eta} - \sum_{\mu=0}^{3} \frac{d}{dx^\mu} \frac{\partial\mathscr{L}}{\partial(\partial_\mu\eta)} \right) \delta\eta = 0 \,,$$

and we obtain the Euler–Lagrange equations

$$\frac{\partial\mathscr{L}}{\partial\eta} = \frac{d}{dx^\mu} \frac{\partial\mathscr{L}}{\partial(\partial_\mu\eta)} \,,$$

with Einstein's summation convention. This we may also write as

$$\frac{d}{dt} \frac{\partial\mathscr{L}}{\partial\dot\eta} = \frac{\partial\mathscr{L}}{\partial\eta} - \frac{d}{dx^k} \frac{\partial\mathscr{L}}{\partial(\partial_k\eta)} \,.$$

The similarity with the usual equation appears more clearly if we use the Lagrange function L instead of the Lagrange density $\mathscr{L}$, but now take it as a functional of the functions η and $\dot\eta$, introducing the *functional derivatives*

$$\frac{\delta L}{\delta\eta} \equiv \frac{\partial\mathscr{L}}{\partial\eta} - \frac{d}{dx^k} \frac{\partial\mathscr{L}}{\partial(\partial_k\eta)} \quad \text{and} \quad \frac{\delta L}{\delta\dot\eta} \equiv \frac{\partial\mathscr{L}}{\partial\dot\eta} \,.$$

If we divide space into N cells and discretize to give $L = \sum_{i=1}^{N} \mathscr{L}_i \, \Delta V_i$, it follows that

$$\delta L = \sum_{i=1}^{N} \left\{ \left(\frac{\partial\mathscr{L}}{\partial\eta} - \frac{d}{dx^k} \frac{\partial\mathscr{L}}{\partial(\partial_k\eta)} \right)_i \delta\eta_i + \frac{\partial\mathscr{L}}{\partial\dot\eta} \delta\dot\eta_i \right\} \Delta V_i \,.$$

Since the variations $\delta\eta_i$ and $\delta\dot\eta_i$ with $i \in \{1, \ldots N\}$ may be performed independently of each other, the limit $\Delta V_i \to 0$ can be considered separately for each cell. The functional derivative $\delta\mathscr{L}/\delta\eta$ still contains a factor $(\Delta V)^{-1}$. Therefore the Lagrange density $\mathscr{L}$ appears on the right. Hence the result reads simply

$$\frac{d}{dt} \frac{\delta L}{\delta\dot\eta} = \frac{\delta L}{\delta\eta} \,,$$

which is similar to the normal Lagrange equation. However, because of the functional derivatives, we are now dealing with a differential equation, from which we must now determine $\eta(t, \mathbf{r})$ rather than $x(t)$—instead of the coordinates x (possibly very many, but nevertheless a finite number), a whole field must now be determined.

The quantity canonically conjugate to the field quantity η_i in the volume ΔV_i is

$$p_i \equiv \frac{\partial L}{\partial \dot{\eta}_i} = \Delta V_i \frac{\partial \mathscr{L}}{\partial \dot{\eta}_i} = \Delta V_i \, \pi_i \,,$$

with the momentum density

$$\pi \equiv \frac{\partial \mathscr{L}}{\partial \dot{\eta}} = \frac{\delta L}{\delta \dot{\eta}} \,,$$

where π_i is its mean value in the volume ΔV_i. If we go over from the Lagrangian to the Hamiltonian mechanics then, with $\mathscr{H}(x^\mu, \eta, \pi, \partial_k \eta)$, we also have

$$dH = \int d^3 x^k \left(\frac{\partial \mathscr{H}}{\partial t} \, dt + \frac{\partial \mathscr{H}}{\partial \eta} \, d\eta + \frac{\partial \mathscr{H}}{\partial \pi} \, d\pi + \frac{\partial \mathscr{H}}{\partial (\partial_l \eta)} \, d(\partial_l \eta) \right).$$

We integrate the last term by parts (without the summation convention):

$$\int dx^l \, \frac{\partial \mathscr{H}}{\partial (\partial_l \eta)} \, d\left(\frac{\partial \eta}{\partial x^l} \right) = \frac{\partial \mathscr{H}}{\partial (\partial_l \eta)} \, d\eta - \int dx^l \, \frac{d}{dx^l} \frac{\partial \mathscr{H}}{\partial (\partial_l \eta)} \, d\eta \,.$$

The integrated term vanishes if the considered system exists only in a finite volume, as we have assumed. Hence,

$$dH = \int d^3 x^k \left\{ \frac{\partial \mathscr{H}}{\partial t} \, dt + \left(\frac{\partial \mathscr{H}}{\partial \eta} - \frac{d}{dx^l} \frac{\partial \mathscr{H}}{\partial (\partial_l \eta)} \right) d\eta + \frac{\partial \mathscr{H}}{\partial \pi} \, d\pi \right\} \,,$$

with the summation convention. Instead of the round bracket, we may also write the functional derivative $\delta H / \delta \eta$. On the other hand, the relation

$$dH = \int d^3 x^k \left(\pi \, d\dot{\eta} + \dot{\eta} \, d\pi - \frac{\delta L}{\delta \eta} \, d\eta - \frac{\delta L}{\delta \dot{\eta}} \, d\dot{\eta} - \frac{\partial \mathscr{L}}{\partial t} \, dt \right)$$

follows from $\mathscr{H} = \pi \, \dot{\eta} - \mathscr{L}$ with $\pi = \partial L / \partial \dot{\eta}$. Here, since $\pi = \delta L / \delta \dot{\eta}$, the first term cancels the fourth. If we also use $\delta L / \delta \eta = \dot{\pi}$, then we may set

$$dH = \int d^3 x^k \left(-\frac{\partial \mathscr{L}}{\partial t} \, dt - \dot{\pi} \, d\eta + \dot{\eta} \, d\pi \right) \,,$$

Comparing with the expression found above, we obtain the *Hamilton equations for a field*, viz.,

$$\frac{\partial \mathscr{H}}{\partial t} = -\frac{\partial \mathscr{L}}{\partial t} \, , \quad \frac{\delta H}{\delta \eta} = -\dot{\pi} \, , \quad \text{and} \quad \frac{\delta H}{\delta \pi} = \dot{\eta} \, ,$$

because $\mathscr{H}$ does not depend on the spatial derivatives of π, and therefore $\delta H/\delta \pi = \partial \mathscr{H}/\partial \pi$.

The Hamilton function H is a conserved quantity if dH/dt vanishes. Clearly,

$$\frac{dH}{dt} = \int d^3 x^k \, \frac{\partial \mathscr{H}}{\partial t} \, ,$$

because $d\eta/dt = \dot{\eta}$ and $d\pi/dt = \dot{\pi}$ cancels the remaining terms of the integrand.

The time dependence of an arbitrary quantity O can be obtained from

$$\frac{dO}{dt} = \int d^3 x^k \left(\frac{\delta O}{\delta \eta} \dot{\eta} + \frac{\delta O}{\delta \pi} \dot{\pi} \right) + \frac{\partial O}{\partial t}$$

$$= \int d^3 x^k \left(\frac{\delta O}{\delta \eta} \frac{\delta \mathscr{H}}{\delta \pi} - \frac{\delta O}{\delta \pi} \frac{\delta \mathscr{H}}{\delta \eta} \right) + \frac{\partial O}{\partial t} = [O, H] + \frac{\partial O}{\partial t} \, .$$

For the last equation we have extended the concept of the Poisson bracket to fields, as an abbreviation for the preceding integral.

The Poisson bracket $[\eta_i, p_i] = 1$ of particle physics has become $[\eta_i, \pi_i] = 1/\Delta V_i$ in field theory. For the limit $\Delta V_i \to 0$,

$$[\eta(t, \mathbf{r}), \pi(t, \mathbf{r}')] = \delta(\mathbf{r} - \mathbf{r}') \, ,$$

and after a Fourier transform $[\eta(t, \mathbf{k}), \pi(t, \mathbf{k}')] = \delta(\mathbf{k} - \mathbf{k}')$.

3.5 Radiation Fields

3.5.1 Solutions of the Inhomogeneous Wave Equations

Now we turn to the potential equations of microscopic electromagnetism from Sect. 3.4.6 (with the Lorentz and Coulomb gauges):

$$\Box A^\mu = \mu_0 \, j^\mu \, , \quad \text{or} \quad \Box A^\mu = \mu_0 \, j^\mu_{\text{trans}} \, .$$

Here the inhomogeneities may also depend on time, because otherwise we just obtain the already known static solutions. We solve both equations with the same Green function, since they involve the same differential operator $\Box$ and differ only in the inhomogeneity. This Green function generalizes the expression (for the Laplace operator Δ) known from statics. In particular, it takes into account the fact that space and time are connected with each other via the velocity of light c_0:

$$\Box \frac{\delta(t' - t \pm |\mathbf{r} - \mathbf{r}'|/c_0)}{4\pi |\mathbf{r} - \mathbf{r}'|} = \delta(t - t')\,\delta(\mathbf{r} - \mathbf{r}')\;.$$

So far we have considered the limit $c_0 \to \infty$ ($\Box \to -\Delta$) and we were therefore allowed to omit the delta function $\delta(t - t')$ on the left- and right-hand sides. We shall use only the Green function with the plus sign: the source at the position $\mathbf{r}'$ acts at the chosen point $\mathbf{r}$ after the lapse of time $t - t' = |\mathbf{r} - \mathbf{r}'|/c_0$. This is called the *retarded solution*. The Green function with the minus sign is known as the *advanced solution*. It is mathematically but not physically allowable, because effects then occur before their cause.

Before proving the validity of these Green functions, we first show their Lorentz invariance. If we use $\delta\{(x_\mu - x_\mu')(x^\mu - x'^\mu)\} = \delta\{(c_0 t - c_0 t')^2 - |\mathbf{r} - \mathbf{r}'|^2\}$ and take into account the equation on p. 20, viz.,

$$\delta\{(c_0 \Delta t)^2 - |\Delta\mathbf{r}|^2\} = \frac{\delta(\Delta t - |\Delta\mathbf{r}|/c_0) + \delta(\Delta t + |\Delta\mathbf{r}|/c_0)}{2\,c_0\,|\Delta\mathbf{r}|}\;,$$

it follows that

$$\frac{\delta(t' - t \pm |\mathbf{r} - \mathbf{r}'|/c_0)}{|\mathbf{r} - \mathbf{r}'|} = 2c_0\,\varepsilon\{\pm(t - t')\}\,\delta\{(x_\mu - x_\mu')(x^\mu - x'^\mu)\}\;.$$

Here, the step function $\varepsilon\{\pm(t - t')\}$ seems to violate Lorentz invariance, but we wish to distinguish uniquely between past and future, and therefore restrict ourselves to the retarded solutions, that is, to *proper Lorentz transformations*. Forwards and backwards light cones then remain separated.

For the actual proof, we use the Fourier representation of the delta function (see p. 21) with $\mathbf{R} \equiv \mathbf{r} - \mathbf{r}'$ and $k = \omega/c_0$, i.e.,

$$\Box \frac{\delta(t' - t \pm R/c_0)}{R} = \Box \frac{1}{2\pi} \int_{-\infty}^{\infty} d\omega\, \frac{\exp\{i\omega(t' - t \pm R/c_0)\}}{R}$$

$$= \frac{1}{2\pi} \int_{-\infty}^{\infty} d\omega\, \exp\{i\omega(t' - t)\}\left(-\frac{\omega^2}{c_0^2} - \Delta\right)\frac{\exp(\pm ikR)}{R}\;.$$

The d'Alembert operator in the "time representation" then becomes $-(\Delta + k^2)$ in the "frequency representation". Now in the general case (the special case $k = 0$ was already considered on p. 26)

$$(\Delta + k^2)\,\frac{\exp(\pm ikR)}{R} = -4\pi\,\delta(\mathbf{R})\;.$$

According to p. 39, for $R \neq 0$, the left-hand side is equal to $R^{-1}(\partial^2/\partial R^2 + k^2)\exp(\pm ikR)$, hence zero. However, for $R = 0$, it is singular, and its volume integral, according to p. 27, is equal to -4π.

Thus, we have for the Lorentz gauge,

$$A^\mu(t, \mathbf{r}) = \frac{\mu_0}{4\pi} \int dt' \, dV' \, j^\mu(t', \mathbf{r}') \, \frac{\delta(t' - t + |\mathbf{r} - \mathbf{r}'|/c_0)}{|\mathbf{r} - \mathbf{r}'|}$$

$$= \frac{\mu_0}{4\pi} \int dV' \, \frac{j^\mu(t - |\mathbf{r} - \mathbf{r}'|/c_0, \mathbf{r}')}{|\mathbf{r} - \mathbf{r}'|}$$

as the retarded solution. The continuity equation already ensures the gauge condition $\partial_\mu A^\mu = 0$. If we use $\int_{-\infty}^{\infty} dt \, \exp(\mathrm{i}\omega t) \, \delta(t' - t + R/c_0) = \exp(\mathrm{i}\omega t') \, \exp(\mathrm{i}kR)$ for the Fourier transform, we obtain the expression

$$A^\mu(\omega, \mathbf{r}) = \frac{1}{\sqrt{2\pi}} \int_{-\infty}^{\infty} dt \, A^\mu(t, \mathbf{r}) \exp(\mathrm{i}\omega t) = \frac{\mu_0}{4\pi} \int dV' \, j^\mu(\omega, \mathbf{r}') \frac{\exp(\mathrm{i}k|\mathbf{r}-\mathbf{r}'|)}{|\mathbf{r}-\mathbf{r}'|} .$$

Note that we take $\exp(\mathrm{i}\omega t)$ and not $\exp(-\mathrm{i}\omega t)$, since that leads us to $\omega t - \mathbf{k} \cdot \mathbf{r} = k_\mu x^\mu$—of course, $j^\mu(\omega, \mathbf{r})$ is related to $j^\mu(t, \mathbf{r})$ via the same Fourier transform. Hence the source density is easy to determine, since $\nabla f(|\mathbf{r} - \mathbf{r}'|) = -\nabla' f(|\mathbf{r} - \mathbf{r}'|)$:

$$\nabla \cdot \mathbf{A}(\omega, \mathbf{r}) = -\frac{\mu_0}{4\pi} \int dV' \, \mathbf{j}(\omega, \mathbf{r}') \cdot \nabla' \frac{\exp(\mathrm{i}k|\mathbf{r} - \mathbf{r}'|)}{|\mathbf{r} - \mathbf{r}'|} .$$

With $\mathbf{j} \cdot \nabla' G = \nabla' \cdot G\mathbf{j} - G\,\nabla' \cdot \mathbf{j}$, we can split the integral into two terms. The first can be converted according to Gauss into a surface integral and does not contribute, since $\mathbf{j}$ vanishes on the surface, while the second can be rewritten with the continuity equation, because using $\rho(t, \mathbf{r}) \propto \rho(\omega, \mathbf{r}) \exp(-\mathrm{i}\omega t)$ and $\mathbf{j}(t, \mathbf{r}) \propto \mathbf{j}(\omega, \mathbf{r}) \exp(-\mathrm{i}\omega t)$, it reads

$$\nabla \cdot \mathbf{j}(\omega, \mathbf{r}) = \mathrm{i}\omega \, \rho(\omega, \mathbf{r}) = \frac{\mathrm{i}\omega}{c_0} \, j^0(\omega, \mathbf{r}) .$$

Consequently,

$$\nabla \cdot \mathbf{A}(\omega, \mathbf{r}) = \frac{\mathrm{i}\omega}{c_0} \, A^0(\omega, \mathbf{r}) ,$$

and hence also $\partial_\mu A^\mu = 0$. In the given expression for A^μ, the continuity equation $\partial_\mu j^\mu = 0$ already ensures the Lorentz gauge.

For the derivation of $\partial_\mu A^\mu = 0$, the current density must vanish on the surface of the integration volume. For the Coulomb gauge, only the transverse current density is of interest (transverse gauge): then it is already sufficient that the current density should not have a normal component there. Then the source freedom for the Fourier transformed $\mathbf{A}(\omega, \mathbf{r})$ is easily checked. As in Sect. 3.2.8, we use Gauss's theorem, assuming no current density at infinity and the source freedom of $\mathbf{j}_{\text{trans}}$. As $\mathbf{A}(\omega, \mathbf{r})$ is solenoidal, this is true also for $\mathbf{A}(t, \mathbf{r})$. In the Coulomb gauge, because $\Delta \Phi = -\rho/\varepsilon_0$, we have

$$\Phi(t, \mathbf{r}) = \frac{1}{4\pi \varepsilon_0} \int dV' \frac{\rho(t, \mathbf{r}')}{|\mathbf{r} - \mathbf{r}'|} ,$$

$$\mathbf{A}(t, \mathbf{r}) = \frac{\mu_0}{4\pi} \int dt' \, dV' \, \mathbf{j}_{\text{trans}}(t', \mathbf{r}') \frac{\delta(t' - t + |\mathbf{r} - \mathbf{r}'|/c_0)}{|\mathbf{r} - \mathbf{r}'|} ,$$

and after a Fourier transform

$$\Phi(\omega, \mathbf{r}) = \frac{1}{4\pi \varepsilon_0} \int dV' \, \rho(\omega, \mathbf{r}') \frac{1}{|\mathbf{r} - \mathbf{r}'|} ,$$

$$\mathbf{A}(\omega, \mathbf{r}) = \frac{\mu_0}{4\pi} \int dV' \, \mathbf{j}_{\text{trans}}(\omega, \mathbf{r}') \frac{\exp(ik|\mathbf{r} - \mathbf{r}'|)}{|\mathbf{r} - \mathbf{r}'|} .$$

We would like to use these expressions for the radiation fields, which is why the Coulomb gauge is often also called *radiation gauge*. The fact that the radiation is transverse is more important for us than Lorentz invariance. For this reason, the radiation gauge is also used in quantum electrodynamics (see Sect. 5.5.1).

3.5.2 Radiation Fields

For the magnetic field, with $\mathbf{B}(\omega, \mathbf{r}) = \nabla \times \mathbf{A}(\omega, \mathbf{r})$, we obtain

$$\mathbf{B}(\omega, \mathbf{r}) = -\frac{\mu_0}{4\pi} \int dV' \, \mathbf{j}_{\text{trans}}(\omega, \mathbf{r}') \times \nabla \frac{\exp(ik|\mathbf{r} - \mathbf{r}'|)}{|\mathbf{r} - \mathbf{r}'|} ,$$

with

$$\nabla \frac{\exp(ik|\mathbf{r} - \mathbf{r}'|)}{|\mathbf{r} - \mathbf{r}'|} = \left(ik - \frac{1}{|\mathbf{r} - \mathbf{r}'|} \right) \frac{\exp(ik|\mathbf{r} - \mathbf{r}'|)}{|\mathbf{r} - \mathbf{r}'|} \frac{\mathbf{r} - \mathbf{r}'}{|\mathbf{r} - \mathbf{r}'|} .$$

We thus have two terms with different position dependence. For time-dependent problems (thus $k \neq 0$), the field decays more weakly with the distance from the source than in the static case. This is also shown by the representation

$$\mathbf{B}(t, \mathbf{r}) = \frac{1}{\sqrt{2\pi}} \int_{-\infty}^{\infty} d\omega \, \mathbf{B}(\omega, \mathbf{r}) \, \exp(-i\omega t) ,$$

because we have

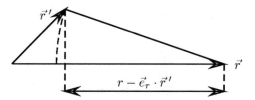

Fig. 3.30 The approximation $|\mathbf{r} - \mathbf{r}'| \approx r - \mathbf{e}_r \cdot \mathbf{r}'$ valid for $r \gg r'$ follows by calculation (using a series expansion) as well as geometrically. The circular radius $|\mathbf{r} - \mathbf{r}'|$ and the double-headed arrow are nearly equally long

$$\frac{1}{\sqrt{2\pi}} \int_{-\infty}^{\infty} d\omega \; \mathbf{j}_{\text{trans}}(\omega, \mathbf{r}') \, e^{-i\omega t} \, e^{ikR} = \mathbf{j}_{\text{trans}}(t - R/c_0, \mathbf{r}') \,,$$

$$\frac{-i}{\sqrt{2\pi}} \int_{-\infty}^{\infty} d\omega \; \omega \, \mathbf{j}_{\text{trans}}(\omega, \mathbf{r}') \, e^{-i\omega t} \, e^{ikR} = \frac{\partial \mathbf{j}_{\text{trans}}(t - R/c_0, \mathbf{r}')}{\partial t} \,,$$

and hence, with $t' = t - |\mathbf{r} - \mathbf{r}'|/c_0$,

$$\mathbf{B}(t, \mathbf{r}) = \frac{\mu_0}{4\pi} \int dV' \left(\frac{\partial \mathbf{j}_{\text{trans}}(t', \mathbf{r}')}{c_0 \, \partial t} + \frac{\mathbf{j}_{\text{trans}}(t', \mathbf{r}')}{|\mathbf{r} - \mathbf{r}'|} \right) \times \frac{\mathbf{r} - \mathbf{r}'}{|\mathbf{r} - \mathbf{r}'|^2} \,,$$

for the magnetic field. Previously, we took the derivative with respect to the position instead of the time and thereby could not account explicitly for the finite propagation velocity.

Since the current density is connected to the velocity, the part with the derivative of $\mathbf{j}$ with respect to time is called the *acceleration field* and the second the *velocity field*. With increasing distance from the source, the acceleration field clearly contributes most to $\mathbf{B}$.

For the electric field, we conclude from $\mathbf{E} = -\nabla\Phi - \partial\mathbf{A}/\partial t$ that

$$\mathbf{E}(\omega, \mathbf{r}) = -\nabla\Phi(\omega, \mathbf{r}) + i\omega\mathbf{A}(\omega, \mathbf{r})$$

$$= -\frac{1}{4\pi\varepsilon_0} \int dV' \left(\rho(\omega, \mathbf{r}') \, \nabla \frac{1}{|\mathbf{r} - \mathbf{r}'|} - \frac{i\omega}{c_0^2} \mathbf{j}_{\text{trans}}(\omega, \mathbf{r}') \, \frac{\exp(ik|\mathbf{r} - \mathbf{r}'|)}{|\mathbf{r} - \mathbf{r}'|} \right) \,,$$

and thus after the Fourier transform $\omega \to t$

$$\mathbf{E}(t, \mathbf{r}) = -\frac{1}{4\pi\varepsilon_0} \int dV' \left(\rho(t, \mathbf{r}') \, \nabla \frac{1}{|\mathbf{r} - \mathbf{r}'|} + \frac{\partial \mathbf{j}_{\text{trans}}(t', \mathbf{r}')/\partial t}{c_0^2 \, |\mathbf{r} - \mathbf{r}'|} \right) \,,$$

noting that we have t in the argument of the charge density, but $t' = t - |\mathbf{r} - \mathbf{r}'|/c_0$ for the current density. Here, we write first the longitudinal and then the acceleration field, even though the acceleration field is more important for greater distances.

For large distances of the chosen point from the source ($r \gg r'$), we may set (see Fig. 3.30)

$$k|\mathbf{r} - \mathbf{r}'| \approx k\,(r - \mathbf{e}_r \cdot \mathbf{r}') = kr - \mathbf{k}' \cdot \mathbf{r}', \quad \text{with} \quad \mathbf{k}' = k\,\mathbf{e}_r .$$

Then, using $\mathbf{e}_r = \mathbf{k}'/k$,

$$4\pi\,\varepsilon_0\,\mathbf{E}(\omega, \mathbf{r}) \approx \frac{ik}{c_0}\,\frac{\exp(ikr)}{r}\int dV'\,\mathbf{j}_{\text{trans}}(\omega, \mathbf{r}')\,\exp(-i\mathbf{k}' \cdot \mathbf{r}') ,$$

$$c_0\,\mathbf{B}(\omega, \mathbf{r}) \approx \mathbf{e}_r \times \mathbf{E}(\omega, \mathbf{r}) .$$

In agreement with Fig. 3.19, the vectors $\mathbf{e}_r$, $\mathbf{E}$, and $\mathbf{B}$ are mutually perpendicular to each other for $r \gg r'$, because with $\mathbf{e}_r = \mathbf{k}'/k$, we have

$$\mathbf{E}(\omega, \mathbf{r}) \cdot \mathbf{e}_r \propto \int dV'\,\mathbf{j}_{\text{trans}}(\omega, \mathbf{r}') \cdot \mathbf{k}'\exp(-i\mathbf{k}' \cdot \mathbf{r}') ,$$

and also $\mathbf{j}_{\text{trans}} \cdot \mathbf{k}'\exp(-i\mathbf{k}' \cdot \mathbf{r}') \propto \mathbf{j}_{\text{trans}} \cdot \nabla'\exp(-i\mathbf{k}' \cdot \mathbf{r}')$, whence generally,

$$\mathbf{j}_{\text{trans}} \cdot \nabla'G = \nabla' \cdot G\mathbf{j}_{\text{trans}} - G\,\nabla' \cdot \mathbf{j}_{\text{trans}} .$$

Thus the volume integral vanishes, because there is no current at the surface and $\mathbf{j}_{\text{trans}}$ is solenoidal.

In the following we shall often use the Fourier transform (see p. 25 and p. 255)

$$\mathbf{j}_{\text{trans}}(\omega, \mathbf{k}) = \frac{1}{\sqrt{2\pi}^3}\int dV\,\exp(-i\mathbf{k} \cdot \mathbf{r})\,\mathbf{j}_{\text{trans}}(\omega, \mathbf{r})$$

$$= \frac{1}{(2\pi)^2}\int dt\,dV\,\exp\{i(\omega t - \mathbf{k} \cdot \mathbf{r})\}\,\mathbf{j}_{\text{trans}}(t, \mathbf{r}) .$$

In particular, we have just obtained the electric field strength at large distances, viz.,

$$\mathbf{E}(\omega, \mathbf{r}) \approx \sqrt{\frac{\pi}{2}}\,\frac{ik}{\varepsilon_0 c_0}\,\frac{\exp(ikr)}{r}\,\mathbf{j}_{\text{trans}}(\omega, \mathbf{k}) , \quad \text{with} \quad \mathbf{k} = k\,\mathbf{e}_r ,$$

and also the magnetic field with $c_0\mathbf{B}(\omega, \mathbf{r}) \approx \mathbf{e}_r \times \mathbf{E}(\omega, \mathbf{r})$.

3.5.3   Radiation Energy

Now the Poynting vector can be related to the properties of the radiation source. To this end, we make a Fourier expansion

$$\mathbf{E}(t, \mathbf{r}) = \frac{1}{\sqrt{2\pi}}\int_{-\infty}^{\infty} d\omega\,\exp(-i\omega t)\,\mathbf{E}(\omega, \mathbf{r}) ,$$

(likewise for $\mathbf{H}$) and obtain for the Poynting vector $\mathbf{S} = \mathbf{E} \times \mathbf{H}$ integrated over the time (according to the Parseval equation on p. 23)

$$\int_{-\infty}^{\infty} dt\, \mathbf{S}(t, \mathbf{r}) = \int_{-\infty}^{\infty} d\omega\, \mathbf{E}^*(\omega, \mathbf{r}) \times \mathbf{H}(\omega, \mathbf{r})\,,$$

because $\mathbf{E}(t, \mathbf{r})$ and $\mathbf{H}(t, \mathbf{r})$ are real functions. Hence, with $\mathbf{E}(\omega, \mathbf{r}) = \mathbf{E}^*(-\omega, \mathbf{r})$ and $\mathbf{H}(\omega, \mathbf{r}) = \mathbf{H}^*(-\omega, \mathbf{r})$, we find

$$\int_{-\infty}^{\infty} dt\, \mathbf{S}(t, \mathbf{r}) = 2\mathrm{Re} \int_{0}^{\infty} d\omega\, \mathbf{E}^*(\omega, \mathbf{r}) \times \mathbf{H}(\omega, \mathbf{r})\,.$$

Far from the radiation source, i.e., beyond any magnetization, so $\mathbf{H} = \mathbf{B}/\mu_0$, the last section now yields

$$\mathbf{E}^*(\omega, \mathbf{r}) \times \mathbf{H}(\omega, \mathbf{r}) \approx \frac{1}{\mu_0 c_0} \mathbf{E}^*(\omega, \mathbf{r}) \cdot \mathbf{E}(\omega, \mathbf{r})\, \mathbf{e}_r \approx \frac{\pi}{2} \frac{k^2}{\varepsilon_0 c_0} \left| \mathbf{j}_{\mathrm{trans}}(\omega, \mathbf{k}) \right|^2 \frac{\mathbf{r}}{r^3}\,.$$

With $k^2/\varepsilon_0 = \mu_0 \omega^2$, the Poynting vector integrated over all times is therefore asymptotically equal to $\pi \mu_0\, \mathbf{r}/(c_0 r^3) \int_0^\infty d\omega\, \omega^2\, |\mathbf{j}_{\mathrm{trans}}(\omega, \mathbf{k})|^2$. The energy (in joule) flowing into the solid angle element $d\Omega = \mathbf{r} \cdot d\mathbf{f}/r^3$ is therefore

$$dW = d\mathbf{f} \cdot \int_{-\infty}^{\infty} dt\, \mathbf{S}(t, \mathbf{r}) = \frac{\pi \mu_0}{c_0} d\Omega \int_0^\infty d\omega\, \omega^2 \left| \mathbf{j}_{\mathrm{trans}}(\omega, \mathbf{k}) \right|^2\,,$$

with $\mathbf{k} = k\, \mathbf{e}_r$. Here $\mathbf{j}_{\mathrm{trans}}$ is the solenoidal part of the current density, for which, according to Sect. 1.1.11, we may also write $\mathbf{j}_{\mathrm{trans}}(\omega, \mathbf{k}) = \mathbf{e}_k \times \{\mathbf{j}(\omega, \mathbf{k}) \times \mathbf{e}_k\}$ with $\mathbf{e}_k = \mathbf{k}/k$. Hence $|\mathbf{j}_{\mathrm{trans}}(\omega, \mathbf{k})|^2 = |\mathbf{j}(\omega, \mathbf{k}) \times \mathbf{k}/k|^2$.

If the frequency range is very sharp, then it is best to work with a single angular frequency $\overline{\omega}$. However, the time integrals then diverge. For a continuous radiation source, we should consider the radiation power averaged over a period: $\mathbf{E}(t, \mathbf{r}) = \mathrm{Re}\,\{\mathbf{E}(\overline{\omega}, \mathbf{r}) \exp(-i\overline{\omega} t)\}$ and the corresponding expression for $\mathbf{H}(t, \mathbf{r})$ lead to

$$\overline{\mathbf{S}} = \frac{\overline{\omega}}{2\pi} \int_0^{2\pi/\overline{\omega}} dt\, \mathbf{E}(t, \mathbf{r}) \times \mathbf{H}(t, \mathbf{r}) = \mathrm{Re}\, \frac{\mathbf{E}^*(\overline{\omega}, \mathbf{r}) \times \mathbf{H}(\overline{\omega}, \mathbf{r})}{2}$$

$$\approx \frac{\pi}{4} \frac{\mu_0 \overline{\omega}\, |\mathbf{j}_{\mathrm{trans}}(\overline{\omega}, \mathbf{k})|^2}{r^2} \mathbf{k}\,.$$

Therefore, for the average radiation power, we obtain

$$d\overline{W} = \overline{\mathbf{S}} \cdot d\mathbf{f} \approx \frac{\pi \mu_0}{4 c_0} \overline{\omega}^2 \left| \mathbf{j}_{\mathrm{trans}}(\overline{\omega}, \mathbf{k}) \right|^2 d\Omega\,.$$

This generally depends on the direction of $\mathbf{k}$—some examples are given in Sect. 3.5.5.

3.5.4   Radiation Fields of Point Charges

For point charges q, the Fourier transform with respect to frequencies does not make sense. Here, it is better to use

$$\int dV' \, j^\mu(t',\mathbf{r}') = \frac{q v^\mu(t',\mathbf{r}')}{\gamma}$$

for the four-potential $A^\mu(t,\mathbf{r})$. The factor γ^{-1} is necessary because of the Lorentz contraction. According to p. 255 and with $\mathbf{r}'$ as a unique function of t', the Lorentz gauge is

$$A^\mu(t,\mathbf{r}) = \frac{\mu_0 q}{4\pi} \int dt' \, \frac{v^\mu(t',\mathbf{r}')}{\gamma} \, \frac{\delta(t'-t+|\mathbf{r}-\mathbf{r}'|/c_0)}{|\mathbf{r}-\mathbf{r}'|}.$$

For the delta function, we use the abbreviation

$$\mathbf{R} \equiv \mathbf{r}-\mathbf{r}', \quad \mathbf{e} \equiv \mathbf{R}/R, \quad \boldsymbol{\beta} \equiv \mathbf{v}/c_0,$$

and set $u \equiv t'-t+R/c_0$. Then $\partial\mathbf{R}/\partial t' = -\mathbf{v}$ and $\partial R/\partial t' = -\mathbf{v}\cdot\mathbf{e}$, so we have $du/dt' = 1 - \boldsymbol{\beta}\cdot\mathbf{e}$, implying $dt' = du/(1-\boldsymbol{\beta}\cdot\mathbf{e})$. Then, because $dt'\,\delta(t'-t+R/c_0) = du\,\delta(u)/(1-\boldsymbol{\beta}\cdot\mathbf{e})$ for the Lorentz gauge, we find the *Liénard–Wiechert potential*

$$A^\mu(t,\mathbf{r}) = \frac{\mu_0 q}{4\pi} \frac{v^\mu(t-R/c_0,\mathbf{r}-\mathbf{R})}{\gamma\,(R-\boldsymbol{\beta}\cdot\mathbf{R})}.$$

For the corresponding equations $\Phi \propto c_0$ and $\mathbf{A} \propto \mathbf{v}$, we have $(A^\mu) \,\hat{=}\, (c_0^{-1}\Phi, \mathbf{A})$ and $(v^\mu) \,\hat{=}\, \gamma\,(c_0, \mathbf{v})$. The factor γ then cancels out. The (retarded) fields spread with finite velocity, and therefore depend on A^μ and v^μ at different times, depending upon the distance R (see Fig. 3.31). Here, with $(\beta^\nu) \,\hat{=}\, \gamma\,(1,\boldsymbol{\beta})$ and $(R^\nu) \,\hat{=}\, (c_0 t - c_0 t', \mathbf{r}-\mathbf{r}')$, $\gamma\,(R-\boldsymbol{\beta}\cdot\mathbf{R})$ can also be written as the scalar product $\beta_\nu R^\nu$. This does not depend on the reference frame—since emitter and receiver move against each other, R alone would not make sense. In fact, information is not radiated evenly in all space directions, but preferentially in the direction of the motion.

For the fields $\mathbf{E} = -\nabla\Phi - \partial\mathbf{A}/\partial t$ and $\mathbf{B} = \nabla\times\mathbf{A}$, we also have to take into account the retardation effect. Instead of the derivative $\partial/\partial t$, we should take the derivative $\partial/\partial t'$, and for ∇, keep t' (but not t) fixed. If, as in Sect. 1.2.7, we indicate the fixed quantity by a subscript on the differential or operator in brackets, we have $(\nabla)_t = (\nabla)_{t'} + (\nabla t')_t\,\partial/\partial t'$. In order to find $(\nabla t')_t$, we determine the action on $R = c_0(t-t')$. This is $(\nabla R)_t = -c_0(\nabla t')_t$ and $(\nabla R)_{t'} = \mathbf{e}$, whence

$$(\nabla)_t = (\nabla)_{t'} - \frac{\mathbf{e}}{(1-\boldsymbol{\beta}\cdot\mathbf{e})\,c_0}\,\frac{\partial}{\partial t'},$$

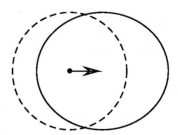

Fig. 3.31 The existence of a point charge ($\bullet$) in the spherical shell becomes noticeable only after the time $\Delta t = R/c_0$. During this time the point charge will already have displaced by $\mathbf{v}\Delta t$, but that becomes observable only later on the spherical shell. The associated Lorentz invariant $\{\gamma\,(R - \boldsymbol{\beta}\cdot\mathbf{R})\}^{-1} = 1/\beta_v R^v$ is sketched here as a *continuous line* for $\beta = \frac{1}{2}$, which is the weight factor in the Liénard–Wiechert potential

and

$$\frac{\partial R}{\partial t} = c_0\left(1 - \frac{\partial t'}{\partial t}\right) = \frac{\partial R}{\partial t'}\frac{\partial t'}{\partial t} = -\mathbf{v}\cdot\mathbf{e}\,\frac{\partial t'}{\partial t} \quad\Longrightarrow\quad \frac{\partial t}{\partial t'} = 1 - \boldsymbol{\beta}\cdot\mathbf{e}\,.$$

From the above expression for Φ and $\mathbf{A}$ (independent of the gauge) and with $\nabla\,(\boldsymbol{\beta}\cdot\mathbf{R}) = \boldsymbol{\beta}$, $\partial R/(c_0\partial t') = -\boldsymbol{\beta}\cdot\mathbf{e}$, and $\partial\,(\boldsymbol{\beta}\cdot\mathbf{R})/(c_0\partial t') = \dot{\boldsymbol{\beta}}\cdot\mathbf{R}/c_0 - \beta^2$, we obtain

$$\mathbf{E}(t,\mathbf{r}) = \frac{q}{4\pi\varepsilon_0}\,\frac{1}{(R - \boldsymbol{\beta}\cdot\mathbf{R})^3}\left(\frac{\mathbf{R} - R\boldsymbol{\beta}}{\gamma^2} + \frac{\mathbf{R}\times\{(\mathbf{R} - R\boldsymbol{\beta})\times\dot{\boldsymbol{\beta}}\}}{c_0}\right),$$

$$c_0\mathbf{B}(t,\mathbf{r}) = \mathbf{e}\times\mathbf{E}(t,\mathbf{r})\,.$$

The second term here decays more weakly by one power of R than the first, but occurs only for accelerated charges: it describes the acceleration field, and the first the velocity field. On the right here, all quantities are to be evaluated at the retarded position of the charge. The magnetic field is always perpendicular to the electric field.

3.5.5 Radiation Fields of Oscillating Dipoles

Let us now investigate a dipole oscillating with the angular frequency ω, with the maximum dipole moment $\widehat{\mathbf{p}}$. In the coordinates t and $\mathbf{r}$, we may then replace $\mathbf{j} = \rho\mathbf{v}$ by $\dot{\mathbf{p}}$. In the equation for $\mathbf{B}(\omega,\mathbf{r})$ from p. 256, we therefore use the expression $-i\omega\widehat{\mathbf{p}}$ as the Fourier component of $\mathbf{j}\,(\omega,\mathbf{r})$:

$$\mathbf{B}(\omega,\mathbf{r}) = \frac{i\mu_0\omega}{4\pi}\left(ik - \frac{1}{r}\right)\widehat{\mathbf{p}}\times\mathbf{e}_r\,\frac{\exp(ikr)}{r}\,.$$

The magnetic field is thus perpendicular to $\mathbf{r}$ and $\widehat{\mathbf{p}}$. For $\widehat{\mathbf{p}} = \widehat{p}\,\mathbf{e}_z$, $\mathbf{B}(\omega, \mathbf{r})$ has only a φ component (proportional to $\widehat{p}\,\sin\theta$). From $\varepsilon_0\,\partial\mathbf{E}/\partial t = \nabla \times \mathbf{B}/\mu_0$, we then also have the associated electric field (outside the origin):

$$\mathbf{E}(\omega, \mathbf{r}) = \frac{i\,c_0^2}{\omega}\,\nabla \times \mathbf{B}(\omega, \mathbf{r})$$

$$= \frac{1}{4\pi\varepsilon_0}\left\{\left(k^2 + \frac{ik}{r} - \frac{1}{r^2}\right)\widehat{\mathbf{p}} - \left(k^2 + \frac{3ik}{r} - \frac{3}{r^2}\right)\widehat{\mathbf{p}}\cdot\mathbf{e}_r\,\mathbf{e}_r\right\}\frac{\exp(ikr)}{r}\,.$$

With $\mathbf{p} = \widehat{p}\,\mathbf{e}_z$, this vector has an r and a θ component.

We derive the picture of the field lines from $\mathrm{d}\mathbf{r} \times \mathbf{E} = 0$, because $\mathrm{d}\mathbf{r}$ must have the direction of $\mathbf{E}$ and may be written $\mathbf{e}_r\,\mathrm{d}r + \mathbf{e}_\theta\,r\,\mathrm{d}\theta$, then express $\mathbf{E} \propto \nabla \times \mathbf{B}$ in spherical coordinates (see p. 39). Since for our choice of the dipole direction, $\mathbf{B}$ has only a φ component, we find (independently of time)

$$\frac{\partial}{\partial r}\,(r\sin\theta\,B_\varphi)\,\mathrm{d}r + \frac{\partial}{\partial\theta}\,(r\sin\theta\,B_\varphi)\,\mathrm{d}\theta = 0\,.$$

This differential equation has the solution $r\sin\theta\,B_\varphi = \mathrm{const.}$, where according to the first equation of this section, B_φ is complex, and so also is the constant. This result is exact and does not rely on approximations—in particular this is not partitioned into near-, middle-, and far-zone, which would be useless in our search for zeros. If we now split into real and imaginary parts and set $\rho = kr$, then, because $r\,B_\varphi \propto (i - \rho^{-1})\sin\theta\,e^{i\rho}$, we find $\sin^{-2}\theta \propto \cos\rho - \sin\rho/\rho$ and $\sin^{-2}\theta \propto \sin\rho + \cos\rho/\rho$. The spherical shells with $\rho = \tan\rho$ or $\rho = -\cot\rho = \tan(\rho + \frac{\pi}{2})$ thus belong to the set of solutions: there the curl densities reverse $\circlearrowright \leftrightarrow \circlearrowleft$ and, according to the induction law, so also does $\partial\mathbf{B}/\partial t$, which means that B is extremal there. It vanishes everywhere on the axis $\theta = 0$ in the direction of the dipoles. Figure 3.32 shows the electric field lines at two times, and Fig. 3.33 the magnetic field lines at the same times.

The distance between the spheres decreases continuously and is only equal to $\lambda/2$ in the far-zone, because the factor $i - \rho^{-1}$ can also be written in the form $i\sqrt{1 + \rho^{-2}}\,\exp(i\arctan\rho^{-1})$, leading to the spherical harmonics of $\rho + \arctan\rho^{-1}$.

3.5.6 Radiation Power for Dipole, Braking, and Synchrotron Radiation

For the radiation power at sufficiently large distances from the source, only the acceleration field is of interest. According to p. 261 for point-like sources, it is given by

$$\mathbf{E} \approx \frac{\mu_0 q}{4\pi}\,\frac{\{\dot{\mathbf{v}} \times (\mathbf{e} - \boldsymbol{\beta})\} \times \mathbf{e}}{(1 - \boldsymbol{\beta}\cdot\mathbf{e})^3\,R}\,, \quad\text{and}\quad \mathbf{B} \approx \frac{\mathbf{e} \times \mathbf{E}}{c_0}\,,$$

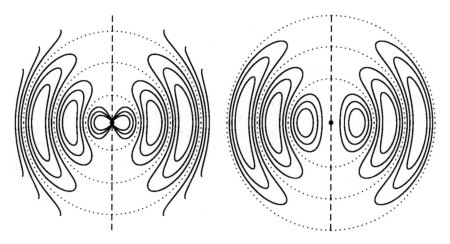

Fig. 3.32 Electric field lines of the Hertz dipole at two times. At the circles (*dotted lines*), the curl densities reverse $\circlearrowleft \leftrightarrow \circlearrowright$, and so also does $\partial \mathbf{B}/\partial t$, implying that B is extremal there

Fig. 3.33 Magnetic field lines of the Hertz dipole at the same times as in Fig. 3.32, here in an inclined view of the central plane. At the *dotted lines*, B is extremal again, while at the *dashed lines*, the field direction is reversed

with $R = |\mathbf{r} - \mathbf{r}'|$. Then also

$$\mathbf{S} \approx \frac{E^2}{\mu_0 c_0} \mathbf{e} \approx \frac{\mu_0}{c_0} \left(\frac{q}{4\pi}\right)^2 \frac{\{(\dot{\mathbf{v}} \times (\mathbf{e} - \boldsymbol{\beta})) \times \mathbf{e}\}^2}{(1 - \boldsymbol{\beta} \cdot \mathbf{e})^6 R^2} \mathbf{e} \, .$$

We shall use this expression (or $d\dot{W}/d\Omega = R^2 \, \mathbf{S} \cdot \mathbf{e}$) for various examples.

In particular, for low velocities ($v \ll c_0$, i.e., $\beta \ll 1$), it follows that

$$\mathbf{E} \approx \frac{\mu_0 q}{4\pi} \frac{(\dot{\mathbf{v}} \times \mathbf{e}) \times \mathbf{e}}{R} \qquad \Longrightarrow \qquad \mathbf{S} \approx \frac{\mu_0}{c_0} \left(\frac{q}{4\pi}\right)^2 \frac{(\dot{\mathbf{v}} \times \mathbf{e})^2}{R^2} \mathbf{e} \, ,$$

and thus for the radiation power into the solid angle element $d\Omega$,

$$\frac{d\dot{W}}{d\Omega} \approx \frac{\mu_0}{c_0} \left(\frac{q}{4\pi}\right)^2 (\dot{\mathbf{v}} \times \mathbf{e})^2 \, .$$

The radiation thus depends on the angle between $\dot{\mathbf{v}}$ and $\mathbf{R}$ through $\sin^2 \theta$. Therefore, with $2\pi \int_0^\pi \sin \theta \, d\theta \, \sin^2 \theta = 2\pi \int_{-1}^1 d\cos\theta \, (1 - \cos^2 \theta) = \frac{2}{3} 4\pi$, the integration over all directions yields the *Larmor formula*

$$\dot{W} = \frac{\mu_0 \, q^2}{6\pi \, c_0} \, \dot{\mathbf{v}} \cdot \dot{\mathbf{v}}$$

for the total radiation power. Here $\dot{\mathbf{v}}$ is to be taken at the retarded time $t' = t - R/c_0$ (as are all the remaining quantities).

Due to this radiation power, the oscillation is damped. This is referred to as the *radiative reaction*. In order to calculate it for a (nearly harmonic) oscillating charge q with mass m, we use the results on p. 99 to relate the decay constant $\gamma = \alpha/(2m)$ to the radiation power $\dot{W} = \alpha \, v^2$, obtaining $\gamma = \dot{W}/(2mv^2)$. Since the ratio of the acceleration and velocity amplitudes for a harmonic oscillation is given by the angular frequency ω, we conclude that the decay constant is

$$\gamma = \frac{\mu_0}{6\pi \, c_0} \frac{q^2}{2m} \, \omega^2 \, .$$

This derivation assumes weak damping, viz., $\gamma \ll \omega$. For electrons, $\omega \ll 3 \times 10^8$ PHz must hold, and this is true even for visible light, where ω is a few PHz. Fourier analysis supplies a not quite sharp frequency: the decay constant leads to a *natural line width*. For heat motion, this is also modified by the *Doppler effect*.

The last equations are valid only for $v \ll c_0$. This condition is always satisfied for the *oscillating dipole* (with sufficiently small displacements). If $\mathbf{p}$ is its dipole moment (at the retarded time), we have to set $q\dot{\mathbf{v}} = \ddot{\mathbf{p}}$. For a harmonic oscillation with angular frequency ω, $\ddot{\mathbf{p}} = -\omega^2\mathbf{p}$, and with the maximum dipole moment $\widehat{\mathbf{p}}$, we find, for the radiation power averaged over a period,

$$\overline{\dot{W}} = \frac{\mu_0}{12\pi \, c_0} \, (\omega^2 \, \widehat{p})^2 \, ,$$

because the square of the spherical harmonics is on average equal to $1/2$. The radiation power (and thus the scattering power) increases as the fourth power of the frequency. Applied to the scattering of sunlight in the air, since $\omega_{\text{blue}} \approx 2\omega_{\text{red}}$, this explains the *blue sky* and the red dawn and dusk.

Dipole radiation is linearly polarized. It oscillates in the plane spanned by $\dot{\mathbf{v}}$ and $\mathbf{e}$, and perpendicularly to $\mathbf{e}$ (transverse).

If we now give up the restriction to small velocities, then for the calculation of the radiation power, we have to account for the retardation. In a time unit, energy is lost at the rate $-\mathrm{d}W/\mathrm{d}t'$, while $\mathbf{S}$ is the energy current density at the position $\mathbf{r}$ and at time t. Therefore, it still depends on $\mathrm{d}t/\mathrm{d}t'$ (see p. 261). Hence, with $\dot{W} = \mathrm{d}W/\mathrm{d}t'$, we find

$$\frac{\mathrm{d}\dot{W}}{\mathrm{d}\Omega} = \frac{\mu_0}{c_0} \left(\frac{q}{4\pi}\right)^2 \frac{\{(\dot{\mathbf{v}} \times (\mathbf{e} - \boldsymbol{\beta})) \times \mathbf{e}\}^2}{(1 - \boldsymbol{\beta} \cdot \mathbf{e})^5} \, .$$

We shall now consider this expression for two special cases.

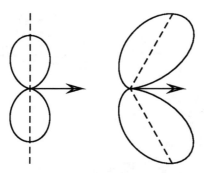

Fig. 3.34 Polar diagram for braking radiation. *Left*: For $\beta_0 \approx 0$. *Right*: For $\beta_0 = \frac{1}{2}$. The *arrow* specifies the direction of the original velocity. The difference in size of the two pictures is to indicate the intensity difference, even though it is not at all to scale, because in total the energy $\frac{1}{2}mv_0^2$ is radiated off *on the left*, and $(\gamma_0 - 1)mc_0^2$ *on the right*, with $v_0^2 \ll c_0^2$. The plane perpendicular to **v** appears as a cone due to aberration, and is indicated by *dashed lines*. $\cos\theta' = 0$ corresponds to $\cos\theta = \beta$

To begin with we assume a longitudinal acceleration (deceleration) $\dot{\mathbf{v}} \parallel \mathbf{v}$. Then $\dot{\mathbf{v}} \times \boldsymbol{\beta}$ vanishes, and we obtain

$$\mathbf{E} \approx \frac{\mu_0 q}{4\pi} \frac{(\dot{\mathbf{v}} \times \mathbf{e}) \times \mathbf{e}}{(1 - \boldsymbol{\beta} \cdot \mathbf{e})^3 R} \quad \text{and} \quad \mathbf{S} = \frac{\mu_0}{c_0} \left(\frac{q}{4\pi}\right)^2 \frac{(\dot{\mathbf{v}} \times \mathbf{e})^2}{(1 - \boldsymbol{\beta} \cdot \mathbf{e})^6 R^2} \mathbf{e} .$$

The electromagnetic field thus differs only from the one for $v \ll c_0$ by the factor $(1 - \boldsymbol{\beta} \cdot \mathbf{e})^{-3}$. Consequently, the radiation into the forwards direction is even stronger than expected non-relativistically.

Let us consider *braking radiation* (also known as deceleration radiation or bremsstrahlung) as an example. Here, $\dot{\mathbf{v}}$ is constant: the velocity decreases at a constant rate from $\mathbf{v}_0$ to zero. From $dv/dt' = \dot{v}$, and hence $\dot{v}^2 dt' = \dot{v}\, dv$, the energy radiated into the solid angle element is (see Fig. 3.34)

$$\frac{dW}{d\Omega} = \frac{\mu_0}{c_0}\left(\frac{q}{4\pi}\right)^2 \dot{v}\sin^2\theta \int_0^{v_0} \frac{dv}{(1 - v\, c_0^{-1}\cos\theta)^5}$$

$$= \frac{\mu_0}{4}\left(\frac{q}{4\pi}\right)^2 \frac{\dot{v}\sin^2\theta}{\cos\theta}\left(\frac{1}{(1 - v_0\, c_0^{-1}\cos\theta)^4} - 1\right) .$$

Of course, this relation holds only for truly constant deceleration.

The linear polarization of braking radiation is given by $\mathbf{E} \propto (\dot{\mathbf{v}} \times \mathbf{e}) \times \mathbf{e}$, as for dipole radiation, thus in the plane spanned by **v** and **e** and perpendicular to **e**.

For *synchrotron radiation*, the acceleration is perpendicular to the velocity, i.e., $\dot{\mathbf{v}} \cdot \boldsymbol{\beta} = 0$. This leads to a radiation power

$$\frac{d\dot{W}}{d\Omega} = \frac{\mu_0}{c_0}\left(\frac{q}{4\pi}\right)^2 \frac{1}{(1 - \boldsymbol{\beta} \cdot \mathbf{e})^3}\left(\dot{\mathbf{v}} \cdot \dot{\mathbf{v}} - \frac{(\dot{\mathbf{v}} \cdot \mathbf{e})^2}{\gamma^2(1 - \boldsymbol{\beta} \cdot \mathbf{e})^2}\right),$$

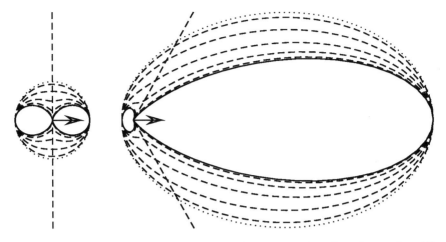

Fig. 3.35 Polar diagrams of the synchrotron radiation. *Left*: For $\beta_0 \approx 0$. *Right*: For $\beta_0 = 1/2$. *Continuous line*: In the plane of the trajectory. *Dotted line*: Perpendicular to the trajectory. *Dashed lines*: In-between in $15°$ steps. The *arrow* specifies the direction of **v**, and the plane perpendicular to **v** is indicated. Here only the line intersecting the plane of the trajectory is important (compare with Fig. 3.32, where the direction of the acceleration is likewise shown *dashed*)

once again more into the forward direction in comparison with the non-relativistic limit.

The linear polarization of synchrotron radiation lies in the plane spanned by $\dot{\mathbf{v}}$ and $\mathbf{e} - \boldsymbol{\beta}$, in particular, perpendicular to **e**, because

$$\mathbf{E} \propto \{\dot{\mathbf{v}} \times (\mathbf{e} - \boldsymbol{\beta})\} \times \mathbf{e} = \dot{\mathbf{v}} \cdot \mathbf{e}\,(\mathbf{e} - \boldsymbol{\beta}) - (1 - \boldsymbol{\beta} \cdot \mathbf{e})\,\dot{\mathbf{v}} \,.$$

The particularly intense radiation in the tangential direction **v** (i.e., for $\mathbf{e} \perp \dot{\mathbf{v}}$) has field strength (see Fig. 3.35)

$$\mathbf{E} \approx -\frac{\mu_0\, q}{4\pi} \frac{\dot{\mathbf{v}}}{(1 - \boldsymbol{\beta} \cdot \mathbf{e})^2\, R} \,.$$

Here, the electric field thus oscillates in the plane of the trajectory.

3.5.7 Summary: Radiation Fields

In this section we have investigated the coupling of the electromagnetic field with its generating sources, and to this end we have appropriately extended the solutions known from the static cases. Here, retardation becomes important. The result has been that the field due to an accelerated charge decreases more weakly by one power of the distance than for a uniformly moving (or resting) charge. At large distances, only the acceleration field is important for the radiation field. Its properties have been considered in the last section for various special cases.

Problems

Problem 3.1 Reformulate $\nabla(\mathbf{a} \cdot \mathbf{b})$ and $\nabla \times (\mathbf{a} \times \mathbf{b})$ such that the operator ∇ has only *one* vector to the right of it (on which it acts). Here the intermediate steps should be taken without components and the differential operator should treat both $\mathbf{a}_c$ and $\mathbf{b}_c$ as constant, so that the product rule reads $\nabla(\mathbf{a} \cdot \mathbf{b}) = \nabla(\mathbf{a} \cdot \mathbf{b}_c) + \nabla(\mathbf{a}_c \cdot \mathbf{b})$, or again $\nabla \times (\mathbf{a} \times \mathbf{b}) = \nabla \times (\mathbf{a} \times \mathbf{b}_c) + \cdots$. The equations $\mathbf{a} \times (\mathbf{b} \times \mathbf{c}) = \mathbf{b}\,(\mathbf{c} \cdot \mathbf{a}) - \mathbf{c}\,(\mathbf{a} \cdot \mathbf{b}) = (\mathbf{c} \cdot \mathbf{a})\,\mathbf{b} - (\mathbf{a} \cdot \mathbf{b})\,\mathbf{c}$ need not be proven. (4 P)

Problem 3.2 Using Cartesian components, determine $\nabla \cdot \mathbf{r}$, $\nabla \times \mathbf{r}$, and $(\mathbf{a} \cdot \nabla)\,\mathbf{r}$. These results will be useful for the following problems. (3 P)

Problem 3.3 Consider an arbitrary (three-times differentiable) scalar function $\psi(\mathbf{r})$ and the three vector fields $\nabla\psi$, $\mathbf{r} \times \nabla\psi$, and $\nabla \times (\mathbf{r} \times \nabla\psi)$. Which of them are source-free and which curl-free? Determine the source and curl strengths as functions of ψ. What is their inversion behavior (parity) if $\psi(-\mathbf{r}) = \psi(\mathbf{r})$? (9 P)

Problem 3.4 Prove $\int_{(V)}(\mathrm{d}\mathbf{f} \cdot \mathbf{a})\,\mathbf{b} = \int_V \mathrm{d}V\,\{\mathbf{b}\,(\nabla \cdot \mathbf{a}) + (\mathbf{a} \cdot \nabla)\,\mathbf{b}\}$ for arbitrary fields $\mathbf{a}(\mathbf{r})$ and $\mathbf{b}(\mathbf{r})$ and show that the volume integral of a source-free vector field $\mathbf{a}$ is always zero, if $\mathbf{a}$ vanishes on the surface (V). (4 P)

Problem 3.5 For which function $\psi(r)$ does the (spatial) central field $\mathbf{a}(\mathbf{r}) = \psi(r)\,\mathbf{r}$ have sources only at the origin? Does it have curls? Investigate this also for a plane central field. Represent the solutions as gradient fields (gradients of scalar fields). (3 P)

Problem 3.6 Let $(\Delta + k^2)\,\psi(\mathbf{r}) = 0$. How can we prove that the three vector fields from Problem 3.3 satisfy the equation $(\Delta + k^2)\,\mathbf{a}(\mathbf{r}) = \mathbf{0}$? Note the sources and curls of the vector fields. (4 P)

Problem 3.7 Determine the vector fields $\nabla(\mathbf{p} \cdot \mathbf{r}/r^3)$ and $\nabla \times (\mathbf{r} \times \mathbf{p}/r^3)$ for constant $\mathbf{p}$ (dipole moment) when $r \neq 0$, and compare them. (5 P)

Problem 3.8 Derive the singular behavior of the two vector fields for $r = 0$ from the volume integral of a sphere around the origin. Express the results in terms of the delta function. (8 P)

Problem 3.9 Prove the representation of the Fourier transform of $f(x) = g(x)\,h(x)$ as a convolution integral given on p. 22. (4 P)

Problem 3.10 For fixed α, β, γ (with $\alpha > 0$, $\beta > 0$, and $0 < \gamma < \pi$), a rectilinear oblique coordinate system x^1, x^2 is given by the two equations $x^1 = \alpha\,(x - y \cot\gamma)$ and $x^2 = \beta\,y$. Which functions $y(x)$ describe the coordinate lines $\{x^1, x^2\}$? At what angle do the coordinate lines cross? How do the basic vectors $\mathbf{g}_i =$ and $\mathbf{g}^i$ read as linear combinations of the Cartesian unit vectors? How do the fundamental tensors g_{ik} and g^{ik} read? (7 P)

Problem 3.11 For spherical coordinates (r, θ, φ), we have to introduce position-dependent unit vectors $\mathbf{e}_r$, $\mathbf{e}_\theta$, and $\mathbf{e}_\varphi$ in the direction of increasing coordinates. Decompose these three vectors in terms of $\mathbf{e}_x$, $\mathbf{e}_y$, and $\mathbf{e}_z$. Determine their partial derivatives with respect to r, θ, φ and express them as multiples of the unit vectors $\mathbf{e}_r$, $\mathbf{e}_\theta$, $\mathbf{e}_\varphi$. (7 P)

Problem 3.12 Determine the covariant and contravariant base vectors $\{\mathbf{g}_i\}$ and $\{\mathbf{g}^i\}$ as multiples of the unit vectors $\mathbf{e}_i$ for spherical coordinates $x^1 = r$, $x^2 = \theta$, and $x^3 = \varphi$. (2 P)

Problem 3.13 With the help the Maxwell construction, draw the force lines of two equally charged parallel lines with charges densities q/l and separated by a distance a. This uses the theorem that, for a source-free field, there is the same flux through any cross-section of a force tube. What changes with this construction for oppositely charged parallel lines, i.e., charge densities $\pm q/l$, separated by a distance a? Why is the construction more precise than the method of drawing trajectories orthogonal to the equipotential lines? (8 P)

Problem 3.14 Determine the equation $f(x, z) = 0$ of the field line of an ideal dipole $\mathbf{p} = p\mathbf{e}_z$ which lies at the origin $(\mathbf{r}' = \mathbf{0})$. Note that, due to the cylindrical symmetry, we may set $y = 0$. (4 P)

Problem 3.15 On the z-axis there are several point charges q_i at the positions z_i. Determine their common potential Φ by Taylor series expansion up to order $(z_i/r)^3$. Examine the result for the potential when $r \gg a$ (write Φ as a multiple of $q_1 = q$) for:

- a dipole $(q_1 = -q_2, z_1 = -z_2 = \frac{1}{2}a)$,
- a linear quadrupole $(q_1 = -\frac{1}{2}q_2 = q_3, z_1 = -z_3 = a, z_2 = 0)$, and
- an octupole $(q_1 = -\frac{1}{3}q_2 = +\frac{1}{3}q_3 = -q_4, z_1 = 3z_2 = -3z_3 = -z_4 = \frac{3}{2}a)$?

Show that the field of a finite dipole may be written approximately as a superposition of a dipole field and an octupole field. How strong is the octupole field compared with the field of a pure quadrupole? Justify with the examples above that an ideal 2^n-pole can be viewed as a superposition of two 2^{n-1}-poles. (8 P)

Problem 3.16 Determine the potential and field strength of a hollow sphere with outer and inner radii R and ηR and a charge Q distributed evenly over its volume. Here $0 \le \eta \le 1$, so a solid sphere has $\eta = 0$ and a surface charge $\eta = 1$. Sketch the results $\Phi(r)$ and $E(r)$ in the limiting cases $\eta = 0$ and $\eta = 1$. In these limiting cases, how much field energy is in the space with $r \le R$? How much is in the external space? (7 P)

Problem 3.17 Express the potential of a metal ring of radius R and charge Q in terms of the *complete elliptic integral of the first kind* $K(m)$ (with $0 \le m \le 1$) of p. 202. Here it will be convenient to replace the spherical coordinate φ by $\pi - 2x$. Determine the potential and the field strength on the axis of the ring. (6 P)

Problem 3.18 Determine the potential and field strength on the axis of a thin metal disc of radius R and charge Q for constant charge density. What is the jump in the field strength at the disc? (3 P)

Problem 3.19 What is obtained for the potential on the axis if the disc has a constant dipole density $\mathbf{p}_A$? What is the jump in the potential? (4 P)

Problem 3.20 On a straight line at distance a from the origin, let there be a point charge $q > 0$, and at distance a' on the same side of the origin, a charge $-q' < 0$. For suitable $q'(q, a, a')$, the potential vanishes on the surface of a sphere about the origin. What is its radius? Use this to determine the charge density ρ_A on a grounded metal sphere of radius R induced by a point charge q at distance a from the center of the sphere. What changes for an ungrounded metal sphere? (6 P)

Problem 3.21 How does the Maxwell stress tensor read for a homogeneous field of strength $\mathbf{E} = E\,\mathbf{e}_z$ in vacuum? How strong is the force on a volume element $dx\,dy\,dz$? Using the stress tensor, determine the force on an area A if its normal is $\mathbf{n} = \mathbf{e}_x \sin\theta + \mathbf{e}_z \cos\theta$.

Hint: Decompose the force into components along $\mathbf{n}$, $\mathbf{t} = \mathbf{e}_x \cos\theta - \mathbf{e}_z \sin\theta$, and $\mathbf{b} = \mathbf{t} \times \mathbf{n}$. Draw the vectors $\mathbf{E}$, $\mathbf{n}$, $\mathbf{t}$, and $\mathbf{F}$ for $\theta = 30^0$. Interpret the result for opposite sides of a cube. (7 P)

Problem 3.22 How does the stress tensor change at the x, y-plane if it carries the charge density ρ_A and is placed in an external (homogeneous) field in the z direction? Can the force on an enclosing layer be related to the mean value of the field strength above and below the plane? Determine the Cartesian components of the Maxwell stress tensor on the plane midway between two equal charges q (each at distance a from this plane)? What force is thus exerted from one of the sides on the plane?

Hint: Express the strength of the field in cylindrical coordinates. (7 P)

Problem 3.23 Determine the electric field around a metal sphere in a homogeneous electric field. Superpose the field of a dipole $\mathbf{p}$ on a suitable homogeneous field $\mathbf{E}_0$ in such a way that the tangential component of the total field vanishes on the surface of the sphere of radius r around the dipole. How large is the normal component (in particular in the direction of $\mathbf{E}_0$, opposite and perpendicular to it)? (4 P)

Problem 3.24 Determine the current density and resistance for half a metal ring with circular cross-section (area πa^2), whose axis forms a semi-circle of radius A (conductivity σ), if there is a voltage U between the faces.[1] Note the special case $a \ll A$. (5 P)

---

[1] Using the substitution $t = \tan\frac{1}{2}x$ with $t' = \frac{1}{2}(1+t^2)$ and $\cos x = (1-t^2)/(1+t^2)$, the integral of $(1 + k\cos x)^{-1}$ for $|k| < 1$ can be transformed into the integral of $2/(1-k)(K^2+t^2)^{-1}$ with $K^2 = (1+k)/(1-k)$. This yields $2\,(1-k^2)^{-1/2}\arctan(t/K)$.

Problem 3.25 In an otherwise homogeneous conductor, there is a spherical void of radius r_0 containing air. Determine the current density $\mathbf{j}$ if it is equal to $\mathbf{j}_0$ for large r. (3 P)

Problem 3.26 Equal currents I flow through two equal coaxial circles (radii R) a distance a apart. For which ratio a/R is the magnetic field strength at the center of the setup *as homogeneous as possible*? What does that mean? Where would we have to place a further pair of loop currents with radius $\frac{1}{2}R$ in order to amplify the homogeneous field? Can the homogeneity be improved by a suitable choice of current strengths in two pairs of loops? (8 P)

Problem 3.27 A closed iron ring with permeability μ and dimensions a and A, as in Problem 3.24, is wrapped around N times with a thin wire. How large is the induction flux $\Phi = \int d\mathbf{f} \cdot \mathbf{B}$ in the ring? How large is the relative error $\delta\Phi = |\Phi - \overline{\Phi}|/\overline{\Phi}$, if we assume a constant magnetic field $\mathbf{H}$ equal to the value $\overline{\Phi}$ at the center of the cross-section? Determine $\overline{\Phi}$ and $\delta\Phi$ for $N = 600$, $\mu = 500\mu_0$, $A = 20$ cm, $\pi a^2 = 10$ cm^2, and $I = 1$ A. The iron ring may have a narrow discontinuity (air gap) of width d. It can be so narrow that no field lines escape from the slit. How does the induction flux depend on the width d if we use a constant magnetic field $\mathbf{H}$ in the cross-section? (7 P)

Problem 3.28 The *mutual inductance* of two coaxial circular rings of radii R and R' a distance a apart is determined as $L = \mu_0\sqrt{RR'}\ \{2\,(K - E) - k^2K\}/k$, with the parameter $k^2 = 4RR'/\{a^2 + (R+R')^2\}$, involving the complete elliptic functions of the first kind, viz., $K(k^2)$ as in Problem 3.17, and the second kind, viz., $E(k^2) = \int_0^{\pi/2} \sqrt{1-k^2\sin^2 z}\ dz$. What is obtained to leading order for L at very large distances $(R \ll a,\ R' \ll a)$?

Hint: Expand K and E in powers of k. (3 P)

Problem 3.29 In the limit of small distances $(R \approx R' \gg a)$, we use the Landen transformation $F(x|k^2) = 2/(1+k)\ F(x'|k'^2)$ for the incomplete elliptic integral of the first kind $F(x|k^2) = \int_0^x dz/\sqrt{1-k^2\sin^2 z}$, with $x' = \frac{1}{2}\{x + \arcsin(k\sin x)\}$ and $k'^2 = 4k/(1+k)^2$.

With $\sin(2z_1 - z) = k\sin z$, we have $\cos(2z_1 - z)(2\,dz_1 - dz) = k\cos z\,dz$, hence also $dz\,\{k\cos z + \cos(2z_1 - z)\} = 2\,dz_1\,\cos(2z_1 - z) = 2\,dz_1\,(1 - k^2\sin^2 z)^{1/2}$. The square of the curly brackets is equal to $k^2\cos^2 z + 2k\cos z\cos(2z_1 - z) + 1 - k^2\sin^2 z$, or again $1 + k^2 + 2k\,\{\cos z\cos(2z_1 - z) - \sin z\sin(2z_1 - z)\}$. The curly bracket may be reformulated as $\cos 2z_1 = 1 - 2\sin^2 z_1$. Then

$$dz/(1 - k^2\sin^2 z)^{1/2} = 2dz_1/\{(1 + k)^2 - 4k\sin^2 z_1\}^{1/2},$$

which is important for the proof of Landen's transformation.

Prove that $K(1 - \varepsilon) \approx \ln(4/\sqrt{\varepsilon})$. What follows for the inductance $L(R, R', a)$? (5 P)

Fig. 3.36 Between two
points of a circuit, a
voltmeter is connected with
thin (loss-free) wires
(resistance R_0), such that the
area A spanned by the circuit
is divided in the ratio $A_1:A_2$

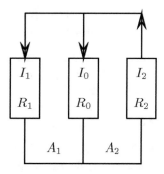

Problem 3.30 Derive from this the *self-inductance* of a thin ring of wire with circular cross-section (abbreviation as in Problem 3.24), which is composed of the
mutual inductances $L = (\pi a^2)^{-2} \int df_1\, df_2\, L_{12}$ of its filaments. Here $\int_0^\pi \ln(A +$
$B \cos\varphi)\, d\varphi = \pi\, \ln\{\frac{1}{2}(A + \sqrt{A^2 - B^2})\}$ for $A \geq |B|$. (For ferromagnetic materials,
there is an additional term, not required here.)   (5 P)

Problem 3.31 For a current strength I, determine the vector potential of a circular
ring of radius R_0 at an arbitrary point $\mathbf{r}$. The circular ring suggests using cylindrical
coordinates (R, φ, z) with $\mathbf{r} = R\mathbf{e}_R + z\mathbf{e}_z$.   (6 P)

Problem 3.32 A very long hollow cylinder with inner radius R_i, outer radius R_a,
and permeability μ is brought into a homogeneous magnetic field $\mathbf{H}_0$ perpendicular
to its axis. Determine $\mathbf{B}$ and $\mathbf{H}$ for all $\mathbf{r}$. How large is the field H_0 compared to its
value on the axis for $\mu \gg \mu_0$?   (9 P)

Problem 3.33 Perpendicular to the circuit shown in Fig. 3.36, made of a thin wire
with resistance $R = R_1 + R_2$, a homogeneous magnetic field changes by equal
amounts in equal time intervals. What voltage does the voltmeter show, and in particular, if the circuit forms a circle and the voltmeter sits at the center of the circle
and is connected with straight wires?   (5 P)

Problem 3.34 An insulating cuboid $(0 \leq x \leq L_x,\ 0 \leq y \leq L_y,\ 0 \leq z \leq L_z)$ of
homogeneous material with scalar permittivity and permeability is enclosed by ide-
ally conducting walls. Investigate the following ansatz for the vector potential:

$$A_x = a_x \cos(\omega t)\cos(k_x x + \varphi_{xx})\cos(k_y y + \varphi_{xy})\cos(k_z z + \varphi_{xz})\,,$$
$$A_y = a_y \cos(\omega t)\cos(k_x x + \varphi_{yx})\cos(k_y y + \varphi_{yy})\cos(k_z z + \varphi_{yz})\,,$$
$$A_z = a_z \cos(\omega t)\cos(k_x x + \varphi_{zx})\cos(k_y y + \varphi_{zy})\cos(k_z z + \varphi_{zz})\,,$$

with the radiation gauge. Can we restrict ourselves here to $0 \leq \varphi_{ik} < \pi$? What is the
relation between ω and $\mathbf{k}$ if all the Maxwell equations are valid? What requirements
follow from the boundary conditions $\mathbf{n} \times \mathbf{E} = \mathbf{0}$ and $\mathbf{n} \cdot \mathbf{B} = 0$?   (7 P)

Problem 3.35 What requirement does the gauge condition $\nabla \cdot \mathbf{A} = 0$ lead to for the
ansatz above? What do we then obtain for the three fields $\mathbf{A}$, $\mathbf{E}$, and $\mathbf{B}$?   (5 P)

Problem 3.36 What do we obtain if **k** is parallel to one of the edges of the cuboid? What is the general ansatz for **A** in Problem 3.34? (3 P)

Problem 3.37 Express the energy density $w(t, \mathbf{r})$ of an electromagnetic wave in terms of its vector potential in the radiation gauge, i.e., with $\Phi = 0$ and $\nabla \cdot \mathbf{A} = 0$. How can Parseval's equation help to re-express the total energy of the wave (integrated over the whole space) as an integral of the square of the absolute value of $\mathbf{A}(t, \mathbf{k})$ and $\partial \mathbf{A}/\partial t$ as weight factors? What is the unknown expression? (5 P)

Problem 3.38 How does the electric field amplitude of the reflected and transmitted waves depend on the incoming amplitude in the limiting cases $\theta = 0^0$ and 90^0 (expressed in terms of the refractive index n)? To what extent are the *parallel* and *perpendicular* components to be distinguished for perpendicular incidence $(\theta = 0^0)$? (4 P)

Problem 3.39 How large is the energy flux $\dot{W}$, averaged over time, for an electromagnetic wave with wave vector **k** passing through an area A perpendicular to **k**? What do we obtain for the reflected and the transmitted waves in the limiting cases investigated above? (4 P)

Problem 3.40 Does the energy conservation law hold true for an electromagnetic wave, incident with the wave vector **k** on the interface between two homogeneous insulators (with an arbitrary angle of incidence)? Investigate this question for arbitrary scalar material constants ε and μ, i.e., also with $\mu \neq \mu_0$. (5 P)

Problem 3.41 For a homogeneous conductor (with scalar σ, ε, and μ), derive the relation between $\overline{w}$, $\mathbf{E}^* \cdot \mathbf{D}$, and $\mathbf{H}^* \cdot \mathbf{B}$ from the Maxwell equations, if only one wave vector is given. How is the time average of the Poynting vector connected to the averaged energy density $\overline{w}$?

Hint: Use the approximation $\alpha^2 \approx \beta^2 \approx \sigma/(2\varepsilon\omega) \gg 1$. (7 P)

List of Symbols

We stick closely to the recommendations of the *International Union of Pure and Applied Physics* (IUPAP) and the *Deutsches Institut für Normung* (DIN). These are listed in *Symbole, Einheiten und Nomenklatur in der Physik* (Physik-Verlag, Weinheim 1980) and are marked here with an asterisk. However, one and the same symbol may represent different quantities in different branches of physics. Therefore, we have to divide the list of symbols into different parts (Table 3.2).

Table 3.2 Symbols used in electromagnetism

	Symbol	Name	Page number
*	Q	Charge	165
	q	Point charge	165
*	ρ	(Space) Charge density	166
a	ρ_A	Surface charge density	167
*	I	Current strength	186
*	$\mathbf{j}$	Current density	186
	$\mathbf{j}_A$	Current density in a surface	195
*	$\mathbf{E}$	Electric field strength	166
*	$\mathbf{D}$	Electric current density (displacement field)	174
*	$\mathbf{B}$	Magnetic current density (magnetic induction field)	181
*	$\mathbf{H}$	Magnetic field strength	193
*	ε	Permittivity (dielectric constant)	176
*	ε_0	Electric field constant (vacuum permittivity)	164, 623
*	μ	Permeability	196
*	μ_0	Magnetic field constant (vacuum permeability)	164, 623
*	$c\ (c_0)$	Light velocity (in vacuum)	164, 216, 623
*	χ_e	Electric susceptibility	175
*	χ_m	Magnetic susceptibility	196
*	$\mathbf{P}$	Electric polarization	174
*	$\mathbf{M}$	Magnetization	191
*	$\mathbf{p}$	Electric dipole moment	171
*	$\mathbf{m}$	Magnetic dipole moment	190
*	U	(Electric) Voltage	169
b	Φ	Electric potential	56
*	$\mathbf{A}$	Vector potential	197
c	E_{pot}	Potential energy	169
*d	W	Work	181
*	w	Energy density	211
e	$\mathbf{N}$	Torque	171
*	C	Capacitance	179
*	R	Electric resistance	187
*	σ	Electric conductivity	187
*f	L	Inductance	201

(continued)

Table 3.2 (continued)

	Symbol	Name	Page number
g	$\mathscr{Z}$	Impedance	213
	S	Poynting vector	211
	T	Stress tensor	184, 215
	$F^{\mu\nu}$	Electromagnetic field tensor	240

[a]The abbreviation σ is actually recommended for this, but it is also used also for the conductivity. The index A reminds us of an area. We also use it for the area divergence and area rotation.
[b]φ is actually recommended, but we use it for the azimuth.
[c] V is needed for the volume.
[d]The abbreviation A, common in mechanics, is needed here for the area.
[e] **M** is recommended, but used here for the magnetization.
[f] L is recommended for the self-inductance. We also use this abbreviation for the mutual inductance.
[g] Z should be taken for the impedance, but $\mathscr{Z}$ stresses the fact that it is a complex quantity: ($\mathscr{Z} = R + iX$, with *resistance* R and *reactance*) X

References

1. J.H. Hannay, Eur. J. Phys. **4**, 141 (1983)
2. I. Brevik, Phys. Rep. **52**, 133 (1979)
3. E.W. Schmid, G. Spitz, W. Lösch, *Theoretical physics with the PC* (Springer, Berlin, 1987)

Suggestions for Textbooks and Further Reading

4. H. Goldstein, J.L. ChP Poole, Safko, *Classical Mechanics*, 3rd edn. (Pearson, 2014)
5. W. Greiner, *Classical Electrodynamics* (Springer, New York, 1998)
6. J.D. Jackson, *Classical Electromagnetism*, 3rd edn. (Wiley, New York, 1998)
7. L.D. Landau, E.M. Lifshitz, *Course of Theoretical Physics Vol. 2 – The Classical Theory of Fields*, 4th edn. (Butterworth–Heinemann, Oxford, 1975)
8. L.D. Landau, E.M. Lifshitz, Course of Theoretical Physics, Vol. 8–Electrodynamics of Continuous Media, 2nd edn. (Butterworth-Heinemann, Oxford, 1984)
9. P. Lorrain, D. Corson, F. Lorrain, *Electromagnetic Fields and Waves*, 3rd edn. (W.H. Freeman, New York, 1988)
10. W. Nolting, *Theoretical Physics 3-Electrodynamics* (Springer, Berlin, 2016)
11. W. Nolting, *Theoretical Physics 4-Special Theory of Relativity* (Springer, Berlin, 2017)
12. W.K.H. Panofsky, M. Phillips, Classical Electricity and Magnetism, 2nd edn. (Addison-Wesley, Reading, 1962)
13. W. Rindler, *Essential Relativity-Special, General, and Cosmological, revised*, 2nd edn. (Springer, New York, 1977)
14. F. Scheck, *Classical Field Theory* (Springer, Berlin, 2018)
15. A. Sommerfeld, *Lectures on Theoretical Physics 3-Electrodynamics* (Academic, London, 1964)
16. A. Sommerfeld, *Lectures on Theoretical Physics 4-Optics* (Academic, London, 1964)
17. W. Thirring, *Classical Mathematical Physics: Dynamical Systems and Field Theories*, 3rd edn. (Springer, New York, 2013)
18. A. Zangwill, Modern Electromagnetics (Cambridge University Press, 2013)

Chapter 4
Quantum Mechanics I

4.1 Wave–Particle Duality

4.1.1 Heisenberg's Uncertainty Relations

A natural law is required to be true without exception: for all observers under equal conditions the same result should be obtained. However, "equal conditions" have to be reproducible and "identical results" can only be ensured within certain error limits. With N measurements, the experimental values x_i in the statistical ensemble scatter around the average value $\bar{x} \equiv \frac{1}{N} \sum_{i=1}^{N} x_i$ with an average error (for the individual measurement)

$$\Delta x \equiv \sqrt{\overline{(x - \bar{x})^2}} = \sqrt{\overline{x^2} - \bar{x}^2} \; .$$

We assume $N \gg 1$ and hence may leave out the factor $\sqrt{N/(N-1)}$ from p. 46. Here, $\overline{x^2}$ is the average value of the squares of the experimental values. These notions have been explained in detail in Sect. 1.3.

A basic feature of quantum physics is that canonically conjugate quantities cannot simultaneously have arbitrarily small error widths: the smaller the one, the larger the other. For example, the momentum $p_k = \partial L / \partial \dot{x}^k$ is canonically conjugate to the coordinate x^k (see p. 99). Since Niels Bohr, such pairs of quantities have been referred to as *complementary*.

In classical physics, this situation does not have the same relevance, even though there are complementary quantities, e.g., for the position x and wavenumber $k = 2\pi/\lambda$ of a wave group, we have $\Delta x \cdot \Delta k \geq 1/2$. The inequality holds in particular for all pairs of quantities connected by Fourier transform. For Gaussian distributions, we

© Springer Nature Switzerland AG 2018
A. Lindner and D. Strauch, *A Complete Course on Theoretical Physics*, Undergraduate Lecture Notes in Physics, https://doi.org/10.1007/978-3-030-04360-5_4

find $\Delta x \cdot \Delta k = 1/2$ (Problem 4.1), and these have the smallest uncertainty product possible for complementary quantities, as will be shown later (p. 321).

However, in classical physics it is often overlooked that canonically conjugate quantities are always complementary to each other, because there the basic error limits may be neglected in comparison to the average values. The situation is different in quantum theory: here the uncertainty relations are indispensable. Hence it must be a statistical theory, as only then do error widths make sense.

According to Heisenberg, for canonically conjugate quantities like position and momentum, we have quantitatively

$$\Delta x^k \cdot \Delta p_{k'} \geq \tfrac{1}{2} \hbar \, \delta^k_{k'} \,,$$

with $\hbar \equiv h/2\pi$. We use h to denote *Planck's action quantum*, but nowadays in quantum theory it is more common to use $\hbar$. This does not occur in the classical relation $\Delta x \cdot \Delta k \geq 1/2$. According to de Broglie, $\mathbf{p} = \hbar \mathbf{k}$ (more on that on p. 319), so the two uncertainty relations are connected to each other. Note, however, that the uncertainty is sometimes defined differently, and then there is a different numerical factor in the uncertainty relations. Note also that Heisenberg [1] calls uncertain quantities "undetermined", but that can be misunderstood.

Thus we cannot, for example, produce an ensemble which is sharp (certain) in the position as well as in the momentum. If we force a ray with sharp direction through a narrow slit, in order to minimize the position uncertainty (perpendicular to the ray direction), then it spreads out because of the diffraction—and this all the more, the narrower the slit. The momentum orthogonal to the old direction of the ray can no longer be neglected and is unsharp (uncertain). (Its average value does not need to change, only the uncertainty.) By eliminating inappropriate parts of the position, we have changed the original ray.

The uncertainty relations are thus already satisfied in the production of a statistical ensemble. The uncertainties can often be attributed to the (then following) measurement, but after the measurement the observation is already finished.

We start from the uncertainty relations as observational facts. As Heisenberg shows for many examples in the above-mentioned book, quantum phenomena only contradict our everyday experience if the uncertainty relations are not considered.

4.1.2 Wave–Particle Dualism

In order to solve Hamilton's equations, unique initial values of position and momentum are necessary, but this requirement can only be satisfied within error limits, because of the uncertainty relations. Hence in quantum theory, we cannot apply the usual notion of determinism to processes—we can only predict how all *possible* states will develop. Classically, we could assign probabilities to the possible orbits, and given what was said above, we should only actually try to find such probabilistic statements.

However, with the probability distributions, interference now occurs, which is the classical proof that waves are involved. On the other hand, other experimental results involve shot noise (granularity) and hence support the idea that it is particles, and not waves, that are involved.

This contradiction shows up clearly in the scattering of monochromatic electrons of sufficient energy from a crystal lattice. Here interference figures result on on the detector screen. This fact is taken as the classical proof of a wave-like nature. With decreasing radiation intensity, the strength of the detection on the screen is reduced, but not continuously—detections appear now here, now there, like shot noise: this is the classical proof of a particle-like nature.

If electrons were classical particles, then they would hit the screen like grains of shot, and the intensity distribution $\rho(\mathbf{r})$ would result—without interference—as a sum of the n intensity distributions $\rho_n(\mathbf{r})$ of the single scattering centers:

$$\rho = \sum_n \rho_n \quad \text{(particle picture)}.$$

The function $\rho(\mathbf{r})$ describes the probability density of the strikes. If we exhaust all possibilities, then we should obtain 1, i.e., $\int d^3\mathbf{r}\, \rho(\mathbf{r}) = 1$. Of course, for discrete possibilities there is a sum instead of an integral.

If electrons were classical waves, then the intensity should decrease continuously, and we should not observe any granularity of the radiation. The intensity distribution $\rho(\mathbf{r})$ would not simply be the sum of the intensity distributions $\rho_n(\mathbf{r})$ of the scattering centers, but would show interference—we would have to work with complex amplitudes $\psi_n(\mathbf{r})$, superpose them, and form the square of the absolute value of the sum:

$$\rho = \left| \sum_n \psi_n \right|^2 = \sum_n |\psi_n|^2 + \sum_{n<m} (\psi_n{}^*\psi_m + \psi_m{}^*\psi_n) \quad \text{(wave picture)}.$$

The mixed terms $(2\,\mathrm{Re}\sum_{n<m}\psi_n{}^*\psi_m)$ describe the interference (see Fig. 4.1).

4.1.3 Probability Waves

Classically, the two pictures (or models), "wave" or "particle", are mutually exclusive. The quantum theory can remove this contradiction: it contains both pictures as limiting cases, but restricts their range of application via the uncertainty relations in such a way that they agree with each other, as we shall soon show.

In the above example, in particular, the point of impact of an individual electron cannot be predicted with certainty. Only the statistical ensemble exhibits equal results

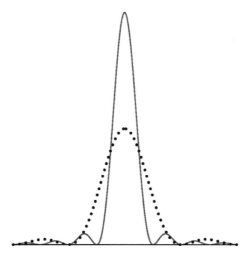

Fig. 4.1 Intensity distribution behind a single slit (*dotted line*) and a double slit (*continuous line*) according to the wave picture. All three slits have width b, and likewise the separation of the double slit. The interference can be explained only with the wave picture—in particular the zeros cannot be explained with the particle picture. For the single slit of width $2a$, we have $\Delta x = a/\sqrt{6}$ and Δk is infinite! For this reason, only the part between the main minima is often considered. Then we obtain $\Delta k = 1/a$ and hence $\Delta x \cdot \Delta k = 1/\sqrt{6}$

under equal conditions, and in particular, always the same interference figure. Only this impact *probability*—or more precisely probability *density*—actually obeys a law, and the theory should only make statements about that. With the observed interference, we need a wave theory.

But it is important that a wave theory only applies for the impact probability, while the classical wave picture is based on "real" waves: it is extrinsic to a wave theory that the field quantity in the statistical ensemble fluctuates, whence deviations from the corresponding classical value may occur. But this statistical error is important for quantum theory: it allows the "granularity" of the radiation, which fits into the particle picture. This granularity remains unnoticed for large particle numbers [2, p. 4]: *The all leveling law of large numbers masks completely the true nature of the single processes.* On the other hand, for large numbers, uncertainties in the particle number barely show up.

In order to capture the granularity for small particle numbers, we have to *quantize* the wave theory, that is, to take the intensities as natural multiples of a basic intensity. We treat *field quantization* in Sect. 4.2.8, and in more detail in Sect. 5.3 on many-particle problems. (Incidentally, the so-called *second quantization* is nothing else than field quantization—a misleading name that can only be understood from the historical perspective. There is in fact only one prescription for quantization, even though it may look different for fields and particles, because it does indeed produce the same result.)

Here we must also consider the fact that the relative phases are important for the superposition of waves. Uncertainty in the phase suppresses the possibility of interference and allows only an incoherent superposition. In fact, as will be shown in Sect. 4.2.9, there is also an uncertainty relation relating particle number and phase, although it cannot generally be written $\Delta N \cdot \Delta \Phi \geq 1/2$, because the phase makes sense only up to a multiple of 2π, and the particle number can only be positive-definite.

We shall speak of particles, as is common practice, and assign probability waves to them. Occasionally, we shall speak of *quanta*, which like particles are natural multiples of an element and can interfere like waves.

Our way to describe the interference of probability waves has already been indicated in the last section: we introduce probability *amplitudes*! These amplitudes are usually called *wave functions*. Several of them can interfere with each other—the amplitudes are added, and the square of the absolute value of the sum yields the probability (or the probability density). Here, the interference phenomenon is expressed in the mixed terms.

These rules are valid, however, only if the individual parts can interfere with each other. (It is well known that, in addition to coherent light, there is also incoherent light, which cannot interfere.) If the phase relations are destroyed by an external manipulation, for example, using light to observe the different paths of the partial waves, then it is no longer the probability amplitudes that are added, but only the probabilities. An "incoherent" superposition arises. For the moment, we restrict ourselves to (interfering) coherent superpositions, which are particularly important for the development of the usual quantum theory. We shall only deal with the general case at the end of the next section (Sect. 4.2.11). This extends the applicability of quantum theory and is important, e.g., in thermodynamics.

Only the probabilities (or probability densities) can be measured (observed), not the associated amplitudes. Hence also only their values and relative phases. In principle, a general phase factor $\exp(i\phi)$ remains arbitrary.

4.1.4 Pure States and Their Superposition (Superposition Principle)

By *probability* we understand the ratio of actual events to the total number of the considered events taken all together. An ensemble with specific attributes (signatures) is investigated for its properties. A *statistical ensemble with attributes* is called a *state* in quantum theory. As already indicated, for the moment we shall not consider incoherent *mixtures*, but only the so-called *pure states*:

> *Objects in the same state have the same properties and cannot be distinguished from each other; if they could be distinguished, then the state would not be characterized with sufficient precision.*

The objects here are intended to be representative of a class of objects. The notion of state serves in a statistical theory. To this end we need a filter which decides whether or not the property exists. Only experience can teach us which attributes are necessary for the complete characterization of a state. For a long time it was believed, for example, that there was only one species of neutrino, whereas three are now distinguished. On the other hand, the subdivision may also be too fine: for several years, the various decay channels of the kaon were assigned to different particles.

As an example, a state is fully specified by the following statement: "There are electrons with polarization direction **s** and momentum **p**." For the sake of simplicity, we shall momentarily take only **p** as a variable. Then the state is considered to be determined by the momentum alone. This is represented by the Dirac symbol $|\mathbf{p}\rangle$, although actually we should write $|electron, \mathbf{s}, \mathbf{p}\rangle$. Instead of the momentum, we could also give the position **r**. The corresponding state is then characterized by $|\mathbf{r}\rangle$. But according to the uncertainty relation we may not take **r** and **p** together, since they cannot be determined simultaneously with such accuracy.

But we may wish to know the probability density of the state $|\mathbf{p}\rangle$ at position **r**. For the corresponding probability amplitude, we write $\langle \mathbf{r}|\mathbf{p}\rangle$ in the Dirac notation. This complex number depends on **r** and **p**.

As a further example, let us consider the *double slit experiment*. Here from Fig. 4.2 we can introduce four different states $|i\rangle$, $|1\rangle$, $|2\rangle$, and $|f\rangle$. Each refers to another position. Here the states $|1\rangle$ and $|2\rangle$ are constructed from $|i\rangle$ and $|f\rangle$ from $|1\rangle$ and $|2\rangle$.

Fig. 4.2 Double slit experiment. The source generates the initial state $|i\rangle$ and the double slit selects the states $|1\rangle$ and $|2\rangle$. The final state $|f\rangle$ is detected. The path between $|i\rangle$ and $|f\rangle$ remains unknown

The corresponding probability amplitudes are clearly $\langle 1|i\rangle$ and $\langle 2|i\rangle$, or $\langle f|1\rangle$ and $\langle f|2\rangle$. The probability amplitude for the formation of $|f\rangle$ from $|i\rangle$ is composed as follows:

$$\langle f|i\rangle = \langle f|1\rangle\langle 1|i\rangle + \langle f|2\rangle\langle 2|i\rangle \,,$$

and the corresponding probability density is

$$|\langle f|i\rangle|^2 = |\langle f|1\rangle\langle 1|i\rangle|^2 + |\langle f|2\rangle\langle 2|i\rangle|^2 + 2\,\mathrm{Re}(\langle f|1\rangle^*\langle 1|i\rangle^*\langle f|2\rangle\langle 2|i\rangle) \,.$$

According to p. 277, we should add the amplitudes for the different paths and then take the square of the absolute value, whence each amplitude will factorize into the product of those amplitudes, to arrive at the slit from the initial state, or at the detector from the slit.

The equation before last suggests that we imagine the initial state $|i\rangle$ at the double slit as decomposed into two states $|1\rangle$ and $|2\rangle$:

$$|i\rangle = |1\rangle\langle 1|i\rangle + |2\rangle\langle 2|i\rangle \,.$$

This *superposition of states* with corresponding probability amplitudes makes good sense (superposition principle of states). Such superpositions can be understood classically with polarized light: we can decompose into either linearly or circularly polarized light (linearly polarized $\{|\,\|\rangle, |\perp\rangle\}$, circularly polarized $\{|+\rangle, |-\rangle\}$):

$$|+\rangle = |\,\|\rangle\,\tfrac{-i}{\sqrt{2}} + |\perp\rangle\,\tfrac{1}{\sqrt{2}} \,,$$
$$|-\rangle = |\,\|\rangle\,\tfrac{i}{\sqrt{2}} + |\perp\rangle\,\tfrac{1}{\sqrt{2}} \,,$$

and conversely,

$$|\,\|\rangle = |+\rangle\,\tfrac{i}{\sqrt{2}} + |-\rangle\,\tfrac{-i}{\sqrt{2}} \,,$$
$$|\perp\rangle = |+\rangle\,\tfrac{1}{\sqrt{2}} + |-\rangle\,\tfrac{1}{\sqrt{2}} \,.$$

Instead of the four states $|+\rangle, |-\rangle, |\,\|\rangle$, and $|\perp\rangle$, we had the four unit vectors $\mathbf{e}_+, \mathbf{e}_-, \mathbf{e}_\|$, and $\mathbf{e}_\perp$ on p. 219.

We may take the states $|\ldots\rangle$ as vectors and, according to the superposition principle, combine them linearly with complex coefficients. As already stressed in the discussion of vector algebra (see p. 3), it is an advantage to state the expansion basis first and then the coefficients. However, this is not true for the dual bra vectors in the next section. We would now like to set up the rules that will be applicable here.

4.1.5 Hilbert Space (Four Axioms)

So far we have denoted states using the Dirac symbol $|\ldots\rangle$ and have written the attribute of the state, e.g., $\mathbf{p}$, between these "brackets" as $|\mathbf{p}\rangle$. If the attributes are countable, then these symbols can be assigned to vectors in a Hilbert space. Here, we list the rules for these (proper) Hilbert vectors. However, for continuous attributes like $\mathbf{p}$, improper Hilbert vectors are necessary, and we shall discuss these in Sect. 4.1.7.

The algebraic and topological structure of the Hilbert space are determined by four axioms. However, in contrast to the usual vector algebra, we shall not restrict ourselves to three dimensions (or even finite dimensions, as is clearly expressed by the fourth axiom):

1. *The Hilbert space is a vector space over the field of complex numbers*, i.e., its elements $|\psi\rangle$, $|\varphi\rangle$, ... can be added and multiplied by complex numbers a, b, ..., where the usual rules of vector algebra apply:

$$|\psi\rangle + |\varphi\rangle = |\varphi\rangle + |\psi\rangle ,$$
$$(|\psi\rangle + |\varphi\rangle) + |\chi\rangle = |\psi\rangle + (|\varphi\rangle + |\chi\rangle) ,$$
$$|\psi\rangle + |o\rangle = |\psi\rangle ,$$
$$|\psi\rangle (a + b) = |\psi\rangle a + |\psi\rangle b ,$$
$$(|\psi\rangle + |\varphi\rangle) a = |\psi\rangle a + |\varphi\rangle a .$$

Occasionally, we will also write $|\psi a + \varphi b\rangle$ for $|\psi\rangle a + |\varphi\rangle b$. In particular, for each $|\psi\rangle$, there is a vector $|-\psi\rangle$ such that $|\psi\rangle + |-\psi\rangle = |o\rangle$ is a "zero vector".

A finite set of vectors $|\psi^1\rangle, \ldots, |\psi^N\rangle$ are said to be *linearly independent* if none of them can be expressed in terms of the remaining ones, so for example, $|\psi^{n'}\rangle \neq \sum_{n \neq n'} |\psi^n\rangle c_n$. An infinite set of vectors are said to be linearly independent of each other if this is true for all finite subsets. N linearly independent vectors span a vector space of *dimension N*. The set of vectors in a one-dimensional Hilbert space forms a *ray*—they differ from each other only by a (complex) number.

2. *The Hilbert space has a Hermitian metric.*[1] This means that a complex number $\langle\psi|\varphi\rangle \equiv \langle\psi\|\varphi\rangle$ is assigned to each pair of vectors $|\psi\rangle$, $|\varphi\rangle$. This is called the *scalar product*, the *inner product*, or the *Dirac bracket*. It has the following properties:

$$\langle\psi|\varphi + \chi\rangle = \langle\psi|\varphi\rangle + \langle\psi|\chi\rangle , \qquad \langle\psi|\varphi\,a\rangle = \langle\psi|\varphi\rangle\,a ,$$
$$\langle\psi|\varphi\rangle = \langle\varphi|\psi\rangle^* , \qquad \langle\psi|\psi\rangle > 0 \;\; \text{if} \;\; |\psi\rangle \neq |o\rangle .$$

It is linear, Hermitian-symmetric and positive-definite. In addition to $\langle\psi|o\rangle = 0 = \langle o|\psi\rangle$, we thus have

---

[1] Charles Hermite (1822–1901) was a French mathematician. Hence the "e" at the end is not pronounced.

$$\langle \psi + \varphi | \chi \rangle = \langle \psi | \chi \rangle + \langle \varphi | \chi \rangle \,, \quad \langle \psi \, a | \varphi \rangle = a^* \, \langle \psi | \varphi \rangle \,.$$

Therefore the scalar product $\langle \psi | \varphi \rangle$ depends linearly on the *ket-vectors* $|\varphi\rangle$, but anti-linearly on the *bra-vectors* $\langle \psi|$. If we multiply the ket-vector by a number a, then likewise the inner product, but if we multiply the bra-vector by a, then the inner product gets multiplied by the conjugate complex a^*. The two types of vectors are in principle different. They cannot be added, because they belong to "dual" spaces, but they can be assigned to each other. A bra-vector is known if its scalar products with all ket-vectors are known. Note that we also distinguish between covariant and contravariant components, and both are needed for a scalar product (see p. 33). Bra- and ket-vectors may also be considered as row and column vectors, respectively, if we allow complex components and do not restrict the dimension.

The quantity $\|\psi\| \equiv \sqrt{\langle \psi | \psi \rangle} \geq 0$ is called the *norm (length)* of the vector $|\psi\rangle$. It vanishes only if $|\psi\rangle$ is the zero vector. We will usually restrict ourselves to vectors of unit norm, whence we always have $|\langle \psi | \varphi \rangle|^2 \leq 1$, as we shall soon see, and that is indispensable for the probability interpretation. For this purpose, there are two further notions we need to consider. Two Hilbert vectors are said to be *orthogonal* to each other if their scalar product vanishes, and they are said to be *parallel* if the two vectors differ only by a numerical factor. From this we can obtain the components of the vector $|\psi\rangle$ parallel and orthogonal to the vector $|\varphi\rangle$ from $|\psi\rangle = |\psi_\|\rangle + |\psi_\perp\rangle$, with $|\psi_\|\rangle = |\varphi\rangle \langle \varphi | \psi \rangle / \langle \varphi | \varphi \rangle$ and $|\psi_\perp\rangle = |\psi\rangle - |\psi_\|\rangle$. Note that we do not assume here that $|\varphi\rangle$ is normalized to 1, and $|\psi_\|\rangle$ and $|\psi_\perp\rangle$ are not generally, even if $|\psi\rangle$ is. Clearly, $\langle \varphi | \psi_\| \rangle = \langle \varphi | \psi \rangle$ and hence $\langle \varphi | \psi_\perp \rangle = \langle \varphi | \psi \rangle - \langle \varphi | \psi_\| \rangle = 0$. It then follows that $\|\psi\|^2 = \|\psi_\|\|^2 + \|\psi_\perp\|^2$ with $\|\psi_\|\|^2 = |\langle \varphi | \psi \rangle|^2 / \|\varphi\|^2$, and because $\|\psi_\perp\| \geq 0$, we thus have $\|\psi\|^2 \geq \|\psi_\|\|^2$ and

$$\text{Schwarz inequality} \qquad \|\psi\| \cdot \|\varphi\| \geq |\langle \psi | \varphi \rangle| \,.$$

In particular, $|\langle \psi | \varphi \rangle|^2 \leq 1$ for vectors normalized to 1. Equality holds only if $|\psi\rangle$ and $|\varphi\rangle$ are parallel to each other and hence $\|\psi_\perp\|$ vanishes. With $\|\psi + \varphi\|^2 = \|\psi\|^2 + 2\text{Re}\langle \psi | \varphi \rangle + \|\varphi\|^2$ and $\text{Re}\langle \psi | \varphi \rangle \leq |\langle \psi | \varphi \rangle|$, Schwarz's inequality also delivers $\|\psi + \varphi\|^2 \leq (\|\psi\| + \|\varphi\|)^2$. This upper limit is true also for $\|\psi - \varphi\|^2 = \|(\psi - \chi) - (\varphi - \chi)\|^2$, where $|\chi\rangle$ can be an arbitrary vector. Hence,

$$\text{Triangle inequality} \qquad \|\psi - \varphi\| \leq \|\psi - \chi\| + \|\varphi - \chi\| \,.$$

For this reason, the norm $\|\psi\|$ is also referred to as the *length* of the Hilbert vector $|\psi\rangle$.

Incidentally, from the Schwarz and the triangle inequalities together, it follows that $\|\psi\| \, \|\varphi\| \geq |\langle \psi | \varphi \rangle| \geq |\text{Re} \, \langle \psi | \varphi \rangle| = \frac{1}{2} |\langle \psi | \varphi \rangle + \langle \varphi | \psi \rangle|$.

We can now investigate *convergence in Hilbert space*. However, we have to distinguish between two notions of convergence:

strong convergence $|\psi^n\rangle \to |\psi\rangle$ if $\lim\limits_{n\to\infty} \||\psi^n - \psi\|| = 0$,

weak convergence $|\psi^n\rangle \rightharpoonup |\psi\rangle$ if $\lim\limits_{n\to\infty} \langle\psi^n|\varphi\rangle = \langle\psi|\varphi\rangle$ for all $|\varphi\rangle$.

With the help of the Schwarz inequality, strong convergence implies weak convergence. For all $|\varphi\rangle$ (with $\|\varphi\| < \infty$), we find

$$|\langle\psi^n|\varphi\rangle - \langle\psi|\varphi\rangle| = |\langle\psi^n - \psi|\varphi\rangle| \le \||\psi^n - \psi\|| \cdot \|\varphi\| \to 0.$$

But weak convergence does not imply strong convergence, unless we also have $\||\psi^n\|| \to \||\psi\||$:

$$\||\psi^n - \psi\||^2 = \||\psi^n\||^2 + \||\psi\||^2 - 2\,\mathrm{Re}\langle\psi^n|\psi\rangle$$
$$\rightharpoonup \||\psi^n\||^2 + \||\psi\||^2 - 2\,\mathrm{Re}\langle\psi|\psi\rangle = \||\psi^n\||^2 - \||\psi\||^2.$$

But there are sequences $\{|\psi^n\rangle\}$ which converge weakly towards the zero vector without their norm tending towards zero, for example, if each $|\psi^n\rangle$ is normalized to 1 and orthogonal to all the others. These issues are investigated in Problem 4.9.

A *Cauchy sequence* of vectors $|\psi^n\rangle$ is understood as a sequence for which $\||\psi^n - \psi^m\||$ becomes smaller than any $\varepsilon > 0$ with increasing n and m. Every strongly convergent sequence is a Cauchy sequence. Conversely, each Cauchy sequence converges strongly if the limit vector also belongs to the Hilbert space. This is taken care of by the third axiom.

3. *The Hilbert space is complete*, in the sense that it contains all its accumulation points. For a finite-dimensional space, this is not in fact an additional requirement. The fourth axiom is then obsolete.

4. *The Hilbert space is of countably infinite dimension* (*separable*), meaning that it contains only a countable infinity of mutually orthogonal unit vectors, $\{|\varepsilon_n\rangle\}$ with $\langle\varepsilon_n|\varepsilon_{n'}\rangle = \delta_{nn'}$ for all natural numbers n and n'. A system of such vectors is referred to as an *orthonormal system*. It consists of vectors which are orthogonal to each other and normalized to 1. We will write $|n\rangle$ for short, instead of $|\varepsilon_n\rangle$.

The Hilbert space vectors are thus abbreviations for states and the scalar products (of vectors normalized to 1) for probability amplitudes. Then, e.g., $\langle\psi|\varphi\rangle$ is the probability amplitude for finding $|\varphi\rangle$ if the system is in the state $|\psi\rangle$. We shall determine such probability amplitudes later, e.g., $\langle\mathbf{r}|\mathbf{p}\rangle = h^{-3/2}\exp(i\mathbf{r}\cdot\mathbf{p}/\hbar)$.

4.1.6 Representation of Hilbert Space Vectors

Every arbitrary (normalizable) Hilbert vector $|\psi\rangle$ can be expanded in terms of a complete basis $\{|n\rangle\}$, where we assume an orthonormal system so that $\langle n|n'\rangle = \delta_{nn'}$ should hold:

$$|\psi\rangle = \sum_n |n\rangle\langle n|\psi\rangle \,, \quad \langle\psi| = \sum_n \langle\psi|n\rangle\langle n| \,.$$

The order of the factors, that is, expansion coefficients *after* ket-vectors and *before* bra-vectors, is actually arbitrary, but will turn out later to be particularly practical. For example, we shall treat $\sum_n |n\rangle\langle n|$ as a unit operator and write equations like $|\psi\rangle = 1|\psi\rangle$ and $\langle\psi| = \langle\psi|1$. Then, according to p. 282, $\langle\psi|n\rangle = \langle n|\psi\rangle^*$ holds. The expansion coefficients

$$\psi_n \equiv \langle n|\psi\rangle$$

are the (complex) vector *components* of $|\psi\rangle$ in this basis. The sequence $\{\psi_n\}$ gives the *representation* of the vector $|\psi\rangle$ in the basis $\{|n\rangle\}$. For the scalar product of two vectors, it then follows that

$$\langle\psi|\varphi\rangle = \sum_n \langle\psi|n\rangle\langle n|\varphi\rangle = \sum_n \psi_n^* \, \varphi_n \,,$$

described as *insertion of intermediate states* or insertion of unity (in Fig. 4.2, we used only two states $|1\rangle$ and $|2\rangle$). The special case $|\varphi\rangle = |\psi\rangle$, viz.,

$$\|\psi\|^2 = \sum_{n=1}^{\infty} |\psi_n|^2 \,,$$

is called the *completeness relation*. It holds only if no basis vector is missing. Finally,

$$\|\psi\|^2 \geq \sum_{n=1}^{N} |\psi_n|^2$$

is *Bessel's inequality*.

The Hilbert vectors which were initially introduced only formally thus become rather simple constructs as soon as a discrete basis is introduced in Hilbert space. Then each vector is given by its (possibly infinitely many) complex components with respect to this basis, i.e., by a sequence of complex numbers. We then speak of *vectors in sequence space*.

If we take the sequence $\{\langle n|\psi\rangle = \psi_n\}$ as a column vector and $\{\langle\psi|n\rangle = \psi_n^*\}$ as a row vector, i.e.,

$$|\psi\rangle \,\hat{=}\, \begin{pmatrix} \psi_1 \\ \psi_2 \\ \vdots \end{pmatrix} \quad \text{and} \quad \langle\psi| \,\hat{=}\, (\psi_1^*, \psi_2^*, \ldots)\,,$$

then the scalar products $\langle\psi|\varphi\rangle$ obey matrix multiplication rules.

Of course, we may introduce a new basis, e.g., $\{|m\rangle\}$ with $\langle m|m'\rangle = \delta_{mm'}$. For the *change of representation* $\{|n\rangle\} \rightarrow \{|m\rangle\}$, we clearly have

$$\psi_m \equiv \langle m|\psi\rangle = \sum_n \langle m|n\rangle\langle n|\psi\rangle = \sum_n m_n{}^* \psi_n \,,$$

i.e., the components in one basis follow from the components in the other basis, where the components of the basis vectors occur as expansion coefficients. This is similar to our procedure for the orthogonal transformation on p. 29, and will be important in Sect. 4.2 as a "unitary transformation".

We now take an obvious step which leads to a new representation of the Hilbert space. We plot the complex numbers $\psi_n \equiv \langle n|\psi\rangle$ versus the natural numbers $\{n\}$ on the number axis, and then, not only to the natural numbers, but to all real numbers x, assign values $\psi(x) \equiv \langle x|\psi\rangle$. This delivers a different representation of the Hilbert space, namely the Hilbert *function space* $\{\psi(x)\}$. It combines all complex functions defined on the real axis for which the square of their absolute value can be integrated in the Lebesgue sense, i.e., they are integrable almost everywhere, in the sense that only a set of arguments of measure zero need be excluded: $\langle \psi|\psi\rangle = \int |\psi(x)|^2 \, \mathrm{d}x < \infty$. Such functions are said to be *normalizable*. With a finite numerical factor, they can be normalized to 1. The range of integration corresponds to the domain of definition, which can be infinite on both sides. This function space is a linear space. It is complete and has a countable infinity of dimensions. The inner product is now given by

$$\langle \psi|\varphi\rangle = \int \psi^*(x)\,\varphi(x)\,\mathrm{d}x \,,$$

i.e., the sum in sequence space becomes an integral in the function space.

With this we can then express a complete orthonormal system of functions $\{g_n(x)\}$ (see p. 21) with $\int g_n{}^*(x)\,g_{n'}(x)\,\mathrm{d}x = \delta_{nn'}$ in the useful form

$$g_n(x) \equiv \langle x|n\rangle \,.$$

An arbitrary (normalizable) function can be *expanded in terms of this orthonormal system* (represented in this basis):

$$\psi(x) = \langle x|\psi\rangle = \sum_n \langle x|n\rangle\langle n|\psi\rangle = \sum_n g_n(x)\,\psi_n \,,$$

with the expansion coefficients ("Fourier coefficients")

$$\psi_n = \langle n|\psi\rangle = \int \langle n|x\rangle\langle x|\psi\rangle\,\mathrm{d}x = \int g_n{}^*(x)\,\psi(x)\,\mathrm{d}x \,.$$

The best-known example is the Fourier expansion, and another the expansion in terms of Legendre polynomials, with domain of definition $-1 \le x \le 1$.

The function $\psi(x)$ in Hilbert function space is then represented as a vector $\{\psi_n\}$ in Hilbert sequence space, or the vector as a function. Depending on the basis, the same Hilbert vector $|\psi\rangle$ appears in different forms—in the sequence space, we obtain Heisenberg's *matrix mechanics*, and in function space, Schrödinger's *wave mechanics*.

4.1.7 Improper Hilbert Vectors

However, $|x\rangle$ and $\langle x|$ are not genuine (*proper*) Hilbert vectors. If we compare the scalar product $\langle\psi|\varphi\rangle = \sum_{nn'}\langle\psi|n\rangle\langle n|n'\rangle\langle n'|\varphi\rangle$ with the expected expression $\iint\langle\psi|x\rangle\langle x|x'\rangle\langle x'|\varphi\rangle\,\mathrm{d}x\,\mathrm{d}x'$, then $\int\langle x|x'\rangle\,\varphi(x')\,\mathrm{d}x'$ must be equal to $\varphi(x)$. The scalar product $\langle x|x'\rangle$ is clearly equal to the Dirac delta function (see Sect. 1.1.10), i.e.,

$$\langle x|x'\rangle = \delta(x - x')\,,$$

and hence is no longer a typical number—Dirac symbols with continuous variables are not proper Hilbert space vectors, but *improper Hilbert vectors*. The normalization to the delta function is called *normalization in the continuum*, and often also *delta-normalization*.

Since x is a continuous variable, $|\langle x|\psi\rangle|^2$ should not be called a probability: it is a probability *density*. Only $|\langle x|\psi\rangle|^2\,\mathrm{d}x$ is a probability, in particular, for the interval $\mathrm{d}x$, and only probabilities can be compared with observed values. For example, there is no particle at the position $\mathbf{r}$, but only in a region d^3r around $\mathbf{r}$. The more certain its position, the more uncertain its momentum! While we may often speak of a particle with the momentum $\mathbf{p}$, e.g., in Sect. 4.1.4, our main interest is not the "small error interval" $\Delta\mathbf{p}$, otherwise its position would have to be totally uncertain.

Continuous variables are often convenient for calculation, and we shall use them repeatedly, even if they are only ever observed in a certain interval. For the same reason we will not be disturbed by the fact that $\langle x|x'\rangle = \delta(x - x')$ is not a standard number (function of x and x'). It is quite sufficient that the delta-function has a definite meaning in an integral.

4.1.8 Summary: Wave–Particle Dualism

Quantum mechanics is more general than its classical limiting case, since quantum theory includes the fact that canonically conjugate quantities cannot simultaneously be sharp (or certain)—they are complementary, in the sense that the more precise one quantity is, the less precise the other will be, a fact overlooked in classical mechanics.

We take Heisenberg's uncertainty relation $\Delta x^k \cdot \Delta p_{k'} \geq \frac{1}{2}\hbar\,\delta^k_{k'}$ as the basic experimental fact. The consequences are far-reaching. In particular, the particle and wave pictures in quantum theory are no longer in contradiction, because all measurable quantities then have "uncertain" values in precisely such a way that the two pictures remain compatible with each other—neither the particle number nor the phase in the statistical ensemble has to have a sharp value.

Uncertainty is a statistical notion and quantum theory a statistical theory for the determination of probabilities and average values. Since interference occurs for these probabilities, we work with probability amplitudes and take these as scalar products of Hilbert vectors. The ket-vector $|\ldots\rangle$ specifies the attributes of the considered ensemble and the bra-vector $\langle\ldots|$ the attributes for the probability. Then the square of the absolute value of the scalar product $\langle\psi|\varphi\rangle$ gives the probability for the attribute ψ in the ensemble φ. The rules for these state vectors have been presented here. We assign proper or improper Hilbert space vectors to them, depending on whether they are valid for countable or continuous variables, respectively.

Concerning the scalar product $\langle\psi|\varphi\rangle$, initially only the square of the absolute value can be measured, i.e., as the associated probability. Only if two amplitudes interfere with each other can the relative phase be determined, and even then, a global phase factor remains free.

Incidentally, as early as 1781, I. Kant wrote in his *Kritik der reinen Vernunft*: "[…] consequently we cannot have knowledge of a matter as a thing as such, but only as much as it is an object of the sensuous perception", something Heisenberg also stressed. Only then can such knowledge be proven as a law, if the experiment is repeated. This leads to statistics. Then the uncertainty relations are valid from the moment the statistical ensemble has been produced, not at the time of the individual measurements. Anyone who does not take this fact into account will very likely find quantum theory *incomplete*.

4.2   Operators and Observables

4.2.1   *Linear and Anti-linear Operators*

The state vectors $|\ldots\rangle$ and $\langle\ldots|$ are mathematical tools to describe pure states in quantum theory. In addition, we need quantities which act on these state vectors, which we call *operators*. We always write them with upper-case letters:

$$|\psi'\rangle = A\,|\psi\rangle\ .$$

Operators assign an image vector $|\psi'\rangle$ to each object vector $|\psi\rangle$. (We can also consider operators which are only defined on a part of space, but we do not wish to deal with those here.) If we know the image vector $|\psi'\rangle$ for each vector $|\psi\rangle$, then we also know the operator A, just as a bra-vector is determined if its scalar products with all ket-vectors are known. If $A\,|\psi\rangle = A'\,|\psi\rangle$ for all $|\psi\rangle$, then the two operators *are equal*, i.e., $A = A'$. The *zero operator* assigns the zero vector to all vectors, i.e.,

$0 \, |\psi\rangle = |o\rangle$, while the *unit operator* assigns the original vectors to all vectors, i.e.,
$1 \, |\psi\rangle = |\psi\rangle$.

In quantum mechanics, only linear and anti-linear operators occur. They are *linear*,
if

$$A \, |\psi + \varphi\rangle = A \, |\psi\rangle + A \, |\varphi\rangle \qquad \text{and} \qquad A \, |\psi \, a\rangle = A \, |\psi\rangle \, a \, ,$$

while for an *anti-linear* operator, $A \, |\psi \, a\rangle = (A \, |\psi\rangle) \, a^*$. In quantum theory, there is
only one important anti-linear operator, namely the time reversal operator $\mathscr{T}$, which
we shall discuss in Sect. 4.2.12 (and also the charge-inversion operator $\mathscr{C}$ for the
Dirac equation). Until then, we shall deal only with linear operators. They can be
added and multiplied by complex numbers:

$$(a \, A + b \, B) \, |\psi\rangle = A \, |\psi\rangle \, a + B \, |\psi\rangle \, b \, .$$

The product of two operators depends on the order of the factors: AB may differ from
BA. We define the commutator and anti-commutator of two operators A and B by

$$
\begin{aligned}
A B - B A &\equiv [A, B] && \text{commutator of } A \text{ and } B \, , \\
A B + B A &\equiv [A, B]_+ \equiv \{A, B\} && \text{anti-commutator of } A \text{ and } B \, .
\end{aligned}
$$

If $A B = B A$ holds, the two operators *commute* with each other. Then also aA and bB
commute with each other. The unit operator and the zero operator commute with all
operators. In quantum theory, it is important to know whether or not two operators
commute with each other, so here are several properties of commutators:

$$
\begin{aligned}
[A, B] &= -[B, A] \, , \\
[A, B + C] &= [A, B] + [A, C] \, , \\
[A, BC] &= [A, B] \, C + B \, [A, C] \, .
\end{aligned}
$$

Hence, with $[A, B^n] = [A, B]B^{n-1} + B[A, B^{n-1}]$ for $n \in \{1, 2, \ldots\}$, it follows that

$$[A, B^n] = \sum_{k=0}^{n-1} B^k \, [A, B] \, B^{n-1-k} \, .$$

In particular, for $[[A, B], B] = 0$, we have $[A, B^n] = n \, [A, B] \, B^{n-1}$. The last expression is sometimes written as $[A, B] \, dB^n / dB$, because we can also differentiate with
respect to operators, as we shall see on p. 316. In addition, we find *Jacobi's identity*

$$[A, [B, C]] + [B, [C, A]] + [C, [A, B]] = 0 \, ,$$

as for the vector product (see p. 4) and the Poisson brackets (p. 124). Note that, together with skew-symmetry $[A, B] = -[B, A]$ and the bilinearity property

$$[aA + bB, C] = a[A, C] + b[B, C] ,$$

in the abstract vector space of the quantities $A, B, C, \ldots$, the Jacobi identity shows that the commutator makes the set of operators into a *Lie algebra*.

There are functions of operators, e.g., polynomials in A like

$$f(A) = c_0 1 + c_1 A + c_2 A^2 + \cdots + c_n A^n ,$$

and the "exponential function" $\exp A \equiv e^A = \sum_{n=0}^{\infty} A^n/(n!)$. Incidentally,

$$e^A\, B\, e^{-A} = B + \frac{[A, B]}{1!} + \frac{[A, [A, B]]}{2!} + \frac{[A, [A, [A, B]]]}{3!} + \cdots$$

is called the *Hausdorff series*. This equation can be proven by considering the function $f(t) = e^{At} B e^{-At}$, for which $\dot{f} = [A, f]$, $\ddot{f} = [A, \dot{f}]$, and similar. With $f(0) = B$, its Taylor series about $t = 0$ for $t = 1$ delivers the Hausdorff series. From $[A, 1] = 0$, it follows in particular that $e^A e^{-A} = 1$, which can be generalized:

$$[A, [A, B]] = 0 = [B, [A, B]] \quad \Longrightarrow \quad e^A\, e^B = e^{\frac{1}{2}[A,B]}\, e^{A+B} .$$

In particular, if we set $f(t) = e^{At} e^{Bt}$, then $\dot{f} = Af + fB = (A + e^{At} B\, e^{-At})f$, and with $[A, [A, B]] = 0$, according to Hausdorff, $\dot{f} = (A + B + [A, B]\, t)\, f$. With $f(0) = 1$ this implies $f(t) = \exp\{(A + B)\, t + \frac{1}{2}[A, B]\, t^2\}$. The claim follows for $t = 1$, since $\exp(\frac{1}{2}[A, B])$ may be factored out, because $[A, B]$ is assumed to commute with A and B.

Numerical factors multiplying Hilbert vectors can be considered as very simple operators. They are multiples of the unit operator and hence commute with every linear operator.

4.2.2   Matrix Elements and Representation of Linear Operators

So far the operators have been acting only on ket-vectors $|\ldots\rangle$. We consider now the scalar product of an arbitrary bra-vector $\langle\psi|$ with the ket-vectors $A\,|\varphi\rangle$, where A is a linear operator. Each scalar product depends linearly on its ket-vector, but now the ket-vector $A\,|\varphi\rangle$ also depends linearly on $|\varphi\rangle$. Thus the scalar product depends linearly on $|\varphi\rangle$. Consequently, a bra-vector can be constructed from the other quantities $\langle\psi|$ and A. Hence, for linear operators, we have

$$\langle \psi | \, (A \, |\varphi\rangle) = ((\langle \psi | A) \, |\varphi\rangle = \langle \psi | A \, |\varphi\rangle \,.$$

This complex number is called the *matrix element of the operator A between the states* $\langle \psi |$ *and* $|\varphi\rangle$.

In order to understand the connection with matrices, we take any discrete complete orthonormal system $\{|n\rangle\}$ and consider $A \, |\psi\rangle = \sum_{n'} A \, |n'\rangle \langle n'|\psi\rangle$ and $A \, |n'\rangle = \sum_n |n\rangle \langle n| A \, |n'\rangle$. If we compare this with the original expression, we have

$$A = \sum_{nn'} |n\rangle \langle n| A \, |n'\rangle \langle n'| \,.$$

The complex numbers $\langle n| A \, |n'\rangle$ form the matrix *of the operator A in the n representation* (possibly with infinitely many matrix elements):

$$\begin{pmatrix} \langle n_1| A \, |n_1\rangle & \langle n_1| A \, |n_2\rangle & \cdots \\ \langle n_2| A \, |n_1\rangle & \langle n_2| A \, |n_2\rangle & \cdots \\ \vdots & \vdots & \ddots \end{pmatrix} \,.$$

If the matrix is known, then so is the operator "in the n representation". Its "diagonal elements" are $\langle n| A \, |n\rangle$ and its "off-diagonal elements" $\langle n| A \, |n'\rangle$ (with $n \neq n'$).

In the Dirac notation, $\langle \ldots | \ldots \rangle$ is thus a number and $| \ldots \rangle \langle \ldots |$ an operator. A particularly important example is the unit operator

$$1 = \sum_n |n\rangle \langle n| \,, \quad \text{with} \quad \langle n| \, 1 \, |n'\rangle = \langle n|n'\rangle = \delta_{nn'} \,.$$

In this sense, the above-mentioned representations $|\psi\rangle = \sum_n |n\rangle \langle n|\psi\rangle$ for states and $A = \sum_{nn'} |n\rangle \langle n|A|n'\rangle \langle n'|$ for operators are to be interpreted as $|\psi\rangle = 1|\psi\rangle$ and $A = 1A1$. This also shows why the notation $\langle n|A|n'\rangle$ is preferred over the abbreviation $A_{nn'}$, even though the symbol takes up more room.

We can take the objects $|n\rangle \langle n|$ as operators projecting onto the states $|n\rangle$. In particular, if $|\psi\rangle$ is normalized to 1, the *projection operator* onto the state $|\psi\rangle$ is

$$P_\psi \equiv |\psi\rangle \langle \psi| \,, \quad \text{with} \quad \langle n| \, P_\psi \, |n'\rangle = \langle n|\psi\rangle \langle \psi|n'\rangle = \psi_n \, \psi_{n'}{}^* \,,$$

because this operator projects an arbitrary vector $|\varphi\rangle$ onto the vector $|\psi\rangle$:

$$P_\psi \, |\varphi\rangle = |\psi\rangle \langle \psi|\varphi\rangle \,.$$

For $\|\psi\| = 1$, we always have $P_\psi{}^2 = P_\psi$, even though $P_\psi \neq 1$ (and $\neq 0$) holds: projection operators are *idempotent*.

For the operator product $A \, B$, we have

$$\langle n| A B |n'\rangle = \sum_{n''} \langle n| A |n''\rangle \langle n''| B |n'\rangle \ ,$$

and hence the usual law of matrix multiplication.

4.2.3   Associated Operators

If the product of two operators is equal to the unit operator, as exemplified by e^A and e^{-A} above, each operator is said to be the *inverse* of the other:

$$A A^{-1} = 1 = A^{-1} A \ .$$

But note that not every operator has an inverse. For *singular* operators, there is a $|\psi\rangle \neq |o\rangle$ with $A |\psi\rangle = |o\rangle$, and from $|o\rangle$ there is no operator leading back to $|\psi\rangle$. For operator products, we have

$$(A B)^{-1} = B^{-1} A^{-1} \ ,$$

because their product with AB gives 1 in both cases.

What the operator A produces for ket-vectors, its (*Hermitian*) *adjoint* operator $A^\dagger$ does for bra-vectors:

$$A |\psi\rangle = |\psi'\rangle \qquad \Longleftrightarrow \qquad \langle \psi| A^\dagger = \langle \psi'| \ .$$

Hence we always have $\langle \psi| A^\dagger |\varphi\rangle = \langle \psi'|\varphi\rangle = \langle \varphi|\psi'\rangle^* = \langle \varphi| A |\psi\rangle^*$, together with

$$A^{\dagger\dagger} = A \qquad \text{and} \qquad (A B)^\dagger = B^\dagger A^\dagger \ ,$$

since $\langle \psi|(AB)^\dagger|\varphi\rangle = \langle\varphi|AB|\psi\rangle^* = \sum_n \langle\varphi|A|n\rangle^* \langle n|B|\psi\rangle^* = \langle\psi|B^\dagger A^\dagger|\varphi\rangle$ for all $\langle \psi|$ and $|\varphi\rangle$. For real matrices, the adjoint is the same as the transpose (reflected in the main diagonal). Incidentally, $(A^\dagger)^{-1} = (A^{-1})^\dagger$ holds, since $A^\dagger (A^{-1})^\dagger = (A^{-1}A)^\dagger = 1^\dagger = 1$. Instead of "Hermitian adjoint", we usually speak of the *Hermitian conjugate* and abbreviate this to h.c., e.g., $A + A^\dagger$ is the same as $A+$ h.c.

An operator would be called *self-adjoint* if

$$A^\dagger = A \ , \qquad \text{i.e.,} \qquad \langle \psi|A|\varphi\rangle = \langle \varphi|A|\psi\rangle^* \ , \qquad \text{for all } |\psi\rangle \text{ and } |\varphi\rangle \ ,$$

which is true, e.g., for any projection operator. Such operators are also said to be *Hermitian*, even though the domains of definition of A and $A^\dagger$ for a Hermitian operator do not have to coincide. Hence, all self-adjoint operators are Hermitian, but not all Hermitian operators are self-adjoint. Note also that Hermitian operators always have real diagonal elements. If all the matrix elements are real, as for the tensor of inertia, then we speak of a *symmetric* matrix rather than a Hermitian matrix. The product of two Hermitian operators is only Hermitian if they commute with each other. But $\{A, B\}$ and $[A, B]/i$ are Hermitian, so we can use

$$AB = \frac{AB + BA}{2} + i\,\frac{AB - BA}{2i} = \frac{\{A, B\}}{2} + i\,\frac{[A, B]}{2i} ,$$

if AB is not Hermitian.

An operator is said to be *unitary*, if

$$U^\dagger = U^{-1} \qquad \Longleftrightarrow \qquad U\,U^\dagger = 1 = U^\dagger U ,$$

whence $\sum_n \langle n'|\,U\,|n\rangle\langle n''|\,U\,|n\rangle^* = \langle n'|n''\rangle = \sum_n \langle n|\,U\,|n'\rangle^* \,\langle n|\,U\,|n''\rangle$. If the matrix is unitary and real, like the rotation matrix on p. 29, then it is also said to be *orthogonal*. Note that any unitary 2×2 matrix can be obtained from 3 real parameters α, β, γ, in particular from

$$U = \begin{pmatrix} \cos\alpha \, \exp(i\beta) & \sin\alpha \, \exp(-i\gamma) \\ -\sin\alpha \, \exp(i\gamma) & \cos\alpha \, \exp(-i\beta) \end{pmatrix} ,$$

if we disregard a common phase factor, hence a fourth parameter. The inverse of a 2 $\times$ 2 matrix is given on p. 71.

Unitary operators U can be derived from self-adjoint operators A via

$$U = \exp(iA) ,$$

since, according to the last section, for $A = A^\dagger$, we find

$$U U^\dagger = \exp(iA - iA)\,\exp(\tfrac{1}{2}[iA, -iA]) = 1 .$$

For *infinitesimal transformations* (with $A = A^\dagger \ll 1$), the approximation $\exp(\pm iA) \approx 1 \pm iA$ is often used. A different relation between a Hermitian and a unitary operator is produced by $U = (1 - iA)(1 + iA)^{-1}$. For the proof, we use the fact that the factors commute.

If all vectors are subjected to a *unitary transformation* U, then their scalar products remain the same:

$$\langle \psi'|\varphi'\rangle = \langle \psi|\,U^\dagger U\,|\varphi\rangle = \langle \psi|\varphi\rangle ,$$

in particular all vectors keep the same norm. Unitary transformations are thus *isometric*. Here, only $U^\dagger U = 1$ is necessary, but not $U U^\dagger = 1$. If U is isometric, then $U U^\dagger$ is a projection operator, since $(U U^\dagger)^2$ is then equal to $U U^\dagger$. For a finite dimension, unitarity follows from isometry.

With a unitary operator U, a complete orthonormal system $\{|n\rangle\}$ can be transformed into a different basis $\{|m\rangle = U\,|n\rangle\}$. A change of representation always corresponds to a unitary transformation. If the vectors are transformed with $|\psi'\rangle = U|\psi\rangle$, then the operators are likewise transformed with

$$A' = UAU^{-1} = UAU^\dagger ,$$

since $|\varphi\rangle = A|\psi\rangle$ and $U|\varphi\rangle = |\varphi'\rangle = A'|\psi'\rangle = A'U|\psi\rangle$ implies $UA = A'U$. Correspondingly, the operator function $f(A)$ turns into $Uf(A)U^{-1} = f(UAU^{-1})$, since all $f(A)$ are to be taken as power series of A, and the unit operator $U^{-1}U$ may be inserted between the individual factors.

The *trace* of an operator, i.e., the sum of its diagonal elements, always remains constant under unitary transformations in finite dimensions, since then tr$(AB) =$ tr(BA) and hence tr$(UAU^{-1}) = $ tr$A$. For infinite dimensions, this is not always true, as we shall see for a counterexample on p. 303.

If A and B commute, then the operator B remains the same for the transformation $U = \exp(iA)$, as follows immediately from the Hausdorff series (p. 290). Here, U does not need to be unitary. This is only necessary if, for self-adjoint A, we also require A' to be self-adjoint, because $A' = UAU^{-1}$ implies $A'^\dagger = U^{-1\dagger}A^\dagger U^\dagger$, and this is equal to A' if $U^{-1} = U^\dagger$.

4.2.4   Eigenvalues and Eigenvectors

These notions are defined as follows (see p. 87). If

$$A|\alpha\rangle = |\alpha\rangle a ,$$

then a is an eigenvalue and $|\alpha\rangle\ (\neq |o\rangle)$ an eigenvector of the operator A. Only the ray specified by $|\alpha\rangle$ is important. For linear operators, $|\alpha\rangle$ is an eigenvector with eigenvalue a if and only if $|\alpha\rangle c$ is an eigenvector with eigenvalue a. Then also $\langle\alpha|A^\dagger = a^*\langle\alpha|$ holds, since $\langle\alpha|A^\dagger|\beta\rangle = \langle\beta|A|\alpha\rangle^* = \langle\beta|\alpha\rangle^* a^* = a^*\langle\alpha|\beta\rangle$, for all $|\beta\rangle$. While the eigenvalues have physical relevance, the state vectors $|\ldots\rangle$ are only mathematical tools. For discrete eigenvalues a_n, the number n is called the *quantum number*, e.g., we speak of the oscillation, angular momentum, direction, and principal quantum numbers in various contexts.

The transformed operator UAU^{-1} has the same eigenvalue a and the eigenvector $U|\alpha\rangle$. From the equation above, it follows in particular that $UAU^{-1} U|\alpha\rangle = UA|\alpha\rangle = U|\alpha\rangle a$.

An important claim is that *Hermitian operators have real eigenvalues*. In particular, if $A = A^\dagger$, then the left-hand side of $\langle\alpha|A|\alpha\rangle = \langle\alpha|\alpha\rangle a$ is real. On the right, the factor $\langle\alpha|\alpha\rangle$ is real, and so therefore is the eigenvalue a.

Unitary operators have only eigenvalues of absolute value 1. If in particular $A^\dagger A = 1$, then we have $\langle\alpha|\alpha\rangle = \langle\alpha|A^\dagger A|\alpha\rangle = a^*\langle\alpha|\alpha\rangle a = |a|^2\langle\alpha|\alpha\rangle$, so $|a|^2 = 1$.

If two eigenvectors of a Hermitian operator $A = A^\dagger$ belong to different eigenvalues, then those eigenvectors are orthogonal to each other. This is because

$A |\alpha_n\rangle = |\alpha_n\rangle a_n$ implies $0 = \langle\alpha_1| A^\dagger - A |\alpha_2\rangle = (a_1 - a_2) \langle\alpha_1|\alpha_2\rangle$, with $a_1 \neq a_2$, so $\langle\alpha_1|\alpha_2\rangle$ must vanish, as required. In fact, we have already shown this for the principal axes of the inertia tensor on p. 88. If all eigenvalues a_n are different, then we may take the normalized eigenvectors $|\alpha_n\rangle$ as an expansion basis in Hilbert space, since they then form a complete orthonormal system, and A will then be diagonal:

$$\langle\alpha_n|\alpha_{n'}\rangle = \delta_{nn'} , \quad 1 = \sum_n |\alpha_n\rangle\langle\alpha_n| , \quad \text{and} \quad A = \sum_n |\alpha_n\rangle a_n \langle\alpha_n| .$$

For this reason, the determination of the eigenvalues and eigenvectors of an operator is referred to as determining the *eigen-representation* (or *diagonalization*) of the operator—it corresponds to a unitary transformation to a more convenient expansion basis for the operator (which gets along without a double sum). For the sum and the product of the eigenvalues, no transformation is necessary, since that changes neither trace nor determinant.

However, an operator can have several linearly independent eigenvectors corresponding to the same eigenvalue, e.g., the unit operator has only the eigenvalue 1. We then speak of *degeneracy*: if there are in total N linearly independent eigenvectors with the same eigenvalue, it is said to be N-fold degenerate. Then N orthonormalized eigenvectors $|\alpha_n\rangle$ can be chosen as basis vectors with this eigenvalue. This is what happens in mechanics when we seek the principal moments of inertia.

The eigenvectors $|\alpha_n\rangle$ of an operator A also diagonalize the powers A^k of the operator A and the operator functions $f(A)$:

$$f(A) = \sum_n |\alpha_n\rangle f(a_n) \langle\alpha_n| .$$

The special case $A^{-1} = \sum_n |\alpha_n\rangle a_n^{-1} \langle\alpha_n|$ shows that none of the eigenvalues can be zero if the inverse exists. If $a_n = 0$ for some n, A is singular.

All functions of the same operator thus have the same eigenvectors, while their eigenvalue spectra can differ. They also commute with each other. Generally, the following claim is true: *two operators A and B commute with each other if and only if they share a complete orthonormal system of eigenvectors*. If they have only common eigenvectors, then they commute, because the order of any product of their eigenvalues is of no importance: $AB = \sum_n |\alpha_n\rangle a_n b_n \langle\alpha_n|$ is equal to $BA = \sum_n |\alpha_n\rangle b_n a_n \langle\alpha_n|$. On the other hand, if initially only $A = \sum_n |\alpha_n\rangle a_n \langle\alpha_n|$ is given with $1 = \sum_n |\alpha_n\rangle\langle\alpha_n|$, then we have

$$AB = \sum_{nn'} |\alpha_n\rangle a_n \langle\alpha_n| B |\alpha_{n'}\rangle\langle\alpha_{n'}| ,$$

$$BA = \sum_{nn'} |\alpha_n\rangle\langle\alpha_n| B |\alpha_{n'}\rangle a_{n'} \langle\alpha_{n'}| .$$

From $AB - BA = 0$, we deduce that $\langle \alpha_n | B | \alpha_{n'} \rangle (a_n - a_{n'}) = 0$, because the zero operator in each basis has only zeros as matrix elements. If there is no degeneracy, then $a_n \neq a_{n'}$ holds for all $n \neq n'$ and hence $\langle \alpha_n | B | \alpha_{n'} \rangle$ is diagonal. Then each $| \alpha_n \rangle$ is also an eigenvector of B. But if eigenvalues a_n are degenerate, then one can make use of the freedom in the choice of the basis vectors to diagonalize the matrix B. When $[A, B] = 0$, there is thus always a complete system of eigenvectors for both operators.

If an operator has degenerate eigenvalues, we must search for further operators which commute with it and lift the degeneracy. Then we can denote the eigenvectors by the associated eigenvalues, e.g., $| \alpha_n \rangle \hateq | a_n, b_n, \ldots \rangle$. Here we may leave out the index n on the right-hand side and write for short $| a, b, \ldots \rangle$. Each eigenvector differs from the others by the order of the values. If there is no degeneracy for A, then the notation $| a \rangle$ suffices. Hence we write in the following

$$ A | a \rangle = | a \rangle a , \quad \text{with} \quad \langle a | a' \rangle = \delta_{aa'} \quad \text{and} \quad \sum_a | a \rangle \langle a | = 1 . $$

Here a is assumed to be discrete. But the operator may also have a continuous eigenvalue spectrum, or even some discrete and some continuous eigenvalues, as happens for the Hamilton operator of the hydrogen atom. For continuous eigenvalues, we have $A = \int | a \rangle a \langle a | \, da$ with $\langle a | a' \rangle = \delta(a - a')$ and $\int | a \rangle \langle a | \, da = 1$. Then sums have to be replaced by integrals and Kronecker symbols by delta functions.

If a Hermitian operator depends on a parameter λ, e.g., $A(\lambda) = \sum_n c_n(\lambda) X^n$, then so do its eigenstates and eigenvalues. For the eigenvalues, we then have the *Hellmann–Feynman theorem*[2]:

$$ A | a \rangle = | a \rangle a \quad \Longrightarrow \quad \langle a | \frac{\partial A}{\partial \lambda} | a \rangle = \frac{\partial a}{\partial \lambda} . $$

For the proof we differentiate $\langle a | A - a | a \rangle = 0$ with respect to λ and make use of A being Hermitian:

$$ \langle a | \frac{\partial A}{\partial \lambda} - \frac{\partial a}{\partial \lambda} | a \rangle + 2 \operatorname{Re} \langle \frac{\partial a}{\partial \lambda} | A - a | a \rangle = 0 . $$

This suffices for the proof because $(A - a) | a \rangle = 0$. The theorem is mainly applied to the Hamilton operator, but we may use it also for other observables. This is connected to the *adiabatic theorem*. If the Hamilton operator $H(t)$ varies sufficiently slowly with time, then a system initially in an eigenstate of $H(t_0)$ remains in the eigenstate developing from it, provided that it always remains non-degenerate. This will be demonstrated in Fig. 4.11 on p. 348 for the time-dependent oscillator.

---

[2]Before these two authors, it was already formulated by Güttinger [3] in his diploma thesis.

4.2.5 Expansion in Terms of a Basis of Orthogonal Operators

Two operators A and B are said to be *orthogonal* to each other if the trace of $A^\dagger B$
vanishes. For matrices of finite dimension, the order of the factors is not important
and nor is it important which factor is chosen as adjoint, since $\mathrm{tr}(A^\dagger B)$ is then equal
to $\mathrm{tr}(BA^\dagger) = \{\mathrm{tr}(A^\dagger B)^\dagger\}^*$ and $(A^\dagger B)^\dagger = B^\dagger A$. In particular, $\mathrm{tr}(A^\dagger A)$ is real, and it is
also non-negative, since $\mathrm{tr}(A^\dagger A) = \sum_{nn'} |\langle n|A|n'\rangle|^2$.

We can introduce an orthogonal system of operators C_n as a common expansion
basis for all operators. If we take Hermitian operators $(C_n{}^\dagger = C_n)$, then that simplifies
the considerations even further, but we shall not do so yet. Thus we only require, for
all n, n',

$$\mathrm{tr}(C_n{}^\dagger C_{n'}) = c\,\delta_{nn'} .$$

Here, $c = c^* > 0$ is a normalization factor, which we can choose at our convenience.
$c = 2$ is often chosen, e.g., for the Pauli matrices in Sect. 4.2.10 and their generalizations to more than two dimensions, the *Gell-Mann matrix*. But occasionally, $c = 1$
is also chosen. In fact, it can also depend on $n = n'$, but we shall not pursue this any
further here.

If the basis $\{C_n\}$ is complete, then for arbitrary operators A, we have

$$A = \sum_n C_n \frac{\mathrm{tr}(C_n{}^\dagger A)}{c}, \quad \text{with} \quad \mathrm{tr}(C_n{}^\dagger A) = \{\mathrm{tr}(C_n A^\dagger)\}^* .$$

For a Hermitian basis $\{C_n{}^\dagger = C_n\}$, all Hermitian operators have real expansion coefficients.

In an N-dimensional vector space, we need N^2 basis operators, one for each matrix
element. But they would all also commute with each other. This is no longer true for
our general basis. Nevertheless, their commutators can be expanded:

$$\mathrm{i}\,[C_{n'}, C_{n''}] = \sum_n C_n \frac{\mathrm{tr}(\mathrm{i}\,C_n{}^\dagger [C_{n'}, C_{n''}])}{c} .$$

If the basis consists of Hermitian operators, then on the right there are only real expansion coefficients, the so-called *structure constants* of the associated Lie algebra (see
[8]), which are antisymmetric in the three indices: symmetric for cyclic permutations
since $\mathrm{tr}(C_n[C_{n'}, C_{n''}]) = \mathrm{tr}(C_{n'}[C_{n''}, C_n])$, and antisymmetric for anti-cyclic permutations since $\mathrm{tr}(C_n[C_{n'}, C_{n''}]) = -\mathrm{tr}(C_{n'}[C_n, C_{n''}])$. Unitary transformations do not
change that.

It is advantageous to start the basis with the unit operator, $C_0 \propto 1$, because this
operator commutes with all other operators and only its trace is non-zero, since the
other operators should be orthogonal to it: $\mathrm{tr}C_n \propto \delta_{n0}$.

A first example will be presented in Sect. 4.2.10. In a two-dimensional vector space, the Pauli operators are useful as an expansion basis. With the Wigner function (Sect. 4.3.5), we also employ an operator basis.

4.2.6 Observables. Basic Assumptions

In the above, we have provided the mathematical tools, and now we turn to physics again. We start with basics. So far we have assumed only that (pure) states can be represented by proper or improper Hilbert vectors and that the scalar product $\langle \psi | \varphi \rangle$ yields the probability amplitude for the state $|\psi\rangle$ to be contained in the state $|\varphi\rangle$. Now we add the following:

> To every measurable quantity (an observable, e.g., position, momentum, energy, angular momentum) is assigned a Hermitian operator. Its real eigenvalues are equal to all possible measurable results of this observable.

If the statistical ensemble is in an *eigenstate*, then the associated eigenvalue is always measured: the measured result is *sharp*. And conversely, if *the same value is always measured*, then it is in this state. In contrast, if the ensemble is not in an eigenstate of the measurable operator, then the measured results scatter about the average value with a non-zero uncertainty.

For dynamical variables, we may take only Hermitian operators, because only they have real eigenvalues; and measured results are real quantities. For a complex quantity, we would have to measure two numbers. As *possible measured results* for the dynamical variable A, only the eigenvalues $\{a\}$ of the assigned operator A occur. This is the physical meaning of the eigenvalues. Furthermore, the orthogonality of two eigenstates can be interpreted physically: the two states always deliver different experimental values. But note that, for degenerate states, we have to consider further properties.

If the system ensemble is in the state $|a\rangle$, the measured results for all variables $f(A)$ are fixed, namely, $f(a)$. In contrast, for all other quantities B with $[A, B] \neq 0$ in the statistical ensemble, generally different values b will be measured. If B does not commute with A, then in most cases $|a\rangle$ cannot be represented by a single eigenvector of B. Then the state $|a\rangle = \sum_b |b\rangle\langle b|a\rangle$ can only be decomposed into several eigenstates $|b\rangle$ of B. This is the physical relevance of the superposition principle.

If the system is prepared in the state $|\psi\rangle$, then generally different values for the variable A are measured—except for the case where $|\psi\rangle$ is an eigenstate of A. We consider now the average value, and in the next section the uncertainty. For the average value, we have to weight the possible measured results a with the associated probabilities $|\langle a|\psi\rangle|^2$. Since we measure the value a with the probability $|\langle a|\psi\rangle|^2$ and the value a' with the probability $|\langle a'|\psi\rangle|^2$,

$$\overline{A} = \sum_a |\langle a|\psi\rangle|^2 \, a = \sum_a \langle\psi|a\rangle \, a \, \langle a|\psi\rangle = \langle\psi|A|\psi\rangle \equiv \langle A\rangle \; .$$

Instead of the average value $\overline{A}$, we also call it the *expectation value* $\langle A\rangle$. The matrix element $\langle\psi|A|\psi\rangle$ delivers the expectation value for the observable A in the state $|\psi\rangle$. The expectation value is determined from the set of possible experimental values. For discrete eigenvalues, it may definitely differ from all possible experimental values, e.g., it may lie between the n th and $(n+1)$ th level.

In the real-space representation $\{|\mathbf{r}\rangle\}$, we have correspondingly

$$\langle\psi|A|\psi\rangle = \int d^3r \, d^3r' \, \langle\psi|\mathbf{r}\rangle\langle\mathbf{r}|A|\mathbf{r}'\rangle\langle\mathbf{r}'|\psi\rangle \; ,$$

with $\langle\mathbf{r}'|\psi\rangle = \psi(\mathbf{r}')$ and $\langle\psi|\mathbf{r}\rangle = \psi^*(\mathbf{r})$, according to Sect. 4.1.6. In most cases, we have to deal with *local operators*. These are diagonal in the real-space representation, so the double integral becomes a single one. For local operators, we thus have $\langle\psi|A|\psi\rangle = \int d^3r \, \psi^*(\mathbf{r}) A(\mathbf{r}) \psi(\mathbf{r}) = \int d^3r \, |\psi(\mathbf{r})|^2 A(\mathbf{r})$.

The general matrix element $\langle\psi|A|\varphi\rangle$ with $\psi \neq \varphi$ cannot be interpreted classically for three reasons: it depends on two states, it is a complex number, and it involves (like $\langle\psi|$ and $|\varphi\rangle$) an arbitrary phase factor. In quantum theory, we deal with the *transition amplitude from $|\varphi\rangle$ with A to $\langle\psi|$*. Note that it is important to get used to reading the expressions in quantum theory from right to left: the operator A acts on the ket-vector and only then is the probability amplitude of this new ket-vector with the bra-vector of importance. These difficulties with the classical meaning do not occur for the diagonal elements $\langle\psi|A|\psi\rangle$: because A is Hermitian, it is real, and if $|\psi\rangle$ is multiplied by $\exp(i\phi)$, then likewise $\langle\psi|$ is multiplied by $\exp(-i\phi)$.

4.2.7 Uncertainty

If the system in an eigenstate of the considered measurable quantity A, then the measured result is known sharply (with certainty) and $\Delta A = 0$. Otherwise, different experimental values occur with their corresponding probabilities. Nevertheless, the average value $\overline{A}$ of the experimental values is equal to the expectation value $\langle\psi|A|\psi\rangle$, and also $\overline{A^2} = \langle\psi|A^2|\psi\rangle$ is known. Hence, according to p. 275, the uncertainty is also known:

$$\Delta A = \sqrt{\langle\psi|A^2|\psi\rangle - \langle\psi|A|\psi\rangle^2} \; .$$

It only vanishes if $|\psi\rangle$ is an eigenstate of A. Otherwise we have $\overline{A^2} > \overline{A}^2$. In particular, if we take a basis with $|\psi\rangle$ as the first vector, then for Hermitian operators A, we have

$$\langle \psi | A^2 | \psi \rangle = \langle \psi | A | \psi \rangle^2 + \langle \psi | A | \psi' \rangle \langle \psi' | A | \psi \rangle + \cdots$$
$$= \langle \psi | A | \psi \rangle^2 + | \langle \psi | A | \psi' \rangle |^2 + \cdots .$$

If $|\psi\rangle$ is not an eigenstate of A, the first term is not the only one to contribute. We then have $\Delta A > 0$.

For the uncertainty relation, we consider two Hermitian operators A and B. Then with $|\alpha\rangle \equiv (A - \overline{A}) |\psi\rangle$ and $|\beta\rangle \equiv (B - \overline{B}) |\psi\rangle$, and because $(\Delta A)^2 = \langle \psi | (A - \overline{A})^2 | \psi \rangle = \|\alpha\|^2$ and $(\Delta B)^2 = \|\beta\|^2$, we obtain for $\|\psi\| = 1$,

$$(\Delta A)^2 \cdot (\Delta B)^2 = \|\alpha\|^2 \|\beta\|^2 \geq |\langle \alpha | \beta \rangle|^2 = |\langle \psi | (A - \overline{A})(B - \overline{B}) | \psi \rangle|^2 ,$$

where we have used Schwarz's inequality (see p. 283). For Hermitian operators C and D, according to p. 293, we now have $|\overline{CD}|^2 = \frac{1}{2}\overline{\{C, D\}}^2 + \frac{1}{2i}\overline{[C, D]}^2$. With $\overline{\{A - \overline{A}, B - \overline{B}\}} = \overline{\{A, B\}} - 2\overline{A}\,\overline{B}$ and $\overline{[A - \overline{A}, B - \overline{B}]} = \overline{[A, B]}$, it thus follows that

$$(\Delta A)^2 \cdot (\Delta B)^2 \geq \langle \psi | \tfrac{1}{2} \{A, B\} - \overline{A}\,\overline{B} | \psi \rangle^2 + \langle \psi | \tfrac{1}{2i} [A, B] | \psi \rangle^2 .$$

If the operators A and B do not commute with each other, but if their commutator is equal to the unit operator up to an imaginary constant, the last term contributes positively and the two quantities A and B cannot both be sharp.

Heisenberg's uncertainty relation for canonically conjugate quantities A and B, viz.,

$$\Delta A \cdot \Delta B \geq \tfrac{1}{2} \hbar ,$$

can thus be guaranteed with non-commuting operators. They only have to obey the requirement

$$[A, \ B] = i\hbar \, 1 .$$

According to Born and Jordan, we can require this of all canonically conjugate quantities and this is the very reason why we actually deal with operators and Hilbert vectors. In the next section, we shall point out connections with these commutation relations.

There are two conditions under which the product of the uncertainties $\Delta A \cdot \Delta B$ is as small as possible. Firstly, we must have $\frac{1}{2}\overline{AB + BA} = \overline{A}\,\overline{B}$, or $\overline{AB} - \frac{1}{2}\overline{[A, B]} = \overline{A}\,\overline{B}$, so that $\overline{(A - \overline{A}) B} = \frac{1}{2}\overline{[A, B]}$. Secondly, according to p. 283, only if the considered vectors $|\alpha\rangle$ and $|\beta\rangle$ are parallel to each other does Schwarz's inequality become an equation, i.e., if

$$(A - \overline{A}) |\psi\rangle = \lambda \, (B - \overline{B}) |\psi\rangle .$$

But then also $\frac{1}{2}\overline{[A, B]} = \langle\psi|(A - \overline{A})B|\psi\rangle = \lambda^* \overline{(B - \overline{B})B} = \lambda^* (\Delta B)^2$. Here, according to the initial equation in the considered extreme case, we have $\Delta A\, \Delta B = \pm(\frac{-\mathrm{i}}{2}\overline{[A, B]})$, where the left-hand expression is ≥ 0 and the one on the right fixes the sign. In short then, $\Delta A\, \Delta B = \mp\mathrm{i}\lambda^*(\Delta B)^2$, or

$$\lambda = \mp\mathrm{i}\,\frac{\Delta A}{\Delta B}\ .$$

For canonically conjugate quantities A and B with $[A, B] = \mathrm{i}\hbar\,1$, we have to take the upper sign, and for $\mathrm{i}\,\overline{[A, B]} > 0$, the lower sign.

4.2.8 Field Operators

Once again, we turn to the wave–particle duality and here restrict ourselves to (many) "quanta in the same state", e.g., with equal momentum. So the following considerations apply only to bosons, but not fermions, e.g., not electrons, because according to the Pauli principle only one fermion may occupy a given state. The discussion here will be useful later for the harmonic oscillator (Sect. 4.5.4), where the transition to neighboring states is always connected with an oscillation quantum of the same energy. Note that sound quanta are also called phonons, and light quanta photons.

The Dirac symbol $|n\rangle$ will now be used to indicate that there are n particles. The numbers $n \in \{0, 1, 2, \dots\}$ are the eigenvalues of the number operator N and $|n\rangle$ its eigenstates, which we shall investigate now in some detail. To this end, we introduce (non-Hermitian) creation and annihilation operators:

$$\Psi^\dagger|n\rangle \propto |n+1\rangle \iff \Psi|n\rangle \propto |n-1\rangle\ .$$

Note that, in many textbooks on quantum mechanics, and also according to the IUPAP recommendations, a or b is used instead of Ψ, which is common practice in field theory though, and indeed this operator has something to do with the state $|\psi\rangle$. $\Psi|\psi\rangle$ results in particular in the vacuum, as we shall soon show. Instead of the state $|\psi\rangle$, we may also speak of the *field* $|\psi\rangle$, if we think of its real-space representation $\langle\mathbf{r}\,|\psi\rangle \equiv \psi(\mathbf{r})$. Since negative eigenvalues n may not occur, $\Psi|0\rangle$ has to deliver the zero vector $|o\rangle$. Note, however, that $|0\rangle$ is not the zero vector $|o\rangle$, but the state with $n = 0$. If n gives the number of "particles", then $|0\rangle$ is the state without particles, the "vacuum", for which $\langle 0|0\rangle = 1$, in contrast to $\langle o|o\rangle = 0$.

Both $\Psi\Psi^\dagger$ and $\Psi^\dagger\Psi$ therefore have the eigenvectors $|n\rangle$. We now require

$$N = \Psi^\dagger\Psi\ .$$

Hence, due to the normalization, it follows from

$$n = \langle n|N|n\rangle = \langle n|\Psi^\dagger\Psi|n\rangle \propto \langle n-1|n-1\rangle\ , \quad \text{for} \quad n > 0\ ,$$

that

$$\Psi\,|n\rangle = |n-1\rangle\,\sqrt{n} \quad\Longleftrightarrow\quad \Psi^{\dagger}\,|n\rangle = |n+1\rangle\,\sqrt{n+1}\,,$$

if we choose the phase factor (arbitrarily) equal to 1. The operator Ψ thus reduces the particle number by one, and is therefore called the *annihilation operator*, while the adjoint operator $\Psi^{\dagger}$ increases it by one and is therefore called the *creation operator*. This leads to

$$|n\rangle = \frac{1}{\sqrt{n!}}\,(\Psi^{\dagger})^{n}\,|0\rangle\,,$$

i.e., all states can be created with this from the "vacuum state" $|0\rangle$. It is special insofar as the annihilation operator Ψ maps only this to the zero vector $|o\rangle$. We have $\Psi^{\dagger}\Psi\,|0\rangle = |o\rangle$, but $\Psi\Psi^{\dagger}\,|0\rangle = |0\rangle$, and generally, $\Psi^{\dagger}\Psi|n\rangle = |n\rangle n$ as well as $\Psi\Psi^{\dagger}|n\rangle = |n\rangle(n+1)$, for all natural numbers n. Hence, we arrive at the basic commutation relation

$$[\Psi,\ \Psi^{\dagger}] \equiv \Psi\Psi^{\dagger} - \Psi^{\dagger}\Psi = 1\,.$$

Thus $\Psi\Psi^{\dagger} = 1 + N$ holds, and we obtain from $\Psi^{\dagger}\Psi\Psi^{\dagger} = \Psi^{\dagger}\,(1+N)$, or the adjoint $\Psi\Psi^{\dagger}\Psi = (1+N)\,\Psi$,

$$[N,\ \Psi^{\dagger}] = \Psi^{\dagger}\,,\qquad [N,\ \Psi] = -\Psi\,.$$

Conversely, from $[\Psi,\ \Psi^{\dagger}] = 1$, we can derive the real eigenvalue spectrum of $\Psi^{\dagger}\Psi$ and the matrix elements of Ψ and $\Psi^{\dagger}$ in the eigenbasis of this Hermitian operator, for an appropriate phase convention. In particular, from $\Psi^{\dagger}\Psi\Psi = (\Psi\Psi^{\dagger} - [\Psi,\ \Psi^{\dagger}])\Psi = \Psi(\Psi^{\dagger}\Psi - 1)$, we conclude that the operator Ψ creates more eigenvectors of $\Psi^{\dagger}\Psi$ from eigenvectors of $\Psi^{\dagger}\Psi$, but with an eigenvalue that is reduced by one. On the other hand, this decrease finally has to lead to the zero vector, and therefore to an end, since $\langle\dots|\Psi^{\dagger}\Psi|\dots\rangle$, being the square of the norm of the Hilbert vector $\Psi|\dots\rangle$, may not become negative and yet is still equal to one of the possible eigenvalues of $\Psi^{\dagger}\Psi$. Hence $\Psi^{\dagger}\Psi$ has the natural numbers as eigenvalues, and we choose the phases such that $\Psi^{\dagger}|n\rangle = |n+1\rangle\sqrt{n+1} \Longleftrightarrow \Psi|n\rangle = |n-1\rangle\sqrt{n}$ holds.

Using the field operators, we can expand the projection operator $|0\rangle\langle 0|$:

$$|0\rangle\langle 0| = \sum_{n} \frac{(-)^{n}}{n!}\,\Psi^{\dagger n}\,\Psi^{n}\,,$$

since, for all natural numbers m, $\langle m|0\rangle\langle 0|m\rangle = \delta_{m0} = (1-1)^{m} = \sum_{n}(-)^{n}\binom{m}{n}$ and $\langle m|\Psi^{\dagger n}\,\Psi^{n}|m\rangle = n!\binom{m}{n}$, and the operator is diagonal in the basis $\{|m\rangle\}$.

Even though the operators $\Psi\Psi^\dagger$ and $\Psi^\dagger\Psi$ have discrete eigenvalues, there are infinitely many of them. Hence the associated basis is of infinite dimension and the traces of both operators diverge. Only then can $\mathrm{tr}(\Psi\Psi^\dagger) = \mathrm{tr}(\Psi^\dagger\Psi)$ hold on the one hand and $\Psi\Psi^\dagger - \Psi^\dagger\Psi = 1$ on the other, whence $\mathrm{tr}[\Psi, \Psi^\dagger] \neq 0$ is valid.

Even if we reject very large eigenvalues $n \gg 1$ as unphysical, a finite basis does of course exist. To investigate this possibility more closely (we shall use it in the next section, but only there), we introduce an upper limit s and require $n \in \{0, \dots, s\}$. Here, s is then assumed large, so that the finite basis comprises all physically necessary states. However, we can then no longer require $[\Psi, \Psi^\dagger] = 1$, since for a finite basis, the trace of the commutators must vanish. But according to Pegg and Barnett [4], there are operators (we shall call them $\widetilde{\Psi}$), which act on physical states like Ψ and nevertheless need only a finite basis:

$$\widetilde{\Psi} = \sum_{n=1}^{s} |n-1\rangle \sqrt{n} \langle n| \quad \Longleftrightarrow \quad \widetilde{\Psi}^\dagger = \sum_{n=1}^{s} |n\rangle \sqrt{n} \langle n-1| .$$

With the finite sum and with $1 = \sum_{n=0}^{s} |n\rangle\langle n|$, we now obtain

$$[\widetilde{\Psi}, \widetilde{\Psi}^\dagger] = 1 - |s\rangle (s+1) \langle s| .$$

The new term ensures that $\mathrm{tr}[\widetilde{\Psi}, \widetilde{\Psi}^\dagger] = 0$, as is appropriate for a finite basis.

Before we make any further use of the field operators for bosons, let us make here a brief mention of the field operators for fermions, even though we shall treat these in more detail in Sect. 4.2.10. We use once again $N = \Psi^\dagger\Psi$, but N will only have the eigenvalues 0 and 1, as required by the *Pauli principle*, and Ψ^2 and hence also $(\Psi^\dagger)^2$ will always be zero. We write the two states as column vectors, with $|0\rangle$ as $\binom{0}{1}$ and $|1\rangle$ as $\binom{1}{0}$, for the number to increase upwards (conversely, for bosons, the state $|n\rangle$ is a column vector with just zeros and a 1 in the nth row). Then all these requirements can be satisfied with

$$N \cong \begin{pmatrix} 1 & 0 \\ 0 & 0 \end{pmatrix}, \qquad \Psi \cong \begin{pmatrix} 0 & 0 \\ 1 & 0 \end{pmatrix},$$

and consequently

$$\Psi^\dagger \cong \begin{pmatrix} 0 & 1 \\ 0 & 0 \end{pmatrix}, \qquad \Psi\Psi^\dagger \cong \begin{pmatrix} 0 & 0 \\ 0 & 1 \end{pmatrix}.$$

For fermions, it is thus the anti-commutator of Ψ and $\Psi^\dagger$ which is equal to 1:

$$\Psi\Psi^\dagger + \Psi^\dagger\Psi = 1 ,$$

and $\Psi^\dagger |0\rangle = |1\rangle$, $\Psi^\dagger |1\rangle = |o\rangle = \Psi |0\rangle$, $\Psi |1\rangle = |0\rangle$. We often find 0 written here instead of $|o\rangle$, even though $A|\psi\rangle$ is a Hilbert vector.

4.2.9 Phase Operators and Wave–Particle Dualism

The natural numbers as eigenvalues fit into the particle picture. But because of the necessary interference, we need an uncertain particle number for wave–particle duality, thus a superposition of different states $|n\rangle$. Then the initial equations $[\Psi, \Psi^\dagger] = 1$ and $N = \Psi^\dagger\Psi$ are still valid further on, but now the phase factors are also important in the wave picture for the superposition of different states $|n\rangle$.

The appropriate determination of the phase operators was long a subject of research. Dirac was himself occupied with this in 1927. Only Pegg and Barnett (see the last section) succeeded in solving the problem: the basis must not be infinite, but only of unmeasurably high dimension. Let us discuss this now, but simply set the phase of the vacuum equal to zero, not leaving it open.

The phases ϕ are unique only between 0 and 2π. In order not to introduce a continuous basis (with improper Hilbert vectors), we set

$$\phi_m = \frac{2\pi}{s+1}\, m\,, \quad \text{with} \quad m \in \{0,\ \ldots s\}\,.$$

Here (as in the last section), we take a very large limit s, but nevertheless finite, since the phase (like any continuous quantity) cannot be measured with arbitrary accuracy. We also introduce a Hermitian operator Φ with eigenvalues ϕ_m, such that $\Phi|\phi_m\rangle = |\phi_m\rangle\,\phi_m$. It is important to show that the states $m = 0$ and $m = s$ are neighboring states. Hence, initially, we search for the unitary operator $E = \exp(\mathrm{i}\Phi)$ with the property $E\,|\phi_m\rangle = |\phi_m\rangle\,\exp(\mathrm{i}\phi_m)$ for $m \in \{0,\ \ldots s\}$ (see Fig. 4.3).

The basis $\{|\phi_m\rangle\}$ is assumed orthonormal and complete. Then we have

$$E = \sum_{m=0}^{s} |\phi_m\rangle\,\exp(\mathrm{i}\phi_m)\,\langle\phi_m| \quad \text{with} \quad \langle\phi_m|\phi_{m'}\rangle = \delta_{mm'}\,.$$

It should be stressed here that s is assumed very large, even though we leave out $\lim_{s\to\infty}$ in front of the sum. In contrast to the last section, however, we may anticipate

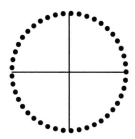

Fig. 4.3 Eigenvalues of the operator $E = \exp(\mathrm{i}\Phi)$ with $\Phi = \Phi^\dagger$. These are evenly distributed over the unit circle in the complex plane. Here $s = 44$ has been chosen. It could be much larger. The only requirement is that it should be finite

that all states $|\phi_m\rangle$ can be important physically, those with $m \approx s$ as much as those with $m \approx 0$, while we shall not take the particle number to be arbitrarily large.

We now relate phase and particle number states. The wave–particle duality allows a sharp phase only for an uncertain particle number, and conversely a sharp phase only for an uncertain particle number. Here we simultaneously require the expansion bases $\{|n\rangle\}$ and $\{|\phi_m\rangle\}$, and use

$$|\phi_m\rangle = \sum_{n=0}^{s} |n\rangle\langle n|\phi_m\rangle \qquad \text{and} \qquad |n\rangle = \sum_{m=0}^{s} |\phi_m\rangle\langle\phi_m|n\rangle ,$$

with $\langle\phi_m|n\rangle = \langle n|\phi_m\rangle^*$. Here the same limit s was deliberately chosen for both expansions, since the last equations are then fully valid—approximations were made previously, in particular, with discrete phases instead of continuous ones and with a finite number of particles. If s is sufficiently large, these assumptions are probably justified.

As the eigenvalues show, E is unitary ($EE^\dagger = E^\dagger E = 1$). With $\widetilde{\Psi}^\dagger\widetilde{\Psi} = N$ and the known decomposition into amplitude and phase factor, we set

$$\widetilde{\Psi} = E\sqrt{N} \qquad \Longleftrightarrow \qquad \widetilde{\Psi}^\dagger = \sqrt{N}\,E^\dagger .$$

Hence, $\widetilde{\Psi} = \sum_{n=1}^{s} |n-1\rangle\sqrt{n}\,\langle n|$ implies that

$$E = \sum_{n=1}^{s} |n-1\rangle\langle n| + |s\rangle\langle 0| .$$

The last term results from the unitarity of E (where we have chosen the phase factor equal to unity). Consequently, we have $\langle n|E = \langle n+1|$, for $0 \le n < s$, and $\langle s|E = \langle 0|$. Hence the eigenvalue equation of E delivers the recursion formula $\langle n+1|\phi_m\rangle = \langle n|E|\phi_m\rangle = \langle n|\phi_m\rangle \exp(i\phi_m)$. If we choose the phase of the vacuum state $|0\rangle$ (arbitrarily) equal to zero, we find $\langle n|\phi_m\rangle = \exp(in\phi_m)/\sqrt{s+1}$ as a solution of the recursion formula, where the normalization factor results from $1 = \langle\phi_m|\phi_m\rangle = \sum_{n=0}^{s}\langle\phi_m|n\rangle\langle n|\phi_m\rangle$.

Hence in the basis $\{|n\rangle\}$, the three matrices N, Ψ (or $\widetilde{\Psi}$), and E read

$$N \cong \begin{pmatrix} 0 & 0 & 0 & \cdots \\ 0 & 1 & 0 & \cdots \\ 0 & 0 & 2 & \cdots \\ \vdots & \vdots & \vdots & \ddots \end{pmatrix}, \; \Psi \cong \begin{pmatrix} 0 & \sqrt{1} & 0 & \cdots \\ 0 & 0 & \sqrt{2} & \cdots \\ 0 & 0 & 0 & \cdots \\ \vdots & \vdots & \vdots & \ddots \end{pmatrix}, \; E \cong \begin{pmatrix} 0 & 1 & 0 & \cdots \\ 0 & 0 & 1 & \cdots \\ 0 & 0 & 0 & \cdots \\ \vdots & \vdots & \vdots & \ddots \\ \underline{1} \end{pmatrix} .$$

The element $\underline{1}$ in the matrix E stands at the end of the first column—then $E = \exp(i\Phi)$ is unitary and Φ cyclic.

From the expression for E in the particle number representation, we have the commutation relation

$$[E, N] = E - |s\rangle \, (s + 1) \, \langle 0| \, ,$$

and also $[E^\dagger, N] = -[E, N]^\dagger = -E^\dagger + |0\rangle \, (s + 1) \, \langle s|$, along with $[\widetilde{\Psi}, \widetilde{\Psi}^\dagger] = 1 - |s\rangle (s + 1) \langle s|$. We now decompose the unitary operator E like $\exp(i\phi)$, using Euler's formula to obtain

$$E \equiv C + iS \, , \quad \text{with} \quad C = \frac{E + E^\dagger}{2} = C^\dagger \, , \quad \text{and} \quad S = \frac{E - E^\dagger}{2i} = S^\dagger \, ,$$

and find

$$[C, N] = +iS + \frac{s+1}{2} \, (|0\rangle \langle s| - |s\rangle \langle 0|) \, ,$$

$$[S, N] = -iC + i \frac{s+1}{2} \, (|0\rangle \langle s| + |s\rangle \langle 0|) \, ,$$

for these Hermitian operators, along with $C^2 + S^2 = \frac{1}{4}\big((E + E^\dagger)^2 - (E - E^\dagger)^2\big) = \frac{1}{2}(E E^\dagger + E^\dagger E) = 1$. Note that, we have $[C, S] = 0$, because $[E, E^\dagger] = 0$.

According to Sect. 4.2.7, we may now derive an uncertainty relation between particle number and phase. In particular, we make use of $\Delta A \cdot \Delta B \geq \frac{1}{2} \, |\overline{[A, B]}|$ for $A = A^\dagger$ and $B = B^\dagger$. Since not all physical states overlap with the state $|s\rangle$, we obtain initially

$$\Delta C \cdot \Delta N \geq \tfrac{1}{2} \, |\overline{S}| \quad \text{and} \quad \Delta S \cdot \Delta N \geq \tfrac{1}{2} \, |\overline{C}| \, ,$$

along with $\Delta C \cdot \Delta S \geq 0$.

If we now associate the phase operator Φ with the unitary operator E, according to p. 293, viz.,

$$E = \exp(i\Phi) \, , \quad \text{with} \quad \Phi = \Phi^\dagger \, ,$$

then, for small phase uncertainty $\Delta\Phi \ll \pi$,

$$\overline{C} \approx \cos\overline{\Phi} \, , \qquad \Delta C \approx |\sin\overline{\Phi}| \, \Delta\Phi \, ,$$
$$\overline{S} \approx \sin\overline{\Phi} \, , \qquad \Delta S \approx |\cos\overline{\Phi}| \, \Delta\Phi \, .$$

Hence the above-mentioned uncertainty relations deliver the inequality $\Delta N \cdot \Delta\Phi \geq \frac{1}{2}$ already announced in Sect. 4.1.3. However, this is not generally valid. If, for example, all phases between 0 and 2π are equally probable (the phase uncertainty thus being as large as possible), then $\overline{C}$ and $\overline{S}$ are both zero. Then $\Delta N = 0$ may hold, even though we have $\Delta\Phi = \pi/\sqrt{3}$. Note that the phase uncertainty depends on the reference phase—this is connected inextricably with the periodicity. In particular, if two neighbouring phase states ϕ_m and ϕ_{m+1} are occupied with equal probability, then for $m < s$, the uncertainty for $s \to \infty$ is negligible, while for $m = s$, it is equal to π. The reference phase is then chosen so that $\Delta\Phi$ becomes as small as possible, or so

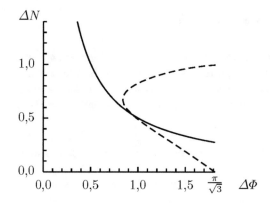

Fig. 4.4 Uncertainty relation between particle number and phase. If all phases are equally probable, then the phase uncertainty is $\Delta\Phi = \pi/\sqrt{3}$ and never greater. The *continuous curve* shows $\Delta N \cdot \Delta\Phi = 1/2$, thus the approximation for $\Delta\Phi \approx 0$. The *dashed curve* shows the approximation for $\Delta N \approx 0$. The two approximations complement each other quite well—only for $\Delta N \approx 1/2$, or $\Delta\Phi \approx 1$, do they differ somewhat from the true curve, and this can be seen only if the image is enlarged. See Phys.Lett. A 218 (1996) 1

that $\overline{\Phi} \approx \pi$. For $\Delta N \approx 0$, it is better to take three neighboring states in the particle number representation with the amplitudes

$$\langle \overline{n} | \psi \rangle = \sqrt{1 - (\Delta N)^2} \quad \text{and} \quad \langle \overline{n} \pm 1 | \psi \rangle = \frac{\exp(\pm i\overline{\phi})}{\sqrt{2}} \Delta N \, ,$$

since, after a Fourier transform, we obtain a phase uncertainty that is as small as possible:

$$(\Delta\Phi)^2 \geq \tfrac{1}{3}\pi^2 - 4\sqrt{2}\, \Delta N \sqrt{1 - (\Delta N)^2} + \tfrac{1}{2} (\Delta N)^2 \, .$$

Here, when calculating $\langle \Phi \rangle$ and $\langle \Phi^2 \rangle$, we replace the sums over ϕ_m by integrals and $\langle \phi | \psi \rangle$ by $(2\pi)^{-1/2} \sum_n \exp(-in\phi) \langle n | \psi \rangle$. The exact limit is shown in Fig. 4.4. It is rather well described by either approximation.

Hence now we understand that the particle and wave pictures—granularity and the capacity to interfere—are not in contradiction if we take into account the uncertainties in particle number and phase.

In the rest of this chapter (Quantum Mechanics I), we will consider only one-particle states (as representative of a statistical ensemble of bosons or fermions), but now these particles will no longer be restricted to a single state.

4.2.10   Doublets and Pauli Operators

The two-dimensional vector space is highly instructive and full of possibilities for applications. It is needed for the spin states of fermions with spin 1/2 (e.g., for electrons), for isospin (neutron and proton states as the two states of the nucleon), and also for the Pauli principle and model calculations of the excitation of atoms (Sect. 5.5.7).

If we call the two states $|\uparrow\rangle$ and $|\downarrow\rangle$ (up and down), then, according to Sect. 4.2.8, the umklapp operator Ψ can be introduced for this system with the property

$$\Psi\Psi^\dagger + \Psi^\dagger\Psi = 1 \; .$$

We now write $|\downarrow\rangle$ instead of $|0\rangle$ and $|\uparrow\rangle$ instead of $|1\rangle$, because up and down are easier to remember than the position of 0 and 1. If we consider these states as column vectors, then with $\Psi^\dagger\,|\downarrow\rangle = |\uparrow\rangle$, $\Psi^\dagger\,|\uparrow\rangle = |o\rangle = \Psi\,|\downarrow\rangle$, and $\Psi\,|\uparrow\rangle = |\downarrow\rangle$, we have

$$\Psi \,\widehat{=}\, \begin{pmatrix} 0 & 0 \\ 1 & 0 \end{pmatrix}, \qquad\qquad \Psi^\dagger \,\widehat{=}\, \begin{pmatrix} 0 & 1 \\ 0 & 0 \end{pmatrix},$$
$$\Psi\Psi^\dagger \,\widehat{=}\, \begin{pmatrix} 0 & 0 \\ 0 & 1 \end{pmatrix}, \qquad\qquad \Psi^\dagger\Psi \,\widehat{=}\, \begin{pmatrix} 1 & 0 \\ 0 & 0 \end{pmatrix}.$$

All other 2×2 matrices can be expressed as linear combinations of these. However, we prefer to have Hermitian matrices as a basis, including among them the unit matrix:

$$C_0 = \Psi^\dagger\Psi + \Psi\Psi^\dagger = 1 \,\widehat{=}\, \begin{pmatrix} 1 & 0 \\ 0 & 1 \end{pmatrix},$$
$$C_1 = \quad \Psi + \Psi^\dagger \quad = \sigma_x \,\widehat{=}\, \begin{pmatrix} 0 & 1 \\ 1 & 0 \end{pmatrix},$$
$$C_2 = \mathrm{i}\,(\Psi - \Psi^\dagger) \; = \sigma_y \,\widehat{=}\, \begin{pmatrix} 0 & -\mathrm{i} \\ \mathrm{i} & 0 \end{pmatrix},$$
$$C_3 = \Psi^\dagger\Psi - \Psi\Psi^\dagger = \sigma_z \,\widehat{=}\, \begin{pmatrix} 1 & 0 \\ 0 & -1 \end{pmatrix}.$$

The notation C_n is taken from Sect. 4.2.5, but the notation with 1 and the *Pauli operator* σ is more often used, where $\sigma_\pm = \frac{1}{2}\,(\sigma_x \pm \mathrm{i}\sigma_y)$ is introduced. Clearly, we also have

$$\Psi = \frac{\sigma_x - \mathrm{i}\,\sigma_y}{2} = \sigma_- \,, \qquad\qquad \Psi^\dagger = \frac{\sigma_x + \mathrm{i}\,\sigma_y}{2} = \sigma_+ \,,$$
$$\Psi\Psi^\dagger = \frac{1 - \sigma_z}{2} \,, \qquad\qquad\quad \Psi^\dagger\Psi = \frac{1 + \sigma_z}{2} \,.$$

The operators of the new basis $\{C_n\}$ are not only Hermitian, but also unitary, this resulting from the necessary normalization in $\mathrm{tr}(C_n{}^\dagger C_{n'}) = \delta_{nn'}\,\mathrm{tr}\,1 = 2\,\delta_{nn'}$, since their squares are equal to the unit operator:

$$C_n = C_n^\dagger = C_n^{-1} , \quad \mathrm{tr} C_n = 2\, \delta_{n0} .$$

Hence, their eigenvalues are real and of absolute value 1. They result from the last equation: C_0 has the two-fold eigenvalue 1, while the other three operators each have the eigenvalues $+1$ and -1. In addition, these three do not commute with each other, but anti-commute. Matrix multiplication delivers $C_1 C_2 = i\, C_3$, and because $C_n^\dagger = C_n$, we thus have $C_2 C_1 = -i\, C_3 = -C_1 C_2$. Cyclic permutation of the indices 1, 2, 3 is allowed, since $C_2 C_3 = C_2 (i C_2 C_1) = i\, C_1 = i\, C_1 C_2 C_2 = -C_3 C_2$, and so on:

$$C_1 C_2 = i\, C_3 = -C_2 C_1 , \quad \text{or} \quad [C_1, C_2] = 2i\, C_3 \quad \text{and cyclic permutations.}$$

Here the notation with the Pauli operator $\boldsymbol{\sigma}$ proves useful because the commutator can then be written as a vector product:

$$\boldsymbol{\sigma} \times \boldsymbol{\sigma} = 2i\, \boldsymbol{\sigma} .$$

The vector product of the Pauli operator $\boldsymbol{\sigma}$ with itself does not vanish, in contrast to what happens with classical vectors, because its components do not commute with each other. Hence apart from 1, only one further component can be diagonalized—in our example, this is $\sigma_z = C_3$. But it could also be σ_x or σ_y. Here only a rotation (unitary transformation) would be necessary, but note that σ_z would then no longer be diagonal.

The four operators are orthogonal to each other:

$$\mathrm{tr}(C_n C_{n'}) = 2\, \delta_{nn'} .$$

Here we recognize why in Sect. 4.2.5 the normalization of the basis operators was left open. With this orthonormalization, according to p. 297, all 2×2-matrices A can be written in the form

$$A = \frac{1 \,\mathrm{tr} A + \boldsymbol{\sigma} \cdot \mathrm{tr}(\boldsymbol{\sigma} A)}{2} .$$

Their eigenvalues follow from $\det (A - a1) = 0$, hence $a^2 - a\,\mathrm{tr} A + \det A = 0$, which implies

$$a_\pm = \frac{\mathrm{tr} A \pm \sqrt{(\mathrm{tr} A)^2 - 4 \det A}}{2} .$$

If we expand the eigenvectors $|\pm\rangle$ in terms of the other basis $\{|\uparrow\rangle, |\downarrow\rangle\}$, then from $A|\pm\rangle = |\pm\rangle\, a_\pm$, we obtain the homogeneous system of equations

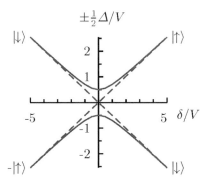

Fig. 4.5 Level repulsion. For fixed interaction $V = 2\langle\downarrow|A|\uparrow\rangle$, the splitting of the eigenvalues $a_\pm$ (i.e., $a_+ - a_-$) is shown here (*red*) as a function of the unperturbed level distance $\delta = \langle\uparrow|A|\uparrow\rangle - \langle\downarrow|A|\downarrow\rangle$. Here the state $|+\rangle$ goes from $|\downarrow\rangle$ to $|\uparrow\rangle$, the state $|-\rangle$ from $-|\uparrow\rangle$ to $|\downarrow\rangle$. Without *anti-crossing*, the *dashed blue lines* would be valid

$$\langle\uparrow|\ A - a_\pm\ |\uparrow\rangle\ \langle\uparrow|\pm\rangle\ + \langle\uparrow|\qquad A\qquad |\downarrow\rangle\ \langle\downarrow|\pm\rangle\ = 0\,,$$
$$\langle\downarrow|\qquad A\qquad |\uparrow\rangle\ \langle\uparrow|\pm\rangle\ + \langle\downarrow|\ A - a_\pm\ |\downarrow\rangle\ \langle\downarrow|\pm\rangle\ = 0\,.$$

This fixes the expansion coefficients only up to a common factor, but because of the normalization condition $|\langle\uparrow|\pm\rangle|^2 + |\langle\downarrow|\pm\rangle|^2 = 1$, only a common phase factor remains open. If (for $A = A^\dagger$), we set

$$\begin{pmatrix}|+\rangle\\|-\rangle\end{pmatrix} = \begin{pmatrix}\cos\alpha & e^{i\beta}\sin\alpha\\-e^{-i\beta}\sin\alpha & \cos\alpha\end{pmatrix}\begin{pmatrix}|\uparrow\rangle\\|\downarrow\rangle\end{pmatrix}\,,$$

with real parameters $0 \le \alpha \le \frac{1}{2}\pi$ and $0 \le \beta < 2\pi$, according to p. 293, and using the abbreviations (for a Hermitian operator A they are real)

$$\delta = \langle\uparrow|A|\uparrow\rangle - \langle\downarrow|A|\downarrow\rangle\qquad\text{and}\qquad \Delta = a_+ - a_- \ge |\delta|\,,$$

since $\Delta = \sqrt{\delta^2 + 4|\langle\downarrow|A|\uparrow\rangle|^2}$, we obtain the equation

$$\exp(i\beta)\,\tan\alpha = \frac{2\,\langle\downarrow|A|\uparrow\rangle}{\Delta + \delta}\,.$$

The phase of $\langle\downarrow|A|\uparrow\rangle$ is thus equal to β and that of $\langle\uparrow|A|\downarrow\rangle$ is equal to $-\beta$, while δ and Δ determine the parameter α:

$$\cos\alpha = \sqrt{\frac{\Delta + \delta}{2\Delta}}\qquad\text{and}\qquad \sin\alpha = \sqrt{\frac{\Delta - \delta}{2\Delta}}\,.$$

If A is the Hamilton operator H, then, with $\Delta \ge |\delta|$, we speak of *level repulsion* or *anti-crossing*. Once the off-diagonal element $\langle\downarrow|A|\uparrow\rangle$ contributes, the separation between the eigenvalues increases (see Fig. 4.5).

4.2.11 Density Operator. Pure States and Mixtures

The properties of a given statistical ensemble can be determined by appropriate measurements. They deliver the expectation values of the corresponding Hermitian operators A, and hence we arrive at conclusions relating to the state of the ensemble.

So far we have dealt only with pure states $|\psi\rangle$. Then we have

$$\langle A \rangle = \langle \psi | A | \psi \rangle = \sum_{nn'} \langle \psi | n' \rangle \langle n' | A | n \rangle \langle n | \psi \rangle ,$$

if we assume a countable basis, otherwise there is a double integral instead of the double sum. The expression on the right can be simplified to $\sum_n \langle n | \psi \rangle \langle \psi | A | n \rangle$. Hence we may also write $\langle A \rangle = \mathrm{tr}(P_\psi A)$, where $P_\psi = |\psi\rangle\langle\psi|$ was introduced on p. 291 as the projection operator acting on the (pure) state $|\psi\rangle$.

A finite number of measurable quantities suffices to determine the given statistical ensemble uniquely. An ensemble of experimental values $\{\langle A_k \rangle\}$ will describe our object. For example, in Sect. 4.1.4, we took the ensemble of electrons with momentum $\langle \mathbf{P} \rangle$ and spin polarization $\langle \mathbf{S} \rangle$. But then the statistical ensemble does not need to form a pure state $|\psi\rangle$. It may also be a mixture thereof, thus an incoherent superposition of pure states $|n\rangle$ (or projectors $|n\rangle\langle n|$) with probabilities ρ_n. Hence, instead of P_ψ, we now take the general density operator

$$\rho = \sum_n |n\rangle\, \rho_n\, \langle n| \quad \Longleftrightarrow \quad \langle A \rangle = \sum_n \rho_n\, \langle n | A | n \rangle = \mathrm{tr}(\rho A) .$$

The ensemble of experimental values $\{\langle A_k \rangle\}$ fixes the density operator ρ, since the matrix elements of A follow from the relevance of this operator (position, momentum, energy, etc.) and hence $\{\langle A_k \rangle\} = \{\mathrm{tr}(\rho A_k)\}$ is an inhomogeneous linear system of equations for the matrix elements of ρ. Here the density operator describes the given system and the Hermitian operators A_k the observables.

The properties of ρ compiled in the following are valid for pure as well as for mixed states (as shown below, the two kinds of states can be distinguished by the easily verifiable attributes of the density operator). We want to fix ρ only by $\{\langle A_k \rangle\} = \{\mathrm{tr}(\rho A_k)\}$ and make use of known properties of the observables.

The density operator is a matrix of finite dimension, determined by a *finite* number of experimental values. Hence the operators commute in $\mathrm{tr}(\rho A)$.

All Hermitian operators A have real expectation values. Hence also the density operator is Hermitian: $\langle A \rangle - \langle A \rangle^* = \mathrm{tr}\{(\rho - \rho^\dagger) A\}$ must always vanish. In addition, all observables with only positive eigenvalues (so-called *positive-definite* operators) have positive expectation values, so the density matrix has to be positive-semidefinite—none of the diagonal elements of ρ can be negative in any representation. Since the unit operator always has unit expectation value, the trace of ρ must be equal to 1. Thus we can list a total of three requirements, viz.,

$$\rho = \rho^\dagger , \quad \langle n| \rho |n\rangle \geq 0 , \quad \mathrm{tr}\rho = 1 ,$$

and this actually results in the fact that ρ is positive-semidefinite, with $\rho = \rho^\dagger$. Here the diagonal element $\langle n|\rho|n\rangle$ gives the probability (or probability density) for the state $|n\rangle$, while the last equation corresponds to our normalization condition for the probabilities. The off-diagonal elements lead to interference and are occasionally referred to as the *coherences* of the system.

Under unitary transformations all operators change, including the density operator, according to the prescription $A' = UAU^\dagger$. But here the expectation values remain the same, because with the finite dimension of ρ, the trace of $\rho'A' = U\rho AU^\dagger$ remains constant, according to p. 294.

For a pure state, we have $\rho = \rho^2$, but not for a mixture. In the eigen-representation of ρ, ρ^2 is also diagonal, and for a pure state only one of these diagonal elements is different from zero (namely 1), but for a mixture at least two are different from zero—and these are then smaller for ρ^2 than for ρ. With $\mathrm{tr}\rho = 1$, it thus follows that

$$\mathrm{tr}\rho^2 = 1 \text{for all pure states,}$$
$$\mathrm{tr}\rho^2 < 1 \text{for all mixtures.}$$

With the trace of ρ^2, we thus have a very simply test of whether we are dealing with a pure state or a mixture, since for the trace, we do not need to search for the eigen-representation, because the diagonal elements suffice.

In particular, for a two-level system with $\mathrm{tr}\rho = 1$ and because $\mathrm{tr}(\sigma\rho) = \langle\sigma\rangle$, according to the last section, we have

$$\rho = \frac{1 + \sigma \cdot \langle\sigma\rangle}{2} \quad \text{and} \quad \mathrm{tr}\rho^2 = \frac{1 + \langle\sigma\rangle \cdot \langle\sigma\rangle}{2} .$$

The quantity $\langle\sigma\rangle$ is called the *polarization*. Since the eigenvalues of the components of σ are equal to ± 1, we have $|\langle\sigma\rangle| \leq 1$. If the equality sign holds here, then we have a pure state, otherwise a mixture, e.g., for an unpolarized state $\langle\sigma\rangle = \mathbf{0}$: unpolarized electrons form a mixture, their two spin states being incoherently superposed.

For an N-state system, the density matrix has N^2 elements, which are determined by equally many real numbers because $\rho^\dagger = \rho$. One of them is known already due to the normalization. Thus $N^2 - 1$ experimental values suffice for this system. In contrast, we could fix a pure state with just $2N - 2$ real numbers, or N complex numbers, but where two real numbers are omitted because of the normalization and the arbitrary common phase. For the density operator, there is no arbitrariness in the phase—its bearing on the bra- and ket-vector cancels.

The smaller $\mathrm{tr}\rho^2$, the less pure the N-state system appears, and $\mathrm{tr}\rho^2$ is smallest when all eigenvalues are equal, in which case it is a *complete mixture*, and then ρ is a multiple of the unit operator, in particular, with the eigenvalues N^{-1}, since $\mathrm{tr}\rho = 1$. Hence we have upper and lower bounds for $\mathrm{tr}\rho^2$:

$$\frac{1}{N} \leq \mathrm{tr}\rho^2 \leq 1 .$$

These only depend on the dimension N of the Hilbert space.

Let us consider these properties for the operator basis $\{C_n\}$ from Sect. 4.2.5. For Hermitian basis operators, the expansion coefficients are real, and we have

$$\rho = \frac{1}{c} \sum_{n=0}^{N^2-1} C_n \langle C_n \rangle \quad \text{and} \quad \mathrm{tr}\rho^2 = \frac{1}{c} \sum_{n=0}^{N^2-1} \langle C_n \rangle^2 .$$

If C_0 is a multiple of the unit operator, then the old requirement $\mathrm{tr}C_0{}^2 = c$ in the N-dimensional Hilbert space leads us to $C_0 = \sqrt{c/N}\, \mathbf{1}$, and hence with $\mathrm{tr}\rho = 1$ to $\langle C_0 \rangle = \mathrm{tr}\rho\, C_0 = \sqrt{c/N}$. Then only the remaining $N^2 - 1$ expectation values $\langle C_n \rangle$ are important, and these can be taken as components of a vector, usually called the *Bloch vector* (more on that in Sect. 4.4.3). The square of its length is

$$\sum_{n=1}^{N^2-1} \langle C_n \rangle^2 = c \left(\mathrm{tr}\rho^2 - \frac{1}{N} \right) ,$$

thus zero for complete mixtures and greatest for pure states, when it is equal to $c\,(1 - N^{-1})$.

4.2.12 Space Inversion and Time Reversal

With a space inversion $\mathscr{P}$, the space directions are reversed, and with a time reversal $\mathscr{T}$, only motions are reversed:

$$\begin{aligned}
\mathscr{P}\, \mathbf{R}\, \mathscr{P}^{-1} &= -\mathbf{R} , & \mathscr{T}\, \mathbf{R}\, \mathscr{T}^{-1} &= +\mathbf{R} , \\
\mathscr{P}\, \mathbf{P}\, \mathscr{P}^{-1} &= -\mathbf{P} , & \mathscr{T}\, \mathbf{P}\, \mathscr{T}^{-1} &= -\mathbf{P} .
\end{aligned}$$

The space inversion is a unitary transformation, but not the time reversal, since unitary transformations do not change algebraic relations between operators—however, $\mathscr{T}[X, P]\, \mathscr{T}^{-1} = [\mathscr{T}X\mathscr{T}^{-1}, \mathscr{T}P\mathscr{T}^{-1}]$ is equal to $-[X, P]$. This can only be inserted into the previous context without contradiction if $\mathscr{T}$ is an *anti-linear* operator, thus changing all numbers into their complex conjugates, and hence $\mathscr{T}\,(i\hbar\,\mathbf{1})\, \mathscr{T}^{-1}$ into $-i\hbar\,\mathbf{1}$.

For anti-linear operators, according to p. 289, we have $\mathscr{T}\,|\psi\, a\rangle = (\mathscr{T}\,|\psi\rangle)\, a^*$. If we set

$$|\overline{\psi}\rangle \equiv \mathscr{T}\,|\psi\rangle ,$$

then $|\overline{\psi}\rangle$ can be expanded with $|\psi\rangle = \sum_n |n\rangle\,\psi_n$, $|\overline{\psi}\rangle = \sum_n |\overline{n}\rangle\,\psi_n{}^*$, and correspondingly $\langle\overline{\varphi}| = \sum_n \varphi_n\langle\overline{n}|$. We obtain generally

$$\langle\overline{\varphi}|\overline{\psi}\rangle = \langle\varphi|\psi\rangle^* \, .$$

Note that $|\overline{\psi}\rangle = \mathscr{T}|\psi\rangle$ depends anti-linearly on $|\psi\rangle$, as does the scalar product $\langle\varphi|\overline{\psi}\rangle$. Consequently, its complex conjugate value depends linearly on $|\psi\rangle$. Correspondingly, from $|\chi\rangle = A\,|\psi\rangle$, we infer $|\overline{\chi}\rangle = \mathscr{T}A\mathscr{T}^{-1}\,|\overline{\psi}\rangle$, and then also

$$\langle\overline{\varphi}|\,\mathscr{T}A\mathscr{T}^{-1}\,|\overline{\psi}\rangle = \langle\varphi\,|A\,|\psi\rangle^* \, .$$

For $A = A^\dagger$, this is equal to $\langle\psi|A|\varphi\rangle$. In particular, we have $\langle\overline{\varphi}\,|\,\mathbf{R}\,|\overline{\psi}\rangle = \langle\psi\,|\,\mathbf{R}\,|\varphi\rangle$ and $\langle\overline{\varphi}\,|\,\mathbf{P}\,|\overline{\psi}\rangle = -\langle\psi\,|\,\mathbf{P}\,|\varphi\rangle$.

If $|\psi\rangle$ is an eigenstate of $\mathscr{T}$, then its phase influences the eigenvalue. In particular, if $\mathscr{T}|\psi\rangle = |\psi\rangle$ holds, then so does $\mathscr{T}\,(|\psi\rangle\,e^{i\phi}) = |\psi\rangle\,e^{-i\phi} = (|\psi\rangle\,e^{i\phi})\,e^{-2i\phi}$. Thus, the two eigenvalues differ by the factor $e^{-2i\phi}$. Hence we can fix each state via the time reversal behavior of the phase, but we cannot assign a quantum number to the time reversal.

For particles without spin, after the choice of a basis with unique phases, the complex conjugation operator $\mathscr{K}$ can be used as the time reversal operator $\mathscr{T}$. Then we have $\mathscr{T}^2 = 1$, independently of the choice of phases.

For particles with half-integer spin, we also have to consider $\mathscr{T}\boldsymbol{\sigma}\,\mathscr{T}^{-1} = -\boldsymbol{\sigma}$. For motion reversal, in particular, the spin becomes inverted along with the angular momentum, since the spin is to be understood as an eigen angular momentum $\mathbf{S}$, as we shall see on p. 329. Now, according to Sect. 4.2.10, $\mathscr{K}\,(\sigma_x, \sigma_y, \sigma_z)\,\mathscr{K}^{-1} = (\sigma_x, -\sigma_y, \sigma_z)$ holds. Hence only $\mathscr{T} = i\sigma_y\,\mathscr{K}$ leads to the final behavior, where the phase factor i is arbitrary, but then the factor in front of $\mathscr{K}$ corresponds to a rotation through the angle π about the y-axis. Independently of this choice of phase, we now have $\mathscr{T}^2 = -1$ (for spin-1/2 particles), a truly astonishing result, since classically the two-fold reversal of the motion leads back to the original state. But note that a $360°$ rotation of a spin-1/2 particle leads to the state with the opposite sign.

For $\mathscr{T}^2 = \pm 1$, we have the equations

$$\langle\phi|\overline{\psi}\rangle = \langle\overline{\phi}|\overline{\overline{\psi}}\rangle^* = \pm\langle\overline{\phi}|\psi\rangle^* = \pm\langle\psi|\overline{\phi}\rangle \, .$$

From this it follows for $\mathscr{T}^2 = -1$ (half-integer spin) that $\langle\psi|\overline{\psi}\rangle = 0$. For half-integer spin, the states $|\psi\rangle$ and $|\overline{\psi}\rangle$ are orthogonal to each other and hence different. Since the Hamilton operator is generally invariant under time reversal, i.e., $H = \mathscr{T}H\,\mathscr{T}^{-1}$, fermions always have pairs of states $(|\psi\rangle, |\overline{\psi}\rangle)$ with equal energy. This is known as *Kramers theorem*. For bound states, $|\psi\rangle$ and $|\overline{\psi}\rangle$ differ by the spin orientation.

The eigenvalue of the space inversion operator $\mathscr{P}$ is the *parity*. Because $\mathscr{P}^2 = 1$, it takes the values ± 1.

4.2.13 Summary: Operators and Observables

In every physical theory, there are observables (measurable quantities). In quantum theory they are described by Hermitian operators, the eigenvalues of which correspond to the possible experimental values. Then the associated eigenvalue always results as the experimental value, and the observable is sharp (certain). Otherwise, the eigenvalue a (of possible experimental values) results with a statistical weight (or probability) given by $\langle a| \rho |a \rangle$, so that, on the average, the expectation value is

$$\langle A \rangle = \sum_a \langle a| \rho |a \rangle \, a \ .$$

For a pure state $|\psi \rangle$, it is $\langle a| \rho |a \rangle = |\langle a|\psi \rangle|^2$, so $\rho = |\psi \rangle \langle \psi|$. For the uncertainty (the average error), we have $\Delta A = \sqrt{\langle A^2 \rangle - \langle A \rangle^2}$ with $\langle A^2 \rangle = \sum_a \langle a| \rho |a \rangle \, a^2$. Hence $\Delta A = 0$ for $\rho = |a \rangle \langle a|$.

Non-commuting operators have no common set of eigenstates. Hence, not all the corresponding observables can be sharp at the same time. In particular, the uncertainty relation $\Delta X \cdot \Delta P \geq \frac{1}{2}\hbar$ follows from the commutation law

$$[X, P] = i \, \hbar \, 1 \ ,$$

with which we shall deal later. Here X and P have continuous eigenvalue spectra which differ from the operators so far considered, and which require improper Hilbert vectors.

4.3 Correspondence Principle

4.3.1 Commutation Relations

According to p. 300, we can ensure Heisenberg's uncertainty relation

$$\Delta X^k \cdot \Delta P_{k'} \geq \tfrac{1}{2}\hbar \, \delta^k_{k'}$$

by assigning Hermitian operators X^k and P_k to the complementary variables position and momentum which obey the commutation relations (of Born and Jordan)

$$[X^k, P_{k'}] = i\hbar \, \delta^k_{k'} \, 1 \ .$$

Since the commutator is proportional to the unit operator, the product $\Delta X^k \cdot \Delta P_k$ of the uncertainties cannot be smaller than $\hbar/2$ for any state $|\psi \rangle$.

Here once again we shall always deal with pairs of canonically conjugate quantities and hence rely on Hamiltonian mechanics. The commutators correspond to the

Poisson brackets, as we shall now show, since we shall use this key idea repeatedly to translate between classical and quantum dynamics.

According to p. 124, all pairs of dynamical quantities u, v have a Poisson bracket defined by

$$[u, v] \equiv \sum_k \left(\frac{\partial u}{\partial x^k} \frac{\partial v}{\partial p_k} - \frac{\partial u}{\partial p_k} \frac{\partial v}{\partial x^k} \right) = -[v, u] \, ,$$

which does not depend on the choice of canonical coordinates x^k and momenta $p_k = \partial L / \partial x^k$ (otherwise it would not be canonical). In particular, classically, we have

$$[x^k, x^{k'}] = 0 = [p_k, p_{k'}] \, , \quad [x^k, p_{k'}] = \delta_{k'}^k \, .$$

If we now require the classical Poisson bracket $[u, v]$ to become the expression $[U, V]/i\hbar$ in quantum theory,

$$[u, v] \overset{!}{\Longrightarrow} \frac{[U, V]}{i\hbar} \, ,$$

where U and V are the Hermitian operators in quantum theory corresponding to the classical u and v, then we do indeed have

$$[X^k, X^{k'}] = 0 = [P_k, P_{k'}] \, , \quad [X^k, P_{k'}] = i\hbar \, \delta_{k'}^k \, 1 \, .$$

If we replace the classical observables by Hermitian operators and the Poisson brackets by commutators divided by $i\hbar$*, then the uncertainty relations are satisfied.*

Since position and momentum operators do not commute with each other, in quantum physics no state can be given which contains position and momentum simultaneously as characterizing items. We have to choose: either the position alone or the momentum alone. But for each additional Cartesian component a new choice can be made. With each new degree of freedom, the state is amended by a new quantum number.

With $[X, P] = i\hbar \, 1$, according to p. 289, we have $[X, P^n] = n i\hbar \, P^{n-1}$. This is also true for negative integers: $[X, P^{-n}] = P^{-n} (P^n X - X P^n) P^{-n} = -n i\hbar \, P^{-n-1}$. Since the operators X and P have continuous eigenvalue spectra, in their eigen-representation, the derivative with respect to X, or indeed P, makes sense—it is simply the derivative with respect to the eigenvalue x, or the eigenvalue p. Hence, we write generally,

$$[X, f(P)] = i\hbar \, \frac{df(P)}{dP} \, , \quad [f(X), P] = i\hbar \, \frac{df(X)}{dX} \, .$$

It follows in particular (see the Hausdorff series on p. 290) that

$$\exp(i\,\mathbf{a}\cdot\mathbf{P})\;\mathbf{R}\;\exp(-i\,\mathbf{a}\cdot\mathbf{P}) = \mathbf{R} + \hbar\,\mathbf{a}\;.$$

According to this, the unitary operator $\exp(i\,\mathbf{a}\cdot\mathbf{P})$ shifts all positions by $\hbar\,\mathbf{a}$, so it is a *displacement operator*. Furthermore, in classical mechanics, the (canonical) momentum is the generating function for infinitesimal displacements (see p. 130). Correspondingly, we have $\exp(i\,\mathbf{a}\cdot\mathbf{R})\;\mathbf{P}\;\exp(-i\,\mathbf{a}\cdot\mathbf{R}) = \mathbf{P} - \hbar\,\mathbf{a}$.

4.3.2 Position and Momentum Representations

In the real-space representation, the position operator X is diagonal. We restrict ourselves initially to one dimension:

$$\langle x|\, X\, |x'\rangle = x\,\delta(x - x')\;.$$

In this representation the momentum operator P follows from the commutation relation $[X, P] = i\hbar\, 1$, since from $i\hbar\,\delta(x - x') = \langle x|\, XP - PX\, |x'\rangle = (x - x')\langle x|\, P\, |x'\rangle$ with $\delta(x) = -x\,\delta'(x)$ (p. 21), we obtain

$$\langle x|\, P\, |x'\rangle = -i\hbar\,\frac{\partial}{\partial x}\,\delta(x - x') = i\hbar\,\frac{\partial}{\partial x'}\,\delta(x - x')\;.$$

Hence we have $\langle x|\, P\, |\psi\rangle = \int dx'\,\langle x|\, P\, |x'\rangle\,\psi(x') = -i\hbar\, d\psi(x)/dx$. This can also be used for higher powers of P in the real-space representation, since for

$$\langle x|\, P^n\, |\psi\rangle = \int dx'\,\langle x|\, P\, |x'\rangle\langle x'|\, P^{n-1}\, |\psi\rangle\;,$$

the integral can be simplified with the delta function to $-i\hbar\,\partial\langle x|\, P^{n-1}\, |\psi\rangle/\partial x = (-i\hbar)^n\,\partial^n\psi/\partial x^n$.

In the real-space representation, we may thus replace $P\,|\psi\rangle$ by $-i\hbar\, d\psi/dx$. This is usually abbreviated as

$$P \mathrel{\widehat{=}} \frac{\hbar}{i}\,\frac{d}{dx}\;,$$

which is of course true only in the real-space representation, if P acts on $\langle x|\psi\rangle \equiv \psi(x)$. Correspondingly, in the displacement operator $U = \exp(iaP)$, all powers of the derivatives with respect to x occur:

$$\exp(iaP) \mathrel{\widehat{=}} \sum_{n=0}^{\infty}\frac{(\hbar a)^n}{n!}\,\frac{d^n}{dx^n}\;,$$

as we also expect for the Taylor series. Note that, with

$$UXU^{-1} = X + \hbar a \quad \text{and} \quad (X + \hbar a)|x - \hbar a\rangle = |x - \hbar a\rangle x \,,$$

$UX|x\rangle = U|x\rangle x$ leads to $U|x\rangle = |x - \hbar a\rangle$, or to $|x + \hbar a\rangle = U^\dagger|x\rangle$, and this in turn to $\psi(x + \hbar a) = \langle x + \hbar a|\psi\rangle = \langle x|U|\psi\rangle$.

In the momentum representation and since $[X, P] = -[P, X]$, we also have

$$\langle p|\, P\, |p'\rangle = p\, \delta(p - p') \,,$$

$$\langle p|\, X\, |p'\rangle = \mathrm{i}\hbar\, \frac{\partial}{\partial p}\, \delta(p - p') = \frac{\hbar}{\mathrm{i}}\, \frac{\partial}{\partial p'}\, \delta(p - p') \,,$$

thus $\langle p|\, P\, |\psi\rangle = p\, \psi(p)$ and $\langle p|\, X\, |\psi\rangle = \mathrm{i}\hbar\, \mathrm{d}\psi/\mathrm{d}p$.

The results are easily extended to three dimensions. With $\mathrm{d}\psi = \nabla_r \psi \cdot \mathrm{d}\mathbf{r} = \nabla_p \psi \cdot \mathrm{d}\mathbf{p}$, we find in particular,

$$
\begin{aligned}
\langle \mathbf{r}|\, \mathbf{R}\, |\mathbf{r}'\rangle &= && \mathbf{r}\, \delta(\mathbf{r} - \mathbf{r}') & \Longrightarrow & \quad \langle \mathbf{r}|\, \mathbf{R}\, |\psi\rangle &= && \mathbf{r}\, \psi(\mathbf{r}) \,, \\
\langle \mathbf{r}|\, \mathbf{P}\, |\mathbf{r}'\rangle &= && \frac{\hbar}{\mathrm{i}}\, \nabla_r\, \delta(\mathbf{r} - \mathbf{r}') & \Longrightarrow & \quad \langle \mathbf{r}|\, \mathbf{P}\, |\psi\rangle &= && \frac{\hbar}{\mathrm{i}}\, \nabla_r\, \psi(\mathbf{r}) \,, \\
\langle \mathbf{p}|\, \mathbf{R}\, |\mathbf{p}'\rangle &= \mathrm{i}\hbar\, \nabla_p\, \delta(\mathbf{p} - \mathbf{p}') & & & \Longrightarrow & \quad \langle \mathbf{p}|\, \mathbf{R}\, |\psi\rangle &= \mathrm{i}\hbar\, \nabla_p\, \psi(\mathbf{p}) \,, \\
\langle \mathbf{p}|\, \mathbf{P}\, |\mathbf{p}'\rangle &= && \mathbf{p}\, \delta(\mathbf{p} - \mathbf{p}') & \Longrightarrow & \quad \langle \mathbf{p}|\, \mathbf{P}\, |\psi\rangle &= && \mathbf{p}\, \psi(\mathbf{p}) \,.
\end{aligned}
$$

This can also be used for the matrix elements of this operator between the states $\langle\varphi|$ and $|\psi\rangle$, if $\varphi(\mathbf{r}) = \langle \mathbf{r}|\varphi\rangle$ or $\varphi(\mathbf{p}) = \langle \mathbf{p}|\varphi\rangle$ are known, since

$$\langle\varphi|A|\psi\rangle = \int \mathrm{d}^3 r\, \varphi^*(\mathbf{r})\, \langle \mathbf{r}|A|\psi\rangle = \int \mathrm{d}^3 p\, \varphi^*(\mathbf{p})\, \langle \mathbf{p}|A|\psi\rangle \,.$$

4.3.3 The Probability Amplitude $\langle \mathbf{r}\,|\,\mathbf{P}\rangle$

We can now determine the Dirac bracket $\langle \mathbf{r}\,|\mathbf{p}\rangle$, i.e., the density of the probability amplitude of the state $|\mathbf{p}\rangle$ at the position $\mathbf{r}$, and then change from the position to the momentum representation. The reverse transformation is possible with $\langle \mathbf{p}\,|\mathbf{r}\rangle = \langle \mathbf{r}\,|\mathbf{p}\rangle^*$. We have in particular $p\, \langle x|p\rangle = \langle x|\, P\, |p\rangle = -\mathrm{i}\hbar\, \partial/\partial x\, \langle x|p\rangle$ and hence $\langle x|p\rangle \propto \exp(\mathrm{i}p\,x/\hbar)$ as a function of x. On the other hand, we also have $x\, \langle x|p\rangle = \langle p|\, X\, |x\rangle^* = -\mathrm{i}\hbar\, \partial/\partial p\, \langle x|p\rangle$, and hence $\langle x|p\rangle \propto \exp(\mathrm{i}x p/\hbar)$ as a function of p. The unknown proportionality factor thus depends neither on x nor on p. We call it temporarily c and determine it from the normalization condition $\delta(p - p') = \int \mathrm{d}x\, \langle p|x\rangle\langle x|p'\rangle = |c|^2 \int \mathrm{d}x\, \exp\{\mathrm{i}\,(p' - p)\,x/\hbar\} = |c|^2\, 2\pi\, \hbar\, \delta(p' - p)$. Hence, it follows that $2\pi\,\hbar\,|c|^2 = 1$, where $2\pi\,\hbar = h$ (so we could just write h here, but $\hbar$ occurs much more often than h, and we shall use it here too). We choose the arbitrary phase factor in the simplest possible way, viz., equal to unity. Then,

$$\langle x|p \rangle = \frac{\exp(ipx/\hbar)}{\sqrt{2\pi\hbar}} \quad\Longleftrightarrow\quad \langle \mathbf{r}\,|\mathbf{p} \rangle = \frac{\exp(i\,\mathbf{p}\cdot\mathbf{r}/\hbar)}{\sqrt{2\pi\hbar}^{\,3}}\,.$$

For time reversal (motion reversal), according to p. 314, we then also have $\langle \bar{x}|\bar{p} \rangle = \langle x|p \rangle^* = (2\pi\hbar)^{-1/2}\exp(-ip\,x/\hbar) = \langle x|-p \rangle$.

The probability of the state $|\mathbf{p}\rangle$ in the space element $d^3\mathbf{r}$ about $\mathbf{r}$ is now given by

$$|\langle \mathbf{r}\,|\mathbf{p} \rangle|^2\, d^3\mathbf{r} = \frac{d^3\mathbf{r}}{(2\pi\hbar)^3}\,.$$

It does not depend on the position, so is equally large everywhere. (For a state with sharp momentum, whence $\Delta P = 0$, ΔX must be infinite!) Note that the integral over the infinite space does not result in 1, as we should require. The improper Hilbert space vector $|\mathbf{p}\rangle$ is not normalizable, so we need an error $\Delta P > 0$.

For the superposition of several states, interference shows up. If, for instance, the state $|\psi\rangle$ contains the momenta $\mathbf{p}_1$ and $\mathbf{p}_2$ with probability amplitudes $\langle \mathbf{p}_1|\psi\rangle$ and $\langle \mathbf{p}_2|\psi\rangle$, respectively, then the associated probability density is

$$|\langle \mathbf{r}\,|\psi \rangle|^2 = \frac{|\exp(i\mathbf{p}_1\cdot\mathbf{r}/\hbar)\,\langle \mathbf{p}_1|\psi\rangle + \exp(i\mathbf{p}_2\cdot\mathbf{r}/\hbar)\,\langle \mathbf{p}_2|\psi\rangle|^2}{(2\pi\hbar)^3}$$

$$= \frac{|\langle \mathbf{p}_1|\psi\rangle|^2}{(2\pi\hbar)^3}\left|\, 1 + \frac{\langle \mathbf{p}_2|\psi\rangle}{\langle \mathbf{p}_1|\psi\rangle}\,\exp\frac{i\,(\mathbf{p}_2-\mathbf{p}_1)\cdot\mathbf{r}}{\hbar}\,\right|^2\,.$$

It now depends on position, in particular in the direction of $\mathbf{p}_2 - \mathbf{p}_1$, and periodically, with the wave vector

$$\mathbf{k} \equiv \frac{\mathbf{p}_2 - \mathbf{p}_1}{\hbar}\,.$$

This we interpret as the interference of probability waves with wave vectors $\mathbf{k}_1$ and $\mathbf{k}_2$. Hence we arrive at the *de Broglie relation*

$$\mathbf{p} = \hbar\,\mathbf{k}\,.$$

It follows therefore from our assumptions.

It is clearly more convenient for the exponential function to use the wave vector $\mathbf{k}$ instead of the momentum $\mathbf{p}$, since then the denominator $\hbar$ drops out, and they are related to each other simply via the de Broglie relation. Hence, $|\mathbf{p}\rangle$ is often replaced by $|\mathbf{k}\rangle$—both states belong to the same ray in (improper) Hilbert space, but are differently normalized. With $\langle \mathbf{p}\,|\mathbf{p}' \rangle = \delta(\mathbf{p}-\mathbf{p}') = \delta\{\hbar(\mathbf{k}-\mathbf{k}')\} = \hbar^{-3}\langle \mathbf{k}|\mathbf{k}' \rangle$, we have

$$|\mathbf{p}\rangle = \frac{|\mathbf{k}\rangle}{\sqrt{\hbar}^{\,3}}\,, \quad \text{and hence} \quad \langle \mathbf{r}\,|\mathbf{k} \rangle = \frac{\exp(i\mathbf{k}\cdot\mathbf{r})}{\sqrt{2\pi}^{\,3}}\,.$$

The transition from the momentum space to the real space representation (or vice versa) is a standard Fourier transform (see p. 22), since we have

$$\langle \mathbf{r} | \psi \rangle = \int d^3k \, \langle \mathbf{r} | \mathbf{k} \rangle \langle \mathbf{k} | \psi \rangle = \frac{1}{\sqrt{2\pi}^3} \int d^3k \, \exp(i\mathbf{k} \cdot \mathbf{r}) \, \langle \mathbf{k} | \psi \rangle \, .$$

Actually, we should have used the term *wave vector representation* instead of *momentum representation*—other authors do not distinguish between these notions and simply state that they could have set $\hbar$ equal to 1.

4.3.4 Wave Functions

The wave function of a state is usually understood to be its real-space representation:

$$\psi(\mathbf{r}) = \langle \mathbf{r} | \psi \rangle \, ,$$

but generally the representation can be in any basis. The real-space representation is often stressed too strongly, since the momentum representation is more suitable for scattering problems and the angular momentum representation for problems with rotation invariance. We shall thus proceed here in a way that is as independent of the representation (as coordinate-free) as possible. The real-space representation is preferred by many, and even if sometimes obvious, it is often rather inconvenient and in principle not superior to the other representations (as emphasized by H. S. Green in the introduction to his textbook, mentioned on p. 396).

If $|\psi\rangle$ is a proper Hilbert space vector, then the function $\psi(\mathbf{r})$ must be normalizable and infinitely differentiable. With the requirement $\langle \psi | \psi \rangle = 1$, we must have

$$\int d^3r \, \psi^*(\mathbf{r}) \, \psi(\mathbf{r}) = 1 \, ,$$

thus in particular $\psi(\mathbf{r}) \to 0$ for $r \to \infty$, and $\psi(\mathbf{r})$ must be differentiable so that the momentum expectation value $\langle \psi | \mathbf{P} | \psi \rangle$ can be calculated. Higher powers of P require higher derivatives, as we have seen in Sect. 4.3.2.

We already had an example of a wave function in the last section, namely the wave function for a given momentum $\mathbf{p}$ (such that ΔP vanishes):

$$\langle \mathbf{r} | \mathbf{p} \rangle = (2\pi\hbar)^{-3/2} \exp(i\mathbf{p} \cdot \mathbf{r}/\hbar) \, .$$

However, $|\mathbf{p}\rangle$ and $|\mathbf{r}\rangle$ are improper Hilbert vectors. Such states are idealizations. An ensemble can only then be characterized by continuous variables when error widths (uncertainties) are included. To each continuous measurable quantity belongs a distribution function (density).

Occasionally, improper Hilbert vectors are required [2]. They are very convenient, and with appropriate distribution functions, a *fuzziness* can still be introduced. For example, a *wave packet* can be formed from $\langle \mathbf{r} | \mathbf{k} \rangle$:

$$\langle \mathbf{r} | \psi \rangle = \int d^3k \, \langle \mathbf{r} | \mathbf{k} \rangle \langle \mathbf{k} | \psi \rangle = \frac{1}{\sqrt{2\pi}^3} \int d^3k \, \exp(i\mathbf{k} \cdot \mathbf{r}) \, \psi(\mathbf{k}) .$$

For $\psi(\mathbf{k}) \neq \delta(\mathbf{k} - \mathbf{k}_0)$, this has a non-vanishing momentum uncertainty, and for $\psi(\mathbf{k}) \neq c$, a position uncertainty.

We may ask which wave function has the smallest possible $\Delta X \cdot \Delta P$, i.e., equal to $\hbar/2$? According to p. 301, we must then have $(X - \overline{X}) | \psi \rangle = -i \, \Delta X / \Delta P \, (P - \overline{P}) | \psi \rangle$ with $1/\Delta P = 2\Delta X / \hbar$. With $\langle x | X | \psi \rangle = x \, \psi(x)$ and $\langle x | P | \psi \rangle = -i\hbar \, \psi'(x)$, we arrive at the differential equation

$$\left(-\frac{d}{dx} + \frac{i}{\hbar} \overline{P} \right) \psi(x) = \frac{x - \overline{X}}{2 (\Delta X)^2} \, \psi(x) .$$

For an appropriate choice of phase, so that no integration constant remains free, its normalized solution reads (for the normalization of the Gauss function, see p. 23)

$$\psi(x) = \frac{1}{\sqrt[4]{2\pi} \sqrt{\Delta X}} \, \exp\left\{ -\left(\frac{x - \overline{X}}{2 \, \Delta X} \right)^2 + \frac{i\overline{P} \, (x - \frac{1}{2}\overline{X})}{\hbar} \right\} .$$

It contains three free parameters, namely $\overline{X}$, ΔX, and $\overline{P}$, but the last drops out for the probability density $|\psi(x)|^2$. This density is a *normal distribution* (Gauss function) with maximum at $\overline{X}$. For the canonically conjugate variable, using a Fourier transform, we find another Gauss function:

$$\psi(p) = \frac{1}{\sqrt[4]{2\pi} \sqrt{\Delta P}} \, \exp\left\{ -\left(\frac{p - \overline{P}}{2 \, \Delta P} \right)^2 - \frac{i\overline{X} \, (p - \frac{1}{2}\overline{P})}{\hbar} \right\} .$$

We shall return to this result in the context of harmonic oscillations (Sect. 4.5.4). The phase factors $\exp(\mp\frac{1}{2} i \overline{P} \overline{X}/\hbar)$ have been added for the sake of the symmetry—then $\psi(x)$ and $\psi(p)$ are really mutually Fourier-transformed quantities.

4.3.5 Wigner Function

In statistical mechanics (p. 523), we introduce the classical density function $\rho_{cl}(\mathbf{r}, \mathbf{p})$ in phase space and use it to determine the average values

$$\overline{A} = \int d^3r \, d^3p \, \rho_{cl}(\mathbf{r}, \mathbf{p}) \, A_{cl}(\mathbf{r}, \mathbf{p}) .$$

In quantum theory this density corresponds to the Wigner function. It follows via Fourier transforms from the density operator ρ in the position or momentum representation.

To show this, we adopt a basis $\{C(\mathbf{r}, \mathbf{p})\}$ of Hermitian, unitary, and orthogonal (in $\mathbf{r}$ and $\mathbf{p}$) operators. For example, the Pauli operators are Hermitian, unitary, and orthogonal in the space of 2×2 matrices, according to p. 309. In the real-space representation,

$$\langle \mathbf{r}_1 | C(\mathbf{r}, \mathbf{p}) | \mathbf{r}_2 \rangle \equiv \delta(2\mathbf{r} - \mathbf{r}_1 - \mathbf{r}_2) \exp \frac{+i\mathbf{p} \cdot (\mathbf{r}_1 - \mathbf{r}_2)}{\hbar} \, ,$$

and in the momentum representation

$$\langle \mathbf{p}_1 | C(\mathbf{r}, \mathbf{p}) | \mathbf{p}_2 \rangle = \delta(2\mathbf{p} - \mathbf{p}_1 - \mathbf{p}_2) \exp \frac{-i\mathbf{r} \cdot (\mathbf{p}_1 - \mathbf{p}_2)}{\hbar} \, .$$

These are practical as an operator basis, according to Sect. 4.2.5, because

$$C(\mathbf{r}, \mathbf{p}) = C^{\dagger}(\mathbf{r}, \mathbf{p}) \ = \ C^{-1}(\mathbf{r}, \mathbf{p}) \, ,$$
$$\mathrm{tr}\{C(\mathbf{r}, \mathbf{p}) \, C(\mathbf{r}', \mathbf{p}')\} = (\tfrac{1}{2}\pi\hbar)^3 \, \delta(\mathbf{r} - \mathbf{r}') \, \delta(\mathbf{p} - \mathbf{p}') \, .$$

As in Sect. 4.2.11, the expectation values of the basis operators are also important. They deliver the *Wigner function*

$$\rho(\mathbf{r}, \mathbf{p}) \ \equiv \ \frac{\langle C(\mathbf{r}, \mathbf{p}) \rangle}{(\pi\hbar)^3} \, ,$$

which is in fact the Fourier transform of the density operator:

$$\langle C(\mathbf{r}, \mathbf{p}) \rangle = \int d^3 r' \, \langle \mathbf{r} - \mathbf{r}' | \rho | \mathbf{r} + \mathbf{r}' \rangle \exp \frac{+2i\,\mathbf{p} \cdot \mathbf{r}'}{\hbar}$$
$$= \int d^3 p' \, \langle \mathbf{p} - \mathbf{p}' | \rho | \mathbf{p} + \mathbf{p}' \rangle \exp \frac{-2i\,\mathbf{r} \cdot \mathbf{p}'}{\hbar} \, .$$

Conversely, we obtain the density operator from the Wigner function (see Figs. 4.6 and 4.7):

$$\langle \mathbf{r} | \rho | \mathbf{r}' \rangle = \int d^3 p \ \rho\left(\frac{\mathbf{r} + \mathbf{r}'}{2}, \mathbf{p}\right) \exp \frac{+i\,(\mathbf{r} - \mathbf{r}') \cdot \mathbf{p}}{\hbar} \, ,$$
$$\langle \mathbf{p} | \rho | \mathbf{p}' \rangle = \int d^3 r \ \rho\left(\mathbf{r}, \frac{\mathbf{p} + \mathbf{p}'}{2}\right) \exp \frac{-i\,(\mathbf{p} - \mathbf{p}') \cdot \mathbf{r}}{\hbar} \, .$$

If we integrate the Wigner function $\rho(\mathbf{r}, \mathbf{p})$ over all momenta or all positions, we obtain the probability densities in position and momentum space, respectively:

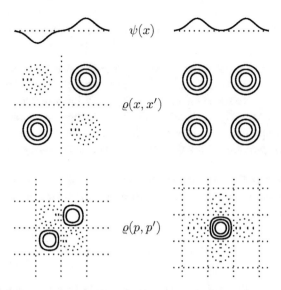

Fig. 4.6 Superposition of the two states $\psi_\pm(x) \propto \exp\{-(x \mp 2)^2\}$: $\psi_+ - \psi_-$ (*left*) and $\psi_+ + \psi_-$ (*right*). Below the wave functions, the density operators $\rho = |\psi\rangle\langle\psi|$ are shown with equal-value lines for $\rho > 0$ (*continuous line*) and for $\rho \leq 0$ (*dotted lines*), in the real-space representation ($\rho(x, x') = \langle x|\rho|x'\rangle$) and the momentum representation ($\rho(p, p') = \langle p|\rho|p'\rangle$). The axes can be recognized as symmetry axes, and ρ is always real here. Along the diagonal $x' = x$, we have $\rho \geq 0$, which corresponds to the "classically expected" density

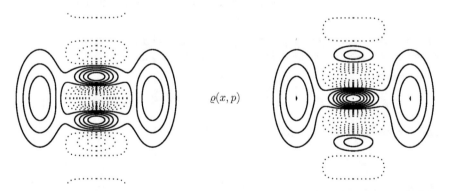

Fig. 4.7 Wigner functions of the superpositions of states from Fig. 4.6. Equal-value lines are shown once again, for $\rho > 0$ (*continuous lines*) and for $\rho \leq 0$ (*dotted lines*). Here, ρ is symmetric with respect to the x- and p-axes. The Wigner function can be negative, which depends sensitively on the phase difference of the superposed states, while in the "classically preferred" phase-space regions (here $x \approx \pm 2$, $p \approx 0$), there is almost no dependence

$$\int d^3 p \, \rho(\mathbf{r}, \mathbf{p}) = \langle \mathbf{r} | \rho | \mathbf{r} \rangle \qquad \text{and} \qquad \int d^3 r \, \rho(\mathbf{r}, \mathbf{p}) = \langle \mathbf{p} | \rho | \mathbf{p} \rangle \, .$$

Hence we have the usual normalization $\int d^3 r \, d^3 p \, \rho(\mathbf{r}, \mathbf{p}) = 1$. Incidentally, a Fourier transform yields $\int d^3 r \, d^3 p \, \rho^2(\mathbf{r}, \mathbf{p}) = (2\pi \hbar)^{-3} \operatorname{tr} \rho^2$. We can also test whether we have a pure state or a mixture using the Wigner function. In addition, the Wigner function, being the expectation value of a Hermitian operator, is real. However, it can also be negative, and this distinguishes it from classical density functions: it is only a *quasi-probability*, but this difference is also necessary for the description of interference.

With the Wigner function, the expectation value of every observable $A(\mathbf{R}, \mathbf{P})$ can be determined. In particular, if we expand the operator A in terms of the basis $\{C(\mathbf{r}, \mathbf{p})\}$, viz.,

$$A = \int d^3 r \, d^3 p \ C(\mathbf{r}, \mathbf{p}) \ (2/\pi \hbar)^3 \operatorname{tr}\{C(\mathbf{r}, \mathbf{p}) A\} \, ,$$

according to p. 297, then using $\langle C(\mathbf{r}, \mathbf{p}) \rangle = (\pi \hbar)^3 \rho(\mathbf{r}, \mathbf{p})$, we can determine $\langle A \rangle$, We set

$$\begin{aligned}
A(\mathbf{r}, \mathbf{p}) &\equiv 2^3 \operatorname{tr}\{C(\mathbf{r}, \mathbf{p}) A\} = 2^3 \int d^3 r' \ \langle \mathbf{r} - \mathbf{r}' | A | \mathbf{r} + \mathbf{r}' \rangle \ \exp \frac{+2\mathrm{i}\, \mathbf{p} \cdot \mathbf{r}'}{\hbar} \\
&= 2^3 \int d^3 p' \ \langle \mathbf{p} - \mathbf{p}' | A | \mathbf{p} + \mathbf{p}' \rangle \ \exp \frac{-2\mathrm{i}\, \mathbf{r} \cdot \mathbf{p}'}{\hbar} \, ,
\end{aligned}$$

because then formally—only formally, since the Wigner function can also be negative—we have the same as in statistical mechanics, that is

$$\langle A \rangle = \int d^3 r \, d^3 \mathbf{p} \ \rho(\mathbf{r}, \mathbf{p}) \, A(\mathbf{r}, \mathbf{p}) \, ,$$

and $A(\mathbf{r}, \mathbf{p})$ is real for a Hermitian operator A.

4.3.6 Spin

So far we have taken the position or momentum representation and then proceeded as if a (pure) state were already defined by $\mathbf{r}$ or $\mathbf{p}$. But for electrons (and nucleons), we must also take into account their *eigen angular momentum* (spin). This degree of freedom must also be determined if the statistical ensemble is to be described uniquely. For this "inner degree of freedom", we only require a Hilbert space of finite dimension. For electrons and nucleons, two dimensions suffice, so here we shall restrict ourselves to that situation and use Sect. 4.2.10. Hence, $|\mathbf{r}, \uparrow\rangle$ and $|\mathbf{r}, \downarrow\rangle$ fix the state, or indeed $|\mathbf{p}, \uparrow\rangle$ and $|\mathbf{p}, \downarrow\rangle$.

Correspondingly, we have to distinguish the operators by the space in which they act. For example, neither **R** nor **P** affects the inner degrees of freedom—they act in the spin space as the unit operator. Conversely, σ does not act on $|\mathbf{r}\rangle$ and $|\mathbf{p}\rangle$. Hence **R** and **P** commute with σ. Of course, there are also operators, which act in two spaces, e.g., the *helicity* $(\mathbf{P} \cdot \mathbf{P})^{-1/2}\,\mathbf{P} \cdot \sigma$, for which the orientation of the spin relative to the momentum is important.

If **A** and **B** do not act in the spin space, then with $\sigma_x{}^2 = 1$ and $\sigma_x\sigma_y = -\sigma_y\sigma_x = \mathrm{i}\sigma_z$ (and cyclic permutations), we have

$$\mathbf{A} \cdot \sigma\,\mathbf{B} \cdot \sigma = \mathbf{A} \cdot \mathbf{B} + \mathrm{i}\,(\mathbf{A} \times \mathbf{B}) \cdot \sigma\,.$$

Since here **A** and **B** may be arbitrary vector operators, we have as special cases of this equation

$$\mathbf{A} \cdot \sigma\,\sigma = \mathbf{A} - \mathrm{i}\,\mathbf{A} \times \sigma \quad \text{and} \quad [\sigma, \mathbf{A} \cdot \sigma] = 2\mathrm{i}\,\mathbf{A} \times \sigma\,.$$

The unit operator in the spin space is not written explicitly, as previously for **R** and **P**. Moreover, on the left of the last equations, we should write $1 \otimes \sigma$ instead of just σ.

If we write σ as a 2×2 matrix, then the Hilbert vectors in the sequence space must also be written as *2-spinors*—for ψ the two elements atop each other, for $\psi^\dagger$ side-by-side and complex-conjugate to those of ψ.

4.3.7 Correspondence Principle

In quantum theory, we describe all observables using Hermitian operators whose eigenvalues correspond to the possible experimental values. So far we have presented only two observables, namely position and momentum, but according to Hamiltonian mechanics, further quantities can be derived. The corresponding observables in quantum theory are in general easy to find—we simply have to take the classical equations as operator equations: *If in classical physics $y = f(x, p)$, where y, x, and p have real values, then usually in quantum theory $Y = f(X, P)$, where Y, X, and P are Hermitian operators.* Hence we have given a mathematical form to Bohr's *correspondence principle*. Classical and quantum mechanical quantities correspond to one another to a large extent, but are distinguished in their mathematical relevance, since instead of classical quantities (number times unit), we now have linear operators.

However, the operators are canonically conjugate quantities and do not commute with each other—for products, the order of the factors is important. This difficulty rarely arises though. Let us take, e.g., the orbital angular momentum

$$\mathbf{L} = \mathbf{R} \times \mathbf{P}\,.$$

In the vector product here, all components commute without posing a problem. Although **L** does not generally commute with **R** and **P**, at least equal components do: $\mathbf{L} \cdot \mathbf{R} = \mathbf{R} \cdot \mathbf{L}$ and $\mathbf{L} \cdot \mathbf{P} = \mathbf{P} \cdot \mathbf{L}$.

If necessary, we can invoke the *Weyl correspondence*. If the Wigner function is used, then a Fourier transform is allowed. In particular, if the classical function $f(x, p)$ is given, its Fourier transform reads

$$f(\alpha, \beta) = \frac{1}{2\pi} \int dx\, dp\ \exp\{-i(\alpha x + \beta p)\} f(x, p) ,$$

and its operator function (where α and β remain real variables)

$$f(X, P) = \frac{1}{2\pi} \int d\alpha\, d\beta\ \exp\{+i(\alpha X + \beta P)\} f(\alpha, \beta) .$$

On p. 290, we already derived the relations (note that $[i\alpha X, i\beta P] = -i\hbar\alpha\beta$)

$$\exp\{i(\alpha X + \beta P)\} = \exp(i\alpha X)\ \exp(i\beta P)\ \exp(+\tfrac{1}{2}i\hbar\,\alpha\beta)$$
$$= \exp(i\beta P)\ \exp(i\alpha X)\ \exp(-\tfrac{1}{2}i\hbar\,\alpha\beta) ,$$

so we can determine $f(X, P)$ from a double Fourier integral, after which we have found $f(\alpha, \beta)$. In this way, $f(x)\, p$ has the Fourier transformed form $f(\alpha, \beta) = \sqrt{2\pi}\, i\, f(\alpha)\, \delta'(\beta)$, and hence, according to Weyl, we have to take $f(X, P) = f(X)\, P - \tfrac{1}{2}i\hbar f'(X)$. According to p. 316, in particular, with $i\hbar f'(X) = [f(X), P]$, this leads to the symmetrized product $\tfrac{1}{2}\{f(X), P\}$. Generally, the power series of the exponential function of $i(\alpha X + \beta P)$ leads to completely symmetrized products of X and P.

If we use quasi-probabilities instead of the Wigner function, we have to order differently, as will be discussed in Sect. 5.5.6.

Let us consider, e.g., the *Hamilton operator* for a particle of mass m and charge q in an electromagnetic field. According to p. 123, the classical Hamilton function is $\frac{1}{2m}(\mathbf{p} - q\mathbf{A}) \cdot (\mathbf{p} - q\mathbf{A}) + q\Phi$. The quantities m and q do not become operators, and in the usual quantum theory neither does the electromagnetic field—this happens only in quantum electrodynamics (see Sect. 5.5). Since **P** does not commute with **A**, we arrive at

$$H = \frac{P^2 - q(\mathbf{P} \cdot \mathbf{A} + \mathbf{A} \cdot \mathbf{P}) + q^2 A^2}{2m} + q\Phi ,$$

thus at the symmetrical product $\{P_k, A^k\}$. Now in the real-space representation, **P** corresponds to the operator $-i\hbar\nabla$, and we find $\nabla \cdot \mathbf{A}\psi = \psi\,\nabla \cdot \mathbf{A} + \mathbf{A} \cdot \nabla\psi$. For the Coulomb gauge, $\nabla \cdot \mathbf{A}$ vanishes, whence $\mathbf{P} \cdot \mathbf{A} = \mathbf{A} \cdot \mathbf{P}$ holds, even though **P** and **R** do not commute with each other. For a homogeneous magnetic field **B**, we have in particular for the vector potential (in the Coulomb gauge) $\mathbf{A} = \tfrac{1}{2}\mathbf{B} \times \mathbf{R}$, and hence $\mathbf{P} \cdot \mathbf{A} + \mathbf{A} \cdot \mathbf{P} = (\mathbf{B} \times \mathbf{R}) \cdot \mathbf{P} = \mathbf{B} \cdot (\mathbf{R} \times \mathbf{P}) = \mathbf{B} \cdot \mathbf{L}$. Here, according to p. 191, a point charge q of mass m with orbital angular momentum $\mathbf{L}$ has magnetic moment $\mu = \frac{1}{2m} q\mathbf{L}$, giving a potential energy $-\mu \cdot \mathbf{B}$ in addition to $q\Phi$.

However, this ansatz does not suffice for electrons in a magnetic field because they have one more inner moment, which is connected to their spin and which has not been accounted for so far. Here it has been shown that the *Pauli equation*, viz.,

$$
\begin{aligned}
H &= \frac{P^2 - q\,\mathbf{B} \cdot (\mathbf{L} + \hbar\boldsymbol{\sigma}) + q^2 A^2}{2m} + q\,\Phi \\
&= \frac{(\mathbf{P} - q\,\mathbf{A}) \cdot (\mathbf{P} - q\,\mathbf{A})}{2m} + q\,\Phi - \frac{q\hbar}{2m}\,\boldsymbol{\sigma} \cdot \mathbf{B},
\end{aligned}
$$

is appropriate. The new feature is the last term, where the factor

$$
\mu_{\text{B}} \equiv \frac{e\hbar}{2m}
$$

is known as the *Bohr magneton*. Due to the factor $\boldsymbol{\sigma}$ in the Pauli equation, H acts on a wave function with two components, a *2-spinor*, which we shall discuss in Sect. 4.5.8. For a homogeneous magnetic field $\mathbf{B}$ (we restrict ourselves to this case), the Pauli equation can be brought into the form

$$
H = \frac{\{(\mathbf{P} - q\,\mathbf{A}) \cdot \boldsymbol{\sigma}\}^2}{2m} + q\,\Phi,
$$

since according to p. 325, we have

$$
\{(\mathbf{P} - q\,\mathbf{A}) \cdot \boldsymbol{\sigma}\}^2 = (\mathbf{P} - q\,\mathbf{A}) \cdot (\mathbf{P} - q\,\mathbf{A}) + i\,\{(\mathbf{P} - q\,\mathbf{A}) \times (\mathbf{P} - q\,\mathbf{A})\} \cdot \boldsymbol{\sigma}.
$$

If $\mathbf{P}$ were to commute with $\mathbf{A}$, then the vector product would vanish, but now for $\mathbf{A} = \frac{1}{2}\mathbf{B} \times \mathbf{R}$, the term $\mathbf{P} \times \mathbf{A} + \mathbf{A} \times \mathbf{P} = -i\hbar\,\mathbf{B}$ remains, since $\mathbf{B}$ commutes with $\mathbf{R}$ and $\mathbf{P}$ and hence $\mathbf{P} \times \mathbf{A} + \mathbf{A} \times \mathbf{P} = \frac{1}{2}[\mathbf{R}, \mathbf{B} \cdot \mathbf{P}] - \frac{1}{2}\mathbf{B}(\mathbf{R} \cdot \mathbf{P} - \mathbf{P} \cdot \mathbf{R})$.

In the form $H = \frac{1}{2m}\{(\mathbf{P} - q\,\mathbf{A}) \cdot \boldsymbol{\sigma}\}^2 + q\,\Phi$, the Pauli equation is the non-relativistic limiting case of the Dirac equation (as will be shown in Sect. 5.6.8). Hence the results here do not describe relativistic effects, even though it is sometimes claimed otherwise. Incidentally, $\hbar\boldsymbol{\sigma}$ will appear as the origin of the doublets of the spin momentum $\mathbf{S}$ in the next section. In the Pauli equation, it thus occurs as the scalar product $(\mathbf{L} + 2\,\mathbf{S}) \cdot \mathbf{B}$. The spin momentum enters with twice the weight (*magneto-mechanical anomaly*). So this factor of 2 is not a relativistic effect.

If the classical equations are valid for operators in quantum theory, this will also apply for the expectation values. However, the expectation value of a product is not generally equal to the product of the expectation values—that would only be true for eigenstates. Hence, generally, we also have $\overline{A}^2 \neq \overline{A^2}$ and then $\Delta A \geq 0$.

4.3.8   Angular Momentum Operator

The orbital angular momentum operator is defined by

$$\mathbf{L} \equiv \mathbf{R} \times \mathbf{P} \,,$$

where the fact that $\mathbf{R}$ and $\mathbf{P}$ do not commute does not create problems, because in the vector product only factors commuting with each other occur together. Hence $\mathbf{L}$ is also Hermitian like $\mathbf{R}$ and $\mathbf{P}$.

From the commutation relations for $\mathbf{R}$ and $\mathbf{P}$, we find

$$[L_x, X] = 0\,, \qquad [L_x, Y] = i\hbar\, Z\,, \qquad [L_x, Z] = -i\hbar\, Y\,,$$
$$[L_x, P_x] = 0\,, \qquad [L_x, P_y] = i\hbar\, P_z\,, \qquad [L_x, P_z] = -i\hbar\, P_y\,,$$
$$[L_x, L_y] = i\hbar\, L_z\,,$$

since we have, e.g., $[L_x, X] = [YP_z - ZP_y, X] = 0$, but

$$[L_x, Y] = -[ZP_y, Y] = Z\,[Y, P_y] = i\hbar Z \,.$$

The above are valid for L_y and L_z, with suitable cyclic permutations. Hence we find the commutator $[L_x, L_y] = [L_x, ZP_x - XP_z] = -i\hbar\, YP_x + i\hbar\, XP_y = i\hbar\, L_z$. Generally, for a vector operator $\mathbf{A}$, we can derive the commutation relation

$$[\mathbf{L} \cdot \mathbf{e}_1 \,,\; \mathbf{A} \cdot \mathbf{e}_2\,] = i\hbar\, \mathbf{A} \cdot (\mathbf{e}_1 \times \mathbf{e}_2)\,,$$

because, according to Hamiltonian mechanics (see p. 130), the angular momentum is the generating function of infinitesimal rotations. In addition, the corresponding equations for the Poisson brackets are valid with $\mathbf{R}$ or $\mathbf{P}$ instead of $\mathbf{A}$ (see Problems 2.44 and 4.30).

The commutation relations $[L_x, L_y] = i\hbar\, L_z$ (and cyclic) mean that there are generally no common eigenvectors for all three components of the angular momentum operator. We can make only one component diagonal. As for the spherical coordinates, we prefer the z-component and choose $\mathbf{e}_z$ as the *quantization direction*. Then the y- and z-components do also have unique expectation values, but with uncertainties. In general, the angular momenta in a state have no sharp direction. They are unsharp (uncertain), as in the time average for each precession, for which only the component along the precession vector is fixed. This is shown in Fig. 4.8.

We have already encountered commutation relations similar to those for the components of the orbital angular momentum $\mathbf{L}$, viz., for the Pauli operators on p. 309. These read $[\sigma_x, \sigma_y] = 2i\, \sigma_z$ and cyclic permutations. Hence with

$$\mathbf{S} = \tfrac{1}{2}\hbar\, \boldsymbol{\sigma} \,,$$

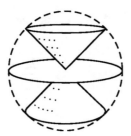

Fig. 4.8 Angular momentum eigenstates. For sharp L^2, all allowable vectors **L** have the same absolute value. They span a sphere (*dashed circle*). Here $l = 1$ is chosen. Then there are three eigenstates $|l, m\rangle$ with sharp L_z, and hence uncertain L_x and L_y. Their angular momentum vectors thus form three cones about the quantization axis. The one for $m = 0$ degenerates to a circle

we conclude

$$[S_x, S_y] = i\hbar S_z \text{ and cyclic permutations.}$$

In fact, we need $\mathbf{S} \equiv \frac{1}{2}\hbar\,\boldsymbol{\sigma}$ for the spin (eigen angular momentum) of electrons and nucleons. But this is easier to treat than the orbital angular momentum, because only two eigenvectors occur. For the three Cartesian components, we have $\sigma_i^2 = 1$, and hence $S^2 \equiv S_x^2 + S_y^2 + S_z^2 = \frac{3}{4}\hbar^2\,1$.

The square of the orbital angular momentum, viz.,

$$L^2 = L_x^2 + L_y^2 + L_z^2 ,$$

is Hermitian and commutes with all components:

$$[L^2, L_z] = 0 = [L^2, L_x] = [L^2, L_y] ,$$

since

$$[L_x^2, L_z] = L_x[L_x, L_z] + [L_x, L_z]L_x$$

is equal to

$$-[L_y, L_z]L_y - L_y[L_y, L_z] = -[L_y^2, L_z] .$$

Hence, there is a complete orthonormal system of eigenvectors of L^2 and L_z.

Since the operators L^2 and L_z are Hermitian, they have real *eigenvalues*, and we shall now seek these, along with a set of common eigenvectors. From the commutation relations, we will determine the eigenvalues $l\,(l+1)\,\hbar^2$ with $l \in \{0, 1, \ldots\}$ of L^2 and the eigenvalues $m\hbar$ with $m \in \{0, \pm 1, \ldots, \pm l\}$ of L_z, where l and m could also take half integer values (1/2, 3/2, etc.). But half-integer values do not lead to a unique real-space representation (see the next section) and are therefore to be discarded. This is different for the inner degree of freedom, where the values $s = 1/2$ and $m = \pm 1/2$ are allowed.

The proof is similar to that for the field operators (see Sect. 4.2.8). We use the non-Hermitian operators

$$L_\pm \equiv L_x \pm iL_y = L_\mp^{\dagger} \quad \Longleftrightarrow \quad L_x = \frac{L_+ + L_-}{2} \;, \quad L_y = \frac{L_+ - L_-}{2i} \;,$$

with the properties

$$[L_z, L_\pm] = \pm\hbar\, L_\pm \;, \quad [L_+, L_-] = 2\hbar\, L_z \;, \quad [L^2, L_\pm] = 0 = [L^2, L_\pm L_\mp] \;.$$

Now let $|a, b\rangle$ be a common eigenvector of L^2 and L_z, so that $L^2 |a, b\rangle = |a, b\rangle\, a\hbar^2$ and $L_z |a, b\rangle = |a, b\rangle\, b\hbar$. Then with the commutation relations, we obtain the following results for $L_\pm |a, b\rangle$:

$$L^2 L_\pm |a, b\rangle = L_\pm |a, b\rangle\, a\hbar^2 \;, \quad L_z L_\pm |a, b\rangle = L_\pm |a, b\rangle\, (b \pm 1)\, \hbar \;.$$

The *ladder operators* $L_\pm$ thus connect eigenstates of L^2 with equal eigenvalue, but with a different eigenvalue of L_z, i.e., $L_\pm |a, b\rangle \propto |a, b \pm 1\rangle$. Hence, we call L_+ a *creation operator* and L_- an *annihilation operator*.

However, the construction method with the ladder operators has to lead to the zero vector after a finite number of steps, and then stop. Otherwise, the norm of the vectors $L_\pm |a, b\rangle$ might become imaginary. From

$$L^2 = L_z^{\,2} + \tfrac{1}{2}\, (L_+ L_- + L_- L_+)$$

and the commutation relation $[L_+,\, L_-] = 2\hbar\, L_z$, it follows that

$$L_\mp L_\pm = L^2 - L_z\, (L_z \pm \hbar) \;,$$

and hence for the squared norm of $L_\pm |a, b\rangle$, which is just the expectation value $\langle a, b| L_\pm^{\dagger} L_\pm |a, b\rangle$, we obtain the value $\{a - b\, (b \pm 1)\}\, \hbar^2$. Hence, the expression must vanish for b_{max} and b_{min}:

$$a = b_{max}\, (b_{max} + 1) = b_{min}\, (b_{min} - 1) \;.$$

We deduce that $b_{min} = -b_{max}$ (or $b_{min} = b_{max} + 1$, but this contradicts $b_{min} \leq b_{max}$). Starting from $|a, b_{min}\rangle$, we must arrive at $|a, b_{max}\rangle$ with the creation operator L_+. Hence, $b_{max} - b_{min} = 2\, b_{max}$ is an integer and the claim is proven. We denote b_{max} by l and usually write m for b. Following the usual practice, we write for short $|l, m\rangle$ instead of $|l\, (l+1), m\rangle$. Incidentally, the orbital angular momentum eigenstates are often not specified by the value of l, but by letters. The first four have historical origin, the rest follow in alphabetical order, without j (see Table 4.1). With the eigenvalue equations

Table 4.1 Coulomb-state "quantum numbers"

l	0	1	2	3	4	5	6	7	...
Name	s	p	d	f	g	h	i	k	...

$$L^2 \,|l, m\rangle = |l, m\rangle \, l\,(l+1)\,\hbar^2 \,, \qquad \text{with } l \in \{0, 1, 2, \ldots\} \,,$$
$$L_z \,|l, m\rangle = |l, m\rangle \, m\,\hbar \,, \qquad \text{with } m \in \{0, \pm 1, \ldots, \pm l\} \,,$$

the phase factors are not yet determined. But since *Condon and Shortley* [5], the phase factor for $L_\pm$ is chosen positive real and the relative phases of the states with equal l are then determined by

$$L_\pm \,|l, m\rangle = |l, m \pm 1\rangle \, \sqrt{l\,(l+1) - m\,(m \pm 1)} \; \hbar$$
$$= |l, m \pm 1\rangle \, \sqrt{(l \mp m)\,(l \pm m + 1)} \; \hbar \,,$$

using $L_\mp L_\pm = L^2 - L_z(L_z \pm \hbar)$. The relative phases of states with unequal l are still free. Hence we can still arrange things so that the matrix elements of all those operators that are invariant under rotations and time-reversal are real. This is possible, e.g., by satisfying the requirement

$$\mathscr{T} \,|l, m\rangle = (-)^{l+m} \,|l, -m\rangle \,.$$

But we shall not deal with this here, because we would then have to investigate the behavior of the states under rotations.

In the states $|l, m\rangle$, the expectation values of $L_\pm$ vanish and so therefore do those of L_x, L_y and $L_+^2 + L_-^2 = 2\,(L_x^2 - L_y^2)$. Consequently, we have $(\Delta L_x)^2 = \langle L_x^2 \rangle = \langle L_y^2 \rangle = (\Delta L_y)^2$:

$$(\Delta L_x)^2 = (\Delta L_y)^2 = \tfrac{1}{2} \langle L^2 - L_z^2 \rangle = \tfrac{1}{2} \{l(l+1) - m^2\}\,\hbar^2 \geq \tfrac{1}{2}\,l\hbar^2 \,.$$

For fixed l, these uncertainties are smallest for $m = \pm l$ and greatest for $m = 0$. Only the s-state is such that all three components of the angular momentum are sharp.

4.3.9 Spherical Harmonics

The spherical harmonics are the real-space representation of the orbital angular momentum eigenstates $|l, m\rangle$. However, it is not the length of the position vector that is important, but only its direction. Hence it is practical to calculate with spherical coordinates (r, θ, φ). With $\langle \mathbf{r} \,|\, \mathbf{R} \,|\mathbf{r}' \rangle = \mathbf{r}\,\delta(\mathbf{r} - \mathbf{r}')$ and $\langle \mathbf{r} \,|\, \mathbf{P} \,|\psi \rangle = -\mathrm{i}\hbar\,\nabla\psi(\mathbf{r})$, we have

$$\langle \mathbf{r} \, | \, \mathbf{L} \, | \psi \rangle = \frac{\hbar}{\mathrm{i}} \, \mathbf{r} \times \nabla \, \psi(\mathbf{r}) \,, \quad \text{with} \ \ \mathbf{r} \times \nabla = -\mathbf{e}_\theta \, \frac{1}{\sin\theta} \, \frac{\partial}{\partial\varphi} + \mathbf{e}_\varphi \, \frac{\partial}{\partial\theta} \,,$$

where (see Fig. 1.12)

$$\begin{aligned}
\mathbf{e}_\theta &= \cos\theta \, (\cos\varphi \, \mathbf{e}_x + \sin\varphi \, \mathbf{e}_y) - \sin\theta \, \mathbf{e}_z \,, \\
\mathbf{e}_\varphi &= \quad\ \ - \sin\varphi \, \mathbf{e}_x + \cos\varphi \, \mathbf{e}_y \,.
\end{aligned}$$

The angular momentum operators thus act only on the angular coordinates $\Omega \equiv (\theta, \varphi)$, not on the length of $\mathbf{r}$. Hence, in the following, we consider

$$\langle \Omega \, | \, lm \rangle \equiv \mathrm{i}^l \, Y_m^{(l)}(\Omega) \,.$$

The factor i^l is a practical phase factor which turns out to be useful for time reversal. In particular, with $\mathscr{T} | \Omega \rangle = | \Omega \rangle$ and $\mathscr{T} | l, m \rangle = (-)^{l+m} | l, -m \rangle$, we find $\langle \Omega | l, m \rangle^* = (-)^{l+m} \langle \Omega | l, -m \rangle$ and hence (with the factor i^l),

$$Y_m^{(l)*}(\Omega) = (-)^m \, Y_{-m}^{(l)}(\Omega) \,.$$

Consequently, all spherical harmonics with $m = 0$ are real—we can even arrange for them all to be positive for Ω in the $\mathbf{z}$-direction, i.e., for $(\theta, \varphi) = (0, 0)$. Without the factor i^l, this would not be possible.

Since L_z in the real-space representation of the operator corresponds to $-\mathrm{i}\hbar \, \partial/\partial\varphi$, and since we also have $L_z | lm \rangle = | lm \rangle \, m\hbar$, the function $\langle \Omega | lm \rangle$ must be connected to φ via the factor $\exp(\mathrm{i}m\varphi)$. It is only unique (mod 2π), if m is an even number—thus also l must be an integer, i.e., $l \in \{0, 1, \ldots\}$. The commutation relations also allow half-integer values, which would be connected with an ambiguity, and this is without contradiction only for unobservable internal coordinates (spin).

We set $Y_m^{(l)}(\Omega) = f_{lm}(\theta) \, \exp(\mathrm{i}m\varphi)$ and determine the unknown function f_{lm} using the ladder operators. With

$$\mathbf{e}_\theta \cdot (\mathbf{e}_x \pm \mathrm{i}\mathbf{e}_y) = \cos\theta \, \exp(\pm \mathrm{i}\varphi) \,, \qquad \mathbf{e}_\varphi \cdot (\mathbf{e}_x \pm \mathrm{i}\mathbf{e}_y) = \pm \mathrm{i} \, \exp(\pm \mathrm{i}\varphi) \,,$$

and consequently also

$$(\mathbf{r} \times \nabla)_\pm = \exp(\pm \mathrm{i}\varphi) \, (- \cot\theta \, \partial/\partial\varphi \pm \mathrm{i} \, \partial/\partial\theta) \,,$$

we have

$$\begin{aligned}
\langle \Omega \, | \, L_\pm \, | \, lm \rangle &= \langle \Omega \, | \, l, m \pm 1 \rangle \, \sqrt{(l \mp m)(l \pm m + 1)} \, \hbar \\
&= \hbar \, \exp(\pm \mathrm{i}\varphi) \left(\pm \frac{\partial}{\partial\theta} + \mathrm{i} \cot\theta \, \frac{\partial}{\partial\varphi} \right) \langle \Omega \, | \, lm \rangle \,.
\end{aligned}$$

Hence, we obtain the differential equation

$$\left(\pm\frac{d}{d\theta} - m \cot\theta\right) f_{lm}(\theta) = f_{l,m\pm1}(\theta) \sqrt{(l\mp m)(l\pm m+1)} \,.$$

In particular, $\langle\Omega| L_\pm | l, \pm l\rangle$ vanishes. Then $(d/d\theta - l \cot\theta) f_{l,\pm l}(\theta) = 0$, and consequently, $f_{l,\pm l}(\theta) \propto \sin^l\theta$. The value of the still missing factors is determined by the normalization condition $\int d\Omega \, |\langle\Omega|lm\rangle|^2 = 1$. From

$$\int_0^\pi \sin^{2l+1}\theta \, d\theta = 2 \, (2^l \, l!)^2/(2l+1)! \,,$$

we deduce an appropriate choice of the phase:

$$Y_{\pm l}^{(l)}(\Omega) = \frac{(\mp)^l}{2^l \, l!} \sqrt{\frac{(2l+1)!}{4\pi}} \, \sin^l\theta \, \exp(\pm i\, l\varphi) \,.$$

The remaining spherical harmonics are now obtained by applying the ladder operators $L_\pm$. However, the operator $\pm d/d\theta - m \cot\theta$ is not quite appropriate here, because it contains two terms. But let us consider the function $\sin^{\mp m}\theta \, f_{lm}$ and take $\cos\theta$ instead of θ as the variable. We only need $0 \le \theta \le \pi$ anyway. Then $d/d\theta = -\sin\theta \, d/d\cos\theta$ leads to

$$\frac{d \sin^{\mp m}\theta \, f_{lm}}{d\cos\theta} = \mp \sin^{\mp m-1}\theta \left(\pm\frac{d}{d\theta} - m\cot\theta\right) f_{lm}$$

$$= \mp \sin^{\mp m-1}\theta \, f_{l,m\pm1} \sqrt{l\mp m}\sqrt{l\pm m+1} \,.$$

After differentiating n times, we have on the right-hand side

$$(\mp)^n \, \sin^{\mp m-n}\theta \, f_{l,m\pm n} \sqrt{\frac{(l\mp m)!}{(l\mp m-n)!} \, \frac{(l\pm m+n)!}{(l\pm m)!}} \,.$$

Hence,

$$f_{l,m\pm n} = (\mp)^n \, \sin^{n\pm m}\theta \, \frac{d^n \sin^{\mp m}\theta \, f_{lm}}{d\cos^n\theta} \sqrt{\frac{(l\pm m)! \, (l\mp m-n)!}{(l\mp m)! \, (l\pm m+n)!}} \,.$$

This *recursion formula* connects all spherical harmonics with equal l to each other. This is achieved by the ladder operators, according to the last section. In particular, with L_- and for $n = m = l$, it leads to $f_{l0} = d^l \sin^l\theta \, f_{ll}/d\cos^l\theta \, (2l)!^{-1/2}$, or (see Fig. 4.9)

$$Y_0^{(l)}(\Omega) = \frac{(-)^l}{2^l \, l!} \sqrt{\frac{2l+1}{4\pi}} \, \frac{d^l \sin^{2l}\theta}{d\cos^l\theta} \equiv \sqrt{\frac{2l+1}{4\pi}} \, P_l(\cos\theta) \,.$$

Fig. 4.9 Spherical harmonics. Their positive real part is shown in *white*, the negative part *hatched*. $l = 0(1)2$ increases upwards from sphere to sphere, m to the right. In addition, there are two frames

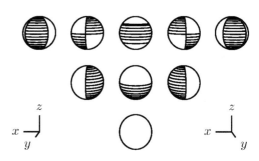

Here $P_l(\cos\theta)$ is a *Legendre polynomial*. We already met them in Sect. 2.2.7, when we considered their *generating function*

$$\frac{1}{\sqrt{1 - 2sz + s^2}} = \sum_{n=0}^{\infty} P_n(z)\, s^n\ , \quad \text{for } |s| < 1\ .$$

They lead to $P_0(z) = 1$, $P_1(z) = z$, and the recursion formula

$$(n + 1)\, P_{n+1}(z) - (2n + 1)\, z\, P_n(z) + n\, P_{n-1}(z) = 0\ .$$

We also proved the orthonormalization condition on p. 82:

$$\int_{-1}^{1} dz\, P_n(z)\, P_{n'}(z) = \frac{2}{2n + 1}\, \delta_{nn'}\ .$$

Hence we can also show the *Rodrigues formula*, viz.,

$$P_n(z) = \frac{1}{2^n\, n!}\, \frac{d^n\, (z^2 - 1)^n}{dz^n}\ .$$

Without this, we would not have met the Legendre polynomials previously at all. If we integrate by parts, where we may assume $n \leq n'$, then we obtain, for $n' > 0$,

$$\int_{-1}^{1} dz\, \frac{d^n(z^2 - 1)^n}{dz^n}\, \frac{d^{n'}(z^2 - 1)^{n'}}{dz^{n'}} = (-)^n \int_{-1}^{1} dz\, \frac{d^{2n}(z^2 - 1)^n}{dz^{2n}}\, \frac{d^{n'-n}(z^2 - 1)^{n'}}{dz^{n'-n}}\ ,$$

with the factor $d^{2n}(z^2 - 1)^n/dz^{2n} = (2n)!$. For $n' > n$, this is zero and otherwise equal to $(2n)! \int_0^\pi d\theta\ \sin^{2n+1}\theta = (2^n n!)^2\, 2/(2n + 1)$. The polynomials defined by Rodrigues' formula are thus also orthonormalized like the Legendre polynomials and are real polynomials of the same degree. Hence, they can differ from each other by at most a sign. But the coefficients for the highest power are positive according to the recursion formula and also according to the Rodrigues formula. This leads to

$$P_n(z) = \frac{1}{2^n} \sum_k (-)^k \frac{(2n-2k)!}{k!\,(n-k)!\,(n-2k)!}\, z^{n-2k} \ .$$

From here, we have $P_n(-z) = (-)^n P_n(z)$.

Clearly, the spherical harmonics with $m = 0$ are real and positive in the z-direction, so the choice of phase for $m = \pm l$ corresponds to our above-mentioned wishes in connection with the factor i^l. Generally, with $m \geq 0$, $f_{l0} = Y_0^{(l)}$ and $f_{l,\pm m} = (\mp)^m \sin^m \theta \ (d^m f_{l0}/d \cos^m \theta) \sqrt{(l-m)!/(l+m)!}$, we obtain the expression

$$Y_{\pm m}^{(l)}(\Omega) = (\mp)^m \sqrt{\frac{2l+1}{4\pi} \frac{(l-m)!}{(l+m)!}} \sin^m \theta \ \frac{d^m P_l(\cos \theta)}{d \cos^m \theta}\, \exp(\pm i\, m\varphi) \ .$$

For spherically symmetric problems, we will often expand the wave functions in terms of these spherical harmonics, beginning in Sect. 4.5.2.

Since $P_l(-\cos \theta) = (-)^l P_l(\cos \theta)$, the spherical harmonics with orbital angular momentum l have parity $(-)^l$, using the standard results $\sin(\pi - \theta) = \sin \theta$, $\cos(\pi - \theta) = -\cos \theta$, and $\exp(\pm i\, m(\varphi + \pi)) = (-)^m \exp(\pm i\, m\varphi)$.

With the spherical harmonics, we know the eigenfunctions of the operator L^2 in the real-space representation (directional representation):

$$\langle \Omega \,|\, L^2 \,|\, lm \rangle = \langle \Omega \,|\, lm \rangle \ l(l+1)\, \hbar^2$$

$$= \left(-\frac{1}{\sin \theta} \frac{\partial}{\partial \theta} \left(\sin \theta \frac{\partial}{\partial \theta} \right) - \frac{1}{\sin^2 \theta} \frac{\partial^2}{\partial \varphi^2} \right) \langle \Omega \,|\, lm \rangle \ \hbar^2 \ .$$

We will need this operator in Sect. 4.5.2 for central fields, because, according to p. 142, the centrifugal potential is proportional to L^2.

4.3.10 Coupling of Angular Momenta

In addition to the orbital angular momentum of electrons and nucleons, we also have to account for their eigen angular momentum (spin). Their total angular momentum involves both. Hence we now consider

$$\mathbf{J} = \mathbf{L} + \mathbf{S} \ .$$

Since $\mathbf{L}$ acts in real space and $\mathbf{S}$ in spin space, the two operators commute. $\mathbf{J}$ is Hermitian like $\mathbf{L}$ and $\mathbf{S}$, so

$$[J_x, J_y] = i\hbar J_z \ , \quad \text{and cyclic permutations.}$$

Hence the considerations in Sect. 4.3.8 deliver

$$J^2 \,|j, m\rangle = |j, m\rangle \, j(j+1)\, \hbar^2 \,, \quad \text{for } j \in \{0, \tfrac{1}{2}, 1, \ldots\}\,,$$

$$J_z \,|j, m\rangle = |j, m\rangle \, m\, \hbar \,, \quad \text{for } m \in \{j, j-1, \ldots, -j\}\,,$$

$$J_\pm \,|j, m\rangle = |j, m\pm 1\rangle \, \sqrt{j(j+1) - m(m\pm 1)}\, \hbar$$

$$\qquad\qquad = |j, m\pm 1\rangle \, \sqrt{(j\mp m)\,(j\pm m+1)}\, \hbar \,,$$

$$\mathscr{T} \,|j, m\rangle = (-)^{j+m} \, |j, -m\rangle \,.$$

We would now like to apply these general equations to the spin $1/2$ case.

Here we could take the *uncoupled representation* $|l, m_l; \tfrac{1}{2}, m_s\rangle$ which diagonal-izes L^2, L_z, S^2, and S_z. But if there is a *spin–orbit coupling*, which we derive from the operator product

$$\mathbf{L} \cdot \mathbf{S} = L_z S_z + \tfrac{1}{2}(L_+ S_- + L_- S_+) \,,$$

then neither L_z nor S_z will be sharp, only their sum J_z. Then the *coupled representation* $|(l, \tfrac{1}{2})j, m\rangle$ is more useful, because it simultaneously diagonalizes L^2, S^2, J^2, and J_z, and hence also $2\mathbf{L} \cdot \mathbf{S} = J^2 - L^2 - S^2$. With $J_z = L_z + S_z$, we then have $m = m_l + m_s$, and for a given l, $m \le l + \tfrac{1}{2} = j$. In fact, $|l, l; \tfrac{1}{2}, \tfrac{1}{2}\rangle$ is also an eigenstate of $J^2 = L^2 + 2\mathbf{L} \cdot \mathbf{S} + S^2$, since with $2\mathbf{L} \cdot \mathbf{S} = 2L_z S_z + L_+ S_- + L_- S_+$, $L_+ |ll\rangle = |o\rangle$, and $S_+ |\tfrac{1}{2}\tfrac{1}{2}\rangle = |o\rangle$, we find that $\{l\,(l+1) + 2\,l\,\tfrac{1}{2} + \tfrac{3}{4}\}\, \hbar^2$ is an eigenvalue, and with $j = l + \tfrac{1}{2}$, this can also be written as $j(j+1)\, \hbar^2$. Hence for $j = l + \tfrac{1}{2}$, we may set the two states $|l, l; \tfrac{1}{2}, \tfrac{1}{2}\rangle$ and $|(l, \tfrac{1}{2})\, l+\tfrac{1}{2}, l+\tfrac{1}{2}\rangle$ equal to each other. Here we finally fix the phase of the coupled state. The remaining states with $j = l + \tfrac{1}{2}$ are obtained from there with the creation operator $J_- = L_- + S_-$. Since we restrict ourselves here to $s = \tfrac{1}{2}$, the operator S_-^2 turns out to be zero, and then we have $J_-^n = L_-^n + n L_-^{n-1} S_-$. For an appropriate choice of the phase and with $J_-^n |jj\rangle = |j, j - n\rangle \sqrt{(2j)!\, n!/(2j - n)!}\, \hbar^n$, it follows that

$$|(l, \tfrac{1}{2})\, l+\tfrac{1}{2}, m\rangle = |l, m+\tfrac{1}{2}; \tfrac{1}{2}, -\tfrac{1}{2}\rangle \sqrt{\frac{l+\tfrac{1}{2}-m}{2l+1}} + |l, m-\tfrac{1}{2}; \tfrac{1}{2}, \tfrac{1}{2}\rangle \sqrt{\frac{l+\tfrac{1}{2}+m}{2l+1}}\,.$$

We then have all $2j + 1 = 2l + 2$ states with $j = l + \tfrac{1}{2}$ in the coupled basis expanded in terms of the uncoupled states. But in the uncoupled basis, there are $(2l + 1) \cdot 2$ states with equal l, thus $2l$ more states. In fact, we can also couple with $j = l - \tfrac{1}{2}$. These states have to be orthogonal to those with equal l and m, so the expansion coefficients are

$$|(l, \tfrac{1}{2})\, l-\tfrac{1}{2}, m\rangle = |l, m+\tfrac{1}{2}; \tfrac{1}{2}, -\tfrac{1}{2}\rangle \sqrt{\frac{l+\tfrac{1}{2}+m}{2l+1}} - |l, m-\tfrac{1}{2}; \tfrac{1}{2}, \tfrac{1}{2}\rangle \sqrt{\frac{l+\tfrac{1}{2}-m}{2l+1}}\,.$$

We may also include a phase factor. The phase of the coupled state remains free to choose—only the relative phases of the states with different m are already fixed by the choice of matrix elements of $J_\pm$. The last equation obeys a second requirement due to *Condon and Shortley*, namely, for $j_1 + j_2 \geq j \geq |j_1 - j_2|$,

$$\langle(j_1, j_2)j, j|j_1, j_1; \ j_2, j-j_1\rangle = \langle j_1, j_1; \ j_2, j-j_1|(j_1, j_2)j, j\rangle > 0 \ ,$$

i.e., all coefficients with $m = j$ and $m_1 = j_1$ are to be positive.

Hence all expansion coefficients of the angular momentum coupling, i.e., all *Clebsch–Gordan coefficients*, are now real. Here we adopt the abbreviation

$$\begin{pmatrix} j_1 & j_2 & j \\ m_1 & m_2 & m \end{pmatrix} \equiv \langle j_1, m_1; \ j_2, m_2|(j_1, j_2)j, m\rangle = \langle(j_1, j_2)j, m|j_1, m_1; \ j_2, m_2\rangle \ ,$$

but other notations do occur. We have now derived, e.g.,

$$\begin{pmatrix} l & \frac{1}{2} & l\pm\frac{1}{2} \\ m+\frac{1}{2} & -\frac{1}{2} & m \end{pmatrix} = \sqrt{\frac{l+\frac{1}{2}\mp m}{2l+1}} = \mp \begin{pmatrix} l & \frac{1}{2} & l\mp\frac{1}{2} \\ m-\frac{1}{2} & \frac{1}{2} & m \end{pmatrix} .$$

Likewise we can now couple two spin-$\frac{1}{2}$ states to *triplet* and *singlet states*. If instead of $|\frac{1}{2}, \frac{1}{2}\rangle$ we write for short $|\uparrow\rangle$ (*spin up*), and instead of $|\frac{1}{2}, -\frac{1}{2}\rangle$ the abbreviation $|\downarrow\rangle$ (*spin down*), it follows that

$$|(\tfrac{1}{2}, \tfrac{1}{2})\,1, +1\rangle = |\uparrow\uparrow\rangle \ , \qquad |(\tfrac{1}{2}, \tfrac{1}{2})\,1, 0\rangle = \frac{|\uparrow\downarrow\rangle + |\downarrow\uparrow\rangle}{\sqrt{2}} \ ,$$

$$|(\tfrac{1}{2}, \tfrac{1}{2})\,1, -1\rangle = |\downarrow\downarrow\rangle \ , \qquad |(\tfrac{1}{2}, \tfrac{1}{2})\,0, 0\rangle = \frac{|\uparrow\downarrow\rangle - |\downarrow\uparrow\rangle}{\sqrt{2}} \ .$$

The triplet states are thus symmetric under exchange of the two uncoupled states, while the singlet state is antisymmetric.

4.3.11 Summary: Correspondence Principle

In the last three sections, we have worked out the basic features of quantum theory. The observables of classical mechanics become Hermitian operators, and relations between measurable quantities become operator equations. Important here is the commutation behavior. The commutator corresponds to the classical Poisson bracket, except for the factor $i\hbar$. The factor i has to occur for a quantity to be Hermitian, while here $\hbar$ introduces Planck's action quantum as a scale factor.

The comparison of the position and momentum representations $\{|\mathbf{r}\rangle\}$ and $\{|\mathbf{p}\rangle\}$ is instructive. These diagonalize the position and momentum operators, respectively. In particular, from the basic commutation relation $[X^k, P_{k'}] = i\hbar\,\delta^k_{k'}\,\mathbf{1}$, we have derived the representation of each operator in the other basis, and also $\langle\mathbf{r}|\mathbf{p}\rangle = \langle\mathbf{p}|\mathbf{r}\rangle^* = (2\pi\hbar)^{-3/2}\exp(i\mathbf{p}\cdot\mathbf{r}/\hbar)$. This probability amplitude is usually called the *wave function of the state with momentum* $\mathbf{p}$. For the derivation we used the equation

Fig. 4.10 Eigenvalues of the angular momentum operator for $m \in \{-j, \ldots, j\}$ and $j \in \{0, \frac{1}{2}, 1, \ldots\}$. Half-integer eigenvalues (*open circles*) occur only for spin momenta, because the real-space representations are then ambiguous

$x\,\delta'(x) = -\delta(x)$ and thus found

$$P_k \mathrel{\widehat{=}} \frac{\hbar}{i}\,\frac{\partial}{\partial x^k} \quad \text{and} \quad X^k \mathrel{\widehat{=}} i\hbar\,\frac{\partial}{\partial p_k}$$

for P_k in the position representation and X^k in the momentum representation. If we do not use Cartesian coordinates, then covariant and contravariant components are different. Note that the metric fundamental tensor generally depends upon the position. For the kinetic energy, which is a scalar, we need, e.g., the quantity $\sum_k P_k P^k \mathrel{\widehat{=}} -\hbar^2 \Delta$. We have already derived the Laplace operator for general coordinates on p. 38:

$$\Delta\psi = \frac{1}{\sqrt{g}}\sum_{ik}\frac{\partial}{\partial x^i}\left(\sqrt{g}\,g^{ik}\,\frac{\partial\psi}{\partial x^k}\right), \quad \text{with} \quad g \equiv \det(g_{ik})\,.$$

We have also investigated the way the non-commutability of operators affects physical laws for the case of the angular momentum. For $l \neq 0$, only one directional component can be sharp, along with the square of the angular momenta, which has eigenvalues $l\,(l+1)\,\hbar^2$ with $l \in \{0, 1, 2, \ldots\}$. The directional quantum number m for a given l can only be an integer between $-l$ and $+l$. Using $\mathbf{L} = \mathbf{R} \times \mathbf{P}$, we derived these properties from those of $\mathbf{R}$ and $\mathbf{P}$ (see Fig. 4.10).

4.4 Time Dependence

4.4.1 Heisenberg Equation and the Ehrenfest Theorem

We now consider time dependence. We shall be guided once again by classical physics.

If a is a function of the canonical position and momentum coordinates, and also of the time, we have

$$\frac{da}{dt} = \sum_k \left(\frac{\partial a}{\partial x^k}\frac{dx^k}{dt} + \frac{\partial a}{\partial p_k}\frac{dp_k}{dt}\right) + \frac{\partial a}{\partial t}\,.$$

As already shown on p. 124, using the Hamilton equations

$$\frac{dx^k}{dt} = \frac{\partial H}{\partial p_k} , \quad \frac{dp_k}{dt} = -\frac{\partial H}{\partial x^k} ,$$

we find classically

$$\frac{da}{dt} = \sum_k \left(\frac{\partial a}{\partial x^k} \frac{\partial H}{\partial p_k} - \frac{\partial a}{\partial p_k} \frac{\partial H}{\partial x^k} \right) + \frac{\partial a}{\partial t} \equiv [a, H] + \frac{\partial a}{\partial t} .$$

The derivative da/dt is thus equal to the Poisson bracket $[a, H]$, if we disregard any explicit time dependence.

Now, in quantum theory, on p. 316 we already assigned the commutator of the corresponding operators (divided by $i\hbar$) to the classical Poisson bracket. This idea for translating between the classical and quantum cases leads us to

$$\frac{dA}{dt} = \frac{[A, H]}{i\hbar} + \frac{\partial A}{\partial t} ,$$

known as the *Heisenberg equation*. Here we have to take any time-independent representation and then differentiate each matrix element of A with respect to time in order to form dA/dt (in this representation). We shall usually restrict ourselves to operators A, which do not depend on time explicitly. Then all operators commuting with H (their eigenvalues are called *good quantum numbers*) are constants of the motion, in particular the Hamilton operator H itself. Hence the energy representation, which diagonalizes H, is particularly important, and we shall consider many examples in the next section. Note that friction effects are beyond the scope of this section and will be treated only in Sect. 4.4.3.

With the Heisenberg equation, we can now determine the derivatives of expectation values with respect to the time, taking time-independent states as the basis:

$$\frac{d \langle A \rangle}{dt} = \frac{i}{\hbar} \langle [H, A] \rangle + \frac{\partial \langle A \rangle}{\partial t} .$$

If we use here $H = P^2/2m + V(\mathbf{R})$ and determine the derivatives of $\langle \mathbf{R} \rangle$ and $\langle \mathbf{P} \rangle$ with respect to time, then $\langle [P^2, \mathbf{R}] \rangle$ is important in the first case and $\langle [V(\mathbf{R}), \mathbf{P}] \rangle$ in the second. Now $[P^2, X] = [P_x^2, X] = -2i\hbar P_x$, and in addition (according to p. 316), $[f(X), P] = i\hbar f'(X)$ holds. Consequently, the following equations are valid:

$$\frac{d \langle \mathbf{R} \rangle}{dt} = \frac{\langle \mathbf{P} \rangle}{m} \quad \text{and} \quad \frac{d \langle \mathbf{P} \rangle}{dt} = \langle -\nabla V \rangle \equiv \langle \mathbf{F} \rangle .$$

Thus the expectation values satisfy the equations of classical physics, which is known as *Ehrenfest's theorem*, although $\langle \mathbf{F}(\mathbf{R}) \rangle$ does not need to be equal to $\mathbf{F}(\langle \mathbf{R} \rangle)$.

In order to see how the uncertainties in **R** and **P** change with time, we determine

$$\frac{d\left(\langle \mathbf{R} \cdot \mathbf{R}\rangle - \langle \mathbf{R}\rangle \cdot \langle \mathbf{R}\rangle\right)}{dt} = \frac{\langle \mathbf{P} \cdot \mathbf{R} + \mathbf{R} \cdot \mathbf{P}\rangle - 2\langle \mathbf{R}\rangle \cdot \langle \mathbf{P}\rangle}{m} ,$$

$$\frac{d\left(\langle \mathbf{P} \cdot \mathbf{P}\rangle - \langle \mathbf{P}\rangle \cdot \langle \mathbf{P}\rangle\right)}{dt} = \langle \mathbf{P} \cdot \mathbf{F} + \mathbf{F} \cdot \mathbf{P}\rangle - 2\langle \mathbf{P}\rangle \cdot \langle \mathbf{F}\rangle .$$

For a constant force (e.g., in the free case), we have $\langle \mathbf{P} \cdot \mathbf{F}\rangle = \langle \mathbf{P}\rangle \cdot \mathbf{F} = \langle \mathbf{F} \cdot \mathbf{P}\rangle$. Thus then the momentum uncertainty remains constant, and for sharp momentum, so does the position uncertainty.

4.4.2 Time Dependence: Heisenberg and Schrödinger Pictures

In the last section, we started from the so-called *Heisenberg picture. In the Heisenberg picture the observables depend on the time, but the states do not*:

$$\frac{d}{dt} A_H = \frac{i}{\hbar} [H_H, A_H] + \frac{\partial}{\partial t} A_H , \quad \frac{d}{dt} |\psi_H\rangle = |o\rangle .$$

To solve the Heisenberg equation, we search for a time-dependent unitary transformation U which connects the operator A_H (A in the Heisenberg picture) with an operator A_S (A in the Schrödinger picture), which does not depend upon time:

$$A_S = U A_H U^\dagger , \quad \text{with} \quad \frac{dA_S}{dt} = 0 .$$

Hence the Heisenberg equation delivers

$$0 = \frac{dU}{dt} A_H U^\dagger + U \left(\frac{i}{\hbar} [H_H, A_H] + \frac{\partial A_H}{\partial t} \right) U^\dagger + U A_H \frac{dU^\dagger}{dt} .$$

If we restrict ourselves to observables which depend on time only implicitly, whence $\partial A_H / \partial t = 0$, then this condition can be satisfied for all operators A_H if the unitary operator U satisfies

$$\frac{dU}{dt} + \frac{i}{\hbar} U H_H = 0 \quad \Longleftrightarrow \quad \frac{dU^\dagger}{dt} - \frac{i}{\hbar} H_H U^\dagger = 0 .$$

Here the zero times coincide in the two pictures: $U(0) = 1$ or $A_H(0) = A_S$. Both requirements are satisfied by the *time-shift operator*

$$U(t) = \exp \frac{-i H_H t}{\hbar} ,$$

if H_H does not depend on time (otherwise we still have to integrate, as we shall see in Sect. 4.4.4). Note that, since p. 317, we already know of similar position- and momentum-shift operators. The Hamilton operator in this situation also commutes with U, and hence $H_H = H_S = H$. We shall now restrict ourselves to this case.

In addition, from $|\psi_S\rangle = U |\psi_H\rangle$, we can say that: *In the Schrödinger picture the states do not depend on time, but the observables (Schrödinger equation) do*:

$$\frac{d}{dt} A_S = 0 , \qquad \frac{d}{dt} |\psi_S\rangle = -\frac{i}{\hbar} H |\psi_S\rangle .$$

In general, differential equations for Hilbert space vectors are easier to integrate than those for operators (the Heisenberg equation). Hence, we shall work mainly in the Schrödinger picture and leave out the subscript S. In particular, we then have, in the real-space representation,

$$\frac{\hbar}{i} \frac{\partial \psi(t, \mathbf{r})}{\partial t} + H \psi(t, \mathbf{r}) = 0 ,$$

where $H(\mathbf{R}, \mathbf{P}) \stackrel{\wedge}{=} H(\mathbf{r}, -i\hbar \nabla)$ is to be taken. This equation is similar to the Hamilton–Jacobi differential equation of p. 135, viz.,

$$\frac{\partial W}{\partial t} + H(\mathbf{r}, \nabla W) = 0 ,$$

if Hamilton's *action function* $W = \int L \, dt$ is replaced by $-i\hbar \psi$, with Planck's *action quantum* $h = 2\pi \hbar$. However, instead of $\nabla W \cdot \nabla W$, we have not $-\hbar^2 \nabla \psi \cdot \nabla \psi$, but rather $-\hbar^2 \nabla \cdot \nabla \psi = -\hbar^2 \Delta \psi$.

If we restrict ourselves to particles of mass m and charge q in an electric potential Φ, then the *time-dependent Schrödinger equation* (in the real-space representation) reads

$$i\hbar \frac{\partial}{\partial t} \psi(t, \mathbf{r}) = \left(-\frac{\hbar^2}{2m} \Delta + V(\mathbf{r}) \right) \psi(t, \mathbf{r}) ,$$

with $V(\mathbf{r}) = q \, \Phi(\mathbf{r})$. If we consider the wave function associated with an eigenstate of H with the sharp energy E_n, where the zero energy may be chosen arbitrarily (a different zero leads only to a new time-dependent phase factor in the wave function, which will not affect the experimental value), we have

$$\psi_n(t, \mathbf{r}) = \exp \frac{-i E_n t}{\hbar} \psi_n(\mathbf{r}) , \quad \text{with} \quad \psi_n(\mathbf{r}) \equiv \psi_n(0, \mathbf{r}) ,$$

and it only remains to solve the *time-independent Schrödinger equation* (in the real-space representation)

$$E_n \, \psi_n(\mathbf{r}) = \left(-\frac{\hbar^2}{2m}\, \Delta + V(\mathbf{r})\right) \psi_n(\mathbf{r}) \, .$$

For a magnetic field, instead of V and in addition to $q\, \Phi$, further terms are still to be considered, as was shown in Sect. 4.3.7. Since for all states with sharp energy, the time appears only in the phase factor, which does not affect the expectation values, they are called *stationary states*.

In the Schrödinger picture, if we transform with any time-dependent unitary operator U, we obtain

$$|\psi'\rangle = U\,|\psi\rangle \, , \quad \text{with} \quad \mathrm{i}\hbar\frac{\mathrm{d}}{\mathrm{d}t}\,|\psi\rangle = H\,|\psi\rangle \quad \text{and} \quad \mathrm{i}\hbar\frac{\mathrm{d}}{\mathrm{d}t}\,|\psi'\rangle = H'\,|\psi'\rangle \, ,$$

and clearly also $\mathrm{i}\hbar\,(\dot{U}\,|\psi\rangle + U\,|\dot{\psi}\rangle) = H'U\,|\psi\rangle$, or

$$H' = \mathrm{i}\hbar\frac{\mathrm{d}U}{\mathrm{d}t}\,U^\dagger + U H\, U^\dagger \, .$$

An example of an application is the unitary transformation to $H' = 0$, which clearly results in $\mathrm{i}\hbar\,\dot{U} = -UH$, or $U = \exp(\mathrm{i}Ht/\hbar)$ (if H does not depend on time). This corresponds to the transition from the Schrödinger to the Heisenberg picture, the states of which do not depend on time.

4.4.3 Time Dependence of the Density Operator

The density operator turns out to be useful also for the time dependence. In particular, we may also use time-independent expansion bases in the Schrödinger picture if the density operator takes care of the time dependence. In the Heisenberg picture, it does not depend on time.

According to p. 312, unitary transformations do not change expectation values. Hence for the time dependence, the notation

$$\langle A \rangle = \mathrm{tr}\{U(t)\, \rho_{\mathrm{H}}\, U^\dagger(t)\, A_{\mathrm{S}}\}$$

is to be preferred, since ρ_{H} and A_{S} do not depend on time, and in addition to $U^\dagger(t)\, A_{\mathrm{S}}\, U(t) = A_{\mathrm{H}}(t)$, we have

$$\rho_{\mathrm{S}}(t) = U(t)\, \rho_{\mathrm{H}}\, U^\dagger(t) \, .$$

We can read off from this that the density operator $\rho_{\mathrm{S}}(t)$ and the observables $A_{\mathrm{H}}(t)$ depend oppositely (contravariantly) on time. With $\rho = U\rho_{\mathrm{H}}U^\dagger$ (leaving out the subscript S), we have the *von Neumann equation*

$$\frac{\mathrm{d}\rho}{\mathrm{d}t} = \frac{[H, \rho]}{\mathrm{i}\hbar} .$$

The equation $\mathrm{d}\rho/\mathrm{d}t = 0$ in the Heisenberg picture corresponds classically to the Liouville equation (see p. 129) $\mathrm{d}\rho/\mathrm{d}t = 0$, which is then reformulated as $\partial\rho/\partial t + [\rho, H] = 0$, because the classical probability density ρ (in phase space) depends upon further variables in addition to t. The density operator depends only on time, the other variables being selected only with their representation. Hence, it does not make sense to write the von Neumann equation (as an operator equation) with the partial derivative $\partial\rho/\partial t = [H, \rho]/\mathrm{i}\hbar$.

In the energy representation, that is, with $H |n\rangle = |n\rangle E_n$, $\langle n|n'\rangle = \delta_{nn'}$, and $\sum_n |n\rangle\langle n| = 1$, the von Neumann equation implies

$$\langle n| \rho (t) |n'\rangle = \langle n| \rho (0) |n'\rangle \, \exp \frac{-\mathrm{i} (E_n - E_{n'}) t}{\hbar} .$$

Only the energy differences are important here—the zero of the energy does not affect the density matrix.

According to the von Neumann equation, none of the expectation values of powers of ρ depend on time, since $\mathrm{d}\langle\rho^n\rangle/\mathrm{d}t \propto \mathrm{tr}(\rho^n[H, \rho])$ always vanishes. This does not lead to arbitrarily many invariants, but to exactly N constants of the motion in an N-dimensional Hilbert space (the normalization condition $\langle\rho^0\rangle = \langle 1\rangle = 1$ counts here). In particular, the purity of a state remains $(\mathrm{tr}\rho^2)$, something that is changed only by dissipation (see Sect. 4.6), and this cannot be described with Hamiltonian mechanics.

The von Neumann equation becomes rather simple for doublets. For these, according to pp. 309 and 312, we have

$$H = \frac{1 \, \mathrm{tr}H + \boldsymbol{\sigma} \cdot \mathrm{tr}(\boldsymbol{\sigma} H)}{2} \quad \text{and} \quad \rho = \frac{1 + \boldsymbol{\sigma} \cdot \langle\boldsymbol{\sigma}\rangle}{2} .$$

We thus search for $\mathrm{d}\langle\boldsymbol{\sigma}\rangle/\mathrm{d}t = \mathrm{tr}(\boldsymbol{\sigma}\,\mathrm{d}\rho/\mathrm{d}t)$. Now $\mathrm{tr}(\boldsymbol{\sigma}\,[H, \rho]) = \langle[\boldsymbol{\sigma}, H]\rangle$. The commutator of $\boldsymbol{\sigma}$ with $\frac{1}{2}\boldsymbol{\sigma} \cdot \mathrm{tr}(\boldsymbol{\sigma} H)$ is thus important, and according to p. 325, this can be derived from the expression $\mathrm{i}\,\mathrm{tr}(\boldsymbol{\sigma} H) \times \boldsymbol{\sigma}$. Hence we obtain in total

$$\frac{\mathrm{d}\langle\boldsymbol{\sigma}\rangle}{\mathrm{d}t} = \boldsymbol{\Omega} \times \langle\boldsymbol{\sigma}\rangle , \quad \text{with} \quad \boldsymbol{\Omega} \equiv \frac{\mathrm{tr}(\boldsymbol{\sigma} H)}{\hbar} ,$$

as for the motion (see p. 92). A well known example is the *Larmor precession* of a magnetic moment in a magnetic field, where $H = -\mu_B \, \boldsymbol{\sigma} \cdot \mathbf{B}$ appears as the Hamilton operator in the Pauli equation (p. 327), whence $\mathrm{tr}(\boldsymbol{\sigma} H) = -2\mu_B \, \mathbf{B}$.

For the Larmor precession, $\langle\boldsymbol{\sigma}\rangle$ denotes the spin polarization. But in general we may also understand $|\uparrow\rangle$ and $|\downarrow\rangle$ as states other than those with $m_s = \pm\frac{1}{2}$. We then speak generally of a *Bloch vector* $\langle\boldsymbol{\sigma}\rangle$. According to p. 308, with $\Psi\Psi^\dagger + \Psi^\dagger\Psi = 1$, we then have

$$H = \Psi^\dagger \Psi \, H_{\uparrow\uparrow} + \Psi^\dagger \, H_{\uparrow\downarrow} + \Psi \, H_{\downarrow\uparrow} + \Psi\Psi^\dagger \, H_{\downarrow\downarrow}$$

$$= 1 \, \frac{H_{\uparrow\uparrow} + H_{\downarrow\downarrow}}{2} + \sigma_x \, \frac{H_{\uparrow\downarrow} + H_{\uparrow\uparrow}}{2} + \sigma_y \, \frac{H_{\downarrow\uparrow} - H_{\uparrow\downarrow}}{2i} + \sigma_z \, \frac{H_{\uparrow\uparrow} - H_{\downarrow\downarrow}}{2} \, .$$

With $\mathrm{tr}\,\boldsymbol{\sigma} = \mathbf{0}$, $\mathrm{tr}\,\boldsymbol{\sigma}\sigma_i = 2\mathbf{e}_i$, and $H_{\uparrow\downarrow} = H_{\downarrow\uparrow}{}^*$, we obtain

$$tr(\boldsymbol{\sigma} \, H) = 2 \left(\mathbf{e}_x \, \mathrm{Re}\, H_{\downarrow\uparrow} + \mathbf{e}_y \, \mathrm{Im}\, H_{\downarrow\uparrow} + \mathbf{e}_z \, \frac{H_{\uparrow\uparrow} - H_{\downarrow\downarrow}}{2} \right) = \hbar\boldsymbol{\Omega} \, ,$$

and this vector determines the precession of the Bloch vectors $\langle \boldsymbol{\sigma} \rangle$ in a space whose z-component contains information about the occupation of the states $|\uparrow\rangle$ and $|\downarrow\rangle$. Here $\hbar\Omega$ tells us how much the two energy eigenvalues differ from each other, which follows from $\det(H - E) = 0$. According to p. 309, we have in particular, $E_\pm = \frac{1}{2}\,(\mathrm{tr}H \pm \hbar\Omega)$, where $\hbar\Omega$ is the square-root of $(\mathrm{tr}H)^2 - 4\det H = (H_{\uparrow\uparrow} - H_{\downarrow\downarrow})^2 + 4|H_{\downarrow\uparrow}|^2$.

The considerations can be transferred from 2 to N dimensions of the Hilbert space, if, according to Sect. 4.2.5, we start from a basis $\{C_n\}$ of time-independent Hermitian operators. In particular, according to the von Neumann equation (see p. 313),

$$\mathrm{tr}\rho^2 = \frac{1}{c} \sum_n \langle C_n \rangle^2$$

is conserved, and for $C_0 = \sqrt{c/N}\,1$, so is $\langle C_0 \rangle$. The Bloch vector with real components $\langle C_1 \rangle, \ldots$ has the same length at all times. Here, according to the von Neumann equation, we have

$$\frac{\mathrm{d}\langle C_n \rangle}{\mathrm{d}t} = \sum_{n'} \Omega_{nn'} \, \langle C_{n'} \rangle \, , \quad \text{with} \quad \hbar\Omega_{nn'} \equiv \frac{\mathrm{tr}(\mathrm{i}H\,[C_n, C_{n'}])}{c} \, ,$$

and for $C_0 \propto 1$, we may restrict ourselves to $n \neq 0 \neq n'$. If H does not depend on time, then neither do any of the coefficients $\Omega_{nn'}$ of the system of linear differential equations. Since they are all real and form a skew-symmetric matrix,

$$\Omega_{nn'} = \Omega_{nn'}{}^* = -\Omega_{n'n} \, ,$$

their eigenvalues are purely imaginary and pairwise complex-conjugate to each other.

The von Neumann equation also yields the time dependence of the Wigner function from Sect. 4.3.5:

$$\rho(t, \mathbf{r}, \mathbf{p}) = \frac{1}{(\pi\hbar)^3} \int \mathrm{d}^3 r' \, \langle \mathbf{r} - \mathbf{r}'| \rho(t) \, |\mathbf{r} + \mathbf{r}' \rangle \, \exp \frac{+2\mathrm{i}\,\mathbf{p} \cdot \mathbf{r}'}{\hbar}$$

$$= \frac{1}{(\pi\hbar)^3} \int \mathrm{d}^3 p' \, \langle \mathbf{p} - \mathbf{p}'| \rho(t) \, |\mathbf{p} + \mathbf{p}' \rangle \, \exp \frac{-2\mathrm{i}\,\mathbf{r} \cdot \mathbf{p}'}{\hbar} \, .$$

With $\langle \mathbf{p} - \mathbf{p}' | [P^2, \rho] | \mathbf{p} + \mathbf{p}' \rangle = -4\mathbf{p} \cdot \mathbf{p}' \langle \mathbf{p} - \mathbf{p}' | \rho | \mathbf{p} + \mathbf{p}' \rangle$, we have in particular $[P^2, \rho]/i\hbar = -2\mathbf{p} \cdot \nabla\rho$, while on the other hand, if V depends upon the position only locally, i.e., if we have $\langle \mathbf{r} | V | \mathbf{r}' \rangle = V(\mathbf{r})\, \delta(\mathbf{r} - \mathbf{r}')$, then

$$\frac{\partial \rho(t, \mathbf{r}, \mathbf{p})}{\partial t} + \frac{\mathbf{p}}{m} \cdot \nabla \rho(t, \mathbf{r}, \mathbf{p})$$

$$= -\frac{i}{\hbar} \frac{1}{(\pi\hbar)^3} \int d^3 r' \, \{V(\mathbf{r} - \mathbf{r}') - V(\mathbf{r} + \mathbf{r}')\} \, \langle \mathbf{r} - \mathbf{r}' | \rho | \mathbf{r} + \mathbf{r}' \rangle \, \exp \frac{2i\mathbf{p} \cdot \mathbf{r}'}{\hbar} \,.$$

For a harmonic oscillation, the right-hand side can be traced back to the expression $\nabla V \cdot \nabla_p \rho(t, \mathbf{r}, \mathbf{p})$, i.e., to the gradient of ρ in momentum space. With $\mathbf{p}/m = \mathbf{v}$, we thus have in the harmonic approximation (and naturally also for the free motion with $\mathbf{F} = 0$),

$$\left(\frac{\partial}{\partial t} + \mathbf{v} \cdot \nabla_r + \mathbf{F} \cdot \nabla_p \right) \rho(t, \mathbf{r}, \mathbf{p}) = 0 \,.$$

This is the *collision-free Boltzmann equation*, which holds quite generally in classical mechanics (and also for other potentials, see Sect. 6.2.3), where $\rho(t, \mathbf{r}, \mathbf{p})$ is then the probability density in phase space.

4.4.4 Time-Dependent Interaction and Dirac Picture

In addition to the Heisenberg and Schrödinger pictures, there is also the Dirac picture, often called the *interaction representation*, used in particular in time-dependent perturbation theory and scattering theory. There the Hamilton operator is split into a free part H_0 and an interaction V, viz.,

$$H = H_0 + V \,,$$

where H_0 does not depend on time—otherwise the following equation would have to be generalized, as will be shown later. If we set

$$U_0(t) = \exp \frac{-i H_0 t}{\hbar} \,,$$

then for $H \approx H_0$, $U \approx U_0$ is also valid, at least for time spans that are not too long. Under the interaction representation, we now understand

$$|\psi_D(t)\rangle = U_0^\dagger(t) \, |\psi_S(t)\rangle = U_0^\dagger(t) \, U(t) \, |\psi_H\rangle \,,$$

and correspondingly

$$A_{\mathrm{D}} = U_0{}^\dagger A_{\mathrm{S}}\, U_0 = U_0{}^\dagger\, U\, A_{\mathrm{H}}\, U^\dagger\, U_0\ .$$

Hence it follows that $\mathrm{d}|\psi_{\mathrm{D}}\rangle/\mathrm{d}t = \mathrm{i}\hbar^{-1}\,(H_0 - U_0{}^\dagger H\, U_0)\,|\psi_{\mathrm{D}}\rangle$, so with $H_0 - H_{\mathrm{D}} = -V_{\mathrm{D}}$, we find

$$\frac{\mathrm{d}}{\mathrm{d}t}\,|\psi_{\mathrm{D}}\rangle = -\frac{\mathrm{i}}{\hbar}\,V_{\mathrm{D}}\,|\psi_{\mathrm{D}}\rangle \qquad \text{and} \qquad \frac{\mathrm{d}}{\mathrm{d}t}\,A_{\mathrm{D}} = \frac{\mathrm{i}}{\hbar}\,[H_0, A_{\mathrm{D}}]\ .$$

In the Dirac picture the time dependence of the observables becomes fixed by H_0 and that of the states by V_{D}.

If we set $|\psi_{\mathrm{D}}(t)\rangle = U_{\mathrm{D}}(t)\,|\psi_{\mathrm{D}}(0)\rangle$ with $|\psi_{\mathrm{D}}(0)\rangle = |\psi_{\mathrm{S}}(0)\rangle = |\psi_{\mathrm{H}}\rangle$, then we obtain

$$U_{\mathrm{D}}(t) = U_0{}^\dagger(t)\, U(t) \qquad \Longleftrightarrow \qquad U(t) = U_0(t)\, U_{\mathrm{D}}(t)\ .$$

Clearly, with $\mathrm{i}\hbar^{-1} U_0{}^\dagger (H_0 - H)U = -\mathrm{i}\hbar^{-1} V_{\mathrm{D}}$, we have the differential equation

$$\frac{\mathrm{d}U_{\mathrm{D}}}{\mathrm{d}t} = -\frac{\mathrm{i}}{\hbar}\,V_{\mathrm{D}}(t)\,U_{\mathrm{D}}(t)\ .$$

To integrate this, we have to respect the order of the operators—an operator at a later time should only act later and thus should stand to the left of operators at earlier times. This requirement is indicated by the special *time-ordering operator* T:

$$U_{\mathrm{D}}(t) = \mathrm{T}\,\exp\!\left(-\frac{\mathrm{i}}{\hbar}\int_0^t \mathrm{d}t'\, V_{\mathrm{D}}(t')\right)\ .$$

The derivative of $\mathrm{T}\exp\int_0^t A(t')\,\mathrm{d}t'$ with respect to t is equal to $A(t)$ times the expression to be differentiated. In addition, we have $\mathrm{T}\exp(0) = 1$. We thus obtain the integral equation

$$\mathrm{T}\,\exp\int_0^t \mathrm{d}t'\, A(t') = 1 + \int_0^t \mathrm{d}t'\, A(t')\,\mathrm{T}\,\exp\int_0^{t'} \mathrm{d}t''\, A(t'')\ ,$$

which can be solved step by step:

$$\mathrm{T}\,\exp\int_0^t \mathrm{d}t'\, A(t') = 1 + \int_0^t \mathrm{d}t'\, A(t') + \int_0^t \mathrm{d}t' \int_0^{t'} \mathrm{d}t''\, A(t')\, A(t'') + \cdots\ .$$

In the term of nth order, there are n time-ordered operators A. This expansion is used in *time-dependent perturbation theory*. Terms higher than the first contribution are usually neglected.

For density operators, we have the equation $\langle A\rangle = \mathrm{tr}(\rho A)$ in each picture. With $A_{\mathrm{S}} = U_0 A_{\mathrm{D}} U_0{}^\dagger$ and $A_{\mathrm{H}} = U^\dagger A_{\mathrm{S}} U = U_{\mathrm{D}}{}^\dagger A_{\mathrm{D}} U_{\mathrm{D}}$, we thus find

$$\rho_D = U_0{}^\dagger \, \rho_S \, U_0 = U_D \, \rho_H \, U_D{}^\dagger \, ,$$

which leads to the differential equation

$$\frac{d\rho_D}{dt} = -\frac{i}{\hbar} \, [V_D, \rho_D] \, .$$

With the series expansion for $U_D(t)$, we thus obtain

$$\rho_D(t) = \rho_D(0) \; - \; \frac{i}{\hbar} \int_0^t dt' \; [V_D(t'), \rho_D(0)]$$

$$- \; \frac{1}{\hbar^2} \int_0^t dt' \int_0^{t'} dt'' \; [V_D(t'), [V_D(t''), \rho_D(0)]] + \cdots \, .$$

Instead of $[V', [V'', \rho]]$, we may also write $[V', V'' \rho] + \text{h.c.}$, where h.c. stands for the Hermitian conjugate, because the operators are Hermitian.

Time-dependent perturbation theory leads to *Fermi's golden rule* for the transition rates. However, the procedure is often superficial. We shall go into more detail when we derive the golden rule in Sect. 4.6.

An exact treatment without approximations can be found for the time-dependent oscillator. This was already done for the classical case in Sect. 2.3.10 and especially in Sect. 2.4.11. In particular, the Hamilton operator

$$H(t) = \frac{1}{\alpha^2(t)} \left(\frac{P^2}{2m} + \frac{m}{2} w^2 X^2 \right)$$

leads to the eigenvalue problem of the usual (time-independent) oscillator of mass m and angular frequency w. The time dependence is contained here in the classical function $\alpha(t)$ and thus involves no time-ordering problem—and the time-independent oscillator has eigenvalues $\hbar w \, (n + \frac{1}{2})$, as will be shown on p. 359. But, as already in classical mechanics (see Sect. 2.4.11), these values for the $-mf(t) X$ with the force f are not the energy

$$E = \frac{(P - mFX)^2}{2m} + \frac{m}{2} \bar{f} X^2 \, , \quad \text{with} \quad \dot{F} = f - \bar{f} \quad \text{and} \quad \bar{F} = 0 \, .$$

In addition to the eigenvalues of $H(t)$, Fig. 4.11 shows the expectation values of the energy with respect to the eigenstates of H. However, the energy uncertainties are very large, and the values actually overlap in the right-hand picture. At least it becomes clear that the eigenvalues do depend on time, although the energy barely does so. In many cases, these properties of a time-dependent interaction are derived only in the *adiabatic approximation* (for sufficiently slow changes).

Fig. 4.11 Eigenvalues of H (*left*) and expectation values $\overline{E} = \langle n|E|n \rangle$ (*right*) for a time-dependent harmonic oscillator (both in the same arbitrary unit). Here $a = 1/2$ and $q = 1/4$ was chosen in the Mathieu equation. For $t = 0$ it is force-free

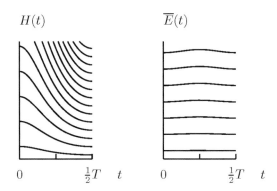

4.4.5 Current Density

For stationary problems, an expression for the probability current follows from the time-dependent Schrödinger equation. Since the total probability is conserved (it is equal to 1), according to p. 187, we have the continuity equation

$$\frac{\partial \rho}{\partial t} + \nabla \cdot \mathbf{j} = 0 \, ,$$

where ρ is the probability density $|\psi(t, \mathbf{r})|^2$. Hence, from the Schrödinger equation, we obtain

$$\frac{\partial \rho}{\partial t} = \psi^* \frac{\partial \psi}{\partial t} + \psi \frac{\partial \psi^*}{\partial t} = \frac{\psi^* H \psi - \psi H \psi^*}{i\hbar} \, ,$$

and with

$$H = \frac{(\mathbf{P} - q\mathbf{A}) \cdot (\mathbf{P} - q\mathbf{A})}{2m} + q \, \Phi \, ,$$

for the Coulomb gauge (i.e., with $\mathbf{P} \cdot \mathbf{A} = \mathbf{A} \cdot \mathbf{P}$) with $\mathbf{P} \cdot \mathbf{P} \,\widehat{=}\, (-i\hbar)^2 \Delta$ and $-\mathbf{A} \cdot \mathbf{P} \,\widehat{=}\, i\hbar \mathbf{A} \cdot \nabla$, we conclude

$$\frac{\partial \rho}{\partial t} = i\hbar \, \frac{\psi^* \Delta \psi - \psi \, \Delta \psi^*}{2m} + q\mathbf{A} \cdot \frac{\psi^* \nabla \psi + \psi \, \nabla \psi^*}{m} \, .$$

Here, according to p. 16, the first numerator is equal to $\nabla \cdot (\psi^* \nabla \psi - \psi \, \nabla \psi^*)$ and the second is equal to $\nabla (\psi^* \psi)$. Hence, assuming the Coulomb gauge, the probability current density is given by

$$\mathbf{j} = \frac{\hbar}{i} \, \frac{\psi^* \nabla \psi - \psi \, \nabla \psi^*}{2m} - \frac{q\mathbf{A} \, \psi^* \psi}{m} \, .$$

Note that, for real wave functions, only the last term contributes. With $\hbar \nabla \psi \cong i \mathbf{P} \psi$ and $\hbar \nabla \psi^* \cong -i(\mathbf{P}\psi)^*$, together with $\mathbf{A} = \mathbf{A}^*$, and in the real-space representation, this is equivalent to

$$\mathbf{j} \cong \text{Re}\left(\psi^* \frac{\mathbf{P} - q\mathbf{A}}{m} \psi\right).$$

Here, classically, $(\mathbf{p} - q\mathbf{A})/m$ is the velocity for a point-like particle of mass m and charge q, and $\psi^*\psi$ is the probability density. For the electric current density, we obtain $q\mathbf{j}$.

For spherically symmetric problems, we prefer to take the wave function

$$\psi_{nlm}(\mathbf{r}) = \frac{u_{nl}(r)}{r} i^l Y_m^{(l)}(\Omega),$$

with the spherical harmonic $Y_m^{(l)}(\Omega)$ of p. 335, which is real up to the factor $\exp(im\varphi)$. Note that the radial functions u_{nl} are real for bound states, but complex for scattering states, as we shall see in Sect. 4.6. If we call the mass m_0, in order not to confuse with the directional quantum number m, and refer to p. 39, then using

$$\nabla = \mathbf{e}_r \frac{\partial}{\partial r} + \mathbf{e}_\theta \frac{1}{r} \frac{\partial}{\partial \theta} + \mathbf{e}_\varphi \frac{1}{r \sin \theta} \frac{\partial}{\partial \varphi},$$

it follows for bound states that

$$\mathbf{j} = \mathbf{e}_\varphi \frac{m\hbar}{m_0} \frac{|\psi_{nlm}(\mathbf{r})|^2}{r \sin \theta}.$$

The term in $\mathbf{A}$ is missing here, because we have restricted ourselves to spherically symmetric potentials. For bound states and eigenstates of the orbital angular momentum $\mathbf{L}$, there is only a probability current along the $\mathbf{L}$-axis, if $m \neq 0$.

For electrons, however, the spin (and magnetic moment) have to be considered. We should take the Pauli equation from p. 327 as the Hamilton operator. Hence, noting that electrons have negative charge $q = -e$, we start with

$$H = H_0 + \mu_B \, \mathbf{B} \cdot \boldsymbol{\sigma}.$$

Since $\boldsymbol{\sigma}$ appears here, we use the spinors

$$\psi \cong \begin{pmatrix} \psi_\uparrow \\ \psi_\downarrow \end{pmatrix} \qquad \Longleftrightarrow \qquad \psi^\dagger \cong (\psi_\uparrow^*, \ \psi_\downarrow^*),$$

and find the equations

$$i\hbar \frac{\partial \psi}{\partial t} = H_0 \psi + \mu_B \mathbf{B} \cdot \boldsymbol{\sigma} \psi \,,$$

$$-i\hbar \frac{\partial \psi^\dagger}{\partial t} = H_0 \psi^\dagger + \mu_B \psi^\dagger \mathbf{B} \cdot \boldsymbol{\sigma} \,.$$

Note that H_0 acts like the unit operator in the spin space, but generally changes ψ in the position space, whence we have $H_0 \psi^\dagger$ and not $\psi^\dagger H_0$ in the last row.

If we multiply the first equation on the left by $\psi^\dagger$ and the second on the right by ψ, then subtract one from the other, it follows that

$$i\hbar \frac{\partial}{\partial t} (\psi^\dagger \psi) = \psi^\dagger H_0 \psi - (H_0 \psi^\dagger) \psi \,.$$

Hence for the probability current density $\mathbf{j}$, we obtain nearly the same expression as previously. Instead of $\psi^* (\mathbf{P} - q\mathbf{A}) \psi$, it now reads

$$\psi^*_\uparrow (\mathbf{P} - q\mathbf{A}) \psi_\uparrow + \psi^*_\downarrow (\mathbf{P} - q\mathbf{A}) \psi_\downarrow \,.$$

For the electric current density, we should now not only take $q\mathbf{j}$ (with $q = -e$ for electrons), but also consider the magnetic moments, and according to p. 192, amend $\nabla \times \mathbf{M}$, using $\mathbf{M} = -\mu_B \psi^\dagger \boldsymbol{\sigma} \psi$ for electrons.

4.4.6 Summary: Time Dependence

The time dependence is determined by the Hamilton operator. Then we distinguish between the Heisenberg and Schrödinger pictures, depending on whether only the observables or only the state vectors depend on time, respectively. In the Schrödinger picture, we have the time-dependent Schrödinger equation

$$i\hbar \frac{d|\psi\rangle}{dt} = H|\psi\rangle \,,$$

and in the Heisenberg picture, the Heisenberg equation

$$\frac{dA}{dt} = \frac{i}{\hbar}[H, A] + \frac{\partial A}{\partial t} \,,$$

which can be looked at as the quantum generalization of the classical equation $da/dt = [a, H] + \partial a/\partial t$ (p. 124). We may also take observables and basis vectors as constant and describe the time dependence by the density operator. This then obeys the von Neumann equation (in the Schrödinger picture), viz.,

$$\frac{d\rho}{dt} = -\frac{i}{\hbar}[H, \rho] \,,$$

which is the generalization of the Liouville equation to quantum theory.

Stationary states have a well-defined energy. Hence, if H does not depend on time, they are eigenstates of the Hamilton operator:

$$H|\psi_n\rangle = |\psi_n\rangle E_n \ ,$$

and in the Schrödinger picture they contain the time factor $\exp(-iE_n t/\hbar)$. This leads from the time-dependent to the time-independent Schrödinger equation (the last equation), which in the real-space representation has the form

$$\left(-\frac{\hbar^2}{2m} \Delta + V(\mathbf{r})\right) \psi_n(\mathbf{r}) = E_n \, \psi_n(\mathbf{r}) \ ,$$

since with $\mathbf{P} \stackrel{\frown}{=} (\hbar/i)\nabla, T = \frac{1}{2m}\mathbf{P} \cdot \mathbf{P}$ turns into $T \stackrel{\frown}{=} -\frac{1}{2m}\hbar^2\Delta$. For particles with spin in a magnetic field, special terms also appear for the potential energy V.

If the problem cannot be solved for the full Hamilton operator H, but for the time-independent approximation $H_0 = H - V$, a perturbation theory is possible, using the Dirac picture. Then H_0 determines the time dependence of the observables and $V_D = U_0^\dagger V U_0$ that of the states.

4.5 Time-Independent Schrödinger Equation

4.5.1 Eigenvalue Equation for the Energy

In this section we search for the eigenvalues E_n and eigenvectors $|n\rangle$ of the *Hamilton operator* H for a given interaction. We deal with the equation $H|n\rangle = |n\rangle E_n$ and assume that H has the form $T + V$ with the (local) potential energy $V(\mathbf{r})$. (We shall treat special cases, and in particular a magnetic field and also particles with spin $\frac{1}{2}$, at the end. The exchange interaction in the Hartree–Fock potential is nonlocal, as we shall see in Sect. 5.4.2.) Actually, V is an operator which is fixed in the real-space representation by $\langle \mathbf{r} | V | \mathbf{r}' \rangle$. But for the local interaction here, we can write $V(\mathbf{r})\,\delta(\mathbf{r} - \mathbf{r}')$ and $V(\mathbf{r})\,\psi(\mathbf{r})$ instead of $\langle \mathbf{r} | V | \psi \rangle = \int d^3 r'\, \langle \mathbf{r} | V | \mathbf{r}' \rangle \langle \mathbf{r}' | \psi \rangle$.

We shall usually take the real-space representation in order to make use of this locality of the interaction. From $(H - E_n)|n\rangle = 0$ and $\langle \mathbf{r} | n \rangle \equiv \psi_n(\mathbf{r})$ with $\mathbf{P} \stackrel{\frown}{=} - i\hbar\,\nabla$, we obtain the differential equation

$$\left(-\frac{\hbar^2}{2m} \Delta + V(\mathbf{r}) - E_n\right) \psi_n(\mathbf{r}) = 0 \ .$$

This is not yet an eigenvalue equation though, but only a partial, linear, and homogeneous differential equation of second order (for which the value and the gradient of the solution at a boundary can still be given arbitrarily in order to fix a special solution).

But $\psi_n(\mathbf{r})$ will now be a probability amplitude, which means that the expression $\int d^3r \, |\psi_n(\mathbf{r})|^2$ will be normalized to 1. However, we have also allowed improper Hilbert vectors, for which we have

$$\int d^3r \, \psi_n^*(\mathbf{r}) \, \psi_{n'}(\mathbf{r}) = \delta(n - n') \, ,$$

with continuous n and n'. But for discrete values we require

$$\int d^3r \, \psi_n^*(\mathbf{r}) \, \psi_{n'}(\mathbf{r}) = \delta_{nn'} \, ,$$

which can only be satisfied for special energies, as will be shown soon.

In order to make that clear, we restrict ourselves to the one-dimensional problem, i.e., to a standard differential equation, and consider

$$\psi''(x) + \frac{2m}{\hbar^2} \{E - V(x)\} \, \psi(x) = 0 \, .$$

If $V(x)$ decreases faster than $|x|^{-1}$ for large $|x|$, so that $E - V \to \hbar^2 k^2/2m$, the asymptotic solutions $\exp(\pm ikx)$ for $k \neq 0$ can be superposed linearly. For $k^2 > 0$, they oscillate and we can normalize in the continuum. But for $k^2 < 0$, we can only take $\exp(-|kx|)$, since $\exp(+|kx|)$ is not normalizable. For $E < V(x)$, all wave functions have to vanish exponentially for $x \to \pm\infty$, with specific dependence according to the differential equation. This is possible only for appropriate (countable) eigenvalues.

These considerations are also valid for the case in which $V(x)$ behaves asymptotically as $|x|^{-1}$ (which requires an amendment $\propto i \ln |kx|$ to the exponent). The sign of $E - V$ is decisive, also in three dimensions.

4.5.2   Reduction to Ordinary Differential Equations

We shall only consider potentials whose variables can be separated, i.e., potentials which can be written as a sum of terms, each of which depends on only one variable. Then the partial differential equation can be separated into three ordinary ones and solved much more easily.

Suppose for example that $V(\mathbf{r}) = V(x) + V(y) + V(z)$. Then the product ansatz, with each term involving just one Cartesian coordinate, i.e., $\langle x|n_x\rangle \langle y|n_y\rangle \langle z|n_z\rangle$ (and energy E_n separating into three terms) provides a way forward. In this way, the given partial differential equation can be reduced to three ordinary ones of the form

$$\left(\frac{d^2}{dx^2} + \frac{2m}{\hbar^2} \{E_{n_x} - V(x)\}\right)\langle x|n_x\rangle = 0 \, .$$

If we multiply this equation by $\langle y|n_y\rangle\langle z|n_z\rangle$ and add the corresponding equations in the variables y and z, then with $E_n = E_{n_x} + E_{n_y} + E_{n_z}$, we have the original partial differential equation. If at least two of these potentials are the same, then degeneracy arises and the different equations result in the same eigenvalues.

For a central potential $V(\mathbf{r}) = V(r)$ spherical coordinates are usually more appropriate than Cartesian ones. As is well known, the Laplace operator in spherical coordinates reads (see p. 39)

$$\Delta\psi = \frac{1}{r}\frac{\partial^2}{\partial r^2}r\psi + \frac{1}{r^2}\left\{\frac{1}{\sin\theta}\frac{\partial}{\partial\theta}\left(\sin\theta\frac{\partial}{\partial\theta}\right) + \frac{1}{\sin^2\theta}\frac{\partial^2}{\partial\varphi^2}\right\}\psi .$$

According to p. 335, the eigenfunctions of the operator in the curly bracket are the spherical harmonics, with the eigenvalue $-l(l+1)$. In classical mechanics, for a central field, we also made use of the angular momentum as a conserved quantity (p. 142). We thus set

$$\psi_{nlm}(r,\theta,\varphi) = \frac{u_{nl}(r)}{r}\, i^l\, Y_m^{(l)}(\Omega) ,$$

where m is the directional quantum number, and obtain the *radial equation*

$$\left(\frac{d^2}{dr^2} - \frac{l(l+1)}{r^2} + \frac{2m}{\hbar^2}\{E_{nl} - V(r)\}\right)u_{nl}(r) = 0 , \quad \text{with} \quad u_{nl}(0) = 0 ,$$

with m the mass once again. This boundary condition requires ψ_{nlm} to be differentiable at the coordinate origin, since we have divided by r. The further boundary condition $u_{nl} \to 0$ for $r \to \infty$ is still required for the normalizability of the bound states. It leads to an eigenvalue equation for the energy. Note that these eigenvalues no longer depend on the directional quantum number m. The spherical symmetry leads to a $2l$-fold degeneracy, i.e., there are $2l+1$ different eigensolutions with equal energy.

Near the origin, for $l \neq 0$, the second term usually outweighs the other ones and we have $u'' - l(l+1)r^{-2}u \approx 0$. This differential equation has the linearly independent solutions r^{-l} and r^{l+1}. Only the second vanishes at the origin (also for $l = 0$). Hence, we usually set u_{nl} in the form $u_{nl}(r) = r^{l+1}f_{nl}(r)$.

4.5.3 Free Particles and the Box Potential

For free particles, the Hamilton operator consists of only the kinetic energy $P^2/(2m)$, so we use the eigenfunctions of the momentum, or indeed of $\mathbf{k} = \mathbf{p}/\hbar$, from Sect. 4.3.3:

$$H = \frac{P^2}{2m} \implies E_k = \frac{\hbar^2k^2}{2m} \quad \text{and} \quad \psi_k(\mathbf{r}) = \frac{\exp(i\mathbf{k}\cdot\mathbf{r})}{\sqrt{2\pi}^3} .$$

There we also saw that $\int d^3r\, \psi_k^*(\mathbf{r})\, \psi_{k'}(\mathbf{r}) = \delta(\mathbf{k} - \mathbf{k}')$ in that case.

The sharp wave vector $\mathbf{k}$ (and the sharp energy E_k) are idealizations. Actually, for these continuous variables, we should consider their uncertainty and hence take a superposition of terms with different wave vectors, a so-called *wave packet*. The energy uncertainty means that we cannot simply split off a factor $\exp(-i\omega t)$, but we only have

$$\psi(t, \mathbf{r}) = \frac{1}{\sqrt{2\pi}^3} \int d^3k\, \psi(\mathbf{k})\, \exp\{i(\mathbf{k} \cdot \mathbf{r} - \omega(\mathbf{k})\,t)\}\,,$$

because $\omega = \hbar k^2/(2m)$ depends upon k. If only wave numbers from the near neighborhood of $\overline{k}$ contribute, then the *group velocity* of this wave packet, viz., $(d\omega/dk)_{\overline{k}} = \hbar\overline{k}/m = \overline{p}/m = \overline{v}$, is twice the *phase velocity* $\overline{\omega}/\overline{k}$. Hence, in the course of time, the wave packet changes shape. If we take, e.g., a Gauss function for $\psi(\mathbf{k})$, as on p. 321 (the smallest possible uncertainty product $\Delta x(0) \cdot \Delta k = 1/2$), then the position uncertainty increases with time:

$$\Delta x(t) = \Delta x(0)\, \sqrt{1 + \{2\hbar\,(\Delta k)^2\,t/m\}^2} = \sqrt{\{\Delta x(0)\}^2 + \{\Delta v\,t\}^2}\,,$$

since $\Delta x(0) \cdot \Delta k = 1/2$, while $\overline{x}$ moves with the velocity $\overline{v} = \hbar\overline{k}/m$.

A further example is that of a box with impermeable walls. Here the probability density may differ from zero only inside the box. Outside the container, the wave function must vanish, since the time-independent Schrödinger equation makes sense only if $V(\mathbf{r})\,\psi(\mathbf{r})$ is finite everywhere. In addition, the wave function must also be differentiable, thus continuous everywhere. This allows only a countable sequence of energies.

In the one-dimensional case, with $V(x) = 0$ for $0 < x < a$, otherwise infinite, the boundary conditions $\psi(0) = 0 = \psi(a)$ and the normalization to 1 fix the eigensolutions up to a phase factor. For $n \in \{1, 2, \ldots\}$ and the abbreviation

$$k_n = \frac{n\pi}{a}\,,$$

and with $\psi_n'' + k_n^2\psi_n = 0$, we have

$$\psi_n(x) = \sqrt{\frac{2}{a}}\, \sin k_n x\,, \quad \text{for } 0 \le x \le a, \text{ otherwise zero}, \quad E_n = \frac{\hbar^2 k_n^2}{2m}\,.$$

There is no normalizable solution for $n = 0$, and negative integers n deliver no further linearly independent solutions (see also Fig. 4.12).

Correspondingly, for a cuboid in three dimensions with side lengths a_x, a_y, a_z, if we have $k_i = n_i\pi/a_i$ with $n_i \in \{1, 2, \ldots\}$,

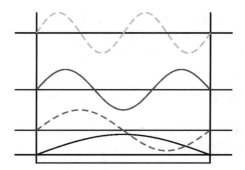

Fig. 4.12 Energy eigenvalues and eigenfunctions of a box potential with infinitely high walls. The figure shows the potential and also the eigenvalues as *horizontal lines*. Each of these lines serves as an axis for the associated eigenfunction, where functions with even n are plotted with *continuous lines*, and those with odd n as *dashed lines*

$$\psi_n(\mathbf{r}) = \sqrt{\frac{8}{V}} \, \sin k_x x \, \sin k_y y \, \sin k_z z \,, \quad \text{for} \;\; 0 \le x \le a_x \,, \; \text{etc.} \,,$$

$$E_n = \frac{\hbar^2 \, (k_x^2 + k_y^2 + k_z^2)}{2m} \,.$$

For a cube ($a_x = a_y = a_z$), there is degeneracy due to the symmetry, since we can permute n_x, n_y, n_z with each other and obtain the same energy value $E_n \propto n^2 = n_x^2 + n_y^2 + n_z^2$. In addition, there are also *accidental degeneracies*. For example, the state $(n_x, n_y, n_z) = (3, 3, 3)$ and the three states $(5,1,1)$, $(1,5,1)$, and $(1,1,5)$ have the same energy, because here n^2 is equal to 27 for each.

The potential discussed here is used for the *Fermi gas model*. In this many-body model, we neglect the interaction between the particles and consider only the quantum conditions, which stem from the inclusion of the particles in the cube volume. In contrast to the classical behavior, only discrete energy values (and wave numbers) are allowed. For such a gas we also need the number of states whose energy is less than an energy bound called the *Fermi energy*:

$$E_F = \frac{\hbar^2}{2m} \, k_F^2 \,.$$

Then clearly, $n^2 \le (a \, k_F / \pi)^2$. Hence contributions come from all points with positive integer Cartesian coordinates inside the sphere of radius $a k_F / \pi$. For sufficiently large $a k_F$, the number of states is

$$N \approx \frac{1}{8} \frac{4\pi}{3} \left(\frac{a k_F}{\pi} \right)^3 = \frac{V}{6\pi^2} k_F^3 = \frac{V}{6\pi^2} \left(\frac{2m E_F}{\hbar^2} \right)^{3/2} \,.$$

According to the Pauli principle, for spin-1/2 particles, each of these states can be occupied by two fermions.

If we search for the bound states, with negative energy eigenvalues $E_n < 0$, in a *box of finite depth* V_0 and width $a = 2\,l$, i.e., with $V(x) = -|V_0|$ for $-l < x < l$, otherwise zero, then with the real abbreviations

$$\kappa_n \equiv \sqrt{\frac{2m}{\hbar^2}\,|E_n|} \quad \text{and} \quad k_n \equiv \sqrt{\frac{2m}{\hbar^2}\,(|V_0| - |E_n|)}\,,$$

the differential equations $\psi'' - \kappa_n^2\,\psi = 0$ for $|x| > l$ and $\psi'' + k_n^2\,\psi = 0$ for $|x| < l$ imply a set of even states $\{\psi_+(x) = \psi_+(-x)\}$:

$$\psi_+(x) \propto \begin{cases} \exp\{\kappa_n(l+x)\} & \text{for} \quad\quad x \le -l\,, \\ \alpha\,\cos(k_n x) & \text{for} \ -l \le x \le +l\,, \\ \exp\{\kappa_n(l-x)\} & \text{for} \quad\quad l \le x\,, \end{cases}$$

and a set of odd states $\{\psi_-(x) = -\psi_-(-x)\}$:

$$\psi_-(x) \propto \begin{cases} +\exp\{\kappa_n(l+x)\} & \text{for} \quad\quad x \le -l\,, \\ \beta\,\sin(k_n x) & \text{for} \ -l \le x \le +l\,, \\ -\exp\{\kappa_n(l-x)\} & \text{for} \quad\quad l \le x\,. \end{cases}$$

The wave functions and their first derivatives have to be continuous everywhere, otherwise a differential equation of second order does not make sense. (In the present case, the second derivative jumps twice by a finite value. For the previously considered infinite potential step, however, the second derivative changes so considerably stepwise that even the first derivative jumps there.) At the limits $x = \pm l$, these properties fix α and β and also require as eigenvalue condition that κ_n/k_n be equal to $\tan(k_n l)$ for the even states and $-\cot(k_n l)$ for the odd states. These requirements with $z \equiv k_n l = \frac{1}{2}\,k_n a$, $\zeta \equiv (2m\hbar^{-2}\,|V_0|)^{1/2}\,l$, and $\kappa_n^2/k_n^2 = \zeta^2/z^2 - 1$ are easier to solve, if we satisfy (starting with $n = 0$)

$$\text{even eigensolution} \quad |\cos z| = \frac{z}{\zeta} \quad \text{for} \quad\quad n\,\pi \le z \le (n+\tfrac{1}{2})\,\pi\,,$$

$$\text{odd eigensolution} \quad\ |\sin z| = \frac{z}{\zeta} \quad \text{for}\ (n+\tfrac{1}{2})\,\pi \le z \le (n+1)\,\pi\,.$$

From $z = k_n l$, it follows that $E_n = -|V_0|\,(1 - z^2/\zeta^2)$. For finite $V_0 a^2$, there are also only finitely many bound-state eigensolutions, namely at most $2\zeta/\pi$ (see Fig. 4.13).

For the unbound solutions ("continuum states" with arbitrary $E > 0$), the potential can be attractive or repulsive:

$$V(x) = V_0\,, \quad \text{for} \quad -a < x < 0\,, \quad \text{otherwise zero}.$$

Here we use the real abbreviations

$$K \equiv \sqrt{2m\hbar^{-2}\,E} \quad \text{and} \quad k \equiv \sqrt{2m\hbar^{-2}\,|E - V_0|}\,,$$

Fig. 4.13 Eigenvalues for the box potential of finite depth. Solutions are the intersections (*full circles*) of the straight line z/ζ with the curves $|\cos z|$ (*continuous lines*) and $|\sin z|$ (*dashed lines*). Here, $\zeta = |2mV_0|^{1/2} a/2\hbar$

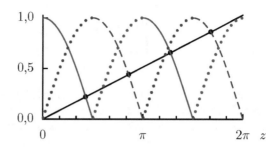

and let a wave come in from the left ($x < -a$). At the potential steps, it is partially reflected and partially refracted. For $E > V_0$, we then have

$$\psi(x) \propto \begin{cases} A \exp\{iK(x+a)\} + B \exp\{-iK(x+a)\}, & \text{for} \quad x \le -a, \\ \cos kx + i\kappa \sin kx, & \text{for} \ -a \le x \le 0, \\ \exp(iKx), & \text{for} \ 0 \le x, \end{cases}$$

with $\kappa = K/k$. Here use has already been made of the continuity of the wave function and of its first derivative at $x = 0$, and the factor for $x > 0$ was set arbitrarily equal to 1, while a common factor is still missing. The continuity conditions for $x = -a$ require

$$A = \cos ka - i\,\frac{\kappa + \kappa^{-1}}{2}\,\sin ka \quad \text{and} \quad B = -i\,\frac{\kappa - \kappa^{-1}}{2}\,\sin ka\,.$$

With the parameter $\zeta = |2mV_0|^{1/2} a/2\hbar$, we have $ka = 2\zeta\,|E/V_0 - 1|^{1/2}$. For $E < V_0$, k is to be replaced by ik (κ by $-i\kappa$) and we note that $\cos iz = \cosh z$ and $\sin iz = i \sinh z$.

If the probability current density j_d is refracted (transmitted), then the probability current density $j_e = j_d\,|A|^2$ comes in and $j_r = j_d\,|B|^2$ is reflected. The *transmittance* $D \equiv j_d/j_e$ and *reflectivity* $R \equiv j_r/j_e$ together sum to 1: $D + R = (1 + |B|^2)/|A|^2 = 1$. We obtain (see Fig. 4.14)

Fig. 4.14 Transmittance D at steps of height V_0 and width a as a function of the energy E for three values of the parameter $\zeta = |2mV_0|^{1/2} a/2\hbar$, namely $1/2$ (*green*), 1 (*blue*), and 2 (*red*). The classical case is shown with a /it dashed line

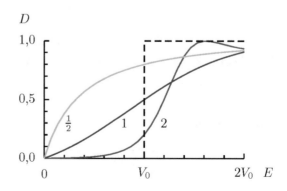

$$D = \begin{cases} \left(1 + \dfrac{V_0{}^2 \sin^2 ka}{4\,E\,|E - V_0|}\right)^{-1} & \text{for } E > V_0\,, \\[2ex] \left(1 + \dfrac{V_0{}^2 \sinh^2 ka}{4\,E\,|E - V_0|}\right)^{-1} & \text{for } E < V_0\,, \end{cases}$$

While for $E < V_0$ nothing is refracted classically, according to quantum theory, the *tunnel effect* occurs because the uncertainty relations have to be observed. Due to the position uncertainty, the finite length a does not "really" act, and because of the momentum uncertainty, neither does the finite potential step height. In particular, for $ka \gg 1$ (and $E < V_0$), we have

$$D \approx \frac{16\,E\,|E - V_0|}{V_0{}^2}\,\exp(-2ka)\,.$$

On the other hand, for $E > V_0$, $D = 1$ classically, but according to quantum theory all is refracted only if $E \gg |V_0|$ or ka is an integer multiple of π. This is also shown in Fig. 4.14.

4.5.4   Harmonic Oscillations

We shall not determine the eigenvalues for linear oscillations here using their differential equation and boundary conditions, but algebraically, using some extremely useful operators. We have $H = \frac{1}{2m}\,P^2 + \frac{m}{2}\,\omega^2\,X^2$. With an energy unit $\hbar\omega$, a momentum unit $p_0 \equiv \sqrt{2\hbar m\omega}$, and a length unit $x_0 \equiv 2\hbar/p_0 = \sqrt{2\hbar/m\omega}$, this leads to the equation

$$\frac{H}{\hbar\omega} = \frac{X^2}{x_0{}^2} + \frac{P^2}{p_0{}^2}\,.$$

If we now set

$$X = x_0\,\frac{\Psi + \Psi^\dagger}{2} \quad \text{and} \quad P = p_0\,\frac{\Psi - \Psi^\dagger}{2i}\,,$$

whence $\Psi = X/x_0 + iP/p_0$ and $\Psi^\dagger = X/x_0 - iP/p_0$, then the commutation relation $[X, P] = i\hbar\,1$ together with $x_0\,p_0 = 2\,\hbar$ imply the equation

$$[\Psi, \Psi^\dagger] = 1\,,$$

and in addition

$$H = \tfrac{1}{2}\,\hbar\omega\,\{\Psi, \Psi^\dagger\} = \hbar\omega\,(\Psi^\dagger\Psi + \tfrac{1}{2})\,.$$

The commutation relation $[\Psi, \Psi^\dagger] = 1$ is known already from p. 302, in particular, for the creation and annihilation operators of bosons. From this commutation relation, we obtained there the eigenvalues of $\Psi^\dagger \Psi$. Hence we already know the energy eigenvalues of the linear oscillator:

$$E_n = \hbar\omega \left(n + \tfrac{1}{2}\right), \quad \text{with} \quad n \in \{0, 1, 2, \ldots\} .$$

The energies of neighboring states all differ by $\hbar\omega$ (see Fig. 4.15). This use of Bose operators makes it possible to treat oscillations as particles. The sound quantum is called a *phonon*, and the quantum of the electromagnetic field (the light quantum) a *photon*.

The energy $\hbar\omega/2$ of the ground state, with $n = 0$, is called the *zero-point energy*. It is not zero, because otherwise position and momentum would both be sharp. But then the product of the uncertainties could be as small as possible. The expectation values of Ψ and $\Psi^\dagger$ vanish in the ground state and so also do $\overline{X}$ and $\overline{P}$. In contrast, for $\overline{X^2}$ and $\overline{P^2}$, it is important to note that $\overline{(\Psi^\dagger \pm \Psi)^2} = \pm \overline{\Psi\Psi^\dagger} = \pm 1$. We thus have $\Delta X = \tfrac{1}{2}x_0$ and $\Delta P = \tfrac{1}{2}p_0$, so their product is equal to $\tfrac{1}{2}\hbar$, and hence as small as possible.

According to p. 128 the Hamilton function of a point charge in a magnetic field can be transformed canonically to that of a linear oscillation with the cyclotron frequency $\omega = qB/m$. Quantum mechanically, we then find the energy eigenvalues (*Landau levels*) with equal distances. However, degeneracy should be noted, as for two-dimensional isotropic oscillations.

According to p. 321, we already know the wave functions of all states with the smallest possible product of the uncertainties $\Delta X \cdot \Delta P$: these are the Gauss functions normalized to 1. Consequently, for the ground state we have

$$\psi_0(x) = \frac{\sqrt[4]{2/\pi}}{\sqrt{x_0}} \exp \frac{-x^2}{x_0{}^2} .$$

Let us now turn to its remaining stationary states, i.e., those with sharp energy. According to p. 302, their eigenfunctions can be can built up with the creation operators $\Psi^\dagger$ from the ground state: $|n\rangle = (n!)^{-1/2} (\Psi^\dagger)^n |0\rangle$. From there, we have $\Psi^\dagger \hat{=} x/x_0 - \tfrac{1}{2}x_0 \, d/dx$. With $s = \sqrt{2}\,x/x_0 = x\sqrt{m\omega/\hbar}$, this becomes $\Psi^\dagger \hat{=} 2^{-1/2} (s - d/ds)$. But we may also replace the operator $s - d/ds$ by $- \exp(\tfrac{1}{2}s^2) \, d/ds \, \exp(-\tfrac{1}{2}s^2)$ and apply n times to ψ_0. Now we have *Rodrigues' formula for Hermite polynomials*:

$$H_n(s) \equiv (-)^n \, \exp(s^2) \, \frac{d^n}{ds^n} \, \exp(-s^2) .$$

With $\delta(s - s') = \delta(x - x') x_0/\sqrt{2}$ and $x_0/\sqrt{2} = \sqrt{\hbar/m\omega}$, which implies $|s\rangle = |x\rangle \sqrt[4]{\hbar/m\omega}$, the result is (see Fig. 4.15)

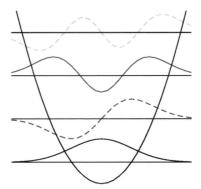

Fig. 4.15 Energy eigenvalues and eigenfunctions for linear oscillations. As in Fig. 4.12, we show the potential and eigenvalues (*horizontal lines*). These lines also serve as axes for the associated eigenfunctions, both even (*continuous lines*) and odd (*dashed lines*). States with sharp energy are stationary. Only for uncertain energy do oscillations occur. This will be discussed in Sect. 5.5.3 (see also Figs. 4.20 and 4.21). As a function of the displacement, the eigenfunctions oscillate in the classically allowed region, while in the classically forbidden regions, they tend to zero monotonically (tunnel effect)

$$\psi_n(s) = \frac{\exp(-\frac{1}{2}s^2)}{\sqrt{2^n\,n!\sqrt{\pi}}}\,H_n(s)\ .$$

So we only need to know the Hermite polynomials.

Clearly, $H_0(s) = 1$ and $H_1(s) = 2s$. The other polynomials can be obtained faster than by differentiation, if we use the *recursion formula*

$$H_{n+1}(s) = 2s\,H_n(s) - 2n\,H_{n-1}(s)\ .$$

Before the proof, we derive the *generating function of the Hermite polynomials*:

$$\exp(2st - t^2) = \sum_{n=0}^{\infty} H_n(s)\,\frac{t^n}{n!}\ .$$

We have $\exp\{-(t - s)^2\} = \sum_n d^n \exp\{-(t - s)^2\}/dt^n|_{t=0}\, t^n/n!$ according to Taylor. Here the derivative up to the factor $(-1)^n$ is equal to the n th derivative with respect to s for $t = 0$, thus equal to $(-1)^n\,d^n \exp(-s^2)/d^n$. Consequently, using the above-mentioned generating function, we may derive further properties of the Hermite polynomials. In particular, we only need to differentiate with respect to t and then compare coefficients in order to prove the formula. For $|s| \gg 1$, we also find $H_n(s) \approx (2s)^n$. If we differentiate the generating function with respect to s, then $H_n' = 2n\,H_{n-1}$

and hence $H_n'' = 2n H_{n-1}'$. If we use the recursion formula in the first derivative, we obtain the differential equation

$$H_n''(s) - 2s\, H_n'(s) + 2n\, H_n(s) = 0 \ .$$

Written as a polynomial, we have

$$H_n(s) = \sum_{k=0}^{[n/2]} a_k\, (2s)^{n-2k} \ , \quad \text{with} \quad \frac{a_{k+1}}{a_k} = -\frac{(n-2k)(n-2k-1)}{k+1} \ ,$$

and $a_0 = 1$. Clearly, $H_n(-s) = (-)^n\, H_n(s)$, so we also know the parities of the states.

According to classical mechanics, there are oscillations only for $T = E - V > 0$. Hence, we would have to require $\frac{1}{2}\hbar\omega\,(2n+1) > \frac{1}{2} m\omega^2 x^2$, or put another way, $s^2 = x^2\, m\omega/\hbar < 2n + 1$. In fact, the Schrödinger equation for linear oscillations can be written in the form $\psi''(s) + (2n + 1 - s^2)\,\psi(s) = 0$. For $s^2 = 2n + 1$, the sign of ψ_n'' therefore changes, without $|\psi_n|^2$ vanishing for larger values of $|s|$. Moreover, in the classically forbidden region (with $T < 0$), there is still a finite probability density. We already met this *tunnel effect* in the last section.

In three dimensions, for the isotropic oscillator, we have

$$E_n = (n + \tfrac{3}{2})\,\hbar\omega \ , \quad \text{with} \quad n = n_x + n_y + n_z \in \{0, 1, 2, \ldots\} \ .$$

Except for the ground state, all states are degenerate: n_x and n_y can be chosen arbitrarily, as long their sum is $\leq n$, while n_z is fixed. There are therefore $\frac{1}{2}(n+2)(n+1)$ different states in the same "oscillator shell". They all have parity $(-1)^n$.

Since a central field is given, we can also express the oscillation quantum number n in terms of the angular momentum quantum number l and the radial quantum number n_r. There are always $2l+1$ degenerate states of equal parity for each value of l. However, the isotropic oscillator is more strongly degenerate. Here n and l are either both even or both odd because of the parity. Their difference is an even number. In fact, we have

$$n = 2\,(n_r - 1) + l \ .$$

Here the radial quantum number n_r starts with the value 1, as is usual in nuclear physics. We then have the following shells: 1s, 1p, 1d-2s, 1f-2p, 1g-2d-3s, and so on.

4.5.5 Hydrogen Atom

In the following we shall investigate only the bound states of a particle with the reduced mass m in an attractive Coulomb potential

$$V(r) = -\frac{e^2}{4\pi\varepsilon_0\, r}\ ,$$

and restrict ourselves therefore to negative energies—we have to consider the scattering off a Coulomb potential ($E > 0$) separately, and we shall do this in Sect. 5.2.3.

The standard example of this potential is the hydrogen atom, but where the magnetic moment μ_B is neglected. If we introduce the charge number Z, we also have the theory for hydrogen-like ions (He$^+$, Li^{++}, etc.). To some approximation, even atoms with one *outer electron* can be treated. If the remaining *core electrons* can be replaced by a point charge at the position of the nucleus, then the considered outer electron is relatively far away from the core (it is said to be in a *Rydberg state*). Then, according to Rydberg, a *quantum defect* δ_l can be introduced, and instead of the principal quantum number n, we have the effective principal quantum number $n^* = n - \delta_l$.

The problem is centrally-symmetric. Hence, according to p. 353, the radial Schrödinger equation

$$\left\{ \frac{d^2}{dr^2} - \frac{l(l+1)}{r^2} + \frac{2m}{\hbar^2}\left(E + \frac{e^2}{4\pi\varepsilon_0\, r}\right) \right\} u_{nl}(r) = 0\ , \quad \text{with} \quad u_{nl}(0) = 0\ ,$$

remains to be solved. We take the *Bohr radius* a_0 and the *Rydberg energy* E_R, which, via the *fine structure constant* (see p. 623)

$$\alpha \equiv \frac{e^2}{4\pi\varepsilon_0}\frac{1}{\hbar c_0} = \frac{1}{137.0\dots}\ ,$$

can be derived from the length unit $\hbar/mc_0$ or the energy unit $mc_0{}^2$, as becomes understandable in the context of the (relativistic) Dirac equation (Sect. 5.6.9):

$$a_0 \equiv \frac{1}{\alpha}\frac{\hbar}{mc_0}\ , \quad E_R \equiv \frac{\alpha^2}{2}\, mc_0{}^2 = \frac{\hbar^2}{2ma_0{}^2}\ .$$

(We shall encounter the fine structure constant in Sect. 4.5.8 for the spin–orbit fine-splitting, which is where it gets its name. For hydrogen-like ions, it is Z times greater.) We set

$$E = -E_R/n^2 \quad \text{and} \quad r = n\, a_0\, \rho\ ,$$

where n will turn out to be the principal quantum number, and obtain the simpler differential equation

$$\left(\frac{d^2}{d\rho^2} - \frac{l(l+1)}{\rho^2} - 1 + \frac{2n}{\rho} \right) u_{nl}(\rho) = 0\ , \quad \text{with} \quad u_{nl}(0) = 0\ .$$

We could already have used the following solution method for the one- and three-dimensional oscillations. It is more cumbersome, but more generally applicable than the methods mentioned so far. Hence I will introduce it here, even though the Coulomb problem can also be solved with operators, which are related to Lenz's vector (see, e.g., [6]).

For large ρ, the differential equation takes the form $u'' - u = 0$, with the two linearly independent solutions $\exp(\pm\rho)$. Only the exponentially decreasing one is normalizable. In contrast, for small ρ, according to p. 353, we have $u \approx \rho^{l+1}$. With these boundary conditions for small and large ρ, we set

$$u(\rho) = \rho^{l+1} \exp(-\rho) F(\rho) , \quad \text{with} \quad F(\rho) = \sum_{k=0}^{n_r} c_k \, \rho^k .$$

For the still unknown function F, the differential equation for u implies

$$\frac{1}{2} \frac{d^2 F}{d\rho^2} + \frac{l+1-\rho}{\rho} \frac{dF}{d\rho} + \frac{n-l-1}{\rho} F = 0 ,$$

and hence for the expansion coefficients c_k, the recursion formula

$$c_k = -\frac{2}{k} \frac{n-l-k}{2l+1+k} c_{k-1} .$$

The coefficient c_0 is not yet fixed by the homogeneous differential equation. Its value is determined from the normalization. But the solution is normalizable only if we are dealing with a polynomial (with $n_r < \infty$), hence if the recursion terminates, otherwise we have in particular $c_k/c_{k-1} \approx 2/k$, which corresponds to the function $\exp(2\rho)$, and despite the remaining factors, it is not normalizable. Hence not only must the *radial quantum number* n_r be a natural number, but so must the *principal quantum number*

$$n = n_r + l + 1 \quad \in \{1, 2, \dots\} .$$

F is thus a polynomial of order n_r, and the energy eigenvalues are (see Fig. 4.16)

$$E_n = -\frac{E_R}{n^2} \quad \text{with} \quad n \in \{1, 2, \dots\} .$$

Except for the ground state, all states are degenerate—and not only like for the centrally symmetric fields (where $2l+1$ states have equal energy), but even more so. A total of

$$\sum_{l=0}^{n-1}(2l+1) = n^2$$

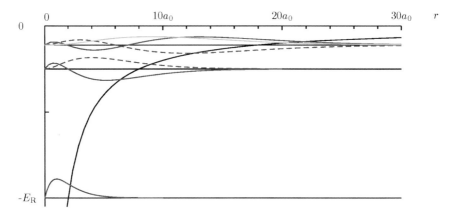

Fig. 4.16 Energy eigenvalues and radial functions of the hydrogen atom. The figure shows the potential, the first (degenerate) eigenvalues, and the associated radial functions, for $l = 0$ (*continuous red lines*), $l = 1$ (*dashed blue lines*), and $l = 2$ (*continuous green lines*)

Table 4.2 Multiplicity of Coulomb states. Note that all these states are to be counted twice because of the spin

States nl	1s	2s-2p	3s-3p-3d	4s-4p-4d-4f	...
$-E_n/E_R$	1	1/4	1/9	1/16	...
Multiplicity	1	4	9	16	...

different states belong to the energy E_n. In atomic physics, it is usual to give the principal quantum number n and the orbital angular momentum, using the letters indicated in Table 4.2.

To determine the polynomials F, we use the variable $s \equiv 2\rho = 2r/na_0$. Then the differential equation reads

$$s \frac{d^2 F}{ds^2} + (2l + 2 - s) \frac{dF}{ds} + (n - l - 1) F = 0 ,$$

the solution of which is the *generalized Laguerre polynomial* $L_{n-l-1}^{(2l+1)}(s)$ (see, e.g., [7]). Other functions also carry this name:

$$L_n^{(m)}(s) \equiv \frac{1}{n!} s^{-m} e^s \frac{d^n(s^{n+m} e^{-s})}{ds^n} = \sum_{k=0}^{n} \binom{n+m}{n-k} \frac{(-s)^k}{k!} ,$$

with the resulting eigenfunctions also shown in Fig. 4.16. As for the Legendre polynomials (p. 334) and the Hermite polynomials (p. 359), the first equation is called *Rodrigues' formula*. It fixes the polynomial by a correspondingly high derivative of a given function. With the *Leibniz formula*

$$\frac{d^n(f\,g)}{dx^n} = \sum_{k=0}^{n} \binom{n}{k} \frac{d^k f}{dx^k} \frac{d^{n-k} g}{dx^{n-k}} \ ,$$

the second expression follows from Rodrigues' formula. It shows that it is indeed a polynomial of n th order. Before we prove that the differential equation is satisfied, let us also deal with the *generating function of the generalized Laguerre polynomials*:

$$\frac{1}{(1-t)^{m+1}} \exp\frac{-st}{1-t} = \sum_{n=0}^{\infty} L_n^{(m)}(s)\, t^n \ , \quad \text{for } |t| < 1 \ .$$

It is easy to prove this. If we differentiate it with respect to s, then the left-hand side leads to $-t \sum_{n=0}^{\infty} L_n^{(m+1)}(s)\, t^n$. Hence, comparing coefficients, we find

$$L_n^{(m+1)} = -\frac{dL_{n+1}^{(m)}}{ds} \qquad \Longleftrightarrow \qquad L_n^{(m)}(s) = (-)^m \frac{d^m L_{m+n}}{ds^m} \ .$$

The generalized Laguerre polynomial $L_n^{(m)}(s)$ is thus equal to the m th derivative of the *Laguerre polynomials* $L_{n+m}(s) \equiv L_{n+m}^{(0)}(s)$, up to the factor $(-1)^m$. In addition, the equation for the generating function holds for $s = 0$, since it is $L_n^{(m)}(0) = \binom{n+m}{n}$, and this *binomial coefficient* also occurs for the Taylor series expansion of $(1-t)^{-m-1}$ in powers of t (for $|t| < 1$), because for arbitrary p and natural number n, we have

$$\binom{p}{n} \equiv \frac{p\,(p-1)\cdots(p-n+1)}{n!} = (-)^n \binom{n-p+1}{n} \ .$$

Hence, $\binom{n+m}{n} = (-)^n \binom{-m-1}{n}$ for $p = n+m$ and the generating function is correct.

If we differentiate it with respect to t and compare the coefficients, we obtain the *recursion formula*

$$(n+1)\, L_{n+1}^{(m)}(s) = (2n+m+1-s)\, L_n^{(m)}(s) - (n+m)\, L_{n-1}^{(m)}(s) \ .$$

Its derivative with respect to s delivers, along with the recursion formula,

$$
\begin{aligned}
L_n^{(m)}(s) &= L_n^{(m-1)}(s) + L_{n-1}^{(m)}(s) \ , \\
(n+1)L_{n+1}^{(m)}(s) &= (n+1-s)L_n^{(m)}(s) + (n+m)L_n^{(m-1)}(s) \ , \\
sL_n^{(m+1)}(s) &= (n+m+1)L_n^{(m)}(s) - (n+1)L_{n+1}^{(m)}(s) \ , \\
sL_n^{(m+1)}(s) &= (n+m)L_{n-1}^{(m)}(s) + (s-n)L_n^{(m)}(s) \ ,
\end{aligned}
$$

and the further *recursion formula*

$$s\, L_n^{(m+1)}(s) - (m+s)\, L_n^{(m)}(s) + (n+m)\, L_n^{(m-1)}(s) = 0 \ ,$$

as well as $s L_{n-1}^{(m+1)} + (s-m) L_n^{(m)} + (n+1) L_{n+1}^{(m-1)} = 0$, which leads to the original differential equation

$$\left(s \frac{d^2}{ds^2} + (m+1-s) \frac{d}{ds} + n \right) L_n^{(m)}(s) = 0 .$$

For the normalization and the matrix elements of R^k, the following equation is important:

$$\int_0^\infty ds \, e^{-s} \, s^k \, L_n^{(m)}(s) L_{n'}^{(m')}(s) = (-)^{n+n'} \sum_l \binom{k-m}{n-l} \binom{k-m'}{n'-l} \frac{(k+l)!}{l!} .$$

It can be derived from the generating function using $\int_0^\infty ds \, e^{-s} s^k = k!$ and $\binom{-x}{n} = (-)^n \binom{x+n-1}{n}$, which is necessary also for $k < m$ or $k < m'$. In particular, the generalized Laguerre polynomials with equal index $m = m'$ in the range $0 \le x \le \infty$ form an orthogonal system for the *weight function* $\exp(-s) \, s^m$:

$$\int_0^\infty ds \, \exp(-s) \, s^m \, L_n^{(m)}(s) L_{n'}^{(m)}(s) = \frac{(m+n)!}{n!} \, \delta_{nn'} .$$

Correspondingly, in the range $-\infty \le x \le \infty$, the Hermite polynomials form an orthogonal system for the weight function $\exp(-s^2)$.

Thus we may set

$$u_{nl}(r) = c \, (\tfrac{1}{2}s)^{l+1} \, \exp(-\tfrac{1}{2}s) \, L_{n-l-1}^{(2l+1)}(s) , \quad \text{with} \quad s \equiv \frac{2r}{n \, a_0} ,$$

with the still unknown normalization factor c, obtaining

$$\langle R^k \rangle = \int r^2 dr \, d\Omega \, |\psi|^2 \, r^k = \int_0^\infty dr \, |u|^2 \, r^k ,$$

according to p. 353 and p. 333 (or Problem 4.35). Hence,

$$|c|^2 = 4^{l+1} \, (n-l-1)!/\{n^2 a_0 \, (n+l)!\} ,$$

and for the ground state

$$u_{10}(r) = \frac{2}{\sqrt{a_0}} \frac{r}{a_0} \exp \frac{-r}{a_0} ,$$

and generally

n

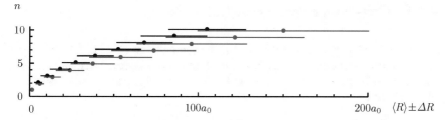

Fig. 4.17 $\langle R \rangle \pm \Delta R$ depends not only on the principal quantum number n, but also on the orbital angular momentum l. Hence the error bars for the lowest l (=0) (*red*) and the highest l (= $n-1$) (*black*) are shown, and the associated $\langle R \rangle$ as a *dot*

$$\langle n, l \, | \, R^k \, | n, l \rangle = \left(\frac{n\,a_0}{2} \right)^k \frac{(n-l-1)!}{2n\,(n+l)!} \sum_m \binom{1+k}{m}^2 \frac{(n+l+1+k-m)!}{(n-l-1-m)!} \, .$$

In particular, we have $a_0 \langle R^{-1} \rangle = n^{-2}$, and hence,

$$\langle V \rangle = -e^2/(4\pi\,\varepsilon_0) \, \langle R^{-1} \rangle = -2\,E_{\mathrm{R}}\,n^{-2} = 2\,E_n \, .$$

With $E_n = \langle T + V \rangle$, we have $\langle T \rangle = -\frac{1}{2} \langle V \rangle$, which also delivers the virial theorem (see p. 79) with a Coulomb field for the time average. For the average distance $\langle R \rangle$, we find $\frac{1}{2} \{3n^2 - l\,(l+1)\}\,a_0$, and in particular, in the ground state, $3a_0/2$. The most probable distance is given by the maximum of $|u(r)|^2$. The states with radial quantum number $n_r = 0$ (and the highest angular momentum in the multiplet of equal principal quantum number) each have only one—at $n^2\,a_0$, in the ground state thus for the Bohr radius—while the probability densities $|u(r)|^2$ of the remaining states have n_r secondary maxima (see p. 367) (Fig. 4.17).

In *Bohr's atomic model*, the centrifugal force cancels the Coulomb force between the electron and nucleus, i.e., $m\,v^2/r = e^2/(4\pi\varepsilon_0 r^2)$. Hence, $T = -\frac{1}{2}\,V$ and $E = \frac{1}{2}\,V = -E_{\mathrm{R}}\,a_0/r$. Here, according to Bohr, not all distances r are allowed, because the orbital angular momentum l_z has to be a multiple of $\hbar$, i.e., $mvr = n\hbar$ with $n \in \{1, 2, \dots\}$. Consequently, according to Bohr's atomic model, we have $r = n^2\,a_0$ and $E_n = -E_{\mathrm{R}}/n^2$. It delivers the same energy values as the Schrödinger equation. However, in Bohr's model, all states have an *orbital angular momentum* $n\hbar$ that differs from zero: s-states are not allowed, and n is not the principal, but the orbital angular momentum quantum number. In addition, Bohr's atomic model assumes a unique orbital curve, and does not incorporate the position and momentum uncertainty.

4.5.6   Time-Independent Perturbation Theory

If, for given H, we cannot solve the eigenvalue equation $(H - E_n)\,|n\rangle = 0$, thus cannot determine the eigenvalues E_n and the eigenvectors $|n\rangle$, then an approximation method often helps. In particular, if $H = \widetilde{H} + \widetilde{V}$ and the eigenvalues and eigenvectors of $\widetilde{H}$ are known,

$$(\widetilde{H} - \widetilde{E}_n)\,|\widetilde{n}\rangle = 0\,, \quad \text{with} \quad \sum_n |\widetilde{n}\rangle\langle\widetilde{n}| = 1 \quad \text{and} \quad \langle\widetilde{n}|\widetilde{n}'\rangle = \delta_{nn'}\,,$$

then we can expand the unknown eigenvector $|\ldots\rangle$ of H for the eigenvalue E with respect to this basis and also determine the matrix elements $\langle\widetilde{n}|\,H - E\,|\widetilde{n}'\rangle$. Using $\langle\widetilde{n}|\,H - E\,|\ldots\rangle = \langle\widetilde{n}|\,\widetilde{H} + \widetilde{V} - E\,|\ldots\rangle = \langle\widetilde{n}|\,\widetilde{E}_n + \widetilde{V} - E\,|\ldots\rangle = 0$, together with $|\ldots\rangle = \sum_n |\widetilde{n}\rangle\langle\widetilde{n}|\ldots\rangle$, we obtain the system of equations

$$\begin{array}{ccccc}
\langle\widetilde{0}|\,\widetilde{E}_0 + \widetilde{V} - E\,|\widetilde{0}\rangle\langle\widetilde{0}|\ldots\rangle + \langle\widetilde{0}| & & \widetilde{V} & |\widetilde{1}\rangle\langle\widetilde{1}|\ldots\rangle + \ldots = 0 \\
\langle\widetilde{1}| & & \widetilde{V} & |\widetilde{0}\rangle\langle\widetilde{0}|\ldots\rangle + \langle\widetilde{1}|\,\widetilde{E}_1 + \widetilde{V} - E\,|\widetilde{1}\rangle\langle\widetilde{1}|\ldots\rangle + \ldots = 0 \\
\vdots & & \vdots & \ddots \quad \vdots
\end{array}$$

Numerical calculations can be performed only for finite basis states $|\widetilde{n}\rangle$, thus only approximately. If we take only two (thus a *doublet*), then we have already determined the eigenvalues on p. 309:

$$E_{\pm} = \tfrac{1}{2}\mathrm{tr}H \pm \tfrac{1}{2}\,\hbar\Omega\,,$$

where now the average value is half of

$$\mathrm{tr}H = \widetilde{E}_0 + \langle\widetilde{0}|\,\widetilde{V}\,|\widetilde{0}\rangle + \widetilde{E}_1 + \langle\widetilde{1}|\,\widetilde{V}\,|\widetilde{1}\rangle\,,$$

and the square of the splitting is

$$(\hbar\Omega)^2 = (\widetilde{E}_0 + \langle\widetilde{0}|\,\widetilde{V}\,|\widetilde{0}\rangle - \widetilde{E}_1 - \langle\widetilde{1}|\,\widetilde{V}\,|\widetilde{1}\rangle)^2 + 4\,|\langle\widetilde{0}|\,\widetilde{V}\,|\widetilde{1}\rangle|^2\,.$$

The two eigenvalues $E_{\pm}$ are always different for $\langle\widetilde{0}|\,\widetilde{V}\,|\widetilde{1}\rangle \neq 0$. With coupling, there is no degeneracy, but the effect of *level repulsion* (see p. 310). Note that, without this coupling, the original eigenvalues $\widetilde{E}_n$ change by the expectation values $\langle\widetilde{n}|\,\widetilde{V}\,|\widetilde{n}\rangle$ to $\widetilde{E}_n + \langle\widetilde{n}|\,\widetilde{V}\,|\widetilde{n}\rangle$. The expansion coefficients $\langle\widetilde{n}|\pm\rangle$ have already been determined in Sect. 4.2.10.

For more than two basis states, the eigenvalue problem can be solved in perturbation theory (or numerically, using the variational method explained in the next section). Here we try to solve for $(E_n - H)\,|n\rangle = 0$ with $(\widetilde{E}_n - \widetilde{H})\,|\widetilde{n}\rangle = 0$. To deal with degeneracy, we take a new basis: if the eigenvalue of $\widetilde{H}$ is, e.g., g-fold, then only the g-dimensional problem $(H - E) |\ldots\rangle = 0$ has to be solved, as was just discussed for $g = 2$.

To derive $|n\rangle$, we avoid cumbersome normalization factors if we now require $\langle \widetilde{n}|n\rangle = 1$. The normalization can be changed again right at the end. Then we have $\langle \widetilde{n}| \widetilde{V} |n\rangle = \langle \widetilde{n}| H - \widetilde{H} |n\rangle = E_n - \widetilde{E}_n$, or

$$E_n = \widetilde{E}_n + \langle \widetilde{n}| \widetilde{V} |n\rangle .$$

The matrix element follows from $|\widetilde{n}\rangle \langle \widetilde{n}| \widetilde{V} |n\rangle = |\widetilde{n}\rangle (E_n - \widetilde{E}_n) = (E_n - \widetilde{H}) |\widetilde{n}\rangle$. Here we use the mutually orthogonal projection operators

$$\mathrm{P} \equiv |\widetilde{n}\rangle \langle \widetilde{n}| \quad \text{and} \quad \mathrm{Q} \equiv 1 - \mathrm{P} .$$

Hence, $\mathrm{P} \widetilde{V} |n\rangle = (E_n - \widetilde{H} |\widetilde{n}\rangle$ and also $\mathrm{P} \widetilde{V} = \widetilde{V} - \mathrm{Q} \widetilde{V} = H - \widetilde{H} - \mathrm{Q}\widetilde{V}$, and consequently $(E_n - \widetilde{H} - \mathrm{Q} \widetilde{V}) |n\rangle = (E_n - \widetilde{H}) |\widetilde{n}\rangle$. If there is no degeneracy, the singular operator $(E_n - \widetilde{H})^{-1}$ can act from the left, since with the projection operator Q, no singular operator appears on the left. The state with $\widetilde{H} |\widetilde{n}\rangle = |\widetilde{n}\rangle \widetilde{E}_n$ is missing and hence the operator $1 - (E_n - \widetilde{H})^{-1}\mathrm{Q} \widetilde{V}$ is regular, while the unit operator appears on the right. Thus with the *propagator*

$$\widetilde{G}(E) \equiv \frac{\mathrm{Q}}{E - \widetilde{H}} ,$$

we find $\{1 - \widetilde{G}(E_n) \widetilde{V}\}|n\rangle = |\widetilde{n}\rangle$, and hence the representation

$$|n\rangle = \{1 - \widetilde{G}(E_n) \widetilde{V}\}^{-1} |\widetilde{n}\rangle ,$$

and the eigenvalue equation

$$E_n = \widetilde{E}_n + \langle \widetilde{n}| \widetilde{V}\{1 - \widetilde{G}(E_n) \widetilde{V}\}^{-1} |\widetilde{n}\rangle .$$

This is the *perturbation theory of Wigner and Brillouin*. Unfortunately, in this result, the unknown quantity E_n also occurs on the right and is not easy to determine. But if we may expand in a geometrical series and the method converges fast enough, we may replace $\widetilde{G}(E_n)$ by $\widetilde{G}(\widetilde{E}_n)$ and can immediately give E_n:

$$E_n = \widetilde{E}_n + \langle \widetilde{n}| \widetilde{V} |\widetilde{n}\rangle + \langle \widetilde{n}| \widetilde{V} \widetilde{G}(\widetilde{E}_n) \widetilde{V} |\widetilde{n}\rangle + \cdots .$$

The expansion is clearly good if the absolute values of the matrix elements of $\widetilde{V}$ are small compared with the energy-level separations $|\widetilde{E}_n - \widetilde{E}_{n'}|$.

By the way, $\widetilde{G}(\widetilde{E}_n)$ is encountered instead of $\widetilde{G}(E_n)$ in the *perturbation theory of Schrödinger and Rayleigh*. With the abbreviation $\Delta_n \equiv E_n - \widetilde{E}_n = \langle \widetilde{n}| \widetilde{V} |n\rangle$, we have $\widetilde{G}(\widetilde{E}_n) = \{1 + \widetilde{G}(\widetilde{E}_n) \Delta_n\} \widetilde{G}(E_n)$, since $A^{-1}(A - B)B^{-1} = B^{-1} - A^{-1}$ delivers

$$A^{-1} = \{1 + A^{-1}(B - A)\}B^{-1} .$$

Hence, $1 - \widetilde{G}(\widetilde{E}_n)\,(\widetilde{V} - \Delta_n)$ factorizes in the form

$$\{1 + \widetilde{G}(\widetilde{E}_n)\,\Delta_n\}\,\{1 - \widetilde{G}(\widetilde{E}_n)\,\widetilde{V}\}\ .$$

For the inverse of $\{1 - \widetilde{G}(\widetilde{E}_n)\,\widetilde{V}\}$, we may also write

$$\{1 - \widetilde{G}(\widetilde{E}_n)\,(\widetilde{V} - \Delta_n)\}^{-1}\,\{1 + \widetilde{G}(\widetilde{E}_n)\,\Delta_n\}\ ,$$

and so avoid $\widetilde{G}(E_n)$. Since $Q\,|\widetilde{n}\rangle$ vanishes, and hence also $\widetilde{G}(\widetilde{E}_n)\,\Delta_n\,|\widetilde{n}\rangle$, we therefore have

$$|n\rangle = \{1 - \widetilde{G}(\widetilde{E}_n)\,(\widetilde{V} - \Delta_n)\}^{-1}\,|\widetilde{n}\rangle\ .$$

With $\{1 - \widetilde{G}(\widetilde{E}_n)\,(\widetilde{V} - \Delta_n)\}^{-1} = 1 + \{1 - \widetilde{G}(\widetilde{E}_n)\,(\widetilde{V} - \Delta_n)\}^{-1}\widetilde{G}(\widetilde{E}_n)\,(\widetilde{V} - \Delta_n)$, this can be reformulated as

$$|n\rangle = \left(1 + \{1 - \widetilde{G}(\widetilde{E}_n)\,(\widetilde{V} - \Delta_n)\}^{-1}\,\widetilde{G}(\widetilde{E}_n)\,\widetilde{V}\right)|\widetilde{n}\rangle\ .$$

The propagator is now taken for the known energy, although Δ_n still contains the unknown energy, so once again there is no explicit expression for it. But at least this equation is easier to solve than the one from the perturbation theory of Wigner and Brillouin. Then we obtain, to third order,

$$(1 + \langle\widetilde{n}|\widetilde{V}\,\widetilde{G}^2\,\widetilde{V}|\widetilde{n}\rangle)\,\Delta_n = \langle\widetilde{n}|\widetilde{V} + \widetilde{V}\widetilde{G}\widetilde{V} + \widetilde{V}\widetilde{G}\widetilde{V}\widetilde{G}\widetilde{V}|\widetilde{n}\rangle\ ,$$

and only encounter nonlinear equations for still higher orders. To second order, we have the same result via both methods.

For $\widetilde{G}\widetilde{V} \ll 1$, the quantum numbers mentioned in $|\widetilde{n}\rangle$ are thus also approximately valid for $|n\rangle$. To next order, however, other states become mixed in. The eigenvalues of operators which commute with $\widetilde{H}$ but not with H are no longer *good quantum numbers*.

4.5.7 Variational Method

If the perturbation theory does not converge fast enough because no good approximation $\widetilde{H}$ is known, then a variational method sometimes helps. It delivers first the ground state and after that also the higher states, if there is no degeneracy.

Each arbitrary approximation $|\psi\rangle$ to the ground state with the energy E_0 delivers an expectation value $\langle\psi|H|\psi\rangle \geq E_0$, since with the eigen representation $\{|n\rangle\}$ of H, we have $\langle\psi|H|\psi\rangle = \sum_n E_n\,|\langle n|\psi\rangle|^2 \geq E_0$,   with $E_0 \leq E_1 \leq \ldots$ and $\sum_n |\langle n|\psi\rangle|^2 = 1$. Consequently, we can take any other basis $\{|\widetilde{n}\rangle\}$, and with $|\psi\rangle = \sum_n^N |\widetilde{n}\rangle\,\langle\widetilde{n}|\psi\rangle$ satisfy

$$\delta\{\langle\psi|\,H\,|\psi\rangle - E\,(\langle\psi|\psi\rangle - 1)\} = 0\,,$$

where E is the Lagrange parameter introduced to deal with the normalization condition. In the framework of the finite basis $\{|\widetilde{n}\rangle\}$, it turns out to be the best approximation to the ground state energy. The expansion coefficients $\langle\widetilde{n}|\psi\rangle$ are to be varied here. Since H is Hermitian, we can trace the variational method back to $\langle\delta\psi|\,H - E\,|\psi\rangle = 0$. Note that this requirement for the matrix elements means that $(H - E)\,|\psi\rangle$ must vanish, since $\langle\delta\psi|$ is arbitrary. Naturally, the method leads more quickly to a useful result the better the basis $\{|\widetilde{n}\rangle\}$ already describes the actual ground state with few states, but it should also be easy to determine $\langle\widetilde{n}|H|\widetilde{n}'\rangle$.

If, in the finite basis $\{|\widetilde{n}\rangle\}$, we find the linear combination which minimizes $\langle\psi|\,H\,|\psi\rangle$ with the additional condition $\langle\psi|\psi\rangle = 1$, then within this framework the ground state $|\psi_0\rangle$ and its energy are determined as well as possible. The proper ground state may still have components orthogonal to the real one. The first excited state then follows with the same variational method and the further additional condition $\langle\psi|\psi_0\rangle = 0$.

4.5.8 Level Splitting

For the coupling of a magnetic moment $\mathbf{m}$ of velocity $\mathbf{v}$ with a centrally symmetric electric field $\mathbf{E} = -\nabla\Phi$, the following expression was derived on p. 372:

$$\widetilde{V} = \frac{1}{r}\frac{d\Phi}{dr}\frac{\mathbf{m}\cdot(\mathbf{r}\times\mathbf{v})}{c_0^2}\,.$$

If we use here the potential $\Phi = e/(4\pi\varepsilon_0 r)$, Weber's equation $c_0^{-2} = \varepsilon_0\mu_0$, the magnetic moment $\mathbf{m} = -e\mathbf{S}/m_0$ for the reduced mass m_0 (see p. 327), and

$$\mathbf{r}\times\mathbf{v} = \mathbf{l}/m_0\,,$$

then according to the correspondence principle—the transition to quantum mechanics with $[\mathbf{L}, R] = 0$ is easy—we find

$$\widetilde{V} = \frac{\mu_0}{4\pi}\frac{e^2}{m_0^2}\frac{\mathbf{L}\cdot\mathbf{S}}{R^3} = \alpha^2\,E_R\,\frac{a_0^3}{R^3}\frac{2\mathbf{L}\cdot\mathbf{S}}{\hbar^2}\,.$$

With the factor $\mathbf{L}\cdot\mathbf{S}$, we speak of *spin–orbit coupling*. (This is stronger in nuclear physics, and leads there, with a box potential, to the "magic nucleon numbers" of the shell model.) The observables L_z and S_z are no longer sharp, but the total angular momentum $\mathbf{J} = \mathbf{L} + \mathbf{S}$ is, as indeed are J^2 and J_z:

$$2\mathbf{L}\cdot\mathbf{S} = 2L_z S_z + L_+ S_- + L_- S_+ = J^2 - L^2 - S^2\,.$$

Fig. 4.18 Fine splitting of the first excited state multiplet of the hydrogen atom. *Left*: Inclusion of the spin–orbit coupling. *Right*: The result of the Dirac theory, with splitting due to a magnetic field of increasing strength. The Landé factor is 2 for $s\frac{1}{2}$ states, $\frac{2}{3}$ for $p\frac{1}{2}$ states, and $\frac{4}{3}$ for $p\frac{3}{2}$ states

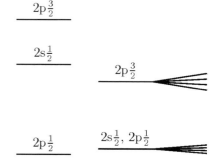

We thus use the coupled basis $\{|(ls)jm\rangle\}$ from Sect. 4.3.10 and find

$$2\mathbf{L}\cdot\mathbf{S}\,|(l\tfrac{1}{2})jm\rangle = |(l\tfrac{1}{2})jm\rangle\,\hbar^2 \begin{cases} l & \text{for } j = l+\tfrac{1}{2}, \\ -l-1 & \text{for } j = l-\tfrac{1}{2}. \end{cases}$$

The degeneracies of the hydrogen levels are thus lifted (for $l > 0$) by the spin–orbit coupling: $\langle\widetilde{V}\rangle = \alpha^2\,E_R\,\langle a_0^3/R^3\rangle\,l$ for the $2l+2$ states with $j = l+\tfrac{1}{2}$ and

$$\langle\widetilde{V}\rangle = -\alpha^2\,E_R\,\langle a_0^3/R^3\rangle\,(l+1)\;,$$

for the $2l$ states with $j = l - 1/2$. The average value of the scalar product $\mathbf{L}\cdot\mathbf{S}$ is thus zero. This is a general *sum rule*.

According to this, the first excited state of the hydrogen atom should split into three. The $2s\frac{1}{2}$ state remains unaltered (as do all s-states), the $2p\frac{3}{2}$ state increases by $\frac{1}{24}\alpha^2 E_R$, and the $2p\frac{1}{2}$ state is lowered by twice that value—the energies given in Sect. 4.5.5 are no longer valid to order $\alpha^2 E_R$. In fact, another fine splitting is found which follows only from the (relativistic) Dirac equation (Sect. 5.6.9). It leads to the result

$$E = -\frac{E_R}{n^2}\left(1 + \frac{\alpha^2}{n}\left(\frac{1}{j+\frac{1}{2}} - \frac{3}{4n}\right) + \cdots\right),$$

and shows that the previously found degeneracy is only partially lifted. It depends on n and j, but not on l and m. The energy of $2p\frac{3}{2}$ is lower than $-\frac{1}{4}E_R$ by $\frac{1}{64}\alpha^2 E_R$ and that of $2p\frac{1}{2}$ and $2s\frac{1}{2}$ even by $\frac{5}{64}\alpha^2 E_R$. According to the Dirac equation, the average value is also lowered, and the splitting amounts to $\frac{1}{16}\alpha^2 E_R$ (see Fig. 4.18).

Incidentally, according to p. 366, we find for the hydrogen atom and $n > l > 0$,

$$\left\langle\frac{a_0^3}{R^3}\right\rangle = \frac{1}{n^3}\,\frac{2}{l\,(l+1)\,(2l+1)}\;.$$

According to this, the classical spin–orbit splitting differs by a factor of 2 from the corresponding splitting due to Dirac.

Even though the spin–orbit coupling in atomic physics is clearly of the same order of magnitude as other "intricacies", it is suitable as an example application, since in nuclear physics it is the spin–orbit coupling which leads to the magic nucleon numbers, as mentioned above. In addition, these considerations support the following chain of thoughts.

The directional degeneracy is lifted by a magnetic field which we would now like to consider in perturbation theory. According to p. 327, we should use the Pauli equation for electrons. We neglect the term proportional to A^2, which leads to diamagnetism, a generally very small effect:

$$\widetilde{V} = -\frac{q}{2m_0}\, \mathbf{B} \cdot (\mathbf{L} + 2\,\mathbf{S}) \ .$$

If we quantize along the magnetic field, then according to perturbation theory, $L_z + 2\,S_z = J_z + S_z$ is important for the state $|(l\frac{1}{2})jm\rangle$. The first term on the right has the eigenvalue $m\,\hbar$, so only the expectation value of S_z in the state $|(l\frac{1}{2})jm\rangle$ is missing. According to Sect. 4.3.10, for this purpose, it follows that

$$\langle\,\widetilde{V}\,\rangle = m\,g\,\mu_B\,B\ , \quad \text{with the } Land\acute{e} \text{ factor} \quad g = \frac{2j+1}{2l+1}\ ,$$

because in the uncoupled basis, we have $S_z|l, m_l; \frac{1}{2}, m_s\rangle = m_s\hbar$ and the Clebsch–Gordan coefficients of p. 337 then deliver

$$\langle (l\frac{1}{2})\, l \pm \frac{1}{2}, m|S_z|(l\frac{1}{2})\, l \pm \frac{1}{2}, m\rangle = \pm m\,\hbar/(2l+1)\ .$$

This result of perturbation theory is true only for small external magnetic fields, such that higher-order terms can be neglected.

4.5.9 Summary: Time-Independent Schrödinger Equation

This Schrödinger equation is a second order differential equation for the unknown wave function. For bound states, the solution must vanish at infinity in order to be normalizable, and only then can it deliver a probability amplitude. On account of these boundary conditions, the time-independent Schrödinger equation becomes an eigenvalue equation for the energy. For unbound states, there is no eigenvalue condition: the energy can change continuously, and the improper Hilbert vectors serve only as an expansion basis for wave packets.

Since, according to the uncertainty relation, position and momentum, and hence also potential and kinetic energy, cannot be sharp simultaneously, there is a tunnel effect in quantum mechanics: for a given energy, there is a finite probability of finding a particle in classically forbidden regions.

Particularly important examples of the application of the time-independent Schrödinger equation are harmonic oscillations (their energy spectrum is equally spaced above the zero-point energy) and the hydrogen atom, or more precisely, the Kepler problem $V(r) \propto r^{-1}$ (with countable energy eigenvalues $E_n = -E_R/n^2$ for bound states and continuous eigenvalues $E > 0$ for scattering states). Free motion and piecewise constant potentials are even simpler to treat.

4.6   Dissipation and Quantum Theory

This section goes beyond the usual scope of a course entitled *Quantum Mechanics I* and, apart from Sect. 4.6.4 on Fermi's golden rule, can be skipped or studied only after the Chaps. 5 and 6.

4.6.1   Perturbation Theory

The Dirac picture is applied in particular to the coupling of atomic structures to their macroscopic surroundings. Without this influence we would not be able to observe atomic objects at all, since all detectors and measuring instruments belong to the macroscopic environment. (Hence this section is indispensable for the *theory of the measurement process*, although we shall not pursue this any further here.) We observe only a few degrees of freedom, but we have to consider their coupling to the many degrees of freedom of the environment. The difference between these two numbers is essential for the following. Hence we shall use the abbreviations "m" and "f" (for many and few) to indicate the two parts. Of course, it would be impossible to follow the many "inner degrees of freedom" of a solid separately. They have to be treated like those of the environment. At any given time, we observe only a few *degrees of freedom of the system*.

Let us consider, e.g., an excited atom, which emits light. In the simplest case we may consider the atom as a two-level system and the environment as the surrounding electromagnetic field. Even if it was initially particularly simple (without photons), the light quantum (photon) can still be emitted from many different states, these being distinguished, e.g., by the propagation direction, but also by the time of arrival at the detector.

For these considerations, pure states alone are not enough. In particular, averaging effects will enhance the degree of "impurity", so we describe everything with density operators. For their time dependence, in the interaction picture (Dirac picture), according to p. 346, we have

$$i\hbar \frac{d\rho_D}{dt} = [V_D, \rho_D] \,,$$

where the operators ρ and V act on both parts. But only the few degrees of freedom of the *open system* are of interest, and hence also only the equation of motion for a simpler *reduced density operator* will concern us, viz.,

$$\rho_f \equiv \mathrm{tr}_m \rho_D \,,$$

since we consider only measurable quantities O_f which do not depend on the many degrees of freedom and hence are unit operators with respect to these degrees of freedom:

$$\langle O_f \rangle = \mathrm{tr}_{vw}\, \rho_D O_f = \mathrm{tr}_f\, \rho_f O_f \,.$$

In particular, we shall derive an equation of motion for ρ_f from the expression for $\dot{\rho}_D$. The result will not be a von Neumann equation: open systems differ in principle from closed systems.

Concerning the experimental conditions, we require that initially the "object" and "environment" should be independent of each other ("uncorrelated"), so that initially ρ_D factorizes into ρ_f and $\rho_m \equiv \mathrm{tr}_f \rho_D$ (more on the notion of correlation in Sect. 6.1.5.) This initial condition is suggestive, because for each repetition of the experiment, we produce the object as identically as possible, but the environment has far too many possibilities of adjustment. Often we simply require that the coupling necessary for the correlation should be *turned on* only at the beginning of the experiment—for the discussion below, both requirements deliver the same result.

Using the product form, the number of independent density matrix parameters is much reduced. If we pay attention only to $\rho = \rho^\dagger$, but not to $\mathrm{tr}\rho = 1$, then an $N \times N$ matrix requires N^2 real parameters, but for the product form, instead of the $(N_m N_f)^2$ parameters, only $N_m{}^2 + N_f{}^2$ are needed. Generally, for uncorrelated systems, we have $\mathrm{tr}\rho^2 = \mathrm{tr}\rho_1^2 \cdot \mathrm{tr}\rho_2^2$, otherwise this is not true—correlated systems form *entangled states*. For example, the singlet state of two electrons in the spin space has $\mathrm{tr}\rho^2 = 1$, but $\mathrm{tr}\rho_i^2 = \frac{1}{2}$.

If the parts are not coupled, then for all times, ρ_D could be split into the product $\rho_m \otimes \rho_f$. But the interaction leads to a correlation. Hence we write

$$\rho_D = \rho_m \otimes \rho_f + \rho_k \,, \quad \text{with} \quad \rho_k(0) = 0 \quad (\text{and } \mathrm{tr}\rho_k(t) = 0, \text{not } 1) \,.$$

Then we obtain $i\hbar\dot{\rho}_f = \mathrm{tr}_m[V_D, \rho_m \otimes \rho_f + \rho_k]$ and a corresponding expression for $i\hbar\dot{\rho}_m$. The term $\mathrm{tr}_m[V_D, \rho_m \otimes \rho_f]$ is equal to the commutator of $\mathrm{tr}_m V_D \rho_m \otimes 1_f$ with ρ_f, where $\mathrm{tr}_m V_D \rho_m \otimes 1_f$ describes the average interaction of the environment with the experimental object. It can be taken as a part of the free Hamilton operator H_f, and correspondingly $\mathrm{tr}_f V_D 1_m \otimes \rho_f$ for H_m. Then these terms for the interaction vanish, and we find

$$\mathrm{i}\hbar \frac{\mathrm{d}\rho_f}{\mathrm{d}t} = \mathrm{tr}_m[V_D, \rho_k] \quad \text{and} \quad \mathrm{i}\hbar \frac{\mathrm{d}\rho_m}{\mathrm{d}t} = \mathrm{tr}_f[V_D, \rho_k] \; .$$

Since ρ_k is of at least first order in the interaction, the changes in ρ_f and ρ_m with time in the Dirac picture are at least of second order in V_D, and this can be exploited in perturbation theory.

The correlation ρ_k changes by one order less:

$$\mathrm{i}\hbar \frac{\mathrm{d}\rho_k}{\mathrm{d}t} = [V_D, \rho_m \otimes \rho_f] \; .$$

Here on the right, the expression $[V_D, \rho_k] - \mathrm{tr}_f[V_D, \rho_k] \otimes \rho_f - \rho_m \otimes \mathrm{tr}_m[V_D, \rho_k]$ is left out, because it depends on a higher order of the coupling. Hence, with regard to the initial value,

$$\rho_k(t) = \frac{-\mathrm{i}}{\hbar} \int_0^t \mathrm{d}t' \; [V_D(t'), \rho_m(t') \otimes \rho_f(t')] \; .$$

The final result is thus a coupled system of integro-differential equations: ρ_k follows from an integral of ρ_m and ρ_f and these quantities from differential equations which depend on ρ_k. In particular, for the unknown ρ_f, we now have the equation

$$\begin{aligned}
\frac{\mathrm{d}\rho_f}{\mathrm{d}t} &= -\frac{1}{\hbar^2} \, \mathrm{tr}_m[V_D(t), \int_0^t \mathrm{d}t' \; [V_D(t'), \rho_m(t') \otimes \rho_f(t')]] \\
&= -\frac{1}{\hbar^2} \int_0^t \mathrm{d}t' \, \mathrm{tr}_m[V_D(t), V_D(t') \, \rho_m(t') \otimes \rho_f(t')] + \mathrm{h.c.}
\end{aligned}$$

Here use was made of the fact that the operators are Hermitian. The double commutator can then be reformulated into two simple commutators.

In order to further simplify the equation, we decompose the coupling V_D into factors which each act only on one of the two parts, although there are several such products, and only their sum delivers the full coupling:

$$V_D = \sum_k C_m^k \otimes V_f^k \; .$$

Then, e.g., for a two-level system, a V_f may occur (even though V_D is Hermitian, the factors on the right-hand side may not be—there are further terms which ensure $V_D = V_D^\dagger$) and both are interconnected with appropriate factors C_m. However, this does not mean that each has only one creation and annihilation operator. In fact, each factor C_m^k embraces a huge set of basis operators (modes) for the environment. But since we are interested only in a few degrees of freedom and, when we form the trace, we average over many degrees of freedom, the notation is rather useful. Here for the time being we shall not fix the normalization of the basis operators C_m^k, so the V_f^k will remain undetermined.

Hence, the integrand splits up into factors for the individual parts:

$$\text{tr}_m[V_D(t), V_D(t')\rho_m(t') \otimes \rho_f(t')] = \sum_{kk'} \text{tr}_m\{C_m^k(t)C_m^{k'}(t')\rho_m(t')\} [V_f^k(t), V_f^{k'}(t')\rho_f(t')] \,.$$

Here the influence of the part with many degrees of freedom is contained in the factors

$$g^{kk'}(t, t') = \text{tr}_m\{C_m^k(t)\, C_m^{k'}(t')\, \rho_m(t')\} \,.$$

If they are determined, then a decoupled integro-differential equation remains for the unknown density operator ρ_f:

$$\frac{d\rho_f}{dt} = \frac{1}{\hbar^2} \sum_{kk'}\left[\int_0^t g^{kk'}(t, t')\, V_f^{k'}(t')\, \rho_f(t')\, dt', V_f^k(t)\right] + \text{h.c.}$$

4.6.2 Coupling to the Environment

So far we have respected the two parts as equivalent terms in a weak coupling and have not yet made use of the fact that they differ essentially by the number of degrees of freedom. This difference allows us to estimate the weight functions $g^{kk'}$ and to simplify the integro-differential equation.

As discussed in Sect. 4.4.4, $\rho(t) = \rho(0) + (i\hbar)^{-1} \int_0^t [V(t'), \rho(t')]\, dt'$ solves the initial equation $i\hbar\,\dot{\rho} = [V, \rho]$, but since the unknown $\rho(t)$ appears on the right, the solution is not found yet. In a perturbation theory, we replace $\rho(t')$ in the integrand by the initial value $\rho(0)$ and then obtain at least an approximate solution. In the given case, we do not need this approximation for $\rho_f(t')$; only for $\rho_m(t')$ will it be necessary. In particular, it will turn out that $g^{kk'}(t, t')$ puts the main weight on $t' \approx t$ and hence the main weight of ρ_f is only for the time t.

Here we start from the fact that *the environment is initially in equilibrium*. Otherwise we would also like to obtain the response of the considered object to new environmental conditions, which is in fact also an important question, but will only be investigated afterwards.

Without coupling of the two parts, the environment would remain in its initial state. We now assume that there is *no feedback*: so the object perturbs its environment (otherwise we could not investigate it at all), but not so strongly that it would be noticed, otherwise we would have to fix the boundary between the two differently. Hence,

$$g^{kk'}(t, t') \approx \text{tr}_m C_m^k(t)\, C_m^{k'}(t')\, \rho_m(0) \,.$$

The "recurrence time" expected for a given closed system depends on the feedback. But with environmental conditions, we shall introduce a damping of the open system which prohibits this feedback.

With $C_m(t) = U_m{}^\dagger(t)\, C_m U_m(t)$ and $U_m(t)\, U_m{}^\dagger(t') = U_m(t - t')$, and because $\rho_m(0)$ is stationary and hence commutes with $U_m(t')$, it follows that

$$g^{kk'}(t, t') = \mathrm{tr}_m C_m^k(t - t')\, C_m^{k'}(0)\, \rho_m(0) = g^{kk'}(t - t') .$$

Thus only the time difference is important for $g^{kk'}$, and the energy representation is therefore particularly useful:

$$g^{kk'}(t - t') = \sum_{n_m n_m'} \langle n_m | C_m^k | n_m' \rangle \langle n_m' | C_m^{k'} | n_m \rangle \langle n_m | \rho_m | n_m \rangle \, \exp \frac{\mathrm{i}(E_{n_m} - E_{n_m'})(t - t')}{\hbar} .$$

Here the many degrees of freedom are reduced to a nearly continuous eigenvalue spectrum of the environmental energy E with the state density $g_m(E)$. We replace the double sum by a double integral,

$$g^{kk'}(t'') = \iint \mathrm{d}E\, \mathrm{d}E'\, g_m(E)\, g_m(E')\, \langle E' | C_m^k | E \rangle \langle E | C_m^{k'} | E' \rangle\, \rho_m(E')\, \exp \frac{\mathrm{i}(E' - E)t''}{\hbar} ,$$

and now make the ansatz $\rho_m(E') = g_m{}^{-1}(E_0)\, \delta(E' - E_0)$. The factor $g_m{}^{-1}(E_0)$ follows from the normalization condition $\int \mathrm{d}E'\, g_m(E')\, \rho_m(E') = 1$. (Actually, we should start from a thermal distribution with a temperature T, but this is not important here.) Hence, we obtain

$$g^{kk'}(t'') = \int \mathrm{d}E\, g_m(E)\, \langle E_0 | C_m^k | E \rangle \langle E | C_m^{k'} | E_0 \rangle\, \exp \frac{\mathrm{i}(E_0 - E)t''}{\hbar} .$$

When forming the trace, we clearly require that an annihilation operator C_m^k always be followed by its adjoint creation operator. Hence the product in front of the exponential function is real (and non-negative). In the last equation of Sect. 4.6.1, in the Hermitian conjugate expression, where $g^{kk'}(t, t')$ is actually to be replaced by $g^{k'k}(t', t)$, we may now also have $g^{k'k*}(t - t')$. If we rephrase $k \leftrightarrow k'$ there, then we arrive at

$$\frac{\mathrm{d}\rho_f}{\mathrm{d}t} = \frac{1}{\hbar^2} \sum_{kk'} \left\{ \left[\int_0^t g^{kk'}(t - t')\, V_f^{k'}(t')\, \rho_f(t')\, \mathrm{d}t',\ V_f^k(t) \right] \right.$$
$$\left. + \left[V_f^{k'}(t),\ \int_0^t g^{kk'*}(t - t')\, \rho_f(t')\, V_f^k(t')\, \mathrm{d}t' \right] \right\} .$$

This integro-differential equation can still be simplified quite decisively using the Markov approximation.

4.6.3 Markov Approximation

Since the environment has many different eigenfrequencies, $g^{kk'}$ changes fast in comparison to ρ_f. The "memory" of the environment is much shorter than that of the atomic object. We therefore expect $g^{kk'}$ to decrease rather fast towards zero with increasing $|t - t'|$. Hence we take $\rho_f(t')$ in the integrand for $t' = t$ (*Markov approximation*) and may then extract it from the integral, whereupon the integro-differential equation becomes a simpler differential equation. The change in ρ_f at time t then depends only on the simultaneous value of ρ_f and no longer on the earlier values. Hence we introduce two dimensionless auxiliary quantities and if $g^{kk'}(t'')$ tends sufficiently fast towards zero, we may also integrate to infinity:

$$A_f^{kk'} \equiv \frac{1}{\hbar} \int_0^\infty g^{kk'}(t'') \, U_f(t'') \, V_f^{k'} \, U_f^\dagger(t'') \, dt'' \, ,$$

$$\bar{A}_f^{k'k} \equiv \frac{1}{\hbar} \int_0^\infty g^{kk'*}(t'') \, U_f(t'') \, V_f^k \, U_f^\dagger(t'') \, dt'' \, .$$

With $A_f^{kk'}(t) = U_f^\dagger(t) A_f^{kk'} U_f(t)$, we then obtain the differential equation

$$\frac{d\rho_f}{dt} = \frac{1}{\hbar} \sum_{kk'} \{[A_f^{kk'}(t) \, \rho_f(t), \, V_f^k(t)] + [V_f^{k'}(t), \, \rho_f(t) \bar{A}_f^{k'k}(t)]\} \, ,$$

where the operators $A_f^{kk'}$ and $\bar{A}_f^{k'k}$ still have to be investigated in more detail.

Hence, we assume that V_f^k changes the energy of the state by δE_f^k, and likewise for $\bar{A}_f^{k'k}$, while $A_f^{kk'}$ changes it by $\delta E_f^{k'}$. If we average now over the fast processes and pay attention only to the contributions of the slower parts, we have $\delta E_f^{k'} = -\delta E_f^k$. For the excitation of an atom by a transverse electromagnetic wave, this procedure is called the *rotating-wave approximation*, because these terms seem to be slowly variable to an observer rotating along with the wave. In each of the two commutators, there is a creation and an annihilation operator V_f. Hence, we have $A_f^{kk'} = \pi \, a^{kk'} \, V_f^{k'}$ and $\bar{A}_f^{k'k} = \pi \, a^{kk'*} \, V_f^k$, using the common abbreviation

$$a^{kk'} \equiv \frac{1}{\pi} \int dE \, g_m(E) \, \langle E_0|C_f^k|E\rangle\langle E|C_m^{k'}|E_0\rangle \, \frac{1}{\hbar} \int_0^\infty dt'' \, \exp \frac{i(E_0 - E + \delta E_f^k) \, t''}{\hbar} \, .$$

The differential equation under consideration then simplifies to

$$\frac{d\rho_f}{dt} = \frac{\pi}{\hbar} \sum_{kk'} \mathrm{Re} \, a^{kk'} \left([V_f^{k'}(t) \, \rho_f(t), \, V_f^k(t)] + [V_f^{k'}(t), \, \rho_f(t) \, V_f^k(t)] \right) \, ,$$

since $\mathrm{Im} \, a^{kk'}$ is multiplied by $[V_f^k(t) V_f^{k'}(t), \rho_f]$. This commutator is not important in the present discussion, because we shall not occupy ourselves with the determination of H_f here; we would only obtain an amendment to the Hamilton operator, e.g., for

the electromagnetic coupling of an atom to the surrounding vacuum, the famous
Lamb shift. According to Sect. 1.1.10, the real part of the integral over t'' is equal to
$\hbar\pi\, \delta(E_0 - E + \delta E_f^k)$:

$$\text{Re}\, a^{kk'} = g_m(E_0 + \delta E_f^k)\, \langle E_0|C_m^k|E_0 + \delta E_f^k\rangle \langle E_0 + \delta E_f^k|C_m^{k'}|E_0\rangle \ .$$

Here, for appropriate normalization of the operators V_f^k, we may take the factors
C_m^k and $C_m^{k'}$ as Bose operators Ψ_k and $\Psi_{k'}{}^\dagger$. This is true for $\delta E_f^k > 0$; for $\delta E_f^k < 0$,
conversely C_m^k is to be replaced by $\Psi_k{}^\dagger$ and $C_m^{k'}$ by $\Psi_{k'}$. In the following we shall
write $\pm\delta E$ instead of δE_f^k and assume $\delta E \geq 0$.

If no degeneracy occurs, then k and k' are uniquely related to each other, and
instead of the double sum, a single sum suffices. Note that, for an isotropic environ-
ment, we have in fact the usual directional degeneracy, but $\text{tr}\, C_m^k(t - t')\, C_m^{k'}(0)\, \rho_m(0)$
then also contributes only as a scalar, and this again relates k and k' to each other
uniquely. For $g_m(E_0 + \delta E)$ stands the factor $\Psi_k\Psi_k{}^\dagger$, for $g_m(E_0 - \delta E)$ the factor
$\Psi_k{}^\dagger\Psi_k = N_k$. With $[\Psi, \Psi^\dagger] = 1$, the factor for $g_m(E_0 + \delta E)$ is therefore greater by
one than that for $g_m(E_0 - \delta E)$. In Sect. 6.5.7, it will be shown that, for thermal
radiation, we have $\bar{n}_k = \{\exp(\hbar\omega_k/k_B T) - 1\}^{-1}$, where the factor k_B in front of the
temperature T is the Boltzmann constant, and the normalization volume V has the
state density $g_m(E) = V E^2/(\pi^2\hbar^3 c^3)$. For the coupling to the vacuum (for *spontaneous emission*), we naturally work with $\bar{n}_k = 0$ (or $T = 0$) so that only the term with
$g_m(E_0 + \delta E)$ appears, and not the term with $g_m(E_0 - \delta E)$. Then there is only forced
absorption, described by H_f, but both forced and spontaneous emission. (For $T > 0$,
there is also spontaneous absorption.) *Spontaneous processes are not described by H_f,
but by the dissipation discussed here.* Taking all this together, if H_f is not degenerate,
we then obtain

$$\frac{d\rho}{dt} = \frac{\pi}{\hbar} \sum_k \{ g_+^k\, [V_-^k(t)\, \rho(t), V_+^k(t)] + g_-^k\, [V_+^k(t)\, \rho(t), V_-^k(t)] \} + \text{h.c.}$$

Here we have left out the subscript f, because all operators now refer to the few
relevant degrees of freedom anyway. In addition, with δE appearing implicitly in k,
we have

$$g_-^k = \bar{n}_k\, g_m(E_0 - \delta E) \qquad \text{and} \qquad g_+^k = (\bar{n}_k + 1)\, g_m(E_0 + \delta E) \ .$$

Note that, if there is no spontaneous absorption, then $\bar{n}_k = 0$ and hence also $g_-^k = 0$.
The Hermitian conjugate of $g_\pm[V_\mp\rho, V_\pm]$ is equal to $g_\pm[V_\mp, \rho V_\pm]$. With

$$g_\pm[V_\mp\rho, V_\pm] + \text{h.c.} = g_\pm\, (2V_\mp\rho V_\pm - \{V_\pm V_\mp, \rho\}) \ ,$$

the equation of motion is often reformulated accordingly.

If we return to the Schrödinger picture (without including the subscript S), then
with time-independent operators H_f and $V_\pm^k$, it follows that

$$\frac{d\rho}{dt} = \frac{[H_f, \rho(t)]}{i\hbar} + \frac{\pi}{\hbar}\sum_k \{g_+^k \; [V_-^k \, \rho(t), V_+^k] + g_-^k \; [V_+^k \, \rho(t), V_-^k] + \text{h.c.}\} \; .$$

We shall apply this *Liouville equation* to different examples. It conserves the trace of ρ, because $d\rho/dt$ can be expressed purely in terms of commutators, but not the purity of the state, since we generally have

$$\frac{d}{dt}\text{tr}\rho^2 = \frac{2\pi}{\hbar}\sum_k \{g_+^k \, \text{tr}([V_-^k \rho, V_+^k] \, \rho) + g_-^k \, \text{tr}([V_+^k \rho, V_-^k] \, \rho)\} \; ,$$

which differs from zero. Hence for dissipation, there is also no unitary operator $U(t)$ with the property $\rho(t) = U(t)\rho(0)U^\dagger(t)$. Incidentally, for real $g_\pm$ and $V_\pm = V_\mp^\dagger$, $\rho \geq 0$ is also conserved, as was proven by Lindblad in 1976.

There may still be amendments $g_0[V\rho, V^\dagger]$ without energy exchange. These destroy the phases of the density operator. For example, for a doublet, we thus have $\rho(t) = \frac{1}{2}(1 + \boldsymbol{\sigma} \cdot \langle\boldsymbol{\sigma}(t)\rangle)$ and $H = \frac{1}{2}\hbar\Omega\sigma_z$, whence

$$\frac{d\rho}{dt} = -i\Omega\frac{[\sigma_z, \rho]}{2} + \left(\gamma_+[\sigma_-\rho, \sigma_+] + \gamma_-[\sigma_+\rho, \sigma_-] + \gamma_0\frac{[\sigma_z\rho, \sigma_z]}{4} + \text{h.c.}\right) \; .$$

Here γ_0 captures the coherence-destroying processes without energy exchange with the environment, γ_+ those with energy delivery to it, and γ_- those with energy intake from it, which is only possible for $T > 0$. Hence with

$$\langle\sigma_z\rangle_\infty = -(\gamma_+ - \gamma_-)/(\gamma_+ + \gamma_-) \; ,$$

we find

$$\langle\sigma_x + i\sigma_y\rangle_t = \langle\sigma_x + i\sigma_y\rangle_0 \, \exp(-i\Omega t) \, \exp\{-(\gamma_+ + \gamma_- + \gamma_0)\, t\} \; ,$$
$$\langle\sigma_z\rangle_t = \langle\sigma_z\rangle_\infty + \{\langle\sigma_z\rangle_0 - \langle\sigma_z\rangle_\infty\} \, \exp\{-2(\gamma_+ + \gamma_-)\, t\} \; .$$

For $\gamma_+/\gamma_- \approx (\bar{n}+1)/\bar{n}$, we have $\langle\sigma_z\rangle_\infty = -(2\bar{n} + 1)^{-1}$, and thermal radiation

$$\bar{n} = \{\exp(\hbar\Omega/k_B T) - 1\}^{-1} \; ,$$

whence the Bloch vector for $k_B T \ll \hbar\Omega$ tends towards $-\mathbf{e}_z$ and for $k_B T \gg \hbar\Omega$ to zero (see Fig. 4.19).

Incidentally, we often see the claim that the dissipation might be describable by a *non-Hermitian* Hamilton operator $H = R - iI$, with $R = R^\dagger$ and $I = I^\dagger$. Then, $H_{nn'}{}^* = R_{n'n} + iI_{n'n}$, and from $i\hbar\dot{\psi}_n = \sum_{n'} H_{nn'}\psi_{n'}$, for $\rho_{nn'} = \psi_n\psi_{n'}{}^*$, the equation $\dot{\rho} = -(i[R, \rho] + \{I, \rho\})/\hbar$ would follow. Here, in contrast to the previously derived equation of motion, the trace of ρ would not be conserved. Thus the ansatz $H = R - iI$ could not be valid generally—at most for special states, e.g., in scattering

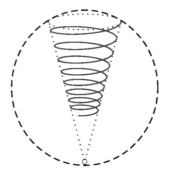

Fig. 4.19 Spiral orbit of the Bloch vector with damping by an environment at temperature $T = 0$. Without damping, according to p. 343, it proceeds on a circle with axis $\mathrm{tr}(\sigma H)$. The damping leads to a spiral orbit. Here $\gamma_+ + \gamma_- = \gamma_0$ is assumed, so the orbit lies on a cone, unless it already starts on the axis. Larger γ_0 narrows the orbit towards the axis and perturbs the coherence even faster. For $T > 0$, the attractor (*open circle*) lies higher, for $kT \gg \hbar\Omega$ in the center

theory, we consider "decaying states" (see Sect. 5.2.5), the probabilities of which decrease in the course of time.

For degeneracy of H_f, the situations are not quite as simple, since the index k actually belongs to V_f^k, while C_m^k embraces many modes, and now for $k \neq k'$ in C_m^k and $C_\mathrm{m}^{k'}$, the same modes may occur, so we may no longer separate the factor $\delta^{kk'}$ from $g_\pm^{kk'}$. Of these, only the mutually degenerate states are captured—instead of the term with k, many terms now occur, corresponding to the degree of degeneracy. We shall discuss this problem in more detail in Sect. 4.6.5.

4.6.4 Deriving the Rate Equation and Fermi's Golden Rule

The Schrödinger and von Neumann equations lead to the time-development operator $U(t) = \exp(-iHt/\hbar)$, and hence immediately after the beginning of the experiment to $U \approx 1 - iHt/\hbar$. If initially only the energy eigenstate $|n_0\rangle$ is occupied, then the occupation probabilities immediately after the beginning of the experiment do not change linearly, but quadratically with the time, i.e., $\langle n|\rho|n\rangle \approx |\langle n|H|n_0\rangle\, t\,/\hbar|^2$ for $n \neq n_0$. Actually, the occupation probability is expected initially (for small t) to increase linearly—the quadratic dependence is so surprising that it is even referred to as the *quantum Zeno paradox*.

But linear behavior follows immediately from the Liouville equation just derived, since for the diagonal elements of the density operator in the energy representation ($H_\mathrm{f}|n\rangle = |n\rangle E_n$), it delivers the *rate equation* (occasionally also called the *Pauli equation*, but which must not be confused with the non-relativistic approximation of the Dirac equation mentioned on p. 327):

$$\frac{\mathrm{d}\langle n|\rho|n\rangle}{\mathrm{d}t} = \sum_{n'} \left(W_{nn'} \langle n'|\rho|n'\rangle - W_{n'n} \langle n|\rho|n\rangle \right) ,$$

with the *transition rate*

$$W_{nn'} \equiv \frac{2\pi}{\hbar} g_{\pm} |\langle n|V_{\mp}|n'\rangle|^2 , \quad \text{for} \quad E_n \lessgtr E_{n'} ,$$

where the index k becomes fixed by n and n'. Note that the transition rate is often referred to as the *transition probability,* but is not normalized to 1. It gives the average number of transitions in the time $\mathrm{d}t$. As for operators, we shall write the initial state after the final state here, even though we are not strictly speaking dealing with matrix elements in the usual sense. If we swap n' and n, we obtain $W_{n'n} = 2\pi g_{\pm} |\langle n'|V_{\mp}|n\rangle|^2/\hbar$ for $E_{n'} \lessgtr E_n$, as is indeed required. In particular, we often also use the abbreviation (without terms $n' = n$)

$$\Gamma_n \equiv \hbar \sum_{n'} W_{n'n} ,$$

which is already useful in the above-mentioned rate equation. We shall discuss such rate equations in Sect. 6.2 and use them to prove in particular the entropy law for "closed systems", i.e., systems separated from their environment, but which also have many internal degrees of freedom in addition to a few observable ones, and hence according to Sect. 4.6.1 fit into the framework considered here. Energy is conserved in such closed systems, so we may set $g_{+}^k = g_{-}^k$ and obtain $W_{nn'} = W_{n'n}$.

Since the change in the atomic system is only relatively slow, this suggests using the initial values on the right-hand side of the rate equation and then determining the derivatives with respect to time, without first integrating the coupled system of equations. If initially we have the pure state $|n_0\rangle$, *Fermi's golden rule* for the determination of the transition rates follows for all states $|n\rangle \neq |n_0\rangle$:

$$\frac{\mathrm{d}\langle n|\rho|n\rangle}{\mathrm{d}t} = W_{nn_0} = \frac{2\pi}{\hbar} g_{\pm} |\langle n|V|n_0\rangle|^2 , \quad \text{for} \quad E_n \lessgtr E_{n_0} .$$

Since the rate equation conserves the trace of ρ, we now also infer

$$\mathrm{d}\langle n_0|\rho|n_0\rangle/\mathrm{d}t = -\sum_{n} W_{nn_0} = -\Gamma_{n_0}/\hbar ,$$

initially, i.e., for $t \ll \hbar/\Gamma_{n_0}$.

For the off-diagonal elements of ρ, the so-called *coherences*, as long as we leave out the terms $g_0 \left(2V\rho V - V^2\rho - \rho V^2\right)$ with energy-conserving $V = V^\dagger$, from the general result of the last section, we obtain

$$\frac{\mathrm{d}\langle n|\rho|n'\rangle}{\mathrm{d}t} = \left(\frac{E_n - E_{n'}}{i\hbar} - \frac{\Gamma_n + \Gamma_{n'}}{2\hbar}\right) \langle n|\rho|n'\rangle , \quad \text{for} \quad n \neq n' .$$

In particular, for $E_n \lessgtr E_{n'}$, from

$$g_+(2V_-\rho V_+ - \{V_+V_-, \rho\}) + g_-(2V_+\rho V_- - \{V_-V_+, \rho\}) \,,$$

only the part $-g_\pm\rho V_\pm V_\mp - g_\mp V_\mp V_\pm\rho$ contributes, because the creation and annihilation operators each connect only two states to each other—only the sum over k comprises all different states. Addition of the term with the factor g_0, viz.,

$$\langle n|2V\rho V - V^2\rho - \rho V^2|n'\rangle = -(\langle n|V|n\rangle - \langle n'|V|n'\rangle)^2 \langle n|\rho|n'\rangle \,,$$

increases the damping in comparison with the expression we have kept here. In this way, the differential equations decouple and lead to

$$\langle n|\rho(t)|n'\rangle = \langle n|\rho(0)|n'\rangle \; \exp\!\left(-\frac{\frac{1}{2}\,(\Gamma_n + \Gamma_{n'}) + \mathrm{i}\,(E_n - E_{n'})}{\hbar}\,t\right),$$

or even more strongly damped. The coherences thus decrease with time. The density operator in the energy representation finally becomes diagonal, and occupation probabilities also become classically understandable. This was discussed for doublets in the last section, and shown in Fig. 4.19. There we had $W_{\downarrow\uparrow} = 2\gamma_+$ and $W_{\uparrow\downarrow} = 2\gamma_-$.

4.6.5   Rate Equation for Degeneracy. Transitions Between Multiplets

When we have degeneracy, we have to consider still further states. We shall denote them with a bar, viz., $|\bar{n}\rangle$ and $|\bar{\bar{n}}\rangle$ will be degenerate with $|n\rangle$, and $|\bar{n}'\rangle$ with $|n'\rangle$. Instead of the rate equation for the occupation probabilities, we have

$$\frac{\mathrm{d}\langle n|\rho|\bar{n}\rangle}{\mathrm{d}t} = \sum_{n'\bar{n}'} W_{n\bar{n}n'\bar{n}'} \,\langle n'|\rho|\bar{n}'\rangle - \sum_{\bar{\bar{n}}} \frac{\Gamma_{\bar{n}n} \,\langle \bar{\bar{n}}|\rho|\bar{n}\rangle + \Gamma_{n\bar{\bar{n}}} \,\langle n|\rho|\bar{\bar{n}}\rangle}{2\hbar} \,,$$

with

$$W_{n\bar{n}n'\bar{n}'} \equiv \frac{2\pi}{\hbar}\, g_\pm \,\langle n|V_\mp|n'\rangle\langle \bar{n}'|V_\pm|\bar{n}\rangle \qquad \text{for } E_n \lessgtr E_{n'} \,,$$

and $\Gamma_{nn'} \equiv \hbar \sum_{n''} W_{n''n''nn'}$. (When there was no degeneracy, we introduced $W_{nn'} = W_{nnn'n'}$ and $\Gamma_n = \Gamma_{nn}$.) In contrast, for the matrix elements of ρ between states of different energy, it follows that

$$\frac{\mathrm{d}\langle n|\rho|n'\rangle}{\mathrm{d}t} = \frac{E_n - E_{n'}}{\mathrm{i}\hbar}\,\langle n|\rho|n'\rangle - \frac{\sum_{\bar{n}} \Gamma_{\bar{n}n} \,\langle \bar{n}|\rho|n'\rangle + \sum_{\bar{n}'} \Gamma_{n'\bar{n}'} \,\langle n|\rho|\bar{n}'\rangle}{2\hbar}\,.$$

Here the sum over $\bar{n}$ also takes the value n, the sum over $\bar{n}'$ takes the value n', and above, the sum over $\bar{\bar{n}}$ also takes the values n and $\bar{n}$.

The directional degeneracy of the angular momentum multiplets delivers an important example. Instead of $|n\rangle$, here it is better to write $|jm\rangle$. In the following, E_j is the energy of the ground state and $E_{j'}$ is the energy of the excited state. If we restrict ourselves to the coupling to the vacuum, with $g_- = 0$ and $g_+ = g_m(E_0 + \delta E)$, then we have

$$W_{jmm',j'm''m'''} = \frac{2\pi}{\hbar}\, g_+ \,\langle jm|V\,|j'm''\rangle\langle j'm'''|V\,|jm'\rangle\,,$$

in addition to $W_{j'm''m''',jmm'} = 0$ with $g_- = 0$. The vacuum does not prefer any direction, and hence leads to a special selection between k and k'. The two interactions couple only to a scalar. We restrict ourselves to radiation of multi-polarity n (usually dipole radiation, i.e., $n = 1$, but in nuclear physics, higher multipole radiation also occurs) and use the *Wigner–Eckart theorem*:

$$\langle jm|\,V_\nu^{(n)}\,|j'm'\rangle = \begin{pmatrix} j' & n & j \\ m' & \nu & m \end{pmatrix} \frac{\langle j\| V^{(n)} \|j'\rangle}{\sqrt{2j+1}}\,.$$

This means that the directional dependence of the matrix elements is included via the Clebsch–Gordan coefficients. Then only one *reduced matrix element* $\langle j\| V^{(n)} \|j'\rangle$ and the factor $(2j+1)^{-1/2}$ remains, split-off in such a way that, for a Hermitian operator, the symmetry $|\langle j\| V^{(n)} \|j'\rangle| = |\langle j'\| V^{(n)} \|j\rangle|$ remains. The above-mentioned isotropy delivers

$$\frac{\hbar\, W_{jmm',j'm''m'''}}{2\pi} = g_+ \sum_\nu \begin{pmatrix} j & n & j' \\ m & \nu & m'' \end{pmatrix}\begin{pmatrix} j & n & j' \\ m' & \nu & m''' \end{pmatrix} \frac{|\langle j\| V^{(n)} \|j'\rangle|^2}{2j'+1}\,,$$

and hence, using the orthogonality of the Clebsch–Gordan coefficients,

$$\Gamma_{j'm''m'''} = \hbar \sum_m W_{jmm,j'm''m'''} = 2\pi\, g_+ \frac{|\langle j\| V^{(n)} \|j'\rangle|^2}{2j'+1}\, \delta_{m''m'''}\,.$$

We note that $m'' = m'''$ has to hold, whence $\Gamma_{j'm''m'''}$ here does not depend on the directional quantum numbers. Hence we set

$$\Gamma_{j'} \equiv 2\pi\, g_+ \frac{|\langle j\| V^{(n)} \|j'\rangle|^2}{2j'+1}\,,$$

and obtain, for the matrix elements of the density operator in the upper multiplet,

$$\frac{d\langle j'm|\rho|j'm'\rangle}{dt} = -\frac{\Gamma_{j'}}{\hbar}\, \langle j'm|\rho|j'm'\rangle\,,$$

for those in the lower multiplet,

$$\frac{d\langle jm|\rho|jm'\rangle}{dt} = \sum_{m''m'''} W_{jmm',j'm''m'''} \, , \, \langle j'm''|\rho|j'm'''\rangle \, ,$$

and for the matrix elements between the two multiplets,

$$\frac{d\langle jm|\rho|j'm'\rangle}{dt} = \left(\frac{E_j - E_{j'}}{i\hbar} - \frac{\Gamma_{j'}}{2\hbar}\right) \langle jm|\rho|j'm'\rangle \, .$$

Here the properties of the Clebsch–Gordan coefficients lead to

$$W_{jmm',j'm''m'''} = \begin{pmatrix} j & n & | & j' \\ m & m'' - m & | & m'' \end{pmatrix} \begin{pmatrix} j & n & | & j' \\ m' & m''' - m' & | & m''' \end{pmatrix} \frac{\Gamma_{j'}}{\hbar} \, \delta_{m-m',m''-m'''} \, ,$$

since all other terms vanish. Consequently, all sub-states of the excited levels decay with the same time constants—and the amplitudes of the coherences $\langle jm|\rho|j'm'\rangle$ decrease exponentially with time, but only with half the time constants. If all sub-states of the excited levels were initially occupied with the same probability and those of the ground state were unoccupied, so that initially $\langle j'm''|\varrho|j'm'''\rangle = \delta_{m''m'''}/(2j' + 1)$, it then follows that

$$\frac{d\langle jm|\rho|jm'\rangle}{dt} = \frac{\Gamma_{j'}}{\hbar} \frac{\delta_{mm'}}{2j + 1} \, ,$$

if we make use of the properties of the Clebsch–Gordan coefficients.

4.6.6 Damped Linear Harmonic Oscillations

An important example is provided by the oscillator coupled to its environment. It is without degeneracy, but has only one creation and one annihilation operator between its states—as long as we neglect multi-quantum processes for the damping (like Ψ^2, but also $\Psi^\dagger\Psi$ for $V_\pm^k$). Hence the index k is superfluous, and we set $V_+ = v\,\Psi^\dagger$ and $V_- = v\,\Psi$ with $[\Psi, \Psi^\dagger] = 1$. The result of Sect. 4.6.3 then takes the form

$$\frac{d\rho}{dt} = -i\omega\,[\Psi^\dagger\Psi, \rho] + \pi v^2 \frac{g_+[\Psi\rho, varPsi^\dagger] + g_-[\Psi^\dagger\rho, \Psi] + \text{h. c.}}{\hbar} \, .$$

Note that expressions like $[\Psi^\dagger\Psi\rho, \Psi^\dagger\Psi] + \text{h. c.}$ lead to pure phase damping, which we shall not pursue here, and multi-quantum processes are still possible. Using the abbreviations

$$\gamma = \pi v^2 \frac{g_+ - g_-}{\hbar} \quad \text{and} \quad \langle \Psi^\dagger \Psi \rangle_\infty = \frac{g_-}{g_+ - g_-} ,$$

we obtain the differential equations

$$\frac{d\langle \Psi^\dagger \Psi \rangle}{dt} = -2\gamma \left\{ \langle \Psi^\dagger \Psi \rangle - \langle \Psi^\dagger \Psi \rangle_\infty \right\} \quad \text{and} \quad \frac{d\langle \Psi^l \rangle}{dt} = -(l\gamma + il\omega) \langle \Psi^l \rangle ,$$

which can be integrated easily:

$$\langle \Psi^\dagger \Psi \rangle_t = \langle \Psi^\dagger \Psi \rangle_\infty + \left\{ \langle \Psi^\dagger \Psi \rangle_0 - \langle \Psi^\dagger \Psi \rangle_\infty \right\} \exp(-2\gamma t) ,$$
$$\langle \Psi^l \rangle_t = \langle \Psi^l \rangle_0 \exp(-il\omega t) \exp(-l\gamma t) .$$

This result is similar to what we found for the two-level system (see Sect. 4.6.3). However, now $\langle \Psi^\dagger \Psi \rangle_\infty = g_-/(g_+ - g_-)$. With $g_+/g_- \approx (\bar{n}+1)/\bar{n}$, the average excitation energy approaches the value $\hbar \omega \bar{n}$, hence the average excitation energy of the environment—for thermal radiation, we have $\bar{n} = \{\exp(\hbar \omega / k_B T) - 1\}^{-1}$, and for the vacuum it is equal to zero.

Since X and P are linear combinations of Ψ and $\Psi^\dagger$, for the damped harmonic oscillation, $\langle X \rangle$ and $\langle P \rangle$ decrease at the rate γ independently of the initial state, while for stationary states the final state is already reached at the outset (see p. 388):

$$\frac{\langle X \rangle}{x_0} + i \frac{\langle P \rangle}{p_0} \equiv \langle \Psi \rangle_t = \langle \Psi \rangle_0 \exp(-i\omega t) \exp(-\gamma t) .$$

Classically (in Sect. 2.3.7), for $\gamma \ll \omega$, i.e., weak coupling to the environment, we have the same result, as Ehrenfest's theorem confirms (Sect. 4.4.1). But classically, we do not have the uncertainties:

$$\left(\frac{\Delta X}{x_0} \right)^2 + \left(\frac{\Delta P}{p_0} \right)^2 = \langle \Psi^\dagger \Psi \rangle_\infty + \tfrac{1}{2} + \left\{ \langle \Psi^\dagger \Psi \rangle_0 - \langle \Psi^\dagger \rangle_0 \langle \Psi \rangle_0 - \langle \Psi^\dagger \Psi \rangle_\infty \right\} \exp(-2\gamma t) ,$$
$$\left(\frac{\Delta X}{x_0} \right)^2 - \left(\frac{\Delta P}{p_0} \right)^2 = \mathrm{Re}\{(\langle \Psi^2 \rangle_0 - \langle \Psi \rangle_0^2) \exp(-2i\omega t)\} \exp(-2\gamma t) .$$

In the course of time, $\Delta X/x_0$ and $\Delta P/p_0$ take the same value, which is determined solely by the environmental temperature and respects the limit $\Delta X \cdot \Delta P = \tfrac{1}{4} x_0 p_0 = \tfrac{1}{2}\hbar$ set by Heisenberg's uncertainty relation.

In addition, the initial values of $\langle X \rangle$, $\langle P \rangle$, ΔX, and ΔP clearly do not yet fix the uncertainties, since $\langle \Psi^2 \rangle_0 - \langle \Psi \rangle_0^2$ is a complex number and therefore requires further input (namely its rate of change, or the direction of the ellipse in Fig. 4.20).

The example shown in Fig. 4.20 comes from an initial "quench state". This will be discussed in Sect. 5.5.4. These are pure states and have $\Delta X/x_0 \neq \Delta P/p_0$ (hence the name), but the smallest possible uncertainty product $\Delta X \cdot \Delta P$, i.e., $\tfrac{1}{4} x_0 p_0 = \tfrac{1}{2}\hbar$. There are of course states for which the product of these uncertainties is greater. For

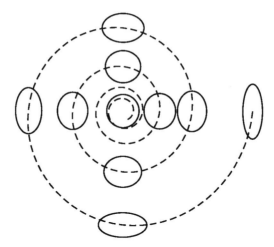

Fig. 4.20 Phase space representation of damped linear oscillations according to quantum theory—with equal damping as in the classical case (see Fig. 2.21). Except for the values (indicated already there) for $\langle X/x_0 \rangle$ and $\langle P/p_0 \rangle$, the uncertainties $\Delta X/x_0$ and $\Delta P/p_0$ can still be read off here. They remain finite, but always become more similar with time. The *circle in the middle* shows the final state. Of course, for the uncertainties, other initial conditions could be valid, as drawn here

Fig. 4.21 Time-dependence of the excitation energy E^* (*dashed green curve*) and its uncertainty for the same damped oscillations as in Fig. 4.20, here relative to the initial energy E^*_0. *Continuous blue curves* show $(E^* \pm \Delta E^*)/E^*_0$ for the initial state there, and *dotted red curves* the same for initially sharp energies

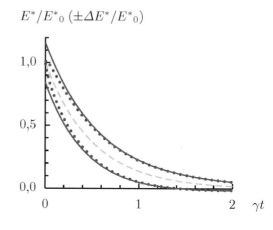

$\langle \Psi^l \rangle$, the above-mentioned phase damping leads to a factor $\exp(-l^2\gamma_0 t)$ which also affects the uncertainties ΔX and ΔP, but not the energy.

Figure 4.21 shows the time dependence of the excitation energy $\langle E^* \rangle = \langle \Psi^\dagger \Psi \rangle\, \hbar\omega$ and its uncertainty. With $(\Delta X/x_0)^2 + (\Delta P/p_0)^2 = \langle \Psi^\dagger \Psi + \frac{1}{2} \rangle - \langle \Psi^\dagger \rangle\langle \Psi \rangle$, the energy is already fixed with the initial values introduced so far, and its uncertainty only by the further initial value $\langle (\Psi^\dagger \Psi)^2 \rangle_0$, which for quench states can be determined using the normal-ordered characteristic function introduced in Sect. 5.5.6 (see p. 481):

$$\langle(\Psi^\dagger\Psi)^2\rangle_t = \tfrac{1}{2}\langle\Psi^\dagger\Psi\rangle_t + \{\langle(\Psi^\dagger\Psi)^2\rangle_0 - \tfrac{1}{2}\langle\Psi^\dagger\Psi\rangle_0\}\exp(-4\gamma t)\ .$$

Thus it can also be zero initially, for $\langle(\Psi^\dagger\Psi)^2\rangle_0 = (\langle\Psi^\dagger\Psi\rangle_0)^2$, but this dependence of the initial uncertainty is rather quickly damped, as Fig. 4.21 shows.

4.6.7 Summary: Dissipation and Quantum Theory

The coupling of an object to unobservable degrees of freedom induces dissipation. The energy does not remain conserved. Classically, this is assigned to friction, which is inaccessible to Hamiltonian mechanics. In quantum theory we also require extensions which go beyond the von Neumann equation (and the Schrödinger equation). Dirac's perturbation theory helps quite a bit here, but further approximations (in particular the Markov approximation) are necessary, until the expressions can be evaluated.

These lead to Fermi's golden rule among other things. The derivative of the occupation probability of an energy state with respect to time, thus the transition rate from the initial to the final state, is equal to the square of the absolute value of its coupling to the initial state times the state density of the relevant reservoir (for finite temperatures, there is one further factor), except for a factor of $2\pi/\hbar$.

But we have also found out how the coherences (the non-diagonal elements of the density operators) depend on time. Their damping ensures *decoherence*: quantum-physical phase effects vanish in favor of classically understandable occupation probabilities. Decoherence leads to a *collapse of the wave function*. It is often overlooked that we always deal with a statistical ensemble, and by selecting a special state, we prepare the old state anew. Decoherence thus leads from quantum physics to classical physics, which is essential for each measurable process, since only then can we arrive at classically realizable situations.

As important as these results are, there remain essential example applications for further chapters (Quantum Mechanics II). We have not yet dealt with many-body problems (where in particular the fact that the particles are indistinguishable has noteworthy consequences), nor with scattering problems and relativistic effects.

Problems

Problem 4.1 Which probability amplitude $\psi(x)$ fits a Gauss distribution $|\psi(x)|^2$ with $\bar{x}=0$ and $\Delta x \neq 0$? What does its Fourier transform

$$\psi(k) = \frac{1}{\sqrt{2\pi}}\int_{-\infty}^{\infty}\exp(-ikx)\,\psi(x)\,dx$$

look like? Show that the factor $1/\sqrt{2\pi}$ here ensures $\int_{-\infty}^{\infty} |\psi(k)|^2 \, dk = 1$. Determine $\Delta x \cdot \Delta k$ for this example.   (6 P)

Problem 4.2 Given a slit of width of $2a$, assume that the probability amplitude $\psi(x) = 1/\sqrt{2a}$ for $|x| \leq a$, otherwise zero. How large is Δx? Determine the Fourier transform. Where are the maximum and the neighboring minima of $|\psi(k)|^2$, and how large are they? Show that the "interference pattern" $|\psi(k)|^2$ becomes more extended with decreasing slit width, but that the product $\Delta x \cdot \Delta k$ is problematic.

(6 P)

Problem 4.3 Consider the Lorentz distribution

$$|\psi(\omega)|^2 \propto 1/\{(\omega - \omega_0)^2 + (\tfrac{1}{2}\gamma)^2\}^{-1} .$$

How large is the uncertainty $\Delta\omega$, and how large is its half-width, i.e., the distance at which $|\psi(\omega)|^2$ has decayed to half the maximum value? Show that $\psi(\omega)$ is the Fourier transform of $\psi(t) \propto \exp\{-i(\omega_0 - i\tfrac{1}{2}\gamma)\,t\}$ for $t \geq 0$, zero for $t < 0$. Can we describe decays with it? How large is the time uncertainty Δt?   (8 P)

Problem 4.4 The transition from the initial state $|i\rangle$ to the final state $|f\rangle$ should be possible via any of the states $|a\rangle$, $|b\rangle$, and $|c\rangle$. How large is the transition probability $|\langle f|i\rangle|^2$ if the states $|a\rangle$ and $|b\rangle$ may interfere, but $|c\rangle$ has to be superposed incoherently?   (2 P)

Problem 4.5 Prove $\int_{-\infty}^{\infty} f(x)\,\delta^{(n)}(x - x')\,dx = (-)^n f^{(n)}(x')$ for   square-integrable functions using integration by parts. Deduce from this that the equation $x\,\delta'(x) = -\delta(x)$ is true for the integrand. Prove $\delta(ax) = \frac{1}{|a|}\,\delta(x)$.   (6 P)

Problem 4.6 A series of functions $\{g_n(x)\}$ forms a complete orthonormal set in the interval from a to b, if $\int_a^b g_n{}^*(x)g_{n'}(x)\,dx = \delta_{nn'}$ and $f(x) = \sum_n g_n(x)f_n$ for all (square-integrable) functions $f(x)$. How can the expansion coefficients f_n be determined? Expand the delta-function $\delta(x - x')$ (with $x' \in [a, b]$) with respect to this basis. Does the sequence $g_n(x) = (2a)^{-1/2} \exp(i\pi nx/a)$ form a complete orthonormal system in the interval $-a \leq x \leq a$?   (6 P)

Problem 4.7 The system of *Legendre polynomials* $P_n(x)$ is complete in the interval $|x| \leq 1$. The *generating function* is $1/\sqrt{1 - 2sx + s^2} = \sum_{n=0}^{\infty} P_n(x)\,s^n$ for $|s| < 1$. How does the associated orthonormal system read? Show that the Legendre polynomials may also be represented by

$$P_n(x) = 1/\{2^n\,n!\}\,d^n(x^2 - 1)^n/dx^n \quad \text{(Rodrigues' formula).}$$

(6 P)

Problem 4.8 The normalized state $|\psi\rangle = |\alpha\rangle\,a + |\beta\rangle\,b$ is constructed from the orthonormalized states $|\alpha\rangle$ and $|\beta\rangle$. What constraint do the coefficients $a \neq 0$ and

$b \neq 0$ satisfy? How do they depend on $|\psi\rangle$? Determine which of the following normalized states $|\psi_i\rangle$ are physically equivalent to $|\psi\rangle$ (disregarding the phase factor): $|\psi_1\rangle = -|\alpha\rangle a - |\beta\rangle b$, $|\psi_2\rangle = |\alpha\rangle a - |\beta\rangle b$, $|\psi_3\rangle = |\alpha\rangle ae^{i\varphi} + |\beta\rangle be^{-i\varphi}$, $|\psi_4\rangle = |\alpha\rangle \cos\varphi \pm |\beta\rangle \sin\varphi$. (6 P)

Problem 4.9 Does the sequence of Hilbert space vectors

$$\begin{pmatrix} 1 \\ 0 \\ 0 \\ \vdots \end{pmatrix}, \begin{pmatrix} 0 \\ 1 \\ 0 \\ \vdots \end{pmatrix}, \begin{pmatrix} 0 \\ 0 \\ 1 \\ \vdots \end{pmatrix}, \ldots$$

converge strongly, weakly, or not at all? If so, give the vector to which the sequence converges. (4 P)

Problem 4.10 Consider the function $\psi(x) = x$ for $-\pi \leq x \leq \pi$. How does it read as a Hilbert vector in the sequence space if we take the basis $\{g_n(x)\}$ of Problem 4.6 (with $a = \pi$)? How does the Hilbert vector in the function space read if it has the components $\psi_n = \delta_{n,1} + \delta_{n,-1}$ in this basis of the sequence space? (4 P)

Problem 4.11 Are the functions $f_0(x) \propto 1$ and $f_1(x) \propto x$ orthogonal to each other for $-\pi \leq x \leq \pi$? Determine their normalization factors. Extend the orthonormalized basis $\{f_0, f_1\}$ so that it is complete for all second-order functions $f(x) = a_0 + a_1 x + a_2 x^2$ in $-\pi \leq x \leq \pi$. (6 P)

Problem 4.12 Determine $\quad [A, [B, C]_\pm] + [B, [C, A]_\pm] + [C, [A, B]_\pm] \quad$ and simplify the expression $[C, [A, B]_\pm]_+ - [B, [C, A]_\pm]_+$. Is

$$(A[B, C]_\pm - [C, A]_\pm B)D + C(A[B, D]_\pm - [D, A]_\pm B)$$

a simple commutator? (6 P)

Problem 4.13 Let the unit operator 1 be decomposed into a projection operator P and its complement Q, viz., $1 = P + Q$. Is Q also idempotent? Are P and Q orthogonal to each other, i.e., is $\text{tr}(PQ) = 0$ true? What are the eigenvalues of P and Q? (4 P)

Problem 4.14 Is the inverse of a unitary operator also unitary? Is the product of two unitary operators unitary? Is $(1 - iA)(1 + iA)^{-1}$ unitary if A is Hermitian? Justify all answers! (4 P)

Problem 4.15 Suppose $(A - a_1 1)(A - a_2 1) = 0$ and let $|\psi\rangle$ be arbitrary, but not an eigenvector of A. Show that $(A - a_1)|\psi\rangle$ and $(A - a_2)|\psi\rangle$ are eigenvectors of A, and determine the eigenvalues. Determine the eigenvalues of the 2×2 matrix A with elements A_{ik}. If the matrix is Hermitian, show that no degeneracy can occur if the matrix is not diagonal. (6 P)

Problem 4.16 Do orthogonal operators remain orthogonal under a unitary transformation?   (2 P)

Problem 4.17 Why is the determinant of the matrix elements of the operator A equal to the product of its eigenvalues?   (4 P)

Problem 4.18 Let the vectors $\mathbf{a}$ and $\mathbf{b}$ commute with the Pauli operator $\boldsymbol{\sigma}$. How can $(\mathbf{a} \cdot \boldsymbol{\sigma})(\mathbf{b} \cdot \boldsymbol{\sigma})$ then be expressed in the basis $\{1, \boldsymbol{\sigma}\}$? What follows for $(\mathbf{a} \cdot \boldsymbol{\sigma})^2$ and what for the anti-commutator $\{\mathbf{a} \cdot \boldsymbol{\sigma}, \mathbf{b} \cdot \boldsymbol{\sigma}\}$? Expand the unitary operator $U = \exp(\mathrm{i}\,\mathbf{a} \cdot \boldsymbol{\sigma})$ in terms of the basis $\{1, \boldsymbol{\sigma}\,\}$.   (6 P)

Problem 4.19 The boson annihilation operator Ψ is in fact not Hermitian and therefore does not necessarily have real eigenvalues, but any complex number ψ may be an eigenvalue of Ψ. Determine (up to the normalization factor) the associated eigenvector in the particle-number basis, and hence the coefficients $\langle n|\psi\rangle$ in $|\psi\rangle = \sum_{n=0}^{\infty} |n\rangle\,\langle n|\psi\rangle$. Why is this not possible for the creation operator $\Psi^\dagger$? For arbitrary complex numbers α and β, consider the scalar product $\langle\alpha|\beta\rangle$ and determine the unknown normalization factor.   (8 P)

Problem 4.20 Show using the method of induction that

$$
\left.\begin{array}{c}\Psi^m\,\Psi^{\dagger\,n} \\ \Psi^{\dagger\,n}\,\Psi^m\end{array}\right\} = \sum_l \frac{(\pm)^l m!\,n!}{l!\,(m-l)!\,(n-l)!}\left\{\begin{array}{c}\Psi^{\dagger\,n-l}\,\Psi^{m-l}\,, \\ \Psi^{m-l}\,\Psi^{\dagger\,n-l}\,.\end{array}\right.
$$

(7 P)

Problem 4.21 Which 2×2 matrices correspond to the Pegg–Barnett operators $\widetilde{\Psi}$, $\widetilde{\Psi}^\dagger$, and $\widetilde{\Psi}\widetilde{\Psi}^\dagger \pm \widetilde{\Psi}^\dagger\widetilde{\Psi}$, if the basis has only two eigenvalues ($s = 1$)? Do these operators behave like field operators for fermions?   (4 P)

Problem 4.22 From $\sigma_x\sigma_y = \mathrm{i}\sigma_z = -\sigma_y\sigma_x$ and $\sigma_x^2 = 1$ (and cyclic permutations), and also $\sigma_\pm = \frac{1}{2}\,(\sigma_x \pm \mathrm{i}\sigma_y)$, determine $\sigma_z\sigma_\pm$, $\sigma_\pm\sigma_z$, $\sigma_\pm\sigma_\mp$ and $\sigma_\pm^2$. What do we obtain therefore for $U\sigma_\pm U^\dagger$ with $U = \exp(\mathrm{i}\alpha\sigma_z)$, according to the Hausdorff series? Simplify the Hermitian operators $\sigma_z\sigma\sigma_z$, $\sigma_\pm\sigma\sigma_\mp$, and $\sigma\sigma_\pm\sigma_\mp + \sigma_\pm\sigma_\mp\sigma$.   (9 P)

Problem 4.23 As is well known, the position and momentum coordinates of a particle span its *phase space*. Show that a classical linear oscillation with angular frequency ω traces an ellipse in phase space, and determine its area as a function of the energy. How large is the probability density for finding the oscillator at the displacement x for oscillations with amplitude $\widehat{x}$, if all phase angles are initially equally probable? (Here we thus consider a statistical ensemble.)   (6 P)

Problem 4.24 Since $\Delta X \cdot \Delta P \geq \frac{1}{2}\hbar$, the phase-space cells may not be chosen arbitrarily small (more finely divided cells would be meaningless). How large is the area if the energy increases by $\hbar\omega$ from cell to cell? Is it possible to associate particles at rest with the cell of lowest energy, which would start oscillating only after gaining energy? What is the mean value of the energy in this cell?   (4 P)

Problem 4.25 Show that the matrix $\langle \psi_1 | P | \psi_2 \rangle = \int_{-\infty}^{\infty} \psi_1^*(x) \frac{\hbar}{i} \frac{d}{dx} \psi_2(x) \, dx$ is Hermitian. What can be concluded from this for the expectation values $\langle P \rangle$ and $\langle P^2 \rangle$ for a real wave function?   (6 P)

Problem 4.26 Derive the 2×2 density matrix of the spin states of unpolarized electrons. Why is it not possible to represent it by a Hilbert vector?   (4 P)

Problem 4.27 Why does the quantum-mechanical expression $\frac{1}{2} \{ f(X) P + P f(X) \}$ correspond to the classical $f(x) p$ according to the Weyl correspondence?

Hint: Use $i\hbar f'(X) = [f(X), P]$.   (6 P)

Problem 4.28 Justify the validity of the following quantum-mechanical expressions—independent of the representation—with a homogeneous magnetic field **B** and Coulomb gauge: $\mathbf{A} = \frac{1}{2} \mathbf{B} \times \mathbf{R}$, $\mathbf{P} \cdot \mathbf{A} + \mathbf{A} \cdot \mathbf{P} = \mathbf{B} \cdot \mathbf{L}$, and $\mathbf{P} \times \mathbf{A} + \mathbf{A} \times \mathbf{P} = -i\hbar \mathbf{B}$.

(4 P)

Problem 4.29 In approximate calculations for motions with high orbital angular momentum, we often replace $\langle L^2 \rangle / \hbar^2$ by the square of a number (as if it were the expectation value of $L/\hbar$). Which number is better than l? How large is the relative error for $l = 3$ and $l = 5$?   (4 P)

Problem 4.30 Is it possible to express the Poisson bracket $[\mathbf{l} \cdot \mathbf{e}_1, \mathbf{a} \cdot \mathbf{e}_2]$ in terms of the triple product $\mathbf{a} \cdot (\mathbf{e}_1 \times \mathbf{e}_2)$ if $\mathbf{a}$ is the position or momentum vector?   (4 P)

Problem 4.31 Derive the uncertainties ΔL_x and ΔL_y for the state $|l, m\rangle$. Hence, determine also $(\Delta L_x)^2 + (\Delta L_y)^2 + (\Delta L_z)^2$.   (2 P)

Problem 4.32 Does **L** commute with R^2 and P^2?   (2 P)

Problem 4.33 For classical vectors **r** and **p**, the equations

$$(\mathbf{r} \times \mathbf{p})^2 = r^2 p^2 - (\mathbf{r} \cdot \mathbf{p})^2 , \qquad \mathbf{p} \times (\mathbf{r} \times \mathbf{p}) = \mathbf{r} p^2 - \mathbf{p}\, \mathbf{r} \cdot \mathbf{p} ,$$

are valid. How do they read for the associated operators?   (4 P)

Problem 4.34 Derive all spherical harmonics for $l = 0, 1$, and 2.   (4 P)

Problem 4.35 Determine the integrals over all directions Ω of $Y_m^{(l)*}(\Omega)$, $Y_{m'}^{(l')}(\Omega)$, and $Y_m^{(l)}(\Omega)$.

Hint: Express the integrals initially with scalar products $\langle \Omega | lm \rangle$.   (2 P)

Problem 4.36 For spherically symmetric problems, the ansatz

$$\psi_{nlm}(\mathbf{r}) = r^{-1} u_{nl}(r) \, i^l \, Y_m^{(l)}(\Omega)$$

turns out to be useful. Using this, reduce $\langle nlm| \, r\cos\theta \, |n'00\rangle$ to a simple integral, given that the integral over the directions is known.

Hint: $r\cos\theta$ corresponds to $\mathbf{R}\cdot\mathbf{e}_z$ in the position representation. (4 P)

Problem 4.37 What do we obtain for $\langle nlm| \, (r\cos\theta)^2 \, |n'00\rangle$ and $\langle nlm| \, \mathbf{P}\cdot\mathbf{e}_z \, |n'00\rangle$ with the ansatz just mentioned? (4 P)

Problem 4.38 The scalar product of two angular momentum operators $\mathbf{J}_1$ and $\mathbf{J}_2$ may be expressed in terms of J_{1z}, $J_{1\pm}$ and J_{2z}, $J_{2\pm}$, viz.,

$$\mathbf{J}_1 \cdot \mathbf{J}_2 = \frac{1}{2}(J_{1+}J_{2-} + J_{1-}J_{2+}) + J_{1z}J_{2z} \; .$$

This helps for the uncoupled basis, but for the coupled basis, the total angular momentum $\mathbf{J} = \mathbf{J}_1 + \mathbf{J}_2$ should be used. Determine the matrix elements of the operator $\boldsymbol{\sigma}_1 \cdot \boldsymbol{\sigma}_2$ in the uncoupled basis $\{|\frac{1}{2}m_1, \frac{1}{2}m_2\rangle\}$ and in the coupled one $\{|(\frac{1}{2}\frac{1}{2})sm\rangle\}$. How can we express the projection operators P_S on the singlet and triplet states (with $S = 0$ and $S = 1$, respectively) using $\boldsymbol{\sigma}_1 \cdot \boldsymbol{\sigma}_2$? (6 P)

Problem 4.39 Represent all $d_{3/2}$ states $|(2\frac{1}{2})\frac{3}{2}m\rangle$ in the uncoupled basis. (4 P)

Problem 4.40 How many p states are there for a spin-$\frac{1}{2}$ particle? Expand in terms of the basis of the total angular momentum. (4 P)

Problem 4.41 Which Ehrenfest equations are valid for the orbital angular momentum? In particular, is the angular momentum a constant on average for a central force? (6 P)

Problem 4.42 Let $\psi(\mathbf{r}) \approx f(\theta)\, r^{-1} \exp(ikr)$ hold for large r. How large is the associated current density for large r? (2 P)

Problem 4.43 How does the position uncertainty for the Gauss wave packet

$$\psi(k) = \exp\{-\frac{1}{4}(\Delta k)^{-2}(k-\bar{k})^2\}/\sqrt[4]{2\pi}\sqrt{\Delta k}$$

depend upon time? In the final result, use $\Delta x(0)$ and Δv instead of Δk. Determine $\bar{x}(t)$ for the case $\bar{x}(0) = 0$. (6 P)

Problem 4.44 Write down the Schrödinger equation for the two-body *hydrogen atom* problem in center-of-mass and relative coordinates. Which (normalized) solution do we have in center-of-mass coordinates? (4 P)

Problem 4.45 For the generalized Laguerre polynomials $L_n^{(m)}(s)$ and for $|t| < 1$, there is a generating function $(1-t)^{-m-1} \exp\{-st/(1-t)\} = \sum_{n=0}^{\infty} L_n^{(m)}(s)\, t^n$. For $\int_0^\infty e^{-s} s^k L_n^{(m)}(s)\, L_{n'}^{(m')}(s)\, ds$, use this to derive the expansion

$$(-)^{n+n'} \sum_{l} \binom{k-m}{n-l}\binom{k-m'}{n'-l} (k+l)!/l! .$$

It is needed for the expectation value $\langle R^k \rangle$ of the hydrogen atom, viz., $\langle R^k \rangle = \int_0^\infty |u|^2 r^k \, dr$, with

$$u_{nl}(r) = \sqrt{\frac{(n-l-1)!}{a_0 \, (n+l)!}} \, \frac{s^{l+1}}{n} \, \exp\left(-\frac{s}{2}\right) L_{n-l-1}^{(2l+1)}(s)$$

and $s \equiv 2r/(na_0)$. How large is $\langle R \rangle$ as a function of n, l, and a_0? (8 P)

List of Symbols

We stick closely to the recommendations of the *International Union of Pure and Applied Physics* (IUPAP) and the *Deutsches Institut für Normung* (DIN). These are listed in *Symbole, Einheiten und Nomenklatur in der Physik* (Physik-Verlag, Weinheim 1980) and are marked here with an asterisk. However, one and the same symbol may represent different quantities in different branches of physics. Therefore, we have to divide the list of symbols into different parts (Table 4.3).

Table 4.3 Symbols used in Quantum Mechanics I

	Symbol	Name	Page number
*	$|\psi\rangle$	Ket-vector (state vector)	282
*	$\langle\psi|$	Bra-vector	283
*	$\langle\varphi|\psi\rangle$	Scalar product,	
		Probability amplitude	282
*	$\langle \mathbf{r}|\psi\rangle \equiv \psi(\mathbf{r})$	Wave function (position representation)	286, 320
*	$\langle \mathbf{p}|\psi\rangle \equiv \psi(\mathbf{p})$	Wave function (momentum representation)	286
	$\langle n| A |n'\rangle \equiv A_{nn'}$	Matrix element of the operator A	290
*	$\langle A\rangle \equiv \bar{A}$	Expectation value of the operator A	298
*	$[A, B] \equiv [A, B]_-$	Commutator of A and B	289

(continued)

Table 4.3 (continued)

	Symbol	Name	Page number
	$\{A,\ B\} \equiv [A,\ B]_+$	Anti-commutator of A and B	289
*	$A^\dagger$	Hermitian adjoint of operator A	292
*	A^{-1}	Inverse of operator A	292
	U	Unitary operator $(U^\dagger = U^{-1})$	293
	Ψ	Annihilation operator	302
	$\Psi^\dagger$	Creation operator	302
	$\mathbf{R}$	Position operator	318
	$\mathbf{P}$	Momentum operator	318
	$\mathbf{L}$	Orbital angular momentum operator	328
	$\mathbf{S}$	Spin (angular momentum) operator	335
*	σ	Pauli operator	308
	H	Hamilton operator	339
	$\mathscr{P}$	Parity operator	313
	$\mathscr{T}$	Time-reversal operator	313
	T	Time-ordering operator	346
	ρ	Density operator	323
	$\rho(\mathbf{r},\ \mathbf{p})$	Wigner function	322
	$Y_m^{(l)}(\Omega)$	Spherical harmonic	332
	$\begin{pmatrix} l & s & j \\ m_l & m_s & m \end{pmatrix}$	Clebsch–Gordan coefficient	337
*	α	Fine structure constant	623
*	a_0	Bohr radius	362
	μ_{B}	Bohr magneton	327

References

1. W. Heisenberg, *The Physical Principles of the Quantum Theory* (Dover, 1930)
2. J. von Neumann, *Mathematische Grundlagen der Quantentheorie* (Springer, Berlin, 1968), p. 4
3. P. Güttinger, Z. Phys. **73**, 169 (1931)
4. D.T. Pegg, S.M. Barnett, Europhys. Lett. **6**(483) (1988). Phys. Rev. A **39**(1665) (1989)
5. E.U. Condon, G.H. Shortley, *The Theory of Atomic Spectra* (Cambridge University Press, 1935)
6. O.L. deLange, R.E. Raab, Phys. Rev. A **34**(1650) (1986)
7. M. Abramowitz, I.A. Stegun, *Handbook of Mathematical Functions* (Dover, New York, 1964)
8. E. Stiefel, A. Fässler, *Group Theoretical Methods and Their Applications (Birkhäuser–Springer* (, Heidelberg, 1992)

Suggestions for Textbooks and Further Reading

9. C. Cohen-Tannoudji, B. Diu, F. Laloè, *Quantum Mechanics 1–2* (Wiley, New York, 1977)
10. R. Dick, *Advanced Quantum Mechanics: Materials and Photons* (Springer, New York, 2012)
11. P.A.M. Dirac: *The Principles of Quantum Mechanics* (Clarendon, Oxford)
12. A.S. Green: *Quantum Mechanics in Algebraic Representation* (Springer, Berlin)
13. W. Greiner, *Quantum Mechanics—An Introduction* (Springer, New York, 2001)
14. G. Ludwig, *Foundations of Quantum Mechanics* (Springer, New York, 1985)
15. L.D. Landau, E.M. Lifshitz: *Course of Theoretical Physics Vol. 3—Quantum Mechanics, Non-Relativistic Theory* 3rd edn. (Pergamon, Oxford, London, 1977)
16. A. Messiah: *Quantum Mechanics I–II* (North-Holland, Amsterdam, 1961–1962)
17. C. Itzykson, J. Zuber, *Quantum Field Theory* (McGraw-Hill, New York, 1980)
18. D. Jackson, *Mathematics for Quantum Mechanics* (Benjamin, New York, 1962)
19. J.M. Jauch, F. Rohrlich, *The Theory of Photons and Electrons. The Relativistic Quantum Field Theory of Charged Particles with Spin One-half* (Springer, Berlin, 1976)
20. W. Nolting, *Theoretical Physics 6—Quantum Mechanics—Basics* (Springer, Berlin, 2017)
21. W. Nolting, *Theoretical Physics 7—Quantum Mechanics—Methods and Approximations* (Springer, Berlin, 2017)
22. P. Roman: *Advanced Quantum Theory* (Addison-Wesley, Reading)
23. J.J. Sakurai, *Advanced Quantum Mechanics* (Addison-Wesley, Reading MA, 1967)
24. J.J. Sakurai, J. Napolitano, *Modern Quantum Mechanics*, 2nd edn. (Addison-Wesley, Boston, 2011)
25. F. Scheck, *Quantum Physics*, 2nd edn. (Springer, Berlin, 2013)
26. F. Schwabl: *Quantum Mechanics* (Springer, Berlin)

Chapter 5
Quantum Mechanics II

5.1 Scattering Theory

5.1.1 Introduction

In simple descriptions of the scattering process, where a sharp energy is assumed and the time factor $\exp(-i\omega t)$ subsequently left out, the obvious result of this chapter can be stated immediately: if a plane wave $\exp(i\mathbf{k} \cdot \mathbf{r})$ falls on a scattering center, then the original wave and the outgoing spherical wave $f(\theta) \exp(ikr)/r$ become superposed, and then the *scattering amplitude* $f(\theta)$ is of decisive importance. Here the center-of-mass system is assumed, and the reduced mass m_0 and kinetic energy $E = \hbar\omega = (\hbar k)^2/2m_0$ are given. As will be shown in the following, for large distances r from the scattering center, we have (see Fig. 5.1)

$$\langle \mathbf{r} \,|\, \mathbf{k} \,\rangle^+ \approx \frac{1}{\sqrt{2\pi}^3} \left(\exp(i\mathbf{k} \cdot \mathbf{r}) + f(\theta)\frac{\exp(ikr)}{r} \right) .$$

Here the scattering amplitude $f(\theta)$ is connected to the *scattering operator* S and the *transition operator* T, these being the important quantities in scattering theory. From the scattering amplitude, we can obtain, e.g., the differential scattering cross-section for the scattering angle θ (as derived on p. 418)

$$\frac{d\sigma}{d\Omega} = |f(\theta)|^2 .$$

With these expressions we can already solve the simplest scattering problems.

To this end, we decompose the plane wave $\exp(i\mathbf{k} \cdot \mathbf{r})$ in terms of *spherical waves*:

$$\exp(i\mathbf{k} \cdot \mathbf{r}) = \frac{4\pi}{kr} \sum_{lm} F_l(kr) \, Y_m^{(l)*}(\Omega_k) \, i^l \, Y_m^{(l)}(\Omega_r) .$$

© Springer Nature Switzerland AG 2018
A. Lindner and D. Strauch, *A Complete Course*
on Theoretical Physics, Undergraduate Lecture Notes in Physics,
https://doi.org/10.1007/978-3-030-04360-5_5

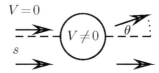

Fig. 5.1 Scattering with scattering angle θ (angle of deflection) and collision parameter s (see Fig. 2.6). If s is too large, there is no scattering force

In order to prove this equation, with $\rho \equiv kr$, we start from $\exp(i\rho \cos \theta)$. We expand this in terms of Legendre polynomials. According to p. 82 (or Problem 4.7), they form an orthogonal system, normalized to $(l+\frac{1}{2})^{-1/2}$, in the variables $\cos \theta$:

$$\exp(i\rho \cos \theta) = \sum_l \frac{2l+1}{\rho} F_l(\rho) \, i^l P_l(\cos \theta) \, ,$$

with the *regular spherical Bessel function* (see Fig. 5.2 top)

$$F_l(\rho) = \frac{\rho}{2 \, i^l} \int_{-1}^{1} d \cos \theta \, P_l(\cos \theta) \, \exp(i\rho \cos \theta) \, .$$

Note that this name usually refers to $j_l(\rho) \equiv \rho^{-1} F_l(\rho) = \sqrt{\pi/(2\rho)} J_{l+1/2}(\rho)$, but for the expansion in terms of spherical harmonics in Sect. 4.5.2, we always wanted to take out a factor $1/r$ from the radial function, and $F_l(\rho)$ actually has more comfortable properties. In particular, $F_0(\rho) = \sin \rho$ and $F_1(\rho) = \rho^{-1} \sin \rho - \cos \rho$ (since $P_0 = 1$ and $P_1 = \cos \theta$), and the higher Bessel functions result from the *recursion relation*

$$F_{l+1}(\rho) = \frac{2l+1}{\rho} F_l(\rho) - F_{l-1}(\rho) \, ,$$

which can themselves be derived from the recursion relations for Legendre polynomials (see p. 82). For the rest of the proof, we still have to expand the Legendre polynomials in terms of spherical harmonics:

$$P_l(\cos \theta) = \frac{4\pi}{2l+1} \sum_{m=-l}^{l} Y_m^{(l)*}(\Omega_k) \, Y_m^{(l)}(\Omega_r) \, .$$

For the proof this *addition theorem for spherical harmonics*, we rotate by a rotational vector $\boldsymbol{\omega}$. We thus have $Y_m^{(l)}(\Omega') = \sum_{m'} Y_{m'}^{(l)}(\Omega) \, \mathscr{D}_{m'm}^{(l)*}(\boldsymbol{\omega})$ and the rotation operator $\mathscr{D}$ is unitary: $\sum_m \mathscr{D}_{m'm}^{(l)}(\boldsymbol{\omega}) \, \mathscr{D}_{m''m}^{(l)*}(\boldsymbol{\omega}) = \delta_{m'm''}$. If we now choose one of the two directions Ω_k or Ω_r as the new z-direction and use Sect. 4.3.9, and in particular, the equations $Y_m^{(l)}(0,0) = \sqrt{(2l+1)/4\pi} \, \delta_{m0}$ and $Y_0^{(l)}(\Omega) = \sqrt{(2l+1)/4\pi} \, P_l(\cos \theta)$, then the addition theorem is proven.

Fig. 5.2 Spherical Bessel functions with l from 0 (*black*) to 3 (*blue*) (*continuous* for l even, *dotted* for l odd). *Top*: regular F_l. *Bottom*: irregular G_l. In addition to these spherical functions, there are also the normal (cylindrical) Bessel functions (see Fig. 5.17)

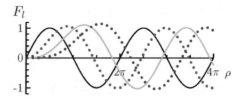

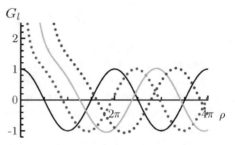

For the regular spherical Bessel functions $F_l(\rho)$, we have asymptotically

$$F_l(\rho) \approx \begin{cases} \rho^{l+1}/(2l+1)!! & \text{for } \rho \approx 0 , \\ \sin(\rho - \tfrac{1}{2}l\pi) & \text{for } \rho \gg l(l+1) , \end{cases}$$

where the *double factorial* $(2l+1)!!$ is the product of all odd integers up to $2l + 1$, viz., $(2l+1)!! = \prod_{n=0}^{l}(2n+1) = (2l+1)!/2^l\, l!$. Then, with $\langle \mathbf{k}'|\mathbf{k}\rangle = (2\pi)^{-3}$ $\int \mathrm{d}^3r\, \exp\{\mathrm{i}(\mathbf{k}-\mathbf{k}')\cdot\mathbf{r}\}$, we have $\int_0^\infty \mathrm{d}r\, F_l(kr)\, F_l(k'r) = \tfrac{1}{2}\pi\, \delta(k-k')$. In addition, it solves the differential equation

$$\left(\frac{\mathrm{d}^2}{\mathrm{d}\rho^2} + 1 - \frac{l(l+1)}{\rho^2}\right)F_l(\rho) = 0 ,$$

as do the remaining spherical Bessel functions, i.e., the *irregular Bessel functions* (*Neumann functions*) (see Fig. 5.2 bottom)

$$G_l(\rho) \approx \begin{cases} (2l-1)!!\,\rho^{-l} , & \text{for } \rho \approx 0 \text{ and } l > 0 \quad (\cos\rho \text{ for } l = 0) , \\ \cos(\rho - \tfrac{1}{2}l\pi) , & \text{for } \rho \gg l(l+1) , \end{cases}$$

the *outgoing Bessel function* (*Hankel function*)

$$O_l(\rho) \equiv G_l(\rho) + \mathrm{i}F_l(\rho) \approx \exp\{\mathrm{i}(\rho - \tfrac{1}{2}l\pi)\} , \quad \text{for } \rho \gg l(l+1) ,$$

and the *incoming Bessel function* (*Hankel function*)

$$I_l(\rho) \equiv G_l(\rho) - \mathrm{i}F_l(\rho) = O_l^*(\rho) .$$

These functions are solutions of the radial Schrödinger equation

$$\left(\frac{\partial^2}{\partial r^2} + k^2 - \frac{l(l+1)}{r^2} - \frac{2m_0}{\hbar^2} V(r)\right) u_l(k, r) = 0 \, ,$$

for large r, because there $V(r)$ will be negligibly small compared to $E > 0$:

$$u_l(k, r) \approx N_l \{F_l(kr) - \pi \, T_l \, O_l(kr)\} \, .$$

Here, we shall actually superpose a plane wave and an outgoing spherical wave. Starting from the boundary condition $u_l(k, 0) = 0$, which is necessary according to p. 353, so that the wave function is differentiable at the origin, and with a convenient slope at the origin which just fixes an inessential factor, we can integrate the differential equation up to the point where the above-mentioned splitting in terms of Bessel functions occurs. Since this is also possible for the first derivative with respect to r, noting that the normalization factor N_l cancels, the unknown *transition amplitude* is given by

$$T_l = \frac{1}{\pi} \frac{W(u_l, F_l)}{W(u_l, O_l)} \, ,$$

with the Wronski determinant

$$W(u_l, F_l) = u_l \frac{\partial F_l}{\partial r} - \frac{\partial u_l}{\partial r} F_l \quad \text{and} \quad W(u_l, O_l) = u_l \frac{\partial O_l}{\partial r} - \frac{\partial u_l}{\partial r} O_l \, .$$

With the normalization factor $N_l = \sqrt{2/\pi}/k$ of u_l in

$$\langle \mathbf{r} | \mathbf{k} \rangle^+ = \sum_{lm} \frac{u_l(k, r)}{r} \, Y_m^{(l)*}(\Omega_k) \, i^l \, Y_m^{(l)}(\Omega_r) \, ,$$

the asymptotic expression for $\langle \mathbf{r} | \mathbf{k} \rangle^+$ with

$$O_l(kr) \approx i^{-l} \exp(ikr)$$

yields the *scattering amplitude*

$$f(\theta) = -\frac{\pi}{k} \sum_l (2l + 1) \, T_l \, P_l(\cos \theta) \, ,$$

and we can derive the scattering cross-section from this. Note that, for low energies, only a few terms contribute to this series. With increasing l the centrifugal potential always dominates the remaining $V(r)$, whence $u_l \to F_l$, and along with it $T_l \to 0$.

Having made this introduction with its prescriptions for proper calculations, we shall now proceed to investigate the scattering process in more detail.

5.1.2 Basics

In order to clarify the basic notions of scattering theory, we restrict ourselves initially to *elastic two-body scattering* and investigate only the change in the motion due to the forces between the two scattering partners. Since the interaction V depends only on the relative distance (and possibly also on the spin) of the scattering partners, it is thus *translation invariant*, and we can disregard the center-of-mass motion. The centre of mass moves unperturbed, with fixed momentum. Therefore, we consider only the relative motion and use the reduced mass m_0—keeping m for the directional (magnetic) quantum number.

As already for classical collisions (Sect. 2.2.3), we assume that the partners before and after the scattering move unperturbed. The coupling V is assumed to have a finite range, i.e., it should decrease more rapidly than r^{-1}. The Coulomb force is an exception, which we consider separately in Sect. 5.2.3. The ray is usually directed toward an uncharged probe, and then there is no Coulomb field, but it is nevertheless important in nuclear physics, because the screening action of the atomic shell may be neglected there, and only the interaction between the nuclei counts. But in the present discussion, we shall assume that the scattering partners act on each other only for a comparably short while—before and after, they are outside the range of the forces and move unperturbed. *Each scattering is a time-dependent situation.* Therefore the unperturbed motion must not be described by a plane wave, since this would be equally probable in the whole space, and there would be no "before" and no "after". Instead we have to take wave packets. This we shall do rather superficially, in the sense that we shall not provide the exact form of the wave packet. We shall then be able to work out basic notions of *time-dependent scattering theory*. The next step will be to go over to *time-independent scattering theory* (with sharp energy) using a Fourier transform, whereupon the calculations become rather simple.

The Schrödinger equation is normally taken as the most important starting equation in any introduction to quantum theory. This is suitable for bound states, because their wave functions are essentially already determined by this differential equation. The boundary conditions are still missing, of course, but these are self-evident for bound states with the required normalizability and lead to the well-known eigenvalue problem. In contrast to the situation for unbound states (scattering states), where the boundary conditions still play an important role in determining the solution, only the asymptotic behavior is significant for many applications. Therefore, we shall struggle with an integral equation which contains the Hamilton operator as well as the boundary conditions, and then of course use the *Lippmann–Schwinger equation* to solve that.

5.1.3 Time Shift Operators in Perturbation Theory

In the Schrödinger picture the development of a state with time t can be given by the unitary time shift operator $U(t, t_0)$:

$$|\psi(t)\rangle = U(t, t_0) \, |\psi(t_0)\rangle \;, \quad \text{with} \quad U(t, t) = 1 \;,$$

and thus $U(t, t_0) = U^{-1}(t_0, t) = U^\dagger(t_0, t)$. Here, according to the Schrödinger equation,

$$i\hbar \, \frac{\partial}{\partial t} \, U(t, t_0) = H \; U(t, t_0) \qquad \Longrightarrow \qquad U(t, t_0) = \exp \frac{-iH \; (t - t_0)}{\hbar} \;,$$

provided that the Hamilton operator H does not depend on time, which we assume.

The time shift operator by itself is not enough for the description of the scattering problem. Initial conditions have to be added. But these refer to states in which there are no forces acting between the scattering partners, so not all of the Hamilton operator H is important, only the free (unperturbed) Hamilton operator H_0:

$$H = H_0 + V \;.$$

We indicate, e.g., the initial state by the relative momentum $\mathbf{p}$ with a suitable distribution function for a wave packet. It remains unaltered only until the interaction V between the scattering partners becomes notable:

$$[H, \mathbf{P}] \neq 0 \;, \quad \text{but} \quad [H_0, \mathbf{P}] = 0 \;.$$

The above-mentioned Hamilton operators do not depend on time, only their effects on the states do.

In addition to the full Hamilton operator H and time shift operator $U(t, t_0)$, it is therefore appropriate to consider also the free operator H_0 or again $U_0(t, t_0)$, and to employ the Dirac picture. According to p. 346, we have $U(t, t_0) = U_0(t, t_0) \, U_D(t, t_0)$ and

$$U_D(t, t_0) = 1 + \int_{t_0}^{t} \mathrm{d}t' \, \frac{V_D(t', t_0) \, U_D(t', t_0)}{i\hbar} \;,$$

with $V_D(t', t_0) = U_0^\dagger(t', t_0) \, V \, U_0(t', t_0)$. Here $U_0(t, t_0)$ can be decomposed into $U_0(t, t') U_0(t', t_0)$, and U_0 is unitary, with $U_0(t, t_0) \, U_0^\dagger(t', t_0) = U_0(t, t')$. From this follows the important equation

$$U(t, t_0) = U_0(t, t_0) + \int_{t_0}^{t} \mathrm{d}t' \, \frac{U_0(t, t') \, V \; U(t', t_0)}{i\hbar} \;,$$

which can be derived from

$$i\hbar \, \frac{\partial}{\partial t'} \, \{U_0(t, t') \, U(t', t_0)\} = U_0(t, t') \, \{-H_0 + H\} \; U(t', t_0)$$

by integrating over t' from t_0 to t. With

$$i\hbar \frac{\partial}{\partial t'} \{U(t, t') U_0(t', t_0)\} = U(t, t') \{-H + H_0\} U_0(t', t_0) \,,$$

we clearly have the equally important result

$$U(t, t_0) = U_0(t, t_0) + \int_{t_0}^{t} dt' \, \frac{U(t, t') \, V \, U_0(t', t_0)}{i\hbar} \,.$$

These two "important" equations form the basis for all that follows. Since $|\psi(t_0)\rangle$ has to be given by the initial conditions, everything worth knowing about the scattering power of the interaction is contained in $U(t, t_0)$. Note that U_0 is known here, but the question remains as to how V affects U.

For stepwise integration, the two forms deliver the same *Neumann series*

$$U(t, t_0) = U_0(t, t_0) + \int_{t_0}^{t} dt' \, \frac{U_0(t, t') \, V \, U_0(t', t_0)}{i\hbar}$$

$$+ \int_{t_0}^{t} dt' \int_{t_0}^{t'} dt'' \, \frac{U_0(t, t') \, V \, U_0(t', t'') \, V \, U_0(t'', t_0)}{(i\hbar)^2} + \cdots \,.$$

It represents the time shift operator $U(t, t_0)$ of the full problem as a sum of time shift operators which feel the potential only at the times t', t'', etc., between t_0 and t and are otherwise determined by H_0, i.e., they are "free" (unperturbed). With the nth term, n interactions occur. If V changes the motion only a little, then this series converges fast. In the *Born approximation*, we terminate after the first term (with one V). This is often a good approximation, but certainly not for resonances.

5.1.4 Time-Dependent Green Functions (Propagators)

We search for the time shift operators for long time spans, because we want to connect the initial and final states. We shall not be concerned with intermediate states that cannot be measured. Therefore, we now set $t_0 = 0$ and investigate the behavior for $t \to \pm\infty$. For these convergence investigations it is better to consider the distant past ($t \to -\infty$) and the far future ($t \to +\infty$) separately.

Using the step function $\varepsilon(x)$ from p. 18 (see Fig. 5.3), whose derivative is the delta function, the following quantities are introduced:

Fig. 5.3 The discontinuity functions $\varepsilon(t)$ (*left*) and $\varepsilon(-t)$ (*right*). Since we have $\varepsilon(-t) = 1 - \varepsilon(t)$, $-\varepsilon(-t)$ has the same derivative as $\varepsilon(t)$, namely $\delta(t)$

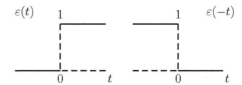

$$G^{\pm}(t) \equiv \frac{\varepsilon(\pm t)\, U(t,0)}{\pm i\hbar} \quad \text{and} \quad G_0^{\pm}(t) \equiv \frac{\varepsilon(\pm t)\, U_0(t,0)}{\pm i\hbar} \,.$$

They satisfy the differential equations

$$\left(i\hbar \frac{d}{dt} - H\right) G^{\pm}(t) = \delta(t), \quad \text{or} \quad \left(i\hbar \frac{d}{dt} - H_0\right) G_0^{\pm}(t) = \delta(t),$$

and are therefore called *Green functions*, since Green functions always solve linear differential equations which have a delta function as the inhomogeneous term. We have encountered other examples of Green functions on pp. 27, 112, and 119. In fact, we are actually dealing with operators, often also called propagators. Clearly, the functions carrying a "+" are unequal to zero only for $t > 0$ and those carrying a "−" only for $t < 0$. Hence we speak of the retarded (+) and advanced (−) Green functions (propagators). We have

$$\text{for } t \gtrless 0, \quad U(t,0) = \pm i\hbar\, G^{\pm}(t), \quad U_0(t,0) = \pm i\hbar\, G_0^{\pm}(t),$$

and use the integral equations of the last sections to derive similar ones for the Green functions:

$$G^{\pm}(t) = G_0^{\pm}(t) + \int_{-\infty}^{\infty} dt'\, G_0^{\pm}(t-t')\, V\, G^{\pm}(t')$$
$$= G_0^{\pm}(t) + \int_{-\infty}^{\infty} dt'\, G^{\pm}(t-t')\, V\, G_0^{\pm}(t').$$

For G^+, the integrand vanishes outside $0 \le t' \le t$, and for G^-, outside $t \le t' \le 0$. With the higher integration limits, we may combine the equations for the retarded and advanced Green functions and obtain integral equations of the Volterra type. Here we find convolution integrals. According to p. 22, we can transform them into products using a Fourier transform and then evaluate the unknown $G^{\pm}$ from $G_0^{\pm}$ and V algebraically.

5.1.5 Energy-Dependent Green Functions (Propagators) and Resolvents

Fourier transforming the integral equations of the time-dependent propagators

$$G^{\pm}(E) \equiv \int_{-\infty}^{\infty} dt\, \exp \frac{iEt}{\hbar}\, G^{\pm}(t),$$

and keeping the factor $\sqrt{2\pi}$, we obtain the*Lippmann–Schwinger equations*

$$G^{\pm}(E) = G_0^{\pm}(E) + G_0^{\pm}(E)\ V\ G^{\pm}(E)$$
$$= G_0^{\pm}(E) + G^{\pm}(E)\ V\ G_0^{\pm}(E)\ ,$$

since, with $\tau = t - t'$,

$$G^{\pm}(E) = G_0^{\pm}(E) + \int_{-\infty}^{\infty} dt \int_{-\infty}^{\infty} dt'\ \exp\frac{iEt}{\hbar}\ G_0^{\pm}(t-t')\ V\ G^{\pm}(t')$$
$$= G_0^{\pm}(E) + \int_{-\infty}^{\infty} d\tau\ \exp\frac{iE\tau}{\hbar}\ G_0^{\pm}(\tau)\ V\ G^{\pm}(E)\ .$$

These equations can be solved formally:

$$G^{\pm}(E) = \frac{1}{1 - G_0^{\pm}(E)V}\ G_0^{\pm}(E) = G_0^{\pm}(E)\ \frac{1}{1 - VG_0^{\pm}(E)}\ .$$

We often write the right-hand side as a Neumann series, viz.,

$$G^{\pm}(E) = G_0^{\pm}(E) + G_0^{\pm}(E)\ V\ G_0^{\pm}(E) + \cdots\ ,$$

and here possibly neglect the higher order terms (Born approximation).

However, before evaluating $G^{\pm}(E)$, we must first determine the simpler propagator $G_0^{\pm}(E)$ of the free motion and here determine the Fourier integral. With $G_0^{\pm}(t) = (\pm i\hbar)^{-1}\ \varepsilon(\pm t)\ U_0(t, 0)$ and $U_0(t, 0) = \exp(-iH_0 t/\hbar)$, we also have

$$G_0^{\pm}(E) = \frac{1}{i\hbar} \int_0^{\pm\infty} dt\ \exp\frac{i\,(E-H_0)\,t}{\hbar} = \frac{1}{\pm i\hbar} \int_0^{\infty} dt\ \exp\frac{\pm i\,(E-H_0)\,t}{\hbar}\ ,$$

where we may use an eigenvalue E_0 of H_0 in the energy representation. We have already investigated these integrals on p. 22 in the context of distributions, and found there

$$\int_0^{\infty} dk\ \exp(\pm ikx) = \frac{\pm i}{x \pm io} = \pm i \left(\frac{P}{x} \mp i\pi\ \delta(x)\right)\ ,$$

where $(x \pm io)^{-1}$ indicates the limiting value $\varepsilon \to +0$ of $(x \pm i\varepsilon)^{-1}$ and P (Cauchy's) principal value:

$$\int_{-\infty}^{\infty} dx\ \frac{P}{x}\ f(x) \equiv P \int_{-\infty}^{\infty} dx\ \frac{f(x)}{x} \equiv \lim_{\varepsilon\to+0}\left(\int_{-\infty}^{-\varepsilon} + \int_{+\varepsilon}^{\infty}\right) dx\ \frac{f(x)}{x}\ .$$

This cuts out a piece around the singular point, with boundaries that converge *symmetrically* towards this position—the cut-out region is investigated by the delta function $\delta(x)$ (as in Fig. 1.6):

$$G_0^\pm(E) = \frac{\mathrm{P}}{E - H_0} \mp i\pi\, \delta(E - H_0) \ .$$

In the following, however, we shall often use

$$G_0^\pm(E) = \frac{1}{E \pm io - H_0} \ ,$$

and correspondingly for $G^\pm$, or even just $G_0 \equiv G_0(\mathscr{E}) \equiv (\mathscr{E} - H_0)^{-1}$, although this is only unique for $\operatorname{Im} \mathscr{E} \neq 0$. The Lippmann–Schwinger equations follow simply from the operator identity

$$\frac{1}{A} = \frac{1}{B} + \frac{1}{B}(B - A)\frac{1}{A} = \frac{1}{B} + \frac{1}{A}(B - A)\frac{1}{B} \ ,$$

if we set $A = E \pm io - H$ and $B = E \pm io - H_0$, then as a consequence we have $B - A = V$, and we replace the limiting value of the product by the product of the limiting values. In addition, we clearly have

$$G^{\pm\dagger} = G^\mp \ , \qquad G_0^{\pm\dagger} = G_0^\mp \ .$$

Retarded and advanced propagators are thus adjoints of one another.

At first glance, it may seem astonishing that we have found an expression for $G_0^\pm(E)$ which makes sense only as a weight function in an integrand. But we describe a time-dependent situation (in particular for each scattering process, we distinguish between before and after) and the Fourier transform $t \to E$ obscures this situation. This procedure is only comprehensible if we calculate with unsharp energies (using wave packets, i.e., integral expressions).

5.1.6 Representations of the Resolvents and the Interactions

The resolvent $G_0^\pm(E) = (E \pm io - H_0)^{-1}$ is diagonal in the *energy representation* $\{|E\Omega\rangle\}$ and also in the *momentum representation* $\{|\mathbf{k}\rangle\}$ (with $E = \hbar^2 k^2/2m_0$), and it is interesting to use both representations for scattering problems:

$$\langle E'\Omega'|\, G_0^\pm(E)\, |E''\Omega''\rangle = \frac{\langle E'\Omega'|E''\Omega''\rangle}{E \pm io - E'} \ ,$$

$$\langle \mathbf{k}'|\, G_0^\pm(E)\, |\mathbf{k}''\rangle = \frac{\langle \mathbf{k}'|\mathbf{k}''\rangle}{E \pm io - \hbar^2 k'^2/2m_0} = \frac{2m_0}{\hbar^2}\, \frac{\langle \mathbf{k}'|\mathbf{k}''\rangle}{k^2 \pm io - k'^2} \ .$$

However, the coupling V is usually given as a function of $\mathbf{r}$. Therefore, we now search for the resolvent in the real-space representation. Using the fact that $\langle \mathbf{r}|\mathbf{k}\rangle = (2\pi)^{-3/2}\exp(i\mathbf{k}\cdot\mathbf{r})$, we find

$$\langle \mathbf{r} \mid G_0^\pm(E) \mid \mathbf{r}' \rangle = \frac{1}{(2\pi)^3} \frac{2m_0}{\hbar^2} \int d^3k' \frac{\exp\{i\mathbf{k}' \cdot (\mathbf{r} - \mathbf{r}')\}}{k^2 \pm io - k'^2} .$$

The integration over the directions of $\mathbf{k}'$ is easy. In particular, if we express the plane wave in terms of spherical harmonics, then introducing $Y_m^{(l)}(\Omega) = i^{-l}\langle \Omega | lm \rangle$ and $Y_0^{(0)}(\Omega) = 1/\sqrt{4\pi}$, the contribution for the integration over all directions comes only from $l = 0$, since $\int d\Omega \langle lm|\Omega\rangle\langle\Omega|00\rangle = \langle lm|00\rangle$:

$$\int d\Omega_k \exp(i\mathbf{k} \cdot \mathbf{a}) = 4\pi \frac{F_0(ka)}{ka} = 4\pi \frac{\sin ka}{ka} .$$

Hence the triple integral is reduced to a single one:

$$\int d^3k' \frac{\exp(i\mathbf{k}' \cdot \mathbf{a})}{k^2 \pm io - k'^2} = \frac{4\pi}{2i} \int_0^\infty \frac{k' \, dk'}{a} \frac{\exp(ik'a) - \exp(-ik'a)}{k^2 \pm io - k'^2}$$

$$= \frac{2\pi}{ia} \int_{-\infty}^\infty k' \, dk' \frac{\exp(ik'a)}{k^2 \pm io - k'^2} .$$

These integrals can be evaluated using complex analysis. The integrands each have two simple poles in the complex k'-plane, with $k_1' = \sqrt{k^2 \pm io}$ and $k_2' = -\sqrt{k^2 \pm io}$. Here, according to the residue theorem $\oint f(z)(z - z_0)^{-1} \, dz = 2\pi i f(z_0)$, the residues in the upper half plane are important because there the integrals over the semi-circle with radius $|k'|$ vanish in the limit $|k'| \to \infty$. Then,

$$\int_{-\infty}^\infty k' \, dk' \frac{\exp(ik'a)}{k^2 \pm io - k'^2} = -2\pi i \, (\pm\sqrt{k^2 \pm io}) \frac{\exp(\pm i\sqrt{k^2 \pm io} \, a)}{\pm 2\sqrt{k^2 \pm io}}$$

$$= -\pi i \, \exp(\pm ika) ,$$

and therefore in the real-space representation, the resolvent becomes

$$\langle \mathbf{r} \mid G_0^\pm(E) \mid \mathbf{r}' \rangle = -\frac{1}{4\pi} \frac{2m_0}{\hbar^2} \frac{\exp(\pm ik|\mathbf{r} - \mathbf{r}'|)}{|\mathbf{r} - \mathbf{r}'|} .$$

It is no accident that we encountered the functions $\exp(\pm ik|\mathbf{r} - \mathbf{r}'|)/|\mathbf{r} - \mathbf{r}'|$ in our discussion of electrodynamics (see p. 255), since we were discussing there the scattering of waves.

Since the momentum representation for scattering problems is actually better than the real-space representation (given that the momenta mark the initial and final states, and the free propagators are diagonal in the momentum representation), we now derive the matrix elements of some popular interactions in the momentum representation. Here we restrict ourselves to couplings, which do not act on the spin, and hence only involve *Wigner forces*, and we shall in fact focus on local and isotropic couplings. Then with $\hbar\mathbf{q}$ as the momentum transfer, we have

Table 5.1 Scattering potentials and their Fourier transforms

Potential	$V(r)/V_0$	$V(q) \cdot (\sqrt{2\pi}/a)^3 / V_0$
Yukawa	$a/r\ \exp(-r/a)$	$4\pi/(1 + a^2 q^2)$
Coulomb	a/r	$4\pi/(a^2 q^2)$
Box	$\varepsilon(a - r)$	$4\pi/(a^2 q^2) \cdot F_1(aq)$
Gauss	$\exp(-r^2/a^2)$	$\sqrt{\pi}^{\,3} \exp(-\tfrac{1}{4}a^2 q^2)$

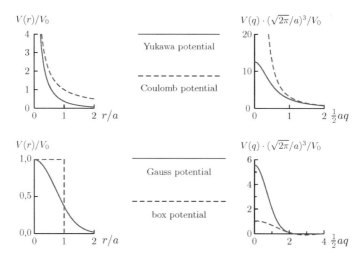

Fig. 5.4 Fourier transforms of several isotropic potentials. *Top*: Yukawa and Coulomb potentials. *Bottom*: Gauss and box potentials. With $\tfrac{1}{2}q = k \sin\tfrac{1}{2}\theta$, $V(q)$ can be used in the Born approximation $T(q) \approx V(q)$ for the differential scattering cross-section, as will be shown

$$\langle \mathbf{k} + \mathbf{q} \,|\, V \,|\, \mathbf{k} \rangle = \int d^3r\ \langle \mathbf{k} + \mathbf{q} \,|\, \mathbf{r} \rangle\ V(r)\ \langle \mathbf{r} \,|\, \mathbf{k} \rangle = \frac{1}{(2\pi)^3} \int d^3r\ V(r)\ \exp(-i\mathbf{q} \cdot \mathbf{r}) \,.$$

This is the Fourier transformed $V(\mathbf{q})$ of the coupling, disregarding the factor $(2\pi)^{-3/2}$. As long as $V(\mathbf{r})$ depends only upon r, as in the present case, we can easily integrate over the directions:

$$\langle \mathbf{k} + \mathbf{q} \,|\, V \,|\, \mathbf{k} \rangle = \frac{V(\mathbf{q})}{(2\pi)^{3/2}} = \frac{4\pi}{(2\pi)^3} \int_0^\infty dr\ r^2\ V(r)\ \frac{\sin qr}{qr} \,.$$

Consequently, this matrix element only depends on the modulus of the momentum transfer: $V(\mathbf{q}) = V(q)$ for each (isotropic) Wigner force. Here $\mathbf{q} = \mathbf{k}_f - \mathbf{k}_i$, and consequently $q^2 = k_f^2 + k_i^2 - 2\mathbf{k}_f \cdot \mathbf{k}_i$, so for elastic scattering $q = 2k \sin\tfrac{1}{2}\theta$, where θ is the scattering angle in the center-of-mass system.

Important examples with two parameters V_0 and a for strength and distance are shown in Table 5.1 and Fig. 5.4, where the spherical Bessel function is

$$F_1(\rho) = \rho^{-1} \sin \rho - \cos \rho \ .$$

Note that the Coulomb potential turns up as the limit $a \to \infty$ of the Yukawa potential, but with aV_0 held fixed. We can thus take

$$\int d^3k \ k^{-2} \ \exp(-i\mathbf{k} \cdot \mathbf{r}) = 4\pi \int_0^\infty dk \ (kr)^{-1} \sin(kr) \ ,$$

because according to Sect. 1.1.10 this is equal to $4\pi \ r^{-1} \ \pi \ \{\varepsilon(r) - \frac{1}{2}\}$, i.e., with $r > 0$, it is equal to $2\pi^2 r^{-1}$. Then k^{-2} is the Fourier transform of $\sqrt{\pi/2} \ r^{-1}$. For the Gauss potential, we can use p. 23.

5.1.7 Lippmann–Schwinger Equations

On p. 407, we derived the Lippmann–Schwinger equations for the propagators $G^\pm = G_0^\pm + G_0^\pm V G^\pm = G_0^\pm + G^\pm V G_0^\pm$. In the following, we shall generally skip the reference to E. Then,

$$G^\pm = G_0^\pm (1 + V G^\pm) = (1 + G^\pm V) G_0^\pm \ ,$$

and also $G_0^\pm = G^\pm (1 - V G_0^\pm) = (1 - G_0^\pm V) G^\pm$. This leads to

$$G_0^\pm = G_0^\pm (1 + V G^\pm) (1 - V G_0^\pm) = (1 - G_0^\pm V) (1 + G^\pm V) G_0^\pm \ ,$$
$$G^\pm = G^\pm (1 - V G_0^\pm) (1 + V G^\pm) = (1 + G^\pm V) (1 - G_0^\pm V) G^\pm \ .$$

Here $G_0^\pm$ acts in the Hilbert space of all states of the unperturbed problem, but $G^\pm$ only in the space of the scattering states: the bound states are missing. Therefore, the projection operator onto the scattering states of H is now useful. Following Feshbach [1], we shall denote this by P. Then it follows that

$$(1 + V G^\pm) (1 - V G_0^\pm) = (1 - G_0^\pm V) (1 + G^\pm V) = 1 \ ,$$
$$(1 - V G_0^\pm) (1 + V G^\pm) = (1 + G^\pm V) (1 - G_0^\pm V) = P \ .$$

We shall return to the fact that the bound states are missing in the next section.

Before that, however, we shall also derive the Lippmann–Schwinger equations for the states. They are superior to the Schrödinger equation for scattering problems, since for a differential equation, we still need boundary conditions in order to determine the solution uniquely.

We denote the free states in the following by $|\psi\rangle$, but the scattering states by $|\psi\rangle^+$ or $|\psi\rangle^-$ (see Fig. 5.5). We take two different ones. In particular, we shall mark the "retarded" solution $|\psi\rangle^+$ of H with the initial momentum—this is not a good quantum number because it is not conserved—and the "advanced" solution $|\psi\rangle^-$ with

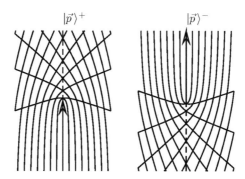

Fig. 5.5 The scattering states $|\mathbf{p}\rangle^{\pm}$ (momentum upwards) with an attractive Coulomb potential, represented by the classical orbital curves (calculated according to Sect. 2.1.6). From orbit to orbit, the collision parameter changes each by one unit. Quantum-mechanically, sharp orbits are not allowed—this is to be noted particularly for the straight orbit through the center

the final momentum. Now t_0 should mean the beginning of the scattering process for $|\psi(t)\rangle^{+}$ and the end for $|\psi(t)\rangle^{-}$. This leads to

$$|\psi(t)\rangle^{\pm} = U(t, t_0)|\psi(t_0)\rangle^{\pm} = \pm i\hbar\, G^{\pm}(t - t_0)|\psi(t_0)\rangle^{\pm} \,,$$

and in both cases $|\psi(t_0)\rangle^{\pm} = |\psi(t_0)\rangle$. In addition, instead of $\pm i\hbar\, G_0^{\pm}(t - t_0)|\psi(t_0)\rangle$, we may also use $|\psi(t)\rangle$. With

$$G^{\pm}(t - t_0) = G_0^{\pm}(t - t_0) + \int_{-\infty}^{\infty} dt'\, G^{\pm}(t - t')\, V\, G_0^{\pm}(t' - t_0) \,,$$

according to p. 406, this leads to the equation

$$|\psi(t)\rangle^{\pm} = |\psi(t)\rangle + \int_{-\infty}^{\infty} dt'\, G^{\pm}(t - t')\, V\, |\psi(t')\rangle \,.$$

Once again, the convolution integral can be transformed into a product via a Fourier transform (in the following, we shall again skip the reference to the energy representation):

$$|\psi\rangle^{\pm} = (1 + G^{\pm} V)\, |\psi\rangle \,.$$

With this the Lippmann–Schwinger equation holds, so $(1 - G_0^{\pm} V)|\psi\rangle^{\pm} = |\psi\rangle$, and hence,

$$|\psi\rangle^{\pm} = |\psi\rangle + G_0^{\pm}\, V\, |\psi\rangle^{\pm} \,.$$

If we use the Born approximation for $G^{\pm}$ or for $|\psi\rangle^{\pm}$,

$$|\psi\rangle^\pm \approx |\psi\rangle + G_0^\pm V\,|\psi\rangle\,,$$

then there are only known quantities on the right.

5.1.8 Möller's Wave Operators

According to the last section, the scattering states $|\psi\rangle^\pm$ are related to the free states $|\psi\rangle$ via operators:

$$|\psi\rangle^\pm = (1 + G^\pm V)\,|\psi\rangle\,.$$

These are *Möller's wave operators* $\Omega^\pm$, with the property

$$\Omega^\pm\,|\psi\rangle = |\psi\rangle^\pm \qquad\Longleftrightarrow\qquad \langle\psi|\Omega^{\pm\dagger} = {}^\pm\langle\psi|\,.$$

Here, in fact, the set $\{|\psi\rangle\}$ forms a complete basis, but the set $\{|\psi\rangle^+\}$ or $\{|\psi\rangle^-\}$ comprises only the scattering states for H. The bound states are missing. If, following Feshbach as before, we introduce the projection operator P onto the scattering states and the projection operator $Q = 1 - P$ onto the bound states, then

$$\Omega^{\pm\dagger}\,\Omega^\pm = 1\,,\quad\text{but}\quad \Omega^\pm\,\Omega^{\pm\dagger} = 1 - Q = P\,.$$

The wave operators are not unitary, but only *isometric*, i.e., they conserve the norm. The wave operators $\Omega^\pm$ do not map onto the whole space, and the adjoints $\Omega^{\pm\dagger}$ from a part of the space onto the whole space. Therefore, in

$$\Omega^\pm = P\,(1 + G^\pm\,V)\,,$$

we should not forget the projection operator P—in any case, in

$$\Omega^\pm\,(1 - G_0^\pm\,V) = P\,,$$

we must not put 1 on the right, because $\Omega^\pm$ does not lead to bound states. On the other hand, with $(1 - G_0^\pm V)\,G^\pm = G_0^\pm$ and $G^{\pm\dagger} = G^\mp$, we have

$$\Omega^\pm\,G_0^\pm = P\,G^\pm \qquad\Longleftrightarrow\qquad G_0^\mp\,\Omega^{\pm\dagger} = G^\mp\,P\,,$$

and with $\Omega^\pm = P\,(1 + G^\pm V)$, the Lippmann–Schwinger equation

$$\Omega^\pm = P\ +\ \Omega^\pm\,G_0^\pm\,V$$

for the wave operators. For the adjoint operators, we then obtain the equations

$$\Omega^{\pm\dagger} = (1 + V\,G^{\mp})\,P = P + V\,G_0^{\mp}\,\Omega^{\pm\dagger}\;,$$

or $(1 - V\,G_0^{\mp})\,\Omega^{\pm\dagger} = P$. While $\Omega^{\pm}$ maps the free states to the scattering states of the full system, conversely, $\Omega^{\pm\dagger}$ maps the scattering states to the unperturbed system, and the bound states $|\psi\rangle^{\mathrm{B}}$ to zero vectors, $\Omega^{\pm\dagger}\,|\psi\rangle^{\mathrm{B}} = |o\rangle$.

Incidentally, we also have

$$H\,\Omega^{\pm} = \Omega^{\pm}\,H_0\;,$$

since for all eigenstates of the energy, we have $H\Omega^{\pm}|\psi\rangle = H|\psi\rangle^{\pm} = E|\psi\rangle^{\pm}$, and the quantum number E commutes with the wave operators $\Omega^{\pm}$, so

$$E\Omega^{\pm}|\psi\rangle = \Omega^{\pm}E|\psi\rangle = \Omega^{\pm}\,H_0|\psi\rangle\;.$$

5.1.9   Scattering and Transition Operators

We shall now look for the transition probability from the initial state $|\psi_{\mathrm{i}}\rangle^{+}$ to the final state $|\psi_{\mathrm{f}}\rangle^{-}$, or more precisely, the amplitude $^{-}\langle\psi_{\mathrm{f}}|\psi_{\mathrm{i}}\rangle^{+} = \langle\psi_{\mathrm{f}}|\,\Omega^{-\dagger}\Omega^{+}\,|\psi_{\mathrm{i}}\rangle$. Note that this does not depend upon time, because $|\psi_{\mathrm{i}}\rangle^{+}$ and $|\psi_{\mathrm{f}}\rangle^{-}$ relate to the same Hamilton operator H. The free states form a complete basis. Therefore, we follow Heisenberg and introduce the *scattering operator*

$$S \equiv \Omega^{-\dagger}\Omega^{+}\;,$$

which relates the initial state directly with the final state:

$$\langle\psi_{\mathrm{f}}|\,S\,|\psi_{\mathrm{i}}\rangle = {}^{-}\langle\psi_{\mathrm{f}}|\psi_{\mathrm{i}}\rangle^{+}\;.$$

If we know its matrix elements, then the scattering problem is essentially solved.

It remains to show that the scattering operator is unitary, even though the wave operators $\Omega^{\pm}$ are only isometric. With $S^{\dagger}S = \Omega^{+\dagger}\Omega^{-}\Omega^{-\dagger}\Omega^{+}$ and $SS^{\dagger} = \Omega^{-\dagger}\Omega^{+}\Omega^{+\dagger}\Omega^{-}$, we therefore investigate $\Omega^{\pm\dagger}\Omega^{\mp}\Omega^{\mp\dagger}\Omega^{\pm} = \Omega^{\pm\dagger}\,P\,\Omega^{\pm}$. Since $\Omega^{\pm}$ maps only onto the space of scattering states, we have $P\Omega^{\pm} = \Omega^{\pm}$, and thus $\Omega^{\pm\dagger}\Omega^{\pm} = 1$ is left over. The scattering operator is thus unitary:

$$S^{\dagger}S = SS^{\dagger} = 1\;.$$

Unitarity guarantees, among other things, that nothing is lost in the scattering process, whence the norm of the original wave remains conserved.

In order to show the influence of the interaction V as clearly as possible, we reformulate the transition amplitude. With

$$|\psi_i\rangle^+ - |\psi_i\rangle^- = (G^+ - G^-)\, V\, |\psi_i\rangle = -2\pi i\, \delta(E-H)\, V\, |\psi_i\rangle\, ,$$
$$^+\langle\psi_f| - ^-\langle\psi_f| = \langle\psi_f|\, V\, (G^- - G^+) = +2\pi i\, \langle\psi_f|\, V\, \delta(E-H)\, ,$$

we have in particular,

$$\langle\psi_f|\, S\, |\psi_i\rangle = {}^-\langle\psi_f|\psi_i\rangle^+ = {}^-\langle\psi_f|\psi_i\rangle^- - 2\pi i\, \delta(E_i - E_f)\, ^-\langle\psi_f|\, V\, |\psi_i\rangle$$
$$= {}^+\langle\psi_f|\psi_i\rangle^+ - 2\pi i\, \delta(E_f - E_i)\, \langle\psi_f|\, V\, |\psi_i\rangle^+\, .$$

Given the isometry of the wave operators, we have $^-\langle\psi_f|\psi_i\rangle^- = \langle\psi_f|\psi_i\rangle = {}^+\langle\psi_f|\psi_i\rangle^+$. Furthermore, the delta function $\delta(E_f - E_i)$ can be extracted and this ensures conservation of the energy:

$$\langle\psi_f|\, S\, |\psi_i\rangle = \delta(E_f - E_i)\, \{\langle\Omega_f|\Omega_i\rangle - 2\pi i\, \langle\psi_f|\, T\, |\psi_i\rangle\}\, ,$$

where the *transition operator* is defined by

$$T \equiv \Omega^{-\dagger}\, V = V\, \Omega^+\, .$$

Here the expressions are only to be evaluated "on the energy shell", i.e., for $E_f = E_i$. Then we have $G_0^+\, T = G_0^+\, \Omega^{-\dagger}\, V = G^+\, P\, V$. Since G^+ acts only in the P-space, we write for short

$$G_0^+\, T = G^+\, V\, , \quad \text{or} \quad T\, G_0^+ = V\, G^+\, .$$

Then for the retarded propagators,

$$G^+ = G_0^+ + G_0^+\, T\, G_0^+$$

from the Lippmann–Schwinger equations. Correspondingly, from $T = V\Omega^+ = V\, P\,(1 + G^+ V)$, we deduce the *Low equation*

$$T = V + V\, G^+\, V\, .$$

According to the above equations, the Lippmann–Schwinger equations are valid for the transition operator T:

$$T = V + V\, G_0^+\, T = V + T\, G_0^+\, V\, .$$

These equations are particularly useful, because the transition operator T is directly connected to the scattering cross-section and indeed other experimental quantities (observables), as we shall see in the next section.

In the Born approximation, we replace T by V and thereby avoid having to compute the resolvents. Then, however, $G^+ V$ must not be too large, which is why the Born approximation fails for resonances. Note finally that, in the Lippmann–Schwinger equation for T, different energies can occur in bra and ket, whereas for two-body scattering, they do not contribute.

5.1.10   The Wave Function $\langle \mathbf{r} \,|\, \mathbf{k} \,\rangle^+$ for Large Distances r

We now consider the real-space representation of the scattering states $|\mathbf{k}\,\rangle^+$ in the relative coordinate $\mathbf{r}$ of the two scattering partners. The limit $r \to \infty$ will be important for the scattering cross-section, with which we shall occupy ourselves subsequently.

Particularly convenient is the starting equation

$$|\mathbf{k}\,\rangle^+ = (1 + G_0^+ T) \,|\mathbf{k}\,\rangle \,,$$

because we have already found the real-space representation of G_0^+ on p. 409:

$$\langle \mathbf{r} \,|\, G_0^\pm \Big(\frac{\hbar^2 k^2}{2m_0}\Big)\, |\mathbf{r}'\rangle = -\frac{1}{4\pi} \frac{2m_0}{\hbar^2} \frac{\exp(\pm \,ik\,|\mathbf{r} - \mathbf{r}'|)}{|\mathbf{r} - \mathbf{r}'|} \,.$$

For $r \gg r'$ and $|\mathbf{r} - \mathbf{r}'| \approx r\sqrt{1 - 2\,\mathbf{r} \cdot \mathbf{r}'/r^2} \approx r - \mathbf{r} \cdot \mathbf{r}'/r$ (see Fig. 3.30), and with the abbreviation

$$\mathbf{k}' \equiv k\,\frac{\mathbf{r}}{r} \,,$$

the last expression goes over into

$$\langle \mathbf{r} \,|\, G_0^\pm \Big(\frac{\hbar^2 k^2}{2m_0}\Big)\, |\mathbf{r}'\rangle \approx -\frac{1}{4\pi} \frac{2m_0}{\hbar^2} \frac{\exp(\pm \,ikr)}{r} \,\exp(\mp \,i\mathbf{k}' \cdot \mathbf{r}') \,.$$

Here, $\exp(-i\mathbf{k}' \cdot \mathbf{r}') = (2\pi)^{3/2}\,\langle \mathbf{k}'|\mathbf{r}'\rangle$. Therefore, we have (see p. 399)

$$\langle \mathbf{r} \,|\mathbf{k}\,\rangle^+ \approx \langle \mathbf{r} \,|\mathbf{k}\,\rangle - \frac{\sqrt{2\pi}\,m_0}{\hbar^2}\, \langle \mathbf{k}'|\,T\,|\mathbf{k}\,\rangle \frac{\exp(ikr)}{r}$$

$$= \frac{1}{\sqrt{2\pi}^{\,3}} \Big(\exp(i\mathbf{k} \cdot \mathbf{r}) + f(\theta)\,\frac{\exp(ikr)}{r}\Big) \,,$$

with *scattering amplitude*

$$f(\theta) \equiv -\Big(\frac{2\pi}{\hbar}\Big)^2 m_0\,\langle \mathbf{k}'|\,T\,|\mathbf{k}\,\rangle = -\frac{(2\pi)^2}{k}\,\langle E\Omega_f|\,T\,|E\Omega_i\rangle \,.$$

For the second formulation here, note that $|\mathbf{k}\,\rangle = |E\Omega\rangle\,\hbar/\sqrt{m_0 k}$, which follows from $\langle \mathbf{k}|\mathbf{k}'\rangle = k^{-2}\delta(k - k')\,\delta(\Omega - \Omega')$ and $\delta(E - E') = 2m_0\hbar^{-2}\,\delta(k^2 - k'^2)$ with $\delta(k^2 - k'^2) = (2k)^{-1}\,\delta(k - k')$ (see p. 20). Here we recognize the difference between the *wave vector* and the *energy representations*. We have already discussed the difference between the *wave vector* and the *momentum representations* on p. 319. Here Ω_i gives the direction before scattering and Ω_f the direction afterwards. If there is a Wigner force—no spin dependence—only the scattering angle θ between the two directions is important, because for rotational invariance the transition operator in the angular momentum representation is diagonal and does not depend upon the directional (magnetic) quantum number:

$$\langle\Omega_f| \, T \, |\Omega_i\rangle = \sum_{lm}\langle\Omega_f|lm\rangle \, T_l \, \langle lm|\Omega_i\rangle = \sum_{lm} Y_m^{(l)}(\Omega_f) \, T_l \, Y_m^{(l)*}(\Omega_i)$$

$$= \sum_l \frac{2l+1}{4\pi} \, T_l \, P_l(\cos\theta) \ .$$

It follows that $f(\theta) = -(\pi/k)\sum_l (2l+1) \, T_l \, P_l(\cos\theta)$, as claimed on p. 402.

5.1.11 Scattering Cross-Section

Scattering cross-sections are not the only observables in scattering processes. For particles with spin, polarizations (i.e., spin distributions) can be measured. But in that case, only the angular momentum algebra need be applied. The basic notions can be explained with the example of spinless particles, and we shall restrict ourselves here to this essentially simple case.

The differential scattering cross-section $d\sigma/d\Omega$ is given by the number of particles scattered into the solid angle element $d\Omega$ relative to the number of incoming particles per area unit and the number of scattering centers. (For stationary currents, we have to refer to equal time spans in the numerator and denominator. In addition, the expression does not hold if the incoming or outgoing particles interact with each other, or if the individual centers scatter coherently, as for the refraction of slow neutrons in crystals.) We can also express the scattering cross-section in terms of the current densities of the scattering wave and the incoming wave:

$$\frac{d\sigma}{d\Omega} = \frac{j_{\text{scat}}(\Omega) \, r^2}{j_i} \ .$$

Here it is well known that, in the real-space representation, we have (see p. 348)

$$\mathbf{j}(\mathbf{r}) = \frac{\hbar}{i} \, \frac{\psi^* \, \nabla\psi - \psi \, \nabla\psi^*}{2m_0} \ ,$$

and from $\psi_{\text{scat}}(\mathbf{r}) \approx (2\pi)^{-3/2} \, \exp(ikr) \, f(\theta)/r$ and $\psi_i(\mathbf{r}) = (2\pi)^{-3/2} \, \exp(i\mathbf{k} \cdot \mathbf{r})$, we obtain the current densities

$$j_\mathrm{i} = \frac{1}{(2\pi)^3}\,\frac{\hbar k}{m_0}\;, \qquad j_\mathrm{scat} \approx \frac{1}{(2\pi)^3}\,\frac{\hbar k}{m_0}\,\frac{|f(\theta)|^2}{r^2}\;.$$

Therefore, the differential scattering cross-section can be evaluated from the scattering amplitude f and the transition matrix T as follows:

$$\frac{\mathrm{d}\sigma}{\mathrm{d}\Omega} = |f(\theta)|^2 = \frac{(2\pi)^4}{k^2}\,|\langle E\Omega_\mathrm{f}|\,T\,|E\Omega_\mathrm{i}\rangle|^2\;,$$

if we also use the last section for the relation between f and T.

Using $\langle E\Omega|\,S\,|E'\Omega'\rangle = \langle E\Omega|E'\Omega'\rangle - 2\pi\mathrm{i}\langle E|E'\rangle\langle E\Omega|T|E\Omega'\rangle$ and the unitarity of the scattering operators, viz., $S^\dagger S = 1$, which expresses current conservation, we obtain

$$\mathrm{i}\langle E\Omega|T|E\Omega'\rangle - \mathrm{i}\langle E\Omega'|T|E\Omega\rangle^* = 2\pi\int\mathrm{d}\Omega''\,\langle E\Omega''|T|E\Omega\rangle^*\langle E\Omega''|T|E\Omega'\rangle\;,$$

after splitting off the factor $2\pi\,\delta(E - E')$. With $\Omega' = \Omega$, this implies

$$-2\mathrm{Im}\langle E\Omega|T|E\Omega\rangle = 2\pi\int\mathrm{d}\Omega'\,|\langle E\Omega'|T|E\Omega\rangle|^2 = \frac{k^2}{(2\pi)^3}\int\mathrm{d}\Omega'\,\frac{\mathrm{d}\sigma}{\mathrm{d}\Omega'}\;,$$

and what is known as the *optical theorem* relating the integrated scattering cross-section and the forward scattering amplitude:

$$\sigma = \frac{(2\pi)^3}{k^2}\,(-2\mathrm{Im}\langle E\Omega|T|E\Omega\rangle) = \frac{4\pi}{k}\mathrm{Im}f(0)\;.$$

To first order in the Born approximation, the forward scattering amplitude is real, which contradicts unitarity. In fact, for the forward scattering amplitude, at least the second order is necessary.

If there are other processes in addition to elastic scattering, such as inelastic or even disorder reactions, then σ in the last equation stands for the sum of all integrated scattering cross-sections, the *total scattering cross-section*, because we have to insert a complete basis in order to arrive at $|T|^2$ when computing $T^\dagger T$.

5.1.12 Summary: Scattering Theory

In the scattering theory, we investigate how an original state is transformed into a new state as a consequence of a perturbation V. In addition to the quantities associated with the unperturbed system, i.e., the Hamilton operator H_0, the time shift operator U_0, the propagators (Green functions) $G_0^\pm$, and the states $|\psi\rangle$, there are quantities associated with the (full) perturbed problem: the Hamilton operator $H = H_0 + V$, the time shift operator U, the propagators $G^\pm$, and the states $|\psi\rangle^\pm$. These quantities are related to each other, in the time-dependent case via integral equations, in the energy-dependent case via the Lippmann–Schwinger equations. The scattering operator S, or again the transition operator T, describe the transition from the unperturbed initial state to the unperturbed final state.

5.2 Two- and Three-Body Scattering Problems

5.2.1 Two-Potential Formula of Gell-Mann and Goldberger

This formula is important for many applications of the generalized scattering theory and starts from

$$V = \widetilde{V} + \delta V \ ,$$

where the approximate scattering problem for $\widetilde{V}$ is considered already solved, so that the propagator for $H_0 + \widetilde{V}$ is known, viz.,

$$\widetilde{G} = G_0 \, (1 + \widetilde{V}\widetilde{G}) = (1 + \widetilde{G} \, \widetilde{V}) \, G_0 \ ,$$

along with the transition operator $\widetilde{T}$:

$$\widetilde{T} = \widetilde{V} \, (1 + G_0 \, \widetilde{T}) = (1 + \widetilde{T} \, G_0) \, \widetilde{V} \ .$$

Note that, from now on, we shall usually skip the indices $\pm$ and the argument E. According to p. 415, we also have $G_0 \widetilde{T} = \widetilde{G} \widetilde{V}$ and $\widetilde{T} G_0 = \widetilde{V}\widetilde{G}$. In addition, using $G = G_0 + G_0 \, (\widetilde{V} + \delta V) \, G$, which implies $(1 - G_0\widetilde{V}) \, G = G_0 \, (1 + \delta V \, G)$, then multiplying by $1 + \widetilde{G}\widetilde{V}$ and using the relation $(1 + \widetilde{G}\widetilde{V})(1 - G_0\widetilde{V}) = P = (1 - \widetilde{V}G_0)(1 + \widetilde{V}\widetilde{G})$ found on p. 411 (with $\delta V = 0$), we deduce that

$$G = \widetilde{G} \, (1 + \delta V \, G) = (1 + G \, \delta V) \, \widetilde{G} \ ,$$

where we just write G instead of PG or GP once again, since we restrict ourselves to scattering states anyway. Another proof this equation follows using $G^{-1} = \mathcal{E} - H$, $\widetilde{G}^{-1} = \mathcal{E} - \widetilde{H}$, and $\delta V = V - \widetilde{V} = \widetilde{G}^{-1} - G^{-1}$:

$$\widetilde{G} \, \delta V \, G = G - \widetilde{G} = G \, \delta V \, \widetilde{G} \ .$$

We thus have a Lippmann–Schwinger equation in which, instead of the full coupling, only the "perturbation" δV appears, but with $\widetilde{G}$ instead of the free propagators G_0.

According to the last equation, we have

$$(1 + G \, \delta V)(1 + \widetilde{G}\widetilde{V}) = 1 + G \, \delta V + \widetilde{G}\widetilde{V} + (G - \widetilde{G}) \, \widetilde{V} = 1 + GV \ .$$

This factorization of $1 + GV$ is useful because then, from $|\psi\rangle^{\pm} = (1 + G^{\pm}V)|\psi\rangle$, with the states deformed by $\widetilde{V}\,|\psi\rangle^{\widetilde{\pm}} = (1 + \widetilde{G}^{\pm}\widetilde{V})|\psi\rangle$, we have the helpful relation

$$|\psi\rangle^{\pm} = (1 + G^{\pm}\,\delta V)\,|\psi\rangle^{\widetilde{\pm}}\,.$$

Note that $1 + VG$ factorizes into $(1 + \widetilde{V}\widetilde{G})(1 + \delta VG)$.

For the Low equation $T = V\,(1 + GV)$, we can also use this kind of factorization. With

$$V\,(1 + G\,\delta V) = \widetilde{V} + (1 + VG)\,\delta V\,, \quad (1 + VG)\,\delta V = (1 + \widetilde{V}\widetilde{G})(1 + \delta V\,G)\,\delta V\,,$$

and the modified Low equation

$$\delta T = (1 + \delta V\,G)\,\delta V\,,$$

along with $\widetilde{T} = \widetilde{V}\,(1 + \widetilde{G}\widetilde{V})$, we obtain the *formula of Gell-Mann and Goldberger*

$$T = \widetilde{T} + (1 + \widetilde{V}\,\widetilde{G})\,\delta T\,(1 + \widetilde{G}\,\widetilde{V}) = \widetilde{T} + (1 + \widetilde{T}\,G_0)\,\delta T\,(1 + G_0\,\widetilde{T})\,,$$

which is extremely useful here.

For the matrix elements of the transition operators, we thus have

$$\langle\psi_f|\,T\,|\psi_i\rangle = \langle\psi_f|\,\widetilde{T}\,|\psi_i\rangle + {}^{\widetilde{}}\langle\psi_f|\,\delta T\,|\psi_i\rangle^{\widetilde{\pm}}\,.$$

If we take the Born approximation $\delta T \approx \delta V$ for δT here, we obtain a better Born approximation known as the distorted-wave Born approximation (DWBA). Whereas all higher order terms in V are left out in the Born approximation, now only those in δV are missing. However, the states $|\psi\rangle^{\widetilde{\pm}}$ (distorted by $\widetilde{V}$) still have to be calculated, as does $\widetilde{T}$.

Note that we also have

$$(1 + G\,\delta V)(1 - \widetilde{G}\,\delta V) = 1 = (1 - \widetilde{G}\,\delta V)(1 + G\,\delta V)\,,$$

since the product is equal to $1 + (G - \widetilde{G} - G\,\delta V\,\widetilde{G})\,\delta V$, and we have already proven $G - \widetilde{G} = G\,\delta V\,\widetilde{G}$. Consequently, multiplying $|\psi\rangle^{\pm} = (1 + G\,\delta V)|\psi\rangle^{\widetilde{\pm}}$ by $1 - \widetilde{G}\,\delta V$, we find $|\psi\rangle^{\widetilde{\pm}} = (1 - \widetilde{G}\,\delta V)|\psi\rangle^{\pm}$, or the Lippmann–Schwinger equation

$$|\psi\rangle^{\pm} = |\psi\rangle^{\widetilde{\pm}} + \widetilde{G}^{\pm}\,\delta V\,|\psi\rangle^{\pm}\,.$$

We shall refer to this in Sect. 5.2.4.

5.2.2 Scattering Phases

This result will now be explained using the methods mentioned in Sect. 5.1.1. There we introduced the spherical Bessel functions

$$F_l \approx \sin(\rho - \tfrac{1}{2}l\,\pi)\,, \qquad O_l \approx \exp\{+\mathrm{i}(\rho - \tfrac{1}{2}l\,\pi)\}\,,$$
$$G_l \approx \cos(\rho - \tfrac{1}{2}l\,\pi)\,, \qquad I_l \approx \exp\{-\mathrm{i}(\rho - \tfrac{1}{2}l\,\pi)\}\,,$$

and expanded the radial function of the Schrödinger equation with respect to two of them in the region with $V = 0$. If V vanishes everywhere (and hence the transition operator along with it), then the function F_l alone suffices, because only this is differentiable at the origin. Generally, $u_l \approx N_l\,(F_l - \pi\,T_l\,O_l)$, where N_l ensures the correct normalization. Given the unitarity of the scattering operators, we set

$$S_l = \exp(2\mathrm{i}\delta_l)\,,$$

and make use of $S_l = 1 - 2\pi\,\mathrm{i}\,T_l$. Then,

$$-\pi\,T_l = \frac{\exp(2\mathrm{i}\delta_l) - 1}{2\mathrm{i}} = \exp(\mathrm{i}\delta_l)\,\sin\delta_l\,,$$

and with $F_l = (O_l - I_l)/(2\mathrm{i})$, it follows that

$$2\mathrm{i}\,u_l/N_l \approx O_l - I_l + \{\exp(2\mathrm{i}\delta_l) - 1\}\,O_l = \exp(2\mathrm{i}\delta_l)\,O_l - I_l\,,$$

so

$$u_l \approx N_l\,\exp(\mathrm{i}\delta_l)\,\sin(\rho - \tfrac{1}{2}l\pi + \delta_l)\,.$$

In order to fix the *scattering phase* δ_l, not only mod π, we also require it to depend continuously on k and vanish for $k \to \infty$, because for $E \to \infty$, the coupling V should be negligible—to the (repulsive) centrifugal force there clearly corresponds the (negative) scattering phase $-\tfrac{1}{2}l\pi$, independent of the energy. Note that, on the other hand, according to the *Levinson theorem*, the phase shift for $k = 0$ is equal to π times the number of bound states.

After these preliminaries, we introduce the scattering phase $\widetilde{\delta}_l$ associated with $\widetilde{V}$, and in addition to $\widetilde{O}_l = \exp(\mathrm{i}\widetilde{\delta}_l)\,O_l$, we use

$$\widetilde{F}_l \approx \cos\widetilde{\delta}_l\,F_l + \sin\widetilde{\delta}_l\,G_l = \cos\widetilde{\delta}_l\,F_l + \sin\widetilde{\delta}_l\,(O_l - \mathrm{i}F_l) = \exp(-\mathrm{i}\widetilde{\delta}_l)\,F_l + \sin\widetilde{\delta}_l\,O_l\,.$$

With this we obtain for u_l asymptotically the expression $N_l\,\{F_l - \pi T_l\,O_l\}$, and instead of the curly brackets, we may also write

$$\exp(\mathrm{i}\widetilde{\delta}_l)\,\{\widetilde{F}_l - \sin\widetilde{\delta}_l\,\exp\,(-\mathrm{i}\widetilde{\delta}_l)\,\widetilde{O}_l - \pi T_l\,\exp(-2\mathrm{i}\widetilde{\delta}_l)\,\widetilde{O}_l\}$$
$$= \exp(\mathrm{i}\widetilde{\delta}_l)\,\{\widetilde{F}_l - \exp(-2\mathrm{i}\widetilde{\delta}_l)\,\widetilde{O}_l\,[\exp(\mathrm{i}\widetilde{\delta}_l)\,\sin\widetilde{\delta}_l + \pi T_l]\}\,.$$

Since we now have to set $\exp(\mathrm{i}\widetilde{\delta}_l) \sin \widetilde{\delta}_l = -\pi \widetilde{T}_l$, we obtain

$$u_l \approx N_l \exp(\mathrm{i}\widetilde{\delta}_l) \{\widetilde{F}_l - \pi(T_l - \widetilde{T}_l) \exp(-2\mathrm{i}\widetilde{\delta}_l) \widetilde{O}_l\} .$$

From this we can conclude that we should take

$$T_l = \widetilde{T}_l + \exp(2\mathrm{i}\widetilde{\delta}_l) \delta T_l ,$$

which corresponds to the two-potential formula. Here, the factor $\exp(2\mathrm{i}\widetilde{\delta}_l)$ originates from the distortion of the states due to the coupling $\widetilde{V}$, because we have used the functions $\widetilde{F}_l$ and $\widetilde{O}_l$.

5.2.3 Scattering of Charged Particles

An important application is to scattering by the Coulomb potential, since it decreases so slowly with increasing r that the previous results cannot simply be carried over. Here we use the *Sommerfeld parameter* (*Coulomb parameter*)

$$\eta \equiv \frac{zZe^2}{4\pi\varepsilon_0} \frac{m_0}{\hbar^2 k} ,$$

together with the *Coulomb scattering phase*

$$\sigma_l(\eta) \equiv \arg \Gamma(l + 1 + \mathrm{i}\eta) \qquad \Longrightarrow \qquad \exp(2\mathrm{i}\sigma_l) = \frac{\Gamma(l + 1 + \mathrm{i}\eta)}{\Gamma(l + 1 - \mathrm{i}\eta)} .$$

The spherical Bessel functions are now replaced by the *Coulomb wave functions*

$$F_l(\eta, \rho) \approx \sin(\rho - \eta \ln 2\rho - \tfrac{1}{2}l\pi + \sigma_l) ,$$
$$O_l(\eta, \rho) \approx \exp\{+\mathrm{i}(\rho - \eta \ln 2\rho - \tfrac{1}{2}l\pi + \sigma_l)\} ,$$

where the logarithm originates from the long range of the potential in the radial Schrödinger equation

$$\left(\frac{\mathrm{d}^2}{\mathrm{d}\rho^2} - \frac{l(l+1)}{\rho^2} + 1 - \frac{2\eta}{\rho}\right) u_l(\rho) = 0 , \quad \text{with } \rho = kr .$$

Note that with the bound states stands -1 instead of $+1$ for the energy and the principal quantum number n instead of $-\eta$ (see p. 362). Despite the long range, we can introduce a *Coulomb scattering amplitude*

$$f_C(\theta) = -\frac{\eta}{2k} \frac{\exp\{2\mathrm{i}(\sigma_0 - \eta \ln \sin \tfrac{1}{2}\theta)\}}{\sin^2 \tfrac{1}{2}\theta} ,$$

and hence determine the *Rutherford cross-section*

$$\frac{d\sigma}{d\Omega} = |f_C(\theta)|^2 = \frac{\eta^2}{4k^2 \sin^4(\frac{1}{2}\theta)} \; .$$

With $f_C(\theta) = -(2\pi)^2 k^{-1} \langle \Omega_f | T_C | \Omega_i \rangle$, the matrix element of the transition operators for the Coulomb problem follows from the scattering amplitude:

$$T_C(\theta) = \frac{\eta}{2} \frac{\exp\{2i\,(\sigma_0 - \eta\,\ln\sin\frac{1}{2}\theta)\}}{(2\pi \sin\frac{1}{2}\theta)^2} \; .$$

Incidentally, its modulus $\frac{1}{2}\eta\,(2\pi \sin\frac{1}{2}\theta)^{-2}$ is equal to $\langle E\Omega_f | V_C | E\Omega_i \rangle$, because with

$$V_C(r) = \frac{zZe^2}{4\pi \varepsilon_0 r} = \frac{\eta\, \hbar^2 k}{m_0\, r} \; ,$$

we have

$$\langle \mathbf{k}_f | V_C | \mathbf{k}_i \rangle = \frac{1}{2}\eta\hbar^2 k\, m_0^{-1}\, (2\pi\, k \sin\frac{1}{2}\theta)^{-2} \; ,$$

according to p. 410, and in addition,

$$\langle E\Omega_f | V_C | E\Omega_i \rangle = m_0 k \hbar^{-2} \langle \mathbf{k}_f | V_C | \mathbf{k}_i \rangle \; ,$$

according to p. 417. Only the phase is missing from the Born approximation!

We thus have $\widetilde{F}_l(\rho) \stackrel{\frown}{=} F_l(\eta, \rho)$ and $\widetilde{O}_l(\rho) \stackrel{\frown}{=} O_l(\eta, \rho)$ for the scattering of charged particles, along with $\widetilde{T} \stackrel{\frown}{=} T_C(\theta)$ and $\widetilde{\delta}_l \stackrel{\frown}{=} \sigma_l(\eta)$. Further forces (e.g., nuclear forces) then contribute in the term δT_l.

5.2.4 Effective Hamilton Operator in the Feshbach Theory

A further important application of the two-potential formula is the *unified theory of nuclear reactions* due to Feshbach (see p. 411). This leads us to a deeper understanding of all resonances and direct reactions (not only in nuclear physics) and embraces several other resonance models.

The decisive point of the Feshbach formalism is the separation of the Hilbert space into two parts, on which we project with the operators P and Q:

$$P = P^\dagger = P^2 \;, \quad Q = Q^\dagger = Q^2 \;, \quad PQ = 0 = QP \;, \quad P + Q = 1 \;.$$

P maps onto those states which do not vanish for large r, viz., the scattering states describing *open channels*, and Q onto the "bound" states, which vanish for large *r* and describe *closed channels*. This division considers only large distances of the scattering partners (*asymptotic boundary conditions*) and allows several cases for short distances. Therefore, different resonance theories are still possible. If we introduce, e.g., a *channel radius R* with the property that the interaction vanishes for larger distances, we may let Q project onto the space $0 \leq r \leq R$ and P onto the space $r > R$: this leads to the scattering matrix of Wigner and Eisenbud [2] (see also [3]). (It differs from the transition matrix of Kapur and Peierls [4] in that the boundary conditions for $r = R$ depend upon the energy.) In the Feshbach formalism, there is no need for the channel radius R.

Along with the division of the Hilbert space into open and closed channels, we also have to decompose the Hamilton operator correspondingly:

$$H = (P + Q)\, H\, (P + Q) \equiv H_{PP} + H_{PQ} + H_{QP} + H_{QQ} \ .$$

For the scattering cross-section, only $P|\psi\rangle^{\pm}$ is important. We now search for the "effective" Hamilton operator acting on these scattering states, and after that derive the associated Lippmann–Schwinger equation.

To begin with, from $(E - H)\,|\psi\rangle^{\pm} = 0$, after projection with P and Q such that $1 = P^2 + Q^2$, we have the general result

$$(E - H_{PP})\, P|\psi\rangle^{\pm} = H_{PQ}\, Q|\psi\rangle^{\pm} \quad \text{and} \quad (E - H_{QQ})\, Q|\psi\rangle^{\pm} = H_{QP}\, P|\psi\rangle^{\pm} \ .$$

Since Q projects onto the closed channels, the inhomogeneous term is missing in its Lippmann–Schwinger equation:

$$Q|\psi\rangle^{\pm} = G_Q\, H_{QP}\, P|\psi\rangle^{\pm} \ , \quad \text{with} \quad G_Q \equiv \frac{1}{E - H_{QQ}} \ .$$

If we insert this into the other relation, we obtain the homogeneous equation

$$(E - H_{PP} - H_{PQ}\, G_Q\, H_{QP})\, P|\psi\rangle^{\pm} = 0 \ .$$

We thus find the *effective Hamilton operator* $H_{PP} + H_{PQ}\, G_Q\, H_{QP}$. Clearly, it can be used for the two-potential formula: H_{PP} plays the role of $H_0 + \widehat{V}$ and $H_{PQ} G_Q H_{QP}$ that of δV. However, from now on, we write $G^{\pm}_P \equiv (\mathscr{E} - H_{PP})^{-1}$ with complex $\mathscr{E}$, instead of $\widetilde{G}^{\pm}$, and according to p. 420, we now have

$$P\,|\psi\rangle^{\pm} = |\psi\rangle^{\widetilde{\pm}} + G^{\pm}_P\, H_{PQ}\, G_Q\, H_{QP}\, P|\psi\rangle^{\pm} \ , \quad \text{with} \quad G^{\pm}_P \equiv \frac{1}{\mathscr{E} - H_{PP}} \ ,$$

as the Lippmann–Schwinger equation for the unknown scattering state.

5.2.5 Separable Interactions and Resonances

The key feature of the new residual interaction $\delta V = H_{PQ} G_Q H_{QP}$ is the product form. Such couplings are said to be *separable*. They can be diagonalized in the space of scattering states and therefore not in real space, and are thus *non-local*: $\langle \mathbf{r} | V | \mathbf{r}' \rangle \neq V_0(\mathbf{r}) \, \delta(\mathbf{r} - \mathbf{r}')$. The transition operator δT now also factorizes, because the relations $\delta T = \delta V (1 + G \, \delta V)$ and $1 + G \, \delta V = (1 - \widetilde{G} \, \delta V)^{-1}$ mentioned on p. 420 deliver $\delta T = \delta V / (1 - \widetilde{G} \, \delta V)$:

$$\delta T = H_{PQ} G_Q H_{QP} \frac{1}{1 - G^+{}_P H_{PQ} G_Q H_{QP}} \; .$$

Here, $A (1 - BA)^{-1} = (1 - AB)^{-1} A$ with $(1 - AB) A = A (1 - BA)$, and thus

$$H_{QP} \frac{1}{1 - G^+{}_P H_{PQ} G_Q H_{QP}} = \frac{1}{1 - H_{QP} G^+{}_P H_{PQ} G_Q} H_{QP} \; .$$

With $\{G_Q (1 - H_{QP} G^+{}_P H_{PQ} G_Q)^{-1}\}^{-1} = (1 - H_{QP} G^+{}_P H_{PQ} G_Q) G_Q^{-1}$, the operator between H_{PQ} and H_{QP} can also be simplified:

$$\delta T = H_{PQ} \frac{1}{E - H_{QQ} - H_{QP} G^+{}_P H_{PQ}} H_{QP} \; .$$

Here, since

$$H_{QP} G^+{}_P H_{PQ} = H_{QP} \left(\frac{P}{E - H_{PP}} - i\pi \, \delta(E - H_{PP}) \right) H_{PQ} = \Delta - \tfrac{1}{2} i \Gamma \; ,$$

it is clear that the poles do not occur at the eigenvalues of H_{QQ}, but are displaced by the *level shift* Δ, and have a *level width* Γ (see Fig. 5.6):

$$|\delta T|^2 \sim \frac{1}{(E - H_{QQ} - \Delta)^2 + \tfrac{1}{4} \Gamma^2} \; ,$$

We will discuss these resonance parameters in the next section. When considering δT, the coupling H_{QP} which leads from the P- to the Q-space is initially important, then the resonance level in Q-space, and finally again the coupling H_{PQ} which leads back from the Q- to the P-space.

Near the resonance,

$$\psi^{(+)}(t) \sim \exp \frac{-i (H_{QQ} + \Delta - \tfrac{1}{2} i \Gamma) \, t}{\hbar} \; ,$$

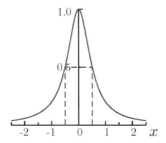

Fig. 5.6 Lorentz curve $\frac{1}{4}/(x^2+\frac{1}{4})$ (*continuous red*), line shape of a scattering resonance about the resonance energy E_R with half-width Γ, where $x = (E - E_R)/\Gamma$. The curve has half the maximum value at two points which have the distance of the half-width (*dashed blue*). For this distribution, $\Delta E = \infty$ and the associated average lifetime is $\hbar/\Gamma$

and consequently,

$$|\psi^{(+)}(t)|^2 \sim \exp\frac{-\Gamma t}{\hbar} = \exp\frac{-t}{\tau} , \quad \text{with} \quad \tau \equiv \frac{\hbar}{\Gamma} ,$$

where τ is the *average lifetime* of the resonance state. We can also view it as the time uncertainty of the state, because it is now $\overline{t^2} - \overline{t}^2 = \tau^2$. The associated distribution function $|\psi^+(E)|^2$ in the energy representation is given by a Lorentz curve (with infinite energy uncertainty according to Problem 4.3). Therefore, the equation $\tau\,\Gamma = \hbar$, which is a lifetime–half-width relation, is *not a time–energy uncertainty relation*, even though this is often claimed—there is no Hermitian time operator in quantum theory and hence there is also no such inequality, even though each finite wave train has a finite time and frequency uncertainty, even classically.

5.2.6   Breit–Wigner Formula

There are various methods for computing

$$\delta T = H_{PQ} \frac{1}{E - H_{QQ} - H_{QP}\, G^+{}_P\, H_{PQ}} H_{QP} .$$

In order to proceed without approximations, we have to diagonalize the denominator, which means searching for the eigen representation of

$$H' \equiv H_{QQ} + H_{QP}\, G^+{}_P\, H_{PQ} , \quad \text{with} \quad G^+{}_P = \frac{P}{E - H_{PP}} - i\pi\,\delta(E - H_{PP}) ,$$

where the last term is not Hermitian. Therefore, we now need two sets of solutions (a *bi-orthogonal system*) in the Q-space,

$$\{\mathscr{E}_\nu - H'(\mathscr{E})\}\,|\Xi_\nu(\mathscr{E})\rangle = 0$$

$$\{\mathscr{E}_\nu^* - H'^\dagger(\mathscr{E})\}\,|\Xi_\nu^A(\mathscr{E})\rangle = 0 \qquad \Longleftrightarrow \qquad \langle\Xi_\nu^A|\,\{\mathscr{E}_\nu - H'(\mathscr{E})\} = 0$$

with $\langle\Xi_\nu^A(\mathscr{E})\mid\Xi_{\nu'}(\mathscr{E})\rangle = \delta_{\nu\nu'}$ and $\sum_\nu|\Xi_\nu(\mathscr{E})\rangle\langle\Xi_\nu^A(\mathscr{E})| = Q$. The eigenvalues $\mathscr{E}_\nu$ of H' are complex, and

$$\frac{Q}{E - H'} = \sum_\nu \frac{|\Xi_\nu(\mathscr{E})\rangle\,\langle\Xi_\nu^A(\mathscr{E})|}{E - \mathscr{E}_\nu}$$

holds. Here $G^+{}_P$ still depends on the energy, and therefore also on H' and the whole bi-orthogonal system. This seriously complicates the computation.

These difficulties can be avoided with an approximation. We take the eigenstates of H_{QQ},

$$(E_n - H_{QQ})\,|n\rangle = 0\,,$$

e.g., those of the box or quadratic potential (see Sects. 4.5.3 and 4.5.4), and obtain the shift and width according to perturbation theory from

$$\langle n|H_{QP}\,G^+{}_P(E)H_{PQ}|n\rangle = P\!\int dE'\,\frac{|\langle n|H_{QP}|\psi(E')\rangle^{\mp}|^2}{E - E'} - i\pi\,|\langle n|H_{QP}|\psi(E)\rangle^{\mp}|^2$$

$$\approx \Delta_n(E) - \tfrac{1}{2}\,i\Gamma_n(E)\,.$$

For elastic scattering, this leads to the *Breit–Wigner formula*

$$\overset{\sim}{}\langle\psi|\,\delta T\,|\psi\rangle^{\mp} \approx \frac{1}{\pi}\sum_n \frac{\tfrac{1}{2}\,\Gamma_n}{E - E_n - \Delta_n + \tfrac{1}{2}i\Gamma_n}\,.$$

With the level width Γ_n, the terms for all real energy values remain finite. This is similar to the result that only finite amplitudes are permitted, as for the damping of a forced oscillation (Sect. 2.3.8) .

5.2.7 *Averaging over the Energy*

So far we have assumed that the energy can be arbitrarily sharp. Actually we should not do this, but rather calculate with mean values. Even disregarding this aspect, it is instructive to given an overview of the average behavior.

We denote the mean values as usual with angular brackets or bars and use suitable weight factors $\rho(E, E')$ to compute them, as in

$$\langle f(E) \rangle \equiv \overline{f(E)} \equiv \int dE' \, \rho(E, E') f(E') \,,$$

where

$$\rho(E, E') = 0 \,, \quad \text{for} \quad |E - E'| \gg I \quad \text{and} \quad \int dE' \, \rho(E, E') = 1 \,.$$

The Lorentz distribution is analytically convenient:

$$\rho(E, E') = \frac{I}{2\pi} \frac{1}{(E - E')^2 + I^2/4} \,.$$

It is symmetric in E and E' and has a maximum for $E' = E$, while the half-width I does not lead to cumbersome boundary effects, as the box distribution does. However, the Lorentz distribution does not have a finite energy uncertainty ΔE—only the half-width is finite. For a test function $f(E)$ which is regular in the upper complex half plane and vanishes sufficiently fast for large $|E|$, we have by the Breit–Wigner formula,

$$f(\mathscr{E}) = \sum_n \frac{a_n}{\mathscr{E} - \mathscr{E}_n} \,, \quad \text{with} \quad \mathscr{E}_n = E_n - \tfrac{1}{2} i \Gamma_n \,, \quad \Gamma_n > 0 \,.$$

The residue theorem then implies

$$\langle f(E) \rangle = \frac{I}{2\pi} \int \frac{dE'}{(E' - E - \tfrac{1}{2} i I)(E' - E + \tfrac{1}{2} i I)} \sum_n \frac{a_n}{E' - \mathscr{E}_n}$$

$$= \frac{I}{2\pi} 2\pi i \sum_n \frac{a_n}{i I (E + \tfrac{1}{2} i I - \mathscr{E}_n)} = f(E + \tfrac{1}{2} i I) \,.$$

While the limit $E + io$ has been necessary so far, the average now already leads to a complex energy: the imaginary part is equal to half of the half-width of the distribution function. In then averaged scattering amplitude, the level widths are thus broadened:

$$\overline{\langle \psi_f | T | \psi_i \rangle} = \langle \psi_f | \widetilde{T} | \psi_i \rangle + \sum_n \frac{\overset{\sim}{\langle \psi_f |} V_{PQ} | \Xi_n \rangle \langle \Xi_n^A | V_{QP} \overset{\sim}{| \psi_i \rangle}}{E - \{ E_n - \tfrac{1}{2} i (\Gamma_n + I) \}} \,.$$

Here we have assumed that $\widetilde{T}$ does not depend strongly on the energy. The interval I of the averaging procedure may be large compared with the resonance widths Γ_n, but it must nevertheless be so small that the average $\widetilde{T}$ is not altered. ($\widetilde{T}$ comprises only the broad "potential resonances", and δT the narrow "compound nucleus (Feshbach) resonances".)

5.2.8 Special Features of Three-Body Problems

In the rest of this section, we shall treat a special aspect of the scattering theory which in fact does not belong to a standard course on Quantum Theory II, although it is nevertheless important and instructive. If three partners 1, 2, and 3 are involved in a reaction, then there are many more reaction possibilities than for only two of them. If initially, e.g., 2 and 3 are bound to each other and form the collision partner for 1, then the following transitions are possible:

$$
\begin{aligned}
1 + (2 + 3) &\rightarrow 1 + (2 + 3) &&\text{elastic (and inelastic) scattering,} \\
&\rightarrow 2 + (3 + 1) &&\text{disorder reaction,} \\
&\rightarrow 3 + (1 + 2) &&\text{disorder reaction,} \\
&\rightarrow 1 + 2 + 3 &&\text{fission reaction.}
\end{aligned}
$$

For fission, one partner can initially also leave the interaction regime, while the others stay together for a while. We then speak of *stepwise decay*, and of a *final-state interaction* between first and second decay, even though this "final state" also decays.

If we trace the reaction back to two-body forces (not including many-body forces), then we must nevertheless be careful to distinguish between genuine three-body operators and those for which the unit operator for one particle may be split-off, then multiplied by a two-body operator for the two remaining particles. For example, for the interaction between particles 2 and 3, we write

$$
V^{23} \equiv V_1 \equiv v_1 \, 1^1 \, .
$$

If the particle is involved, then its number appears up, if it is not involved, then it appears down. Lower-case letters now indicate two-body operators. For two-body forces, we then have $V = V^{23} + V^{31} + V^{12} = V_1 + V_2 + V_3$.

Since for the disorder reaction $1 + (2 + 3) \rightarrow 2 + (3 + 1)$, initially V_1 and then V_2 leads to a bound state of the corresponding pair, instead of the free Hamilton operators H_0, we clearly now need the *channel Hamilton operators*

$$
H_\alpha \equiv H_0 + V_\alpha \, ,
$$

and the "residual interaction" is

$$
V^\alpha \equiv V - V_\alpha = H - H_\alpha \, .
$$

V^α thus contains all two-body interactions involving α, e.g., then $V^1 = V^{12} + V^{13} = V_3 + V_2$.

In order to capture the fission, let us also allow $\alpha = 0$, i.e., $\alpha \in \{0, 1, 2, 3\}$, and require $V_0 \equiv 0$ or $V^0 \equiv V$. In addition to the full resolvent G, we also introduce *channel resolvents* G_α:

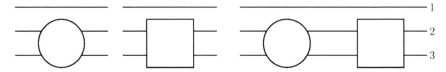

Fig. 5.7 Unconnected graphs for three-body scattering. Here partner 1 is not involved and delivers useless factors. *Left*: V_1 *Center*: T_1. *Right*: $V_1 G_0 T_1$. Partners 2 and 3 participate in the two-body scattering

$$G(\mathcal{E}) \equiv \frac{1}{\mathcal{E} - H} \;, \quad G_\alpha(\mathcal{E}) \equiv \frac{1}{\mathcal{E} - H_\alpha} \;.$$

Then according to Sect. 5.2.1, the Lippmann–Schwinger equations are valid:

$$G_\alpha = (1 + G_\alpha V_\alpha)\, G_0 = G_0\, (1 + V_\alpha\, G_\alpha) \;,$$
$$G = (1 + G\, V^\alpha)\, G_\alpha = G_\alpha\, (1 + V^\alpha\, G) \;.$$

These equations are in fact correct, but the last row does not fix the unknown resolvent G uniquely. Here we would have to invert the operator

$$1 - G_\alpha\, V^\alpha = 1 - G_0\, (1 + V_\alpha\, G_\alpha)\, V^\alpha = 1 - G_0\, V^\alpha - G_0\, V_\alpha\, G_\alpha\, V^\alpha \;.$$

But with $V^1 = V_2 + V_3$ (with $\alpha = 1$), it contains the parts $G_0 V_2$ and $G_0 V_3$, and hence different unit operators (the "non-involved part", *unconnected graphs*) (see Fig. 5.7). In the energy and momentum representations, this leads to delta functions, and in the real-space representation to divergent integrals, which requires another approach. Note that such problems do not occur for $V_\alpha G_\alpha V^\alpha$, because all parts are involved.

5.2.9 The Method of Kazaks and Greider

One possibility for solution is a method due to Kazaks and Greider [5]. As for the two-potential formula, we deal initially only with parts of the interaction. In particular, we take the transition operators for the two-body scattering to v_α (with $\alpha \neq 0$),

$$t_\alpha = v_\alpha\, (1 + g_0\, t_\alpha) = (1 + t_\alpha\, g_0)\, v_\alpha \;,$$

and use the energy $E - E_\alpha$ in g_0. We leave the particles α untouched and begin by solving the scattering problem for the two remaining partners. Then we may also use

$$T_\alpha = t_\alpha\, 1^\alpha = V_\alpha\, (1 + G_0\, T_\alpha) = (1 + T_\alpha\, G_0)\, V_\alpha \;,$$

with

$$(1 - G_0 V_\alpha)(1 + G_0 T_\alpha) = 1 \quad \text{and} \quad T_\alpha G_0 = V_\alpha G_\alpha ,$$

and we need T_1, T_2, and T_3. Then with $\alpha \neq \beta \neq \gamma \neq \alpha$, and thus $V^\alpha = V_\beta + V_\gamma$, we obtain

$$1 - G_0 V^\alpha = 1 - G_0 V_\beta - G_0 V_\gamma = (1 - G_0 V_\beta)\{1 - (1 + G_0 T_\beta) G_0 V_\gamma\} .$$

The last factor is equal to $1 - G_0 V_\gamma - G_0 T_\beta G_0 V_\gamma$, and with $T_\gamma = (1 + T_\gamma G_0) V_\gamma$, or $V_\gamma = T_\gamma (1 - G_0 V_\gamma)$, it may also be factorized:

$$1 - G_0 V_\gamma - G_0 T_\beta G_0 V_\gamma = (1 - G_0 T_\beta G_0 T_\gamma)(1 - G_0 V_\gamma) .$$

Consequently, $(1 - G_0 V^\alpha)^{-1}$ can be decomposed into three factors:

$$(1 - G_0 V^\alpha)^{-1} = (1 + G_0 T_\gamma)(1 - G_0 T_\beta G_0 T_\gamma)^{-1}(1 + G_0 T_\beta) .$$

Here β and γ may be exchanged. Therefore, for the transition operator T^α associated with $V^\alpha = V_\beta + V_\gamma$ (with $\alpha \neq 0$), we obtain

$$T^\alpha = (1 + T^\alpha G_0) V^\alpha = V^\alpha (1 + G_0 T^\alpha) = V^\alpha (1 - G_0 V^\alpha)^{-1} ,$$

along with $V_\beta (1 + G_0 T_\beta) = T_\beta$ and $V_\gamma (1 + G_0 T_\gamma) = T_\gamma$, and hence the expression (see Fig. 5.8)

$$
\begin{aligned}
T^\alpha = {} & T_\beta (1 - G_0 T_\gamma G_0 T_\beta)^{-1}(1 + G_0 T_\gamma) \\
& + T_\gamma (1 - G_0 T_\beta G_0 T_\gamma)^{-1}(1 + G_0 T_\beta) .
\end{aligned}
$$

The initially non-invertible operator $1 - G_\alpha V^\alpha$ with $V^\alpha = T^\alpha (1 - G_0 V^\alpha)$ may now be split-up into a product:

$$1 - G_\alpha V^\alpha = 1 - G_0 V^\alpha - G_0 T_\alpha G_0 V^\alpha = (1 - G_0 T_\alpha G_0 T^\alpha)(1 - G_0 V^\alpha) .$$

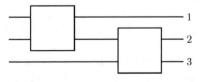

Fig. 5.8 Connected graphs for three-body scattering. Here we consider the example $T_3 G_0 T_1$. These arise for the method of Kazaks and Greithe and also for the iterated Faddeev equation. The scattering problem is therefore soluble

Both factors are invertible. In particular, $(1 - G_0 V^\alpha) (1 + G_0 T^\alpha) = 1$. Therefore, for the unknown resolvent G from $(1 - G_\alpha V^\alpha) G = G_\alpha$, we have the unique result

$$G = (1 + G_0 T^\alpha) (1 - G_0 T_\alpha G_0 T^\alpha)^{-1} G_\alpha .$$

The operators T_α, T_β, and T_γ are extremely useful for solving the problem. Only by controlling the two-body scattering can we treat the three-body scattering.

5.2.10 Faddeev Equations

In the last equation, G may be decomposed into three parts:

$$G = G^1 + G^2 + G^3 ,$$

where (with $\alpha = 1$)

$$\begin{aligned}
G^1 &= G_1 + G_0 T_1 (G^2 + G^3) \\
G^2 &= G_0 T_2 (G^1 + G^3) \\
G^3 &= G_0 T_3 (G^1 + G^2) .
\end{aligned}$$

Hence,

$$\begin{aligned}
G^2 &= G_0 T_2 G^1 + G_0 T_2 G_0 T_3 (G^1 + G^2) \\
&= (1 - G_0 T_2 G_0 T_3)^{-1} G_0 T_2 (1 + G_0 T_3) G^1 .
\end{aligned}$$

Using $(1 - A B)^{-1} A = A (1 - B A)^{-1}$, this is equivalent to

$$\begin{aligned}
G^2 &= G_0 T_2 (1 - G_0 T_3 G_0 T_2)^{-1} (1 + G_0 T_3) G^1 , \\
G^3 &= G_0 T_3 (1 - G_0 T_2 G_0 T_3)^{-1} (1 + G_0 T_2) G^1 .
\end{aligned}$$

We then also have $G^2 + G^3 = G_0 T^1 G^1$ and thus $G^1 = G_1 + G_0 T_1 G_0 T^1 G^1$. If we solve with respect to G^1, then we find $G^1 = (1 - G_0 T_1 G_0 T^1)^{-1} G_1$. Consequently, the initial equation is equivalent to

$$G = (1 + G_0 T^1) G^1 = (1 + G_0 T^1) (1 - G_0 T_1 G_0 T^1)^{-1} G_1 .$$

This expression for the resolvent G was also derived in the last section. Hence, if the initial state has $\alpha = 1$, we have proven the *Faddeev equations*

$$\begin{pmatrix} G^1 \\ G^2 \\ G^3 \end{pmatrix} = \begin{pmatrix} G_1 \\ 0 \\ 0 \end{pmatrix} + G_0 \begin{pmatrix} 0 & T_1 & T_1 \\ T_2 & 0 & T_2 \\ T_3 & T_3 & 0 \end{pmatrix} \begin{pmatrix} G^1 \\ G^2 \\ G^3 \end{pmatrix} ,$$

which deliver $G = G^1 + G^2 + G^3$. After an iteration, they have a unique solution, because then only connected graphs occur:

$$\begin{pmatrix} G^1 \\ G^2 \\ G^3 \end{pmatrix} = \begin{pmatrix} G_1 \\ G_0\,T_2\,G_1 \\ G_0\,T_3\,G_1 \end{pmatrix} + G_0 \begin{pmatrix} T_1G_0(T_2+T_3) & T_1G_0T_3 & T_1G_0T_2 \\ T_2G_0T_3 & T_2G_0(T_1+T_3) & T_2G_0T_1 \\ T_3G_0T_2 & T_3G_0T_1 & T_3G_0(T_1+T_2) \end{pmatrix} \begin{pmatrix} G^1 \\ G^2 \\ G^3 \end{pmatrix}.$$

More details can be found in the book by Schmid and Ziegelmann [6].

5.2.11 Summary: Two- and Three-Body Scattering Problems

Here we presented the generalized framework for scattering theory, followed by several important applications. These made use of the two-potential formula ($V = \widetilde{V} + \delta V$) due to Gell-Mann and Goldberger: $T = \widetilde{T} + (1 + \widetilde{T}G_0)\,\delta V\,(1 + G_0\widetilde{V})$. This helps, e.g., with the scattering of charged particles, because the Coulomb potential has too long a range for a simple scattering theory, but also for resonances, where the coupling of the scattering states to bound states becomes important.

5.3 Many-Body Systems

5.3.1 One- and Many-Body States

Since generally n is taken as the occupation number for many-particle problems, we shall now write $|v\rangle$ to indicate a one-particle basis, instead of $|n\rangle$ as used so far. We start from a complete orthonormal set of one-particle states $|v\rangle$, whence

$$\sum_v |v\rangle\langle v| = 1 \quad \text{and} \quad \langle v|v'\rangle = \delta_{vv'} .$$

For continuous quantum numbers v, there will be an integral here instead of the sum, and the delta function instead of the Kronecker symbol. Here we order the states $|v\rangle$ with respect to their energy. This is not actually important for the time being, but the notation $v < v'$ should always make sense, and later it is mainly states with low energy that will be occupied.

N particles have N times as many degrees of freedom as a single particle, and the Hilbert space has correspondingly more quantum numbers and dimensions. As long as they do not interact with each other, for each individual particle, we can identify the one-particle state it is in—if there are pure states, the case to which we restrict ourselves here. Let the first particle be in the state $|v_1\rangle$, the second in $|v_2\rangle$, and so on. Then we may consider a *product of one-particle states*

$$|\nu_1 \nu_2 \ldots \nu_N\rangle \equiv |\nu_1\rangle \otimes |\nu_2\rangle \otimes \cdots \otimes |\nu_N\rangle$$

for the corresponding N-particle state.

One basic assumption in the following is now that these N-particle states always form a complete and orthonormalized basis, even if the particles interact with each other. Then any possible N-particle state $|N \ldots\rangle$ may be built from these states:

$$|N \ldots\rangle = \sum_{\nu_1 \ldots \nu_N} |\nu_1 \ldots \nu_N\rangle \langle \nu_1 \ldots \nu_N | N \ldots\rangle \,,$$

since the states $|\nu_1 \ldots \nu_N\rangle$ form a complete basis, i.e.,

$$\sum_{\nu_1 \ldots \nu_N} |\nu_1 \ldots \nu_N\rangle \langle \nu_1 \ldots \nu_N| = 1 \,,$$

and are orthonormalized, i.e.,

$$\langle \nu_1 \ldots \nu_N | \nu_1' \ldots \nu_N' \rangle = \langle \nu_1 | \nu_1' \rangle \cdots \langle \nu_N | \nu_N' \rangle \,.$$

Here we shall also allow for improper Hilbert vectors, where integrals occur instead of sums.

This framework is generally unnecessary, however, for identical particles. For indistinguishable particles, we cannot state which is the first or which is the last, because the interchange of two particles does not change the expectation value of an arbitrary observable—otherwise the particles were not identical. Since we shall now occupy ourselves with such indistinguishable particles, it is clear that we should only have superpositions of states with an exchange symmetry: if the order of the particles changes, at most the phase factor of the states can change.

5.3.2 Exchange Symmetry

Let the transposition operator $\mathscr{P}_{kl} = \mathscr{P}_{lk}$ exchange the particles labelled k and l:

$$\mathscr{P}_{kl} \, | \ldots \nu_k \ldots \nu_l \ldots\rangle = | \ldots \nu_l \ldots \nu_k \ldots\rangle \,.$$

Since $\mathscr{P}_{kl}^2$ leads back to the old state, the operator $\mathscr{P}_{kl}$ has the eigenvalues ± 1. Its eigenstates for the particles k and l are said to be *symmetric* ($p_{kl} = +1$) or *anti-symmetric* ($p_{kl} = -1$). Let us now consider all $N!$ different permutations $\mathscr{P}$ of an N-particle state $|\nu_1 \ldots \nu_N\rangle$. They can be built from products of pair-exchange operators $\mathscr{P}_{kl}$, although not uniquely. The only thing that is fixed is whether an even or an odd number of pair exchanges is necessary. We speak of *even* and *odd permutations* (see Fig. 5.9).

Fig. 5.9 The 3! = 6 different permutations of three objects. The even permutations are the identity, the cyclic, and the anti-cyclic permutations, the odd ones are the three transpositions. The last shows three transpositions, even though it can also be understood as a single transposition with particle 2 remaining unchanged

Fig. 5.10 Representation of $\mathscr{P}_{km}\mathscr{P}_{lm}\mathscr{P}_{km}\mathscr{P}_{kl} = 1$. As $p_{km}^2 = 1$, it is clear that $p_{lm} = p_{lk}$ for all $k \neq l \neq m$

For identical particles, the eigenvalues p_{kl} have to be either all $+1$ or all -1, because the exchange symmetry is a characteristic of the considered particles: they form either symmetric or antisymmetric states. The state cannot have one exchange symmetry in the pairs (k, l) and (k, m), but the other in the pair (l, m), as Fig. 5.10 shows. Therefore we may restrict ourselves to either completely symmetric or completely antisymmetric states.

In the following, we label symmetric states with an s on the Dirac symbol and anti-symmetric ones with an a:

$$|\ldots v_k \ldots v_l \ldots\rangle_s = +|\ldots v_l \ldots v_k \ldots\rangle_s, \qquad \text{for all } k \text{ and } l,$$
$$|\ldots v_k \ldots v_l \ldots\rangle_a = -|\ldots v_l \ldots v_k \ldots\rangle_a, \qquad \text{for all } k \text{ and } l,$$

or, with $\delta_{\mathscr{P}} = +1$ for even permutations and $\delta_{\mathscr{P}} = -1$ for odd,

$$\mathscr{P}|v_1 \ldots v_N\rangle_s = |v_1 \ldots v_N\rangle_s,$$
$$\mathscr{P}|v_1 \ldots v_N\rangle_a = \delta_{\mathscr{P}}|v_1 \ldots v_N\rangle_a.$$

Symmetric states describe *bosons*, and antisymmetric ones *fermions*.

Hence, two fermions cannot occupy the same one-particle state, because upon transposition of the two particles, the many-body state has to change sign. We now have the basic ingredient for the famous *Pauli exclusion principle*. For symmetric states (bosons), this restriction does not exist. If n_v gives the particle number in the state $|v\rangle$, then for bosons, $n_v \in \{0, 1, 2, \ldots\}$ holds, while for fermions $n_v \in \{0, 1\}$. The sum of all occupation numbers n_v yields the total number N of particles, viz., $N = \sum_v n_v$.

The permutation operators $\mathscr{P}$ all have an inverse:

$$\mathscr{P}\mathscr{P}^{-1} = 1 = \mathscr{P}^{-1}\mathscr{P}.$$

In addition, nothing changes if all bra- and all ket-vectors undergo the same permutation,

$$\mathscr{P}^{\dagger}\mathscr{P} = 1 \quad \Longrightarrow \quad \mathscr{P}^{\dagger} = \mathscr{P}^{-1} \, .$$

Permutation operators are thus unitary.

All observables O of an N-particle system have to commute with permutations, as long as we are dealing with identical particles:

$$O = \mathscr{P}^{\dagger}O\mathscr{P} \quad \Longrightarrow \quad [O, \mathscr{P}] = 0 \, .$$

Therefore, no perturbation can alter the symmetry: $O = \mathscr{P}^{\dagger}_{kl}\mathscr{P}_{kl}O = \mathscr{P}^{\dagger}_{kl}O\mathscr{P}_{kl}$ delivers $_s\langle v_1 \ldots v_N | O | v'_1 \ldots v'_N \rangle_a = -_s\langle v_1 \ldots v_N | O | v'_1 \ldots v'_N \rangle_a = 0$. In particular, symmetric and antisymmetric states are orthogonal to each other, which follows by inserting $O = 1$, and the symmetry does not change with time, because the Hamilton operator is invariant under permutations.

5.3.3   Symmetric and Antisymmetric States

In order to form arbitrary many-body states $|v_1 \ldots v_N\rangle$ from symmetric and antisymmetric states, we take the symmetrizing and anti-symmetrizing operators

$$\mathscr{S} = \frac{1}{N!}\sum_{\mathscr{P}}\mathscr{P} \quad \text{and} \quad \mathscr{A} = \frac{1}{N!}\sum_{\mathscr{P}}\delta_{\mathscr{P}}\mathscr{P} \, .$$

Here the sums run over all $N!$ different permutations. The two expressions can be proven together. For this we set

$$\Lambda = \frac{1}{N!}\sum_{\mathscr{P}}\lambda(\mathscr{P})\,\mathscr{P} \, , \quad \text{with} \quad \begin{cases} \Lambda = \mathscr{S}, \; \lambda(\mathscr{P}) = 1 \quad \text{for bosons,} \\ \Lambda = \mathscr{A}, \; \lambda(\mathscr{P}) = \delta_{\mathscr{P}} \; \text{for fermions.} \end{cases}$$

In particular, with

$$\mathscr{P}_{kl}\mathscr{S} \, |v_1 \ldots v_N\rangle = \mathscr{S} \, |v_1 \ldots v_N\rangle \, , \quad \mathscr{P}_{kl}\mathscr{A} \, |v_1 \ldots v_N\rangle = -\mathscr{A} \, |v_1 \ldots v_N\rangle \, ,$$

we find $\mathscr{P}_{kl}\,\Lambda = \lambda(\mathscr{P}_{kl})\,\Lambda$ and therefore also

$$\mathscr{P}\Lambda = \lambda(\mathscr{P})\,\Lambda = \Lambda\mathscr{P} \, .$$

It remains to show that Λ is idempotent, to be sure that it is a projection operator. But now $N!\,\Lambda^2 = \sum_{\mathscr{P}}\lambda(\mathscr{P})\,\mathscr{P}\Lambda = \sum_{\mathscr{P}}\lambda^2(\mathscr{P})\,\Lambda$ holds and $\sum_{\mathscr{P}}1 = N!$, so we do indeed have $\Lambda^2 = \Lambda$. In addition, Λ is Hermitian, because $\mathscr{P}$ is unitary, $\lambda(\mathscr{P}) = \lambda(\mathscr{P}^{-1})$, and the sum over all $\mathscr{P}$ is equal to the sum over all $\mathscr{P}^{-1}$:

$$\Lambda = \Lambda^2 = \Lambda^{\dagger} \, , \quad \text{for} \; \Lambda = \mathscr{S} \; \text{and} \; \Lambda = \mathscr{A} \, .$$

The operator Λ is a linear combination of the unitary operators $\mathscr{P}$, and hence itself not unitary. Furthermore, although we have already found the projection operators $\mathscr{S}$ and $\mathscr{A}$, we must nevertheless also normalize the unknown symmetric and antisymmetric states correctly. If n_ν gives the number of bosons in the one-particle state $|\nu\rangle$, then we have

$$|\nu_1 \ldots \nu_N\rangle_s = \sqrt{\frac{N!}{n_1! n_2! \ldots}}\, \mathscr{S}\, |\nu_1 \ldots \nu_N\rangle\,,$$
$$|\nu_1 \ldots \nu_N\rangle_a = \sqrt{N!}\quad \mathscr{A}\, |\nu_1 \ldots \nu_N\rangle\,.$$

For fermions, the last equation with $\Lambda^\dagger \Lambda = \Lambda$ delivers

$$_a\langle \nu_1 \ldots \nu_N | \nu_1 \ldots \nu_N\rangle_a = N!\, \langle \nu_1 \ldots \nu_N | \mathscr{A}\, | \nu_1 \ldots \nu_N\rangle$$
$$= \sum_{\mathscr{P}} \delta_{\mathscr{P}}\, \langle \nu_1 \ldots \nu_N | \mathscr{P}\, | \nu_1 \ldots \nu_N\rangle\,.$$

But here only $\mathscr{P} = 1$ contributes—with $\langle \nu_1 \ldots \nu_N | \nu_1' \ldots \nu_N'\rangle = \langle \nu_1|\nu_1'\rangle \ldots \langle \nu_N|\nu_N'\rangle$ and because for fermions all ν_i have to be different. Thus $|\nu_1 \ldots \nu_N\rangle_a$ is normalized correctly. In contrast, in the expression $\langle \nu_1 \ldots \nu_N | \mathscr{P}\, | \nu_1 \ldots \nu_N\rangle$ for bosons, the $n_1! n_2! \ldots$ terms contribute a 1, for which $\mathscr{P}|\nu_1 \ldots \nu_N\rangle$ is equal to $|\nu_1 \ldots \nu_N\rangle$. This implies

$$|\nu_1 \ldots \nu_N\rangle_s = \frac{1}{\sqrt{N!}} \frac{1}{\sqrt{n_1! n_2! \ldots}} \sum_{\mathscr{P}} \mathscr{P}\, |\nu_1 \ldots \nu_N\rangle\,,$$
$$|\nu_1 \ldots \nu_N\rangle_a = \frac{1}{\sqrt{N!}} \sum_{\mathscr{P}} \delta_{\mathscr{P}} \mathscr{P}\, |\nu_1 \ldots \nu_N\rangle\,,$$

where both sums run over all $N!$ permutations. The first sum has $n_1! n_2! \ldots$ equal terms and can be summed up correspondingly:

$$|\nu_1 \ldots \nu_N\rangle_s = \sqrt{\frac{n_1! n_2! \ldots}{N!}} \sum_{\mathscr{P}'} \mathscr{P}'|\nu_1 \ldots \nu_N\rangle\,,$$

if we take only the permutations $\mathscr{P}'$ which lead to different states.

To compute matrix elements with $\Lambda^\dagger O \Lambda = \Lambda O = O\Lambda$, it is sufficient to symmetrize only in the bra- or ket-vectors. But we then have to normalize correctly:

$$_s\langle \nu_1 \ldots \nu_N | O | \nu_1' \ldots \nu_N'\rangle_s = \sqrt{\frac{N!}{n_1! n_2! \ldots}}\, \langle \nu_1 \ldots \nu_N | O | \nu_1' \ldots \nu_N'\rangle_s\,,$$
$$_a\langle \nu_1 \ldots \nu_N | O | \nu_1' \ldots \nu_N'\rangle_a = \sqrt{N!}\, \langle \nu_1 \ldots \nu_N | O | \nu_1' \ldots \nu_N'\rangle_a\,.$$

Note that the completeness relation for the N-fermion system (hence considering only antisymmetric states) with

$$|\nu_1' \ldots \nu_N'\rangle_a = \frac{1}{N!} \sum_{\nu_1 \ldots \nu_N} |\nu_1 \ldots \nu_N\rangle_{aa}\langle \nu_1 \ldots \nu_N|\nu_1' \ldots \nu_N'\rangle_a$$

$$= \sum_{\nu_1 < \cdots < \nu_N} |\nu_1 \ldots \nu_N\rangle_{aa}\langle \nu_1 \ldots \nu_N|\nu_1' \ldots \nu_N'\rangle_a$$

may be written in two ways, viz.,

$$\sum_{\nu_1 \ldots \nu_N} |\nu_1 \ldots \nu_N\rangle_{aa}\langle \nu_1 \ldots \nu_N| = N! \ ,$$

or with far fewer terms

$$\sum_{\nu_1 < \cdots < \nu_N} |\nu_1 \ldots \nu_N\rangle_{aa}\langle \nu_1 \ldots \nu_N| = 1 \ .$$

In the real-space representation, the N-fermion state $|\nu_1 \ldots \nu_N\rangle_a$ reads

$$\langle \mathbf{r}_1 \ldots \mathbf{r}_N|\nu_1 \ldots \nu_N\rangle_a = \frac{1}{\sqrt{N!}} \sum_{\mathscr{P}} \delta_{\mathscr{P}} \langle \mathbf{r}_1 \ldots \mathbf{r}_N|\mathscr{P}|\nu_1 \ldots \nu_N\rangle$$

$$= \frac{1}{\sqrt{N!}} \begin{vmatrix} \langle \mathbf{r}_1|\nu_1\rangle & \ldots & \langle \mathbf{r}_N|\nu_1\rangle \\ \vdots & \ddots & \vdots \\ \langle \mathbf{r}_1|\nu_N\rangle & \ldots & \langle \mathbf{r}_N|\nu_N\rangle \end{vmatrix} \ .$$

The last expression is called the (normalized) *Slater determinant*,

Calculations with symmetric or antisymmetric states can be greatly simplified with creation and annihilation operators: the former increase the particle number by one, while the latter lower them by one. Therefore, in the following we have the particle vacuum $|0\rangle$, one-particle states $|\nu\rangle$, and N-particle states $|\nu_1 \ldots \nu_N\rangle_s$ and $|\nu_1 \ldots \nu_N\rangle_a$ for $N \geq 2$. The set of their Hilbert spaces forms the *Fock space*. Here states with different particle number N are orthogonal to each other. The Hilbert vector $|0\rangle$ of the vacuum state should not be confused with the zero vector $|o\rangle$, where $\langle 0|0\rangle = 1$, but $\langle o|o\rangle = 0$.

We begin with fermions, because the states are easier to normalize than those of bosons.

5.3.4 *Creation and Annihilation Operators for Fermions*

Let the operator $A_\nu^\dagger$ create a fermion in the one-particle state $|\nu\rangle$ from the vacuum $|0\rangle$:

$$A_\nu^\dagger|0\rangle = |\nu\rangle \ .$$

Let generally $A_\nu^\dagger$ make the state $|\nu_1 \ldots \nu_N \nu\rangle_a$ from the N-fermion state $|\nu_1 \ldots \nu_N\rangle_a$, if this is possible at all—the state $|\nu\rangle$ has to be unoccupied previously, so that an antisymmetric $(N+1)$-particle state can be constructed:

$$A_\nu^\dagger |\nu_1 \ldots \nu_N\rangle_a = \begin{cases} |o\rangle & \text{if } \nu \in \{\nu_1 \ldots \nu_N\} \,, \\ |\nu_1 \ldots \nu_N \nu\rangle_a & \text{if } \nu \notin \{\nu_1 \ldots \nu_N\} \,. \end{cases}$$

Note the *phase convention* employed here: if ν is arranged differently, the state may differ in sign. For example, $|\nu\nu_1 \ldots \nu_N\rangle_a$ requires another creation operator. This will be discussed separately on p. 442. It follows that

$$|\nu_1 \ldots \nu_N\rangle_a = A_{\nu_N}^\dagger \cdots A_{\nu_1}^\dagger |0\rangle \,,$$

and the antisymmetry requires

$$A_\nu^\dagger A_{\nu'}^\dagger = -A_{\nu'}^\dagger A_\nu^\dagger \,, \quad \text{in particular } (A_\nu^\dagger)^2 = 0 \text{ (Pauli principle)} \,.$$

States with different particle number should be orthogonal to each other. Therefore,

$$\langle 0|A_\nu^\dagger = \langle o| \quad \text{and} \quad \langle v'|A_\nu^\dagger = \langle v'|\nu\rangle\langle 0| \,.$$

For the operator A_ν Hermitian conjugate to $A_\nu^\dagger$, it follows that

$$A_\nu A_{\nu'} = -A_{\nu'} A_\nu \,, \quad {}_a\langle \nu_1 \ldots \nu_N| = \langle 0|A_{\nu_1} \cdots A_{\nu_N} \,,$$

together with

$$A_\nu |0\rangle = |o\rangle \,, \quad \text{and} \quad A_\nu |v'\rangle = |0\rangle\langle \nu|v'\rangle \,.$$

We thus have

$$A_\nu |\nu_1 \ldots \nu_N\rangle_a = \begin{cases} |\nu_1 \ldots \nu_{N-1}\rangle_a & \text{if } \nu = \nu_N \,, \\ -|\nu_1 \ldots \nu_{N-2}\nu_N\rangle_a & \text{if } \nu = \nu_{N-1} \,, \\ \vdots & \vdots \\ (-)^{N-1}|\nu_2 \ldots \nu_N\rangle_a & \text{if } \nu = \nu_1 \,, \\ |o\rangle & \text{if } \nu \notin \{\nu_1 \ldots \nu_N\} \,. \end{cases}$$

The operator A_ν thus removes a fermion from the state $|\nu\rangle$. Therefore, A_ν is an annihilation operator of the fermion in the state $|\nu\rangle$.

With these creation and annihilation operators, many-fermion states can be treated very conveniently—also if the particle number does not change at all, e.g., if equally many creation and annihilation operators occur in operator products. In particular, there is no longer any need to pay attention to the antisymmetry. We merely follow the calculation rules for the new operators:

$$A_\nu^\dagger A_{\nu'}^\dagger + A_{\nu'}^\dagger A_\nu^\dagger = 0 = A_\nu A_{\nu'} + A_{\nu'} A_\nu \,,$$
$$A_\nu A_{\nu'}^\dagger + A_{\nu'}^\dagger A_\nu = \langle \nu | \nu' \rangle \,.$$

The first two commutation laws have already been proven. In addition, we have $\langle 0 | A_\nu A_{\nu'}^\dagger | 0 \rangle = \langle \nu | \nu' \rangle$ and $\langle 0 | A_{\nu'}^\dagger A_\nu | 0 \rangle = 0$. For more particles, we first consider the case $\nu \neq \nu'$: $A_\nu A_{\nu'}^\dagger$ and $A_{\nu'}^\dagger A_\nu$ create one fermion in the state $|\nu'\rangle$ and destroy one in the state $|\nu\rangle$, but the new states have opposite sign, e.g., $A_{\nu_N}^\dagger A_{\nu_{N-1}}$ changes from the state $|\nu_1 \ldots \nu_{N-1}\rangle_a$ to the state $|\nu_1 \ldots \nu_{N-2} \nu_N\rangle_a$, but $A_{\nu_{N-1}} A_{\nu_N}^\dagger$ results in $A_{\nu_{N-1}} |\nu_1 \ldots \nu_N\rangle_a = -|\nu_1 \ldots \nu_{N-2} \nu_N\rangle_a$. Therefore, it only remains to show that $A_\nu A_\nu^\dagger + A_\nu^\dagger A_\nu = 1$:

$$A_\nu A_\nu^\dagger |\nu_1 \ldots \nu_N\rangle_a = \begin{cases} |o\rangle & \text{if } \nu \in \{\nu_1 \ldots \nu_N\} \,, \\ |\nu_1 \ldots \nu_N\rangle_a & \text{if } \nu \notin \{\nu_1 \ldots \nu_N\} \,, \end{cases}$$

$$A_\nu^\dagger A_\nu |\nu_1 \ldots \nu_N\rangle_a = \begin{cases} |\nu_1 \ldots \nu_N\rangle_a & \text{if } \nu \in \{\nu_1 \ldots \nu_N\} \,, \\ |o\rangle & \text{if } \nu \notin \{\nu_1 \ldots \nu_N\} \,. \end{cases}$$

With this our claim is proven: simple anti-commutation relations are valid for fermion field operators.

An additional result is that the eigenvalue of

$$N_\nu = A_\nu^\dagger A_\nu$$

gives the occupation number of the state $|\nu\rangle$, and hence that N_ν is the occupation number operator. For the particle number operator, we must therefore take

$$N = \sum_\nu A_\nu^\dagger A_\nu \,.$$

Its eigenvalue gives the total number of fermions.

5.3.5 Creation and Annihilation Operators for Bosons

So far we have associated each particle with a one-particle state. For bosons, such a state may be repeated very often, because many of them can be in the same one-particle state. As already shown on p. 437, the occupation numbers n_ν are important for the normalization. A particularly suitable representation for bosons is the so-called *occupation-number representation*. We fix an order for the one-particle states $|\nu\rangle$ and then give only the occupation numbers n_ν, thus writing $|n_1 \ldots n_\nu \ldots\rangle$ with $N = \sum_\nu n_\nu$. If unoccupied states also occur, we may leave those out of the representation. Therefore, we order the one-particle states with respect to their energy (see p. 433). We consider here only the (symmetric) boson states $| \ldots n_\nu \ldots \rangle_s$.

The boson creation operators $B_\nu^\dagger$ with the property

$$B_\nu^\dagger | \ldots n_\nu \ldots \rangle_s = | \ldots n_\nu + 1 \ldots \rangle_s \, c(n_\nu + 1) \, ,$$

whose normalization factors $c(n_\nu + 1)$ remain unknown for the moment, have to commute with each other so that a symmetric state is produced:

$$B_\nu^\dagger B_{\nu'}^\dagger - B_{\nu'}^\dagger B_\nu^\dagger = 0 \, .$$

For the Hermitian adjoint annihilation operator, we have

$$B_\nu B_{\nu'} - B_{\nu'} B_\nu = 0 \, .$$

With $_s\langle \ldots n_\nu \ldots | B_\nu | \ldots n_{\nu'} \ldots \rangle_s^* = {_s}\langle \ldots n_{\nu'} \ldots | B_\nu^\dagger | \ldots n_\nu \ldots \rangle_s = \delta_{n_{\nu'}, n_\nu + 1} \, c(n_\nu + 1)$, we may infer

$$B_\nu | \ldots n_{\nu'} \ldots \rangle_s = | \ldots n_{\nu'} - 1 \ldots \rangle_s \, c^*(n_{\nu'}) \, .$$

Hence $c(0) = 0$ holds, because no particles can be destroyed in the vacuum:

$$B_\nu | 0 \rangle = | o \rangle \, .$$

For $\nu \neq \nu'$, it follows that $B_\nu B_{\nu'}^\dagger | \ldots n_\nu \ldots n_{\nu'} \ldots \rangle_s = B_{\nu'}^\dagger B_\nu | \ldots n_\nu \ldots n_{\nu'} \ldots \rangle_s$, so both are equal to $| \ldots n_\nu - 1 \ldots n_{\nu'} + 1 \ldots \rangle_s \, c^*(n_\nu) \, c(n_{\nu'} + 1)$, in contrast to the case for $\nu = \nu'$, where

$$B_\nu B_\nu^\dagger | \ldots n_\nu \ldots \rangle_s = | \ldots n_\nu \ldots \rangle_s \, |c(n_\nu + 1)|^2 \, ,$$
$$B_\nu^\dagger B_\nu | \ldots n_\nu \ldots \rangle_s = | \ldots n_\nu \ldots \rangle_s \, |c(n_\nu)|^2 \, .$$

If we choose therefore $c(n_\nu) = \sqrt{n_\nu}$, so that $|c(n_\nu + 1)|^2 - |c_\nu|^2 = 1$, then for boson field operators, the following simple commutator relations are valid:

$$B_\nu B_{\nu'}^\dagger - B_{\nu'}^\dagger B_\nu = \langle \nu | \nu' \rangle \, ,$$
$$B_\nu^\dagger B_{\nu'}^\dagger - B_{\nu'}^\dagger B_\nu^\dagger = 0 = B_\nu B_{\nu'} - B_{\nu'} B_\nu \, .$$

In addition then,

$$N_\nu = B_\nu^\dagger B_\nu$$

is the occupation-number operator for the one-particle state $| \nu \rangle$ and

$$N = \sum_\nu B_\nu^\dagger B_\nu$$

is the operator for the total number of bosons. We also have the equation

$$|n_1 n_2 \ldots\rangle_s = \frac{(B_1{}^\dagger)^{n_1}}{\sqrt{n_1!}} \frac{(B_2{}^\dagger)^{n_2}}{\sqrt{n_2!}} \cdots |0\rangle .$$

The field operators in Sect. 4.2.8 have the same properties—there we sought operators N whose eigenvalues were the natural numbers n. And the ladder operators for harmonic oscillations in Sect. 4.5.4 also had those properties.

5.3.6   General Properties of Creation and Annihilation Operators

We now summarize the previous considerations. Except for the all important sign in the commutation relations, the creation and annihilation operators for bosons and fermions are very similar. Therefore, we now write, with the upper sign for bosons and the lower sign for fermions,

$$[\Psi_\nu, \Psi_{\nu'}^\dagger]_\mp = \langle\nu|\nu'\rangle \quad \text{and} \quad [\Psi_\nu, \Psi_{\nu'}]_\mp = 0 = [\Psi_\nu^\dagger, \Psi_{\nu'}^\dagger]_\mp .$$

With these field operators, the many-body states $|n_1 \ldots\rangle_s$ and $|n_1 \ldots\rangle_a$ can be created from the vacuum state $|0\rangle$:

$$|n_1 \ldots\rangle_{s,a} = \prod_{\nu=1}^{\infty} \frac{(\Psi_\nu^\dagger)^{n_\nu}}{\sqrt{n_\nu!}} |0\rangle .$$

For fermions, the occupation numbers n_ν are only equal to zero or one, because for them $(\Psi_\nu)^2 = -(\Psi_\nu)^2$ vanishes, and hence $n_\nu! = 1$. In addition, with $\Sigma(i) \equiv \sum_{k=i+1}^{\infty} n_k$, we have

$$\Psi_{\nu_i}^\dagger |\ldots n_i \ldots\rangle_{s,a} = (\pm)^{\Sigma(i)} |\ldots n_i+1 \ldots\rangle_{s,a} \sqrt{1 \pm n_i} ,$$
$$\Psi_{\nu_i} |\ldots n_i \ldots\rangle_{s,a} = (\pm)^{\Sigma(i)} |\ldots n_i-1 \ldots\rangle_{s,a} \sqrt{n_i} .$$

The other phase convention $\Sigma(i) = \sum_{k=1}^{i} n_k$, already mentioned in Sect. 5.3.4, is often used. It seems simple, because then k only runs over a finite number of values, but it is less convenient because the states of higher energy are all unoccupied and usually only the states near the Fermi energy are important.

For a change of representation, viz.,

$$|\mu\rangle = \sum_\nu |\nu\rangle\langle\nu|\mu\rangle ,$$

new creation and annihilation operators are necessary. From

$$|\mu\rangle = \Psi_\mu^\dagger |0\rangle = \sum_\nu \Psi_\nu^\dagger |0\rangle \langle \nu|\mu\rangle \,,$$

it follows generally that

$$\Psi_\mu^\dagger = \sum_\nu \langle \mu|\nu\rangle^* \, \Psi_\nu^\dagger \quad\Longleftrightarrow\quad \Psi_\mu = \sum_\nu \langle \mu|\nu\rangle \, \Psi_\nu \,.$$

For example, we can go over to the real-space representation. Let $\Psi^\dagger(\mathbf{r})$ create a particle at the position $\mathbf{r}$ and $\Psi(\mathbf{r})$ destroy one there. Then we have

$$\Psi^\dagger(\mathbf{r}) = \sum_\nu \langle \mathbf{r}|\nu\rangle^* \, \Psi_\nu^\dagger \quad\Longleftrightarrow\quad \Psi(\mathbf{r}) = \sum_\nu \langle \mathbf{r}|\nu\rangle \, \Psi_\nu \,,$$

with $\langle \mathbf{r}|\nu\rangle = \psi_\nu(\mathbf{r})$. Conversely,

$$\Psi_\nu^\dagger = \int d^3r \, \langle \nu|\mathbf{r}\rangle^* \, \Psi^\dagger(\mathbf{r}) \quad\Longleftrightarrow\quad \Psi_\nu = \int d^3r \, \langle \nu|\mathbf{r}\rangle \, \Psi(\mathbf{r}) \,.$$

The commutation laws are transferred:

$$[\Psi_\mu, \Psi_{\mu'}^\dagger]_\mp = \sum_{\nu\nu'} \langle \mu|\nu\rangle \langle \mu'|\nu'\rangle^* \, [\Psi_\nu, \Psi_{\nu'}^\dagger]_\mp = \sum_\nu \langle \mu|\nu\rangle \langle \nu|\mu'\rangle = \langle \mu|\mu'\rangle \,,$$

and correspondingly $[\Psi_\mu, \Psi_{\mu'}]_\mp = 0$. The particle number operator reads

$$N = \sum_\nu \Psi_\nu^\dagger \Psi_\nu = \sum_\nu \iint d^3r\,d^3r' \, \langle \nu|\mathbf{r}\rangle^* \langle \nu|\mathbf{r}'\rangle \, \Psi^\dagger(\mathbf{r}) \Psi(\mathbf{r}')$$
$$= \iint d^3r\,d^3r' \, \langle \mathbf{r}|\mathbf{r}'\rangle \, \Psi^\dagger(\mathbf{r}) \Psi(\mathbf{r}') = \int d^3r \, \Psi^\dagger(\mathbf{r}) \Psi(\mathbf{r}) \,.$$

Thus $\Psi^\dagger(\mathbf{r}) \Psi(\mathbf{r})$ is the particle density operator. Note that the expectation value of the particle-number operator $\Psi_\nu^\dagger \Psi_\nu$ does not need to be an integer: if it is so in the basis $\{|\nu\rangle\}$, then it will not generally be so in the basis $\{|\mu\rangle\}$.

5.3.7 The Two-Body System as an Example

Here there are only the permutations $|\nu_1\nu_2\rangle$ and $|\nu_2\nu_1\rangle$:

$$|\nu_1\nu_2\rangle_s = \frac{|\nu_1\nu_2\rangle + |\nu_2\nu_1\rangle}{\sqrt{2(1+\langle\nu_1|\nu_2\rangle)}} \quad \text{and} \quad |\nu_1\nu_2\rangle_a = \frac{|\nu_1\nu_2\rangle - |\nu_2\nu_1\rangle}{\sqrt{2}} \,.$$

For the matrix elements of an arbitrary operator O and with

$$\langle v_1 v_2 | O | v_1{}' v_2{}' \rangle = \langle v_2 v_1 | O | v_2{}' v_1{}' \rangle \,,$$

it follows generally that

$$_{s,a}\langle v_1 v_2 | O | v_1{}' v_2{}' \rangle_{s,a} = \frac{\langle v_1 v_2 | O | v_1{}' v_2{}' \rangle \pm \langle v_1 v_2 | O | v_2{}' v_1{}' \rangle}{\sqrt{1 + \langle v_1 | v_2 \rangle} \sqrt{1 + \langle v_1{}' | v_2{}' \rangle}} \,.$$

We shall only be concerned with one-particle operators T and two-body operators V. Here, for a one-particle operator,

$$\langle v_1 v_2 | T | v_1{}' v_2{}' \rangle = \langle v_1 | T | v_1{}' \rangle \langle v_2 | v_2{}' \rangle + \langle v_1 | v_1{}' \rangle \langle v_2 | T | v_2{}' \rangle \,.$$

Its bra- and ket-states are thus distinguished at most in one particle, otherwise the matrix element vanishes. (For a two-body operator, two states are distinguished in at most two particles.) For fermions, we must also have $v_1 \neq v_2$ and $v_1{}' \neq v_2{}'$, otherwise the parts cancel each other.

The last equation also follows if we build T up from bilinear products of creation and annihilation operators and take the matrix elements of T in the chosen representation as expansion coefficients:

$$T = \sum_{v v'} \langle v | T | v' \rangle \, \Psi_v^\dagger \Psi_{v'} \,.$$

We have in particular,

$$_{s,a}\langle v_1 v_2 | \Psi_v^\dagger \Psi_{v'} | v_1{}' v_2{}' \rangle_{s,a} = \frac{\langle 0 | \Psi_{v_1} \Psi_{v_2} \Psi_v^\dagger \Psi_{v'} \Psi_{v_2{}'}^\dagger \Psi_{v_1{}'}^\dagger | 0 \rangle}{\sqrt{1 + \langle v_1 | v_2 \rangle} \sqrt{1 + \langle v_1{}' | v_2{}' \rangle}} \,,$$

and with this, the factor

$$\Psi_{v_1} \Psi_{v_2} \Psi_v^\dagger = \Psi_{v_1} \left(\langle v_2 | v \rangle \pm \Psi_v^\dagger \Psi_{v_2} \right) = \langle v_2 | v \rangle \Psi_{v_1} \pm \left(\langle v_1 | v \rangle \pm \Psi_v^\dagger \Psi_{v_1} \right) \Psi_{v_2} \,,$$

and also the other factor $\Psi_{v'} \Psi_{v_2{}'}^\dagger \Psi_{v_1{}'}^\dagger$ in the expectation value of the adjoint of $\Psi_{v_1{}'} \Psi_{v_2{}'} \Psi_{v'}^\dagger$. With $\langle 0 | \Psi_v^\dagger = \langle o |$ and $\Psi_v | 0 \rangle = | o \rangle$, together with

$$\langle 0 | \Psi_\mu \Psi_{\mu'}^\dagger | 0 \rangle = \langle \mu | \mu' \rangle \,,$$

we thus find

$$_{s,a}\langle v_1 v_2 | \Psi_v^\dagger \Psi_{v'} | v_1{}' v_2{}' \rangle_{s,a} = \langle v_2 | v \rangle \frac{\langle v' | v_2{}' \rangle \langle v_1 | v_1{}' \rangle \pm \langle v' | v_1{}' \rangle \langle v_1 | v_2{}' \rangle}{\sqrt{1 + \langle v_1 | v_2 \rangle} \sqrt{1 + \langle v_1{}' | v_2{}' \rangle}}$$
$$+ \langle v_1 | v \rangle \frac{\langle v' | v_1{}' \rangle \langle v_2 | v_2{}' \rangle \pm \langle v' | v_2{}' \rangle \langle v_2 | v_1{}' \rangle}{\sqrt{1 + \langle v_1 | v_2 \rangle} \sqrt{1 + \langle v_1{}' | v_2{}' \rangle}} \,,$$

which is what was to be shown.

If we now consider a two-body operator, e.g., the interaction $V(\mathbf{r}_1, \mathbf{r}_2) = V(\mathbf{r}_2, \mathbf{r}_1)$, then

$$_{s,a}\langle \nu_1 \nu_2 | V | \nu_1' \nu_2' \rangle_{s,a} = \frac{\langle \nu_1 \nu_2 | V | \nu_1' \nu_2' \rangle \pm \langle \nu_1 \nu_2 | V | \nu_2' \nu_1' \rangle}{\sqrt{1 + \langle \nu_1 | \nu_2 \rangle} \sqrt{1 + \langle \nu_1' | \nu_2' \rangle}} \; .$$

With μ and μ' from the same basis as ν and ν', we may also write

$$V = \tfrac{1}{2} \sum_{\nu \mu \nu' \mu'} \langle \nu \mu | V | \nu' \mu' \rangle \; \Psi_\mu^\dagger \Psi_\nu^\dagger \Psi_{\nu'} \Psi_{\mu'} \; ,$$

because if we use

$$\Psi_{\nu_1} \Psi_{\nu_2} \Psi_\mu^\dagger \Psi_\nu^\dagger = \Psi_{\nu_1} \left(\langle \nu_2 | \mu \rangle \pm \Psi_\mu^\dagger \Psi_{\nu_2} \right) \Psi_\nu^\dagger$$

$$= \langle \nu_2 | \mu \rangle \left(\langle \nu_1 | \nu \rangle \pm \Psi_\nu^\dagger \Psi_{\nu_1} \right) \pm \left(\langle \nu_1 | \mu \rangle \pm \Psi_\mu^\dagger \Psi_{\nu_1} \right) \left(\langle \nu_2 | \nu \rangle \pm \Psi_\nu^\dagger \Psi_{\nu_2} \right)$$

in the previous equation and its adjoint for $\Psi_\nu \Psi_\mu \Psi_{\nu_2}^\dagger \Psi_{\nu_1}^\dagger$, along with $\langle 0 | \Psi^\dagger = \langle o |$ and $\Psi | 0 \rangle = | o \rangle$, then it follows that

$$_{s,a}\langle \nu_1 \nu_2 | V | \nu_1' \nu_2' \rangle_{s,a} = \frac{\langle \nu_1 \nu_2 | V | \nu_1' \nu_2' \rangle + \langle \nu_2 \nu_1 | V | \nu_2' \nu_1' \rangle}{2\sqrt{1 + \langle \nu_1 | \nu_2 \rangle} \sqrt{1 + \langle \nu_1' \nu_2' \rangle}}$$

$$\pm \frac{\langle \nu_1 \nu_2 | V | \nu_2' \nu_1' \rangle + \langle \nu_2 \nu_1 | V | \nu_1' \nu_2' \rangle}{2\sqrt{1 + \langle \nu_1 | \nu_2 \rangle} \sqrt{1 + \langle \nu_1' \nu_2' \rangle}} \; ,$$

and hence, with $\langle \nu \mu | V | \nu' \mu' \rangle = \langle \mu \nu | V | \mu' \nu' \rangle$, the original equation.

The expectation value of the symmetry-independent first terms $\langle \nu_1 \nu_2 | V | \nu_1 \nu_2 \rangle$ is called the *direct term* while that of the symmetry-dependent term $\langle \nu_1 \nu_2 | V | \nu_2 \nu_1 \rangle$ is known as the *exchange term*. The expansion of one- and two-body operators with respect to products of creation and annihilation operators turns out to be useful for all N-particle states, as we shall now show.

5.3.8 Representation of One-Particle Operators

One-particle operators such as the kinetic energy and the one-particle potential act on the degrees of freedom of only one particle. Clearly,

$$T = \sum_{\nu \nu'} \langle \nu | T | \nu' \rangle \; \Psi_\nu^\dagger \Psi_{\nu'} \; ,$$

where $\Psi_\nu^\dagger \Psi_{\nu'}$ is the *one-particle density operator*. In the occupation-number representation, it has the matrix elements

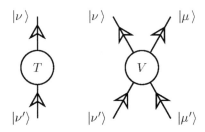

Fig. 5.11 The two Feynman diagrams are to be read from bottom to top, i.e., from initial to final state(s). One-particle operators (T) act on one particle, two-body operators (V) on two. Uninvolved partners do not change their state and would appear here as simple straight arrows. Such diagrams are useful for compositions of operations, similar to those in Figs. 5.7 (right) and 5.8

$$_{\text{s,a}}\langle \ldots n_i n_j \ldots | \Psi_i^\dagger \Psi_j | \ldots n_i' n_j' \ldots \rangle_{\text{s,a}}$$
$$= (\pm)^{\Sigma(i) - \Sigma(j)} \sqrt{n_i n_j'} \; _{\text{s,a}}\langle \ldots n_i - 1 n_j \ldots | \ldots n_i' n_j' - 1 \ldots \rangle_{\text{s,a}} \, .$$

We have in particular,

$$_{\text{s,a}}\langle \nu_1 \nu_2 \ldots \nu_N | T | \nu_1 \nu_2 \ldots \nu_N \rangle_{\text{s,a}} = \sum_{n=1}^{N} \langle \nu_n | T | \nu_n \rangle \, ,$$
$$_{\text{s,a}}\langle \nu_1 \nu_2 \ldots \nu_N | T | \nu_1' \nu_2 \ldots \nu_N \rangle_{\text{s,a}} = \langle \nu_1 | T | \nu_1' \rangle \sqrt{n_1 n_1'} \, , \quad \text{for } \nu_1 \neq \nu_1' \, ,$$
$$_{\text{s,a}}\langle \nu_1 \nu_2 \ldots \nu_N | T | \nu_1' \nu_2' \ldots \nu_N \rangle_{\text{s,a}} = 0 \, , \quad \text{for } \nu_1 \text{ and } \nu_2 \notin \{\nu_1' \nu_2'\} \, .$$

A one-particle operator can change the quantum numbers of at most one particle. Therefore, its matrix elements do not depend on the symmetry of the many-body state. Its expectation value is the sum of the expectation values of all occupied one-particle states (see Fig. 5.11 left).

5.3.9   Representation of Two-Body Operators

Two-body operators like the interaction V between two particles can alter the quantum numbers of two particles. Here we have

$$V = \tfrac{1}{2} \sum_{\nu\mu\nu'\mu'} \langle \nu\mu | V | \nu'\mu' \rangle \, \Psi_\mu^\dagger \Psi_\nu^\dagger \Psi_{\nu'} \Psi_{\mu'} \, .$$

The expression

$$V = \sum_{\nu \leq \mu, \nu' \leq \mu'} \frac{1}{\sqrt{1 + \langle \nu|\mu\rangle}\sqrt{1 + \langle \nu'|\mu'\rangle}} \, _{\text{s,a}}\langle \nu\mu | V | \nu'\mu' \rangle_{\text{s,a}} \, \Psi_\mu^\dagger \Psi_\nu^\dagger \Psi_{\nu'} \Psi_{\mu'}$$

is equivalent, where, as in Sect. 5.3.7,

$$_{s,a}\langle \nu\mu|V|\nu'\mu'\rangle_{s,a} = \frac{\langle \nu\mu|V|\nu'\mu'\rangle \pm \langle \nu\mu|V|\mu'\nu'\rangle}{\sqrt{1+\langle \nu|\mu\rangle}\sqrt{1+\langle \nu'|\mu'\rangle}} \ .$$

With the commutation behavior of the annihilation operators, we obtain

$$(\langle \nu\mu|V|\nu'\mu'\rangle \pm \langle \nu\mu|V|\mu'\nu'\rangle)\,\Psi_\mu^\dagger \Psi_\nu^\dagger \Psi_{\nu'}\Psi_{\mu'}$$
$$= (\langle \nu\mu|V|\mu'\nu'\rangle \pm \langle \nu\mu|V|\nu'\mu'\rangle)\,\Psi_\mu^\dagger \Psi_\nu^\dagger \Psi_{\mu'}\Psi_{\nu'} \ ,$$

and in the sum $\nu' \le \mu'$, we may swap these two indices (with $\mu' \le \nu'$) without any consequences. Instead of the claimed equation, we then have

$$\begin{aligned}
V &= \tfrac{1}{2}\sum_{\nu\le\mu,\nu'\mu'} \frac{\langle \nu\mu|V|\nu'\mu'\rangle \pm \langle \nu\mu|V|\mu'\nu'\rangle}{1+\langle \nu|\mu\rangle}\,\Psi_\mu^\dagger \Psi_\nu^\dagger \Psi_{\nu'}\Psi_{\mu'} \\
&= \tfrac{1}{4}\sum_{\nu\mu,\nu'\mu'}(\langle \nu\mu|V|\nu'\mu'\rangle \pm \langle \nu\mu|V|\mu'\nu'\rangle)\,\Psi_\mu^\dagger \Psi_\nu^\dagger \Psi_{\nu'}\Psi_{\mu'} \\
&= \tfrac{1}{4}\sum_{\nu\mu,\nu'\mu'}\langle \nu\mu|V|\nu'\mu'\rangle\,\Psi_\mu^\dagger \Psi_\nu^\dagger (\Psi_{\nu'}\Psi_{\mu'} \pm \Psi_{\mu'}\Psi_{\nu'}) \ ,
\end{aligned}$$

where the summation indices have been renamed at the end. As above, the upper sign holds for bosons, the lower one for fermions. Therefore, $\pm\Psi_{\mu'}\Psi_{\nu'} = \Psi_{\nu'}\Psi_{\mu'}$. Thus the two expressions are indeed equal.

The expectation value of a two-body operator consists of the direct and exchange terms. It depends on the symmetry of the many-body state and, according to Sect. 5.3.6, this yields

$$\begin{aligned}
_{s,a}\langle n_1 \dots | \Psi_\mu^\dagger \Psi_\nu^\dagger \Psi_{\nu'}\Psi_{\mu'}|n_1 \dots\rangle_{s,a} \\
= \sum_{i\ne j} n_i n_j \,(\langle \nu_i|\nu\rangle\langle \nu_j|\mu\rangle \pm \langle \nu_j|\nu\rangle\langle \nu_i|\mu\rangle)\,\langle \nu'|\nu_i\rangle\langle \mu'|\nu_j\rangle \\
+ \sum_i n_i(n_i-1)\,\langle \nu_i|\nu\rangle\langle \nu_i|\mu\rangle\langle \nu'|\nu_i\rangle\langle \mu'|\nu_i\rangle \ .
\end{aligned}$$

Then quite generally,

$$\begin{aligned}
_{s,a}\langle n_1 \dots |V|n_1 \dots\rangle_{s,a} = \tfrac{1}{2}\sum_{i\ne j} n_i n_j\,(\langle \nu_i\nu_j|V|\nu_i\nu_j\rangle \pm \langle \nu_j\nu_i|V|\nu_i\nu_j\rangle) \\
+ \tfrac{1}{2}\sum_i n_i(n_i-1)\,\langle \nu_i\nu_i|V|\nu_i\nu_i\rangle \ .
\end{aligned}$$

For fermions, the second sum does not contribute—none of the states is doubly occupied. Therefore, the result may also be reformulated as

$$_{s,a}\langle \nu_1 \ldots \nu_N | V | \nu_1 \ldots \nu_N\rangle_{s,a} = \frac{1}{2} \sum_{\substack{n \neq m}}^{N} \frac{\langle \nu_n \nu_m | V | \nu_n \nu_m \rangle \pm \langle \nu_n \nu_m | V | \nu_m \nu_n \rangle}{1 + \langle \nu_n | \nu_m \rangle}$$

$$= \frac{1}{2} \sum_{\substack{n \neq m}}^{N} {}_{s,a}\langle \nu_n \nu_m | V | \nu_n \nu_m \rangle_{s,a} \;,$$

where we also use p. 444. For the expectation value, it thus follows that

$$_{s,a}\langle \nu_1 \ldots \nu_N | V | \nu_1 \ldots \nu_N\rangle_{s,a} = \sum_{n<m}^{N} {}_{s,a}\langle \nu_n \nu_m | V | \nu_n \nu_m \rangle_{s,a} \;.$$

Apart from this, we must consider the off-diagonal matrix elements of V. A two-body operator can alter the quantum numbers of at most two particles (see Fig. 5.11 right). For $\nu_1 \neq \nu_1'$, it follows that (compare with the result for one-particle operators in the last section)

$$_{s,a}\langle \nu_1 \nu_2 \ldots \nu_N | V | \nu_1{}' \nu_2 \ldots \nu_N\rangle_{s,a}$$

$$= \sum_{n=2}^{N} {}_{s,a}\langle \nu_1 \nu_n | V | \nu_1{}' \nu_n \rangle_{s,a} \sqrt{\frac{n_1}{1 + \langle \nu_1 | \nu_n \rangle} \; \frac{n_1{}'}{1 + \langle \nu_n | \nu_1{}' \rangle}} \;,$$

and for ν_1 and $\nu_2 \notin \{\nu_1{}' \nu_2{}'\}$,

$$_{s,a}\langle \nu_1 \nu_2 \nu_3 \ldots \nu_N | V | \nu_1{}' \nu_2{}' \nu_3 \ldots \nu_N\rangle_{s,a}$$

$$= {}_{s,a}\langle \nu_1 \nu_2 | V | \nu_1{}' \nu_2{}' \rangle_{s,a} \sqrt{\frac{n_1 \,(n_2 - \langle \nu_1 | \nu_2 \rangle)}{1 + \langle \nu_1 | \nu_2 \rangle} \; \frac{n_1{}' \,(n_2{}' - \langle \nu_1{}' | \nu_2{}' \rangle)}{1 + \langle \nu_1{}' | \nu_2{}' \rangle}} \;.$$

As before the particle numbers n_i refer to the bra-vector and the particle numbers $n_i{}'$ to the ket-vector. If the two vectors differ in more than two particles, the matrix element is zero.

5.3.10   Time Dependence

So far all our considerations of many-body systems are valid for a fixed time. Now we ask how the creation and annihilation operators behave under exchange at different times. We assume that the time-dependence is determined by a Hermitian Hamilton operator consisting only of one- and two-body operators, which neither depends on time explicitly nor changes the particle number:

$$H = \sum_{\nu\nu'} \langle \nu | T | \nu' \rangle \, \Psi_\nu^\dagger \Psi_{\nu'} + \frac{1}{2} \sum_{\nu\mu\,\nu'\mu'} \langle \nu\mu | V | \nu'\mu' \rangle \, \Psi_\mu^\dagger \Psi_\nu^\dagger \Psi_{\nu'} \Psi_{\mu'} \;.$$

Here, in addition to the kinetic energy, T may also include other one-particle operators. We shall return to this shortly.

The Schrödinger picture is less suitable for field operators than the Heisenberg picture (and the Dirac picture), because in field theory, we also trace the states back to operators, and acting on the vacuum, which does not depend on time. Therefore, we now transfer the relation $A_H(t) = U^\dagger(t) A_H(0) U(t)$ with $U(t) = \exp(-iHt/\hbar)$ from Sect. 4.4.2 to field theory (without reference to the Heisenberg picture):

$$\Psi(t) = U^\dagger(t)\,\Psi(0)\,U(t) \qquad \Longleftrightarrow \qquad \Psi^\dagger(t) = U^\dagger(t)\,\Psi^\dagger(0)\,U(t)\,.$$

We thus take over the equations valid for observables and apply them to field operators. Using $\dot{U} = -iHU/\hbar = -iUH/\hbar$, this yields

$$\frac{d\Psi}{dt} = \frac{i}{\hbar}\,[H, \Psi] \qquad \Longleftrightarrow \qquad \frac{d\Psi^\dagger}{dt} = \frac{i}{\hbar}\,[H, \Psi^\dagger]\,.$$

With $[AB, C] = A[B, C]_\mp - [C, A]_\mp B$, we obtain $[\Psi_\nu^\dagger \Psi_{\nu'}, \Psi_\kappa] = -\langle \kappa|\nu\rangle\,\Psi_{\nu'}$, and also

$$[\Psi_\mu^\dagger \Psi_\nu^\dagger \Psi_{\nu'} \Psi_{\mu'}, \Psi_\kappa] = [\Psi_\mu^\dagger \Psi_\nu^\dagger, \Psi_\kappa]\,\Psi_{\nu'}\Psi_{\mu'}\,,$$

with

$$[\Psi_\mu^\dagger \Psi_\nu^\dagger, \Psi_\kappa] = \mp\langle \kappa|\nu\rangle\,\Psi_\mu^\dagger - \langle \kappa|\mu\rangle\,\Psi_\nu^\dagger\,.$$

For bosons and also for fermions, this leads to

$$[\Psi_\nu^\dagger \Psi_{\nu'}, \Psi_\kappa] = -\langle \kappa|\nu\rangle\,\Psi_{\nu'}\,,$$
$$[\Psi_\mu^\dagger \Psi_\nu^\dagger \Psi_{\nu'}\Psi_{\mu'}, \Psi_\kappa] = -\langle \kappa|\mu\rangle\,\Psi_\nu^\dagger \Psi_{\nu'}\Psi_{\mu'} - \langle \kappa|\nu\rangle\,\Psi_\mu^\dagger \Psi_{\mu'}\Psi_{\nu'}\,.$$

Therefore, for both sorts of particles, the Heisenberg equation with the chosen Hamilton operator can be reformulated as

$$i\hbar\,\frac{d\Psi_\kappa}{dt} = \sum_\nu \langle \kappa|T|\nu\rangle\,\Psi_\nu + \sum_{\nu\nu'\mu'} \langle \nu\kappa|V|\nu'\mu'\rangle\,\Psi_\nu^\dagger \Psi_{\nu'}\Psi_{\mu'}\,,$$

if we use $\langle \nu\mu|V|\nu'\mu'\rangle = \langle \mu\nu|V|\mu'\nu'\rangle$.

Later on we shall introduce the *average two-body interaction* $\overline{V}$, a one-particle operator which can be combined with T to give $H_0 = T + \overline{V}$. Subtracting from V, this yields the *residual interaction* $V - \overline{V}$ as a two-body operator, which is often small and unimportant. If we neglect it and take the one-particle basis which diagonalizes H_0, we obtain

$$i\hbar\,\frac{d\Psi_\nu}{dt} \approx \langle \nu|H_0|\nu\rangle\,\Psi_\nu\,, \quad \text{e.g.,} \quad i\hbar\,\frac{\partial \Psi(t, \mathbf{r})}{\partial t} \approx \langle \mathbf{r}|H_0|\mathbf{r}\rangle\,\Psi(t, \mathbf{r})\,.$$

This is similar to the usual Schrödinger equation. However, Ψ_ν is not a state, but an operator, and a matrix element of the Hamilton operator H_0 is to be taken. We discuss this further in the next section.

Before that we shall set $\langle \nu | H_0 | \nu \rangle = \hbar \omega_\nu$, whence

$$\Psi_\nu(t) = \Psi_\nu(0) \, \exp(-\mathrm{i}\omega_\nu t) \quad \text{and} \quad \Psi_\nu^\dagger(t) = \Psi_\nu^\dagger(0) \, \exp(+\mathrm{i}\omega_\nu t) \; .$$

In the last paragraph the two-body interaction was neglected. There the equations are valid (without neglecting this) in the Dirac picture, which includes the time dependence of the states due to V. So far H_0 should be sufficiently simple for mathematical treatment, but now we distinguish between H_0 and V through physical properties, namely, whether one- or two-body operators are involved.

5.3.11   Wave–Particle Dualism

In this chapter we have always started from (several) particles, but we would also have arrived at creation and annihilation operators if we had quantized the wave picture, i.e., if we had taken each field strength as a Hermitian operator (e.g., the electromagnetic field, as will be shown in Sect. 5.5). In fact, such a field quantization would have taken us beyond the usual scope of a course on quantum mechanics, but otherwise would also have had many advantages. Note that the term *second quantization* instead of field quantization is misleading: we quantize either once in the field picture or once in the particle picture, with several particles.

So far we have investigated the laws governing the behavior of particles and looked at them as representatives of a class of identical particles. For single particles, there are only statements about probabilities. Therefore, we always take a very large number N of equal particles and use them to repeat the same experiment. The more often the same particle attribute appears, the higher its probability. But this probability now shows interference effects and therefore requires a wave theory. We assume for these considerations that the particles do not act on each other, which we also had to do when deriving the generalized Schrödinger equation in the last section.

For $N \gg 1$, it is of no importance whether N is a natural number: even a (small) uncertainty in the particle number might occur. It is just then that a sharp probability statement (appropriate for the wave picture) holds for single particles! On the other hand, if an uncertainty in the wave quantities (phase) is not important, then sharp statements in the particle picture are possible. This has already been pointed out for the uncertainty relation between particle number and phase (Sect. 4.2.9).

Considering the relative frequency, it is easily overlooked how the particle picture is contained in such a seemingly unimportant constraint: N had to be a natural number—other values were meaningless. This *granularity* is foreign to the wave picture. Field intensities and wave functions are appropriate for there, but classically these distributions can be arbitrarily normalized. Clearly, the wave picture then has to be modified in such a way that the arbitrary values for N become restricted to natural numbers by quantum conditions—and the observables of the wave picture (field strengths and intensities) have to become operators. We have become acquainted with the commutation laws for the field operators in this chapter.

As long as wave functions are taken as classical field quantities, not as probability amplitudes normalized to 1, the Schrödinger equation is not an equation of quantum mechanics, but of classical physics.

5.3.12 Summary: Many-Body Systems

In the quantum mechanics of many-particle problems, bosons and fermions behave differently: bosons form symmetric states and fermions antisymmetric states. Such states with special exchange symmetry are easily treated by introducing creation and annihilation operators $\Psi_\nu^\dagger$ and Ψ_ν which satisfy the commutation laws (upper sign for bosons, lower for fermions):

$$\Psi_\nu \Psi_{\nu'}^\dagger \mp \Psi_{\nu'}^\dagger \Psi_\nu = \langle \nu | \nu' \rangle \qquad \text{and} \qquad \Psi_\nu \Psi_{\nu'} \mp \Psi_{\nu'} \Psi_\nu = 0 \, .$$

Both are called *field operators*. They also arise when quantizing the wave picture. Here it is best to work in the Heisenberg or Dirac picture, where $\Psi_\nu(t) = \Psi_\nu(0) \exp(-i\omega_\nu t)$.

It is also important to distinguish between one- and two-body operators:

$$T = \sum_{\nu\nu'} \langle \nu | T | \nu' \rangle \, \Psi_\nu^\dagger \Psi_{\nu'} \quad \text{and} \quad V = \tfrac{1}{2} \sum_{\nu\mu\nu'\mu'} \langle \nu\mu | V | \nu'\mu' \rangle \, \Psi_\mu^\dagger \Psi_\nu^\dagger \Psi_{\nu'} \Psi_{\mu'} \, .$$

Here T can also contain "average one-particle potentials".

5.4   Fermions

5.4.1   Fermi Gas in the Ground State

As a first application, we shall evaluate the one- and two-body densities

$$\rho(\mathbf{r}) \quad = {}_a\langle \nu_1 \ldots \nu_N | \quad \Psi^\dagger(\mathbf{r}) \, \Psi(\mathbf{r}) \quad\quad\quad |\nu_1 \ldots \nu_N\rangle_a \, ,$$
$$\rho(\mathbf{r}, \mathbf{r}') = {}_a\langle \nu_1 \ldots \nu_N | \quad \Psi^\dagger(\mathbf{r}') \, \Psi^\dagger(\mathbf{r}) \, \Psi(\mathbf{r}) \, \Psi(\mathbf{r}') \, |\nu_1 \ldots \nu_N\rangle_a \, ,$$

of a Fermi gas in the ground state, i.e., the probability density for a particle at the position $\mathbf{r}$ or for two particles at the positions $\mathbf{r}$ and $\mathbf{r}'$. Actually, in a Fermi gas, the individual fermions do not interact with each other, but the antisymmetry nevertheless correlates them. Since only one fermion may be in any one-particle state, in the ground state of the Fermi gas, the particles are distributed over the various states with as low a total energy as possible.

The following calculation would be more complicated if we enclosed the fermions in a cube of volume a^3, as in Sect. 4.5.3. It is easier with *periodic boundary conditions*, i.e., with $\psi(x, y, z) = \psi(x+a, y, z) = \psi(x, y+a, z) = \psi(x, y, z+a)$. They lead to the eigenfunctions

$$\psi_\nu(\mathbf{r}) = \frac{\exp(i\mathbf{k}_\nu \cdot \mathbf{r})}{\sqrt{V}} \; ,$$

if we leave out the spin functions $\chi_\nu(\mathbf{s})$. The wave vector $\mathbf{k}_\nu$ and the energy E_ν are then determined by the constraints

$$\mathbf{k}_\nu = \frac{2\pi}{a}\,\mathbf{n}_\nu \quad \text{and} \quad E_\nu = \frac{(\hbar k_\nu)^2}{2m} \; ,$$

where each Cartesian component of $\mathbf{n}_\nu$ has to be an integer $(0, \pm1, \ldots)$. With the cube and impenetrable walls (see p. 355), each component of $\mathbf{n}_\nu$ takes only the values $1, 2, 3, \ldots$. However, then for $\mathbf{k}_\nu \propto \mathbf{n}_\nu$, there is a factor π/a instead of $2\pi/a$. Moreover, for periodic boundary conditions, all states apart from those with a vanishing $\mathbf{n}_\nu$ component are more strongly degenerate by the factor $2^3 = 8$, but lie further apart from each other by the same factor than for the cube. The density of states is the same in both cases. But the boundary conditions are not important for our problem. In particular, as in Sect. 4.5.3, for the number of states with $E \leq E_F$, a factor of 2 accounts for the two spin states of spin-1/2 particles and we have

$$N \approx 2\,\frac{V}{6\pi^2}\,k_F^3 \; .$$

With the above-mentioned wave functions, we shall now calculate the one- and two-particle densities $\rho(\mathbf{r})$ and $\rho(\mathbf{r}, \mathbf{r}')$. Generally, we have

$$ {}_a\langle \nu_1 \ldots \nu_N | \; \Psi^\dagger(\mathbf{r})\,\Psi(\mathbf{r})\;|\nu_1 \ldots \nu_N\rangle_a = \sum_{n=1}^N |\psi_n(\mathbf{r})|^2 \; , $$

and using $|\chi_n(\mathbf{s})|^2 = 1 = |\chi_m(\mathbf{s})|^2$,

$$ {}_a\langle \nu_1 \ldots \nu_N | \; \Psi^\dagger(\mathbf{r}')\,\Psi^\dagger(\mathbf{r})\,\Psi(\mathbf{r})\,\Psi(\mathbf{r}')\;|\nu_1 \ldots \nu_N\rangle_a $$

$$ = \tfrac{1}{2} \sum_{n,m=1}^N \Big\{ |\psi_n(\mathbf{r})|^2\,|\psi_m(\mathbf{r}')|^2 - \psi_n{}^*(\mathbf{r})\,\psi_m{}^*(\mathbf{r}')\,\psi_n(\mathbf{r}')\,\psi_m(\mathbf{r})|\langle \mathbf{s}_n|\mathbf{s}_m\rangle|^2 $$

$$ + |\psi_m(\mathbf{r})|^2\,|\psi_n(\mathbf{r}')|^2 - \psi_m{}^*(\mathbf{r})\,\psi_n{}^*(\mathbf{r}')\,\psi_m(\mathbf{r}')\,\psi_n(\mathbf{r})|\langle \mathbf{s}_n|\mathbf{s}_m\rangle|^2 \Big\} \; . $$

Here we see immediately the advantage of periodic boundary conditions: the one-particle density is then constant and given by

$$\rho(\mathbf{r}) = \frac{N}{V} = \frac{k_F^3}{3\pi^2} = \rho_0 \,,$$

and the two-particle density simplifies to

$$\rho(\mathbf{r}, \mathbf{r}') = \frac{1}{V^2} \sum_{n,m=1}^{N} \left\{ 1 - \tfrac{1}{2} \exp\{i\,(\mathbf{k}_n - \mathbf{k}_m) \cdot (\mathbf{r}' - \mathbf{r})\} \,|\langle s_n|s_m\rangle|^2 \right.$$
$$\left. - \tfrac{1}{2} \exp\{i\,(\mathbf{k}_n - \mathbf{k}_m) \cdot (\mathbf{r} - \mathbf{r}')\} \,|\langle s_n|s_m\rangle|^2 \right\}.$$

The double sum can be approximated by a double integral, integrating over $d^3 n = 2V/(2\pi)^3\,d^3k$ and including a factor of 2 for spin-1/2 particles. For the latter, we have on average $|\langle s_n|s_m\rangle|^2 = 1/2$. Using this, the double integral factorizes:

$$\rho(\mathbf{r}, \mathbf{r}') \approx \frac{N^2}{V^2} - \frac{1}{2V^2} \left(\frac{2V}{(2\pi)^3} \int d^3k \, \exp\{i\mathbf{k} \cdot (\mathbf{r} - \mathbf{r}')\} \right)^2 .$$

Here we have to integrate over all directions of $\mathbf{k}$ and the modulus from 0 to $k_F = (3\pi^2\rho_0)^{1/3}$. According to p. 409, we obtain $\int d\Omega_k \exp(i\mathbf{k} \cdot \mathbf{a}) = 4\pi \sin ka/(ka)$, and consequently, using $F_1(x) = x^{-1} \sin x - \cos x$ from p. 400,

$$\rho(\mathbf{r}, \mathbf{r}') = \rho_0^2 \left\{ 1 - \tfrac{1}{2} \left(\frac{3F_1(x)}{x^2} \right)^2 \right\} \qquad \text{with } x = k_F|\mathbf{r} - \mathbf{r}'| .$$

Of course, the factor of $1/2$ comes from the two spin states.

The antisymmetry thus correlates fermions of equal spin. They tend to avoid each other, each fermion being surrounded by an *exchange hole* (see Fig. 5.12). This anti-correlation shows up only for short distances.

Therefore, the boundary conditions are not important for this consideration, and the function $(3F_1/x^2)^2$ may be approximated by a Gauss function.

Fig. 5.12 Exchange hole around a fermion. Two-body density as a function of $x = k_F|\mathbf{r} - \mathbf{r}'|$ (*continuous red*) and the approximation $1 - \tfrac{1}{2}\exp(-x^2/5)$ (*dashed blue*)

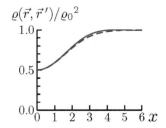

5.4.2 Hartree–Fock Equations

Each N-fermion state can be expanded in the basis $\{|\nu_1 \ldots \nu_N\rangle_a\}$. We shall use this freedom in the choice of one-particle states $\{|\nu\rangle\}$ to diagonalize the Hamilton operator

$$H = \sum_{\nu\nu'} \langle \nu|H_0|\nu'\rangle \, \Psi_\nu^\dagger \, \Psi_{\nu'} + \tfrac{1}{2} \sum_{\nu\mu\nu'\mu'} \langle \nu\mu|V|\nu'\mu'\rangle \, \Psi_\mu^\dagger \Psi_\nu^\dagger \Psi_{\nu'} \Psi_{\mu'}$$

as well as possible with a single state $|\nu_1 \ldots \nu_N\rangle_a$. Here, in addition to the kinetic energy, H_0 also contains the potential energy which originates from external forces, while V describes the coupling of the single fermions among each other.

The diagonal elements of the Hamilton operator $H = H_0 + V$ just mentioned, viz.,

$$_a\langle \nu_1 \ldots \nu_N|H|\nu_1 \ldots \nu_N\rangle_a = \sum_{n=1}^{N} \langle \nu_n|H_0|\nu_n\rangle + \sum_{n<m}^{N} {}_a\langle \nu_n\nu_m|V|\nu_n\nu_m\rangle_a \, ,$$

supply the energy eigenvalues in zeroth order perturbation theory. Of the remaining matrix elements $_a\langle \nu_1 \ldots \nu_N|H|\nu_1' \ldots \nu_N'\rangle_a$, all of those whose bra- and ket-states differ in more than two particles will in fact vanish, but generally neither

$$_a\langle \nu_1\nu_2\nu_3 \ldots \nu_N|H|\,\nu_1'\nu_2'\nu_3 \ldots \nu_N\rangle_a = {}_a\langle \nu_1\nu_2|V|\nu_1'\nu_2'\rangle_a \, , \quad \text{with } \{\nu_1, \nu_2\} \notin \{\nu_1', \nu_2'\},$$

nor the matrix elements which are not diagonal with respect to just one particle will vanish. For example,

$$_a\langle \nu_1\nu_2 \ldots \nu_N|H|\nu_1'\nu_2 \ldots \nu_N\rangle_a = \langle \nu_1|H_0|\nu_1'\rangle + \sum_{n=1}^{N} {}_a\langle \nu_1\nu_n|V|\nu_1'\nu_n\rangle_a \, ,$$

if $\nu_1 \neq \nu_1'$. At least this second kind of non-diagonal element vanishes if we determine the basis $\{|\nu\rangle\}$ (one-particle states) from the *Hartree–Fock equations*:

$$\langle \nu|H_0|\nu'\rangle + \sum_{n=1}^{N} {}_a\langle \nu\nu_n|V|\nu'\nu_n\rangle_a = e_\nu \, \langle \nu|\nu'\rangle \, .$$

We shall thus derive the one-particle states $|\nu\rangle$ and the one-particle energies e_ν from the Hartree–Fock equations and take approximately $|\nu_1 \ldots \nu_N\rangle_a$ for the N-fermion states. With this in fact generally the non-diagonal elements of H will not all vanish yet, but a better approximation can be obtained using just a superposition of several states—we shall come back to this in the next section.

In the Hartree–Fock equations, the test particle is coupled to the remaining fermions. The sum of the one-particle energies e_ν of the occupied states counts this coupling twice, so this is not equal to the ground-state energy E_0 of the N fermions:

$$E_0 = {}_a\langle v_1 \ldots v_N | H | v_1 \ldots v_N \rangle_a = \sum_{n=1}^{N} \langle v_n | H_0 | v_n \rangle + \frac{1}{2} \sum_{n,m=1}^{N} {}_a\langle v_n v_m | V | v_n v_m \rangle_a$$

$$= \sum_{n=1}^{N} e_n - \frac{1}{2} \sum_{n,m=1}^{N} {}_a\langle v_n v_m | V | v_n v_m \rangle_a \ .$$

But we have *Koopman's theorem*: the last particle has the energy

$$e_N = E_0(N) - E_0(N-1) \ ,$$

since

$$e_N = \langle v_N | H_0 | v_N \rangle + \sum_{n=1}^{N-1} {}_a\langle v_N v_n | V | v_N v_n \rangle_a \ .$$

We now consider the Hartree–Fock equations in the real-space representation, where we restrict ourselves to local Wigner forces. Then the following abbreviations are useful:

$$V_H(\mathbf{r}) \equiv \sum_{n=1}^{N} \int d^3 r' \ \psi_n{}^*(\mathbf{r}') \ V(\mathbf{r}, \mathbf{r}') \ \psi_n(\mathbf{r}') \qquad \text{(Hartree term)},$$

$$V_F(\mathbf{r}, \mathbf{r}') \equiv \sum_{n=1}^{N'} \psi_n{}^*(\mathbf{r}') \ V(\mathbf{r}, \mathbf{r}') \ \psi_n(\mathbf{r}) \qquad \text{(Fock term)}.$$

Note that only the states $|v_n\rangle$ with the same spin orientation as the unknown solution (indicated by N') contribute to the Fock term. For spin-independent operators H_0 and V, we have

$$\big(H_0 + V_H(\mathbf{r})\big) \psi_v(\mathbf{r}) - \int d^3 r' \ V_F(\mathbf{r}, \mathbf{r}') \psi_v(\mathbf{r}') = \psi_v(\mathbf{r}) \ e_v \ .$$

Here the direct term (Hartree term) and the exchange term (Fock term) contain the wave functions to be determined. The Hartree–Fock equations can be solved only iteratively. We first use a suitable ansatz for V_H and V_F, solve the eigenvalue equation, and then use the eigenfunctions found in this way to get a better approximation for V_H and V_F, and so on. This method has to be repeated until the solutions of the Hartree–Fock equations do not change within given limits (until they are *self-consistent*).

The exchange term is non-local and impedes the calculations. If we neglect it, we have the simple *Hartree equations*, but their solutions are not orthogonal to each other, because they belong to a wrong Hamilton operator [7]. The exchange hole is less effective for repulsive (Coulomb) forces than for attractive (nuclear) forces. The Hartree equations are therefore essentially more appropriate for atomic physics than for nucleus physics.

5.4.3 Rest Interaction and Pair Force

The Hartree–Fock method thus delivers the best one-particle states—it diagonalizes the Hamilton operator as well as possible with a single antisymmetric product of such states. However, there are still off-diagonal elements originating from the two-body coupling. In fact, only

$$H_{\mathrm{HF}} = \sum_{\nu\nu'} \left(\langle \nu| \, H_0 \, |\nu'\rangle + \sum_{n=1}^{N} {}_{\mathrm{a}}\langle \nu\nu_n| \, V \, |\nu'\nu_n\rangle_{\mathrm{a}} \right) \Psi_\nu^\dagger \, \Psi_{\nu'}$$

becomes diagonalized—there remains a *residual interaction*

$$V_{\mathrm{R}} = H - H_{\mathrm{HF}} \ .$$

In order to include this term, we have to superpose several product states whose components differ from each other by the quantum numbers of at last two particles. This *configuration mixture* delivers a further correlation, in addition to the symmetry condition, which is related to the exchange hole.

We thus ask which parts of the coupling V are already well approximated by a one-particle operator and which remain as the residual interaction. Clearly, the parts with longer range change only weakly with the distance from the remaining partners. These can be well described by an average one-particle potential. Consequently, the residual interaction describes the parts of short range.

In order to study those effects, we could investigate the limit of a delta force $\propto \delta(\mathbf{r} - \mathbf{r}')$. But since the matrix elements

$$\langle \nu\mu| \, \delta \, |\nu'\mu'\rangle = \int \mathrm{d}^3 r \ \psi_\nu^*(\mathbf{r}) \, \psi_\mu^*(\mathbf{r}) \, \psi_{\nu'}(\mathbf{r}) \, \psi_{\mu'}(\mathbf{r})$$

very often differ from zero, the corresponding problem is still too involved. We take the so-called *pair force*, which, to exaggerate somewhat, has even shorter range: it acts only between fermions in mutually time-reversed states (and which are therefore equally probable everywhere). For a Hamilton operator with time-reversal symmetry, they have the same energy according to Kramers' theorem (see p. 314).

Thus as residual interaction we take

$$V_{\mathrm{pair}} = \sum_{\nu\nu'} {}_{\mathrm{a}}\langle \nu\bar{\nu}| \, V \, |\nu'\bar{\nu}'\rangle_{\mathrm{a}} \, \Psi_{\bar{\nu}}^\dagger \, \Psi_{\nu}^\dagger \, \Psi_{\nu'} \, \Psi_{\bar{\nu}'} \ ,$$

without summation over $\bar{\nu}$ and $\bar{\nu}'$. The two states $|\nu\rangle$ and $|\bar{\nu}\rangle$ have opposite momentum and angular momentum. Therefore, it is often assumed that ν and $\bar{\nu}$ differ only in the sign ($\nu = -\bar{\nu} > 0$) and we then require $\nu, \nu' > 0$ for the sum. Since the matrix elements of the delta function are always positive, we shall also assume that the matrix elements of the pair force all have the same sign, which, for an attractive pair force, will be negative.

For such a pair force, close to the Hartree–Fock ground state, it is particularly convenient for the energy if the fermion level is pairwise occupied or empty. If $|\nu\rangle$ is occupied, then so is $|\bar{\nu}\rangle$. If the ground state according to the Hartree–Fock method (with even particle number) is of the form $|\nu_1 \bar{\nu}_1 \dots \nu_{N/2} \bar{\nu}_{N/2}\rangle_a$, it now also contains (superposed) states which differ by pairs $\nu\bar{\nu}$. These have neither momentum nor angular momentum. In excited states, these pairs can also *break up*.

5.4.4 Quasi-Particles in the BCS Formalism

Despite all the simplifications which result from the pair force (compared to the actually expected residual interaction), the eigenvalue problem is still too difficult. Bardeen, Cooper, and Schrieffer proposed an approximating ansatz for the ground state which allows the pair force to be diagonalized rather easily:

$$|\text{BCS}\rangle = \prod_{\nu>0}(u_\nu + v_\nu \, \Psi_\nu^\dagger \Psi_{\bar{\nu}}^\dagger)|0\rangle \ ,$$

where instead of u_ν and v_ν we could also take $\cos\varphi_\nu$ and $\sin\varphi_\nu$ (see p. 310):

$$u_\nu{}^2 + v_\nu{}^2 = 1 \ , \quad u_\nu = u_\nu{}^* \geq 0 \ , \quad v_\nu = v_\nu{}^* \ .$$

The occupation probabilities of the states $|\nu\rangle$ and $|\bar{\nu}\rangle$ are thus equal and easy to remember. With probability u_ν^2, they are *unoccupied* (empty), and with probability v_ν^2, they are occupied (filled). However, the ansatz has the disadvantage that the particle number is not sharp. In fact, we require the *expectation value* to deliver the correct particle number n, i.e.,

$$\langle\text{BCS}| \, N \, |\text{BCS}\rangle = \sum_{\nu>0} 2v_\nu{}^2 = n \ ,$$

but the particle number is not sharp, as will be shown later:

$$(\Delta N)^2 = \langle\text{BCS}| \, N^2 \, |\text{BCS}\rangle - \langle\text{BCS}| \, N \, |\text{BCS}\rangle^2 = 4 \sum_{\nu>0} u_\nu{}^2 v_\nu{}^2 \ .$$

In fact, for most terms, we find either $u_\nu{}^2 = 0$ or $v_\nu{}^2 = 0$, and hence $(\Delta N)^2 \ll 4\sum_\nu v_\nu{}^2 = 2n$, but this uncertainty is nevertheless irritating for the smaller particle numbers—hence particularly in atomic and nuclear physics, but less in solid-state physics. Clearly, in this approximation, we cannot describe any properties varying quickly with the particle number, only the slowly varying ones.

The state $|\text{BCS}\rangle$ may be taken as a *quasi-vacuum*. It has neither momentum nor angular momentum, but energy. Moreover, its particle number is not zero. Acting on this quasi-vacuum are *quasi-particle operators* $\Phi_\nu^{(\dagger)}$, which again obey the Fermi exchange rule. Carrying out the *Bogoliubov transformation*,

$$\Phi_\nu \equiv u_\nu\,\Psi_\nu - v_\nu\,\Psi_{\bar\nu}^\dagger \quad\Longleftrightarrow\quad \Phi_\nu^\dagger = u_\nu\,\Psi_\nu^\dagger - v_\nu\,\Psi_{\bar\nu}\,,$$

the commutation rule for the operators Ψ implies

$$\Phi_\nu\,\Phi_{\nu'}^\dagger + \Phi_{\nu'}^\dagger\,\Phi_\nu = \langle\nu|\nu'\rangle \quad\text{and}\quad \Phi_\nu\,\Phi_{\nu'} + \Phi_{\nu'}\,\Phi_\nu = 0\,.$$

Whether a particle is annihilated in the state $|\nu\rangle$ or created in the state $|\bar\nu\rangle$ makes no difference to the momentum and angular momentum—only for the particle number and the energy. For this reason, the Bogoliubov transformation is not as peculiar as it may appear at first sight.

According to p. 314, for fermions, we have $|\bar{\bar\nu}\rangle = -|\nu\rangle$. This yields

$$\Phi_{\bar\nu} = u_\nu\,\Psi_{\bar\nu} + v_\nu\Psi_\nu^\dagger \quad\Longleftrightarrow\quad \Phi_{\bar\nu}^\dagger = u_\nu\,\Psi_{\bar\nu}^\dagger + v_\nu\Psi_\nu\,,$$

$$\Psi_\nu = u_\nu\,\Phi_\nu + v_\nu\,\Phi_{\bar\nu}^\dagger \quad\text{and}\quad \Psi_{\bar\nu} = u_\nu\,\Phi_{\bar\nu} - v_\nu\,\Phi_\nu^\dagger\,.$$

Now we may deduce that

$$\Phi_\nu\,|\text{BCS}\rangle = |o\rangle \quad\text{and}\quad \Phi_\nu{}^\dagger|\text{BCS}\rangle = \Psi_\nu^\dagger \prod_{\nu'(\neq\nu)>0} (u_{\nu'} + v_{\nu'}\,\Psi_{\nu'}^\dagger\,\Psi_{\bar{\nu}'}{}^\dagger)|0\rangle\,,$$

as well as $\Phi_\nu\,\Phi_{\nu'}^\dagger|\text{BCS}\rangle = |\text{BCS}\rangle\,\langle\nu|\nu'\rangle$. We can see that the particle number operator $\sum_{\nu>0}(\Psi_\nu^\dagger\,\Psi_\nu + \Psi_{\bar\nu}^\dagger\,\Psi_{\bar\nu})$ is generally no longer diagonal by considering

$$N = \sum_{\nu>0} 2v_\nu{}^2 + (u_\nu{}^2 - v_\nu{}^2)(\Phi_\nu^\dagger\,\Phi_\nu + \Phi_{\bar\nu}^\dagger\,\Phi_{\bar\nu}) + 2u_\nu v_\nu\,(\Phi_\nu^\dagger\,\Phi_{\bar\nu}^\dagger + \Phi_{\bar\nu}\,\Phi_\nu)\,.$$

With this result, we can also prove the above-mentioned expression for $(\Delta N)^2$.

5.4.5 Hartree–Fock–Bogoliubov Equations

Using the *Bogoliubov transformation*, we can go over from particle to quasi-particle operators and generalize the Hartree–Fock equations in such a way that, within the framework of the BCS ansatz, pair correlations are also included. Then the quasi-particle energies e_ν and the occupation probabilities $v_\nu{}^2 = 1 - u_\nu{}^2$ of the ground state are conserved.

As in the Hartree–Fock method, we also diagonalize the one-particle parts here, but now in the quasi-particle formalism. Since the particle number is no longer sharp, we want to obtain at least its mean value correctly, and therefore introduce the *chemical potential* μ as a Lagrangian parameter (see p. 560). The *Hartree–Fock–Bogoliubov equations* read

$$\langle \nu | H_0 + V - \mu N | \nu' \rangle = \langle \nu | \nu' \rangle \, e_\nu \, .$$

Here ν and ν' are either both positive or both negative, time-reversed states being orthogonal to each other in any case. The Hamilton operator should be Hermitian and invariant under time reversal. Then according to p. 314, we have $\langle \overline{\nu}' | H_0 | \overline{\nu} \rangle = \langle \overline{\nu} | H_0 | \overline{\nu}' \rangle^* = \langle \nu | H_0 | \nu' \rangle$, so

$$H_0 = \sum_{\nu\nu'>0} \langle \nu | H_0 | \nu' \rangle \, (\Psi_\nu^\dagger \Psi_{\nu'} + \Psi_{\overline{\nu}'}^\dagger \Psi_{\overline{\nu}}) + \langle \overline{\nu} | H_0 | \nu' \rangle \Psi_{\overline{\nu}}^\dagger \Psi_{\nu'} + \langle \nu | H_0 | \overline{\nu}' \rangle \Psi_\nu^\dagger \Psi_{\overline{\nu}'} \, .$$

In order to determine e_ν, we need only the part with the factors

$$\begin{aligned}
\Psi_\nu^\dagger \Psi_{\nu'} + \Psi_{\overline{\nu}'}^\dagger \Psi_{\overline{\nu}} = 2v_\nu^{\,2} \langle \nu | \nu' \rangle &+ \left(u_\nu u_{\nu'} - v_\nu v_{\nu'} \right) \left(\Phi_\nu^\dagger \Phi_{\nu'} + \Phi_{\overline{\nu}'}^\dagger \Phi_{\overline{\nu}} \right) \\
&+ \left(u_\nu v_{\nu'} + v_\nu u_{\nu'} \right) \left(\Phi_\nu^\dagger \Phi_{\overline{\nu}'}^\dagger + \Phi_{\overline{\nu}} \Phi_{\nu'} \right) \, .
\end{aligned}$$

The remaining terms of H_0 do not contribute to the matrix element above, because they have opposite signs of ν and ν'.

Only the terms with pairwise positive or negative $\nu\mu\nu'\mu'$ are important for $V = \sum_{\nu<\mu,\nu'<\mu'} {}_a\langle \nu\mu | V | \nu'\mu' \rangle_a \Psi_\mu^\dagger \Psi_\nu^\dagger \Psi_{\nu'} \Psi_{\mu'}$, viz.,

$$\begin{aligned}
&+\tfrac{1}{4} \sum_{\nu\mu\nu'\mu'>0} {}_a\langle \nu\mu | V | \nu'\mu' \rangle_a \, (\Psi_\mu^\dagger \Psi_\nu^\dagger \Psi_{\nu'} \Psi_{\mu'} + \Psi_{\overline{\mu}'}^\dagger \Psi_{\overline{\nu}'}^\dagger \Psi_{\overline{\nu}} \Psi_{\overline{\mu}}) \\[4pt]
&+\tfrac{1}{2} \sum_{\nu\mu\nu'\mu'>0} {}_a\langle \overline{\nu}\mu | V | \overline{\nu}'\mu' \rangle_a \, (\Psi_\mu^\dagger \Psi_{\overline{\nu}}^\dagger \Psi_{\overline{\nu}'} \Psi_{\mu'} + \Psi_{\overline{\mu}'}^\dagger \Psi_{\nu'}^\dagger \Psi_\nu \Psi_{\overline{\mu}}) \\[4pt]
&= +\sum_{\nu\mu>0} \left({}_a\langle \nu\mu | V | \nu\mu \rangle_a + {}_a\langle \overline{\nu}\mu | V | \overline{\nu}\mu \rangle_a \right) v_\nu^{\,2} v_\mu^{\,2} \\[4pt]
&\quad +\sum_{\nu\mu>0} {}_a\langle \overline{\nu}\nu | V | \overline{\mu}\mu \rangle_a \, u_\nu v_\nu u_\mu v_\mu \\[4pt]
&\quad +\sum_{\nu\mu\nu'>0} \left({}_a\langle \nu\mu | V | \nu'\mu \rangle_a + {}_a\langle \nu\overline{\mu} | V | \nu'\overline{\mu} \rangle_a \right) v_\mu^{\,2} \\
&\qquad\qquad\{ +(u_\nu u_{\nu'} - v_\nu v_{\nu'})(\Phi_\nu^\dagger \Phi_{\nu'} + \Phi_{\overline{\nu}'}^\dagger \Phi_{\overline{\nu}}) \\
&\qquad\qquad +(u_\nu v_{\nu'} + v_\nu u_{\nu'})(\Phi_\nu^\dagger \Phi_{\overline{\nu}'}^\dagger + \Phi_{\overline{\nu}} \Phi_{\nu'}) \} \\[4pt]
&\quad +\sum_{\nu\mu\nu'>0} {}_a\langle \overline{\nu}\nu' | V | \overline{\mu}\mu \rangle_a \, v_\mu u_\mu \\
&\qquad\qquad\{ -(u_\nu v_{\nu'} + v_\nu u_{\nu'})(\Phi_\nu^\dagger \Phi_{\nu'} + \Phi_{\overline{\nu}'}^\dagger \Phi_{\overline{\nu}}) \\
&\qquad\qquad +(u_\nu u_{\nu'} - v_\nu v_{\nu'})(\Phi_\nu^\dagger \Phi_{\overline{\nu}'}^\dagger + \Phi_{\overline{\nu}} \Phi_{\nu'}) \} + \cdots \, ,
\end{aligned}$$

where we have left out terms with 4 quasi-particle operators, because we do not need them in the following. The quasi-particle representation of the particle-number operator N was already given at the end of the last section. With this we have all the terms necessary for the Hartree–Fock–Bogoliubov equations. In particular, with the abbreviations

$$\langle v | \Gamma | v' \rangle = \sum_{\mu > 0} \left({}_a\langle v\mu | V | v'\mu \rangle_a + {}_a\langle v\overline{\mu} | V | v'\overline{\mu} \rangle_a \right) v_\mu^{\,2} \, ,$$

$$\Delta_{\overline{v}v'} = - \sum_{\mu > 0} {}_a\langle \overline{v}v' | V | \overline{\mu}\mu \rangle_a \, u_\mu \, v_\mu \, ,$$

the expectation value of the energy in the ground state is

$$\langle \mathrm{BCS} | H | \mathrm{BCS} \rangle = \sum_{v > 0} 2v_v^{\,2} \, \langle v | H_0 + \Gamma | v \rangle - u_v \, v_v \, \Delta_{\overline{v}v} \, .$$

New compared with the Hartree–Fock expression are the terms $\Delta_{\overline{v}v'}$, i.e., they are no longer neglected in the Hartree–Fock–Bogoliubov method. However, in addition to the one-particle energies, the occupation probabilities $v_v^{\,2} = 1 - u_v^{\,2}$ must now also be determined. They follow from the Hartree–Fock–Bogoliubov equations

$$e_v \langle v | v' \rangle = (u_v \, u_{v'} - v_v \, v_{v'}) \, \langle v | H_0 + \Gamma - \mu | v' \rangle + (u_v \, v_{v'} + v_v \, u_{v'}) \, \Delta_{\overline{v}v'} \, ,$$

$$0 = (u_v \, v_{v'} + v_v \, u_{v'}) \, \langle v | H_0 + \Gamma - \mu | v' \rangle - (u_v \, u_{v'} - v_v \, v_{v'}) \, \Delta_{\overline{v}v'} \, .$$

The states $|v\rangle$ are required to diagonalize the operator $H_0 + \Gamma$, whose eigenvalues $\varepsilon_v + \mu$ are the Hartree–Fock one-particle energies:

$$\langle v | H_0 + \Gamma | v' \rangle = \langle v | v' \rangle \, (\varepsilon_v + \mu) \, .$$

In addition, we restrict ourselves to the pair force as the residual interaction and assume an attractive pair force (see p. 456):

$$\Delta_{\overline{v}v'} = \langle v | v' \rangle \, \Delta_v \, , \quad \text{with } \Delta_v \geq 0 \, .$$

Then the Hartree–Fock–Bogoliubov equations read

$$e_v = + (u_v^{\,2} - v_v^{\,2}) \, \varepsilon_v + 2u_v v_v \, \Delta_v \, ,$$

$$0 = - (u_v^{\,2} - v_v^{\,2}) \, \Delta_v + 2u_v v_v \, \varepsilon_v \, .$$

With $u_v^{\,2} + v_v^{\,2} = 1$, we set $u_v \to \cos \varphi_v$, $v_v \to \sin \varphi_v$ and make use of the properties of the trignonometric functions: $u_v^{\,2} - v_v^{\,2} \to \cos(2\varphi_v)$, $2u_v v_v \to \sin(2\varphi_v)$. The second Hartree–Fock–Bogoliubov equation then delivers $\cot(2\varphi_v) = \varepsilon_v / \Delta_v$: φ_v decreases from $\pi/2$ to 0 between $\varepsilon_v \ll -\Delta_v$ and $\varepsilon_v \gg \Delta_v$. According to the first Hartree–Fock–Bogoliubov equation, the quasi-particle energies e_v are never negative, and with $\sin \alpha = (1 + \cot^2 \alpha)^{-1/2}$ and $\cos \alpha = \cot \alpha \cdot \sin \alpha$, it follows that (see Fig. 5.13)

Fig. 5.13 Effects of the pair
force. Quasi-particle
energies e_μ for equidistant
one-particle energies as a
function of the gap
parameter Δ (*left*) and
occupation probability of the
BCS ground state as a
function of ε_μ / Δ (*right*)

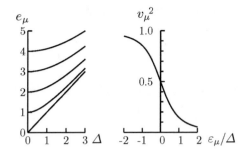

$$e_\nu = +\sqrt{\varepsilon_\nu{}^2 + \Delta_\nu{}^2}, \quad u_\nu = \sqrt{\frac{e_\nu + \varepsilon_\nu}{2\,e_\nu}}, \quad v_\nu = \sqrt{\frac{e_\nu - \varepsilon_\nu}{2\,e_\nu}}.$$

For $\Delta_\nu = 0$, we do not find pair effects, but the usual Hartree–Fock result: either
$u_\nu = 0$, $v_\nu = 1$, and $e_\nu = -\varepsilon_\nu$ or $u_\nu = 1$, $v_\nu = 0$, and $e_\nu = +\varepsilon_\nu$. While the Hartree–
Fock one-particle energies ε_ν, evaluated at the Fermi energy μ, can be positive or
negative, the Hartree–Fock–Bogoliubov eigenvalues e_ν are always positive.

Generally, the pair potential satisfies $\Delta_\nu \neq 0$. Then the Fermi edge is not sharp,
and that alters the states close to it. Thus there, the quasi-particle energies $e_\nu =
(\varepsilon_\nu{}^2 + \Delta_\nu{}^2)^{1/2}$ are different from the Hartree–Fock energies ε_ν. An *energy gap* Δ_ν
appears, and only above this gap are there quasi-particle levels. Note that the energy
gap corresponds to the rest energy mc^2 in the expression $E/c = \sqrt{p^2 + (mc)^2}$ for the
energy of free particles according to special relativity theory (see p. 245).

The gap parameter Δ_ν with $\Delta_\nu = -\sum_{\mu>0} {}_a\langle \bar{\nu}\nu | V | \bar{\mu}\mu \rangle_a \, u_\mu v_\mu$ has to satisfy the
so-called *gap condition* (or gap equation)

$$\Delta_\nu = -\sum_{\mu>0} {}_a\langle \bar{\nu}\nu | \, V \, | \bar{\mu}\mu \rangle_a \frac{\Delta_\mu}{2 e_\mu}.$$

It is mainly the terms with $\mu \approx \nu_F$ that contribute to the sum, because $u_\mu v_\mu$ is
only different from zero close to the Fermi edge. In addition, the matrix elements
${}_a\langle \bar{\nu}\nu | \, V \, | \bar{\mu}\mu \rangle_a$ are particularly large for $\nu \approx \mu$, hence so is the gap parameter Δ_ν for
$\nu \approx \nu_F$. The pair interaction can thus only be felt close to the Fermi edge. As long as
we are only interested in states close to Fermi edge, we may use an average matrix
element

$$G \equiv -\overline{{}_a\langle \bar{\nu}\nu | \, V \, | \bar{\mu}\mu \rangle_a}, \quad \text{for } \nu \approx \nu_F, \text{ otherwise zero}.$$

Then the gap parameter Δ_ν no longer depends on the state $|\nu\rangle$, and the gap condition
simplifies to

$$\Delta = \frac{G\Delta}{2} \sum_{\mu>0} \frac{1}{\sqrt{\varepsilon_\mu{}^2 + \Delta^2}}.$$

In addition to the trivial solution $\Delta = 0$, there is another if

$$\frac{G}{2} \sum_{\mu>0} \frac{1}{|\varepsilon_\mu|} > 1 \ .$$

The pair correlations grow stepwise with increasing pair force G, and hence every perturbation theory fails.

5.4.6 Hole States

So far we have described the transition of a fermion from $|v\rangle$ to $|v'\rangle$ using the operator $\Psi^\dagger_{v'} \Psi_v$. Such "particle scattering" occurs for small excitation energies only close to the Fermi edge, and in fact preferably with $e_v < e_{\mathrm{F}}$ and $e_{v'} > e_{\mathrm{F}}$. Here we assume a unique Fermi edge—in atomic and nuclear physics (for non-deformed nuclei), we take closed shells, otherwise at least an even fermion number so that the ground state is not degenerate. We denote this "normal state" by $|\widehat{0}\rangle$.

   Removing a particle from the state $|\bar{v}\rangle$ turns this normal state into a *hole state* $|v^{-1}\rangle$. It behaves with respect to momentum and angular momentum like the state $|v\rangle$. Instead of particle scattering, we may thus also speak of particle–hole *pair generation* $|\widehat{0}\rangle \to |v^{-1}v'\rangle$. Below the Fermi edge, we also use *hole operators* Φ, and above the Fermi edge, the particle operators Ψ as before, with

$$\Phi^\dagger_v |\widehat{0}\rangle = |v^{-1}\rangle \ , \quad \Phi_v \Phi^\dagger_{v'} + \Phi^\dagger_{v'} \Phi_v = \langle v|v'\rangle \ , \quad \Phi_v \Phi_{v'} = -\Phi_{v'} \Phi_v \ .$$

With $\Phi^\dagger_v = \Psi_{\bar{v}}$, whence $\Phi_v = \Psi^\dagger_{\bar{v}}$, $\Phi^\dagger_{\bar{v}} = -\Psi_v$, and $\Phi_{\bar{v}} = -\Psi^\dagger_v$, they barely differ from the BCS quasi-particle operators. Here, for states below the Fermi edge, we carry out a Bogoliubov transformation of all field operators with $u_v = 0$, $v_v = -1$ (see Fig. 5.14).

5.4.7 Summary: Fermions

The treatment of many-body systems with fermion creation and annihilation operators was explained using the example of the Fermi gas. The best one-particle basis derives from the Hartree–Fock equations. For pair forces, it is better to use quasi-particles and the Hartree–Fock–Bogoliubov equations to introduce pair correlations.

Fig. 5.14 Feynman graphs
with hole states. Hole arrows
point downwards (time
reversal of the particle arrow
in Fig. 5.11). *Upper row:*
The four diagrams for a
one-particle operator (T),
viz., pair creation, pair
annihilation, hole scattering,
and the vacuum expectation
value $\langle \widehat{0}|T|\widehat{0}\rangle$. *Lower row:* A
selection of two-particle
operators, viz., particle–hole
and hole–hole scattering,
particle scattering with pair
creation, and the one-particle
potential

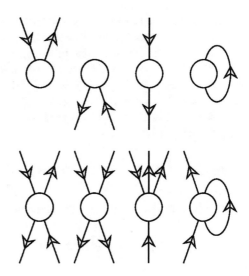

5.5 Photons

5.5.1 Preparation for the Quantization of Electromagnetic Fields

The electromagnetic field is described classically by the Maxwell equations. According to p. 215, for homogeneous non-conductors, they deliver wave equations for the electric field strength **E** and the magnetic flux density **B**, and likewise for the scalar potential Φ and the vector potential **A**. In the following, we restrict ourselves to homogeneous and isotropic media, hence constant scalar ε and μ.

According to quantum theory, we have to alter our notion of waves to permit a particle interpretation—radiation may exhibit interference effects, but it may also be granular. This can be obtained only via uncertainties: the experimental quantities have to be replaced by Hermitian operators with suitable commutation behavior.

For the wave function we prefer the four-potential instead of the field strengths **E** and **B**, because, from the relations $\partial \mathbf{B}/\partial t = -\nabla \times \mathbf{E}$ and $\nabla \cdot \mathbf{B} = 0$, we see that their components are not independent of each other. These two equations are already automatically satisfied with the ansatz $\mathbf{E} = -\partial \mathbf{A}/\partial t - \nabla \Phi$ and $\mathbf{B} = \nabla \times \mathbf{A}$. However, the potentials cannot be measured and also depend on the gauge, but then the wave functions for electrons are not measurable and contain an arbitrary phase.

It is better to characterize free particles by their momentum (wave vector) than by their position. Therefore, we now consider the Fourier transform of the fields and take the Coulomb or radiation gauge $\mathbf{k} \cdot \mathbf{A}(t, \mathbf{k}) = 0$. Then the transverse parts of the field strengths $\mathbf{E} = -\partial \mathbf{A}/\partial t - \mathrm{i}\mathbf{k}\Phi$ and $\mathbf{B} = \mathrm{i}\mathbf{k} \times \mathbf{A}$ are $-\partial \mathbf{A}/\partial t$ and $\mathrm{i}\mathbf{k} \times \mathbf{A}$ and their longitudinal parts are $-\mathrm{i}\mathbf{k}\Phi$ and **0**. For any other gauge the vector potential also has a longitudinal part. Note, however, that the Coulomb gauge is not Lorentz invariant.

If we do adopt the Lorentz gauge, we encounter other difficulties in quantum theory, because the Lorentz condition cannot be transferred to operators. Then we have to introduce longitudinal and scalar photons, which are not easily normalized (see, e.g., [8]). Here $\mathbf{E}_{\text{long}} = -\mathrm{i}\mathbf{k}\rho/(\varepsilon k^2)$ holds, according to the third Maxwell equation.

We now consider the energy $W = \frac{1}{2}\int d^3r\,(\mathbf{E}\cdot\mathbf{D}+\mathbf{H}\cdot\mathbf{B})$ (see p. 211) and the momentum $\mathbf{P} = \int d^3r\,\mathbf{D}\times\mathbf{B}$ (see p. 215) in a non-conductor, i.e., with $\rho = 0$ and $\mathbf{j}=\mathbf{0}$, as well as $\mathbf{D}=\varepsilon\mathbf{E}$ and $\mathbf{H}=\mathbf{B}/\mu$. According to Parseval's equation (p. 23), for the energy

$$W(t) = \frac{\varepsilon}{2}\int d^3k\,(\mathbf{E}^*\cdot\mathbf{E}+c^2\mathbf{B}^*\cdot\mathbf{B}) = \frac{\varepsilon}{2}\int d^3k\,\left(\frac{\partial\mathbf{A}^*}{\partial t}\cdot\frac{\partial\mathbf{A}}{\partial t}+\omega^2\mathbf{A}^*\cdot\mathbf{A}\right),$$

with transverse gauge, and for the momentum

$$\mathbf{P}(t) = \varepsilon\int d^3k\,(\mathbf{E}^*\times\mathbf{B}) = -\mathrm{i}\varepsilon\int d^3k\,\mathbf{k}\,\frac{\partial\mathbf{A}^*}{\partial t}\cdot\mathbf{A}.$$

According to p. 216, we have

$$\mathbf{A}(t,\mathbf{k}) = \frac{\mathbf{A}(\mathbf{k})\exp(-\mathrm{i}\omega t)+\mathbf{A}^*(-\mathbf{k})\exp(+\mathrm{i}\omega t)}{2} = \mathbf{A}^*(t,-\mathbf{k}),$$

and thus $\partial\mathbf{A}(t,\mathbf{k})/\partial t = -\frac{\mathrm{i}}{2}\omega\,\{\mathbf{A}(\mathbf{k})\exp(-\mathrm{i}\omega t)-\mathbf{A}^*(-\mathbf{k})\exp(+\mathrm{i}\omega t)\}$. For the energy, we may replace the integrand $\mathbf{A}^*(-\mathbf{k})\cdot\mathbf{A}(-\mathbf{k})$ by $\mathbf{A}^*(\mathbf{k})\cdot\mathbf{A}(\mathbf{k})$ and for the momentum, $\mathbf{k}\,\mathbf{A}^*(-\mathbf{k})\cdot\mathbf{A}(-\mathbf{k})$ by $-\mathbf{k}\,\mathbf{A}^*(\mathbf{k})\cdot\mathbf{A}(\mathbf{k})$ (a variable transformation), to deduce the time-independent expressions

$$W = \frac{\varepsilon}{2}\int d^3k\,\omega^2\,\mathbf{A}^*(\mathbf{k})\cdot\mathbf{A}(\mathbf{k}),$$
$$\mathbf{P} = \frac{\varepsilon}{2}\int d^3k\,\omega\mathbf{k}\,\mathbf{A}^*(\mathbf{k})\cdot\mathbf{A}(\mathbf{k}),$$

since the oscillating factors cancel for the energy and the momentum—in the latter case, for the symmetry under $\mathbf{k}\leftrightarrow-\mathbf{k}$. This distinguishes the results calculated with potentials from those calculated with field strengths.

Because of the spins, we also have to consider the *angular momentum*:

$$\mathbf{J} = \varepsilon\int d^3r\,\mathbf{r}\times(\mathbf{E}\times\mathbf{B}).$$

Here we replace only $\mathbf{B}$ by $\nabla\times\mathbf{A}$, but not $\mathbf{E}$ by $-\partial\mathbf{A}/\partial t$ for the time being. If $\mathbf{E}_c$ and $\mathbf{r}_c$ are now treated as constant, then

$$\mathbf{E}\times(\nabla\times\mathbf{A}) = \nabla\,\mathbf{E}_c\cdot\mathbf{A}-\mathbf{E}\cdot\nabla\,\mathbf{A},$$

according to Sect. 1.1.8,

$$\mathbf{r} \times \{\mathbf{E} \times (\nabla \times \mathbf{A})\} = -\nabla \times (\mathbf{E}_c \cdot \mathbf{A}\, \mathbf{r}) - \mathbf{E} \cdot \nabla\, \mathbf{r}_c \times \mathbf{A}$$
$$= -\nabla \times (\mathbf{E}_c \cdot \mathbf{A}\, \mathbf{r}) - \mathbf{E} \cdot \nabla\, \mathbf{r} \times \mathbf{A} + \mathbf{E} \times \mathbf{A}\,.$$

The volume integrals of $\nabla \times (\mathbf{E}_c \cdot \mathbf{A}\, \mathbf{r})$ and $\mathbf{E} \cdot \nabla\, \mathbf{r} \times \mathbf{A}$ can be changed into surface integrals. Then there is initially only one more volume integral, of $\mathbf{r} \times \mathbf{A}\, \nabla \cdot \mathbf{E}$, but the electric field is source-free here. These surface integrals $\int \mathrm{d}\mathbf{f} \times \mathbf{r}\, \mathbf{E} \cdot \mathbf{A}$ and $\int \mathrm{d}\mathbf{f} \cdot \mathbf{E}\, \mathbf{r} \times \mathbf{A}$ pick up the orbital angular momentum of the fields. They depend on where the origin of the position vectors lies and do not have a component in the direction of propagation. This is different with the volume integral of $\mathbf{E} \times \mathbf{A} = \mathbf{A} \times \partial \mathbf{A}/\partial t$. Here, using Parseval's equation, we arrive at the eigen angular momentum

$$\mathbf{S} = \varepsilon \int \mathrm{d}^3 k\, \mathbf{A}^*(t, \mathbf{k}) \times \frac{\partial \mathbf{A}(t, \mathbf{k})}{\partial t}\,.$$

Since only terms even in $\mathbf{k}$ contribute to the integral, the parts oscillating at 2ω cancel again, and we find

$$\mathbf{S} = \frac{-\mathrm{i}\varepsilon}{2} \int \mathrm{d}^3 k\, \omega\, \mathbf{A}^*(\mathbf{k}) \times \mathbf{A}(\mathbf{k})\,.$$

The result $\mathbf{S}(\mathbf{k}) = -\frac{\mathrm{i}}{2}\varepsilon\, \omega\, \mathbf{A}^*(\mathbf{k}) \times \mathbf{A}(\mathbf{k})$ is useful for the helicity $\mathbf{S}(\mathbf{k}) \cdot \mathbf{e}_k$.

Because of the transversality, on p. 218, we already introduced two mutually orthogonal unit vectors $\mathbf{e}_\parallel$ and $\mathbf{e}_\perp$ with $\mathbf{e}_\parallel \times \mathbf{e}_\perp = \mathbf{e}_k$, and shortly after that also complex unit vectors $\mathbf{e}_\pm \propto (\mathbf{e}_\parallel \pm \mathrm{i}\mathbf{e}_\perp)/\sqrt{2}$. There, however, we did not determine the phase factor, which we now adjust to the spherical harmonics $Y_m^{(l)}(\Omega)$. Let $\mathbf{e}(\Omega) = \mathbf{e}_r$ be the unit vector in the direction of $\Omega = (\theta, \varphi)$. Then, with $\mathbf{e}_0 = \mathrm{i}\,\mathbf{e}_k$, we require

$$\mathbf{e}_m \cdot \mathbf{e}(\Omega) = \sqrt{\frac{4\pi}{3}}\, \mathrm{i} Y_m^{(1)}(\Omega) = \begin{cases} \mathrm{i}\, \cos\theta & \text{for } m = 0\,, \\ \frac{\mp\mathrm{i}}{\sqrt{2}}\, \sin\theta\, \exp(\pm\mathrm{i}\varphi) & \text{for } m = \pm 1\,. \end{cases}$$

According to p. 332, we always took the factor i^l for the expansions of functions $f(\mathbf{r})$ in terms of spherical harmonics. If, for $\mathbf{k}$ in the z-direction, we choose $\mathbf{e}_\parallel = \mathbf{e}_x$ and $\mathbf{e}_\perp = \mathbf{e}_y$, then we have $\mathbf{e}_\pm \cdot \mathbf{e}_\parallel = \mp\mathrm{i}/\sqrt{2}$ and $\mathbf{e}_\pm \cdot \mathbf{e}_\perp = 1/\sqrt{2}$. Therefore, for the expansion in terms of circularly polarized light, we take

$$\mathbf{e}_\pm \equiv \mp\mathrm{i}\, \frac{\mathbf{e}_\parallel \pm \mathrm{i}\mathbf{e}_\perp}{\sqrt{2}} = \mathbf{e}_\mp^*\,,$$

with the properties

$$\mathbf{e}_\pm^* \cdot \mathbf{e}_\pm = 1\,, \qquad \mathbf{e}_\pm^* \times \mathbf{e}_\pm = \pm\mathrm{i}\mathbf{e}_k\,,$$
$$\mathbf{e}_\pm^* \cdot \mathbf{e}_\mp = 0\,, \qquad \mathbf{e}_\pm^* \times \mathbf{e}_\mp = \mathbf{0}\,.$$

The amplitudes for the two helicities are then

$$A_\pm(\mathbf{k}) = \mathbf{e}_\pm{}^* \cdot \mathbf{A}(\mathbf{k}) \qquad \Longleftrightarrow \qquad \mathbf{A}(\mathbf{k}) = \mathbf{e}_+ A_+(\mathbf{k}) + \mathbf{e}_- A_-(\mathbf{k}) ,$$

and hence we deduce the two equations

$$\begin{aligned}
\mathbf{A}^*(\mathbf{k}) \cdot \mathbf{A}(\mathbf{k}) &= |A_+(\mathbf{k})|^2 + |A_-(\mathbf{k})|^2 , \\
\mathbf{A}^*(\mathbf{k}) \times \mathbf{A}(\mathbf{k}) &= (|A_+(\mathbf{k})|^2 - |A_-(\mathbf{k})|^2)\, \mathrm{i}\, \mathbf{e}_k .
\end{aligned}$$

We can also give the contribution of the respective helicities to the energy and the momentum, as soon as we know the amplitude of $A_\pm$.

Actually, for $\mathbf{e}_\pm$, we should also include the argument $\mathbf{k}$, because we need to note that $\mathbf{e}_\pm{}^*(-\mathbf{k}) \times \mathbf{e}_\pm(-\mathbf{k}) = \pm\mathrm{i}\,(-\mathbf{e}_k)$, and $\mathbf{e}_\pm(-\mathbf{k}) = \mathbf{e}_\pm^*(\mathbf{k}) = \mathbf{e}_\mp(\mathbf{k})$. With this we deduce

$$\mathbf{A}(t, \mathbf{k}) = \sum_{\lambda=\pm} \mathbf{e}_\lambda \, \frac{A_\lambda(\mathbf{k})\, \exp(-\mathrm{i}\omega t) + A_\lambda^*(-\mathbf{k})\, \exp(+\mathrm{i}\omega t)}{2} ,$$

or $A_\lambda(t, \mathbf{k}) = \frac{1}{2}\{A_\lambda(\mathbf{k})\, \exp(-\mathrm{i}\omega t) + A_\lambda{}^*(-\mathbf{k})\, \exp(+\mathrm{i}\omega t)\}$. Here we also have

$$A_\lambda(-\mathbf{k}) = \mathbf{e}_\lambda(\mathbf{k}) \cdot \mathbf{A}(-\mathbf{k}) .$$

5.5.2   Quantization of Photons

Clearly, the two quantities $|A_\pm(\mathbf{k})|^2$ depend on the intensity of the radiation field. Classically, in the wave picture, they may take arbitrary values ≥ 0, but in quantum physics, only natural numbers. There are only integer light quanta, no fractions of them. We usually speak of *photons* rather than *light quanta*.

The properties of these photons can be read off from the previous expressions for energy, momentum, and helicity densities in $\mathbf{k}$-space:

$$\begin{aligned}
W(\mathbf{k}) &= \tfrac{1}{2}\varepsilon\, \omega^2\, \{|A_+(\mathbf{k})|^2 + |A_-(\mathbf{k})|^2\} , \\
\mathbf{P}(\mathbf{k}) &= \tfrac{1}{2}\varepsilon\, \omega\mathbf{k}\, \{|A_+(\mathbf{k})|^2 + |A_-(\mathbf{k})|^2\} , \\
\mathscr{H}(\mathbf{k}) &= \tfrac{1}{2}\varepsilon\, \omega\, \{|A_+(\mathbf{k})|^2 - |A_-(\mathbf{k})|^2\} .
\end{aligned}$$

The ratio of their energy to their momentum is thus $\omega/k = c$. According to relativity theory (see p. 245), for all massless particles we can state that photons do not have mass and therefore move with the velocity of light.

If we now assume the known Planck–de Broglie relations for single photons, viz.,

$$E = \hbar\omega \quad \text{and} \quad \mathbf{p} = \hbar\mathbf{k} \ ,$$

then the density of the quanta with helicity $\lambda = \pm 1$ is obtained as

$$\rho_\lambda(\mathbf{k}) = \frac{\varepsilon\omega}{2\hbar} \, |A_\lambda(\mathbf{k})|^2 \ .$$

The angular momentum in the motional direction thus yields $\pm\hbar$. We distinguish between two helicities or two sorts of photons. In fact, they all have spin one, but it is oriented only in or opposite to the direction of motion, not orthogonal to it—this is a relativistic effect, which relates to the Lorentz contraction. With integer spins, they are therefore bosons. (Electrons also have only two spin states, but they are fermions.)

The integral $\int d^3k \, \rho_\lambda(\mathbf{k})$ in the classical calculation does not need to be an even number. But in the particle picture, we have to enforce this by a special quantum condition, viz., for photons we have to take creation and annihilation operators satisfying the Bose commutation law:

$$[\Psi_\lambda(\mathbf{k}), \ \Psi_{\lambda'}^\dagger(\mathbf{k}')] = \langle \mathbf{k}, \lambda | \mathbf{k}', \lambda' \rangle \quad \text{and} \quad [\Psi_\lambda(\mathbf{k}), \ \Psi_{\lambda'}(\mathbf{k}')] = 0 \ .$$

According to p. 450, in the Heisenberg picture, the time dependence is given by

$$\Psi_\lambda(t, \mathbf{k}) = \Psi_\lambda(\mathbf{k}) \, \exp(-i\omega t) \ .$$

Since we are dealing with bosons, several photons can be in the same state $|\mathbf{k}, \lambda\rangle$. From the expression for the particle density, which we understand as the expectation value of $\Psi^\dagger \Psi$, we deduce the assignment

$$\Psi_\lambda(\mathbf{k}) \mathrel{\widehat{=}} \sqrt{\frac{\varepsilon\omega}{2\hbar}} \, A_\lambda(\mathbf{k}) \quad \text{and} \quad \Psi_\lambda^\dagger(\mathbf{k}) \mathrel{\widehat{=}} \sqrt{\frac{\varepsilon\omega}{2\hbar}} \, A_\lambda{}^*(\mathbf{k}) \ .$$

The Hamilton, momentum operator, and helicity operator then follow with $\omega = ck$:

$$H = \int d^3k \ \hbar\omega \ \{\Psi_+^\dagger(\mathbf{k}) \, \Psi_+(\mathbf{k}) + \Psi_-^\dagger(\mathbf{k}) \, \Psi_-(\mathbf{k})\} \ ,$$

$$\mathbf{P} = \int d^3k \ \hbar\mathbf{k} \ \{\Psi_+^\dagger(\mathbf{k}) \, \Psi_+(\mathbf{k}) + \Psi_-^\dagger(\mathbf{k}) \, \Psi_-(\mathbf{k})\} \ ,$$

$$\mathscr{H} = \int d^3k \qquad \{\Psi_+^\dagger(\mathbf{k}) \, \Psi_+(\mathbf{k}) - \Psi_-^\dagger(\mathbf{k}) \, \Psi_-(\mathbf{k})\} \ .$$

The vector potential has now become an operator:

$$\mathbf{A}(t, \mathbf{k}) = \sqrt{\frac{2\hbar}{\varepsilon\omega}} \sum_{\lambda=\pm} \mathbf{e}_\lambda \, \frac{\Psi_\lambda(t, \mathbf{k}) + \Psi_\lambda^\dagger(t, -\mathbf{k})}{2} \ .$$

Hence it follows that $\mathbf{A}^{\dagger}(t, \mathbf{k}) = \mathbf{A}(t, -\mathbf{k})$, with $\mathbf{e}_\lambda{}^*(\mathbf{k}) = \mathbf{e}_\lambda(-\mathbf{k})$. The transverse electric and magnetic field operators are then obtained from $\mathbf{E} = -\partial\mathbf{A}/\partial t$ and $\mathbf{B} = \mathrm{i}\mathbf{k} \times \mathbf{A}$ (as well as from Weber's equation):

$$\mathbf{E}(t, \mathbf{k}) = \mathrm{i}\sqrt{2\hbar\omega/\varepsilon} \sum_{\lambda=\pm} \mathbf{e}_\lambda \; \frac{\Psi_\lambda(t, \mathbf{k}) - \Psi_\lambda^{\dagger}(t, -\mathbf{k})}{2} \; ,$$

$$\mathbf{B}(t, \mathbf{k}) = \mathrm{i}\sqrt{2\hbar\omega\mu} \sum_{\lambda=\pm} \mathbf{e}_k \times \mathbf{e}_\lambda \; \frac{\Psi_\lambda(t, \mathbf{k}) + \Psi_\lambda^{\dagger}(t, -\mathbf{k})}{2} \; ,$$

where we can also use $\mathrm{i}\mathbf{e}_k \times \mathbf{e}_\lambda = \lambda\,(\mathbf{e}_\lambda{}^* \times \mathbf{e}_\lambda) \times \mathbf{e}_\lambda = \lambda\mathbf{e}_\lambda$, although this does not always help.

In order to make the transition from $\mathbf{k}$ to $\mathbf{r}$, we consider arbitrary Cartesian components n, unrelated to $\mathbf{k}$, instead of the helicities, and investigate $[\Psi_n(\mathbf{k}),\ \Psi_{n'}^{\dagger}(\mathbf{k}')]$. This is equal to $\sum_{\lambda\lambda'} \mathbf{e}_n \cdot \mathbf{e}_\lambda \, \mathbf{e}_{\lambda'}^* \cdot \mathbf{e}_{n'} \, \langle \mathbf{k}, \lambda | \mathbf{k}', \lambda' \rangle$. Because of the last factor, we may restrict ourselves to $\lambda = \lambda'$. Here $\sum_\lambda \mathbf{e}_\lambda \, \mathbf{e}_\lambda^* \cdot \mathbf{e}_{n'}$ is the part of $\mathbf{e}_{n'}$ perpendicular to $\mathbf{k}/k \equiv \mathbf{e}_k$, which, according to p. 4, we may thus write as $\mathbf{e}_{n'} - \mathbf{e}_k \, \mathbf{e}_k \cdot \mathbf{e}_{n'}$. Therefore we deduce

$$[\Psi_n(\mathbf{k}),\ \Psi_{n'}^{\dagger}(\mathbf{k}')] = (\delta_{nn'} - \mathbf{e}_n \cdot \mathbf{e}_k \, \mathbf{e}_k \cdot \mathbf{e}_{n'}) \, \langle \mathbf{k} | \mathbf{k}' \rangle \; ,$$

as a generalization of $[\Psi_\lambda(\mathbf{k}),\ \Psi_{\lambda'}^{\dagger}(\mathbf{k}')] = \langle \mathbf{k}, \lambda | \mathbf{k}', \lambda' \rangle$.

For the fields $\mathbf{A}$ and $\mathbf{B}$, there is the sum of Ψ and $\Psi^{\dagger}$, and for $\mathbf{E}$, their difference. Therefore, the commutation laws are different. In fact,

$$0 = [A_n(\mathbf{k}),\ A_{n'}(\mathbf{k}')] = [E_n(\mathbf{k}),\ E_{n'}(\mathbf{k}')] = [B_n(\mathbf{k}),\ B_{n'}(\mathbf{k}')]$$

and $[A_n(\mathbf{k}),\ B_{n'}(\mathbf{k}')] = 0$, but also, using $\langle \mathbf{k}\lambda | - \mathbf{k}'\lambda' \rangle = \langle \mathbf{k} | - \mathbf{k}' \rangle \langle \lambda | - \lambda' \rangle$ and $\mathbf{e}_{-\lambda} = \mathbf{e}_\lambda^*$ as well as Weber's equation,

$$[A_n(\mathbf{k}),\ E_{n'}(\mathbf{k}')] = \frac{\hbar}{\mathrm{i}\varepsilon} \, (\delta_{nn'} - \mathbf{e}_n \cdot \mathbf{e}_k \, \mathbf{e}_k \cdot \mathbf{e}_{n'}) \, \langle \mathbf{k} | - \mathbf{k}' \rangle \; ,$$

$$[E_n(\mathbf{k}),\ B_{n'}(\mathbf{k}')] = \frac{\hbar}{\varepsilon} \, (\mathbf{e}_n \times \mathbf{e}_{n'}) \cdot \mathbf{k} \; \langle \mathbf{k} | - \mathbf{k}' \rangle \; .$$

Here we have at last made use of $\sum_\lambda \mathbf{e}_\lambda \, \mathbf{e}_\lambda^* \cdot (\mathbf{e}_{n'} \times \mathbf{k}) = \mathbf{e}_{n'} \times \mathbf{k}$.

After a Fourier transform $\mathbf{k} \to \mathbf{r}$, the corresponding operator functions of $\mathbf{r}$, rather than $\mathbf{k}$, are

$$\mathbf{A}(t, \mathbf{r}) = \frac{1}{\sqrt{2\pi}^{\,3}} \int \mathrm{d}^3k \; \exp(\mathrm{i}\mathbf{k} \cdot \mathbf{r}) \, \mathbf{A}(t, \mathbf{k}) = \mathbf{A}^{\dagger}(t, \mathbf{r}) \; ,$$

where the last equation corresponds to the classical relation $\mathbf{A}(t, \mathbf{r}) = \mathbf{A}^*(t, \mathbf{r}) \Longleftrightarrow \mathbf{A}(t, \mathbf{k}) = \mathbf{A}^*(t, -\mathbf{k})$. With $\Psi(t) = \Psi(0) \exp(-\mathrm{i}\omega t)$ and $\Psi^{\dagger}(t) = \Psi^{\dagger}(0) \exp(\mathrm{i}\omega t)$, it is often useful to decompose the fields into the so-called *positive-frequency part* $\mathbf{A}^+(t, \mathbf{r})$ and *negative-frequency part* $\mathbf{A}^-(t, \mathbf{r}) = \mathbf{A}^{+\dagger}(t, \mathbf{r})$:

$$\mathbf{A}(t, \mathbf{r}) = \mathbf{A}^+(t, \mathbf{r}) + \mathbf{A}^-(t, \mathbf{r}) ,$$

where

$$\mathbf{A}^+(t, \mathbf{r}) = \frac{1}{\sqrt{2\pi}^3} \int d^3k \sqrt{\frac{\hbar}{2\varepsilon\omega}} \sum_{\lambda=\pm} \mathbf{e}_\lambda \, \Psi_\lambda(\mathbf{k}) \, \exp\{i\,(\mathbf{k}\cdot\mathbf{r} - \omega t)\} ,$$

and likewise for the electric and magnetic fields.

In real space, with the *transverse delta function*

$$\delta_{nn'}^{\text{trans}}(\mathbf{r}) = \frac{1}{(2\pi)^3} \int d^3k \, (\delta_{nn'} - \mathbf{e}_n \cdot \mathbf{e}_k \, \mathbf{e}_k \cdot \mathbf{e}_{n'}) \, \exp(i\mathbf{k}\cdot\mathbf{r}) ,$$

we have the not so simple commutation laws

$$[A_n(\mathbf{r}), E_{n'}(\mathbf{r}')] = \frac{\hbar}{i\varepsilon} \, \delta_{nn'}^{\text{trans}}(\mathbf{r} - \mathbf{r}') ,$$

$$[E_n(\mathbf{r}), B_{n'}(\mathbf{r}')] = \frac{\hbar}{i\varepsilon} \, (\mathbf{e}_n \times \mathbf{e}_{n'}) \cdot \nabla\delta(\mathbf{r} - \mathbf{r}') .$$

Integration of $[E_n(\mathbf{r}), B_{n'}(\mathbf{r}')]$ over a space element around $\mathbf{r}'$ yields zero. Electric and magnetic field strengths at the same position commute, and equal components ($n = n'$) of $\mathbf{E}$ and $\mathbf{B}$ also commute everywhere. Note that $\mathbf{A}(\mathbf{r})$ and $-\varepsilon\mathbf{E}^{\text{trans}}(\mathbf{r})$ may be taken as canonical conjugates, provided that we have ensured that the fields are transverse.

The transverse delta function clearly has the following symmetries:

$$\delta_{nn'}^{\text{trans}}(\mathbf{r}) = \delta_{n'n}^{\text{trans}}(\mathbf{r}) = \delta_{nn'}^{\text{trans}}(-\mathbf{r}) .$$

In addition, it is source-free, i.e., $\sum_n \partial\delta_{nn'}^{\text{trans}}/\partial x_n = 0$, because

$$\sum_n k_n \, (\delta_{nn'} - k_n k_{n'}/k^2) = 0 , \quad \text{with } k^2 = \sum_n k_n^2 .$$

To relate this to the usual delta function, we consider

$$-\int \frac{d^3k}{(2\pi)^3} \frac{k_n k_{n'}}{k^2} \, \exp(i\mathbf{k}\cdot\mathbf{r}) = \frac{\partial^2}{\partial x_n \partial x_{n'}} \int \frac{d^3k}{(2\pi)^3} \frac{\exp(i\mathbf{k}\cdot\mathbf{r})}{k^2} .$$

According to p. 410, the right-hand integral is equal to $(4\pi r)^{-1}$. We thus have

$$\delta_{nn'}^{\text{trans}}(\mathbf{r}) = \delta_{nn'} \, \delta(\mathbf{r}) + \frac{\partial^2}{\partial x_n \partial x_{n'}} \frac{1}{4\pi r} ,$$

where, according to p. 172,

$$\frac{\partial^2}{\partial x_n \partial x_{n'}} \frac{1}{4\pi r} = \frac{3x_n x_{n'}/r^2 - \delta_{nn'}}{4\pi r^3} - \frac{\delta_{nn'} \, \delta(\mathbf{r})}{3} \, .$$

Thus it has to be accounted for even when $n \neq n'$, otherwise we would also have to split off the factor $\delta_{nn'}$.

All commutation laws have been derived here for equal times—and in the Schrödinger picture, the field operators do not depend upon time. To avoid integrals and improper Hilbert vectors, we must consider a finite volume V and periodic boundary conditions, as in Sect. 5.4.1.

5.5.3   Glauber States

According to Sect. 4.2.8, the commutation law $[\Psi, \Psi^\dagger] = 1$ leads to the eigenvalues $n \in \{0, 1, 2, \ldots\}$ of the operators $\Psi^\dagger \Psi$, and for a suitable phase convention to

$$\Psi|n\rangle = |n{-}1\rangle \sqrt{n} \qquad \Longleftrightarrow \qquad \Psi^\dagger|n\rangle = |n{+}1\rangle \sqrt{n+1} \, .$$

If n is the particle number, $|0\rangle$ corresponds to the vacuum state, Ψ is an annihilation operator, and $\Psi^\dagger$ is a creation operator.

In Sect. 4.5.4, we used these operators for linear oscillations and set $X = x_0 (\Psi + \Psi^\dagger)/2$, $P = p_0 (\Psi - \Psi^\dagger)/(2i)$. Since we are dealing here with canonically conjugate quantities, for which the scale factors are not essential, we now consider the components

$$A_1 \equiv \frac{\Psi + \Psi^\dagger}{2} = A_1{}^\dagger \quad \text{and} \quad A_2 \equiv \frac{\Psi - \Psi^\dagger}{2i} = A_2{}^\dagger \, .$$

If in particular the mean value (expectation value) $\overline{A}_1$ oscillates harmonically, then so does $\overline{A}_2$, but with the phase shifted by $\pi/2$. The commutation law $[\Psi, \Psi^\dagger] = 1$ delivers

$$[A_1, A_2] = \tfrac{1}{2}\mathrm{i} \, ,$$

and thus, according to p. 300, the uncertainty relation $\Delta A_1 \cdot \Delta A_2 \geq 1/4$.

In this and the next section, we shall consider in detail those states whose uncertainty product $\Delta A_1 \cdot \Delta A_2$ is as small as possible, thus "as classical as possible". Then, according to p. 300, we must have

$$(A_1 - \overline{A}_1)\,|\psi\rangle = -\mathrm{i}\frac{\Delta A_1}{\Delta A_2}\,(A_2 - \overline{A}_2)\,|\psi\rangle \, .$$

In this section, we restrict ourselves to $\Delta A_1 = \Delta A_2 = 1/2$ and hence to *Glauber states* (which were in fact introduced by Schrödinger much earlier [9]), also called *coherent states*, although this is somewhat misleading, because all pure states can be superposed coherently. They are particularly important for the electromagnetic field (the "photon states") of lasers. In the next section we shall consider the more general case $\Delta A_1 \neq \Delta A_2$, and in particular, *quenched states*.

With the field operators $\Psi |\psi\rangle = |\psi\rangle \overline{\Psi}$ from above, the constraint $(A_1 - \overline{A}_1) |\psi\rangle = -i(A_2 - \overline{A}_2) |\psi\rangle$ reads: *Glauber states are eigenstates of the annihilation operator* Ψ. This operator is not Hermitian. Therefore, we need a complex number in order to label the eigenvalue. α is normally used, and we shall follow that here:

$$\Psi |\alpha\rangle = |\alpha\rangle \, \alpha \, , \quad \text{with} \quad \langle \alpha | \alpha \rangle = 1 \, .$$

Then $\langle \alpha | \Psi^\dagger = \alpha^* \langle \alpha |$, and consequently,

$$\langle \alpha | A_1 | \alpha \rangle = \text{Re}\,\alpha \quad \text{and} \quad \langle \alpha | A_2 | \alpha \rangle = \text{Im}\,\alpha \, ,$$

or $\alpha = \overline{A}_1 + i \overline{A}_2$. Note that, when $X = x_0 A_1$ and $P = p_0 A_2$, we also have $\alpha = \overline{X}/x_0 + i\overline{P}/p_0$, so we take the two real phase-space components of the one-dimensional oscillation as a complex number.

We can create the Glauber state $|\alpha\rangle$ with a unitary operator $D(\alpha)$ (the exponent is anti-Hermitian) from the ground state $|0\rangle$:

$$D(\alpha) \equiv \exp(\alpha \Psi^\dagger - \alpha^* \Psi) \, , \quad \text{with} \quad D^\dagger(\alpha) = D(-\alpha) = D^{-1}(\alpha) \, .$$

Using the property $D^\dagger(\alpha) \Psi D(\alpha) = \Psi + \alpha \, 1$ (the Hausdorff series, see p. 290, only contains two terms here), $D(\alpha)$ is called the *displacement operator*. It leads to

$$\Psi D(\alpha)|0\rangle = D(\alpha) \, (\Psi + \alpha)|0\rangle = D(\alpha)|0\rangle \, \alpha \, ,$$

so

$$|\alpha\rangle = D(\alpha)|0\rangle \, .$$

Here, according to p. 290, we may factorize, hence,

$$D(\alpha) = \exp(\alpha \Psi^\dagger) \, \exp(-\alpha^* \Psi) \, \exp(-\tfrac{1}{2}|\alpha|^2) \, ,$$

and use $\exp(-\alpha^* \Psi)|0\rangle = |0\rangle$ along with $\Psi^{\dagger n}|0\rangle = |n\rangle \sqrt{n!}$:

$$|\alpha\rangle = \exp(-\tfrac{1}{2}|\alpha|^2) \sum_{n=0}^{\infty} \frac{\alpha^n}{\sqrt{n!}} \, |n\rangle \, .$$

Incidentally, $D(\alpha + \beta)$ does not simply factorize into $D(\alpha)\,D(\beta)$, because a phase factor also occurs: $D(\alpha + \beta) = \exp\{i\mathrm{Im}(\alpha^*\beta)\}\,D(\alpha)\,D(\beta)$. This yields

$$D(\alpha)\,D(\beta) = \exp(\alpha\beta^* - \alpha^*\beta)\,D(\beta)\,D(\alpha)\ .$$

Consequently, we also have

$$\langle\alpha|\alpha'\rangle = \exp\{-\tfrac{1}{2}|\alpha - \alpha'|^2 + i\mathrm{Im}(\alpha^*\alpha')\}\ .$$

The eigenstates of the non-Hermitian operators Ψ are thus neither countable nor orthogonal to each other. They nevertheless form a complete basis. We only have to integrate over the whole complex plane. Instead of $\mathrm{dRe}\alpha\ \mathrm{dIm}\alpha$, however, we write for short $\mathrm{d}^2\alpha$ and take α and α^* to be independent of each other. Then

$$\int \frac{\mathrm{d}^2\alpha}{\pi}\ |\alpha\rangle\langle\alpha| = 1\ .$$

If we expand the left-hand side in terms of the complete basis $\{|n\rangle\}$, then we obtain $\langle n|\alpha\rangle\langle\alpha|n'\rangle = \exp(-|\alpha|^2)\,\alpha^n\,\alpha^{*n'}/\sqrt{n!\,n'!}$, with $\alpha = a\exp(i\varphi)$, or $\mathrm{d}^2\alpha = a\,\mathrm{d}a\,\mathrm{d}\varphi$ and

$$\int \frac{\mathrm{d}^2\alpha}{\pi}\ \langle n|\alpha\rangle\langle\alpha|n'\rangle = \int_0^\infty \frac{a^{n+n'+1}e^{-a^2}}{\sqrt{n!\,n'!}}\ \mathrm{d}a\ \frac{1}{\pi}\int_0^{2\pi} e^{i(n-n')\varphi}\ \mathrm{d}\varphi\ .$$

The last integral is equal to $2\pi\,\delta_{nn'}$ and, for $n = n'$, the one to the left of it is equal to $1/2$ (if set we $x = a^2$, so that $\mathrm{d}x = 2a\,\mathrm{d}a$, then $\int_0^\infty x^n\exp(-x)\,\mathrm{d}x = n!$ leads to the result). The double integral is equal to $\langle n|n'\rangle$.

So far we have always taken orthogonal bases and, for continuous variables, have arrived at simple integrals. But now the states are no longer orthogonal to each other and we require double integrals. The basis $\{|\alpha\rangle\}$ is said to be *over-complete*. An arbitrary state can be decomposed in terms of these, but no longer uniquely, because the basis states now depend linearly on each other. Hence, e.g., for all $n \in \{1, 2, \ldots\}$,

$$|o\rangle = \Psi^n|0\rangle = \int \frac{\mathrm{d}^2\alpha}{\pi}\ \Psi^n|\alpha\rangle\langle\alpha|0\rangle = \int \frac{\mathrm{d}^2\alpha}{\pi}\ |\alpha\rangle\,\alpha^n\exp(-\tfrac{1}{2}|\alpha|^2)\ .$$

Consequently, there are even infinitely many linear combinations of states $|\alpha\rangle$ which may result in the zero vector $|o\rangle$.

In the Glauber state $|\alpha\rangle$, the operator $N = \Psi^\dagger\Psi$ has the expectation value

$$\langle\alpha|\,N\,|\alpha\rangle = |\alpha|^2\ ,$$

and, with $N^2 = \Psi^\dagger(\Psi^\dagger\Psi + 1)\Psi$, the uncertainty $\Delta N = |\alpha|$. Note that this increases with $|\alpha|$, but the relative uncertainty $\Delta N/\overline{N} = |\alpha|^{-1}$ decreases, as expected for the transition to classical mechanics. For a harmonic oscillation, we obtain the result $\overline{H} = \hbar\omega\,(|\alpha|^2 + 1/2)$ and $\Delta H = \hbar\omega\,|\alpha|$.

Furthermore, the probability for the *Fock state* $|n\rangle$, with sharp particle number and unsharp phase, depends only on the modulus of α:

$$|\langle n|\alpha\rangle|^2 = \exp(-|\alpha|^2)\,\frac{|\alpha|^{2n}}{n!}\ .$$

This is a *Poisson distribution* ρ_n with mean value $\langle n\rangle = |\alpha|^2$ and uncertainty $\Delta n = |\alpha|$ (see p. 519).

For the time dependence, using $H|n\rangle = |n\rangle\,\hbar\omega\,(n+\frac{1}{2})$ and $|\alpha(0)\rangle = |\alpha_0\rangle$, we obtain

$$|\alpha(t)\rangle = \exp\frac{-\mathrm{i}Ht}{\hbar}\,|\alpha_0\rangle = \exp\frac{-|\alpha_0|^2 - \mathrm{i}\omega t}{2}\,\sum_{n=0}^{\infty}\frac{(\alpha_0\,\mathrm{e}^{-\mathrm{i}\omega t})^n}{\sqrt{n!}}\,|n\rangle\ ,$$

whence $\langle\alpha(t)|\,\Psi\,|\alpha(t)\rangle = \alpha_0\,\exp(-\mathrm{i}\omega t)$, and its complex conjugate for the expectation value of $\Psi^\dagger$. Consequently, we have

$$\overline{X(t)} = x_0\,\mathrm{Re}(\alpha_0\,\mathrm{e}^{-\mathrm{i}\omega t})\quad\text{and}\quad\overline{P(t)} = p_0\,\mathrm{Im}(\alpha_0\,\mathrm{e}^{-\mathrm{i}\omega t})\ .$$

The Glauber states oscillate harmonically with angular frequency ω and with fixed position, momentum, and energy uncertainties. Ehrenfest's equations are also valid here.

5.5.4 Quenched States

We now allow for $\Delta A_1 \neq \Delta A_2$, but keep searching for further states with an uncertainty product $\Delta A_1 \Delta A_2$ as small as possible. The necessary equation mentioned in the last section can be reformulated as the eigenvalue equation

$$(A_1 + \mathrm{i}\,\frac{\Delta A_1}{\Delta A_2}\,A_2)\,|\psi\rangle = |\psi\rangle\,(\overline{A}_1 + \mathrm{i}\,\frac{\Delta A_1}{\Delta A_2}\,\overline{A}_2)\ ,\quad\text{with}\ A_n = A_n^\dagger\ .$$

But the non-Hermitian operator is now composed linearly of the annihilation operator Ψ and the creation operator $\Psi^\dagger$. Therefore, we consider the *Bogoliubov transformation*, but now for boson operators, with $u = u^* > 0$ and $v = v^*$:

$$\Phi = u\Psi + v\Psi^\dagger\quad\Longleftrightarrow\quad\Phi^\dagger = u\Psi^\dagger + v\Psi\ .$$

Note that a common phase factor is unimportant, so we may choose $u = u^* > 0$, and $v \neq v^*$ would then lead to $\Delta A_1 \cdot \Delta A_2 > 1/4$.

With $[\Phi,\,\Phi^\dagger] = (u^2 - v^2)\,[\Psi,\,\Psi^\dagger]$, we also require

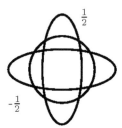

Fig. 5.15 Uncertainties of the squared components (A_1 horizontal and A_2 vertical) for the Glauber state ($z = 0$) and quenched states ($z = \pm 1/2$). As in Fig. 4.20, instead of the ΔA values, we now plot ellipses of the same area with corresponding principal axes, where the product is $\Delta A_1 \cdot \Delta A_2 = 1/4$

$$u^2 - v^2 = 1 \quad \Longleftrightarrow \quad [\Phi, \Phi^\dagger] = 1 \;.$$

For $u = \cosh z$ and $v = \sinh z$, this is possible with a single real parameter z (see Fig. 5.15). Recall that, for fermions, we had $u^2 + v^2 = 1$ (see p. 457), and therefore circular instead of hyperbolic functions. Note also that, with $u \geq 1$, we are no longer allowed to choose $u = 0$ and then replace Ψ by $\Phi^\dagger$. Conversely, then $\Psi = u\Phi - v\Phi^\dagger$ and $\Psi^\dagger = u\Phi^\dagger - v\Phi$.

The Bogoliubov transformation can be carried out by a unitary operator S: $\Phi = S\,\Psi S^\dagger$. In particular, if we set $S = \exp A$ with $A^\dagger = -A$ with $S^\dagger = S^{-1}$, then according to Hausdorff (see p. 290), A follows from

$$\Phi = u\Psi + v\Psi^\dagger = \cosh z \;\Psi + \sinh z \;\Psi^\dagger$$
$$= S\,\Psi S^\dagger = \Psi + \tfrac{1}{1!}\,[A,\,\Psi] + \tfrac{1}{2!}\,[A,\,[A,\,\Psi]] + \cdots \;,$$

since here only $[A,\,\Psi] = z\Psi^\dagger$ has to hold, and thus $[A,\,\Psi^\dagger] = z\Psi$. Consequently, $A = \tfrac{1}{2}z\,(\Psi^2 - \Psi^{\dagger 2})$ up to an arbitrary phase factor in S. The *quench operator* (or "squeeze operator")

$$S(z) = \exp\frac{z\,(\Psi^2 - \Psi^{\dagger 2})}{2}$$

affects the ratio $\Delta A_1 / \Delta A_2$, as will now be shown.

Corresponding to the Glauber state $|\alpha\rangle$ (eigenstate of Ψ) is a *quenched state* $S|\alpha\rangle$, an eigenstate of Φ. With $\Phi S = S\Psi$, we have in particular,

$$\Phi S|\alpha\rangle = S|\alpha\rangle\,\alpha \;.$$

In order to be able to employ $|\alpha\rangle = D(\alpha)|0\rangle$, we investigate the product $SD(\alpha)$ with $D(\alpha) = \exp(\alpha\Psi^\dagger - \alpha^*\Psi) = f(\Psi, \Psi^\dagger)$. Since

$$Sf\,(\Psi,\,\Psi^\dagger)S^\dagger = f\,(S\Psi S^\dagger,\,S\Psi^\dagger S^\dagger) = f\,(\Phi,\,\Phi^\dagger)\;,$$

we also have $Sf(\Psi, \Psi^\dagger) = f(\Phi, \Phi^\dagger)S$. Here,

$$\alpha\Phi^\dagger - \alpha^*\Phi = (\alpha u - \alpha^* v)\,\Psi^\dagger - (\alpha u - \alpha^* v)^*\,\Psi\;.$$

If we therefore set

$$\beta \equiv u\alpha - v\alpha^* \qquad \Longleftrightarrow \qquad \alpha = u\beta + v\beta^*\;,$$

we find that $SD(\alpha) = D(\beta)\,S$, so for the eigenstate of Φ with eigenvalue α,

$$S\,|\alpha\rangle = S\,D(\alpha)\,|0\rangle = D(\beta)\,S\,|0\rangle\;.$$

For the quasi-vacuum ("quenched vacuum"), we have $S|0\rangle$, hence $\Phi S|0\rangle = |o\rangle$.

The expectation value of the operator $F(\Psi, \Psi^\dagger)$ in the quenched state $S|\alpha\rangle$ is thus the vacuum expectation value of

$$D^\dagger(\alpha)\,S^\dagger\,F(\Psi, \Psi^\dagger)\,S\,D(\alpha) = S^\dagger\,D^\dagger(\beta)\,F(\Psi, \Psi^\dagger)\,D(\beta)\,S\;.$$

In the term $D^\dagger(\beta)\,F(\Psi, \Psi^\dagger)\,D(\beta) = F(\Psi + \beta, \Psi^\dagger + \beta^*)$, it is now useful to change the representation $\Psi \to u\Phi - v\Phi^\dagger$ and $\Psi^\dagger \to u\Phi^\dagger - v\Phi$, with $\langle 0|\Phi^\dagger S^\dagger = \langle o|$ and $\Phi S|0\rangle = |o\rangle$. Thus, it follows in particular that

$$\langle\Psi\rangle = \beta\;, \qquad \langle\Psi^2\rangle = \beta^2 - uv\;, \qquad \langle\Psi^\dagger\Psi\rangle = |\beta|^2 + v^2\;,$$

and for the Hermitian operator $F = f\Psi + f^*\Psi^\dagger$,

$$\langle F\rangle = f\beta + f^*\beta^* \quad\text{and}\quad \Delta F = |fu - f^*v| = |f|\sqrt{\cosh 2z - \cos 2\varphi\,\sinh 2z}\;,$$

where $f = e^{i\varphi}|f|$. For the two components in the quenched state $S|\alpha\rangle$ with $f = 1/2$ or $f = -1/2i$ and $u \geq v$, we find

$$\Delta A_1 = \tfrac{1}{2}(u - v) \qquad\text{and}\qquad \Delta A_2 = \tfrac{1}{2}(u + v)\;.$$

Here we have $u \mp v = \exp(\mp z)$ and hence $\Delta A_1 \cdot \Delta A_2 = \tfrac{1}{4}$ and $\Delta A_1/\Delta A_2 = \exp(-2z)$.

Quenched states are appropriate, e.g., when comparing two oscillations of different frequency, because their ground states have $\Delta X = \tfrac{1}{2}x_0$ and $\Delta P = \tfrac{1}{2}p_0$, with $x_0 p_0 = 2\hbar$, but x_0 and $p_0 = \sqrt{2\hbar m\omega}$ depend upon the given frequency. In an "inappropriate basis", the oscillation appears compressed (or expanded) (see Fig. 5.16).

The quenched states are formed under *parametric amplification* (in this context, see the discussion of parametric resonance in Sects. 2.3.10 and 2.4.11), with the Hamilton operator

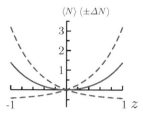

Fig. 5.16 Influence of the quench parameter z on the particle number (*continuous red*) and its uncertainty $\Delta N = \sqrt{|u\beta - v\beta^*|^2 + u^2v^2}$ (*dashed blue*), here shown for the quenched vacuum. The average particle number is in fact then as small as possible for a given z, but greater than zero for $z \neq 0$

$$H = \hbar\omega \, \Psi^\dagger\Psi - i\hbar\kappa \, \frac{\exp(i\omega_p t)\,\Psi^2 - \exp(-i\omega_p t)\,\Psi^{\dagger 2}}{2} \,,$$

or with $H = \hbar\omega \, \Psi^\dagger\Psi + \frac{1}{2}\hbar\kappa \, \{\exp(i\omega_p t)\,\Psi^2 + \exp(-i\omega_p t)\,\Psi^{\dagger 2}\}$, which can lead back to the former under the phase transformation $\Psi \to \exp(-i\pi/4)\,\Psi$. Here ω is the angular frequency of the considered light and $\omega_p = 2\omega$ that of the pump light, while κ gives the (real) coupling constant. The pump light is described classically here, with fixed intensity, and in this sense, the above Hamilton operator is "semi-classical". This will be discussed in more detail in Sect. 5.5.6. We thus have the Heisenberg equation

$$\frac{d\Psi}{dt} = \frac{i[H,\,\Psi]}{\hbar} = -i\omega\,\Psi + \kappa\,\exp(-2i\omega t)\,\Psi^\dagger \,.$$

It can be solved by carrying out the time-dependent Bogoliubov transformation

$$\Psi(t) = \exp(-i\omega t)\,\{\cosh(\kappa t)\,\Psi(0) + \sinh(\kappa t)\,\Psi^\dagger(0)\}\,.$$

The phase factor is unimportant here. This therefore leads to quenched states. For the photon number operator $N(t) = \Psi^\dagger(t)\,\Psi(t)$, we have

$$N(t) = \sinh^2(\kappa t)\,\,1 + \cosh(2\kappa t)\,\,N(0) + \sinh(2\kappa t)\,\frac{\Psi^{\dagger 2}(0) + \Psi^2(0)}{2}\,.$$

If there is no light initially so that $N(0) = 0$, then the average photon number increases as $\sinh^2(\kappa t)$, although the result for long times is certainly not correct, because the pump light cannot supply energy inexhaustibly.

5.5.5 *Expansion in Terms of Glauber States*

In Sect. 5.5.2, we gave different observables (e.g., $\mathbf{E}, \mathbf{B}, H, \mathbf{P}$) as functions of the field operators $\Psi, \Psi^\dagger$. If we now expand the operator $f(\Psi, \Psi^\dagger)$ in terms of Glauber states,

$$f(\Psi, \Psi^\dagger) = \int \frac{d^2\alpha}{\pi} \frac{d^2\alpha'}{\pi} |\alpha\rangle\langle\alpha| f(\Psi, \Psi^\dagger) |\alpha'\rangle\langle\alpha'| ,$$

then we may evaluate the coefficients, if $f(\Psi, \Psi^\dagger)$ is *normal ordered*, i.e., in all products, the creation operators occur to the left of the annihilation operators:

$$f(\Psi, \Psi^\dagger) = \sum_{rs} f_{rs}^{(n)} \Psi^{\dagger r} \Psi^s .$$

Then $\langle\alpha| f(\Psi, \Psi^\dagger) |\alpha'\rangle = \sum_{rs} f_{rs}^{(n)} \alpha^{*r}\alpha'^s \langle\alpha|\alpha'\rangle$. With the abbreviation

$$f^{(n)}(\alpha', \alpha^*) = \sum_{rs} f_{rs}^{(n)} \alpha^{*r}\alpha'^s ,$$

it follows that $\langle\alpha| f(\Psi, \Psi^\dagger) |\alpha'\rangle = f^{(n)}(\alpha', \alpha^*) \langle\alpha|\alpha'\rangle$ with

$$\langle\alpha|\alpha'\rangle = \exp\{-\frac{1}{2}|\alpha-\alpha'|^2 + i\mathrm{Im}(\alpha^*\alpha')\} ,$$

according to p. 472.

The operator $f(\Psi, \Psi^\dagger)$ may also be *anti-normal-ordered*, with the creation operators to the right of the annihilation operators:

$$f(\Psi, \Psi^\dagger) = \sum_{rs} f_{rs}^{(a)} \Psi^r \Psi^{\dagger s} .$$

Then for the function $f(\Psi, \Psi^\dagger)$, just one double integral (over $d^2\alpha$) suffices. If we insert the unit operator between Ψ^r and $\Psi^{\dagger s}$, then we obtain the important relation

$$f(\Psi, \Psi^\dagger) = \int \frac{d^2\alpha}{\pi} f^{(a)}(\alpha, \alpha^*) |\alpha\rangle\langle\alpha| ,$$

with

$$f^{(a)}(\alpha, \alpha^*) = \sum_{rs} f_{rs}^{(a)} \alpha^r \alpha^{*s} .$$

Here we have $f^{(a)}(\alpha, \alpha^*) \neq f^{(n)}(\alpha, \alpha^*)$, as can be recognized, e.g., from $f(\Psi, \Psi^\dagger) = \Psi\Psi^\dagger = \Psi^\dagger\Psi + 1$, because then $f^{(a)}(\alpha, \alpha^*) = |\alpha|^2$, $f^{(n)}(\alpha, \alpha^*) = |\alpha|^2 + 1$, and so $f^{(n)}(\alpha, \alpha^*) = f^{(a)}(\alpha, \alpha^*) + 1$. More general than $\Psi\Psi^{\dagger n} = \Psi^{\dagger n}\Psi + n\Psi^{\dagger n-1}$ is

$$\left.\begin{matrix} \Psi^m \Psi^{\dagger n} \\ \Psi^{\dagger n} \Psi^m \end{matrix}\right\} = \sum_l \frac{(\pm)^l m! \, n!}{l! \, (m-l)! \, (n-l)!} \begin{cases} \Psi^{\dagger n-l} \Psi^{m-l} , \\ \Psi^{m-l} \Psi^{\dagger n-l} , \end{cases}$$

as can be shown by induction (see Problem 4.20).

Note that, from $f^{(a)}(\alpha, \alpha^*)$, we can also determine $f^{(n)}(\alpha, \alpha^*)$, but we cannot determine $f^{(n)}(\alpha', \alpha^*)$ for $\alpha' \neq \alpha$:

$$f^{(n)}(\alpha', \alpha'^*) = \int \frac{d^2\alpha}{\pi} \, \exp(-|\alpha-\alpha'|^2) \, f^{(a)}(\alpha, \alpha^*) \,,$$

with $f^{(n)}(\alpha', \alpha'^*) = \langle \alpha' | f(\Psi, \Psi^\dagger) | \alpha' \rangle$ and $|\langle \alpha | \alpha' \rangle|^2 = \exp(-|\alpha-\alpha'|^2)$.

Generally, we may set

$$f(\Psi, \Psi^\dagger) = \int \frac{d^2\xi}{\pi} \, \exp(\xi\Psi^\dagger) \, \exp(-\xi^*\Psi) \, F^{(n)}(\xi, \xi^*)$$

$$= \int \frac{d^2\xi}{\pi} \, \exp(-\xi^*\Psi) \, \exp(\xi\Psi^\dagger) \, F^{(a)}(\xi, \xi^*) \,,$$

with the expansion coefficients

$$F^{(n)}(\xi, \xi^*) = \mathrm{tr}\{\exp(\xi^*\Psi) \, \exp(-\xi\Psi^\dagger) \, f(\Psi, \Psi^\dagger)\} \,,$$
$$F^{(a)}(\xi, \xi^*) = \mathrm{tr}\{\exp(-\xi\Psi^\dagger) \, \exp(\xi^*\Psi) \, f(\Psi, \Psi^\dagger)\} \,.$$

If we replace $f(\Psi, \Psi^\dagger)$ in $\mathrm{tr}\{\exp(\xi^*\Psi) \exp(-\xi\Psi^\dagger) f(\Psi, \Psi^\dagger)\}$ by the normal-ordered double integral $\int d^2\xi' \, \pi^{-1} \exp(\xi'\Psi^\dagger) \exp(-\xi'^*\Psi) F^{(n)}(\xi', \xi'^*)$, then we arrive at $\int d^2\xi' \, \pi^{-1} F^{(n)}(\xi', \xi'^*) \, \mathrm{tr}\{\exp[(\xi - \xi')^*\Psi] \exp[-(\xi-\xi')\Psi^\dagger]\}$. If we insert the unit operator $\int d^2\alpha \, \pi^{-1} |\alpha\rangle\langle\alpha|$ between the two exponential functions in the trace, then we obtain $\int d^2\alpha \, \pi^{-1} \exp\{(\xi-\xi')^*\alpha - (\xi-\xi')\alpha^*\}$, and the exponent is $2\mathrm{i}\,\mathrm{Im}\{(\xi-\xi')^*\alpha\}$, thus equal to $2\mathrm{i}\,\mathrm{Re}(\xi-\xi')\mathrm{Im}\alpha - 2\mathrm{i}\,\mathrm{Im}(\xi-\xi')\mathrm{Re}\alpha$. In this way, we arrive at the Fourier expansions of delta functions of the real and imaginary parts of $2(\xi-\xi')$. This is easily integrated over ξ', and we arrive at $F^{(n)}(\xi, \xi^*)$. The proof for $F^{(a)}(\xi, \xi^*)$ is very similar. We thus obtain the Fourier transforms

$$F^{(n)}(\xi, \xi^*) = \int \frac{d^2\alpha}{\pi} \, \exp(\xi^*\alpha - \xi\alpha^*) \, f^{(n)}(\alpha, \alpha^*) \,,$$

$$f^{(n)}(\alpha, \alpha^*) = \int \frac{d^2\xi}{\pi} \, \exp(\xi\alpha^* - \xi^*\alpha) \, F^{(n)}(\xi, \xi^*) \,.$$

Note that we usually require the normalization factor 2π for the Fourier transform. Here π suffices, because the factor of 2 is already contained in the expression $2\,\mathrm{Im}(\xi^*\alpha) = \mathrm{Im}(2\xi^*\alpha)$. Of course, the relation between $F^{(a)}(\xi, \xi^*)$ and $f^{(a)}(\alpha, \alpha^*)$ is also a Fourier transform.

We have the trace of the anti-normal-ordered products $\exp(\xi^*\Psi) \exp(-\xi\Psi^\dagger)$ for $F^{(n)}(\xi, \xi^*)$, and that of the normal-ordered products for $F^{(a)}(\xi, \xi^*)$. In both cases, the product of the exponential functions and of $f(\Psi, \Psi^\dagger)$ can be reformulated as a normal-ordered product of powers of Ψ and $\Psi^\dagger$, and the unit operator inserted between the two factors.

5.5.6 Density Operator in the Glauber Basis

If we set likewise

$$\rho(\Psi, \Psi^\dagger) = \sum_{rs} \rho_{rs}^{(n)} \, \Psi^{\dagger r} \Psi^s = \sum_{rs} \rho_{rs}^{(a)} \, \Psi^r \Psi^{\dagger s} ,$$

for the density operator $\rho(\Psi, \Psi^\dagger)$, then

$$\mathrm{tr}\,(\Psi^r \Psi^{\dagger s} \Psi^{\dagger u} \Psi^v) = \int d^2\alpha \, \pi^{-1} \alpha^{*s+u} \alpha^{r+v}$$

implies the equations

$$\langle f(\Psi, \Psi^\dagger) \rangle = \int \frac{d^2\alpha}{\pi} \, \rho^{(a)}(\alpha, \alpha^*) f^{(n)}(\alpha, \alpha^*) = \int \frac{d^2\alpha}{\pi} \, \rho^{(n)}(\alpha, \alpha^*) f^{(a)}(\alpha, \alpha^*) ,$$

where one normal and one anti-normal-ordered operator always occur, like covariant and contravariant components for the scalar product. Since $f(\Psi, \Psi^\dagger) = 1$, we therefore also have $\int d^2\alpha \, \rho^{(n)}(\alpha, \alpha^*) = \int d^2\alpha \, \rho^{(a)}(\alpha, \alpha^*) = \pi$.

As for the Wigner function (see Fig. 4.7), the different representations of $\langle f \rangle$ suggest introducing *quasi-probability densities*, and in particular, the *P-function*

$$P(\alpha) \equiv \frac{\rho^{(a)}(\alpha, \alpha^*)}{\pi} , \quad \text{with} \quad \int d^2\alpha \, P(\alpha) = 1 ,$$

and the *Q-function* (or *Husimi function*)

$$Q(\alpha) \equiv \frac{\rho^{(n)}(\alpha, \alpha^*)}{\pi} , \quad \text{with} \quad \int d^2\alpha \, Q(\alpha) = 1 .$$

It then follows that

$$\langle f(\Psi, \Psi^\dagger) \rangle = \int d^2\alpha \, P(\alpha) f^{(n)}(\alpha, \alpha^*) = \int d^2\alpha \, Q(\alpha) f^{(a)}(\alpha, \alpha^*) .$$

Since $\rho = \int d^2\alpha \, P(\alpha) \, |\alpha\rangle\langle\alpha|$, but $|\alpha\rangle\langle\alpha|$ does not project on orthogonal states, the *P*-function is only a quasi-probability density. The *Q*-function does in fact have the properties of a probability density, i.e., it is real and never negative, with $\rho^{(n)}(\alpha, \alpha^*) = \langle\alpha|\rho|\alpha\rangle$, but does not lead to the full density operator.

Very useful here are also the *normal-ordered characteristic function*

$$C^{(n)}(\xi, \xi^*) \equiv \langle \exp(\xi \Psi^\dagger) \, \exp(-\xi^* \Psi) \rangle$$

and the *anti-normal-ordered characteristic function*

$$C^{(a)}(\xi, \xi^*) \equiv \langle \exp(-\xi^* \Psi) \exp(\xi \Psi^\dagger) \rangle .$$

These can be used to derive the *moments* at the position $\xi = \xi^* = 0$:

$$\langle \Psi^{\dagger r} \Psi^s \rangle = (-)^s \frac{\partial^{r+s} C^{(n)}}{\partial \xi^r \partial \xi^{*s}} \quad \text{and} \quad \langle \Psi^r \Psi^{\dagger s} \rangle = (-)^r \frac{\partial^{r+s} C^{(a)}}{\partial \xi^{*r} \partial \xi^s} .$$

The two functions are related, because, according to p. 290,

$$\exp(\xi \Psi^\dagger - \xi^* \Psi) = \exp(\xi \Psi^\dagger) \exp(-\xi^* \Psi) \exp(-\tfrac{1}{2}|\xi|^2)$$
$$= \exp(-\xi^* \Psi) \exp(\xi \Psi^\dagger) \exp(+\tfrac{1}{2}|\xi|^2) ,$$

so $C^{(n)}(\xi, \xi^*) = C^{(a)}(\xi, \xi^*) \exp(|\xi|^2)$. The characteristic functions are the Fourier transforms of $\rho(\alpha, \alpha^*)$, so

$$C^{(a)}(\xi, \xi^*) = \int \frac{d^2 \alpha}{\pi} \exp(\xi \alpha^* - \xi^* \alpha) \, \rho^{(n)}(\alpha, \alpha^*) ,$$

$$\rho^{(n)}(\alpha, \alpha^*) = \int \frac{d^2 \xi}{\pi} \exp(\xi^* \alpha - \xi \alpha^*) \, C^{(a)}(\xi, \xi^*) ,$$

and likewise $C^{(n)}(\xi, \xi^*)$ and $\rho^{(a)}(\alpha, \alpha^*)$ are Fourier transforms of one another.

Let us consider some useful examples:

(1) Clearly, for the Glauber state $|\alpha\rangle$, we have

$$C^{(n)}(\xi, \xi^*) = \exp(\xi \alpha^* - \xi^* \alpha) , \quad \text{with} \quad \rho = |\alpha\rangle\langle\alpha| .$$

(2) For the laser, a superposition of these states with equal amplitude and unknown phase $\arg \alpha = \varphi$ is important. We have to average over φ to obtain $\rho = \overline{|\alpha\rangle\langle\alpha|}$. Then we arrive at $C^{(n)}(\xi, \xi^*) = \frac{1}{2\pi} \int_0^{2\pi} d\varphi \, \exp\{|\xi \alpha|(e^{-i\varphi} - e^{i\varphi})\}$. With $z = |2\xi\alpha|$ and $t = \exp(-i\varphi)$, the integrand can be expanded in terms of *regular Bessel functions* $J_n(z)$, because they have the *generating function*

$$\exp\left(z \frac{t - t^{-1}}{2}\right) = \sum_{n=-\infty}^{\infty} J_n(z) \, t^n , \quad \text{for } t \neq 0 ,$$

with the symmetry $J_{-n}(z) = (-)^n J_n(z)$. If we expand $\exp(\tfrac{1}{2}zt)$ and $\exp(-\tfrac{1}{2}z/t)$ in series, we obtain the regular Bessel functions

$$J_n(z) = \sum_{k=0}^{\infty} \frac{(-)^k \, (z/2)^{n+2k}}{k! \, (n+k)!} ,$$

Fig. 5.17 Regular Bessel functions and irregular Bessel functions, also called Neumann functions, for n from 0 (*black*) to 3 (*blue*) (*continuous* for n even, *dotted* for n odd). Asymptotically, $J_n(x) \approx$

$$\sqrt{\frac{2}{\pi x}} \; \cos\{x - (n + \tfrac{1}{2})\tfrac{1}{2}\pi\}$$

and $N_n(x) \approx$

$$\sqrt{\frac{2}{\pi x}} \; \sin\{x - (n + \tfrac{1}{2})\tfrac{1}{2}\pi\}$$

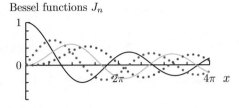

Bessel functions J_n

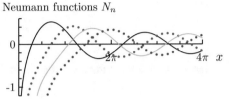

Neumann functions N_n

as shown in Fig. 5.17. Note that the *spherical Bessel functions* $F_l(z)$ mentioned on p. 401 are *Bessel functions of half-integer index*, viz., $F_l(z) = \sqrt{\pi z/2}\, J_{l+1/2}(z)$. From the last equation,

$$\exp(iz \sin \varphi) = \sum_{n=-\infty}^{\infty} J_n(z) \; \exp(in\varphi) \;.$$

With this we obtain

$$C^{(\mathrm{n})}(\xi, \xi^*) = J_0(2|\xi \alpha|) \;, \quad \text{with} \quad \rho = \overline{|\alpha\rangle\langle\alpha|} \;.$$

The anti-normal-ordered function $C^{(\mathrm{a})}$ also contains the factor $\exp(-|\xi|^2)$.

(3) The quenched state $S|\alpha\rangle$ has the normal-ordered characteristic function

$$\langle\alpha|S^\dagger \exp(\xi\Psi^\dagger)\exp(-\xi^*\Psi)S|\alpha\rangle = \exp(\tfrac{1}{2}|\xi|^2)\, \langle 0|S^\dagger D^\dagger(\beta)D(\xi)D(\beta)S|0\rangle \;,$$

with $\beta = u\alpha - v\alpha^*$. Here, according to p. 472, we have

$$D^\dagger(\beta)\, D(\xi)\, D(\beta) = \exp(\xi\beta^* - \xi^*\beta)\, D(\xi) \;,$$

whence

$$C^{(\mathrm{n})}(\xi, \xi^*) = \exp(\tfrac{1}{2}|\xi|^2 + \xi\beta^* - \xi^*\beta)\, \langle 0|\, S^\dagger \exp(\xi\Psi^\dagger - \xi^*\Psi)\, S\, |0\rangle \;.$$

As on p. 475, we replace $\xi\Psi^\dagger - \xi^*\Psi \to (u\xi + v\xi^*)\, \Phi^\dagger - (u\xi + v\xi^*)^*\, \Phi$, and the vacuum expectation value is found to be $\exp(-\tfrac{1}{2}|u\xi + v\xi^*|^2)$. So in total, for the quenched state,

$$C^{(n)}(\xi, \xi^*) = \exp\left(\beta^*\xi - \beta\xi^* - v^2\xi\xi^* - uv\,\frac{\xi^2 + \xi^{*\,2}}{2}\right) .$$

This leads, e.g., to the expressions mentioned in connection with Fig. 4.21.

(4) According to p. 580, the canonical density operator

$$\rho = \sum_n \frac{\langle N \rangle^n}{\langle N + 1 \rangle^{n+1}}\, |n\rangle\langle n|$$

with $\langle N \rangle = \{\exp(\hbar\omega/kT) - 1\}^{-1}$ and thus $\langle N + 1 \rangle = \{1 - \exp(-\hbar\omega/kT)\}^{-1}$ is associated with the temperature T. Hence, $|\langle\alpha|n\rangle|^2 = \exp(-|\alpha|^2)\,|\alpha|^{2n}/n!$ implies

$$\rho^{(n)}(\alpha, \alpha^*) = \langle\alpha|\,\rho\,|\alpha\rangle = \frac{1}{\langle N + 1 \rangle}\,\exp\,\frac{-|\alpha|^2}{\langle N + 1 \rangle} .$$

This means that $C^{(a)}(\xi, \xi^*)$ is the Fourier component of a Gauss function, thus also a Gauss function, according to p. 23:

$$C^{(a)}(\xi, \xi^*) = \exp\{-\langle N + 1 \rangle\,|\xi|^2\} .$$

The normal-ordered function $C^{(n)}(\xi, \xi^*)$ also requires the factor $\exp(|\xi|^2)$:

$$C^{(n)}(\xi, \xi^*) = \exp\{-\langle N \rangle\,|\xi|^2\} \qquad \Longleftrightarrow \qquad \rho^{(a)}(\alpha, \alpha^*) = \frac{1}{\langle N \rangle}\,\exp\,\frac{-|\alpha|^2}{\langle N \rangle}$$

for the canonical distribution.

5.5.7 Atom in a Light Field

We consider an atom with two eigenstates $\{|\uparrow\rangle, |\downarrow\rangle\}$ at the energies $\pm\frac{1}{2}\hbar\omega_A$ and a light field with the energy quantum $\hbar\omega_L$. The atom can be described using Pauli operators σ and the field using Bose operators Ψ, $\Psi^\dagger$, and for the coupling $-\mathbf{p}\cdot\mathbf{E}$, the dipole moment with $\sigma_x = \sigma_+ + \sigma_-$ and the field strength with $i(\Psi - \Psi^\dagger)$, if we combine all remaining factors into the real factor $\frac{1}{2}\hbar g$. The phase transformation $\Psi \to i\Psi$ changes $-i(\Psi - \Psi^\dagger)$ into $\Psi + \Psi^\dagger$. In comparison to $\sigma_+\Psi + \sigma_-\Psi^\dagger$, the parts $\sigma_+\Psi^\dagger + \sigma_-\Psi$ couple to states of much higher frequency, viz., $\omega_L + \omega_A$ instead of $\omega_L - \omega_A$, and therefore do not contribute to the time average. Note that $\sigma_+\Psi$ describes induced or *forced absorption*, and $\sigma_-\Psi^\dagger$ induced or *forced emission*. With this we arrive at the Hamilton operator of the *Jaynes–Cummings model*:

$$H = \tfrac{1}{2}\hbar\omega_A\,\sigma_z + \hbar\omega_L\,\Psi^\dagger\Psi + \tfrac{1}{2}\hbar g\,(\sigma_+\Psi + \sigma_-\Psi^\dagger) .$$

Fig. 5.18 Eigenfrequencies $\omega_\pm$ in the Jaynes–Cummings model as a function of the detuning $\Delta \equiv \omega_L - \omega_A$, each relative to ω_A, and here for $g/\omega_A = 0.1$

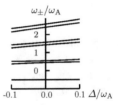

Above the ground state $|\downarrow, 0\rangle$, with energy $-\frac{1}{2}\hbar\omega_A$, it couples the state pair $|\uparrow, n\rangle$ and $|\downarrow, n+1\rangle$, where n is the photon number:

$$H|\uparrow, n\rangle = \hbar\{(n+\tfrac{1}{2})\,\omega_L - \tfrac{1}{2}\Delta\}\,|\uparrow, n\rangle + \tfrac{1}{2}\hbar g\sqrt{n+1}\,|\downarrow, n+1\rangle ,$$
$$H|\downarrow, n+1\rangle = \tfrac{1}{2}\hbar g\sqrt{n+1}\,|\uparrow, n\rangle + \hbar\{(n+\tfrac{1}{2})\,\omega_L + \tfrac{1}{2}\Delta\}\,|\downarrow, n+1\rangle ,$$

with *detuning* $\Delta = \omega_L - \omega_A$ between the light field and the atom. According to p. 309 ($\frac{1}{2}\mathrm{trH} \pm \frac{1}{2}\sqrt{(\mathrm{trH})^2 - 4\det H}$), the eigenvalues of H are (see Fig. 5.18)

$$\omega_\pm = \omega_L\,(n+\tfrac{1}{2}) \pm \tfrac{1}{2}\Omega_n ,$$

with the generalized (to $\Delta \neq 0$) *Rabi frequency*

$$\Omega_n = \sqrt{(n+1)\,g^2 + \Delta^2} .$$

According to p. 310, the eigenstates associated with this doublet are

$$|+, n\rangle = |\uparrow, n\rangle \cos\theta_n + |\downarrow, n+1\rangle \sin\theta_n ,$$
$$|-, n\rangle = -|\uparrow, n\rangle \sin\theta_n + |\downarrow, n+1\rangle \cos\theta_n ,$$

where $\cos\theta_n = \sqrt{1-\Delta/\Omega_n}/\sqrt{2}$ and $\sin\theta_n = \sqrt{1+\Delta/\Omega_n}/\sqrt{2}$. They are thus eigenstates of $\Psi^\dagger\Psi + \sigma_+\sigma_-$ with eigenvalue $n+1$. For the remaining expectation values, we can use

$$\cos(2\theta_n) = -\frac{\Delta}{\Omega_n} \quad \text{and} \quad \sin(2\theta_n) = \frac{\sqrt{n+1}\,g}{\Omega_n} .$$

For example, the matrix elements of $\Psi^\dagger\Psi$ and $\sigma_z = 2\sigma_+\sigma_- - 1$ between the basis states with

$$(\Psi^\dagger\Psi - \sigma_+\sigma_-)|\pm, n\rangle = |\pm, n\rangle(n \mp \cos(2\theta_n)) + |\mp, n\rangle \sin(2\theta_n)$$

are easy to evaluate, and their time dependence is known to be $\exp(-i\omega_\pm t)$. If initially either the state $|\uparrow, n\rangle$ was occupied (upper sign) or the state $|\downarrow, n+1\rangle$ (lower sign), it follows that

$$\langle \sigma_z \rangle = \pm \left(\cos^2(2\theta_n) + \sin^2(2\theta_n) \, \cos \Omega_n t \right) .$$

In particular, only for a resonance ($\Delta = 0$) do all atoms end up in the other state. This is of course true for other initial light fields, e.g., for the Glauber state $|\alpha\rangle$. If initially the state $|\uparrow, \alpha\rangle$ was occupied (upper sign) or the state $|\downarrow, \alpha\rangle$ (lower sign), we arrive at the weight factors $\exp(-|\alpha|^2) \, |\alpha|^{2n+1\mp 1} \{(n + \tfrac{1}{2} \mp \tfrac{1}{2})!\}^{-1}$. We shall restrict ourselves to the case $|\alpha| \gg 1$. Then the weight factors for $n + \tfrac{1}{2} \mp \tfrac{1}{2} \approx |\alpha|^2 - \tfrac{1}{2}$ are particularly large (Stirling's formula on p. 518 is used in the proof), and therefore for an approximate calculation we use the generalized Rabi frequency

$$\Omega_\alpha = \sqrt{\left(|\alpha|^2 \pm \tfrac{1}{2} \right) g^2 + \Delta^2}$$

in $\cos^2(2\theta_\alpha) = (\Delta / \Omega_\alpha)^2 = 1 - \sin^2(2\theta_\alpha)$. But for the sum over $\cos(\Omega_n t)$, we have to calculate more precisely by one order. Here the abbreviations

$$\omega = \frac{g^2}{2\Omega_\alpha} \ll \Omega_\alpha \quad \text{and} \quad \kappa = |\alpha|^2 + \frac{2\Delta^2}{g^2}$$

are useful, because then for the important terms we have $\Omega_n \approx (\kappa + n)\,\omega$, and this leads to the approximation

$$\langle \sigma_z \rangle = \pm [\cos^2(2\theta_\alpha)$$
$$+ \sin^2(2\theta_\alpha) \, \exp\{-|\alpha|^2 \, (1 - \cos(\omega t))\} \, \cos\{\kappa \omega t\} + |\alpha|^2 \sin(\omega t)\}] ,$$

with the upper sign for the initial state $|\uparrow, \alpha\rangle$ and the lower sign for $|\downarrow, \alpha\rangle$. Here, in the time $\tfrac{1}{2}\pi/\omega$, the factor $\exp\{-|\alpha|^2 \, (1 - \cos \omega t)\}$ decreases from one to a negligibly small value. The oscillations observed for the Fock state stop after this time, and set in again at the time $2\pi/\omega$ (see Fig. 5.19).

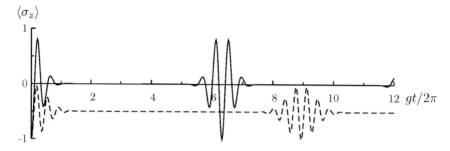

Fig. 5.19 Absence and return of the excitation of an atom in a light field described by a Glauber state. Initially, $|\alpha|^2 = 10$ and the atom was in the ground state. *Continuous curve*: resonance. *Dashed curve*: Detuning $\Delta = |\alpha g|$. Here $\langle \sigma_z \rangle = 0$ indicates that on average there are equally many atoms in excited states as in the ground state

This absence and return ("collapse" and "revival") occurs only with the unsharp Rabi frequency $\{\Omega_n\}$, as can be seen by comparing with the *semi-classical ansatz*: only the atom is treated according tos quantum physics, but the field classically. Its Hamilton operator describes an *illuminated atom* (quasi-atom or *dressed atom*), which is the expectation value of H with respect to a Glauber state $|\alpha\rangle$:

$$\tilde{H} \equiv \langle \alpha | H - \hbar\omega_L \Psi^\dagger \Psi | \alpha \rangle = \tfrac{1}{2}\hbar\omega_A \sigma_z + \tfrac{1}{2}\hbar g \, (\alpha \sigma_+ + \alpha^* \sigma_-) \, .$$

Note that we have taken $\hbar\omega_L |\alpha|^2$ as the zero energy and subtracted, as usual. Here, according to p. 473, the quantity $\alpha = |\alpha| \exp(-i\omega_L t)$, and consequently also $\tilde{H}$, depends on time. But this can be eliminated by a unitary transformation $U(t) = \exp(\tfrac{1}{2}i\omega_L t \, \sigma_z)$. Here we go over to a reference frame rotating with the light wave. The the *rotating-wave approximation* (RWA) neglects the terms $\sigma_+ \Psi^\dagger + \sigma_- \Psi$, and with the new axes, we arrive likewise at the time-independent Hamilton operator $\overline{H}$. Since U depends on time, using Problem 4.22, we find

$$\overline{H} = U\tilde{H}U^\dagger + i\hbar \dot{U} U^\dagger = \tfrac{1}{2}\hbar\mu \, \sigma_x - \tfrac{1}{2}\hbar\Delta \, \sigma_z \, , \quad \text{with} \quad \mu = |\alpha| \, g \, .$$

Its eigenvalues are $\overline{E}_\pm = \pm\tfrac{1}{2}\hbar\Omega_\alpha$ with $\Omega_\alpha = \sqrt{\mu^2 + \Delta^2}$. Using

$$\boldsymbol{\Omega}_\alpha = \frac{\text{tr}(\boldsymbol{\sigma}\overline{H})}{\hbar} = \mu \mathbf{e}_x - \Delta \, \mathbf{e}_z \, ,$$

in the equation for the Bloch vector $\langle \boldsymbol{\sigma} \rangle \equiv \text{tr}(\rho\boldsymbol{\sigma})$ deduced from the von Neumann equation $i\hbar\dot{\rho} = [\overline{H}, \rho]$ on p. 343, we find

$$\frac{d\langle \boldsymbol{\sigma} \rangle}{dt} = \boldsymbol{\Omega}_\alpha \times \langle \boldsymbol{\sigma} \rangle \, .$$

According to the semi-classical ansatz, the Bloch vector thus rotates about the vector $\boldsymbol{\Omega}_\alpha$ in the reference frame rotating with the light frequency, and with this a complete change from $\langle \sigma_z \rangle = \pm 1$ to $\langle \sigma_z \rangle = \mp 1$ is only possible for resonance. But since the Bloch vector rotates about $\boldsymbol{\Omega}_\alpha$ for arbitrarily long times, there is no absence and return semi-classically.

So far we have not considered *spontaneous emission* (the coupling to the remaining modes)—and this is often more apparent than the absence or the neglected terms $\sigma_+ \Psi^\dagger + \sigma_- \Psi$. For a two-level system, it is easy to write down the differential equation, according to p. 381 for $T = 0$:

$$\frac{d\rho}{dt} = \frac{[H, \rho]}{i\hbar} + \frac{[\sigma_- \rho, \sigma_+] + [\sigma_-, \rho\sigma_+]}{2\tau} + \frac{[\sigma_z \rho, \sigma_z] + \text{h.c.}}{2\tau_0} \, .$$

This implies the *Bloch equation* (see also Problem 4.22)

$$\frac{d\langle\boldsymbol{\sigma}\rangle}{dt} = \boldsymbol{\Omega}_\alpha \times \langle\boldsymbol{\sigma}\rangle - \frac{(1+4\tau/\tau_0)\langle\sigma_x\,\mathbf{e}_x + \sigma_y\,\mathbf{e}_y\rangle + 2\langle\sigma_z+1\rangle\,\mathbf{e}_z}{2\tau},$$

or again, writing γ^{-1} instead of 2τ and setting $\beta = 1+4\tau/\tau_0$,

$$\frac{d\langle\boldsymbol{\sigma}\rangle}{dt} = O\,\langle\boldsymbol{\sigma}\rangle - 2\gamma\,\mathbf{e}_z\,, \quad \text{with } O = \begin{pmatrix} -\beta\gamma & \Delta & 0 \\ -\Delta & -\beta\gamma & -\mu \\ 0 & \mu & -2\gamma \end{pmatrix}.$$

The previously skew-symmetric operator O thus obtains some diagonal elements: its eigenvalues are not purely imaginary, and its real part leads to damping. The inverse of O is

$$O^{-1} = \frac{-1}{\gamma\,(2\beta^2\gamma^2 + 2\Delta^2 + \beta\mu^2)} \begin{pmatrix} \mu^2 + 2\beta\gamma^2 & 2\gamma\,\Delta & -\mu\Delta \\ -2\gamma\,\Delta & 2\beta\gamma^2 & -\beta\mu\gamma \\ -\mu\Delta & \beta\mu\gamma & \beta^2\gamma^2 + \Delta^2 \end{pmatrix}.$$

Using $\langle\boldsymbol{\sigma}\rangle \approx 2\gamma\,O^{-1}\mathbf{e}_z$, the z-component of the stationary final state is

$$\langle\sigma_z\rangle \approx -\frac{\beta^2\gamma^2 + \Delta^2}{\beta^2\gamma^2 + \Delta^2 + \frac{1}{2}\beta\mu^2} = \frac{-1}{1 + \frac{1}{2}\beta\mu^2/(\beta^2\gamma^2 + \Delta^2)} = \frac{-1}{1 + I/I_S}\,,$$

since $\mu^2 = |g\alpha|^2$ is proportional to the light intensity I. The *saturation intensity* I_S is clearly proportional to $\beta^2\gamma^2 + \Delta^2$, so at *resonance* ($\Delta = 0$), it is particularly small and increases quadratically with the detuning Δ. For $I \ll I_S$, $\langle\sigma_z\rangle$ approaches -1, which corresponds to the lower energy state, but for $I \gg I_S$, it tends towards 0, the two states then being equally probable. For the rotating-wave transformation, the z-component is conserved, while

$$\langle\sigma_x + i\sigma_y\rangle \approx \mu\,\frac{\Delta + i\beta\gamma}{\beta^2\gamma^2 + \Delta^2 + \frac{1}{2}\beta\mu^2}$$

becomes constant due to this transformation (see Fig. 5.20).

We have considered spontaneous emission only semi-classically. In a full quantum mechanical treatment, we would also have to describe the electromagnetic field using operators (Ψ, $\Psi^\dagger$), and hence assume the Jaynes–Cummings model. In addition to the considered damping, we would also have to include terms $[\Psi\rho, \Psi^\dagger] + [\Psi, \rho\,\Psi^\dagger]$. This damping couples the Jaynes–Cummings doublets and can be solved analytically only with further approximations.

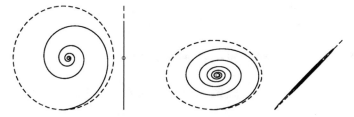

Fig. 5.20 Motion of the Bloch vector for the illuminated two-level atom (from out of the ground state) using the rotating-wave approximation in the (y, z) plane (*top view*) and (x, z) plane (*side view*): *left* for resonance ($\Delta = 0$) and *right* with detuning (here $\Delta = \mu$). *Dashed curves* indicate without spontaneous emission (dissipation, $\gamma = 0$), and *continuous curves* with spontaneous emission (here $\gamma = \mu/10$ and $\beta = 1$). Without dissipation, a circle is obtained, otherwise a spiral with the attractor indicated by the *open circle*. For resonance, the quantization axis lies in the plane of the circle, otherwise not (so the right-hand circle for detuning is inclined, and smaller)

5.5.8 Summary: Photons

As an example of a many-boson system, we have considered the light field and quantized the classical Maxwell equations, thereby investigating the quantum properties of a classical field. Instead of the occupation-number representation, we prefer to take Glauber states, which are "as classical as possible". Then as polar coordinate we have the amplitude and phase of the field and we do indeed find oscillations, in contrast to states with sharp energy.

5.6 Dirac Equation

5.6.1 Relativistic Invariance

The Dirac equation is a relativistic equation. Therefore, we use the notation with four-vectors known from electrodynamics (Sect. 3.4). The position vector with its Cartesian components

$$x^k : \qquad (x^1, x^2, x^3) = (x, y, z) , \quad \text{with} \ \ k \in \{1, 2, 3\} ,$$

is amended with a further component $x^0 = ct$ (the "light path"), to yield the four-vector x with contravariant components

$$x^\mu : \qquad (x^0, x^1, x^2, x^3) \mathrel{\widehat{=}} (ct, x^k) , \quad \text{with} \ \ \mu \in \{0, 1, 2, 3\} .$$

Correspondingly, the components of the *mechanical momentum p* are (see p. 245)

$$p^\mu : \qquad (p^0, p^1, p^2, p^3) \triangleq \left(\frac{E}{c}, p^k\right),$$

and those of the vector potential A are (see p. 239)

$$A^\mu : \qquad (A^0, A^1, A^2, A^3) \triangleq \left(\frac{\Phi}{c}, A^k\right).$$

If we consider a particle with charge q in the electromagnetic field, its mechanical momentum differs from the *canonical momentum P*, which has components (see p. 247)

$$P^\mu = p^\mu + q\, A^\mu \ .$$

Apart from the contravariant components (upper index), we also need the *covariant components* (lower index). These can be derived for the pseudo-Euclidean metric of special relativity theory using the metric tensor

$$(g^{\mu\nu}) = \begin{pmatrix} 1 & 0 & 0 & 0 \\ 0 & -1 & 0 & 0 \\ 0 & 0 & -1 & 0 \\ 0 & 0 & 0 & -1 \end{pmatrix} = (g_{\mu\nu})\ .$$

We shall always use Einstein's summation convention from now on, and thus leave out the summation sign whenever the summation index in a product occurs once up, once down. For the present case, $x_\mu = g_{\mu\nu}\, x^\nu$ and hence $x_0 = x^0$, $x_k = -x^k$.

The Lorentz invariant scalar products are sums over products of covariant and contravariant components. In particular, for free particles, we have

$$v^\mu\, v_\mu = c^2\ , \quad \text{and with} \quad p^\mu = m\, v^\mu\ , \quad \text{also} \quad p^\mu\, p_\mu = m^2 c^2\ .$$

Here m is the mass of the particle under consideration. With $p^\mu\, p_\mu = (p^0)^2 - \mathbf{p} \cdot \mathbf{p}$, we thus have for free particles

$$(E/c)^2 = (m\, c)^2 + \mathbf{p} \cdot \mathbf{p}\ .$$

However, we shall generally use the equation $p^\mu\, p_\mu = m^2 c^2$.

5.6.2   Quantum Theory

In the following we have to replace the observables by Hermitian operators, but we shall use the same letters. In particular, p should mean the mechanical momentum and P the canonical momentum. Here we have to account for the fact that P does not commute with A. Therefore, for all bilinear equations, we shall restrict ourselves initially to the case $qA = 0$ and treat the generalized case only in Sect. 5.6.8.

The Dirac equation is a relativistic equation for a wave field ψ which we shall interpret as a probability amplitude. For the superposition principle to remain valid, the equation has to be linear in ψ. In addition, if $\psi(t_0)$ is given, everything at later times should be fixed. Consequently, it must also be a first order differential equation in time, and relativistic covariance then allows only first derivatives with respect to the position. We note that the Schrödinger equation also contains only first derivatives with respect to time, but second derivatives with respect to the position.

According to the correspondence principle, we have to obtain classical mechanics in the classical limit of special relativity theory. However, we cannot use the equation $p^\mu p_\mu = m^2 c^2$, because taking into account

$$P_\mu \mathrel{\hat{=}} i\hbar \, \frac{\partial}{\partial x^\mu} = i\hbar \, \partial_\mu \, , \quad \text{or} \quad p_\mu \mathrel{\hat{=}} i\hbar \, \partial_\mu - q A_\mu \, ,$$

it leads to a differential equation of second order, i.e., the Klein–Gordon equation [10, 11], derived also by [12] and [13]. According to Dirac [14], we should make an ansatz with a linear expression in p_μ:

$$(\gamma^\mu p_\mu - mc) \, \psi = 0 \, , \quad \text{or} \quad (i\,\gamma^\mu \, \partial_\mu - \frac{q}{\hbar} \, \gamma^\mu A_\mu - \kappa) \, \psi(x) = 0 \, ,$$

where $\kappa \equiv mc/\hbar$. (It is common practice to set $\hbar = c = 1$ and put the mass m instead of κ, even though the Compton wavelength $2\pi/\kappa$ is a well known quantity.) Note that, setting

$$\gamma^\mu \mathrel{\hat{=}} (\gamma^0, \gamma^k) \, ,$$

together with $p_\mu \mathrel{\hat{=}} (E/c, -p^k)$ and $A_\mu \mathrel{\hat{=}} (\Phi/c, -A^k)$, we have on the one hand,

$$\gamma^\mu p_\mu = \gamma^0 \, \frac{E}{c} - \boldsymbol{\gamma} \cdot \mathbf{p} \, ,$$

$$\gamma^\mu A_\mu = \gamma^0 \, \frac{\Phi}{c} - \boldsymbol{\gamma} \cdot \mathbf{A} \, ,$$

but on the other,

$$\gamma^\mu \, \partial_\mu = \gamma^0 \, \frac{1}{c} \frac{\partial}{\partial t} + \boldsymbol{\gamma} \cdot \boldsymbol{\nabla} \, ,$$

where $\partial_\mu \mathrel{\hat{=}} (\partial/(c\partial t), \nabla^k)$.

We could also have written the Dirac equation in the form $(\gamma^\mu p_\mu + mc) \, \psi = 0$, because the only restriction is $p^\mu p_\mu = (mc)^2$. In this bilinear equation, we would have to restrict ourselves to $qA = 0$—the generalization to $qA \neq 0$ will follow in Sect. 5.6.8.) We must now deal with this ambiguity.

5.6.3   Dirac Matrices

The novel feature in Dirac's ansatz is to take the square root of $p^\mu p_\mu$, i.e., to require $p^\mu p_\mu = (\gamma^\mu p_\mu)^2$. This equation requires $\gamma^\mu \gamma^\nu + \gamma^\nu \gamma^\mu = 0$ for $\mu \neq \nu$ and $\gamma^\mu \gamma^\mu = g^{\mu\mu}$, if we assume that all the γ^μ commute with the operators considered so far. The four quantities γ^μ must therefore anti-commute, i.e., they cannot be normal numbers. If we make an ansatz with matrices, then ψ must have correspondingly many components. We combine the last equations to give

$$\gamma^\mu \gamma^\nu + \gamma^\nu \gamma^\mu = 2\, g^{\mu\nu} ,$$

which is the basic relation defining a *Clifford algebra*. On the right, we should write the unit operator, but we shall leave it out for many of the following equations.

If only three such operators were necessary, then we could take the Pauli matrices discussed on p. 308, viz.,

$$\sigma^1 = \begin{pmatrix} 0 & 1 \\ 1 & 0 \end{pmatrix}, \qquad \sigma^2 = \begin{pmatrix} 0 & -i \\ i & 0 \end{pmatrix}, \qquad \sigma^3 = \begin{pmatrix} 1 & 0 \\ 0 & -1 \end{pmatrix} .$$

Note that, for $\mu \in \{1, 2, 3\}$, we should also have a factor $\pm i$ for $\sigma^{\mu\, 2} = -1$ to hold. Together with the unit matrix, these form a complete basis for 2×2 matrices. Consequently, the Dirac matrices must have a higher dimension.

Since the squares of the γ^μ are equal to $+1$ or -1, we can form a total of 16 different products. These include unity and the four operators γ^μ, plus six 2-products $\gamma^\mu \gamma^\nu$ with $\mu < \nu$, four 3-products $\gamma^\lambda \gamma^\mu \gamma^\nu$ with $\lambda < \mu < \nu$, and finally, the 4-product

$$\gamma^5 = i\gamma^0 \gamma^1 \gamma^2 \gamma^3 .$$

The index 5 is commonly used, since μ is sometimes allowed to run from 1 to 4 instead of 0–3. In contrast, authors vary in the use of the factor i. In any case, the abbreviation for the four-product is suggested because $\gamma^\mu \gamma^5 + \gamma^5 \gamma^\mu = 0$ and $(\gamma^5)^2 = 1$. Therefore we shall also set $g^{\mu 5} = g^{5\mu} = 0$ for $\mu \neq 5$ and $g^{55} = 1$, which is not common practice, and then generalize the starting equation $[\gamma^\mu, \gamma^\nu]_+ = 2g^{\mu\nu}$.

As basis operators, we prefer to use

$$\sigma^{\mu\nu} = \tfrac{i}{2}\, [\gamma^\mu, \gamma^\nu]$$

in the following, instead of the six 2-products and the four 3-products,

$$\sigma^{\mu 5} = \tfrac{i}{2}\, [\gamma^\mu, \gamma^5] ,$$

and this is also not standard practice. Given that $\sigma^{\mu\nu} = -\sigma^{\nu\mu}$, this introduces 10 new quantities for which we have included a factor of i. For $\mu \neq \nu$ (including 5), we then have $\sigma^{\mu\nu} = i\gamma^\mu \gamma^\nu$. We also have (again including 5)

$$\gamma^\mu \gamma^\nu = g^{\mu\nu} - i\sigma^{\mu\nu} \,,$$
$$\gamma^\mu \gamma^\nu \gamma^\kappa = g^{\mu\nu}\,\gamma^\kappa + g^{\nu\kappa}\,\gamma^\mu - g^{\kappa\mu}\,\gamma^\nu + \sum_{\lambda<\rho} \varepsilon^{\mu\nu\kappa}{}_{\lambda\rho}\,\sigma^{\lambda\rho} \,,$$

where $\varepsilon^{\mu\nu\kappa}{}_{\lambda\rho} = \varepsilon^{\mu\nu\kappa\lambda'\rho'}\,g_{\lambda'\lambda}g_{\rho'\rho}$, and $\varepsilon^{\mu\nu\kappa\lambda\rho}$ is the totally antisymmetric Levi-Civita symbol. In particular, $\varepsilon^{01235} = 1$. For the commutators, we now have (also with 5)

$$[\gamma^\mu,\ \gamma^\nu] = -2i\,\sigma^{\mu\nu} \,,$$
$$[\gamma^\kappa,\ \sigma^{\mu\nu}] = -2i\,(g^{\kappa\nu}\,\gamma^\mu - g^{\kappa\mu}\,\gamma^\nu) \,,$$
$$[\sigma^{\kappa\lambda},\ \sigma^{\mu\nu}] = -2i\,(g^{\kappa\mu}\,\sigma^{\lambda\nu} + g^{\lambda\nu}\,\sigma^{\kappa\mu} - g^{\kappa\nu}\,\sigma^{\lambda\mu} - g^{\lambda\mu}\,\sigma^{\kappa\nu}) \,,$$

and for the anti-commutators

$$[\gamma^\mu,\ \gamma^\nu]_+ = 2\,g^{\mu\nu} \,,$$
$$[\gamma^\kappa,\ \sigma^{\mu\nu}]_+ = 2i \sum_{\lambda<\rho} \varepsilon^{\kappa\mu\nu}{}_{\lambda\rho}\,\sigma^{\lambda\rho} \,,$$
$$[\sigma^{\kappa\lambda},\ \sigma^{\mu\nu}]_+ = 2\,(i\,\varepsilon^{\kappa\lambda\mu\nu}{}_\rho\,\gamma^\rho + g^{\kappa\mu}\,g^{\lambda\nu} - g^{\kappa\nu}\,g^{\lambda\mu}) \,.$$

In the last equation, the Einstein convention implies a sum over ρ. For $\kappa \neq \lambda \neq \mu \neq \kappa$, we now obtain

$$\sigma^{\kappa\lambda}\,\sigma^{\lambda\mu} = -\sigma^{\lambda\mu}\,\sigma^{\kappa\lambda} = i\,g^{\lambda\lambda}\,\sigma^{\kappa\mu} \qquad \text{and} \qquad \sigma^{\kappa\lambda}\sigma^{\kappa\lambda} = g^{\kappa\kappa}\,g^{\lambda\lambda} \,.$$

For the space-like components, we have $\sigma^{12}\sigma^{23} = -\sigma^{23}\sigma^{12} = -i\sigma^{13} = i\sigma^{31}$ (and cyclic permutations in 1, 2, 3), as is usual for the three Pauli operators, whence the letter σ has been adopted.

We have thus introduced 16 operators γ^A. Of these, only the unit operator commutes with all the others, while each of the others commutes with eight and anti-commutes with the remaining eight. Only the unit operator commutes with all four operators γ^μ.

The traces of the 15 operators γ^A without the unit operator all vanish. This is immediately clear for the ten products $\sigma^{\mu\nu}$, because tr[A, B] always vanishes. For the five remaining ones, using $2i\,g^{\kappa\kappa}\,\gamma^\mu = [\gamma^\kappa,\ \sigma^{\kappa\mu}]$, we arrive likewise at a commutator and can therefore infer vanishing traces here.

Each product of two operators γ^A and γ^B is (except for the sign and possibly a factor of i) equal to one of the 16 operators. These 16 operators are linearly independent, because if $\sum_A a_A\,\gamma^A = 0$ were to hold, then any one of them could be multiplied by some γ^B, and by forming the trace, we could conclude that $a_B = 0$. Clearly, the linear combination gives zero only if all coefficients vanish. Therefore, the 16 operators are indeed linearly independent.

All 16 matrices γ^A are unitary for a Hermitian Hamilton operator. To show this, we multiply the starting equation $(\gamma^\mu p_\mu - mc)\psi = 0$ from the left by $c\,\gamma^0$ and use $(\gamma^0)^2 = 1$. We have $\gamma^0\,(\gamma^\mu p_\mu - mc) = p_0 - \gamma^0\,(\boldsymbol{\gamma}\cdot\mathbf{p} + mc)$, and with $cp_0 \cong i\hbar\,\partial/\partial t - q\Phi$, it then follows that

$$i\hbar\,\frac{\partial\psi}{\partial t} = H\,\psi\,, \quad \text{with}\;\; H = q\,\Phi + \gamma^0\,(\boldsymbol{\gamma}\cdot c\mathbf{p} + mc^2)\,.$$

The Hamilton operator H is only Hermitian for $\gamma^0 = \gamma^{0\,\dagger}$ and $\gamma^0\,\gamma^k = (\gamma^0\,\gamma^k)^\dagger = \gamma^{k\,\dagger}\,\gamma^0$, so $\gamma^{k\,\dagger} = -\gamma^k$ and $\gamma^{5\,\dagger} = \gamma^5$. In what follows, we shall instead use the equation

$$\gamma^{\mu\,\dagger} = \gamma^0\gamma^\mu\gamma^0\,, \quad \text{for}\;\; \mu \in \{0, 1, 2, 3\}\,.$$

Consequently, we have $\gamma^{\mu\,\dagger}\gamma^\mu = 1$ for $\mu \in \{0, 1, 2, 3, 5\}$. With this, the remaining operators $\sigma^{\mu\nu}$ are also unitary:

$$\gamma^{A\,-1} = \gamma^{A\,\dagger} = \gamma_A\,.$$

This is what we set out to prove, and with this we have also derived the Hamilton operator of the Dirac theory.

5.6.4   Representations of the Dirac Matrices

Since we have arrived at 16 linearly independent operators γ^A, we are dealing with at least 4×4 matrices, which may all be written as (super-)matrices of the Pauli matrices, the 2×2 zero matrix, and the 2×2 unit matrix. In the *standard representation* γ^0 is diagonal, in the *Weyl representation* γ^5 is. For each, we set

$$\sigma^{kl} = \begin{pmatrix} \sigma^m & 0 \\ 0 & \sigma^m \end{pmatrix}, \quad \text{thus in particular}\;\; \sigma^{12} = \begin{pmatrix} \sigma^3 & 0 \\ 0 & \sigma^3 \end{pmatrix},$$

with $(k, l, m) = (1, 2, 3)$ or a cyclic permutation thereof, and σ^m the 2×2 Pauli matrix. Except for these three matrices (and the unit matrix), the two representations are different (see Table 5.2).

These representations can be unitarily transformed into each other using the operator $U = (\gamma^0 + \gamma^5)/\sqrt{2} = U^\dagger = U^{-1}$. It relates the first and third components, or again the second and fourth components:

$$(\psi_1, \psi_2, \psi_3, \psi_4) \leftrightarrow (\psi_1 + \psi_3, \psi_2 + \psi_4, \psi_1 - \psi_3, \psi_2 - \psi_4)/\sqrt{2}\,.$$

With $\gamma^0\gamma^k = -i\sigma^{0k}$, the Hamilton operator reads $H = q\Phi + \gamma^0\,(\boldsymbol{\gamma}\cdot c\mathbf{p} + mc^2)$ in the standard representation

$$H_{\mathrm{D}} = q\,\Phi + \boldsymbol{\alpha}\cdot c\mathbf{p} + \beta\,mc^2 = \begin{pmatrix} q\,\Phi + mc^2 & \boldsymbol{\sigma}\cdot c\mathbf{p} \\ \boldsymbol{\sigma}\cdot c\mathbf{p} & q\,\Phi - mc^2 \end{pmatrix},$$

and in the Weyl representation

Table 5.2 Standard and Weyl representations of the γ matrices	4×4 matrix	Standard representation	Weyl representation
	$\begin{pmatrix} 1 & 0 \\ 0 & -1 \end{pmatrix} = \beta$	γ^0	γ^5
	$\begin{pmatrix} 0 & 1 \\ 1 & 0 \end{pmatrix}$	γ^5	γ^0
	$\begin{pmatrix} 0 & -1 \\ 1 & 0 \end{pmatrix}$	$+\mathrm{i}\sigma^{05}$	$-\mathrm{i}\sigma^{05}$
	$\begin{pmatrix} \sigma^k & 0 \\ 0 & -\sigma^k \end{pmatrix}$	$-\mathrm{i}\sigma^{k5}$	$-\mathrm{i}\sigma^{0k}$
	$\begin{pmatrix} 0 & \sigma^k \\ \sigma^k & 0 \end{pmatrix} = \alpha^k$	$-\mathrm{i}\sigma^{0k}$	$-\mathrm{i}\sigma^{k5}$
	$\begin{pmatrix} 0 & -\sigma^k \\ \sigma^k & 0 \end{pmatrix} =$ $\alpha^k \beta = -\beta \alpha^k$	$-\gamma^k$	γ^k

$$H_{\mathrm{W}} = \begin{pmatrix} q\,\Phi + \boldsymbol{\sigma} \cdot c\mathbf{p} & mc^2 \\ mc^2 & q\,\Phi - \boldsymbol{\sigma} \cdot c\mathbf{p} \end{pmatrix} .$$

The standard representation is in fact convenient for low energies, i.e., for $|\boldsymbol{\sigma} \cdot c\mathbf{p}| \ll mc^2$, but otherwise the Weyl representation is to be preferred, not only for neutrinos and quarks which may have very small masses, but because H_{W} is easier to diagonalize. The *helicity* $\boldsymbol{\sigma} \cdot \mathbf{p}/p$ is a good quantum number for massless Dirac particles (even for $q\Phi \neq 0$)—neutrinos are left-handed and anti-neutrinos right-handed.

Later, we shall also need the complex-conjugate Dirac matrices for the anti-linear operators used to describe time reversal and charge conjugation. Therefore, for $\mu = 0$ to 3, we now consider

$$\gamma^{\mu\,*} = \mathscr{B}\,\gamma^\mu\,\mathscr{B}^{-1} \qquad \Longrightarrow \qquad \gamma^{5\,*} = -\mathscr{B}\,\gamma^5\,\mathscr{B}^{-1} .$$

This fixes $\mathscr{B}$ only up to a numerical factor. We may choose $\mathscr{B}$ unitary:

$$\mathscr{B}^\dagger = \mathscr{B}^{-1} .$$

This fixes the modulus of the numerical factor, while its phase remains free to be chosen.

The operator $\mathscr{B}$ depends on the representation of the operators γ^μ. $\mathscr{B}$ has to commute with the real γ^μ (for $\mu \in \{0, 1, 2, 3\}$) and to anti-commute with the imaginary ones. Then conversely for γ^5, e.g., $\gamma^5 \mathscr{B} = -\mathscr{B}\gamma^5$ for real γ^5. In both the representations considered above, γ^0, γ^1, γ^3, and γ^5 are real and γ^2 is imaginary, thus in both cases $\mathscr{B} \propto \sigma^{25}$, and only the phase factor remains open. We choose $\mathscr{B} = \sigma^{25}$ and $\mathscr{B}$ real: $\mathscr{B} = \mathscr{B}^*$ and thus $\mathscr{B}^{-1} = \mathscr{B}^\dagger = \widetilde{\mathscr{B}}$.

In any case, $\mathscr{B}$ is antisymmetric in both representations:

$$\widetilde{\mathscr{B}} = -\mathscr{B} \, .$$

This actually holds in all representations, because each unitary transformation leaves $\mathscr{B}$ unitary and antisymmetric. If $\gamma' = \mathscr{U} \gamma \, \mathscr{U}^\dagger$ holds with $\gamma^* = \mathscr{B} \gamma \, \mathscr{B}^{-1}$ and $\gamma'^* = \widetilde{\mathscr{U}}^{-1} \gamma^* \widetilde{\mathscr{U}} = \mathscr{B}' \gamma' \mathscr{B}'^{-1}$, then also $\gamma' = \mathscr{U} \mathscr{B}^{-1} \widetilde{\mathscr{U}} \mathscr{B}' \gamma' (\mathscr{U} \mathscr{B}^{-1} \widetilde{\mathscr{U}} \mathscr{B}')^{-1}$. Therefore, $\mathscr{U} \mathscr{B}^{-1} \widetilde{\mathscr{U}} \mathscr{B}'$ commutes with the four γ'^μ, and consequently, according to p. 491, it is a multiple of the unit. Except for a phase factor, $\mathscr{B}' = \mathscr{U}^* \mathscr{B} \mathscr{U}^\dagger$, so $\widetilde{\mathscr{B}}' = -\mathscr{B}'$ for $\widetilde{\mathscr{B}} = -\mathscr{B}$.

The complex conjugation operator $\mathscr{K}$ also depends on the representation. In contrast to $\mathscr{B}$, it acts on all degrees of freedom, and according to Sect. 4.2.12, it is anti-linear and anti-unitary:

$$\mathscr{K} \, c \, \mathscr{K}^{-1} = c^* \quad \text{and} \quad \mathscr{K}^\dagger = \mathscr{K}^{-1} = \mathscr{K} \, .$$

But the product $\mathscr{K} \mathscr{B} = \mathscr{B} \mathscr{K}$ does not depend on the representation. Here,

$$(\mathscr{K} \mathscr{B})^2 = \mathscr{K} \mathscr{B} \mathscr{K}^{-1} \mathscr{B} = \mathscr{B}^* \mathscr{B} \, ,$$

and since $\mathscr{B}^* = -\mathscr{B}^{-1}$, we have generally

$$(\mathscr{K} \mathscr{B})^2 = -1 \quad \text{and} \quad (\mathscr{K} \mathscr{B})^\dagger = (\mathscr{K} \mathscr{B})^{-1} \, .$$

For $\mu \in \{0, 1, 2, 3\}$, we have in addition $\mathscr{K} \mathscr{B} \gamma^\mu = \mathscr{K} \gamma^{\mu*} \mathscr{B} = \gamma^\mu \mathscr{K} \mathscr{B}$, while

$$\mathscr{K} \mathscr{B} \gamma^5 = -\mathscr{K} \gamma^{5*} \mathscr{B} = -\gamma^5 \mathscr{K} \mathscr{B} \, .$$

5.6.5 Behavior of the Dirac Equation Under Lorentz Transformations

The equation

$$(\gamma^\mu p_\mu - mc) \, \psi = 0 \, , \quad \text{with} \quad \gamma^\mu \gamma^\nu + \gamma^\nu \gamma^\mu = 2 \, g^{\mu\nu} \, ,$$

is written in a relativistically covariant way. For a change of coordinates $x^\mu \to x'^\mu$, we have

$$(\gamma'^\mu p'_\mu - mc) \, \psi = 0 \, .$$

The notation $\gamma^\mu p_\mu$ indicates a relativistic invariant (a scalar). Hence for a homogeneous Lorentz transformation (see p. 232),

$$x'^\mu = a^\mu{}_\nu\, x^\nu \,, \quad \text{with } a^\mu{}_\nu\, a_\mu{}^\lambda = g_\nu{}^\lambda = a_\nu{}^\mu\, a^\lambda{}_\mu \ \text{ and } \ a^\mu{}_\nu{}^* = a^\mu{}_\nu \,,$$

the Dirac matrices have to transform as a 4-vector, viz.,

$$\gamma'^\mu = a^\mu{}_\nu\, \gamma^\nu \,,$$

and with $\gamma^\kappa \gamma^\lambda + \gamma^\lambda \gamma^\kappa = 2\, g^{\kappa\lambda}$, it follows that

$$\gamma'^\mu \gamma'^\nu + \gamma'^\nu \gamma'^\mu = a^\mu{}_\kappa\, a^\nu{}_\lambda\, (\gamma^\kappa \gamma^\lambda + \gamma^\lambda \gamma^\kappa) = 2\, a^{\mu\lambda}\, a^\nu{}_\lambda = 2\, g^{\mu\nu} \,.$$

We deduce that $\sigma^{\mu\nu}$ transforms as a tensor of second rank, and the unit as a scalar.

We now prove that $\gamma^5 = i\,\gamma^0 \gamma^1 \gamma^2 \gamma^3$ transforms as a pseudo-scalar if $\gamma'^\mu = a^\mu{}_\nu \gamma^\nu$. It suffices to show that

$$\gamma'^5 = \gamma^5\, \det a \,,$$

because, according to p. 228, all proper Lorentz transformations have $\det a = +1$, while for a space inversion, we have $\det a = -1$. The properties of the determinant can be described with the totally antisymmetric tensor $\varepsilon_{\kappa\lambda\mu\nu}$:

$$\varepsilon_{\kappa'\lambda'\mu'\nu'}\, \det a = \varepsilon_{\kappa\lambda\mu\nu}\, a^\kappa{}_{\kappa'}\, a^\lambda{}_{\lambda'}\, a^\mu{}_{\mu'}\, a^\nu{}_{\nu'} \,.$$

The matrix γ^5 can be taken as $\frac{i}{4!}\,\varepsilon_{\kappa\lambda\mu\nu}\,\gamma^\kappa\,\gamma^\lambda\,\gamma^\mu\,\gamma^\nu$ and γ'^5 as $\frac{i}{4!}\,\varepsilon_{\kappa\lambda\mu\nu}\,\gamma'^\kappa\,\gamma'^\lambda\,\gamma'^\mu\,\gamma'^\nu$, whence

$$\gamma'^5 = \frac{i}{4!}\,\varepsilon_{\kappa\lambda\mu\nu}\, a^\kappa{}_{\kappa'}\, a^\lambda{}_{\lambda'}\, a^\mu{}_{\mu'}\, a^\nu{}_{\nu'}\,\gamma^{\kappa'}\gamma^{\lambda'}\gamma^{\mu'}\gamma^{\nu'} = \gamma^5\,\det a \,.$$

The claim is therefore proven, and $\sigma^{\mu 5} \propto \gamma^\mu \gamma^5$ is an axial vector, or pseudo-vector.

With $[\gamma^\mu, \gamma^\nu]_+ = 2g^{\mu\nu} = [\gamma'^\mu, \gamma'^\nu]_+$, it is usual to take the same Gamma matrices and transfer the transformation to the states ψ. After multiplication from the left by $\mathscr{L}$, the transformed Dirac equation $(\gamma'^\mu p'_\mu - mc)\,\psi = 0$, with

$$\gamma'^\mu = \mathscr{L}^{-1}\,\gamma^\mu\,\mathscr{L} \,,$$

becomes

$$(\gamma^\mu p'_\mu - mc)\,\psi' = 0 \,, \quad \text{with } \psi' = \mathscr{L}\,\psi \,.$$

We may thus always calculate with the same Gamma matrices, if we transform the states suitably.

In order to determine the form of $\mathscr{L}$, we start with $\mathscr{L}^{-1}\gamma^\mu\mathscr{L} = a^\mu{}_\nu\,\gamma^\nu$. If we take the Hermitian conjugate and multiply on the left and right by γ^0, then it follows that $\gamma^0\mathscr{L}^\dagger\gamma^0\gamma^\mu\gamma^0\mathscr{L}^{-1\,\dagger}\gamma^0 = a^\mu{}_\nu\gamma^\nu = \mathscr{L}^{-1}\gamma^\mu\mathscr{L}$, and with $\gamma^0 = \gamma^{0\,-1}$ and $\mathscr{L}^{-1\,\dagger} = \mathscr{L}^{\dagger\,-1}$, we also have $\mathscr{L}\gamma^0\mathscr{L}^\dagger\gamma^0\gamma^\mu = \gamma^\mu\mathscr{L}\gamma^0\mathscr{L}^\dagger\gamma^0$, i.e., $\mathscr{L}\gamma^0\mathscr{L}^\dagger\gamma^0$ commutes with all four γ^μ and therefore, according to p. 491, is a multiple of the unit: $\mathscr{L}\gamma^0\mathscr{L}^\dagger = b\,\gamma^0$. Here b has to be real, because γ^0 and hence also the left-hand side are Hermitian. Its sign, with $\mathscr{L}^\dagger = \gamma^0\mathscr{L}^{-1}b\,\gamma^0$ or $\mathscr{L}^\dagger\mathscr{L} = b\,\gamma^0\,a^0{}_\nu\gamma^\nu$, is determined by $a^0{}_0$, because $\gamma^0\,a^0{}_\nu\gamma^\nu = a^0{}_0 - \mathrm{i}\,a^0{}_k\sigma^{0k}$ and $\mathrm{tr}\,\sigma^{\mu\nu} = 0$ leads to $4b\,a^0{}_0 = \mathrm{tr}\,\mathscr{L}^\dagger\mathscr{L} > 0$. For *orthochronous Lorentz transformations*, the time direction remains unchanged, so $a^0{}_0 > 0$ (see p. 228) and also $b > 0$, while for time reversal, $b < 0$. Here, $|b| = 1$, if we impose the group property that the product of two Lorentz transformations is another Lorentz transformation, and hence (as for the canonical transformations in Sect. 2.4.3) that $\det\mathscr{L} = 1$ has to be valid. Taking this together then, only $b = \pm 1$ remains possible, so

$$\mathscr{L}^\dagger = \pm\gamma^0\,\mathscr{L}^{-1}\,\gamma^0\;,$$

with the plus sign for orthochronous Lorentz transformations and the minus sign for time reversal. This means that $\mathscr{L}$ is not always unitary, and in fact $\psi^\dagger\psi$ transforms as the time-like component of a four-vector, as will be shown in the next section.

Let us now consider an infinitesimal Lorentz transformation

$$a_{\mu\nu} \approx g_{\mu\nu} + \omega_{\mu\nu}\;,\quad\text{with}\quad\omega_{\mu\nu} = -\omega_{\nu\mu}\;,\quad\text{and}\quad|\omega_{\mu\nu}| \ll 1\;,$$

and make the ansatz $\mathscr{L} \approx 1 - \frac{\mathrm{i}}{2}\,\omega^{\mu\nu}S_{\mu\nu}$, whence $\mathscr{L}^{-1} \approx 1 + \frac{\mathrm{i}}{2}\,\omega^{\mu\nu}S_{\mu\nu}$. Then $S_{\mu\nu} = -S_{\nu\mu}$ remains to be determined. Since on the one hand,

$$a^\mu{}_\nu\,\gamma^\nu = \mathscr{L}^{-1}\,\gamma^\mu\,\mathscr{L} \approx \gamma^\mu - \frac{\mathrm{i}}{2}\,\omega^{\kappa\lambda}\,(\gamma^\mu S_{\kappa\lambda} - S_{\kappa\lambda}\,\gamma^\mu)\;,$$

and on the other,

$$a^{\mu\nu}\gamma_\nu = (g^{\mu\nu} + \omega^{\mu\nu})\,\gamma_\nu \approx \gamma^\mu + \frac{1}{2}\,\omega^{\kappa\lambda}\,(g^\mu{}_\kappa\,\gamma_\lambda - g^\mu{}_\lambda\,\gamma_\kappa)\;,$$

we infer that $-\mathrm{i}[\gamma^\mu, S_{\kappa\lambda}] = g^\mu{}_\kappa\,\gamma_\lambda - g^\mu{}_\lambda\,\gamma_\kappa$. Here, according to p. 491, the quantity $g^\mu{}_\kappa\,\gamma_\lambda - g^\mu{}_\lambda\,\gamma_\kappa$ is equal to $-\frac{\mathrm{i}}{2}[\gamma^\mu, \sigma_{\kappa\lambda}]$. This suggests

$$S_{\mu\nu} = \tfrac{1}{2}\,\sigma_{\mu\nu}\;.$$

However, a term can be added which commutes with the Dirac matrices, hence a multiple of the unit. But that contradicts the constraint $\det\mathscr{L} = 1$. Consequently, for infinitesimal transformations,

$$\mathscr{L} \approx 1 - \tfrac{\mathrm{i}}{4}\,\omega^{\mu\nu}\,\sigma_{\mu\nu}$$

holds uniquely, and, e.g., for a rotation through the small angle ε about the z-axis, i.e., with $\omega_{21} = -\omega_{12} = \varepsilon$, all others being zero, $\mathscr{L}(\varepsilon) = 1 + \frac{1}{2}\,\varepsilon\,\sigma_{12}$. With $\sigma_{12}{}^2 = 1$, this can be generalized for a finite rotation $\mathscr{L}(\phi) = \mathscr{L}^{\phi/\varepsilon}(\varepsilon)$ to

$$\mathscr{L} = \cos\frac{\phi}{2} + i\,\sigma_{12}\,\sin\frac{\phi}{2}\;.$$

Here we recognize that these particles have spin $1/2$. In particular, for a rotation through 2π, the sign switches, and only after two full rotations does the system return to its original state.

From infinitesimal Lorentz transformations, we can obtain all proper Lorentz transformations. For the improper ones, we may restrict ourselves to time reversal and space inversion, possibly combined with a proper Lorentz transformation, and we shall discuss these in detail in Sect. 5.6.7. There we shall also consider charge inversion (charge conjugation). We may then understand why the solutions ψ have four rather than two components.

5.6.6 Adjoint Spinors and Bilinear Covariants

So far we have been considering the Dirac equation $(\gamma^\mu p_\mu - mc)\,\psi = 0$. Then, with $\gamma^{\mu\,\dagger} = \gamma^0\gamma^\mu\gamma^0$, for $\mu \in \{0, 1, 2, 3\}$, and $(\gamma^0)^2 = 1$, the Hermitian *adjoint Dirac equation* is

$$\bar{\psi}\,(\gamma^\mu p_\mu - mc) = 0\;, \quad \text{with} \quad \bar{\psi} \equiv \psi^\dagger\,\gamma^0\;.$$

Instead of the Hermitian conjugate spinors $\psi^\dagger$, it is better then to consider the *adjoint* $\bar{\psi}$ of ψ, because the same operator acts on ψ and $\bar{\psi}$, once on the right, once on the left. Here, in the standard representation, we have $\bar{\psi} = (\psi_1{}^*, \psi_2{}^*, -\psi_3{}^*, -\psi_4{}^*)$, but in the Weyl representation, $\bar{\psi} = (\psi_3{}^*, \psi_4{}^*, \psi_1{}^*, \psi_2{}^*)$, where we have set $\psi^\dagger = (\psi_1{}^*, \psi_2{}^*, \psi_3{}^*, \psi_4{}^*)$ in both cases.

In the real-space representation of $p_\mu = P_\mu - qA_\mu$, according to p. 489, P_μ corresponds to the operator $i\hbar\,\partial_\mu$. With $\langle\psi|P_\mu^\dagger|x^\mu\rangle = \langle x^\mu|P_\mu|\psi\rangle^* = -i\hbar\partial_\mu\psi^*$, $p_\mu^\dagger$ acts like the operator $-i\hbar\,\partial_\mu - qA_\mu$ acting on the left. Consequently, we may write the adjoint Dirac equation in the real-space representation in the form

$$(i\hbar\,\partial_\mu + q\,A_\mu)\,\bar{\psi}\,\gamma^\mu + mc\,\bar{\psi} = 0\;,$$

or free of any representation, $(P_\mu + qA_\mu)\,\bar{\psi}\,\gamma^\mu + mc\,\bar{\psi} = 0$.

For an orthochronous transformation $\psi \to \psi' = \mathscr{L}\,\psi$ with $\mathscr{L}^\dagger = +\gamma^0\,\mathscr{L}^{-1}\gamma^0$, we have

$$\bar{\psi}' = \psi'^\dagger\,\gamma^0 = \psi^\dagger\,\mathscr{L}^\dagger\,\gamma^0 = \psi^\dagger\,\gamma^0\,\mathscr{L}^{-1} = \bar{\psi}\,\mathscr{L}^{-1}$$

and $\bar{\psi}' \gamma^\mu \psi' = \bar{\psi} \mathscr{L}^{-1} \gamma^\mu \mathscr{L} \psi = \bar{\psi} \gamma'^\mu \psi$. Thus,

$$
\begin{array}{lll}
\bar{\psi} & 1 & \psi \text{ scalar}, \\
\bar{\psi} & \gamma^\mu & \psi \text{ vector}, \\
\bar{\psi} & \sigma^{\mu\nu} & \psi \text{ tensor}, \\
\bar{\psi} & \sigma^{\mu 5} & \psi \text{ axial vector}, \\
\bar{\psi} & \gamma^5 & \psi \text{ pseudo-scalar},
\end{array}
$$

as was to be expected for the operators γ^A according to the last section.

From the differential equations for $\psi(x)$ and $\bar{\psi}(x)$, viz.,

$$
\gamma^\mu \left(i\hbar\, \partial_\mu - q A_\mu \right) \psi = +mc\, \psi, \quad \text{i.e.,} \quad \gamma^\mu \partial_\mu\, \psi = -\tfrac{i}{\hbar} \left(q A_\mu\, \gamma^\mu\, \psi + mc\, \psi \right),
$$
$$
\left(i\hbar\, \partial_\mu + q A_\mu \right) \bar{\psi}\, \gamma^\mu = -mc\, \bar{\psi}, \quad \text{i.e.,} \quad \partial_\mu\, \bar{\psi}\, \gamma^\mu = +\tfrac{i}{\hbar} \left(q A_\mu\, \bar{\psi}\, \gamma^\mu + mc\, \bar{\psi} \right),
$$

we deduce the "continuity equation"

$$
\partial_\mu \left(\bar{\psi}\, \gamma^\mu\, \psi \right) = 0,
$$

and according to p. 239, a conservation law for $\int d^3r\, \bar{\psi}\, \gamma^0\, \psi = \int d^3r\, \psi^\dagger \psi \geq 0$. Therefore, we relate the time-like component $\bar{\psi}\, \gamma^0 \psi$ to a "density", in fact the charge density, as will be shown in the next section. However, the different components of γ^μ do not commute with each other, and therefore the probability current is not sharp. This is worth noting for a plane wave, which solves the Dirac equation in the field-free space (with $A^\mu = 0$). Therefore, we often speak here of *Zitterbewegung* (trembling motion), but we should nevertheless explain the fact that ψ has four components, not just two, as would have been expected for spin-1/2 particles. Hence we consider improper Lorentz transformations and then treat the phenomenon of Zitterbewegung on p. 505.

5.6.7 Space Inversion, Time Reversal, and Charge Conjugation

For these three improper Lorentz transformations, the Dirac equation keeps the same form. However, for time reversal and charge conjugation, we also need here the anti-linear complex conjugation operator $\mathscr{K}$, which already appeared for time reversal in non-relativistic quantum mechanics (see p. 313). Since the operator $\mathscr{K}$ does not act only on the Dirac matrices, but also on the remaining quantities, we shall now give the full transformation operator, differently from the proper Lorentz transformations considered so far.

Under a the space inversion, all polar three-vectors change their sign, while the axial vectors do not—so all time-like components remain conserved. Consequently, $(P'_0, P'_k) = (P_0, -P_k)$ and also $(\Phi'(t', \mathbf{r}'), \mathbf{A}'(t', \mathbf{r}')) = (\Phi(t, -\mathbf{r}), -\mathbf{A}(t, -\mathbf{r}))$.

The Dirac equation thus keeps the same form if (γ^0, γ^k) are transformed into $(\gamma^0, -\gamma^k)$. This can be done with

$$\mathscr{P} = \gamma^0 \mathscr{P}_0 \,,$$

where $\mathscr{P}_0$ is the inversion in the usual space, which we already need in non-relativistic quantum mechanics. The sign remains undetermined, because a rotation by 2π changes the sign of ψ without changing any measurement values. The phase factor has been chosen such that

$$\mathscr{P}^2 = 1 \,,$$

as in the non-relativistic case. We then also have $\mathscr{P} = \mathscr{P}^\dagger = \mathscr{P}^{-1}$ and

$$(\gamma^\mu p'_\mu - mc) \, \mathscr{P}\psi = 0 \,,$$

as claimed.

Under time reversal, $(t, \mathbf{r})$ has to change into $(t', \mathbf{r}') = (-t, \mathbf{r})$ and $(\Phi, \mathbf{A})$ changes into $(\Phi'(t', \mathbf{r}'), \mathbf{A}'(t', \mathbf{r}')) = (\Phi(-t, \mathbf{r}), -\mathbf{A}(-t, \mathbf{r}))$, because the magnetic field switches sign for motion reversal. The position vectors remain the same for time reversal, but not the momentum vectors. We thus need an anti-linear transformation, as was shown already on p. 313.

In fact, the time reversal operator $\mathscr{T}$ in real space has the same properties as the anti-linear complex conjugation operator $\mathscr{K}$, but the latter also changes the Dirac matrices, as we have seen in Sect. 5.6.4. Only the operator $\mathscr{K}\mathscr{B}$ commutes with them. $\mathscr{B}$ acts like a unit operator in real space.

For the invariance of the Dirac equation under time reversal (motion reversal), we need an anti-linear operator which changes the sign of the three space-like Dirac matrices. This we can do with

$$\mathscr{T} = \gamma^0 \mathscr{K}\mathscr{B} \,,$$

where the sign is arbitrary. Since $(\gamma^0 \mathscr{K}\mathscr{B})^2 = (\gamma^0)^2(\mathscr{K}\mathscr{B})^2$ with $(\gamma^0)^2 = 1$ and $(\mathscr{K}\mathscr{B})^2 = -1$, we thus have

$$\mathscr{T}^2 = -1 \quad \text{and} \quad \mathscr{T}^\dagger = \mathscr{T}^{-1} \,.$$

These two properties do not depend on the representation. In both the standard and the Weyl representation (with $\mathscr{B} = \sigma^{25}$), we have to take $\mathscr{T} = i\sigma^{31}\mathscr{K}$.

Starting with the adjoint Dirac equation $(P_\mu + qA_\mu) \, \bar{\psi} \, \gamma^\mu + mc\bar{\psi} = 0$ of the last sections, we can construct the charge-conjugate solution. In particular, if we take the space-inverted matrices of this equation and set

$$\widetilde{\gamma^\mu} = -\mathscr{U}^{-1} \gamma^\mu \mathscr{U} \,, \quad \text{for } \mu \in \{0, 1, 2, 3\} \quad (\Longrightarrow \quad \widetilde{\gamma^5} = +\mathscr{U}^{-1} \gamma^5 \mathscr{U}) \,,$$

and if we multiply $\widetilde{\gamma^\mu} \, (P_\mu + qA_\mu) \, \widetilde{\tilde{\psi}} + mc\widetilde{\tilde{\psi}} = 0$ by $-\mathcal{U}$, we obtain

$$\{\gamma^\mu \, (P_\mu + qA_\mu) - mc\} \, \mathcal{U}\widetilde{\tilde{\psi}} = 0 \; .$$

The sign of the charge q has been reversed here, relative to that in the original Dirac equation, and this is the required *charge conjugation*. Hence we infer the *charge conjugation operator*

$$\mathscr{C} = \gamma^0 \mathcal{U} \mathscr{K} \; ,$$

since with $\bar{\psi} = \psi^\dagger \gamma^0$, we have $\widetilde{\tilde{\psi}} = \widetilde{\gamma^0}\psi^*$ and therefore $\mathcal{U}\widetilde{\tilde{\psi}} = -\gamma^0 \mathcal{U} \mathscr{K} \psi$. Note that the phase factor remains arbitrary.

The properties of the operators $\mathcal{U}$ follow from $\widetilde{\gamma^\mu} = -\mathcal{U}^{-1}\gamma^\mu \mathcal{U}$, but only up to a factor, which allows us to choose $\mathcal{U}$ unitary, i.e., $\mathcal{U}^{-1} = \mathcal{U}^\dagger$. Since $\widetilde{\gamma^\mu} = (\gamma^{\mu\,\dagger})^* = (\gamma^0\gamma^\mu\gamma^0)^* = \mathscr{B}\gamma^0\gamma^\mu\gamma^0\mathscr{B}^{-1}$, we must still require $\gamma^0\gamma^\mu\gamma^0 = -\mathscr{B}^{-1}\mathcal{U}^{-1}\gamma^\mu\mathcal{U}\mathscr{B}$. Thus the three operators γ^k commute with $\mathcal{U}\mathscr{B}$, while γ^0 and γ^5 anti-commute with it. Therefore, $\mathcal{U}\mathscr{B}$ is proportional to σ^{05}, independently of the representation, and consequently $\mathcal{U}$ with $\mathscr{B}^{-1} = -\mathscr{B}$ is proportional to $\sigma^{05}\mathscr{B}$. The still missing factor has to have the absolute value one, because $\mathcal{U}$, σ^{05}, and $\mathscr{B}$ should be unitary. We can thus write $\mathcal{U} = u\sigma^{05}\mathscr{B}$ with $|u| = 1$.

For the charge conjugation operator $\mathscr{C}$, we thus have $u\gamma^0\sigma^{05}\mathscr{K}\mathscr{B} = iu\gamma^5\mathscr{K}\mathscr{B}$. In the following, we choose $u = -i$, whence the charge conjugation operator is

$$\mathscr{C} = \gamma^5 \mathscr{K} \mathscr{B} \; .$$

Independently of the representation, we thus find

$$\mathscr{C}^\dagger \mathscr{C} = (\gamma^5 \mathscr{K} \mathscr{B})^\dagger \, \gamma^5 \mathscr{K} \mathscr{B} = (\mathscr{K} \mathscr{B})^{-1} \gamma^{5\,\dagger} \gamma^5 \mathscr{K} \mathscr{B} = 1 \; ,$$

along with $\mathscr{C}^2 = (\gamma^5 \mathscr{K} \mathscr{B})^2 = -(\gamma^5)^2 (\mathscr{K} \mathscr{B})^2 = 1$, and hence,

$$\mathscr{C}^\dagger = \mathscr{C}^{-1} = \mathscr{C} \; .$$

The charge conjugation operator is thus unitary and anti-commutes with all Gamma matrices except for the unit: $\mathscr{C}\gamma^A = -\gamma^A \mathscr{C}$, for $\gamma^A \neq 1$. Due to the factor $\mathscr{K}$, it is anti-linear and therefore $\mathscr{C}P_\mu = -P_\mu \mathscr{C}$, but $\mathscr{C}A_\mu = A_\mu \mathscr{C}$. In both the standard and the Weyl representation, we have $\mathscr{C} = -i\gamma^2\mathscr{K}$.

The common transition $\mathscr{T}\mathscr{P}$ $(= \mathscr{P}\mathscr{T})$ is thus described by $\mathscr{K}\mathscr{B}\mathscr{P}_0$, and the transition $\mathscr{T}\mathscr{P}\mathscr{C}$ $(= -\mathscr{C}\mathscr{P}\mathscr{T})$ by $\gamma^5\mathscr{P}_0$. In the next section, we will see how important the operator $\gamma^5 = \gamma^{5\,\dagger}$ is. But let us already recognize a noteworthy property of the *CPT* transformation: with $(\gamma^5)^2 = 1$ and $\gamma^\mu\gamma^5 = -\gamma^5\gamma^\mu$, it leaves scalars, pseudo-scalars, and tensors of second rank unaltered, while for vectors and pseudo-vectors, the sign changes—such statements form the object of the *CPT theorem*.

If we denote the charge-conjugate state of $|\psi\rangle$ by $|\psi\rangle_c$, then by p. 314,

$$|_c\langle\varphi|\psi\rangle_c = \langle\varphi|\psi\rangle^* \quad \text{and} \quad |_c\langle\varphi| \mathscr{C} O \mathscr{C}^{-1} |\psi\rangle_c = \langle\psi| O^\dagger |\varphi\rangle .$$

With $\mathscr{C}\gamma^A\mathscr{C}^{-1} = -\gamma^A$, for $\gamma^A \neq 1$, and $\gamma^{A\,\dagger} = \gamma_A$, this leads to the expectation values

$$\langle\gamma^A\rangle_c = -\langle\gamma_A\rangle , \quad \text{thus} \quad \langle\gamma^0\rangle_c = -\langle\gamma^0\rangle , \quad \langle\gamma^k\rangle_c = +\langle\gamma^k\rangle ,$$
$$\langle\sigma^{0k}\rangle_c = +\langle\sigma^{0k}\rangle , \quad \langle\sigma^{ik}\rangle_c = -\langle\sigma^{ik}\rangle ,$$

and

$$\langle 1\rangle_c = +\langle 1\rangle , \quad \langle X^\mu\rangle_c = +\langle X^\mu\rangle , \quad \langle P^\mu\rangle_c = -\langle P^\mu\rangle , \quad \langle A^\mu\rangle_c = +\langle A^\mu\rangle .$$

Moreover, $H = q\Phi + \gamma^0\{\boldsymbol{\gamma}\cdot c\,(\mathbf{P} - q\mathbf{A}) + mc^2\}$ yields $\mathscr{C}H = -H\mathscr{C}$, or

$$\langle H(q)\rangle_c = -\langle H(-q)\rangle .$$

Thus *the eigenvalues of the Hamilton operator change their sign along with the charge*. If we take them as energy eigenvalues, then we necessarily arrive at negative energy values and find no ground state. Thus an arbitrary amount of energy could be emitted. (However, for time-dependent forces the Hamilton operator and the energy operator agree only for a suitable gauge, so we can also require $E = |H|$ here.) We can repair this difficulty if we quantize the field and attach zero energy to the vacuum. Every particle creation should cost energy, independently of the charge. Due to charge conservation, particles can only be created from the vacuum in pairs of opposite charge, and with a supply of energy. Then twice the energy is necessary compared with what would be required for one particle (disregarding the binding energy between the two).

Here we recall the non-relativistic Fermi gas. In its ground state, all one-particle states below the Fermi edge are occupied, while all those above are empty. Adding energy raises a fermion from an occupied state to an unoccupied one. The excited state differs from the ground state by a particle–hole pair. This picture is also suitable for the Dirac theory. We only have to choose the Fermi edge as the zero energy, i.e., as a quasi-vacuum. If a particle is missing from this quasi-vacuum, then we have a hole, i.e., an *anti-particle*, which is a particle of opposite charge and energy (see Fig. 5.21).

As the quantity adjoint to $\psi_c = \mathscr{C}\psi$, we take $\overline{\psi_c} = \psi^\dagger\gamma^0\mathscr{C}$. This implies $\overline{\psi_c}\gamma^0\psi_c = \psi^\dagger\gamma^0\mathscr{C}\gamma^0\mathscr{C}\psi = -\psi^\dagger(\gamma^0)^2\mathscr{C}^2\psi = -\overline{\psi}^\dagger\gamma^0\psi$, as expected, with $\langle\gamma^0\rangle_c = -\langle\gamma^0\rangle$.

5.6.8 Dirac Equation and Klein–Gordon Equation

We now turn to the problem mentioned in Sect. 5.6.2 that $\mathbf{P}$ does not commute with $\mathbf{A}$, and therefore additional terms occur for $q\mathbf{A} \neq \mathbf{0}$ compared to the *Klein–Gordon equation*

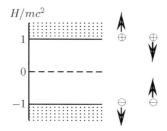

Fig. 5.21 Charge symmetry. For charge inversion, the signs of $\langle H \rangle$, $\langle \mathbf{P} \rangle$, $\langle \mathbf{p} \rangle$, and $\langle \boldsymbol{\sigma} \rangle$ are all reversed. The continuum of H eigenvalues of free particles is indicated by *dotted lines* (*left*). The eigenvalues of $\mathbf{p}$ are shown next to it: *top* for $\oplus$ and *bottom* for anti-particles $\ominus$, and right next to it the same after time reversal (*right*)

$$(p^\mu p_\mu - m^2 c^2)\,\psi = 0 \ .$$

To this end, it can be advantageous to use the projection operators

$$\mathrm{P}_\pm \equiv \tfrac{1}{2}\left(1 \pm \gamma^5\right) = \mathrm{P}_\pm{}^\dagger = \mathrm{P}_\pm{}^2 \ , \quad \mathrm{P}_\pm \mathrm{P}_\mp = 0 \ , \quad \mathrm{P}_+ + \mathrm{P}_- = 1 \ .$$

They commute with p_μ, but not with γ^μ, for $\mu \in \{0, 1, 2, 3\}$:

$$\mathrm{P}_\pm \gamma^\mu = \gamma^\mu \mathrm{P}_\mp \ , \quad \text{but} \quad \mathrm{P}_\pm \gamma^5 = \gamma^5 \mathrm{P}_\pm = \pm \mathrm{P}_\pm \ .$$

Therefore, $\mathrm{P}_\pm \gamma^\mu p_\mu\,\psi = \gamma^\mu p_\mu \mathrm{P}_\mp\,\psi$ also holds. On the other hand, the Dirac equation implies $\gamma^\mu p_\mu \psi = mc\,\psi$, and mc commutes with $\mathrm{P}_\pm$, so from $\mathrm{P}_\mp mc\,\psi = \mathrm{P}_\mp \gamma^\mu p_\mu \psi = \gamma^\mu p_\mu \mathrm{P}_\pm \psi$, we may infer

$$(\mathrm{P}_\mp + \mathrm{P}_\pm)\,mc\,\psi = (\gamma^\mu p_\mu + mc)\,\mathrm{P}_\pm \psi \ ,$$

where $\mathrm{P}_\mp + \mathrm{P}_\pm = 1$ and division by mc is allowed for $m \neq 0$. From a component $\mathrm{P}_+\psi$ or $\mathrm{P}_-\psi$, we thus obtain the total solution ψ which has to satisfy the Dirac equation. Consequently,

$$(\gamma^\mu p_\mu - mc)\,(\gamma^\nu p_\nu + mc)\,\mathrm{P}_\pm \psi = (\gamma^\mu \gamma^\nu p_\mu p_\nu - m^2 c^2)\,\mathrm{P}_\pm \psi = 0 \ ,$$

i.e., each component $\mathrm{P}_\pm \psi$ satisfies the same equation. With $\gamma^\mu \gamma^\nu = g^{\mu\nu} - i\sigma^{\mu\nu}$ and $\sigma^{\mu\nu} p_\mu p_\nu = -\sigma^{\nu\mu} p_\mu p_\nu = -\sigma^{\mu\nu} p_\nu p_\mu = \tfrac{1}{2}\sigma^{\mu\nu}[p_\mu, p_\nu]$, together with

$$[p_\mu, p_\nu] = [P_\mu - q A_\mu,\ P_\nu - q A_\nu] = q\,([P_\nu,\ A_\mu] + [A_\nu,\ P_\mu])$$
$$\mathrel{\widehat{=}} iq\hbar\,(\partial_\nu A_\mu - \partial_\mu A_\nu) = -iq\hbar\,F_{\mu\nu} \ ,$$

the equation for the components can be reformulated as

$$(p^\mu p_\mu - m^2 c^2 - \tfrac{1}{2}\,q\hbar\,\sigma^{\mu\nu}\,F_{\mu\nu})\,\mathrm{P}_\pm \psi = 0 \ .$$

The operator $p^\mu p_\mu - m^2 c^2$ of the Klein–Gordon equation must therefore be amended by the term $-\frac{1}{2} q\hbar\, \sigma^{\mu\nu} F_{\mu\nu}$. This couples the different components of $P_\pm \psi$ via the operators $\sigma^{\mu\nu}$, and in the standard representation, disregarding a factor $q\hbar$, it reads

$$-\tfrac{1}{2} (\sigma^{\mu\nu} F_{\mu\nu})_{\mathrm{D}} = \boldsymbol{\sigma} \cdot \mathbf{B} - i\boldsymbol{\alpha} \cdot \mathbf{E}/c = \begin{pmatrix} \boldsymbol{\sigma} \cdot \mathbf{B} & -i\boldsymbol{\sigma} \cdot \mathbf{E}/c \\ -i\boldsymbol{\sigma} \cdot \mathbf{E}/c & \boldsymbol{\sigma} \cdot \mathbf{B} \end{pmatrix} ,$$

while in the Weyl representation, it reads

$$-\tfrac{1}{2} (\sigma^{\mu\nu} F_{\mu\nu})_{\mathrm{W}} = \begin{pmatrix} \boldsymbol{\sigma} \cdot (\mathbf{B} - i\mathbf{E}/c) & 0 \\ 0 & \boldsymbol{\sigma} \cdot (\mathbf{B} + i\mathbf{E}/c) \end{pmatrix} .$$

Since the projection operators $P_\pm$ are also diagonal in the Weyl representation,

$$P_+ = \begin{pmatrix} 1 & 0 \\ 0 & 0 \end{pmatrix} , \qquad P_- = \begin{pmatrix} 0 & 0 \\ 0 & 1 \end{pmatrix} ,$$

this leads us to 2-spinors $\psi_\pm \equiv (P_\pm \psi)_{\mathrm{W}}$, an advantage over the standard representation:

$$\{p^\mu p_\mu - m^2 c^2 + q\hbar\, \boldsymbol{\sigma} \cdot (\mathbf{B} \mp i\mathbf{E}/c)\}\, \psi_\pm = 0 .$$

Generally, we have

$$p^\mu p_\mu = (P^\mu - qA^\mu)\, (P_\mu - qA_\mu) = P^\mu P_\mu - q\, (P^\mu A_\mu + A^\mu P_\mu) + q^2\, A^\mu A_\mu .$$

Now P^μ commutes with A_μ for the Lorentz gauge $\partial_\mu A^\mu = 0$, so it follows that $P^\mu A_\mu + A^\mu P_\mu = 2 A^\mu P_\mu$. In the scalar product, the order of the operators P and A is thus irrelevant for the Lorentz gauge and we obtain

$$p^\mu p_\mu = (E - q\Phi)^2/c^2 - (\mathbf{P} - q\mathbf{A})^2 ,$$

whereupon

$$\{(E - q\Phi)^2 - c^2\, (\mathbf{P} - q\mathbf{A})^2 - (mc^2)^2 + q\hbar c\, \boldsymbol{\sigma} \cdot (c\mathbf{B} \mp i\mathbf{E})\}\, \psi_\pm = 0 .$$

In this way we have reformulated the Dirac equation as two similar equations for 2-spinors, each being an equation for spin-1/2 particles. (In the standard representation, the same goal is pursued with the *Foldy–Wouthuysen transformation*, but this proceeds only stepwise and approximations have to be made.)

How are the components ψ_+ and ψ_- to be interpreted? To find out, we consider the equation $\mathcal{K}\mathcal{B}\gamma^5 = -\gamma^5 \mathcal{K}\mathcal{B}$. It yields $\mathcal{C}P_\pm = P_\mp \mathcal{C}$. If $P_\pm \psi$ describes a particle, then $\mathcal{C}P_\pm \psi$ describes its anti-particle, which we find as $P_\mp \mathcal{C}\psi$ in the complementary space of the particle. We may thus interpret ψ_+ as a particle and ψ_- as its anti-particle.

In the non-relativistic limit, we have $E - q\Phi \approx mc^2$ and consequently,

$$(E - q\Phi)^2 - (mc^2)^2 \approx 2\,mc^2\,(E - q\Phi - mc^2)\ .$$

In addition, we may then neglect $\hbar\boldsymbol{\sigma} \cdot \mathbf{E}$ compared to $2mc\,\Phi$, since for $\mathbf{E} = -\nabla\Phi$ with $\Delta x \cdot \Delta P \geq \frac{1}{2}\hbar$, we have

$$\left| \frac{\hbar\boldsymbol{\sigma} \cdot \mathbf{E}}{2mc} \right| \approx \frac{\hbar}{2mc}\,\frac{\Delta\Phi}{\Delta x} \leq \frac{\Delta P\,\Delta\Phi}{mc} \ll \overline{|\Phi|}\ .$$

Therefore, in the non-relativistic limit, we find the *Pauli equation* (see p. 327)

$$\left\{ E - \left(mc^2 + \frac{(\mathbf{P} - q\,\mathbf{A})^2}{2m} + q\,\Phi - \frac{q\hbar}{2m}\,\boldsymbol{\sigma} \cdot \mathbf{B} \right) \right\} \psi_\pm = 0\ .$$

Hence there is a real magnetic dipole moment $q\hbar\,\boldsymbol{\sigma}/2m$, and according to the preceding equation, there is also an electric dipole moment, although this is imaginary and therefore not observable, as Dirac himself stressed [14].

5.6.9   Energy Determination for Special Potentials

For free motion ($qA^\mu = 0$), we arrive at the equation

$$E^2 - c^2 P^2 - (mc^2)^2 = 0 \qquad \Longrightarrow \qquad E = c\sqrt{(mc)^2 + P^2}\ .$$

Here, the energy does not depend on the spin (degeneracy). In addition to the momentum, the helicity $\boldsymbol{\sigma} \cdot \mathbf{p}/p$ also commutes with the free Hamilton operator (in both the standard and the Weyl representation). Therefore the free 2-spinors can be decomposed in terms of their helicity ($\eta = \pm 1$). If $\mathbf{p}$ has the direction (θ, φ), then the helicity states, i.e., the eigenstates of $(\sigma_x \cos\varphi + \sigma_y \sin\varphi) \sin\theta + \sigma_z \cos\theta$, can be represented by

$$|+\rangle \mathrel{\widehat{=}} \begin{pmatrix} c \\ s \end{pmatrix} \qquad \text{and} \qquad |-\rangle \mathrel{\widehat{=}} \begin{pmatrix} -s^* \\ c \end{pmatrix}\ ,$$

where we use the abbreviations $c \equiv \cos(\frac{1}{2}\theta)$ and $s \equiv \sin(\frac{1}{2}\theta)\,\exp(i\varphi)$, along with $\langle +| \mathrel{\widehat{=}} (c, s^*)$ and $\langle -| \mathrel{\widehat{=}} (-s, c)$. The directions of $\mathbf{p}$ and $\boldsymbol{\sigma}$ are reversed under charge conjugation, so the helicity is conserved.

So far we have had to write the Hamilton operator for the free motion as a 4×4 matrix

$$H_{\mathrm{D}} = \gamma^0 mc^2 + c\,\gamma^0 \boldsymbol{\gamma} \cdot \mathbf{P}\ ,$$

but now we can decompose it into two 2×2 matrices, viz.,

$$H_\pm = \pm c\sqrt{(mc)^2 + P^2} \,,$$

where we choose H_+ for particles and H_- for anti-particles.

The advantage of this separation can be illustrated by considering, e.g., the velocity. We determine the derivative of the position operator $\mathbf{R}$ with respect to time via the Heisenberg equation. In the standard representation, this yields

$$\frac{d\mathbf{R}_D}{dt} \equiv \frac{[\mathbf{R},\, H_D]}{i\hbar} = c\,\gamma^0 \boldsymbol{\gamma} \,.$$

Hence not all three Cartesian velocity components—each with modulus c—can be sharp simultaneously, because they do not commute with each other. This is often interpreted as *Zitterbewegung*. But with $[\mathbf{R},\, f(P^2)] = 2i\hbar\,(\partial f/\partial P^2)\,\mathbf{P}$, we also have the equation

$$\frac{d\mathbf{R}_\pm}{dt} \equiv \frac{[\mathbf{R},\, H_\pm]}{i\hbar} = c\,\frac{c\mathbf{P}}{H_\pm} \,,$$

which does indeed make sense, because according to p. 245, for free particles, we have $\mathbf{p} = c^{-2}E\mathbf{v}$. The split into particles and anti-particles clarifies this matter. The anti-particles move against their momenta here.

For free motion, the associated 4×4 matrix H_W (see p. 492) can also be decomposed into two 2×2 matrices, one for each of the two helicities $\eta = \pm1$. If we now use the parameter τ to distinguish particles ($\tau = 1$) and anti-particles ($\tau = -1$), then we obtain the eigenvalue equation

$$\begin{pmatrix} \eta\,cp - \tau E & mc^2 \\ mc^2 & -\eta\,cp - \tau E \end{pmatrix} \begin{pmatrix} \psi_{\tau\eta} \\ \varphi_{\tau\eta} \end{pmatrix} = 0 \,.$$

This leads to the above-mentioned energy eigenvalue (with $E > 0$) and to

$$\frac{\psi_{\tau\eta}}{\varphi_{\tau\eta}} = \frac{\tau E + \eta\,cp}{mc^2} = \frac{mc^2}{\tau E - \eta\,cp} \,.$$

For the normalization, we invoke the invariant $\bar{\psi}\psi = 2\mathrm{Re}\,\psi_{\tau\eta}{}^*\varphi_{\tau\eta}$, noting that $|\psi_{\tau\eta}|^2 + |\varphi_{\tau\eta}|^2$ is not suitable here. Then the expressions just found deliver

$$\bar{\psi}\psi = 2|\psi_{\tau\eta}|^2 \,\mathrm{Re}\frac{\varphi_{\tau\eta}}{\psi_{\tau\eta}} = 2|\psi_{\tau\eta}|^2 \,\frac{mc^2}{\tau E + \eta\,cp}$$

$$= 2|\varphi_{\tau\eta}|^2 \,\mathrm{Re}\frac{\psi_{\tau\eta}}{\varphi_{\tau\eta}} = 2|\varphi_{\tau\eta}|^2 \,\frac{mc^2}{\tau E - \eta\,cp} \,.$$

With $E > cp$, $\bar{\psi}\psi$ is positive for particles ($\tau = 1$) and negative for anti-particles ($\tau = -1$). We therefore require $\bar{\psi}\psi = \tau$ and infer

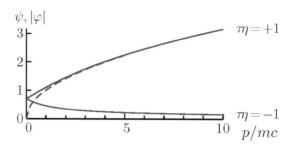

Fig. 5.22 Large amplitude (*red*) and small amplitude (*magenta*). These differ by the product $\tau\eta = \pm 1$ and depend on p/mc. In the case of free motion considered here, we have $p = m\gamma v$, with $v/c = \beta$ and hence $p/mc = \gamma\beta$ and the approximation $\sqrt{p/mc}$ (*dashed blue*)

$$|\psi_{\tau\eta}|^2 = \frac{E + \tau\eta\, cp}{2mc^2} \qquad \text{and} \qquad |\varphi_{\tau\eta}|^2 = \frac{E - \tau\eta\, cp}{2mc^2}\ .$$

We choose $\psi_{\tau\eta}$ and $\varphi_{\tau\eta}$ real, and $\psi_{\tau\eta} \geq 0$. With this and with $2\mathrm{Re}\,\psi_{\tau\eta}{}^*\varphi_{\tau\eta} = \tau$, the sign of $\varphi_{\tau\eta}$ is the same as that of τ:

$$\psi_{\tau\eta} = \sqrt{\frac{E + \tau\eta\, cp}{2mc^2}} \qquad \text{and} \qquad \varphi_{\tau\eta} = \tau\sqrt{\frac{E - \tau\eta\, cp}{2mc^2}}\ ,$$

again with $E > 0$. For high energies, $E \approx cp$ and therefore one or the other amplitude is negligible—but $|\mathrm{Re}\psi_{\tau\eta}{}^*\varphi_{\tau\eta}| = \frac{1}{2}$. We speak here of *large* and *small amplitudes* (see Fig. 5.22).

For the Weyl representation, these expressions are then also to be multiplied by the helicity amplitudes mentioned above:

$$\begin{pmatrix} \psi_{++}\, c \\ \psi_{++}\, s \\ \varphi_{++}\, c \\ \varphi_{++}\, s \end{pmatrix}, \qquad \begin{pmatrix} -\psi_{+-}\, s^* \\ \psi_{+-}\, c \\ -\varphi_{+-}\, s^* \\ \varphi_{+-}\, c \end{pmatrix}, \qquad \begin{pmatrix} \psi_{-+}\, c \\ \psi_{-+}\, s \\ \varphi_{-+}\, c \\ \varphi_{-+}\, s \end{pmatrix}, \qquad \begin{pmatrix} -\psi_{--}\, s^* \\ \psi_{--}\, c \\ -\varphi_{--}\, s^* \\ \varphi_{--}\, c \end{pmatrix}.$$

The momentum eigenfunction must be included with all these "internal" wave functions.

In a homogeneous magnetic field $\mathbf{B} = \mathbf{B}_0$, the Coulomb gauge is $\mathbf{A} = \frac{1}{2}\mathbf{B}_0 \times \mathbf{R}$, with Φ and $\mathbf{E}$ equal to zero, and we have $\mathbf{P} \cdot \mathbf{A} = \mathbf{A} \cdot \mathbf{P} = \frac{1}{2}\hbar\mathbf{B}_0 \cdot \mathbf{L}$, where we introduce the dimensionless quantity $\mathbf{L} = \mathbf{R} \times \mathbf{P}/\hbar$. This yields

$$(\mathbf{P} - q\,\mathbf{A}) \cdot (\mathbf{P} - q\,\mathbf{A}) = P^2 - q\hbar\,\mathbf{B}_0 \cdot \mathbf{L} + q^2 A^2$$

and

$$\frac{E^2}{c^2} = (mc)^2 + P^2 - q\hbar\,\mathbf{B}_0 \cdot (\mathbf{L} + \boldsymbol{\sigma}) + \frac{q^2}{4}\,(\mathbf{B}_0 \times \mathbf{R})^2\ .$$

For charge conjugation, q is to be replaced by $-q$ and $\langle \mathbf{L} + \boldsymbol{\sigma} \rangle$ by $-\langle \mathbf{L} + \boldsymbol{\sigma} \rangle$, and we thus arrive at the same value of $|E|$, despite the charge (a)symmetry.

For the hydrogen problem with $q\Phi = -e^2/(4\pi\varepsilon_0\,r)$ and $\mathbf{B} \equiv \mathbf{0}$, it is advantageous to take the fine-structure constant

$$\alpha \equiv \frac{e^2}{4\pi\varepsilon_0}\frac{1}{\hbar c} = \frac{1}{137.\ldots} \, ,$$

and use the further abbreviation $\mathbf{r}' = \mathbf{r}/r$. With $P^2 \,\hat{=}\, -\hbar^2(\mathrm{d}^2/\mathrm{d}r^2 - L^2/r^2)$, we arrive at the differential equation

$$\left(\frac{\mathrm{d}^2}{\mathrm{d}r^2} - \frac{m^2c^4 - E^2}{\hbar^2 c^2} + \frac{\alpha E}{\hbar c}\frac{2}{r} - \frac{L^2 \mp \mathrm{i}\alpha\,\mathbf{r}'\cdot\boldsymbol{\sigma} - \alpha^2}{r^2} \right) \psi(r) = 0 \, .$$

It is similar to the non-relativistic radial equation of the hydrogen problem, investigated in more detail in [15] (see also p. 422):

$$\left(\frac{\mathrm{d}^2}{\mathrm{d}\rho^2} - 1 + \frac{2\eta}{\rho} - \frac{l(l+1)}{\rho^2} \right) u_l(\rho) = 0 \, ,$$

with the Coulomb parameter η (not to be confused with the helicity, which we shall no longer consider). Normalizability requires $\eta - l$ to be a natural number (1, 2, 3, ...). We shall denote it by $n_r + 1$, whence n_r gives the number the nodes of the radial function. (The zeros at the boundaries 0 and ∞ do not constitute nodes.)

To exploit this well known result, we now have to express the eigenvalues of $L^2 \mp \mathrm{i}\alpha\,\mathbf{r}'\cdot\boldsymbol{\sigma} - \alpha^2$ in terms of $\lambda(\lambda + 1)$. In fact, λ is somewhat smaller than l, as will now be shown.

The dipole–field coupling $\propto \boldsymbol{\sigma}\cdot\mathbf{r}'/r^2$ does not commute with the orbital angular momentum, but like any scalar, it does commute with the total angular momentum $(\mathbf{L} + \mathbf{S})\,\hbar$, so for the spin angular momentum, we split off the factor $\hbar$ and write $\mathbf{S} = \frac{1}{2}\boldsymbol{\sigma}$. It is therefore appropriate to take the coupled representation $|(l\frac{1}{2})jm\rangle$ of p. 336. In particular, the operator $\mathbf{L}\cdot\boldsymbol{\sigma}$ is diagonal, and with $\mathbf{L}\times\mathbf{L} = \mathrm{i}\mathbf{L}$, according to p. 325, we have $(\mathbf{L}\cdot\boldsymbol{\sigma})^2 = L^2 + \mathrm{i}(\mathbf{L}\times\mathbf{L})\cdot\boldsymbol{\sigma} = L^2 - \mathbf{L}\cdot\boldsymbol{\sigma}$, so $L^2 = \mathbf{L}\cdot\boldsymbol{\sigma}\,(\mathbf{L}\cdot\boldsymbol{\sigma} + 1)$.

The term $\mp\mathrm{i}\alpha\,\mathbf{r}'\cdot\boldsymbol{\sigma} - \alpha^2$ with $(\mathbf{r}'\cdot\boldsymbol{\sigma})^2 = 1$ may also be written

$$\mp\mathrm{i}\alpha\,\mathbf{r}'\cdot\boldsymbol{\sigma}\,(1 \mp \mathrm{i}\alpha\,\mathbf{r}'\cdot\boldsymbol{\sigma}) \, .$$

Therefore, it follows that

$$L^2 \mp \mathrm{i}\alpha\,\mathbf{r}'\cdot\boldsymbol{\sigma} - \alpha^2 = \Lambda\,(\Lambda + 1) \, , \quad \text{with} \quad \Lambda = -\mathbf{L}\cdot\boldsymbol{\sigma} \mp \mathrm{i}\alpha\,\mathbf{r}'\cdot\boldsymbol{\sigma} - 1 \, ,$$

if we can prove that $\mathbf{L}\cdot\boldsymbol{\sigma}\,\mathbf{r}'\cdot\boldsymbol{\sigma} + \mathbf{r}'\cdot\boldsymbol{\sigma}\,\mathbf{L}\cdot\boldsymbol{\sigma} = -2\mathbf{r}'\cdot\boldsymbol{\sigma}$. Here, according to p. 325, the left-hand side is equal to $(\mathbf{L}\cdot\mathbf{r}' + \mathbf{r}'\cdot\mathbf{L}) + \mathrm{i}\,(\mathbf{L}\times\mathbf{r}' + \mathbf{r}'\times\mathbf{L})\cdot\boldsymbol{\sigma}$, and with this the first bracket vanishes, because $\mathbf{L}$ and $1/R$ commute and we have $\mathbf{R}\cdot(\mathbf{R}\times\mathbf{P}) = -\mathbf{R}\cdot(\mathbf{P}\times\mathbf{R}) = -(\mathbf{R}\times\mathbf{P})\cdot\mathbf{R}$. For the second, we may use $\mathbf{R}\cdot\mathbf{P} = 3\mathrm{i}\hbar + \mathbf{P}\cdot$

$\mathbf{R}$ along with $[\mathbf{R}, \mathbf{P} \cdot \mathbf{R}] = i\hbar\mathbf{R}$ and $[\mathbf{P}, R^2] = -2i\hbar\mathbf{R}$. This leads to $\mathbf{L} \times \mathbf{R} + \mathbf{R} \times \mathbf{L} = 2i\mathbf{R}$, whence $\mathbf{L} \cdot \boldsymbol{\sigma}\, \mathbf{r}' \cdot \boldsymbol{\sigma} + \mathbf{r}' \cdot \boldsymbol{\sigma}\, \mathbf{L} \cdot \boldsymbol{\sigma}$ is indeed equal to $-2\mathbf{r}' \cdot \boldsymbol{\sigma}$. We thus obtain

$$\Lambda^2 = (\mathbf{L} \cdot \boldsymbol{\sigma} + 1)^2 - \alpha^2 .$$

The eigenvalues of this Hermitian operator depend only on α^2, not on the sign $\mp i\alpha$. But in our further calculations, we have to distinguish between $j = l \pm \frac{1}{2}$ and we also need the different signs now for another purpose.

In particular, by p. 372,

$$(\mathbf{L} \cdot \boldsymbol{\sigma} + 1)\, |(l\tfrac{1}{2})jm\rangle = \pm |(l\tfrac{1}{2})jm\rangle\, (j + \tfrac{1}{2}) , \quad \text{for } j = l \pm \tfrac{1}{2} \ (\in \{\tfrac{1}{2}, \ldots\}) ,$$

and consequently,

$$\Lambda^2\, |(l\tfrac{1}{2})jm\rangle = |(l\tfrac{1}{2})jm\rangle\, \{(j + \tfrac{1}{2})^2 - \alpha^2\} ,$$

as well as

$$\Lambda\, |(l\tfrac{1}{2})jm\rangle = \mp |(l\tfrac{1}{2})jm\rangle\, \{(j + \tfrac{1}{2})^2 - \alpha^2\}^{1/2} , \quad \text{for } j = l \pm \tfrac{1}{2} .$$

Note that the sign follows from the limit $\alpha \to 0$, whence Λ tends towards $-\mathbf{L} \cdot \boldsymbol{\sigma} - 1$. If we now denote the eigenvalue of $\Lambda\,(\Lambda + 1)$ by $\lambda(\lambda + 1)$, we have

$$\lambda = \{(j + \tfrac{1}{2})^2 - \alpha^2\}^{1/2} - (j + \tfrac{1}{2} - l) ,$$

and hence,

$$\lambda = l - \varepsilon_j , \quad \text{with } \varepsilon_j \equiv j + \tfrac{1}{2} - \sqrt{(j + \tfrac{1}{2})^2 - \alpha^2} \approx \frac{\alpha^2}{2j + 1} \ll 1 .$$

With this we may now return to the known result of the non-relativistic calculation. Comparing the two radial equations with $(m^2 c^4 - E^2)/(\hbar c)^2 = k^2$ and $\alpha E/(\hbar c) = \eta k$ leads to

$$\eta \;\widehat{=}\; \frac{\alpha E}{\sqrt{m^2 c^4 - E^2}} \qquad \Longrightarrow \qquad E = \frac{mc^2}{\sqrt{1 + (\alpha/\eta)^2}} .$$

Normalizability now requires

$$\eta = n_r + 1 + \lambda = n - \varepsilon_j ,$$

with the *principal quantum number* $n \equiv n_r + l + 1$ (see p. 363). Finally, we obtain

$$E = \frac{mc^2}{\sqrt{1 + \alpha^2/(n - \varepsilon_j)^2}} = mc^2 - \frac{E_R}{n^2}\left\{1 + \frac{\alpha^2}{n}\left(\frac{1}{j + \frac{1}{2}} - \frac{3}{4n}\right) + \cdots\right\},$$

where $j \in \{\frac{1}{2}, \ldots, n - \frac{1}{2}\}$, so that $1/(j + \frac{1}{2}) > 3/4n$, and the Rydberg energy (see p. 362)

$$E_R \equiv \frac{1}{2}\alpha^2 mc^2 .$$

As can already be seen from Fig. 4.18, there is now no degeneracy with respect to the angular momentum j (only with respect to the orbital angular momentum l), in contrast to the non-relativistic calculation. The terms indicated by dots in the above may be left out, being smaller than the effects neglected in the Dirac theory (like the Lamb shift mentioned on p. 380).

5.6.10 Difficulties with the Dirac Theory

In fact, the Dirac equation describes electrons (and neutrinos) better than the Schrödinger equation, because it accounts for relativistic effects and spin (although it is still not the end of the story). In particular, it also holds for anti-particles (positrons), and their energy spectrum is reflected at $E = 0$. There are thus infinitely many states of negative energy, with no lower bound. In particular, the free Dirac equation allows any energy above mc^2 and below $-mc^2$, but none in-between.

Dirac suggested viewing the vacuum as a many-body state, where all states of negative energy are occupied and all states of positive energy empty. If this vacuum is excited by more than $2mc^2$ (through photon absorption), then a particle switches from a state of negative energy into a state of positive energy. This creates a *particle–hole pair*, which may be interpreted as electron–positron pair creation. Conversely, there may also be pair annihilation, where a particle makes a transition to a *hole state* and emits electromagnetic radiation.

Even though pair generation and annihilation may be described with the hole theory, the Dirac equation leaves some questions open. In particular, it cannot be a one-particle theory. The many particles of negative energy should interact with each other. In addition it remains to be clarified whether electrons or positrons have negative energy. These problems can only be tackled by field quantization.

List of Symbols

We stick closely to the recommendations of the *International Union of Pure and Applied Physics* (IUPAP) and the *Deutsches Institut für Normung* (DIN). These are listed in *Symbole, Einheiten und Nomenklatur in der Physik* (Physik-Verlag,

Table 5.3 Symbols used in quantum mechanics II

	Symbol	Name	Page number
	H	Full Hamilton operator	404
	H_0	Free Hamilton operator	404
	V	Interaction operator	404
	G	Propagator for H	405
	G_0	Propagator for H_0	405
	S	Scattering operator	414
	T	Transition operator	415
		One-particle operator	444
	Ω	Solid angle	417
	$\Omega^{\pm}$	Möller's wave operators	413
	P, Q	Projection operators	413
	$\mid\,\rangle^{\pm}$	Scattering states	412
*	σ	Scattering cross-section	418
	δ	Scattering phase	421
*	Γ	Level width	425
	Ψ	Annihilation operator	442
	$\Psi^{\dagger}$	Creation operator	442
	N	Particle number operator	443
	γ^{μ}	Dirac matrix	489
	$\sigma^{\mu\nu}$	Dirac matrix	490

Weinheim 1980) and are marked here with an asterisk. However, one and the same symbol may represent different quantities in different branches of physics. Therefore, we have to divide the list of symbols into different parts (Table 5.3).

References

1. H. Feshbach, Ann. Phys. **19**, 287 (1962)
2. E.P. Wigner, L. Eisenbud, Phys. Rev. **72**, 29 (1947)
3. A.M. Lane, R.G. Thomas, Rev. Mod. Phys. **30**, 257 (1958)
4. P.I. Kapur, R. Peierls, Proc. Roy. Soc. A **166**, 277 (1937)
5. P.A. Kazaks, K.R. Greider, Phys. Rev. C **1**, 856 (1970)
6. E.W. Schmid, H. Ziegelmann, *The Quantum Mechanical Three-Body Problem* (Vieweg, Braunschweig, 1974)
7. G.E. Brown, *Many-Body Problems* (North-Holland, Amsterdam, 1972), p. 22
8. C. Cohen Tannoudji, J. Dupont-Roc, G. Grynber, *Photons and Atoms* (Wiley, New York, 1989), Chap. 5

9. E. Schrödinger, Naturwissenschaften **14**, 664 (1926)
10. O. Klein, Z. Phys. **37**, 895 (1926)
11. W. Gordon, Z. Phys. **40**, 117 (1926)
12. E. Schrödinger, Ann. Physics **79**, 489 (1926)
13. V. Fock, Z. Phys. **38**, 242; **39**, 226 (1926)
14. P.A.M. Dirac, Proc. Roy. Soc. A **117**, 610 (1928)
15. R.A. Swainson, G.W.F. Drake: J. Phys. A **24**, 79, 95 (1991)

Suggestions for Textbooks and Further Reading

16. W. Greiner, J. Reinhardt: *Field Quantization* (Springer, New York 1996)
17. W. Greiner: *Relativistic Quantum Mechanics: Wave Equations* (Springer, New York 2000)
18. V.B. Berestetskii, E.M. Lifshitz, L.P. Pitaevskii, *Course of Theoretical Physics—Quantum Electrodynamics*, vol. 4 ,2nd edn. (Butterworth–Heinemann, Oxford 1982)

Chapter 6
Thermodynamics and Statistics

6.1 Statistics

6.1.1 Introduction

Although this chapter is announced in the usual way as being about *thermodynamics and statistics*, we shall nevertheless begin with statistics. Then we shall be able to justify thermodynamics with quantum theory, and present the entropy S in a more logical way.[1] The entropy is a key basic notion in the theory of heat and must otherwise be introduced axiomatically. In such a representation, thermodynamics starts with the following main theorems, where the notion of *state variable* appears three times and, as an observable, is associated with the instantaneous state of the considered system, e.g., position, momentum, and energy in particle mechanics:

Zeroth main theorem (R. H. Fowler): There is a state variable called *temperature T* (in kelvin K). Two systems (or two parts of a systems) are only in thermal equilibrium if they have equal temperature.

First main theorem (R. Mayer, H. v. Helmholtz): There is a state variable called the *internal energy U* of the system. It increases by the (reversible or irreversible) addition of an amount of heat δQ and addition of work δA:

---

[1]It is interesting to quote Carathéodory [1] in his inaugural address to the Prussian Academy as cited in [2]: "It is possible to ask the question as to how to construct the phenomenological science of thermodynamics when it is desired to include only directly measurable quantities, that is volumes, pressures, and the chemical composition of systems. The resulting theory is logically unassailable and satisfactory for the mathematician because, starting solely with observed facts, it succeeds with a minimum of hypotheses. And yet, precisely these merits impede its usefulness to the student of nature, because, on the one hand, temperature appears as a derived quantity, and on the other, and above all, it is impossible to establish a connection between the world of visible and tangible matter and the world of atoms through the smooth walls of the all too artificial structure."

© Springer Nature Switzerland AG 2018
A. Lindner and D. Strauch, *A Complete Course*
on Theoretical Physics, Undergraduate Lecture Notes in Physics,
https://doi.org/10.1007/978-3-030-04360-5_6

$$dU \equiv \delta Q + \delta A .$$

Note that dU is a complete differential, while the terms on the right-hand side are not necessarily so. For a cycle, $\oint dU = 0$ holds, while not all closed integrals of the individual quantities on the right would vanish. Therefore, U is a state variable, but heat and work are not, as already stressed in Fig. 2.1 (more on that in Sect. 6.4.2). Symbols containing δ are common in variational calculus (see Sects. 2.1.2 and 2.1.3). For a closed system the energy conservation law holds, i.e., $dU = 0$. Generally, there are no conservation laws for heat or work alone.

Second main theorem (R. Clausius, W. Thomson/Lord Kelvin): There is a state variable called *entropy* S. This increases by the reversibly added quantity $\delta Q_{rev}/T$,

$$dS \equiv \frac{\delta Q_{rev}}{T} ,$$

and for a closed system it can only increase with time:

$$\frac{dS}{dt} \geq 0 , \quad \text{for a closed system.}$$

This inequality is called the *entropy law*.

Third main theorem (W. Nernst): At the absolute zero of the temperature $T = 0$, the entropy depends only on the degree of degeneracy of the ground state. There we can set $S = 0$.

The entropy seems to many people like a mysterious auxiliary quantity. What is important for its measurement is the amount of heat added *reversibly*, and this depends only on the entropy, whether or not a process can be reversed in a closed system. It may possibly break time-reversal invariance.

On the other hand, if we begin with statistics and derive the phenomena associated with heat from the disordered motion of particles like molecules, atoms, or photons, as described in [3, 4], for example, then we can begin by introducing the information entropy (the many different possibilities of expression). This can be used to justify the main theorems of thermodynamics. On the other hand, in statistics we rely on "sensible" assumptions.

Therefore, we can already clarify the notion of entropy in this section. In Sect. 6.2, we introduce the time dependence and justify the entropy law. After that we will consider equilibrium distributions and use this to understand what entropy can do for us. In Sect. 6.4, we can then deal with the main theorems of thermodynamics and subsequently turn to applications.

In the following, we consider *systems* with very many degrees of freedom (very many "particles"), whose individual characteristics neither can nor shall be pursued in detail. If we take a mole of some gas (i.e., nearly a septillion molecules), then we can neither solve the coupled equations of motion, nor set the initial conditions for all the individual particles correctly and actually follow their time evolution. In fact, we do not want to observe the single molecules, but only a few properties (parameters) of the system as a whole. We can fix the *macro state* through a handful of collective (macroscopic) parameters and follow its evolution, but not the basic *micro state*, which contains far too many microscopic parameters. (Even if only a few particles appear to be important, their coupling to the environment with its many degrees of freedom cannot be switched off completely, and this environment is continually changing.)

A truly enormous number of different micro states belong to any given macro state, specified by its particle number and type, its energy and volume, etc. We shall treat these many states using statistical methods. All the micro states belonging to the same macro state form a *statistical ensemble*.

6.1.2 Statistical Ensembles and the Notion of Probability

A statistical ensemble is described by a small number of parameters, while many other parameters vary from member to member within the ensemble. As an example, we have already mentioned a gas of molecules, whose energy, volume, and particle number have been given. Another statistical ensemble need not even assume a fixed number of particles. So the local particle densities in the considered gases may differ significantly from the mean value N/V. The different values occur with different sub-ensembles of particles in the ensembles.

From the occurrence of an attribute (signature) in a sequence (of micro states), we may infer its probability, i.e., its relative occurrence in the limit of large sequences. If we consider, e.g., the results of tossing a dice, then the "6" will not always occur exactly once for very six throws (sometimes not at all, sometimes repeatedly), but for a fair dice every number $z \in \{1, \ldots, 6\}$ will occur on the average equally often. The probabilities ρ_z for a fair dice do not depend on z. Summed over all possibilities, we must therefore obtain unity, i.e., $\sum_z \rho_z = 1$, because the probability that the event z_1 *or* z_2 occurs, is generally equal to $\rho_1 + \rho_2$ (to be contrasted with the probability that *only* z_1, *then* z_2 appears, which is equal to $\rho_1 \cdot \rho_2$, and that *once* z_1 and *once* z_2 appears, equal to $\rho_1 \cdot \rho_2 + \rho_2 \cdot \rho_1 = 2\rho_1 \cdot \rho_2$). With $\rho_1 = \cdots = \rho_6$ and $\sum_{z=1}^{6} \rho_z = 1$, we conclude that the probability ρ_z for each number of spots is equal to 1/6, for a fair dice.

If z is generally a natural number which may assume Z values, and ρ_z the associated probability (relative occurrence in the statistical ensemble), then ρ_z is real, non-negative, and normalized:

$$\rho_z = \rho_z^* \geq 0 , \qquad \sum_{z=1}^{Z} \rho_z = 1 .$$

If z is a continuous variable, then $\rho(z)$ will be a probability *density*, and instead of the sum, there will be an integral. Only $\rho(z)\,\mathrm{d}z$ is then a probability, namely that the variable takes a value between z and $z + \mathrm{d}z$.

The mean value of a quantity A_z in an ensemble given by $\{\rho_z\}$ is clearly

$$\overline{A} \equiv \langle A \rangle = \sum_{z=1}^{Z} \rho_z A_z \, ,$$

since each value A_z is weighted here with its associated probability. In quantum mechanics (see Sect. 4.2.11), ρ and A are Hermitian operators, which we may represent in a basis $\{|n\rangle\}$ as matrices. Then,

$$\langle A \rangle = \sum_{nn'} \langle n| \rho |n'\rangle \langle n'| A |n\rangle = \sum_{n} \langle n| \rho A |n\rangle = \text{tr}(\rho A) \, .$$

In the eigen representation of ρ or A, only the diagonal elements of ρ and A are necessary, and thus only a sum over $\rho_n A_n$. Therefore, in the following we shall often write $\langle A \rangle = \text{tr}(\rho A)$, even though we think mostly of ρ and A as the classical quantities.

The mean value is a linear functional, i.e., for arbitrary constant α and β, we have

$$\langle \alpha\, A + \beta\, B \rangle = \alpha \langle A \rangle + \beta \langle B \rangle \, ,$$

since $\text{tr}\{\rho\,(\alpha A + \beta B)\} = \alpha\text{tr}(\rho A) + \beta\text{tr}(\rho B)$. With $\langle 1 \rangle = 1$, the mean value of the deviations from the mean value vanishes:

$$\langle A - \langle A \rangle \rangle = \langle A \rangle - \langle A \rangle \langle 1 \rangle = 0 \, ,$$

but generally the square of the *fluctuation* (the *variance* or *dispersion*) will not be zero:

$$(\Delta A)^2 \equiv \langle (A - \langle A \rangle)^2 \rangle = \langle A^2 \rangle - \langle A \rangle^2 \, ,$$

where $\Delta A \geq 0$. We call ΔA the *standard deviation* (*error width* or *average square deviation*, and in quantum theory, the *uncertainty*) and $\Delta A / \langle A \rangle$ the *relative fluctuation*: the smaller it is, the less frequently the members of the ensemble are in states with a value A_z which deviates essentially from $\langle A \rangle$. Results of measurements will be given in the form $\langle A \rangle \pm \Delta A$.

6.1.3 Binomial Distribution

For the probability distribution $\{\rho_z\}$ of a statistical ensemble of Z mutually independent experiments, we are often led to ask whether z of them have turned out to be convenient (positive), with the remaining ones being inconvenient (negative), i.e., there are essentially two outcomes. This is similar to the problem of a one-dimensional random walk with fixed step length. Here, a body can move forward or backward, and in fact with the probability p for each step ahead and probability $q = 1 - p$ for each step back: then we ask where it will be after Z steps? If there are z steps ahead and $Z - z$ steps back, then finally it will have made $z - (Z - z) = 2z - Z$ steps ahead—for $z < \frac{1}{2} Z$, it will thus have moved back.

The probability that out of the Z steps the first z were ahead and the remaining ones back is clearly equal to $p^z q^{Z-z}$. Here the order is not important at all, since we ask only for the probability that in total there were z steps ahead. As is well known, there is a total of $Z!$ different ways to order Z distinguishable objects—the last can be placed at Z sites and therefore increases the number of ways by a factor of Z, while for $Z = 1$, there is only one site. But here only two outcomes are distinguished (ahead or back), and therefore $Z!$ is to be divided by $z! (Z - z)!$ Thus $\binom{Z}{z}$ different combinations deliver the same result. The unknown probability is therefore equal to the probability $p^z q^{Z-z}$ for the first-mentioned possibility times this number of equivalent series. We find the *binomial distribution* (*Bernoulli distribution*)

$$\rho_z = \binom{Z}{z} p^z q^{Z-z} .$$

With $\sum_{z=0}^{Z} \binom{Z}{z} p^z q^{Z-z} = (p + q)^Z$ and $p + q = 1$, we do indeed obtain $\sum_z \rho_z = 1$.

The mean value of "convenient" occurrences is $\langle z \rangle = \sum_z \rho_z z$. Such mean values can often be evaluated as derivatives with respect to suitable parameters. For the binomial distribution, for instance,

$$\langle z \rangle = \sum_z \binom{Z}{z} p^z q^{Z-z} z = p \frac{\partial}{\partial p} \sum_z \binom{Z}{z} p^z q^{Z-z} ,$$

so we can say $\langle z \rangle = p \, \partial (p + q)^Z / \partial p$ at the site $p = 1 - q$. This yields

$$\langle z \rangle = p Z ,$$

as expected, because the probability p is the ratio $\langle z \rangle / Z$. We can also find the mean value of z^2 for this distribution in a similar way, by noting that $\langle z^2 \rangle$ is equal to

$$(p \, \partial / \partial p)^2 (p + q)^Z |_{p=1-q} = p \, \partial / \partial p \, \{pZ (p + q)^{Z-1}\}|_{p=1-q} ,$$

which is equal to

$$pZ + p^2 Z (Z - 1) = p^2 Z^2 + p (1 - p) Z .$$

Hence the binomial distribution yields the standard deviation and relative fluctuation (see Fig. 6.1)

$$\Delta z = \sqrt{pq Z} \quad \text{and} \quad \frac{\Delta z}{\langle z \rangle} = \sqrt{\frac{q}{p}} \frac{1}{\sqrt{Z}} .$$

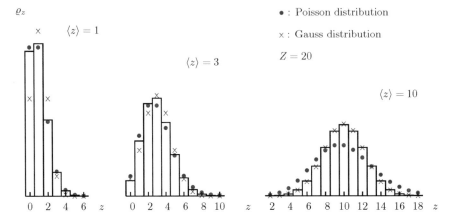

Fig. 6.1 Binomial distributions (Bernoulli distributions), represented by bars, with 20 possibilities and the mean values $\langle z \rangle \in \{1, 3, 10\}$. For comparison, we also show the values for the associated Poisson and Gauss distributions

With increasing Z, we find that $\Delta z / \langle z \rangle$ becomes ever smaller, the maximum of $\{\rho_z\}$ becoming sharper. For example, with $Z = 10^{20}$ and $p \approx q$, the measurement value is uncertain only in the tenth digit.

6.1.4 Gauss and Poisson Distributions

For very large Z the binomial coefficients are difficult to evaluate. Then it is better to use approximation formulas for the factorials of Z and $Z - z$, and in particular, *Stirling's formula*

$$Z! \approx (Z/e)^Z \sqrt{2\pi Z} \ .$$

This can be proven using $n! = \int_0^\infty x^n\, e^{-x}\, dx$, if the exponent $n \ln x - x$ can be expanded in a power series about the maximum $x = n$ (Problem 6.2). For very large Z, we may even leave out the square-root factor, because $\ln \sqrt{2\pi Z} \ll Z\, (\ln Z - 1) = \ln (Z/e)^Z$, as also represented in Fig. 6.2. The logarithmic scale is very appropriate here.

Let us start by considering the case $p \ll 1$ (or equivalently $q \ll 1$, because then we need to interchange only p and q). We have $\langle z \rangle \ll Z$, implying that only $z \ll Z$ is important. Therefore, we may now approximate the binomial coefficients $\binom{Z}{z}$ with $(1 - z/Z)^z \approx 1$, but $(1 - z/Z)^Z \approx e^{-z}$ as follows:

$$\binom{Z}{z} \approx \frac{1}{z!}\, \frac{(Z/e)^Z}{\{(Z - z)/e\}^{Z - z}} = \frac{1}{z!} \left(\frac{Z}{e}\right)^z \frac{(1 - z/Z)^z}{(1 - z/Z)^Z} \approx \frac{Z^z}{z!}\ .$$

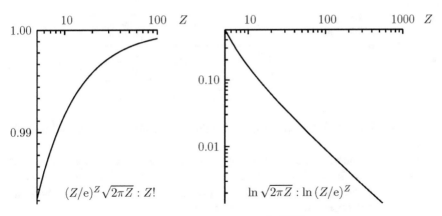

Fig. 6.2 Quality of the Stirling formula. The ratio $(Z/e)^Z \sqrt{2\pi Z}/Z!$ (*left*) and the ratio of the logarithms of $\sqrt{2\pi Z}$ and $(Z/e)^Z$ (*right*) versus Z

In addition, with $\ln(1 - p) \approx -p$, we may set $q^{Z-z} \approx e^{-p(Z-z)}$. For $z \ll Z$, the factor e^{pz} can be neglected in comparison to e^{-pZ}. Consequently, for $Z \gg 1$ and $p \ll 1$, with $\langle z \rangle = pZ$, the binomial distribution goes over into the *Poisson distribution*

$$\rho_z = \exp(-\langle z \rangle) \, \frac{\langle z \rangle^z}{z!} \; .$$

Since $\sum_{z=0}^{Z} \langle z \rangle^z/z!$ tends to $e^{\langle z \rangle}$ for $Z \gg \langle z \rangle$, the normalization is conserved, despite the approximations. In addition, for $q \approx 1$, from the standard deviation of the binomial distribution, we now obtain $(\Delta z)^2 = pZ = \langle z \rangle$ and likewise from the Poisson distribution.

The Poisson distribution always occurs if there are a great many possibilities, but only a few are actually realized, e.g., for the probability of weakly coupled quanta striking the atoms of a multi-layered lattice, or for the clump probability in a beam of mutually independent particles, where we may ask how soon one quantum is followed by the next and we refer to the average distance. (The sequence is no longer independent, if the quanta occur preferably in pairs or single.)

So far, with $p \ll 1$, only $z \ll Z$ was important, or equivalently, for $q \ll 1$, only $z \approx Z \gg 1$ was important. If neither p nor q are very small, then these boundary values are no longer relevant. We may then take z as a continuous variable and expand $\ln \rho(z)$ in a Taylor series about the maximum $\langle z \rangle$. If we use the Stirling formula for the factorials in $\binom{Z}{z}$, then we obtain (Problem 6.4) the *Gauss distribution*, also called the *normal distribution* (see Figs. 1.15 and 6.1):

$$\rho(z) = \frac{1}{\sqrt{2\pi}\,\Delta z} \, \exp \frac{-(z - \langle z \rangle)^2}{2(\Delta z)^2} \; .$$

Here we always have $\langle z \rangle = pZ$ and $\Delta z = \sqrt{pq Z}$. Instead of the error width Δz, the *Lorentz distribution* is sometimes taken, i.e., the interval, in which $\rho(z)$ is greater

	Size	Light	Average weight	Heavy
Table 6.1 Correlation between people's size and weight in terms of thin (○) and thick (●)	Short	○	●	
	Tall		○	●

than or equal to half the maximum value. For the Gauss distribution, it is $2\sqrt{\ln 4}\,\Delta z \approx 2.35\Delta z$.

What is important in all these examples of probability distributions is the result that the relative deviation from the mean value with increasing Z becomes ever smaller, and in the limit $Z \gg 1$, the uncertainty Δz becomes negligible, because we can only give the mean value $\langle z \rangle$ to a few significant figures.

6.1.5 Correlations and Partial Systems

We usually consider several observables and investigate how they are connected to each other. We restrict ourselves here to two quantities A and B. Their deviations from the average value may be correlated to each other, e.g., people's height and weight (see Table 6.1).

A measure for such correlations is clearly

$$K_{AB} \equiv \langle (A - \langle A \rangle)(B - \langle B \rangle) \rangle = \langle AB \rangle - \langle A \rangle \langle B \rangle \,,$$

which can be usefully related to the fluctuations ΔA and ΔB. A better measure is the normalized correlation or *correlation coefficient*

$$\kappa_{AB} \equiv \frac{K_{AB}}{\Delta A \cdot \Delta B} = \frac{\langle AB \rangle - \langle A \rangle \langle B \rangle}{\Delta A \cdot \Delta B} \,.$$

The fluctuation (squared) $(\Delta A)^2$ is thus equal to the *auto-correlation* K_{AA}, and κ_{AA} is 1. While K_{AB} may be negative, in which case we speak of an *anti-correlation*, this is not possible for the auto-correlation. Note that, in quantum mechanics, we have $\langle AB \rangle \neq \langle BA \rangle$, if the operators A and B do not commute. Then, according to p. 326, for the correlation, we often use the symmetrized product $\frac{1}{2}(AB+BA)$ and takes $K_{AB} = \frac{1}{2}\langle AB+BA \rangle - \langle A \rangle \langle B \rangle$ as the correlation coefficient.

If several independent variables $z^{(1)}, \ldots, z^{(n)}$ occur, we combine them into a vector $\mathbf{z}$ and consider $\rho(\mathbf{z})$. We shall soon see that $\rho(\mathbf{z})$ may be written exactly as a product $\rho^{(1)}(z^{(1)}) \cdots \rho^{(n)}(z^{(n)})$ if there are no correlations between observables, which are only related to different variables.

In particular, if we take a property $A^{(i)}$, for which only the ith variable $z^{(i)}$ is important, then we may immediately sum over all other variables $z^{(k \neq i)}$ in $\langle A^{(i)} \rangle = \sum_{\mathbf{z}} \rho(\mathbf{z}) A^{(i)}(\mathbf{z})$, because $A^{(i)}(\mathbf{z})$ does not depend on them. With this sum,

$\rho(\mathbf{z})$ becomes a function of $z^{(i)}$ alone, and in fact $\rho^{(i)}(z^{(i)})$, if $\rho(\mathbf{z})$ factorizes and use is made of $\sum \rho^{(k)}(z^{(k)}) = 1$.

For $i \neq k$, we then have

$$\langle A^{(i)} A^{(k)} \rangle = \sum_{z^{(i)}, z^{(k)}} \rho^{(ik)}(z^{(i)}, z^{(k)}) A^{(i)}(z^{(i)}) A^{(k)}(z^{(k)}) \ .$$

If $\rho^{(ik)}(z^{(i)}, z^{(k)})$ factorizes, this is equal to $\langle A^{(i)} \rangle \langle A^{(k)} \rangle$, so the mean value of the products is the product of the mean values. Conversely, if all correlations vanish, then the probability factorizes.

If a system can be decomposed into mutually independent parts, then there are no correlations between them, and its probability can be broken down into the products of the individual probabilities, one for each part.

6.1.6 Information Entropy

To each probability distribution $\{\rho_z\}$, we assign an "information measure" $I \geq 0$. It vanishes if the same thing always happens, i.e., if only a "boring" case z' is always realized, thus if $\rho_z = \delta_{zz'}$. The more there are other possibilities that can be realized, the more information can be transmitted, the sooner there will be a rare message, and the greater will be the uncertainty concerning the present event. As *information measure*, we take *the number of yes–no decisions with which, for a given distribution* $\{\rho_z\}$, *on the average, one of the possibilities can be determined*. This information measure is

$$I \equiv -\sum_z \rho_z \mathrm{lb}\rho_z \ ,$$

where lb denotes the *binary logarithm*, i.e., to the base 2, defined by

$$2^{\mathrm{lb}x} \equiv x \ , \quad \text{whence} \quad \mathrm{lb}x \equiv \log_2 x = \frac{\ln x}{\ln 2} \ .$$

Occasionally, ld$x$ is used instead of lb$x$, referred to as the *logarithmus dualis*. The unit of information is the *bit* (binary number). For example, a set of $32 = 2^5$ playing cards contains 5 bit of information, as we shall see soon.

However, the information measure I only evaluates how rarely an event occurs, but does not account for its worthiness, in the sense of how much it is worth to us. The playing cards have different values for the different team members, but each contributes an information content of 5 bit, and a row with 100 arbitrarily chosen letters (and punctuation marks) has the same information content as an equally long piece of prose or verse. Since there may be overwhelmingly many "misprints", I is often called a measure of *disorder*. (It is interesting to note that, in written texts, the letters do not not all occur with equal probability. In German, the information content of one of the 26 letters of the alphabet, together with the space, is not $\mathrm{lb}27 = 4.76$ bit, but only approximately 4 bit.)

In order to understand that the given sum achieves what is required of it, we proceed stepwise. First, we restrict ourselves to Z equally probable possibilities, whence $\rho_z = 1/Z$. For $Z = 2^m$ possibilities, clever questioners after each response drop half the remaining possibilities: after m responses, they know which of the 2^m possibilities actually exist. Here we thus have $I = \mathrm{lb}Z$. If Z is not a power of 2, we do not always need the same number of questions. For $Z = 3$, in a third of all cases, one question suffices. Then with the second question, we could already check the next attribute. Here the information measure for the questions for two attributes has to be additive: if the first attribute Z_1 has equally probable possibilities and the second Z_2 likewise, then in total there are $Z = Z_1 Z_2$ equally probable possibilities, and we must have $I(Z_1 Z_2) = I(Z_1) + I(Z_2)$. This requirement is fulfilled only by the function $I(Z) = c \ln Z$, where the factor c cannot depend on Z, and clearly has to be equal to $1/\ln 2$, so that everything is correct for $Z = 2^m$. For $\rho_z = 1/Z$, we do indeed find the above-mentioned expression, because $-Z(1/Z)\mathrm{lb}(1/Z) = \mathrm{lb}Z$.

The additivity of the information measure for independent attributes must also be valid for other distributions $\{\rho_z \neq 1/Z\}$. For these, we take the largest common divisor $1/Z'$ of all fractions ρ_z and start from a total of Z' equally probable events which we combine into Z groups, each with $N_z = \rho_z Z'$ members (see Figs. 6.3 and 6.4). The information measure $\mathrm{lb}Z'$ may then be composed of two terms: one is the unknown auxiliary quantity I_z and measures the information which is related to the characterization of the group z, while the other rates the information in this group and clearly has the value $\mathrm{lb}N_z$. Then $\mathrm{lb}Z' = I_z + \mathrm{lb}N_z$, and with $N_z/Z' = \rho_z$, this delivers the expression $I_z = -\mathrm{lb}\rho_z$, while its mean value gives the unknown variable. Thus on the average $I = -\sum_z \rho_z \mathrm{lb}\rho_z$ questions are indeed necessary before reaching the final decision.

As $\rho_z \to 0$, the quantity $\mathrm{lb}\rho_z$ increases beyond all bounds, but nevertheless so slowly that $\rho_z \mathrm{lb}\rho_z \to 0$. We do not need to question completely improbable possibilities as they do not contribute to the uncertainty—physicists often speak of *frozen degrees of freedom*.

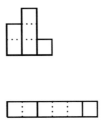

Fig. 6.3 Information measure for $\{\rho_1 = \frac{1}{3}, \rho_2 = \frac{1}{2}, \rho_3 = \frac{1}{6}\}$. This is indicated here by the upper probability distribution. The problem can be mapped onto $Z' = 6$ equally probable cases, whence the steps turn into a single bar of equal area. With the additivity of the information measure, it then follows that $\mathrm{lb}6 = I_1 + \mathrm{lb}2 = I_2 + \mathrm{lb}3 = I_3 + \mathrm{lb}1$ and hence $I_z = -\mathrm{lb}\rho_z$

Fig. 6.4 Information entropy I for the binary system. It has only two states, and therefore $\rho_1 + \rho_2 = 1$. Hence, I may be represented here as a function of ρ_1. Note the steep slope for $\rho_1 \approx 0$ and $\rho_2 \approx 0$. The uncertainty is greatest when the two states are occupied with equal probability (ρ_1 and ρ_2 both equal to 1/2)

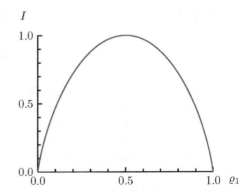

In thermodynamics, instead of the information measure I, we use the *information entropy*

$$S \equiv (k \ln 2)\, I = -k \sum_z \rho_z \ln \rho_z \, ,$$

where k is the *Boltzmann constant*, already mentioned in the list of fundamental constants on p. 623. Note that we prefer the natural logarithm $\ln x$, because it can be differentiated with respect to x more easily than $\mathrm{lb}\,x$. With $0 \le \rho_z \le 1$, the entropy is never negative. It vanishes if only one state is occupied, and takes its largest value if all possible states are equally probable (Problem 6.6).

6.1.7 Classical Statistics and Phase Space Cells

The notion of entropy just introduced is useful only for countable attributes z. This is because ρ_z has to be dimensionless, given that we cannot take a logarithm of a probability *density*. This means that continuous variables have to be discretized. We shall investigate this more precisely for the probability density $\rho(x, p)$.

According to Hamiltonian mechanics, a system of N point masses is completely determined if their positions and momenta are given. This therefore means specifying $6N$ quantities. Classical N-particle systems will be represented by a point (x, p) in the $6N$-dimensional *phase space*. (This is also called the *larger phase space* or Γ-*space*, the generalization of the 6-dimensional phase space of a single particle, which is also called μ-*space*. In μ-space, N points are occupied.) The vectors x and p each have $3N$ components.

We are concerned here with statistical ensembles and therefore assign a probability density $\rho(x, p)$ with the following properties to each phase space point:

$$\rho(x, p) = \rho^*(x, p) \ge 0 \, , \qquad \int \rho(x, p)\, \mathrm{d}^{3N}x \, \mathrm{d}^{3N}p = 1 \, ,$$

i.e., $\rho(x, p)$ is real, non-negative, and normalized.

Using this, the mean values of the quantities $A(x, p)$ may be evaluated from

$$\langle A \rangle \equiv \int \rho(x, p) \, A(x, p) \, \mathrm{d}^{3N}x \, \mathrm{d}^{3N}p \,.$$

Here arbitrary canonical transformations $(x, p) \leftrightarrow (x', p')$ are allowed, i.e., those ensuring $\mathrm{d}x \, \mathrm{d}p = \mathrm{d}x' \, \mathrm{d}p'$, because we require $\rho(x, p) = \rho'(x', p')$ and $A(x, p) = A'(x', p')$, according to Sect. 2.4.4.

In quantum theory, this is true if we take the Wigner function as the density (see p. 324). However, it is sometimes negative. This disadvantage can be avoided with the density operator (see Sect. 4.2.11), hence with $\langle A \rangle = \mathrm{tr}(\rho A)$. (In the position representation, this is equal to $\int \langle x|\rho|x' \rangle \langle x'|A|x \rangle \, \mathrm{d}^{3N}x \, \mathrm{d}^{3N}x'$, and in the momentum representation, to $\int \langle p|\rho|p' \rangle \langle p'|A|p \rangle \, \mathrm{d}^{3N}p \, \mathrm{d}^{3N}p'$. In contrast, the Wigner function uses x and p, even though they cannot be sharp simultaneously.) The density operator is Hermitian, non-negative, and normalized. Here, unitary transformations U are also permitted, so instead of the position representation, the momentum or any other representation may be used. If we have $\rho' = U\rho \, U^{-1}$ and $A' = U A \, U^{-1}$, then $\mathrm{tr}(AB) = \mathrm{tr}(BA)$ implies $\mathrm{tr}(\rho'A') = \mathrm{tr}(\rho A)$.

We shall now divide each continuous variable x, p into equal sections δx and δp, so that the phase space is divided into cells of size $(\delta x \, \delta p)^{3N}$. The smaller these cells, the more precisely the states are determined. Here, in the classical description, the cell size may be arbitrarily small, while in quantum physics, according to Heisenberg's uncertainty relation, position and momentum cannot both be arbitrarily sharp, because $\Delta x \cdot \Delta p \geq \frac{1}{2} \hbar$. In fact, only for

$$\delta x \cdot \delta p = h \equiv 2\pi \, \hbar$$

do classical and quantum mechanics yield the same number of states. We shall now show this for free particles in a cube. Another example is given in Fig. 6.5 (or Problem 6.7), namely for harmonic oscillators.

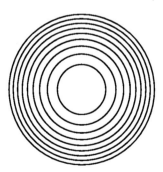

Fig. 6.5 Phase space cells partition the action variable (with the phase integrals $J = \oint p \, \mathrm{d}x$, as discussed on p. 136). They lead us to the action quantum. In addition to linear cell boundaries (as in Fig. 2.28), curved ones are also possible. Thus polar coordinates are appropriate for an oscillation. If the phase angle is completely unsharp (as on the time average), then for suitable scale factors, the phase space cells are concentric circular rings of equal area

In a cube of side L, according to quantum theory (p. 355), the Cartesian components of the wave vector have eigenvalues $k_n = n\,\pi/L$ with $n \in \{1, 2, \ldots\}$. Only the wave function $\sqrt{2/L}\,\sin(k_n x)$ vanishes at the container walls $x = 0$ and $x = L$. The number of one-particle states with momenta $p \le p_F = \hbar k_F = \hbar\,n_F\,\pi/L = \frac{1}{2}h\,n_F/L$ is thus equal to the number of unit cubes in the octant with radius $n_F = 2h^{-1}L\,p_F$:

$$\Omega = \frac{1}{8}\frac{4\pi}{3}\,n_F{}^3 = \frac{4\pi}{3}\frac{V}{h^3}\,p_F{}^3 .$$

If we divide the phase space volume $\frac{4}{3}\pi\,p_F{}^3\,V$ into cells of size h^3, then we have as many cells as states according to quantum theory, and we shall exploit this in the following.

We recognize here the meaning of the Planck constant h for thermodynamics. While the classical description for discretization remains completely undetermined, the quantum-mechanical uncertainty relation supplies a unique cell size in phase space. So we are not dealing with an *uncertainty relation*, a name that can be considered less appropriate.

6.1.8 Summary: Statistics

In statistics, we consider ensembles in which the Z possibilities occur with probabilities ρ_z. The probability distribution $\{\rho_z\}$ satisfies the constraints $\rho_z = \rho_z{}^* \ge 0$ and $\sum_z \rho_z = 1$. (For continuous z, integrals occur instead of sums. Nevertheless, according to quantum theory, the phase space cells have size $\delta x\,\delta p = h$ and we may discretize.) The observable A in the statistical ensemble has the average value $\langle A \rangle = \mathrm{tr}(\rho A)$ and the uncertainty (error width) $\Delta A = \sqrt{\langle A^2 \rangle - \langle A \rangle^2}$. Two quantities A and B have the correlation $K_{AB} = \langle AB \rangle - \langle A \rangle\langle B \rangle$. With such correlations, we can determine whether mutually independent variables occur in the statistical ensemble. If this is the case, then the probability distribution may be factorized into a product whose factors each depend only on one of the variables. Important for the following is also the information entropy $S = -k\,\mathrm{tr}\,(\rho \ln \rho)$. Disregarding the factor $k \ln 2$, this gives the average number of yes–no decisions with which one of the possibilities for the given probability distribution $\{\rho_z\}$ may be determined. This entropy is one of the most important parameters characterizing the statistical ensemble.

6.2 Entropy Theorem

6.2.1 Entropy Law and Rate Equation

The information entropy must satisfy the extremely important *entropy law*:

$$\frac{\mathrm{d}S}{\mathrm{d}t} \ge 0 , \quad \text{for all closed systems.}$$

This is also called *Boltzmann's H-theorem*, because instead of the entropy, Boltzmann used the upper-case Greek letter Eta, which resembles the Latin letter H, and he defined $H \equiv \mathrm{tr}(\rho \ln \rho) = -S/k$. We shall avoid this quantity here, because it could be confused with the enthalpy, which, according to international recommendations, should be abbreviated with the Latin H. Here a system is called *closed*, if it is not in contact with the environment, whence it exchanges neither energy nor particles, nor anything else. Therefore, in addition to invariable macro parameters, only its entropy depends on the time (or the probability distribution, which for its part does depend on the external parameters, and the time). We shall only allow for changes in other macro parameters at the end of the next section.

As will be shown in Sect. 6.2.3, this inequality follows from the *rate equation* for the probability (also called the *balance* or *master equation*), demonstrated in Sect. 4.6.4:

$$\frac{d\rho_z}{dt} = \sum_{z' \neq z} (W_{zz'}\, \rho_{z'} - W_{z'z}\, \rho_z) \,,$$

where $W_{z'z}$ (≥ 0) gives the *transition rate* from the state z into the state z'. Note that, as in quantum theory, the final state is also on the left of the initial state here. On p. 383, $W_{z'z} \propto |\langle z'|\, H\, |z\rangle|^2$ for $z' \neq z$ was already determined. Such rate equations are often set as an ansatz (further examples in the second part of this section), which should not be confused with the (entropy conserving) Liouville or von Neumann equation, which we shall discuss in Sect. 6.2.3. The term $\sum_{z'} W_{zz'}\rho_{z'}$ is the yield rate and $\sum_{z'} W_{z'z}\rho_z$ the loss rate for the state z, and the balance depends on both.

As a rate equation, we may also take the *diffusion equation*

$$\frac{\partial \rho}{\partial t} = D\, \Delta \rho \,,$$

as we shall now show in one dimension, in particular, with $\partial^2 \rho / \partial z^2$ instead of $\Delta \rho$. Thus we discretize the position parameter z of the cell with size δz and obtain a connection with the neighboring cells:

$$\frac{d\rho_z}{dt} = D\, \frac{\rho_{z+1} - 2\rho_z + \rho_{z-1}}{(\delta z)^2} \,.$$

The transition rate W and the diffusion constant D are related by

$$W_{zz'} = \delta_{z',z\pm 1}\, D/(\delta z)^2 = W_{z'z} \,,$$

and from $W_{z'z} \geq 0$, it then follows that $D \geq 0$.

While open systems may prefer the transition in one direction (for example, they transmit energy to a colder environment), for closed systems, $W_{zz'} = W_{z'z}$. Therefore, the rate equation for closed systems simplifies to $\dot{\rho}_z = \sum_{z'} W_{zz'} (\rho_{z'} - \rho_z)$. Hence,

$$\frac{dS}{dt} = -k \sum_z \frac{d(\rho_z \ln \rho_z)}{dt} = -k \sum_z \frac{d\rho_z}{dt} (\ln \rho_z + 1) = k \sum_{zz'} W_{zz'} (\rho_z - \rho_{z'})(\ln \rho_z + 1) \,.$$

If we now swap the summation indices ($z \leftrightarrow z'$) and add the expressions, we obtain

$$2 \frac{dS}{dt} = k \sum_{zz'} W_{zz'} (\rho_z - \rho_{z'})(\ln \rho_z - \ln \rho_{z'}) \ .$$

With $\rho_z > \rho_{z'}$, we also have $\ln \rho_z > \ln \rho_{z'}$, so there are no negative terms here. The entropy law thus follows from the rate equation if the transition rates $W_{z'z}$ and $W_{zz'}$ are equal, and this applies to closed systems.

The entropy increases until it has taken the largest value compatible with the remaining constraints. In particular, the rate equation does not change at all if $W_{zz'} \rho_{z'} = W_{z'z} \rho_z$ holds for all pairs (z, z'). In this situation, the system is said to be in *detailed equilibrium*.

6.2.2 Irreversible Changes of State and Relaxation-Time Approximation

If the entropy of a closed system has increased in the course of time, then according to the entropy law, it never ever returns to the initial state by itself, because the entropy would have to decrease again. The change of state is thus not reversible, and is said to be *irreversible*.

We take a two-level system as the simplest example. We already investigated its rate equation in Sect. 4.6.4. With $\rho_1 + \rho_2 = 1$, it may be decoupled to yield

$$\dot{\rho}_1 = W_{12} \rho_2 - W_{21} \rho_1 = W_{12} - (W_{12} + W_{21}) \rho_1 \ ,$$

whence it has the solution

$$\rho_1(t) = W_{12}\tau + \{\rho_1(0) - W_{12}\tau\} \exp \frac{-t}{\tau} \ ,$$

with the *relaxation time*

$$\tau = \frac{1}{W_{12} + W_{21}} \ .$$

In quantum physics, τ is called the *average lifetime*. It is occasionally replaced by the *decay time* $T_{1/2} = \tau \ln 2$, because $\frac{1}{2} = \exp(t_{1/2}/\tau)$. It is a measure of how fast equilibrium is reached. The more strongly the two states are coupled to each other, the faster this happens. The solution ρ_1 approaches the limiting value $W_{12}\tau$ monotonically, and $\rho_2 = 1 - \rho_1$, the value $W_{21}\tau$. The value with the highest entropy (here 1/2) is reserved for a closed system, in particular, when $W_{12} = W_{21}$.

For Z states, we modify the rate equation into a linear system of equations $\dot{\rho}_z = \sum_{z'} a_{zz'} \rho_{z'}$, where

$$a_{zz'} = \begin{cases} W_{zz'} \ , & \text{for} \quad z \neq z' \ , \\ -\sum_{z'' \neq z} W_{z''z} \ , & \text{for} \quad z = z' \ . \end{cases}$$

The sums of the columns in its coefficient matrix $(a_{zz'})$ are all zero, whence two important properties follow. Firstly, the determinant of this matrix must be zero, and therefore there is a zero eigenvalue, hence a stationary eigen solution. The second property follows because only the diagonal elements are actually negative: all eigenvalues have a non-positive real part. According to *Gerschgorin's theorem* [5], the position (in the complex plane) of the (suitably ordered) kth eigenvalue of a complex matrix has a distance from the kth diagonal element which is less than the sum of the moduli of the non-diagonal elements of the kth column. If the transition rates for inverse processes are equal (as for each closed system), then the matrix is Hermitian and thus has only real eigenvalues, which we set equal to $-\tau_k^{-1}$ (each τ_k is then a relaxation time). We presume in the following that the eigenvalue 0 is not degenerate, otherwise there may be different final states. Then the solutions $\rho_z(t)$ of the rate equation each consist of a constant term $\rho_z(\infty)$ and $Z-1$ terms $c_{zk}\exp(-t/\tau_k)$. After a sufficiently long time, only the largest value of the τ_k is important, which we now denote by τ:

$$\rho_z(t) \approx \rho_z(\infty) + c_z \exp\frac{-t}{\tau}.$$

In this *relaxation-time approximation*, the factors c_z are determined by the initial state. If it differs only little from the final state, we may approximate by setting $c_z \approx \rho_z(0) - \rho_z(\infty)$.

The stationary final state is given by $\dot{\rho}_z = 0$ (for all z). With

$$\sum_{z'} a_{zz'}\,\rho_{z'}(\infty) = 0 , \qquad \sum_z \rho_z = 1 ,$$

it may be traced back to the adjoint $A_{zz'}$ of the matrix $(a_{zz'})$ of coefficients. Then, up to the sign $(-)^{z+z'}$, the adjoint $A_{zz'}$ is the sub-determinant (or first minor) generated by eliminating the zth row and z'th column, and therefore $\det a = \sum_{z'} a_{zz'}A_{zz'}$:

$$\rho_z(\infty) = \frac{A_{z'z}}{\sum_{z''} A_{z'z''}} ,$$

where here z' may be chosen arbitrarily. For $W_{zz'} = W_{z'z}$, the matrix $(a_{zz'})$ is also symmetric, and therefore $\sum_z a_{zz'} = 0$ implies that $\sum_{z'} a_{zz'} = 0$ and all $\rho_z(\infty)$ are equally large.

Radioactive decay corresponds to an open system. The decay products move away from each other and never recombine. Therefore, there is in fact a transition mother $\rightarrow$ daughter, but not vice versa. From the differential equation $\dot{\rho} = -\rho/\tau$ for the probability of the mother state, we obtain the solution $\rho(t) = \rho(0)\exp(-t/\tau)$. Note that the solution for the final state can be broken up into three factors: two for the decay products and one for the relative motion. According to p. 525, a great many possible states with energies between $E_F - \frac{1}{2}dE$ and $E_F + \frac{1}{2}dE$ belong to this third factor, in fact, $4\pi V h^{-3} m\sqrt{2m E_F}\,dE$, implying therefore a high entropy.

Fig. 6.6 Time dependence
of the stepwise decay
$1 \to 2 \to 3$, and in fact here
for $\tau_2 = 3\tau_1$

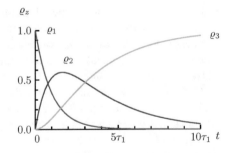

For a *stepwise decay*, we have to set up the following system of equations for
the probabilities of the radiating substances and the final products, once again with
$W_{zz'} \neq W_{z'z}$:

$$\dot{\rho}_1 = -\frac{\rho_1}{\tau_1} , \qquad \dot{\rho}_2 = \frac{\rho_1}{\tau_1} - \frac{\rho_2}{\tau_2} , \qquad \dot{\rho}_3 = \frac{\rho_2}{\tau_2} ,$$

with the solutions

$$\rho_1(t) = \rho_1(0) \, \exp \frac{-t}{\tau_1} ,$$

$$\rho_2(t) = \frac{\tau_2}{\tau_1 - \tau_2} \left(\rho_1(t) - \rho_1(0) \, \exp \frac{-t}{\tau_2} \right) ,$$

$$\rho_3(t) = \rho_1(0) - \rho_1(t) - \rho_2(t) ,$$

if we restrict ourselves to $\rho_1(0) = 1$, and therefore $\rho_2(0) = \rho_3(0) = 0$ (see Fig. 6.6).
(But note that, with $\tau_1 = \tau_2$, we have $\rho_2(t) = \rho_1(t) \, t/\tau_1$.) According to the above,
we have $\rho_3(\infty) = \rho_1(0)$. But this does not mean that the entropies of the initial and
final states were equal, because once again the relative motion is missing, and this
would lead to an increase in the entropy.

6.2.3 Liouville and Collision-Free Boltzmann Equation

In classical mechanics, we label each N-particle system by a point in the (larger)
phase space and a statistical ensemble of such systems by a swarm of points with the
probability density $\rho(t, x, p)$. The single points move in this space as time goes by,
but their total number remains constant. We then have the *Liouville equation*

$$\frac{\mathrm{d}\rho}{\mathrm{d}t} = \frac{\partial \rho}{\partial t} + \sum_{k=1}^{3N} \left(\frac{\partial \rho}{\partial x^k} \dot{x}^k + \frac{\partial \rho}{\partial p_k} \dot{p}_k \right) = 0 .$$

We proved this in Sect. 2.4.4: a volume element in the phase space keeps its probability if it follows the equations of motion (by swimming along the particle trajectories,

as it were)—as for an incompressible liquid. Its shape can in fact change, but not its content. Recall also that, according to p. 342, the *von Neumann equation* is the quantum theoretical counterpart of the Liouville equation. This is even more general than the (time-dependent) Schrödinger equation, because it holds not only for pure states, but also for mixtures.

Under special conditions, the Liouville equation is also called the collision-free Boltzmann equation, in particular, if there is a swarm of interaction-free molecules, which cannot therefore collide. Then the probability distribution $\rho(t, \mathbf{r}, \mathbf{p})$ of one molecule suffices, because any other will have the same distribution. Note that, since there are no correlations, the probability distribution of the gas factorizes even if not all the molecules have the same mass, although then that will appear differently in $\rho(t, \mathbf{r}, \mathbf{p})$. Momentum changes may be traced back to an external force $\mathbf{F} = \dot{\mathbf{p}}$ via

$$\left(\frac{\partial}{\partial t} + \mathbf{v} \cdot \nabla_r + \mathbf{F} \cdot \nabla_p\right) \rho(t, \mathbf{r}, \mathbf{p}) = 0 .$$

Note, however, that the canonical momentum then has to be equal to the mechanical one, but charged particles would also interact with each other. If we take the velocity $\mathbf{v}$ instead of the momentum $\mathbf{p}$, then setting $\dot{\mathbf{v}} = \mathbf{a}$, we obtain the *collision-free Boltzmann equation*

$$\left(\frac{\partial}{\partial t} + \mathbf{v} \cdot \nabla_r + \mathbf{a} \cdot \nabla_v\right) \rho(t, \mathbf{r}, \mathbf{v}) = 0 ,$$

which in plasma physics is also called the *Vlasov equation*. For $\mathbf{a} \equiv \mathbf{0}$, it is solved by any function $\rho(\mathbf{r} - \mathbf{v}t, \mathbf{v})$.

For all these examples, the total entropy is conserved if there is no friction force. (Actually, as mentioned before, we cannot take a logarithm of a density, because it carries a dimension. But we may divide the phase space into cells and associate probabilities with them.) With $\partial(\rho \ln \rho)/\partial t = (\ln \rho + 1) \partial \rho/\partial t$, the collision-free Boltzmann equation delivers $\partial(\rho \ln \rho)/\partial t = -(\mathbf{v} \cdot \nabla_r + \mathbf{a} \cdot \nabla_v)\rho \ln \rho$, and therefore

$$\frac{\mathrm{d}S}{\mathrm{d}t} = -k \int \frac{\partial(\rho \ln \rho)}{\partial t} \, \mathrm{d}^3r \, \mathrm{d}^3v = k \int (\mathbf{v} \cdot \nabla_r + \mathbf{a} \cdot \nabla_v) \, \rho \ln \rho \ \mathrm{d}^3r \, \mathrm{d}^3v .$$

Since the velocity cannot be arbitrarily high, the surface integral of $\mathbf{a} \, \rho \ln \rho$ in the velocity space vanishes. Therefore, Gauss's theorem supplies

$$\int \mathbf{a} \cdot \nabla_v \, \rho \ln \rho \, \mathrm{d}^3v = - \int \rho \ln \rho \, \nabla_v \cdot \mathbf{a} \, \mathrm{d}^3v .$$

Since the external force, and hence the acceleration $\mathbf{a}$, should not depend on the velocity, the last expression vanishes. For a friction force, the situation is different, but this can be traced back to collisions which we will account for only in the next section.

In order to determine a local change in the entropy, we integrate the further terms only over the velocity. Since $\mathbf{r}$ and $\mathbf{v}$ are mutually independent variables, we find

$$\int \mathbf{v} \cdot \nabla_r \, \rho \ln \rho \; \mathrm{d}^3 v = \nabla_r \cdot \int \mathbf{v} \, \rho \ln \rho \; \mathrm{d}^3 v \; .$$

The entropy may thus change locally, but not globally, because then according to Gauss's theorem, the surface integral would have to be investigated. But for $r \to \infty$, the factor $\rho \ln \rho$ is zero.

6.2.4 Boltzmann Equation

We now consider an example in which the entropy can increase with time. If molecules of equal mass collide, then further terms appear in the Boltzmann equation mentioned above, which describe the collision-induced gain and loss of the probability density $\rho(t, \mathbf{r}, \mathbf{v})$:

$$\left(\frac{\partial}{\partial t} + \mathbf{v} \cdot \nabla_r + \mathbf{a} \cdot \nabla_v \right) \rho(t, \mathbf{r}, \mathbf{v}) = R_+ - R_- \; .$$

This relation is also more general than the rate equation initially considered, because in fact $\mathrm{d}\rho/\mathrm{d}t$ stands on the left, while on the right the gain and loss have not been split-up into transition rate and density. This will be done later.

We evaluate the new terms using the following approximations. Firstly, we account for collisions between only two particles and restrict ourselves to time spans during which a molecule collides at most once. Both assumptions presume a sufficiently low density. Secondly, we neglect the influence of the container walls, which is justified for sufficiently large systems. Thirdly, we restrict ourselves to elastic scattering (point-like collision partners without internal degrees of freedom). The differential scattering cross-section σ may depend only on the velocities. Finally, in addition to energy and momentum conservation, we also make use of space-inversion and time-reversal invariance:

$$\sigma(\mathbf{v}_1, \mathbf{v}_2 \to \mathbf{v}_1', \mathbf{v}_2') = \sigma(-\mathbf{v}_1, -\mathbf{v}_2 \to -\mathbf{v}_1', -\mathbf{v}_2') \; , \qquad \mathbf{r} \to -\mathbf{r} \; ,$$
$$= \sigma(-\mathbf{v}_1', -\mathbf{v}_2' \to -\mathbf{v}_1, -\mathbf{v}_2) \; , \qquad t \to -t \; .$$

Then the scattering cross-sections for *inverse collisions* are equal,

$$\sigma(\mathbf{v}_1, \mathbf{v}_2 \to \mathbf{v}_1', \mathbf{v}_2') = \sigma(\mathbf{v}_1', \mathbf{v}_2' \to \mathbf{v}_1, \mathbf{v}_2) \; ,$$

something we shall use to establish the relation between R_+ and R_-, or to establish $W_{zz'} = W_{z'z}$. Due to energy and momentum conservation, $\mathbf{v}_1$ and $\mathbf{v}_2$ already fix $\mathbf{v}_1'$ and $\mathbf{v}_2'$, except for the direction of the relative velocity. For the proof in the next section, this is of no help. Instead of $\int \sigma \, \mathrm{d}\Omega$, it is better to write $\int \sigma(\mathbf{v}_1, \mathbf{v}_2 \rightarrow \mathbf{v}_1', \mathbf{v}_2') \, \mathrm{d}^3 v_1' \, \mathrm{d}^3 v_2'$. Then σ is not actually an area, but it is probably not appropriate to use another letter.

Here the decrease in the probability density $\rho(t, \mathbf{r}, \mathbf{v}_1)$ is the product of the scattering cross-section and the current strength, which themselves may be calculated from the probability density and the relative velocity:

$$R_-(t, \mathbf{r}, \mathbf{v}_1) = \int \sigma(\mathbf{v}_1, \mathbf{v}_2 \rightarrow \mathbf{v}_1', \mathbf{v}_2') \, \rho(t, \mathbf{r}, \mathbf{v}_1, \mathbf{v}_2) \, |\mathbf{v}_1 - \mathbf{v}_2| \, \mathrm{d}^3 v_2 \, \mathrm{d}^3 v_1' \, \mathrm{d}^3 v_2' \; .$$

For the gain in the probability density, on the other hand, we obtain

$$R_+(t, \mathbf{r}, \mathbf{v}_1) = \int \sigma(\mathbf{v}_1', \mathbf{v}_2' \rightarrow \mathbf{v}_1, \mathbf{v}_2) \, \rho(t, \mathbf{r}, \mathbf{v}_1', \mathbf{v}_2') \, |\mathbf{v}_1' - \mathbf{v}_2'| \, \mathrm{d}^3 v_2 \, \mathrm{d}^3 v_1' \, \mathrm{d}^3 v_2' \; .$$

Since the scattering cross-sections for inverse collisions are equal and the energy is conserved, whence also $|\mathbf{v}_1' - \mathbf{v}_2'| = |\mathbf{v}_1 - \mathbf{v}_2|$, this may be reformulated as

$$R_+(t, \mathbf{r}, \mathbf{v}_1) = \int \sigma(\mathbf{v}_1, \mathbf{v}_2 \rightarrow \mathbf{v}_1', \mathbf{v}_2') \, \rho(t, \mathbf{r}, \mathbf{v}_1', \mathbf{v}_2') \, |\mathbf{v}_1 - \mathbf{v}_2| \, \mathrm{d}^3 v_2 \, \mathrm{d}^3 v_1' \, \mathrm{d}^3 v_2' \; .$$

Finally, we obtain

$$\left(\frac{\partial}{\partial t} + \mathbf{v}_1 \cdot \boldsymbol{\nabla}_r + \mathbf{a} \cdot \boldsymbol{\nabla}_{v_1} \right) \rho(t, \mathbf{r}, \mathbf{v}_1)$$

$$= \int |\mathbf{v}_1 - \mathbf{v}_2| \, \sigma(\mathbf{v}_1, \mathbf{v}_2 \rightarrow \mathbf{v}_1', \mathbf{v}_2') \, \{ \rho(t, \mathbf{r}, \mathbf{v}_1', \mathbf{v}_2') - \rho(t, \mathbf{r}, \mathbf{v}_1, \mathbf{v}_2) \}$$

$$\mathrm{d}^3 v_2 \, \mathrm{d}^3 v_1' \, \mathrm{d}^3 v_2' \; .$$

On the left is the unknown probability distribution for a single particle, and on the right the unknown probability distribution for two particles. This equation is soluble only by a further approximation, derived from the assumption of *molecular chaos*: the probability distribution of two particles (at time t and at the same position $\mathbf{r}$) is assumed to factorize, the velocities of the colliding molecules being assumed not to be correlated (such a factorization was already assumed in Sect. 4.6.1 in order to arrive at a calculable expression for the dissipation in quantum-mechanical systems):

$$\rho(t, \mathbf{r}, \mathbf{v}_1, \mathbf{v}_2) = \rho(t, \mathbf{r}, \mathbf{v}_1) \cdot \rho(t, \mathbf{r}, \mathbf{v}_2) \; .$$

In this situation, we obtain a non-linear integro-differential equation known as the *Boltzmann equation* (*Boltzmann transport equation*)

$$\left(\frac{\partial}{\partial t} + \mathbf{v}_1 \cdot \nabla_r + \mathbf{a} \cdot \nabla_{v_1}\right) \rho(t, \mathbf{r}, \mathbf{v}_1)$$

$$= \int |\mathbf{v}_1 - \mathbf{v}_2| \, \sigma(\mathbf{v}_1, \mathbf{v}_2 \to \mathbf{v}_1', \mathbf{v}_2') \, \{\rho(t, \mathbf{r}, \mathbf{v}_1') \, \rho(t, \mathbf{r}, \mathbf{v}_2') - \rho(t, \mathbf{r}, \mathbf{v}_1) \, \rho(t, \mathbf{r}, \mathbf{v}_2)\}$$

$$\mathrm{d}^3 v_2 \, \mathrm{d}^3 v_1' \, \mathrm{d}^3 v_2' \, .$$

The *collision integral* on the right-hand side can usually be further simplified by exploiting energy and momentum conservation (see the previous page). We have thus derived a balance equation and traced the transition rates back to known notions.

Note that the Boltzmann equation may be used to describe a range of different *transport processes*, e.g., in reactors, superfluids, or stars [6].

6.2.5 Proof of the Entropy Law Using the Boltzmann Equation

In order to investigate the influence of the collision integrals on the entropy, we begin by excluding external forces ($\mathbf{a} = \mathbf{0}$) and assume that the probability density does not depend upon the position, so that only $\rho(t, \mathbf{v})$ appears. We then have

$$S(t) = -k \int \rho(t, \mathbf{v}) \, \ln \rho(t, \mathbf{v}) \, \mathrm{d}^3 v$$

and

$$-\frac{1}{k} \frac{\mathrm{d}S}{\mathrm{d}t} = \int \frac{\partial \rho}{\partial t} \, \{\ln \rho + 1\} \, \mathrm{d}^3 v$$

$$= \int |\mathbf{v}_1 - \mathbf{v}_2| \, \sigma(\mathbf{v}_1, \mathbf{v}_2 \to \mathbf{v}_1', \mathbf{v}_2') \, \{\rho(t, \mathbf{v}_1') \, \rho(t, \mathbf{v}_2') - \rho(t, \mathbf{v}_1) \, \rho(t, \mathbf{v}_2)\}$$

$$\{\ln \rho(t, \mathbf{v}_1) + 1\} \, \mathrm{d}^3 v_1 \, \mathrm{d}^3 v_2 \, \mathrm{d}^3 v_1' \, \mathrm{d}^3 v_2' \, .$$

With the symmetry of the collision partners 1 and 2, this may also be written as

$$-\frac{2}{k} \frac{\mathrm{d}S}{\mathrm{d}t} = \int |\mathbf{v}_1 - \mathbf{v}_2| \, \sigma(\mathbf{v}_1, \mathbf{v}_2 \to \mathbf{v}_1', \mathbf{v}_2') \, \{\rho(t, \mathbf{v}_1') \, \rho(t, \mathbf{v}_2') - \rho(t, \mathbf{v}_1) \, \rho(t, \mathbf{v}_2)\}$$

$$\{\ln (\rho(t, \mathbf{v}_1) \, \rho(t, \mathbf{v}_2)) + 2\} \, \mathrm{d}^3 v_1 \, \mathrm{d}^3 v_2 \, \mathrm{d}^3 v_1' \, \mathrm{d}^3 v_2' \, .$$

Since inverse collisions have scattering cross-sections equal to the original ones and since the modulus of the relative velocity remains conserved, we may swap the primed and the unprimed velocities, and then, as on p. 527, infer $\mathrm{d}S/\mathrm{d}t \geq 0$.

If the probability density also depends on the position, we have to respect the additional term $\int \mathbf{v} \cdot \nabla_r \, \rho \ln \rho \, \mathrm{d}^3 r \, \mathrm{d}^3 v$. As shown in the section before last, the entropy may then change locally, but not globally. Likewise, an external force $\mathbf{F}(\mathbf{r})$ would change nothing in the result.

The Boltzmann equation can be used, not only to prove the entropy law, but even to evaluate the entropy gain, provided that the scattering cross-section is known. It originates uniquely from the change in the states under collisions. There can be no entropy gain without collisions.

It is well known that the usual basic equations of mechanics and electromagnetism do not change under time reversal. To each solution of the basic equations belongs a "time-reversed" solution, for which everything proceeds in the reverse order, i.e., t is replaced by $-t$. In particular, elastic scattering is invariant under time reversal, and this has even been used explicitly. Nevertheless, the entropy of a closed system may only increase with time, never decrease.

In reality, there is no contradiction. In fact, we evaluate the entropy using another distribution function than the one actually planned for the (time-reversal invariant) Liouville equation. We describe the system with its vast number of degrees of freedom using only a small number of variables, average over the remaining ones, and thereby lose the time-inversion symmetry. This shows up, e.g., in the derivation of the Boltzmann equation. Here the entropy changes, because we have assumed molecular chaos—by doing this, we have *averaged out possible correlations* and lost information! Actually, the one-particle density is related to the two-particle density, this with the three-particle density, and so on. Collisions couple the one- to the many-body densities. But in order to be able to proceed at all, we have to terminate this sequence somewhere and come back to molecular chaos.

Although these considerations were initially applied only to the calculated entropy, the question remains as to whether they might not also apply to the experimental quantity, if the entropy is used as a state variable like, e.g., energy or volume. In fact, we always adopt only a few state parameters, far too few to be able to describe a system microscopically. This will become clear in the next section.

If the allowed states are all equally probable, the return probabilities of a many-body system ($N \gg 1$) are unbelievably small. If, for example, each particle is independent of the others and equally probable in both halves of a container, then all N particles are in the one half only with the probability 2^{-N}, thus for $N = 100$ only with the probability 10^{-30} (see Problem 6.10).

6.2.6 Molecular Motion and Diffusion

In order to investigate the influence of correlations in more detail, we consider a gas at rest, consisting of molecules of the same kind. Then $\langle \mathbf{v} \rangle = \mathbf{0}$ holds as the ensemble average and also as the time average. Note that an ensemble is said to be *ergodic*, if its ensemble average is equal to its time-average value. But $\langle v^2 \rangle$ is not zero. According to the equidistribution law on p. 559, the average kinetic energy per degree of freedom for the absolute temperature T is $\frac{1}{2}kT$. We shall allow for motions along a straight line, in a plane, or in space. Therefore, let n be the number of dimensions. Consequently, $\langle v^2 \rangle = nkT/m$.

Collisions alter the velocity of a test particle and lead to an irregularly fluctuating acceleration $\mathbf{a}$ around the mean value zero. Then the *auto-correlation function* of the velocity $\langle \mathbf{v}(t) \cdot \mathbf{v}(t') \rangle$ for $t = t'$ is in fact equal to $\langle v^2 \rangle > 0$, but for $|t - t'| \to \infty$, it surely approaches $\langle \mathbf{v}(t) \rangle \cdot \langle \mathbf{v}(t') \rangle$, i.e., it must approach zero. We set

$$\langle \mathbf{v}(t) \cdot \mathbf{v}(t') \rangle = \langle v^2 \rangle \, \chi(t - t') \, ,$$

with $\chi(t - t') = \chi(t' - t)$, $\chi(0) = 1$, and $\chi(\infty) = 0$. Up to the first collision, χ keeps the same value, because until then the velocity does not change. Thus we shall assume now that each individual collision proceeds very fast (an assumption we drop in the section after next), and the initial and final velocities will no longer be correlated. The probability of a collision is (supposedly) equally large for equal timespans. If we call the average time up to a collision τ, then we have

$$\chi(t) = \exp \frac{-|t|}{\tau} \, .$$

This τ does indeed correspond to a relaxation time. On average, in each time span τ, the same fraction of the original attributes is removed.

If we choose the origin at $\mathbf{r}(0)$, then from $\mathbf{r}(t) = \int_0^t dt' \, \mathbf{v}(t')$ and $\chi(t - t') = \chi(t' - t) = \exp(-|t - t'|/\tau)$, we find for the squared fluctuation

$$\langle r^2(t) \rangle = \int_0^t dt' \int_0^t dt'' \, \langle \mathbf{v}(t') \cdot \mathbf{v}(t'') \rangle = 2 \langle v^2 \rangle \int_0^t dt' \int_0^{t'} dt'' \, \chi(t' - t'')$$

$$= 2 \langle v^2 \rangle \, \tau^2 \left(\frac{t}{\tau} - 1 + \exp \frac{-t}{\tau} \right) .$$

For $|t| \ll \tau$, this *Ornstein–Fürth relation* goes over into $\langle r^2 \rangle \approx \langle v^2 \rangle \, t^2$, and for $t \gg \tau$, into $\langle r^2 \rangle \approx 2 \langle v^2 \rangle \, \tau t$, and both are easy enough to understand: up to the first collision, we have $\mathbf{r} = \mathbf{v} t$ and thus $\langle r^2 \rangle = \langle v^2 \rangle \, t^2$, but after many collisions $\langle r^2 \rangle$ increases only in proportion to t (see Fig. 6.7). This is the same for random walks and for diffusion, as we shall now show.

Fig. 6.7 Ornstein–Fürth relation. Distance of a gas molecule from its initial position as a function of time (*continuous red curve*). For $t \gg \tau$, the approximation $\sqrt{t/\tau}$ holds, represented by the *dashed blue parabola*

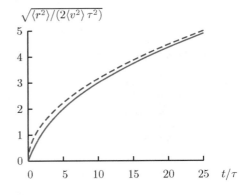

For the *random walk*, we assume that the test body after each collision moves along a new direction which is not correlated with the direction prior to the collision. For N collisions therefore, using $\mathbf{r} = \sum_{i=1}^{N} s_i\,\mathbf{e}_i$ with $\langle \mathbf{e}_i \cdot \mathbf{e}_k \rangle = \delta_{ik}$, we obtain the expression $\langle r^2 \rangle = \sum_{i=1}^{N} \langle s_i{}^2 \rangle$. Here $\langle s_i{}^2 \rangle = \langle v^2 \rangle \langle t_i{}^2 \rangle$ and $\langle t_i{}^2 \rangle = 2\tau^2$ is independent of i, so $\langle r^2 \rangle \propto N$ and hence proportional to the total time.

This squared fluctuation also increases in accordance with the *diffusion equation*

$$\frac{\partial \rho}{\partial t} = D\,\Delta\rho \,,$$

hence linearly with time. In particular, if we set the initial value $\rho(0, \mathbf{r}) = \delta(\mathbf{r})$, then for n dimensions, the solution of this differential equation (Problem 6.9) reads

$$\rho(t, \mathbf{r}) = \frac{\exp\{-r^2/(4Dt)\}}{\sqrt{4\pi\,Dt}^{\,n}} \,,$$

and with $\rho(0, \mathbf{r}) = f(\mathbf{r})$, $\rho(t, \mathbf{r}) = \int d^3 r' f(\mathbf{r}')\exp\{-|\mathbf{r}-\mathbf{r}'|^2/4Dt\}/\sqrt{4\pi\,Dt}^{\,n}$ then solves the diffusion equation (see Fig. 6.8).

From this we obtain $\langle r^2 \rangle = 2n\,Dt$. Comparing with the expression $\langle r^2 \rangle \approx 2\langle v^2 \rangle\,\tau t$ derived above, we arrive at $nD = \langle v^2 \rangle\,\tau$. The relation $\langle v^2 \rangle = nkT/m$ is generally used:

$$D = \frac{\langle v^2 \rangle}{n}\,\tau = \frac{kT}{m}\,\tau \,.$$

The diffusion constant D is thus related to the relaxation time τ, where the mass of the test particle and the temperature of its environment are also involved.

As already mentioned, the result $\langle r^2 \rangle \propto t$ can be valid only for sufficiently long times, because up to the first collision, $\langle r^2 \rangle \propto t^2$ has to hold. We could also have derived the relation $\langle r^2 \rangle = 2n\,Dt$ for all $t \geq 0$ by using the ansatz $\langle \mathbf{v}(t) \cdot \mathbf{v}(t') \rangle = 2nD\,\delta(t-t')$. Although we also make the ansatz for the auto-correlation function as a delta function, it is only an approximation. The diffusion equation has to be improved at the outset. Only the differential equation (*improved diffusion equation*)

Fig. 6.8 One-dimensional diffusion. Shown is the distribution function $\sqrt{D\tau}\,\rho(t, x)$ as a function of $x/\sqrt{D\tau}$ at the times $t = \frac{1}{10}\tau$ (*red curve*), $\frac{1}{3}\tau$ (*blue curve*), and τ (*green curve*). For $t \to \infty$, we find $\rho \to 0$

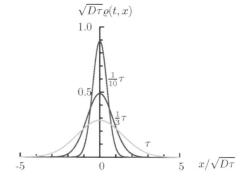

$$\frac{\partial \rho}{\partial t} = (1 - e^{-t/\tau})\, D\, \Delta\rho$$

is solved, for the initial condition $\rho(0, \mathbf{r}) = \delta(\mathbf{r})$, by

$$\rho(t, \mathbf{r}) = \frac{\exp(-r^2/4Dt')}{\sqrt{4\pi Dt'}^n}, \quad \text{with } t' \equiv t - \tau\,(1 - e^{-t/\tau}),$$

and this leads to the Ornstein–Fürth relation and to $\langle v^2 \rangle = 2nD\, t' \approx \langle v^2 \rangle\, t^2$.

These considerations are also valid for *Brownian molecular motion*, where an inert particle is struck by much faster ones. However, its velocity in this collision does not change as much as above and its relaxation time τ is therefore very much longer than the average time between two collisions.

6.2.7 Langevin Equation

In the preceding section, we determined $\langle r^2(t) \rangle$ with a time-dependent probability density $\rho(t, \mathbf{r})$. This corresponds to the Schrödinger picture (Sect. 4.4.2) in quantum theory. There we also used the Heisenberg picture—then the probability density does not depend on time, but rather on the observable $\mathbf{r}$. This picture has the advantage that derivatives of mean values with respect to time are equal to mean values of derivatives with respect to time.

If we differentiate the Ornstein–Fürth relation

$$\langle r^2(t) \rangle = 2\,\langle v^2 \rangle\, \tau\, \{t - \tau\,(1 - e^{-t/\tau})\}$$

with respect to time, we obtain

$$\langle \mathbf{r} \cdot \mathbf{v} \rangle = \langle v^2 \rangle\, \tau\, \left(1 - \exp\frac{-t}{\tau}\right).$$

If we differentiate this once more with respect to time, then we obtain $\langle v^2 \rangle + \langle \mathbf{r} \cdot \dot{\mathbf{v}} \rangle$ on the left and $\langle v^2 \rangle\, e^{-t/\tau} = \langle v^2 \rangle - \langle \mathbf{r} \cdot \mathbf{v} \rangle/\tau$ on the right. It therefore follows that

$$\langle \mathbf{r} \cdot \dot{\mathbf{v}} \rangle = -\frac{\langle \mathbf{r} \cdot \mathbf{v} \rangle}{\tau} = -\langle v^2 \rangle\, \left(1 - \exp\frac{-t}{\tau}\right).$$

At the beginning, when $|t| \ll \tau$, it is clear that $\langle \mathbf{r} \cdot \mathbf{v} \rangle \approx \langle v^2 \rangle\, t$ and $\langle \mathbf{r} \cdot \dot{\mathbf{v}} \rangle \approx -\langle v^2 \rangle\, t/\tau$, while later, when $t \gg \tau$, the two correlation functions $\langle \mathbf{r} \cdot \mathbf{v} \rangle \approx \langle v^2 \rangle\tau > 0$ and $\langle \mathbf{r} \cdot \dot{\mathbf{v}} \rangle \approx -\langle v^2 \rangle < 0$ are constant. These properties, including the sign, are easily understood for diffusion: initially, $\mathbf{r}$, $\mathbf{v}$, and $\dot{\mathbf{v}}$ are independent of each other, but then a correlation is established, and collisions hinder the diffusion, rather as for a frictional force.

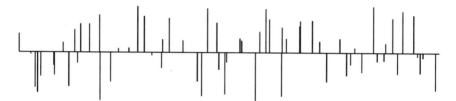

Fig. 6.9 Stochastic force as a function of time. This acts irregularly in time, strength, and direction (only one component is shown here)

This is taken into account by the*Langevin equation*:

$$\frac{d\mathbf{v}}{dt} = \mathbf{a} - \frac{\mathbf{v}}{\tau} , \quad \text{with} \quad \langle \mathbf{a} \rangle = \mathbf{0} .$$

It is generally set in the form

$$\mathbf{F} = \mathbf{F}' - \alpha\,\mathbf{v} , \quad \text{with} \quad \langle \mathbf{F}' \rangle = \mathbf{0} ,$$

and $\alpha \equiv m/\tau$ is referred to as a *frictional constant*. We have already investigate a Stokes frictional force $-\alpha\mathbf{v}$ on p. 99. The *stochastic force* $\mathbf{F}'$ fluctuates irregularly to and fro (see Fig. 6.9), and cancels out in the ensemble and the time average. Likewise the *stochastic acceleration* $\mathbf{a}(t)$, which differs from the derivative of the velocity with respect to time.

The Langevin equation actually yields the required properties of $\langle \mathbf{r} \cdot \dot{\mathbf{v}} \rangle$ and $\langle \mathbf{r} \cdot \mathbf{v} \rangle$. Since no correlations are to be expected between $\mathbf{r}$ and $\mathbf{a}$ (at equal times), and since $\langle \mathbf{r} \cdot \mathbf{a} \rangle$ vanishes, we deduce $\langle \mathbf{r} \cdot \dot{\mathbf{v}} \rangle = -\langle \mathbf{r} \cdot \mathbf{v} \rangle/\tau$ and in addition

$$\frac{d\langle \mathbf{r} \cdot \mathbf{v} \rangle}{dt} = \langle v^2 \rangle - \frac{\langle \mathbf{r} \cdot \mathbf{v} \rangle}{\tau} .$$

Since $\langle v^2 \rangle$ does not depend on time and $\langle \mathbf{r} \cdot \mathbf{v} \rangle$ vanishes initially,

$$\langle \mathbf{r} \cdot \mathbf{v} \rangle = \langle v^2 \rangle\, \tau \left(1 - \exp\frac{-t}{\tau} \right)$$

solves the problem. Then all requirements are satisfied, and the Ornstein–Fürth relation follows (with $\langle r^2(0) \rangle = 0$) by integrating over time.

We know the solution of the Langevin equation, because in Sect. 2.3.8 we treated the forced damped oscillation and solved a still more general inhomogeneous differential equation via a Laplace transformation. The solution to

$$\ddot{x}(t) + 2\gamma\,\dot{x}(t) + \omega_0{}^2\, x(t) = a(t)$$

is $x(t) = x_0(t) + \int_0^t dt'\, g(t - t')\, a(t')$, where $x_0(t)$ and $g(t)$ satisfy the homogeneous differential equation and have the initial values $x_0(0) = x(0)$, $\dot{x}_0(0) = \dot{x}(0)$, and $g(0) = 0$, $\dot{g}(0) = 1$. We are only interested in the first derivative $\dot{x}$, for which, using $g(0) = 0$, we find the expression

$$\dot{x}(t) = \dot{x}_0(t) + \int_0^t dt'\, \dot{g}(t - t')\, a(t') \, .$$

The average force is also absent ($\omega_0 = 0$). Now, we have the simple differential equation $\ddot{g} + \dot{g}/\tau = 0$ with $\dot{g}(0) = 1$, which leads to $\dot{g}(t) = \exp(-t/\tau)$. Therefore, the solution of the Langevin equation for $t \geq 0$ reads

$$\mathbf{v}(t) = \mathbf{v}(0) \exp\frac{-t}{\tau} + \int_0^t dt'\, \exp\frac{-(t - t')}{\tau}\, \mathbf{a}(t') \, ,$$

and from $\langle \mathbf{v}(0) \rangle = \mathbf{0}$, it follows that $\langle \mathbf{v}(t) \rangle = \mathbf{0}$. After many collisions, the initial velocity $\mathbf{v}(0)$ is thus "forgotten", and likewise the acceleration, the longer back it lies. For $\tau \to \infty$, nothing is forgotten, but then the diffusion constant from the last section was much too large.

6.2.8 Generalized Langevin Equation and the Fluctuation–Dissipation Theorem

So far we have assumed that the collisions are so fast that we could have taken the correlation to be $\langle \mathbf{a}(t) \cdot \mathbf{a}(t') \rangle \propto \delta(t - t')$. We now drop this approximation, assuming that the collisions last for a while. We set

$$\langle \mathbf{a}(t) \cdot \mathbf{a}(t') \rangle = \langle v^2 \rangle\, \gamma(|t - t'|) \, ,$$

because for an equilibrium distribution, only the time difference $|t - t'|$ may be of importance, and we leave open the way γ may be affected, although it will surely be monotonically decreasing towards zero. It is convenient to factorize the fixed factor $\langle v^2 \rangle$.

In fact, we only need to modify the solution of the Langevin equation considered above, viz.,

$$\mathbf{v}(t) = \mathbf{v}(0)\, \chi(t) + \int_0^t dt'\, \chi(t - t')\, \mathbf{a}(t') \, ,$$

insofar as the *linear response function* χ to the perturbation $\mathbf{a}$ is no longer equal to the old function $\dot{g}(t) = e^{-t/\tau}$. In particular, it is determined by $\langle \mathbf{a}(t) \cdot \mathbf{a}(t') \rangle$. Therefore, we have to generalize the Langevin equation. Note that the linear response function χ is sometimes called the *generalized susceptibility*.

As before, we assume $\langle \mathbf{a} \rangle = \mathbf{0}$ and for the equilibrium distribution, i.e., with $\langle \mathbf{v}(0) \cdot \mathbf{v}(0) \rangle = \langle v^2 \rangle$ and $\langle \mathbf{v}(0) \cdot \mathbf{a}(t) \rangle = 0$, we obtain

$$\frac{\langle \mathbf{v}(t) \cdot \mathbf{v}(t') \rangle}{\langle v^2 \rangle} = \chi(t)\, \chi(t') + \int_0^t dt'' \int_0^{t'} dt''' \, \chi(t-t'')\, \chi(t'-t''')\, \gamma(|t''-t'''|) \,.$$

This expression has to be a function of $|t - t'|$. But how does γ depend on χ?

This may be answered by doing a Laplace transform. Instead of $\mathcal{L}\{\gamma\}$ as in Sect. 2.3.8, we now write $\widehat{\gamma}$ for the Laplace transform of γ:

$$\widehat{\gamma}(s) \equiv \int_0^\infty dt\; e^{-st}\, \gamma(t) \,.$$

Because γ depends only on $|t - t'|$, we now consider the *double Laplace transform*

$$\widehat{\widehat{\gamma}}(s, s') \equiv \int_0^\infty dt \int_0^\infty dt'\; e^{-st-s't'}\, \gamma(|t - t'|) \,,$$

and relate it to the single Laplace transform of γ. In particular, using $st + s't' = (s+s')\,t + s'\,(t'-t)$ and $t'' = t' - t$, it follows that

$$\widehat{\widehat{\gamma}}(s, s') = \int_0^\infty dt\; e^{-(s+s')t} \int_{-t}^\infty dt''\; e^{-s't''}\, \gamma(|t''|) \,.$$

We split the last integral into two, one from 0 to ∞ and one from $-t$ to 0, and then set $t' = -t''$:

$$\widehat{\widehat{\gamma}}(s, s') = \frac{\widehat{\gamma}(s')}{s + s'} + \int_0^\infty dt\; e^{-(s+s')t} \int_0^t dt'\; e^{s't'}\, \gamma(t') \,.$$

Since $\exp\{-(s + s')\,t\}$ is the derivative of $-\exp\{-(s + s')\,t\}/(s + s')$ with respect to t, we may integrate by parts:

$$(s + s')\, \widehat{\widehat{\gamma}}(s, s') = \widehat{\gamma}(s') - e^{-(s+s')t} \int_0^t dt'\; e^{s't'}\, \gamma(t') \bigg|_{t=0}^{t=\infty} + \int_0^\infty dt\; e^{-(s+s')t}\, e^{s't}\, \gamma(t) \,.$$

Clearly, the "boundary values" do not contribute—the factor $\exp\{-(s + s')\,t\}$ kills the integral for $t \to \infty$, and for $t = 0$ the integral does not contribute. Since all the functions γ depend only on $|t - t'|$, we have the "noteworthy property"

$$\widehat{\widehat{\gamma}}(s, s') = \frac{\widehat{\gamma}(s) + \widehat{\gamma}(s')}{s + s'} \,.$$

The double Laplace transform of $\langle \mathbf{v}(t) \cdot \mathbf{v}(t') \rangle$ reads accordingly, because for this, too, only $|t - t'|$ is of importance. It contains the expression

$$\mathcal{L} \equiv \int_0^\infty dt \int_0^\infty dt' \, e^{-st-s't'} \int_0^t dt'' \int_0^{t'} dt''' \, \chi(t-t'') \, \chi(t'-t''') \, \gamma(|t''-t'''|) \, .$$

If we interchange the order of integration here, i.e., swapping t with t'' and t' with t''', then t'' is integrated from 0 to ∞ and t from t'' to ∞, etc. If we then replace $t - t'' \to t$ and $t' - t''' \to t'$, all four integrals have the limits 0 and ∞ and are easily reformulated:

$$\mathcal{L} = \int_0^\infty dt'' \int_0^\infty dt''' \int_0^\infty dt \int_0^\infty dt' \, e^{-s(t+t'')-s'(t'+t''')} \, \chi(t) \, \chi(t') \, \gamma(|t''-t'''|)$$
$$= \widehat{\chi}(s) \, \widehat{\chi}(s') \, \widehat{\widehat{\gamma}}(s, s') \, .$$

The double Laplace transform of $\langle \mathbf{v}(t) \cdot \mathbf{v}(t') \rangle / \langle v^2 \rangle$ is thus equal to

$$\widehat{\chi}(s) \, \widehat{\chi}(s') \, \{1 + \widehat{\widehat{\gamma}}(s, s')\} = \frac{\widehat{\chi}(s) \, \widehat{\chi}(s') \, \{s' + \widehat{\gamma}(s')\} + \widehat{\chi}(s') \, \widehat{\chi}(s) \, \{s + \widehat{\gamma}(s)\}}{s + s'} \, .$$

This has to apply to a function which depends only on $|t - t'|$ and therefore has the above-mentioned "noteworthy property". Consequently, $\widehat{\chi}(s) \, \{s + \widehat{\gamma}(s)\}$ cannot depend on s at all, and so has to be a constant. Its value is determined by the requirement $\chi(0) = 1$, with $\mathbf{v}(t)$ equal to $\mathbf{v}(0)$ for $t = 0$, and is in fact independent of γ. If we use this for $\widehat{\chi}(s)$ in the limit $s \to \infty$ from $\widehat{\chi} \approx \chi(0)/s$, we arrive at the desired relation

$$\widehat{\chi}(s) = \frac{1}{s + \widehat{\gamma}(s)} \, ,$$

and hence also obtain the correlation function of the velocities, viz.,

$$\langle \mathbf{v}(t) \cdot \mathbf{v}(t') \rangle = \langle v^2 \rangle \, \chi(|t - t'|) \, .$$

The auto-correlation functions of the acceleration and velocity are thus related to each other uniquely, and so also the fluctuations are related to the diffusion. This important discovery is called the *fluctuation–dissipation theorem*. Instead of the pair of notions *reversible–irreversible* (with respect to time), we take the pair *conservative–dissipative* with regard to the energy.

For a correlation function

$$\gamma(t) = \Gamma^2 \exp(-2\mu t) \, ,$$

with $\widehat{\gamma}(s) = \Gamma^2/(s + 2\mu)$, the fluctuation–dissipation theorem leads to the function $\widehat{\chi}(s) = (s + 2\mu)/\{(s + \mu)^2 - (\mu^2 - \Gamma^2)\}$. Since we normally set $\gamma(t) \propto \delta(t)$, we should expect $\mu \gg \Gamma$. Using this and the abbreviation $\nu = \sqrt{\mu^2 - \Gamma^2} < \mu$, we obtain the correlation function

$$\chi(t) = \exp(-\mu t) \, [\cosh(\nu t) + \frac{\mu}{\nu} \, \sinh(\nu t)] \, .$$

For $t \gg \nu^{-1}$, it takes the form $\frac{\mu+\nu}{2\nu} \exp\{-(\mu-\nu)t\}$.

The connection between χ and γ is also useful for the derivative of $\mathbf{v}$ with respect to t. The starting equation leads to $\widehat{v} = v(0)\,\widehat{\chi} + \widehat{\chi}\,\widehat{a}$ with $\widehat{\chi} = 1/(s+\widehat{\gamma})$, and hence to the equation $s\widehat{v} - v(0) = \widehat{a} - \widehat{\gamma}\,\widehat{v}$. This expression is equal to the Laplace transform of $\dot{v}$. Therefore, we infer the *generalized Langevin equation*, for which the history of the object is important:

$$\frac{d\mathbf{v}}{dt} = \mathbf{a}(t) - \int_0^t dt'\,\gamma(t-t')\,\mathbf{v}(t') ,$$

if $\langle \mathbf{a} \rangle = \mathbf{0}$ and $\langle \mathbf{a}(t) \cdot \mathbf{a}(t') \rangle = \langle v^2 \rangle\,\gamma(|t-t'|)$.

In the last section, we found $\chi(t) \approx \exp(-t/\tau)$ for $t \geq 0$, which yields $\widehat{\chi}(s) \approx 1/(s+\tau^{-1})$. According to the fluctuation–dissipation theorem, $\widehat{\gamma}(s) \approx \tau^{-1}$ was obtained, i.e., $\gamma(t) \approx 2\tau^{-1}\,\delta(t)$. This also implies

$$\int_0^\infty dt\,\langle \mathbf{a}(0) \cdot \mathbf{a}(t) \rangle \approx \frac{\langle v^2 \rangle}{\tau} ,$$

which, according to p. 536, is equal to nD/τ^2. With $\langle v^2 \rangle = nkT/m$ and $\alpha = m/\tau$, we also have

$$\int_0^\infty dt\,\langle \mathbf{F}'(0) \cdot \mathbf{F}'(t) \rangle \approx nkT\,\alpha ,$$

where $\mathbf{F}' = m\mathbf{a}$ is again the statistically fluctuating force.

Even if we avoid the approximation of the last section, viz., $\gamma(t) \propto \delta(t)$, we may nevertheless generally rely on $\gamma(t)$ decreasing almost to zero with increasing t. Then it seems worthwhile considering a Taylor series expansion of $\mathbf{v}(t')$ about $t' \approx t$ in the integrand of the generalized Langevin equation. With t' instead of $t-t'$, this leads to

$$\frac{d\mathbf{v}}{dt} = \mathbf{a} - \mathbf{v}(t) \int_0^t dt'\,\gamma(t') + \frac{d\mathbf{v}}{dt} \int_0^t dt'\,t'\,\gamma(t') + \cdots .$$

This takes the form of the usual Langevin equation if the first integral does not depend upon t at all (and may be set equal to τ^{-1}) and the remaining integrals do not contribute. These requirements are satisfied if only the average changes in $\mathbf{v}$ are important, averaged over the collision time, so that γ has already decreased to its final value.

6.2.9 Fokker–Planck Equation

We now consider the distribution function $\rho(t, \mathbf{v})$ for the velocity. We expect to obtain a diffusion equation $\partial\rho/\partial t = D_v\,\Delta_v\rho$ with $D_v \geq 0$. The Fokker–Planck equation [7] also contains a *drift term*, since it reads

$$\frac{\partial \rho}{\partial t} = D_v \Delta_v \rho + \frac{\boldsymbol{\nabla}_v \cdot (\rho \mathbf{v})}{\tau} , \quad \text{with} \quad D_v = \frac{D}{\tau^2} = \frac{kT}{m\tau} \geq 0 .$$

To derive this, we proceed in two steps. To begin with, we consider the *Kramers–Moyal expansion* (in one dimension):

$$\frac{\partial \rho}{\partial t} = \sum_{k=1}^{\infty} \left(-\frac{\partial}{\partial v} \right)^k D^{(k)}(v) \, \rho .$$

We then justify the claim that it is mainly the first two terms that contribute. Here the *general Fokker–Planck equation* assumes neither that the *drift coefficient* is $D^{(1)} \propto v$, nor that the *diffusion coefficient* $D^{(2)}$ has to be constant—it may even also depend upon t, not only on v. However, $D^{(2)} \geq 0$ has to hold.

If, in the short time Δt, the velocity changes by $\mathbf{w}$ with the probability density $P(t, \mathbf{v} \leftarrow t - \Delta t, \mathbf{v} - \mathbf{w})$, then

$$\rho(t, \mathbf{v}) = \int d^3\mathbf{w} \, P(t, \mathbf{v} \leftarrow t - \Delta t, \mathbf{v} - \mathbf{w}) \, \rho(t - \Delta t, \mathbf{v} - \mathbf{w}) .$$

If we restrict ourselves for the time being only to motion along a straight line, then a Taylor expansion about $w = 0$ delivers

$$P(t, v \leftarrow t - \Delta t, v - w) \, \rho(t - \Delta t, v - w)$$
$$= \sum_{k=0}^{\infty} \frac{(-w)^k}{k!} \left(\frac{\partial}{\partial v} \right)^k P(t, v + w \leftarrow t - \Delta t, v) \, \rho(t - \Delta t, v) .$$

Therefore, we introduce the moments

$$\langle w^k \rangle \equiv \int dw \, P(t, v + w \leftarrow t - \Delta t, v) \, w^k .$$

They depend upon v, t, and Δt. With $P(t, v \leftarrow t, v - w) = \delta(w)$, all moments with $k > 0$ have to vanish for $\Delta t = 0$. In contrast, $\langle w^0 \rangle$ is always equal to 1. For the determination of $\partial \rho / \partial t$, we may restrict ourselves to the linear terms in Δt (the term $k = 0$ does not contribute), and using

$$\frac{\langle w^k \rangle}{k!} = D^{(k)}(t, v) \, \Delta t + \cdots , \quad \text{with} \ k \in \{1, 2, 3, \ldots\} ,$$

we arrive at the above-mentioned Kramers–Moyal expansion

$$\frac{\partial \rho}{\partial t} = \sum_{k=1}^{\infty} \left(-\frac{\partial}{\partial v} \right)^k D^{(k)} \, \rho .$$

Here it is clear that none of the coefficients $D^{(k)}$ with even k are negative, because the probability density P has this property.

To derive the Fokker–Planck equation, we now have to consider the expansion coefficients

$$D^{(k)}(t, v) \equiv \frac{1}{k!} \frac{\partial \langle w^k \rangle}{\partial \Delta t} .$$

They can be determined from the Langevin equation $\dot{\mathbf{v}} = \mathbf{a} - \mathbf{v}/\tau$ with $\langle \mathbf{a} \rangle = \mathbf{0}$. If in the time Δt, the collision acceleration averages out, and on the other hand Δt nevertheless remains so small that we may restrict ourselves to the linear term, we may conclude that $\langle w \rangle = -v \, \Delta t / \tau$, while for short times, only the auto-correlation of the collision accelerations contributes to $\langle w^2 \rangle$:

$$\langle w^2 \rangle \approx \iint_0^{\Delta t} dt' \, dt'' \, \langle a(t') \cdot a(t'') \rangle \approx \Delta t \int_{-\infty}^{\infty} dt \, \langle a(0) \cdot a(t) \rangle = 2 D_v \, \Delta t .$$

Here the expansion coefficients $D^{(k)}$ vanish for $k > 2$ if for even k, we start from

$$\langle a(t_1) \cdots a(t_k) \rangle = \sum_{\text{all pairs}} \langle a(t_i) \, a(t_j) \rangle \cdots \langle a(t_l) \, a(t_k) \rangle ,$$

and a similar sum for $k + 1$, where each term also contains a further factor $\langle a \rangle$. This ensures that $\langle w^{2\kappa+1} \rangle$ vanishes for $\kappa > 0$. In addition, it then follows that $\langle w^{2\kappa} \rangle \propto (\Delta t)^{\kappa}$, so only $D^{(1)}$ and $D^{(2)}$ actually remain different from zero.

With this we can now derive the Fokker–Planck equation (in the three-dimensional space, correlations between the different directions are not expected):

$$\frac{\partial \rho}{\partial t} = \frac{\boldsymbol{\nabla}_v \cdot \rho \, \mathbf{v}}{\tau} + D_v \, \Delta_v \, \rho = \left(3 + \mathbf{v} \cdot \boldsymbol{\nabla}_v + \tau D_v \, \Delta_v \right) \frac{\rho}{\tau} .$$

Reformulation can help us find solutions. The average term vanishes if we introduce the variable $\mathbf{u} = \mathbf{v} \exp(t/\tau)$ instead of $\mathbf{v}$ (p. 43 is useful for such reformulations):

$$\left(\frac{\partial \rho}{\partial t} \right)_v = \left(\frac{\partial \rho}{\partial t} \right)_u + \left(\boldsymbol{\nabla}_u \, \rho \right) \cdot \left(\frac{\partial \mathbf{u}}{\partial t} \right)_v = \left(\frac{\partial \rho}{\partial t} \right)_u + \frac{\mathbf{u}}{\tau} \cdot \left(\boldsymbol{\nabla}_u \, \rho \right)_t = \left(\frac{\partial \rho}{\partial t} \right)_u + \frac{\mathbf{v}}{\tau} \cdot \boldsymbol{\nabla}_v \, \rho .$$

Therefore, with ρ now a function of t and $\mathbf{u}$, and with $\Delta_v = \exp(2t/\tau) \, \Delta_u$, we arrive at

$$\frac{\partial \rho}{\partial t} = \left(3 + \tau D_v \, \exp \frac{2t}{\tau} \, \Delta_u \right) \frac{\rho}{\tau} .$$

The first term on the right-hand side disappears if we consider the differential equation for $f = \rho \exp(-3t/\tau)$:

$$\frac{\partial f}{\partial t} = \left(\frac{\partial \rho}{\partial t} - \frac{3\rho}{\tau} \right) \exp \frac{-3t}{\tau} = \exp \frac{2t}{\tau} \, D_v \, \Delta_u f .$$

Fig. 6.10 Fokker–Planck
equation. Diffusion equation
with a drift term (see Fig. 6.8
for the situation without this
term). Also represented are
initially sharp solutions for
the times $\frac{1}{10}\tau$ (*red curve*), $\frac{1}{3}\tau$
(*blue curve*), and τ (*green
curve*). At the beginning,
$\langle v \rangle = -3\sqrt{D_v\tau}$ holds. The
stationary final distribution is
the *dashed curve*

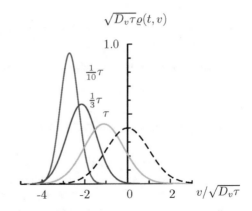

Finally, we also set $t' = \frac{1}{2}\tau\,\{\exp(2t/\tau) - 1\}$, and with $dt' = \exp(2t/\tau)\,dt$, we obtain
the diffusion equation (in the velocity space)

$$\frac{\partial f}{\partial t'} = D_v\,\Delta_u f\;.$$

According to p. 536, its solution is $f = \sqrt{4\pi D_v t'}^{\,-3}\exp\{-(\mathbf{u} - \mathbf{u}_0)^2/4D_v t'\}$. Using
this, and if the initial velocity $\mathbf{v}_0$ is given as sharp, the desired solution of the Fokker–
Planck equation reads (see Fig. 6.10)

$$\rho(t, \mathbf{v}) = \frac{1}{\sqrt{2\pi\tau D_v\{1 - \exp(-2t/\tau)\}}^{\,3}}\;\exp\frac{-\{\mathbf{v} - \mathbf{v}_0\exp(-t/\tau)\}^2}{2\tau D_v\,\{1 - \exp(-2t/\tau)\}}\;.$$

Consequently, the mean value $\langle \mathbf{v} \rangle = \mathbf{v}_0\exp(-t/\tau)$ decreases down to the equilibrium value $\mathbf{0}$. But the drift term also limits the squared fluctuation, viz.,

$$(\Delta v)^2 = 3\tau D_v\left(1 - \exp\frac{-2t}{\tau}\right),$$

which then approaches the equilibrium value $3\tau D_v$ twice as fast (with half the relaxation time $\frac{1}{2}\tau$)—otherwise, with τ very large compared to the observation time
t, it would have increased permanently with $(\Delta v)^2 = 6D_v t$. This time-dependent
squared fluctuation helps us even for the distribution function:

$$\rho(t, \mathbf{v}) = \frac{1}{\left(\sqrt{2\pi/3}\,\Delta v(t)\right)^3}\;\exp\left\{-\frac{3}{2}\left(\frac{\mathbf{v} - \mathbf{v}_0\exp(-t/\tau)}{\Delta v(t)}\right)^2\right\}\;.$$

For $t \gg \tau$ with $(\Delta v)^2 \rightarrow 3\tau D_v = 3kT/m$, it goes over into the equilibrium distribution

$$\rho(\mathbf{v}) = \frac{\exp(-\frac{1}{2}mv^2/kT)}{\sqrt{2\pi kT/m}^{\,3}} \; .$$

We shall derive this *Maxwell distribution* again in a different way in Sect. 6.3.1.

6.2.10 Summary: Entropy Law

Our aim here was to justify the thermodynamically important entropy law. The entropy of a closed system can only increase as time goes by, never decrease. This holds for macroscopic systems with many degrees of freedom if we describe them with only a small number of variables, and in any case, we could by no means account for all of them. If the entropy of a closed system increases, it changes irreversibly, even though all the basic equations of mechanics and electromagnetism remain the same under time reversal. The entropy law follows from the rate equation. A particularly impressive example of a rate equation is supplied by the Boltzmann equation. It holds for a gas of colliding molecules, as long as their probability distributions are uncorrelated (the assumption of molecular chaos).

   The increase in the entropy in closed systems does not contradict the observation of biological systems, which always become more intricate, and hence less probable. They are not closed systems.

6.3 Equilibrium Distribution

6.3.1 Maxwell Distribution

The collision integral in the Boltzmann equation vanishes for collisions of identical molecules, if (see p. 527 for detailed equilibrium)

$$\rho(t, \mathbf{r}, \mathbf{v}_1)\, \rho(t, \mathbf{r}, \mathbf{v}_2) = \rho(t, \mathbf{r}, \mathbf{v}_1')\, \rho(t, \mathbf{r}, \mathbf{v}_2') \; .$$

Energy and momentum conservation also impose the constraints

$$v_1{}^2 + v_2{}^2 = v_1'{}^2 + v_2'{}^2 \,, \quad \mathbf{v}_1 + \mathbf{v}_2 = \mathbf{v}_1' + \mathbf{v}_2' \; .$$

Consequently, for elastic collisions, $(\mathbf{v}_1 - \mathbf{v}_0)^2 + (\mathbf{v}_2 - \mathbf{v}_0)^2$ is conserved for arbitrary $\mathbf{v}_0$. The first equation may be brought into this form:

$$\ln \rho(t, \mathbf{r}, \mathbf{v}_1) + \ln \rho(t, \mathbf{r}, \mathbf{v}_2) = \ln \rho(t, \mathbf{r}, \mathbf{v}_1') + \ln \rho(t, \mathbf{r}, \mathbf{v}_2') \; .$$

Note that the sum of two one-particle quantities is conserved. Since both $\mathbf{v}_1$ and $\mathbf{v}_2$ may be chosen quite arbitrarily, the general solution is

$$\ln \rho = -A \, (\mathbf{v} - \mathbf{v}_0)^2 + \ln C \, ,$$

and this yields the *local Maxwell distribution*

$$\rho(t, \mathbf{r}, \mathbf{v}) = C(t, \mathbf{r}) \, \exp\{-A(t, \mathbf{r}) \, (\mathbf{v} - \mathbf{v}_0(t, \mathbf{r}))^2\} \, ,$$

with initially arbitrary functions $C(t, \mathbf{r})$, $A(t, \mathbf{r})$, and $v_0(t, \mathbf{r})$, provided that it is normalized correctly, i.e., $\int d^3r \, d^3v \, \rho(t, \mathbf{r}, \mathbf{v}) = 1$.

Let us take here the special case in which the probability density depends only on $\mathbf{v}$. We then have $\rho(\mathbf{v}) = C \, \exp\{-A \, (\mathbf{v} - \mathbf{v}_0)^2\}$ with $\int d^3v \, \rho(\mathbf{v}) = 1$. This Gauss distribution is symmetric with respect to $\mathbf{v}_0$. Therefore,

$$\langle \mathbf{v} \rangle = \mathbf{v}_0 \, .$$

Consequently, $\mathbf{v}_0$ is the average velocity of a molecule. The normalization requires $C = (A/\pi)^{3/2}$, and the parameter A is related to the squared fluctuation in the velocity by $(\Delta v)^2 = \frac{3}{2} A^{-1}$. We thus obtain

$$\rho(\mathbf{v}) = \frac{1}{\left(\sqrt{2\pi/3} \, \Delta v\right)^3} \, \exp\left(-\frac{3}{2} \frac{(\mathbf{v} - \mathbf{v}_0)^2}{(\Delta v)^2}\right) \, .$$

This is the famous *Maxwell distribution*, if we take $(\Delta v)^2$ as a measure of the disorderliness of the motion and relate the associated kinetic energy to the temperature according to

$$\tfrac{1}{2}m \, (\Delta v)^2 = \tfrac{3}{2} \, kT \, ,$$

by setting $(\Delta v)^2 = 3kT/m$, as discussed on p. 534.

If we restrict ourselves to gases which are on the average at rest (something that can always be realized with suitable coordinates), then $\mathbf{v}_0 = \mathbf{0}$, and the distribution is isotropic. Only the modulus of $\mathbf{v}$ is important in this case. Using $d^3v = v^2 \, dv \, d\Omega_v$, if we require $\int_0^\infty dv \, \rho(v) = 1$, then

$$\rho(v) = \frac{4\pi \, v^2 \, \exp(-\tfrac{1}{2}mv^2/kT)}{\sqrt{2\pi kT/m}^{\,3}} \, .$$

Clearly, the maximum of $\rho(v)$ is at $\hat{v} = \sqrt{2kT/m}$, and thus $\tfrac{1}{2}m \, \hat{v}^2 = kT$. The mean value of the modulus of v lies somewhat higher, namely at $\langle v \rangle = (2/\sqrt{\pi}) \, \hat{v}$.

But instead of $\rho(v)$, we often consider $\rho(E)$, the distribution with respect to the kinetic energy E, and use $dE = mv \, dv$:

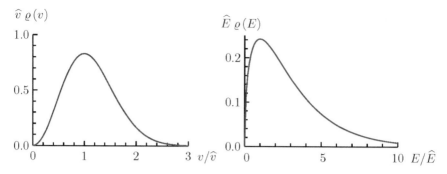

Fig. 6.11 Maxwell distributions. $\rho(v)$ (*left*), $\rho(E)$ (*right*) in suitable temperature-independent units: $\widehat{v} = \sqrt{2kT/m}$ and $\widehat{E} = \frac{1}{2}kT$

$$\rho(E) = \frac{2\,\sqrt{E/kT}\,\exp(-E/kT)}{\sqrt{\pi}\,kT}\,.$$

The maximum of this distribution lies at $\widehat{E} = \frac{1}{2}kT$, and its mean value is $\langle E \rangle = \frac{3}{2}kT = 3\widehat{E}$. The uncertainty is $\Delta E = \sqrt{3/2}\,kT$ (see Fig. 6.11).

6.3.2   Thermal Equilibrium

The Maxwell distribution is an equilibrium distribution, because it was expressly assumed that collisions do not alter anything. Therefore, in particular, the entropy is also conserved, despite the collisions.

Generally, *thermal* (*thermodynamic* or also *statistical*) *equilibrium* exists if the entropy does not change with time by itself. Such an equilibrium always exists if we consider closed systems with an entropy as high as possible. Of course, all parameters which characterize our statistical ensemble must then be given as fixed.

In the Schrödinger picture, a sufficient equilibrium condition is

$$\frac{\partial \rho}{\partial t} = 0 \quad \Longrightarrow \quad \text{equilibrium,}$$

since then neither ρ nor the mean values $\{\langle A_i \rangle\}$ depend upon time, including the entropy. With the Liouville equation, the constraint $\partial \rho/\partial t = 0$ may also be replaced by the requirement

$$[H, \rho] = 0 \,, \quad \text{or} \quad \sum_k \frac{\partial H}{\partial x^k}\frac{\partial \rho}{\partial p_k} - \frac{\partial H}{\partial p_k}\frac{\partial \rho}{\partial x^k} = 0 \,.$$

This is satisfied if, instead of the distribution function ρ with its $6N$ variables, we take the distribution function $\rho(H)$ with the energy as its only variable. Then,

$$\frac{\partial \rho}{\partial p_k} = \frac{\partial \rho}{\partial H} \frac{\partial H}{\partial p_k} \quad \text{and} \quad \frac{\partial \rho}{\partial x^k} = \frac{\partial \rho}{\partial H} \frac{\partial H}{\partial x^k} ,$$

and the Poisson bracket $[H, \rho(H)]$ always vanishes.

In quantum theory, stationary states are eigenstates of the Hamilton operator: their density operator ρ commutes with H. Conversely, from $[H, \rho] = 0$, in the energy representation, it follows that $(E_z - E_{z'}) \langle z| \rho |z'\rangle = 0$. If there is no degeneracy, i.e., $E_z \neq E_{z'}$ for $z \neq z'$, then the density operator of an equilibrium state is diagonal: $\langle z| \rho |z'\rangle = \rho(E_z) \langle z|z'\rangle$, or $\rho = \sum_z |z\rangle \rho(E_z) \langle z|$. Here $\rho(E_z)$ is the probability of the state $|z\rangle$ with energy E_z. (We divide possible degeneracies into two classes, namely those which spring from special symmetries of the Hamilton operator, and those which are merely accidental. We account for symmetries by further quantum numbers, or simply multiply $\rho(E_z)$ by the number of degenerate states. However, we shall disregard accidental degeneracies here. We assume that accidental degeneracies occur so rarely that they have no statistical weight.)

The above-mentioned equilibrium condition $\partial \rho / \partial t = 0$ may also be replaced by the sufficient constraint that ρ depend only on the energy. (However, this is not necessary, because according to the Liouville equation, for degenerate states there may also be entropy-increasing exchanges without energy change.) In the following, we shall determine several *canonical distributions* for different equilibrium conditions. Here we must always make an assumption concerning the energy with reference to the equilibrium conditions.

6.3.3 Micro-canonical Ensemble

Closed systems belong to a micro-canonical ensemble if they have the same external parameters, their energy lies in the interval between E and $E + dE$, and they are in equilibrium. Their entropy is then as high as possible, otherwise it would not be an equilibrium. According to Problem 6.6 (Sect. 6.1.6), all Z_{MC} permitted (accessible) states have the same probability, the values resulting from the normalization of ρ:

$$\rho_{MC}(E_z) = \begin{cases} Z_{MC}^{-1} , & \text{for } E \leq E_z \leq E + dE , \\ 0 , & \text{otherwise} . \end{cases}$$

The constant Z_{MC}, which is the number of states in the considered energy regime, is the *partition function*. Note that, since the letter Z is the generally accepted notation for the partition function, we count the states with z and the upper boundary is called Z. Here the partition functions are related to the various ensembles, which is why we append the subscript "MC" for *micro-canonical*.

The energy values E_z depend on the given problem. We shall take care of this later. Here we are interested primarily in the question of the probabilities with which the single energies occur in the ensemble, in order to make the entropy as high as possible, since this determines the equilibrium.

The idea of requiring equal "a priori probabilities" is suggestive even without considering the entropy. It is the only sensible assumption, as long as there are no reasons to prefer certain states over others in the considered regime. For any other distribution, there are irreversible transitions between the states until equilibrium is reached, at which point the entropy is maximal. According to Sect. 6.1.6, this highest entropy is $S = k \ln Z_{MC}$. It belongs to Z_{MC} states with equal probabilities $\rho_z = Z_{MC}^{-1}$.

It is often claimed that the entropy S may be expressed in terms of the *thermodynamic probability* W in the form $S = k \ln W$, even though it is admitted that this "probability" might be greater than one, which contradicts the notion of probability. In contrast, there is a corresponding equation with the micro-canonical partition function Z_{MC} rather than the thermodynamic probability W. In some sense though, this partition function may be connected to an occurrence, and relative occurrences do lead to probabilities. In this context, we compare two micro-canonical ensembles: the original one with the partition function Z_{MC} and another, which is less restricted and also contains other states. Then its partition function $Z_{MC>}$ is greater than Z_{MC}. According to the basic assumption of equal a priori probabilities, the probability of a state of the original ensemble in this larger ensemble is given by $Z_{MC}/Z_{MC>}$. Here $Z_{MC>}$ is in fact not uniquely fixed, but this freedom "only" relates to the zero of the entropy: the denominator necessary for the normalization in fact shifts the origin of the entropy, but what is important are usually only differences in entropy.

The relation $S = k \ln W$ is called *Boltzmann's principle*. From $W = \exp(S/k)$ and $\dot{S} \geq 0$, it follows that $\dot{W} \geq 0$, which tells us that the "disorder" in an isolated system can only increase as time goes by.

6.3.4 Density of States in the Single-Particle Model

For macroscopic bodies, the density of the energy eigenvalues E_z increases approximately exponentially with the energy, as we shall now show with a particularly simple example.

We consider a system of very many distinguishable particles which all feel the same average force, but no rest interaction—thus without correlations between the particles. (As long as the rest interaction can be treated with perturbation theory, the results barely change. The levels may move relative to each other, but this affects neither the partition function nor the average level density.) According to quantum theory, the one-particle potential fixes the one-particle energies and hence also the number of states below the energy E, which for the N-particle system we shall denote by $\Omega(E, N)$. Note that, on p. 525, we wrote Ω for $\Omega(E, 1)$. We now have

$$Z_{MC} = \Omega(E + dE, N) - \Omega(E, N) ,$$

and instead of summing over z, we may also integrate over the energy, if we take the *density of states* $\partial\Omega/\partial E$ as weight factor: $\sum_z \hat{=} \int dE\, \partial\Omega/\partial E$.

Since we have assumed only particles that are independent of each other, and therefore neglect correlations, for this "number of states", we have

$$\Omega(E, N) \approx \Omega^N(E/N, 1) .$$

Here the approximation consists in saying that not all particles have to have the same energy—only the total energy is given. But we shall soon see that for sufficiently large N, $\Omega(E, N)$ depends so strongly on the energy that other energy separations barely contribute to the density of states. The number of one-particle states does not in fact depend particularly strongly on the energy, e.g., according to p. 525, for a gas of interaction-free molecules, we find $p_F^3 \propto E^{3/2}$. But the huge power N leads to a very strong energy dependence of $\Omega(E, N)$ for the N-particle system. In particular, if $\Omega(\frac{1}{2}E, \frac{1}{2}N) = a\, E^M$ holds with $M \gg 1$, then the product is

$$\Omega\left(\frac{1}{2}(E + \varepsilon), \frac{1}{2}N\right) \cdot \Omega\left(\frac{1}{2}(E - \varepsilon), \frac{1}{2}N\right) = a^2\,(E^2 - \varepsilon^2)^M .$$

Even for $\varepsilon/E = \sqrt{\alpha/M}$, this is smaller than $a^2\, E^{2M}$ by the factor $e^{-\alpha}$, e.g., with a millimol and $\varepsilon/E = 10^{-9}$, whence $\alpha = \frac{3}{2} \times 6 \times 10^{20} \times 10^{-18} = 900$ by nearly 400 orders of magnitude. Therefore, only $\Omega(E/N, 1)$ is actually important. An example is shown in Fig. 6.12.

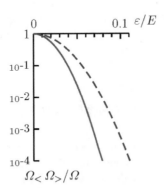

Fig. 6.12 The number $\Omega(E, N)$ of states up to the energy E of an N-particle system decreases rapidly if the energy is not distributed evenly over all particles. Here, one half has the energy $E_< = \frac{1}{2}(E - \varepsilon)$ and the other half the energy $E_> = \frac{1}{2}(E + \varepsilon)$. We plot the ratio $\Omega(E_<, \frac{1}{2}N) \cdot \Omega(E_>, \frac{1}{2}N)/\Omega(E, N)$ against ε/E for $N = 1000$ (*dashed curve*) and for $N = 2000$ (*continuous curve*)

Fig. 6.13 Probability distribution $\rho(E_z)$ of a micro-canonical ensemble of 100 particles in a cube as a function of E_z. Here the density of states increases with E_z. The higher energies in the allowed regime contribute more strongly than the lower ones

For an energy shift $E \rightarrow E + \delta E$, the function $\Omega(E, N)$ changes so much that a Taylor series makes sense only for its logarithm:

$$\ln \Omega(E + \delta E, N) \approx \ln \Omega(E, N) + \frac{\partial \ln \Omega(E, N)}{\partial E} \, \delta E \ .$$

Here the factor in front of δE is huge, namely $\frac{3}{2} N/E$ for $\Omega \propto (E^{3/2})^N$. Even for one millimol and $\delta E/E = 10^{-9}$, $\ln \Omega$ increases by nearly a trillion—and the number of states increases in this approximation exponentially with the energy δE to

$$\Omega(E + \delta E, N) \approx \Omega(E, N) \ \exp\left(\frac{\partial \ln \Omega(E, N)}{\partial E} \, \delta E\right) .$$

This property of the partition function or of the density of states $\partial \Omega/\partial E$ leads us to a new problem: for all mean values of the micro-canonical ensemble, the upper energy regime is much more important than the lower one. Here, only the mean value of the energy is accessible to us macroscopically, so we should give $\langle E \rangle$ and not start from the micro-canonical ensemble (see Fig. 6.13).

Note that the density of states also increases with the particle number N and the volume V as strongly as with the energy E, because the above considerations may be transferred to all other *extensive parameters*. By an extensive parameter, we understand a macroscopic parameter which is proportional to the size of the system, like the particle number, the energy, and the volume. In contrast, *intensive parameters* keep their value under subdivision of the system, e.g., the temperature T and the pressure p.

6.3.5 Mean Values and Entropy Maximum

For all "canonical ensembles" except for the micro-canonical one, we always fix average values: for the canonical ensemble, the energy $\langle E \rangle$, for the grand canonical ensemble, also the particle number $\langle N \rangle$, and for the generalized grand canonical ensemble also other mean values, such as the volume $\langle V \rangle$, which for the other ensembles is given precisely, just as the particle number N is given precisely for the canonical ensemble.

We now search for the general distribution $\{\rho_z\}$ with the highest entropy which is consistent with the constraints given by the mean values $\langle A_i \rangle$. Here we take only mean values of extensive quantities, such that the error widths remain as negligible as possible.

An indispensable constraint is $\mathrm{tr}\rho \equiv \langle 1 \rangle = 1$. Therefore, we begin with $i = 0$ and set $A_0 = 1$. For n further constraints, i runs up to n. With the Lagrangian parameters $-k\lambda_i$ for the unknown ρ_z, we have the variation problem

$$\delta \left(S - k \sum_{i=0}^{n} \lambda_i \langle A_i \rangle \right) = 0 , \quad \text{or} \quad \delta \sum_{z} \rho_z \left(\ln \rho_z + \sum_{i=0}^{n} \lambda_i A_{iz} \right) = 0 .$$

The extremum is obtained from $\ln \rho_z + \sum_{i=0}^{n} \lambda_i A_{iz} + 1 = 0$ and leads to

$$\rho_z = \exp \left(-1 - \sum_{i=0}^{n} \lambda_i A_{iz} \right) = \exp \left(- \sum_{i=1}^{n} \lambda_i A_{iz} \right) / \exp(1 + \lambda_0) .$$

The Lagrangian parameter λ_0 follows from the norm $\mathrm{tr}\rho = 1$. If no further mean values are given, then the highest entropy belongs to $\rho_z = 1/Z$ with $Z = \sum_z 1$, as we know already from Sect. 6.1.6. Otherwise, taking the *partition function*

$$Z \equiv \sum_{z} \exp \left(- \sum_{i=1}^{n} \lambda_i A_{iz} \right) ,$$

with $\sum_z \rho_z = 1$, we have the equation $\exp(1 + \lambda_0) = Z$. Hence, $\lambda_0 = \ln Z - 1$ and

$$\rho_z = \frac{1}{Z} \exp \left(- \sum_{i=1}^{n} \lambda_i A_{iz} \right) .$$

The remaining Lagrangian parameters λ_i are related to the corresponding mean values:

$$\langle A_i \rangle = \frac{1}{Z} \sum_{z} A_{iz} \exp \left(- \sum_{j=1}^{n} \lambda_j A_{jz} \right) = -\frac{1}{Z} \frac{\partial Z}{\partial \lambda_i} = -\frac{\partial \ln Z}{\partial \lambda_i} .$$

The mean values $\langle A_i \rangle$ thus follow from derivatives of the partition function Z, so we have to determine $Z(\lambda_1, \ldots, \lambda_n)$ such that, for all $i \in \{1, \ldots, n\}$, the equations $\langle A_i \rangle = -\partial \ln Z/\partial \lambda_i$ are satisfied, where the remaining Lagrangian parameters λ_j with $j \neq i$ are to be kept fixed.

We have thus found the constraints for the extremum of $S[\rho]$. It is a maximum, because $-k \rho_z (\ln \rho_z + \sum_{i=0}^{n} \lambda_i A_{iz})$ differentiated twice with respect to ρ_z is equal to $-k/\rho_z < 0$. We shall investigate the physical meaning of the Lagrangian parameters $\lambda_1, \ldots, \lambda_n$ in Sect. 6.3.8. These are adjustable parameters and lead us among other things to the temperature and the pressure.

Note that the partition function also yields the squared fluctuation of $\langle A_i \rangle$, because from

$$\langle A_i^2 \rangle = \frac{1}{Z} \frac{\partial^2 Z}{\partial \lambda_i^2} = \frac{\partial}{\partial \lambda_i} \left(\frac{1}{Z} \frac{\partial Z}{\partial \lambda_i} \right) + \frac{1}{Z^2} \left(\frac{\partial Z}{\partial \lambda_i} \right)^2 = -\frac{\partial \langle A_i \rangle}{\partial \lambda_i} + \langle A_i \rangle^2 \ ,$$

we deduce

$$(\Delta A_i)^2 = -\frac{\partial \langle A_i \rangle}{\partial \lambda_i} \ .$$

Since the squared fluctuation is non-negative, the partial derivative must not be positive. If it is zero, then there is no unique relation $\langle A_i \rangle \to \lambda_i$. Otherwise $\langle A_i \rangle$ is a monotonically decreasing function of λ_i, and so λ_i is a monotonically decreasing function of $\langle A_i \rangle$. Clearly, also $(\Delta A_i)^2 = \partial^2 \ln Z / \partial \lambda_i^2$ holds.

If the mixed derivatives $\partial^2 \ln Z / (\partial \lambda_i \, \partial \lambda_j)$ are continuous, then the order of the derivatives may be interchanged. Then we arrive at the equations

$$\frac{\partial \langle A_i \rangle}{\partial \lambda_j} = \frac{\partial \langle A_j \rangle}{\partial \lambda_i} \ .$$

These are *Maxwell's integrability conditions (Maxwell relations)*, which will turn out to be useful later on.

6.3.6 Canonical and Grand Canonical Ensembles

For the *canonical ensemble*, the mean value of the energy $\langle E \rangle$ is given, in addition to the norm $\langle 1 \rangle$. According to the last section, we then have the *canonical partition function*:

$$Z_C \equiv \sum_z \exp(-\lambda_E E_z) = \mathrm{tr}[\exp(-\lambda_E E)]$$

and the probability distribution

$$\rho_C = \frac{1}{Z_C} \exp(-\lambda_E E) \ .$$

Note that the Lagrangian parameter λ_E is related to the energy, but the letter β is usually used, even though β will be used for the pressure coefficients (see p. 619). Here, for brevity, we have left out the index z for ρ_C and E. For the same reason, the trace notation is convenient for the partition function. If states are *degenerate*, we have to multiply by their degree of degeneracy.

For canonical ensembles, what is important is thus to know how the given mean value $\langle E \rangle$ depends upon the adjustable parameter λ_E. According to the last section, we have (see Fig. 6.14)

$$\langle E \rangle = -\frac{\partial \ln Z_C}{\partial \lambda_E} \quad \text{and} \quad (\Delta E)^2 = -\frac{\partial \langle E \rangle}{\partial \lambda_E} \ .$$

Fig. 6.14 The level density increases approximately as $E^{3N/2}$, and the occupation probability decreases as $e^{-\lambda E}$. Hence $x^n e^{-x}$ is important, with maximum for $x = n$. Here this function is shown for $n \in \{2, 4, 8, 16\}$ relative to its maximum, and therefore as a function of x/n

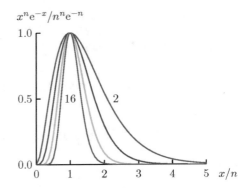

We shall later relate the temperature to the parameter λ_E. Indeed, we shall find that λ_E is the reciprocal of kT.

For any canonical ensemble of macroscopic bodies, only a small energy range δE is of importance. If we approximate its partition function Z_C by an integral of the energy with the integrand $f(E, N) = \exp(-\lambda_E E)\, \partial\Omega(E, N)/\partial E$, then for large N, $f(E, N)$ has a very sharp maximum at $\widehat{E}$. For the density of states of a gas of interaction-free molecules, for example, the integrand $f(E) \propto \exp(-\lambda_E E)\, E^{3N/2}$ is to be considered near its maximum at $\widehat{E} = \frac{3}{2} N/\lambda_E$, and after a Taylor series expansion,

$$f(\widehat{E} + \delta E) \approx f(\widehat{E})\, \exp\{-\tfrac{3}{4}N\, (\delta E/\widehat{E})^2\}\,,$$

we find a Gauss distribution with the tiny width $\widehat{E}/\sqrt{3N/2} \ll \widehat{E}$. (Who would ever determine the energy up to twelve digits for one mole?) Consequently, we have $\widehat{E} \approx \langle E \rangle$, and for such a sharp maximum, only the states from the nearest neighborhood are important. Therefore, the canonical and micro-canonical ensembles are very similar—the energy uncertainty (via λ_E) is given instead of the energy range dE. Therefore, a distribution parameter λ_E may even be assigned to a micro-canonical ensemble, and with this a *temperature*, as will be shown in Sect. 6.3.8. The requirement is that $\exp(-\lambda_E E)\, \partial\Omega/\partial E$ should have its maximum at $\widehat{E}$, which requires $\lambda_E\, \partial\Omega/\partial E = \partial^2\Omega/\partial E^2$, or $\lambda_E = \partial \ln(\partial\Omega/\partial E)/\partial E$ for each $E = \widehat{E}$, i.e., $(kT)^{-1} = \partial \ln(\partial\Omega/\partial E)/\partial E|_{\widehat{E}}$.

For the *grand canonical ensemble*, in addition to $\langle 1 \rangle$ and $\langle E \rangle$, we also fix the particle number $\langle N \rangle$ only on the average. Then we have

$$\rho_{GC} = \frac{1}{Z_{GC}}\, \exp(-\lambda_E E - \lambda_N N)\,,$$

with

$$Z_{GC} \equiv \mathrm{tr}\, \exp(-\lambda_E E - \lambda_N N) = \sum_N \exp(-\lambda_N N)\, Z_C(N)\,.$$

Even more mean values characterize the *generalized grand canonical ensemble*. For this, in addition to $\langle 1 \rangle$, $\langle E \rangle$, and $\langle N \rangle$, further quantities $\langle V_i \rangle$ are given, e.g., the average volume. Then we have

$$\rho = \frac{1}{Z} \, \exp(-\lambda_E E - \lambda_N N - \textstyle\sum_i \lambda_i V_i) \,,$$

with

$$Z \equiv \mathrm{tr} \, \exp(-\lambda_E E - \lambda_N N - \textstyle\sum_i \lambda_i V_i)$$

and

$$\langle E \rangle = -\frac{\partial \ln Z}{\partial \lambda_E} \,, \qquad (\Delta E)^2 = -\frac{\partial \langle E \rangle}{\partial \lambda_E} \geq 0 \,,$$

$$\langle N \rangle = -\frac{\partial \ln Z}{\partial \lambda_N} \,, \qquad (\Delta N)^2 = -\frac{\partial \langle N \rangle}{\partial \lambda_N} \geq 0 \,,$$

$$\langle V_i \rangle = -\frac{\partial \ln Z}{\partial \lambda_i} \,, \qquad (\Delta V_i)^2 = -\frac{\partial \langle V_i \rangle}{\partial \lambda_i} \geq 0 \,.$$

In the following, we shall imagine as the other quantities V_i only the volume and then, instead of $\sum_i \lambda_i V_i$, take only $\lambda_V V$. Here we shall sometimes fix the particle number, thus give only $\langle 1 \rangle$, $\langle E \rangle$, and $\langle V \rangle$ as mean values. This ensemble has no special name.

According to the last section, the entropy is

$$S = -k \mathrm{tr} \, (\rho \ln \rho) = -k \sum_z \rho_z \ln \frac{\exp(-\sum_{i=1}^{n} \lambda_i A_{iz})}{Z} = k \left(\ln Z + \sum_{i=1}^{n} \lambda_i \langle A_i \rangle \right) \,.$$

Using this, for generalized grand canonical ensembles, we obtain

$$S = k \, (\ln Z + \lambda_E \langle E \rangle + \lambda_N \langle N \rangle + \lambda_V \langle V \rangle) \,,$$

with somewhat simpler expressions for canonical and grand canonical ensembles, which are not so important at the moment, because we also wish to investigate the dependence on $\langle N \rangle$ and $\langle V \rangle$. Here Z is a function of the Lagrangian parameters λ_E, λ_N, and λ_V (see, e.g., Fig. 6.15). We investigate the canonical partition function $Z_C(\lambda_E, N, V)$ on p. 575 and the grand canonical partition function $Z_{GC}(\lambda_E, \lambda_N, V)$ on p. 579.

In the following, we shall usually drop the bracket symbols $\langle \; \rangle$, because we consider only mean values anyway, if not explicitly stated otherwise. In addition, we adopt the common practice in thermodynamics of writing U for the energy E. It is referred to as the *internal energy*, bearing in mind that there are also other forms of energy. In Sect. 6.3.1, for the Maxwell distribution, we divided the kinetic energy into the collective part $\frac{m}{2} \langle v \rangle^2$ and the disordered part $\frac{m}{2} (\Delta v)^2$, since we have $\langle v^2 \rangle = \langle v \rangle^2 + (\Delta v)^2$. For such an ideal gas, only the disordered motion counts for

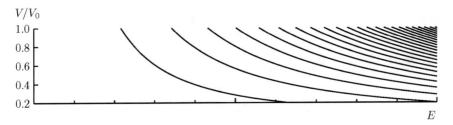

Fig. 6.15 If the volume V or another extensive parameter changes, then every energy eigenvalue E_z also changes, and so therefore does the *density of states* $\partial \Omega/\partial E$, here shown for the same example as in Fig. 6.13, viz., 100 molecules in a cube

the internal energy, the collective center-of-mass motion being considered as one of the macroscopic parameters.

6.3.7 Exchange Equilibria

For *inhibited (partial) equilibria* with special "constraints" (inhibitions), only parts of the system are in equilibrium (each with an entropy as high as possible), which for the total system without the inhibition would have a higher entropy. It is not in *total (global) equilibrium*. We are not interested in the exact description of the transition from partial to total equilibrium under removal of the inhibition—for that, we would have to solve rate equations. Here the initial and final state suffice: the new equilibrium is reached by suitable alterations of the partial systems—an *exchange equilibrium* (total equilibrium) then develops.

We exemplify by considering two separate closed systems, each of which is in equilibrium and has the average energy U_n and entropy S_n ($n \in \{1, 2\}$). If the two systems come into contact, in most cases, the total system will not yet be in equilibrium: then the two parts exchange energy, as long as the total entropy increases, i.e., $S_f \geq S_i$. Here it is assumed that the coupling is so weak and the energy exchange so slow that, for the total energy, $U = U_1 + U_2$ always holds and the probability distribution always factorizes (and thus $S = S_1 + S_2$ holds). In the new equilibrium state, the total entropy is then as high as possible: $\delta S = \delta S_1 + \delta S_2 = 0$ under the constraint $\delta U = \delta U_1 + \delta U_2 = 0$. Exchange equilibrium with respect to the energy leads to the requirement $\delta S = \sum_n (\partial S_n/\partial U_n)\, \delta U_n = 0$, thus to

$$\frac{\partial S_1}{\partial U_1} = \frac{\partial S_2}{\partial U_2}, \quad \text{or} \quad \lambda_{E1} = \lambda_{E2},$$

since we have $S = k\,(\ln Z + \lambda_E U)$—because Z is a function of λ_E and hence of U, thus $Z(\lambda_E(U))$—and this implies that

$$\frac{1}{k}\frac{\partial S}{\partial U} = \left(\frac{\partial \ln Z}{\partial \lambda_E} + U\right)\frac{\partial \lambda_E}{\partial U} + \lambda_E$$

and also, for $U = -\partial \ln Z/\partial \lambda_E$ and $\partial U/\partial \lambda_E = -(\Delta U)^2$,

$$\frac{1}{k}\frac{\partial S}{\partial U} = \frac{-U + U}{-(\Delta U)^2} + \lambda_E = \lambda_E .$$

Note that, in the partial derivatives, N and V, or λ_N and λ_V, are held constant. The equilibrium state of systems in thermal contact can thus be recognized by all parts having equal distribution parameter λ_{En}.

These considerations are clearly valid not only for the energy U, but also for the particle number and the volume. Under the constraint $\delta N = 0$, $\delta S = 0$ delivers

$$\frac{\partial S_1}{\partial N_1} = \frac{\partial S_2}{\partial N_2} , \quad \text{or} \quad \lambda_{N1} = \lambda_{N2} ,$$

and under the constraint $\delta V = 0$, $\delta S = 0$ delivers

$$\frac{\partial S_1}{\partial V_1} = \frac{\partial S_2}{\partial V_2} , \quad \text{or} \quad \lambda_{V1} = \lambda_{V2} .$$

The exchange equilibrium is only reached if the Lagrangian parameters in all parts agree with each other.

Now we can better understand how *reversible* and *irreversible changes of state* are distinguished. In the last section, we removed closed systems of inhibitions and local differences were then equalized, e.g., by diffusion or temperature adjustment. Such a change of state proceeds by itself and is not reversible, but irreversible—and the entropy increases.

However, we may also modify external parameters, e.g., supply energy. This may also happen reversibly, or one part reversibly and another part irreversibly. *The change is then reversible if it proceeds solely through equilibrium states.* However, this constraint is only satisfied if no internal equalization is necessary.

6.3.8 Temperature, Pressure, and Chemical Potential

According to p. 513, the *zeroth main theorem* of thermodynamics states that: *There is a state variable called temperature T and two parts of a system are only in thermal equilibrium if they have the same temperature.* This equilibrium depends in particular on the possibility that energy may be exchanged. Like λ_E, the temperature is the same in all parts—the two parameters describe the same situation. The larger λ_E, the more important are the states of low energy, and the cooler the considered body: λ_E is inversely proportional to the temperature. They are related by the Boltzmann constant k according to

$$\lambda_E = \frac{1}{kT} , \quad \text{or} \quad T = \frac{1}{k\lambda_E} ,$$

as we shall show. This also implies

$$\frac{\partial}{\partial \lambda_E} = -kT^2 \frac{\partial}{\partial T} .$$

If the average energy is given, then the (thermodynamic) temperature T characterizes the equilibrium distribution, but if the energy has to be sharp, then the notion of temperature is useless.

However, the zeroth main theorem only states when two temperatures are equal. We could also take another function $f(T)$ as the temperature. In this sense any uncalibrated mercury thermometer serves its purpose within its measurable range, but without a gauge, not even temperature differences can be given uniquely. Thus for a canonical distribution, the thermodynamic temperature is uniquely determined by $T = (k\lambda_E)^{-1}$. Then the behavior of macroscopic models, e.g., of an ideal gas, can be determined as a function of the temperature (or of the parameter λ_E), and hence a gas thermometer can be constructed as a measuring device. In Sect. 6.5.4, we shall prove the thermal equation of state for ideal gases (the Gay-Lussac law), viz.,

$$pV = NkT ,$$

from which the gas thermometer gauge may be derived. And we shall actually prove $pV = N/\lambda_E$ there!

It is immediately clear that, for $T = 0$, special situations occur, since then $\lambda_E = \infty$. Now for all equilibria with a finite energy uncertainty (with $T > 0$),

$$(\Delta U)^2 = -\frac{\partial U}{\partial \lambda_E} > 0 \quad \Longrightarrow \quad \frac{\partial U}{\partial T} > 0 .$$

With decreasing temperature T, the internal energy U thus also decreases, implying that the states of low energy are preferentially occupied. In this limit, only the ground state is occupied, if it is not degenerate. Correspondingly, the equilibrium distribution for $T = 0$ only depends on whether or not the ground state is degenerate, and likewise the entropy. If there is no degeneracy, then ρ_z is different from 0 for only one z and hence $S = 0$. This property is called the *third main theorem of thermodynamics*.

In classical statistical mechanics, the following *equidistribution law* can be derived: *All canonical variables (positions, momenta) which occur in only one term in the Hamilton function, and there as squared, contribute the value $\frac{1}{2} kT$ to the internal energy in a canonical ensemble.* For the proof, we take the Hamilton function $H = H_0 + cx^2$, where H_0 and c do not depend upon the coordinate x. In a canonical ensemble, this variable x contributes

$$\int_{-\infty}^{\infty} dx \ \rho(x) \ cx^2 = \frac{\int dx \ \exp(-\lambda_E \ cx^2) \ cx^2}{\int dx \ \exp(-\lambda_E \ cx^2)} = -\frac{\partial}{\partial \lambda_E} \ln \int_{-\infty}^{\infty} dx \ \exp(-\lambda_E cx^2)$$

to the internal energy. The integral has the value $\sqrt{\pi/(\lambda_E \, c)}$, whence $\frac{1}{2} \ln \lambda_E$ has to be differentiated with respect to λ_E, which results in $\frac{1}{2}/\lambda_E = \frac{1}{2} kT$. This proves the equidistribution law.

Then, for example, for force-free motion, the squares of the components of the momentum for the three space directions enter as separate terms—a single free particle thus has the energy $\frac{1}{2}m \, (\Delta v)^2 = \frac{3}{2} kT$, as claimed on p. 547 for the Maxwell distribution, and now proven. Consequently, ideal gases with N atoms (without internal degrees of freedom) have $U = \frac{3}{2} NkT$. Correspondingly, for the linear harmonic oscillator, the internal energy is $\frac{2}{2} kT$. The virial theorem in mechanics (see p. 79) then shows that $\langle E_{\text{pot}} \rangle = \langle E_{\text{kin}} \rangle$. It thus also holds in quantum theory, but it should be noted, however, that it often delivers discrete energy eigenvalues so the above-mentioned integrals are then sums $\sum_z \rho(E_z) \, E_z$, which for low temperatures leads to deviations from classical statistics. This shows up quite clearly in connection with the *freezing of degrees of freedom*.

If two parts exchange not only energy, but also volume, then not only do their temperatures become equal, but also their values of the parameter λ_V. It is common to set

$$\lambda_V = p \, \lambda_E = \frac{p}{kT} \ ,$$

because pV is then an energy. This means that p is an energy/volume $\hat{=}$ force/area and has the unit $N/m^2 = Pa = 10^{-5}$ bar. In addition, for fixed λ_E (> 0), $(\Delta V)^2 = -\partial V/\partial \lambda_V > 0$ implies the relations $\partial V/\partial p < 0$ and $\partial p/\partial V < 0$. If the volume decreases, then p increases, provided that no other parameters change: p is the *pressure* with which the system acts on the container walls. It is only when it is the same in all parts that any volume exchange will cease.

Correspondingly, the Lagrangian parameter λ_N becomes the same in all parts of a system if particles can be exchanged. We assume that the temperature becomes equal and set

$$\lambda_N = -\mu \, \lambda_E = -\frac{\mu}{kT} \ .$$

Then μN is an energy, and so is μ, the *chemical potential*. Like temperature and pressure, it is a distribution parameter and important for chemical reactions, as will be shown below. Since $(\Delta N)^2 > 0$, we have $\partial N/\partial \mu > 0$ for fixed λ_E (> 0). As observed, e.g., in Figs. 6.19 and 6.22, the chemical potential is often, but not always negative.

For materials involving different types of particles, the expression μN in the exchange equilibrium is replaced by $\sum_i \mu_i N_i$, as will be proven in Sect. 6.5.5. However, *chemical equilibria* have to be treated separately, because the molecules are counted as particles, but in chemical reactions, only the number of atoms is constant, and not necessarily the number of molecules, e.g., not for 2 $H_2O \rightarrow 2$ $H_2 + O_2$. If we take X_i as a symbol for the ith sort of molecule, then we have

$$\sum_i \nu_i \, X_i = 0 \,,$$

where the *stoichiometric coefficients* ν_i are positive for reaction products, negative for reaction partners (and then integers as small as possible)—in the above-mentioned example, they take the values -2, 2, and 1. After dn reactions, we have $dN_i = \nu_i \, dn$ (actually n is a natural number, but we may go over to a continuum by referring to the very large total number). This implies $\delta S = \sum_i (\partial S / \partial N_i) \, \nu_i \, dn = 0$ as equilibrium condition. Then, according to the last section, $\sum_i \lambda_{Ni} \nu_i = 0$, and hence,

$$\sum_i \nu_i \, \mu_i = 0 \,.$$

We shall use this equation on p. 588 for the law of mass action for chemical reactions.

6.3.9 Summary: Equilibrium Distributions

Equilibrium distributions do not change with time—the entropy is as high as possible given the constraints. This happens if the probability distribution depends only on the energy. For the micro-canonical ensemble, all states in the energy range from E to $E + dE$ are occupied with equal probability. For the other canonical ensembles, some parameters are given only as average values (for macroscopic systems, the fluctuations about the mean value are normally extremely small). To each mean value there is a distribution parameter which, in the exchange equilibrium, is the same for all parts. To the energy corresponds the temperature T, to the volume the pressure p, and to the particle number the chemical potential μ. Here the Lagrangian parameter $\lambda_E = 1/kT$, $\lambda_V = p/kT$, and $\lambda_N = -\mu/kT$ were initially introduced as distribution parameters. For n given mean values $\{\langle A_i \rangle\}$, the partition function $Z = \mathrm{tr}[\exp(-\sum_{i=1}^n \lambda_i A_i)]$ turns out to be useful because $\langle A_i \rangle = -\partial \ln Z / \partial \lambda_i$ and $(\Delta A_i)^2 = \partial^2 \ln Z / \partial \lambda_i^2$.

6.4 General Theorems of Thermodynamics

6.4.1 The Basic Relation of Thermodynamics

From the relation for the entropy of a generalized grand canonical ensemble, we shall now derive the following important equation of macroscopic thermodynamics:

$$dU = T \, dS - p \, dV + \mu \, dN \,.$$

Since we take the equilibrium expression for S, it holds only for reversible changes of state, or at least for *changes* of state in which so far all external parameters have been kept fixed, so $dV = 0$ and $dN = 0$.

In Sect. 6.3.6, we derived the equation

$$S = k \ (\ln Z + \lambda_E U + \lambda_V V + \lambda_N N)$$

for the entropy. Here the partition function Z is a function of the three Lagrangian parameters λ_E, λ_V, and λ_N, and according to the same discussion, $\langle A_i \rangle = -\partial \ln Z / \partial \lambda_i$ implies $d \ln Z = -U \ d\lambda_E - V \ d\lambda_V - N \ d\lambda_N$, and hence

$$dS = k \ (\lambda_E \ dU + \lambda_V \ dV + \lambda_N \ dN) \ .$$

According to Sect. 6.3.8, the Lagrangian parameters λ_E, λ_V, and λ_N are related to the temperature T, the pressure p, and the chemical potential μ:

$$\lambda_E = \frac{1}{kT} \ , \qquad \lambda_V = p \, \lambda_E \ , \quad \text{and} \quad \lambda_N = -\mu \, \lambda_E \ .$$

Consequently, for $T \neq 0$,

$$dS = \frac{dU + p \, dV - \mu \, dN}{T} \ ,$$

and we have thus proven the claim that $dU = T \ dS - p \ dV + \mu \ dN$.

For the grand canonical ensemble, the term $-p \, dV$ does not occur, because the volume is to be kept constant, and for the canonical ensemble, the term $\mu \, dN$ is also missing, because the particle number is then also fixed. Particularly often, the equation is used with $dN = 0$, namely, in the form $dU = T \ dS - p \ dV$.

If the changes in the state quantities do not proceed purely through equilibrium states, but nevertheless begin and end with such states, then, in addition to the reversible change of state just treated, there will also be an irreversible one. According to the entropy law—and from now on we always assume $dt > 0$—the entropy increases without a change in the other macroscopic parameters. This can be accounted for by

$$dS \geq \frac{dU + p \, dV - \mu \, dN}{T} \ ,$$

or again, for $T > 0$,

$$dU \leq T \ dS - p \ dV + \mu \ dN \ .$$

The equations for reversible processes become inequalities for irreversible ones, if we stay with fixed $dt > 0$.

6.4.2 Mechanical Work and Heat

For fixed particle number ($dN = 0$), we now consider the inequality

$$dU \leq T\,dS - p\,dV$$

somewhat more deeply, thus allowing for irreversible changes of state. We think, for example, of a gas with pressure p in a cylinder with (friction-free) mobile pistons. In order to reduce the volume ($dV < 0$), we have to do work $\delta A = -p\,dV$ *on* the system. This energy is buffered in the gas—its pressure increases, because the molecules hit the walls more often. Alternatively, a spring might be extended or compressed. Instead of $\delta A = -p\,dV$, we may also take $\delta A = (\pm)\sum_k F_k\,dx^k$ with generalized coordinates x^k and associated generalized forces F_k. The sign has to be adjusted to the relevant notion.

The work δA is not generally a complete differential, because heat is also transferred. Even in a *cycle process*, i.e., going through different states before returning to the initial state, $\oint \delta A$ does not generally vanish. If it did, this would be a sign of a complete differential dA, or a *state variable A*, whence the integral $\int dA$ would depend only on the initial and final points of the path and not on the path in-between.

We know this situation already from mechanics (p. 56). Only for $\oint \mathbf{F} \cdot d\mathbf{r} = 0$ can we introduce a potential energy—Lorentz and frictional forces are situations where this is not possible. At least the Lorentz force (see Sect. 2.3.4) can be derived from a generalized potential energy $q\,(\Phi - \mathbf{v} \cdot \mathbf{A})$, or $q\,v_\mu A^\mu$ if, in addition to the position, we also allow the velocity as a variable, provided that there is no frictional force. As is well known, this leads to heat, our subject here.

The internal energy U also increases if we supply energy without changing the volume V. Here the temperature does not even need to increase notably (*latent heat*). Then, e.g., at the normal freezing temperature of water, we need a *melting heat* of 6 kJ/mole to melt ice. This is often written in the form $(H_2O) = [H_2O] + 6$ kJ. If the solid phase is set in angular brackets, the liquid in round, and the gaseous in curly, then we have (per mole)

$$
\begin{aligned}
(\ldots) &= [\ldots] + \text{melting heat}\,, \\
\{\ldots\} &= (\ldots) + \text{vaporization heat}\,, \\
\{\ldots\} &= [\ldots] + \text{sublimation heat}\,.
\end{aligned}
$$

Here, we may neglect the volume change for melting, but not of course for vaporization, which is why there are tables, e.g., [8], listing the *vaporization enthalpy*, i.e., the energy difference for constant pressure. We shall return to this in Sect. 6.4.4.

If we set δQ for the amount of heat in an infinitesimal process, the energy conservation law for $dN = 0$ takes the form

$$dU = \delta Q + \delta A\,, \quad \text{with} \quad dU = 0\,, \quad \text{for closed systems}\,.$$

This important equation is called the *first main theorem of thermodynamics*. Here, irreversible processes are also permitted. The essentially new aspect compared to mechanics is the kind of energy, i.e., "heat".

If we restrict ourselves to reversible processes, the comparison with the first mentioned equation $dU = T\,dS + \delta A$ supplies the *second main theorem of thermodynamics*, viz.,

$$\delta Q_{\text{rev}} = T\,dS\,, \quad \text{or} \quad dS = \frac{\delta Q_{\text{rev}}}{T}\,.$$

After our rather detailed investigation of the entropy, this is almost self-evident, as soon as the notion of the amount of heat has been clarified by the first main theorem.

While the entropy for reversible δQ_{rev} may increase or decrease, depending on its sign, for irreversible processes it always increases. We have already investigated in detail the entropy law "$dS/dt \geq 0$ for closed systems" as a further constituent of the second main theorem. Therefore, all the main theorems of thermodynamics have been explained sufficiently—we have already discussed the zeroth and third in Sect. 6.3.8.

Note that, using the second main theorem, a thermometer can be gauged, which is a problem, according to p. 559. In particular, by the second main theorem, the equation

$$\oint dS = \oint \frac{\delta Q_{\text{rev}}}{T} = 0$$

holds for a cycle.

The *Carnot process* appears in the (S, T) diagram in Fig. 6.16 as a rectangle with

$$0 = \oint dS = \frac{Q_+}{T_+} - \frac{Q_-}{T_-} \quad \Longrightarrow \quad \frac{T_-}{T_+} = \frac{Q_-}{Q_+}\,.$$

Hence, via the reversibly exchanged amounts of heat, the temperature can be measured in arbitrary units—the discussion in Sect. 6.3.8 did not reach this far.

Fig. 6.16 In the Carnot cycle, the amount of heat Q_+ is reversibly taken in at the temperature T_+ and the amount of heat Q_- is reversibly taken out at the temperature T_-. No heat is exchanged in-between, and the total work taken in is $Q_+ - Q_-$, equal to the enclosed area in the (S, T) diagram. For a more general cycle, see Problem 6.25

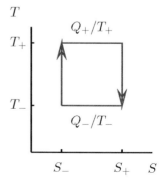

The Carnot cycle is the ideal of a steam engine. In the combustion chamber, an amount of heat Q_+ is taken in at the temperature T_+, and in the condenser, an amount of heat Q_- is taken out at the temperature T_- and given off to the cooling water (usually also at intermediate temperatures, which is not convenient). The difference $Q_+ - Q_- = \oint \delta Q$ can at most be converted to exploitable work $- \oint \delta A$, the energy remaining conserved for cyclic systems on the time average, and always for closed systems. The ratio of this work to the gained (input) energy Q_+ is the *thermodynamic efficiency* η of the machine. (Modern power plants can reach $\eta > 45\%$, James Watt had $\eta \approx 3\%$, and its predecessors, e.g., Thomas Savery, a tenth of it.) According to Carnot, this efficiency has an upper limit $\eta_C < 1$, because $\eta = (Q_+ - Q_-)/Q_+ = 1 - T_-/T_+$ and the cooling water (without energy input) cannot be cooler than the environment (and the fire cannot be arbitrarily hot). In reality, the efficiency is less, because heat is exchanged for intermediate temperatures and everything should go quickly, so changes are not only quasi-stationary.

In essence, the steam engine converts a part of the disordered motion (at high temperature) into ordered motion (work)—the energy is thereby changed from many degrees of freedom to a few. Nevertheless, the total entropy does not decrease, because it moves heat from the fire into the cooling water, and there the entropy increases more notably.

6.4.3 State Variables and Complete Differentials

State variables characterize a state, e.g., energy U, particle number N, and volume V are state variables in thermodynamics. They may be taken as functions of other state variables $(x_1, \ldots) \,\widehat{=}\, \mathbf{x}$. Then,

$$\mathrm{d}f \equiv f(\mathbf{x} + \mathrm{d}\mathbf{x}) - f(\mathbf{x}) = \sum_i \frac{\partial f}{\partial x^i}\, \mathrm{d}x^i \quad \text{and} \quad \oint \mathrm{d}f = 0 \,.$$

This quantity $\mathrm{d}f$ is called a *complete* (or *total* or *exact*) *differential*.

But not every infinitesimal quantity δf is a complete differential $\mathrm{d}f$. We shall write δf for all differential forms of the kind encountered in the variational calculus, while many use only $\mathrm{d}f$, even for non-exact differentials. Then,

$$\delta f = \sum_i a_i\, \mathrm{d}x^i$$

is a complete differential only if $a_i = \partial f/\partial x^i$ for all i, and on all simply-connected regions,

$$\frac{\partial a_i}{\partial x^k} = \frac{\partial a_k}{\partial x^i} \,, \quad \text{for all } i \text{ and } k \,.$$

Thus $\partial^2 f/\partial x^k \partial x^i = \partial^2 f/\partial x^i \partial x^k$ is required, but the partial derivatives only commute if they are continuous. If this necessary and sufficient constraint on a complete differential is violated, then the infinitesimal quantity δf is a "non-exact differential". Then the path becomes decisively important for the integration. For example, $\delta f = \alpha x^{-1}\, dx + \beta x\, dy$ is not exact, since $\partial a_x/\partial y = 0$, but $\partial a_y/\partial x = \beta$. If we integrate here from $(1, 1)$ to $(2, 2)$, going parallel to each axis in turn, then the path via $(2, 1)$ yields $\int \delta f = \alpha \ln 2 + 2\beta$, while the path via $(1, 2)$ yields $\int \delta f = \beta + \alpha \ln 2$, whence $\oint \delta f \neq 0$.

In three dimensions, this necessary and sufficient constraint for a complete differential can also be expressed by

$$\nabla \times \mathbf{a} = \mathbf{0} \,.$$

In mechanics, therefore, a potential can only be introduced for curl-free forces (see p. 56).

Note that, always in two dimensions, and in special cases in higher dimensions, an incomplete differential can be made into a complete differential by multiplying by a suitable function (the *integrating factor*, also called *Euler's integrating factor*), which then becomes a state variable. The integrating factor for Q_{rev} is T^{-1}.

Changes of state are named after the conserved variable:

$$
\begin{array}{llll}
dS = 0 & \text{isentropic}\,, & dV = 0 & \text{isochoric}\,,\\
dT = 0 & \text{isothermal}\,, & dp = 0 & \text{isobaric}\,.
\end{array}
$$

For reversible processes, *isotropic* means the same as *adiabatic*, i.e., without heat exchange. With the ideal Carnot process, the states change either isotropically or isothermally, so in the (S, T) diagram, it is easier to represent than in the (V, p) diagram.

6.4.4 Thermodynamical Potentials and Legendre Transformations

For the internal energy U, on p. 561, we derived the differential form

$$dU = T\, dS - p\, dV + \mu\, dN$$

for reversible processes. Consequently, the state variables S, V, and N, the so-called *natural variables*, are particularly well suited as independent variables for the internal energy. We can in particular obtain the associated intensive quantities T, p, and μ from the internal energy U by differentiation:

$$\left(\frac{\partial U}{\partial S}\right)_{V,N} = T\,, \quad \left(\frac{\partial U}{\partial V}\right)_{S,N} = -p\,, \quad \left(\frac{\partial U}{\partial N}\right)_{S,V} = \mu\,.$$

Likewise, the potential energy E_{pot} may be differentiated with respect to the general-ized coordinates x^k, which delivers generalized forces $\partial E_{\text{pot}}/\partial x^k = -F_k$. Therefore, the internal energy U is one of the *thermodynamic potentials*.

As already mentioned for the vaporization heat on p. 563, it is often appropriate to replace the extensive variables S, V, or N by their associated intensive parameters T, p, or μ, respectively, if, e.g., the temperature and pressure are kept fixed, but not the entropy and volume.

We have already encountered such transformations of variables in mechanics, where we replaced the Lagrange function $L(t, x, \dot{x})$ by the Hamilton function $H(t, x, p)$ by $p = \partial L/\partial \dot{x}$. This is made possible using a *Legendre transformation*:

$$\frac{\partial A}{\partial B} = C, \quad \text{or} \quad \mathrm{d}A = C \,\mathrm{d}B,$$

$$\Longrightarrow \quad \mathrm{d}\,(BC - A) = B \,\mathrm{d}C, \quad \text{or} \quad \frac{\partial\,(BC - A)}{\partial C} = B.$$

If we thus want to replace the variable B by $C = \partial A/\partial B$, then we take $BC - A$ instead of A. So, when $H = \dot{x} p - L$ was chosen, we obtained $\partial H/\partial p = \dot{x}$.

We now introduce the following *thermodynamic potentials*:

U	internal energy,
$H \equiv U + pV$	enthalpy,
$F \equiv U - TS$	(Helmholtz) free energy,
$G \equiv H - TS = F + pV$	free enthalpy (Gibbs free energy),

to obtain new *natural variables* with their differentials:

$$\begin{aligned}
\mathrm{d}U &= +T \,\mathrm{d}S - p \,\mathrm{d}V + \mu \,\mathrm{d}N, \\
\mathrm{d}H &= +T \,\mathrm{d}S + V \,\mathrm{d}p + \mu \,\mathrm{d}N, \\
\mathrm{d}F &= -S \,\mathrm{d}T - p \,\mathrm{d}V + \mu \,\mathrm{d}N, \\
\mathrm{d}G &= -S \,\mathrm{d}T + V \,\mathrm{d}p + \mu \,\mathrm{d}N.
\end{aligned}$$

Clearly, we could also introduce four further *grand canonical potentials* $U - \mu N$, $H - \mu N$, $F - \mu N \equiv J$, and $G - \mu N$. Of these, we shall also need

$$\mathrm{d}J = -S \,\mathrm{d}T - p \,\mathrm{d}V - N \,\mathrm{d}\mu,$$

from Sect. 6.5.2 onward. However, we often consider systems with a given particle number. Then we have $\mathrm{d}N = 0$, the four equations are simplified (the chemical potential no longer plays a role), and the grand canonical potential becomes obsolete. If, on the other hand, further variables are important, then additional terms appear, e.g., with electric or magnetic fields.

The expression *thermodynamic potential* is, however, only justified if it is taken as a function of its natural variables, thus, e.g., $U(S, V, N)$. Otherwise, simple partial derivatives do not result. Then according to p. 43 and this section,

$$\left(\frac{\partial U}{\partial V}\right)_{T,N} = \left(\frac{\partial U}{\partial V}\right)_{S,N} + \left(\frac{\partial U}{\partial S}\right)_{V,N}\left(\frac{\partial S}{\partial V}\right)_{T,N} = -p + T\left(\frac{\partial S}{\partial V}\right)_{T,N},$$

and the last mentioned derivative has still to be determined. We shall return to this in the next section.

From the Legendre transformation equations above with $(\partial C/\partial B)(\partial B/\partial C) = 1$ for $C = \partial A/\partial B$, it is clear that $\partial^2 A/\partial B^2 \cdot \partial^2(BC-A)/\partial C^2 = 1$. Taking the first equation, e.g., with $A = U$, $B = V$, and $C = -p$ for fixed S, this delivers

$$-1 = \left(\frac{\partial^2 H}{\partial p^2}\right)_S\left(\frac{\partial^2 U}{\partial V^2}\right)_S = \left(\frac{\partial^2 F}{\partial T^2}\right)_V\left(\frac{\partial^2 U}{\partial S^2}\right)_V$$
$$= \left(\frac{\partial^2 G}{\partial p^2}\right)_T\left(\frac{\partial^2 F}{\partial V^2}\right)_T = \left(\frac{\partial^2 G}{\partial T^2}\right)_p\left(\frac{\partial^2 H}{\partial S^2}\right)_p,$$

each for fixed particle number N. Here we have written first the negative and then the positive factor, and we shall encounter such sign rules in the next section.

6.4.5 Maxwell's Integrability Conditions and Thermal Coefficients

The thermodynamic potentials are state variables, and therefore integrability conditions are valid: their mixed derivatives do not depend upon the sequence of differentiations (except for phase transitions). We shall use this now and always keep the particle number fixed. Then, with $f(x, y)$ instead of $\partial^2 f/\partial x\,\partial y = \partial^2 f/\partial y\,\partial x$, we write more precisely

$$\left(\frac{\partial}{\partial x}\right)_y\left(\frac{\partial f}{\partial y}\right)_x = \left(\frac{\partial}{\partial y}\right)_x\left(\frac{\partial f}{\partial x}\right)_y.$$

These imply four integrability conditions, depending on which pair of S, T, V, and p is taken as the natural variables:

$$dU = + T\,dS - p\,dV \qquad -\left(\frac{\partial p}{\partial S}\right)_V = +\left(\frac{\partial T}{\partial V}\right)_S,$$

$$dH = + T\,dS + V\,dp \qquad +\left(\frac{\partial V}{\partial S}\right)_p = +\left(\frac{\partial T}{\partial p}\right)_S,$$

$$dF = - S\,dT - p\,dV \qquad -\left(\frac{\partial p}{\partial T}\right)_V = -\left(\frac{\partial S}{\partial V}\right)_T,$$

$$dG = - S\,dT + V\,dp \qquad +\left(\frac{\partial V}{\partial T}\right)_p = -\left(\frac{\partial S}{\partial p}\right)_T.$$

Here derivatives of p and V with respect to S and T are related to the "inverse derivatives" of S and T with respect to p and V. Here the partner is always kept fixed:

p and V form one pair, S and T the other. For the derivative $\partial p/\partial S = (\partial S/\partial p)^{-1}$, there occurs a minus sign. For all four derivative pairs, we shall now introduce abbreviations.

The derivative $(\partial p/\partial T)_V$ is the *pressure coefficient*. It is denoted by β, but note that β is often used for $(kT)^{-1}$. It is related to p by the *thermal stress coefficient* $\alpha_p = \beta/p$, and related to the volume derivative $(\partial V/\partial T)_p$ by the *thermal expansion coefficient* α:

$$\alpha \equiv \frac{1}{V}\left(\frac{\partial V}{\partial T}\right)_p = -\frac{1}{V}\left(\frac{\partial S}{\partial p}\right)_T \quad \text{expansion coefficient},$$

$$\beta \equiv \left(\frac{\partial p}{\partial T}\right)_V = \left(\frac{\partial S}{\partial V}\right)_T \quad \text{pressure coefficient}.$$

The derivative $(\partial T/\partial V)_S$ in the first pair $-(\partial p/\partial S)_V = (\partial T/\partial V)_S$, now referring to p. 43, can be traced back to

$$\left(\frac{\partial T}{\partial V}\right)_S = -\left(\frac{\partial T}{\partial S}\right)_V\left(\frac{\partial S}{\partial V}\right)_T = -\beta \Big/ \left(\frac{\partial S}{\partial T}\right)_V,$$

and the second in a corresponding manner to

$$\left(\frac{\partial T}{\partial p}\right)_S = -\left(\frac{\partial T}{\partial S}\right)_p\left(\frac{\partial S}{\partial p}\right)_T = \alpha\, V \Big/ \left(\frac{\partial S}{\partial T}\right)_p.$$

Here the derivatives $\partial S/\partial T$ are related to the *heat capacities*. We avoid the notion of *specific heat* (heat capacity/mass), because in the next section we divide by the particle number N instead of the mass, which is theoretically more convenient:

$$C_p \equiv T\left(\frac{\partial S}{\partial T}\right)_p = \left(\frac{\partial H}{\partial T}\right)_p \quad \text{isobaric heat capacity},$$

$$C_V \equiv T\left(\frac{\partial S}{\partial T}\right)_V = \left(\frac{\partial U}{\partial T}\right)_V \quad \text{isochoric heat capacity}.$$

Besides these, we also introduce the *compressibilities*:

$$\kappa_T \equiv -\frac{1}{V}\left(\frac{\partial V}{\partial p}\right)_T \quad \text{isothermal compressibility},$$

$$\kappa_S \equiv -\frac{1}{V}\left(\frac{\partial V}{\partial p}\right)_S \quad \text{adiabatic (isentropic) compressibility}.$$

The signs for the heat capacities and compressibilities were chosen such that none of the four coefficients is negative. According to p. 559, we have in particular $(\partial U/\partial T)_V > 0$ with $(\Delta U)^2 > 0$, and according to p. 560, $(\partial V/\partial p)_T < 0$ with $(\Delta V)^2 > 0$, whence $C_V \geq 0$ and $\kappa_T \geq 0$. In addition, we shall soon see that $C_p \geq C_V$ and $\kappa_S = (C_V/C_p)\,\kappa_T$.

The expansion coefficient α and the pressure coefficient β are mostly positive, but they can both be negative (e.g., in water at the freezing temperature). However, at least their product is always positive.

The adiabatic compressibility can be determined from the *sound velocity* c and the mass density ρ. In the case of sound, there is a force density $-\nabla p$, and therefore the impulse density has the modulus dp/c. It is equal to the momentum density $c\,d\rho$. Consequently, $c^2 = dp/d\rho$ holds. Here the entropy is conserved, because there is no time for heat exchange. With $\rho\,(\partial p/\partial\rho)_S = V^{-1}\,(\partial p/\partial V^{-1})_S = -V\,(\partial p/\partial V)_S = \kappa_S^{-1}$, we see that κ_S, ρ, and c^2 are actually connected:

$$\kappa_S = \frac{1}{\rho\,c^2}\ .$$

The thermal coefficients for fixed intensive quantities are thus rather easy to measure, including the expansion coefficient α and the heat capacity C_p for fixed pressure, as well as the isothermal compressibility κ_T. However, the pressure coefficient β and the heat capacity C_V for fixed volume are not. Therefore, the following three relations are helpful:

- Firstly the equation

$$\beta = \frac{\alpha}{\kappa_T}\ .$$

For its proof in $(\partial p/\partial T)_V$, we need only swap the fixed and the altered variable, according to p. 43.
- Secondly, the equation

$$\frac{C_p}{C_V} = \frac{\kappa_T}{\kappa_S}\ .$$

The left-hand side is equal to $(\partial S/\partial T)_p\,(\partial T/\partial S)_V$, and, according to p. 44, we may swap the pair $(S,\,T)$ with the pair $(p,\,V)$ to obtain the right-hand side.
- The third equation

$$C_p - C_V = TV\alpha\beta$$

follows immediately (as a product $T \cdot \beta \cdot V\alpha$), according to p. 43, from

$$\left(\frac{\partial S}{\partial T}\right)_p - \left(\frac{\partial S}{\partial T}\right)_V = \left(\frac{\partial S}{\partial V}\right)_T\left(\frac{\partial V}{\partial T}\right)_p\ .$$

With $\alpha\beta = \alpha^2/\kappa_T \geq 0$, we see that α and β have equal sign. Independently of this sign, we clearly have $C_p \geq C_V$ and $\kappa_T \geq \kappa_S$. Ten derivatives of the potentials can be traced back to expansion and pressure coefficients in addition to T, S, p, and V (the remaining thermal coefficients also occur in other derivatives):

$$\left(\frac{\partial U}{\partial V}\right)_T = -p + \beta T = \beta\left(\frac{\partial U}{\partial S}\right)_T = -\frac{\beta}{\alpha V}\left(\frac{\partial U}{\partial p}\right)_T,$$

$$\left(\frac{\partial H}{\partial p}\right)_T = (1 - \alpha T)V = -\alpha V\left(\frac{\partial H}{\partial S}\right)_T = -\frac{\alpha V}{\beta}\left(\frac{\partial H}{\partial V}\right)_T,$$

$$\left(\frac{\partial F}{\partial T}\right)_p = -S - \alpha p V = \alpha V\left(\frac{\partial F}{\partial V}\right)_p,$$

$$\left(\frac{\partial G}{\partial T}\right)_V = -S + \beta V = \beta\left(\frac{\partial G}{\partial p}\right)_V.$$

The first of these equations was already discussed on p. 567. The remaining ones follow in a similar way (Problem 6.34).

6.4.6 Homogeneous Systems and the Gibbs–Duhem Relation

How do the different quantities depend on the number of particles N? To answer this question we restrict ourselves now to particles of one sort and always assume *homogeneous systems*: all adjustable parameters have the same value everywhere, such that everything is in local equilibrium.

As mentioned on p. 552, state variables are said to be *extensive* if they are proportional to the number of particles, e.g., S, V, and the thermodynamic potentials U, H, F, and G. In contrast, in equilibrium, *intensive* state variables have the same value everywhere, e.g., T, p, and μ are intensive state variables. Except for the temperature, all extensive quantities will be denoted with upper case letters and all intensive ones with lower case letters.

Of course, we can also divide the extensive quantities by the particle number and then arrive at intensive quantities. We denote them by the corresponding lower case letters—the only exception is the temperature—and then we have no other extensive quantities than N:

$$v = \frac{V}{N}, \quad s = \frac{S}{N}, \quad u = \frac{U}{N}, \quad h = \frac{H}{N}, \quad f = \frac{F}{N}, \quad g = \frac{G}{N}.$$

This separation is particularly convenient, if in addition to N only the intensive quantities T and p occur as independent variables, hence the natural variables of the free enthalpy G.

If the weight of a particle (molecule) or the molecular weight M_r is known, then a scale suffices for the determination of the particle number $N = M/(M_r u)$ of a macroscopic probe, where $u = \frac{1}{12}$ of the mass of ^{12}C is the *atomic mass unit* (*atomic mass constant*) (see Table A.3). Therefore, "specific" quantities, i.e., divided by the mass, are normally preferred, e.g., the *specific heats* rather than the heat capacities/particle. (But note that the *specific weight* gives the ratio M/V.)

It is common to refer to a special particle number, namely the *Loschmidt number* N_L. It corresponds to a *mole*, i.e., M_r gram of the substance. Note that the *Avogadro constant* N_A differs only by dimension: $N_A = N_L$/mole. (This constant was introduced by Avogadro in 1811, but the value of this number was first determined by Loschmidt in 1865.) Then, for example, on p. 563, the melting heat was given in kJ/mole. It is necessary for N_L molecules. The product of N_A and the Boltzmann constant k is called the *gas constant* and denoted by

$$R \equiv N_A k .$$

Quantities referring to one mole are common in physical chemistry and are called *molar* quantities. To obtain these, we multiply the quantities valid for a single molecule by the Avogadro constant N_A.

The chemical potential μ is the adjustable parameter corresponding to the particle number N. According to p. 567, it is obtained from any of the four thermodynamic potentials by differentiation with respect to N, if the other natural variables are kept fixed. The free enthalpy is particularly suitable, because it depends otherwise only on intensive quantities: $\mu = (\partial G/\partial N)_{Tp}$. Hence, for homogeneous systems in equilibrium, $G = N \, g(T, p)$ clearly implies $\mu = g(T, p)$, and thus the famous *Gibbs–Duhem relation*

$$G = \mu N ,$$

which will prove to be extremely useful. For homogeneous systems, with

$$G = H - TS = F + pV = U - TS + pV ,$$

it yields

$$H = T S + \mu N , \qquad F = -p V + \mu N , \qquad U = T S - p V + \mu N .$$

For homogeneous mixtures of different sorts of particles, μN is to be replaced by $\sum_i \mu_i N_i$, as shown on p. 587.

Note that the chemical potential always decreases with increasing temperature, because $\mathrm{d}F = -S \, \mathrm{d}T - p \, \mathrm{d}V + \mu \, \mathrm{d}N$ implies the integrability condition

$$(\partial \mu/\partial T)_{V,N} = -(\partial S/\partial N)_{T,V} = -s \, (T, \, V) ,$$

and the fact that the entropy is never negative.

6.4.7 Phase Transitions and the Clausius–Clapeyron Equation

We shall now investigate the equilibrium condition for the exchange of particles, energy, or volume, in particular the phase equilibrium. As is well known, the same molecules may exist in different *phases* (*aggregation states*): solid, liquid, gaseous,

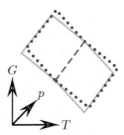

Fig. 6.17 For first order phase transitions, the first derivative of the free enthalpy $G(T, p)$ makes a jump, here indicated by the *dashed red line*. The structure indicated by the *dotted blue lines* would have higher G than the stable phase (*continuous green lines*). Here $\partial G/\partial T = -S < 0$ and $\partial G/\partial p = V > 0$ always hold

etc. In Sect. 6.3.8 we derived the constraints $T_+ = T_-$, $p_+ = p_-$, and $\mu_+ = \mu_-$. According to the Gibbs–Duhem relation, we thus also have

$$g_+(T, \ p) = g_-(T, \ p) \ .$$

This equation defines a *coexistence curve* $p\,(T)$ in the (T, p) plane, where the two phases are in equilibrium (see Fig. 6.17). Away from this curve, there is only the one or the other phase, namely the one with the lower free enthalpy, as will be shown in Sect. 6.4.9. Three phases may exist in simultaneous equilibrium only at the *triple point* T_{tr}, p_{tr}. This is the meeting point of the three branches corresponding to the phase equilibria for *melting, vaporization,* and *sublimation,* or those of other phase transitions.

For the coexistence curve $p\,(T)$, the differential equation of Clausius and Clapeyron holds. Along this curve, we have $dg_+ = dg_-$. Hence $dg = -s\,dT + v\,dp$ leads to

$$-s_+\,dT + v_+\,dp = -s_-\,dT + v_-\,dp \ ,$$

and this in turn implies the *Clausius–Clapeyron equation*:

$$\frac{dp}{dT} = \frac{s_+ - s_-}{v_+ - v_-} \ .$$

The entropy change $S_+ - S_-$ times the transition temperature T is equal to the transition heat for the phase change: melting, vaporization, or sublimation heat (see p. 563). For these heats, we are dealing with transition *enthalpies*, since we then have to care for $\Delta p = 0$ and have therefore $T\,\Delta S = \Delta H$:

$$\frac{dp}{dT} = \frac{1}{T}\frac{\Delta H}{\Delta V} \ .$$

We usually have $dp/dT > 0$, but there are nevertheless also counter-examples, for instance, for the transition ice $\rightarrow$ water with $\Delta H = 6.007$ kJ/mol and $\Delta V = -0.0900$ cm^3/g.

The different substances in a mixture do not usually transform at the same temperature. If we have, for example, two metals mixed in a melt and then cool it down, without altering the pressure, then often only one of the metals will freeze, or at least with a mixing ratio different from the one given for the melt. The mixing ratio of the melt also changes, and along with it its transition temperature. On further cooling, the two metals do not necessarily segregate. The lowest melting temperature may occur for a certain mixing ratio of the two metals, hence higher for neighboring mixing processes. This special mixture is called *eutecticum*: it freezes (at the eutectic temperature) like a pure metal, while for other compositions, inhomogeneities are formed in the *alloy*.

The *mixing entropy* is important for such mixtures, where we are concerned, for example, by things like the *lowering of the freezing point* and *raising of the boiling point* of water by addition of salts. This will be discussed in Sect. 6.5.5, because only there will we be able to determine the temperature change.

6.4.8 Enthalpy and Free Energy as State Variables

The last two sections have shown the utility of the notion of free enthalpy G for homogeneous systems and for phase transitions. In particular, it is conserved for isobaric–isothermal processes, just as the internal energy is for isochoric–isentropic processes. In contrast, for phase transitions with volume changes, and fixed pressure, the *enthalpy* H (not the free enthalpy) is important for the transition heat, in addition to the internal energy and also the (mechanical) *work* $p \, dV$.

The enthalpy is also important for the *isentropic flow* of frictionless liquids through tube narrowings and widenings: here neither work nor heat is exchanged through the wall of the tube, but pressure and temperature vary with the tube cross-section. The idea is to follow a mass element M in a stationary flow, and in addition to its internal energy U, to account also for its collective kinetic energy $\frac{1}{2} M \bar{v}^2$, work pV, and potential energy Mgh in the gravitational field of the Earth. Only the sum of the enthalpy $H = U + pV$ and the center-of-mass energy $\frac{1}{2} M \bar{v}^2 + Mgh$ is conserved along the path. Here the pressure changes with the tube cross-section, as is easy to see for incompressible liquids because the continuity equation requires $\nabla \cdot \mathbf{v} = 0$. The smaller the tube cross-section, the higher the collective velocity $\bar{v}$ parallel to the wall, and the lower the pressure on the wall. The *Bernoulli equation* (Daniel Bernoulli, 1738) can be applied here. According to this, $\frac{1}{2} \rho \bar{v}^2 + p + \rho g h$ is conserved along the path, where the pressure dependence of the internal energy (for fixed volume) is neglected compared to the other contributions, along with the friction (*viscosity*).

The enthalpy is conserved in the *throttling experiment of Joule and Thomson*. Here a suitable penetrable obstacle ("a piece of cotton wool") ensures a pressure difference between the high and low pressure regions, and here again there is no heat exchange with the environment. The kinetic energy of the center-of-mass is negligible ($\bar{v} = 0$), and therefore the enthalpy is conserved.

For real gases in the throttling experiment, the temperature changes (*Joule–Thomson effect*). According to p. 43, we have

$$\left(\frac{\partial T}{\partial p}\right)_H = -\left(\frac{\partial T}{\partial H}\right)_p \left(\frac{\partial H}{\partial p}\right)_T .$$

Then, according to Sect. 6.4.5 with $dH = T\,dS + V\,dp$, we have

$$\left(\frac{\partial T}{\partial H}\right)_p = \frac{1}{C_p} \quad \text{and} \quad \left(\frac{\partial H}{\partial p}\right)_T = (1 - \alpha\,T)\,V .$$

Note that C_p and V are extensive quantities and for the Joule–Thomson coefficients only their ratio is important. Ideal gases have $\alpha T = 1$ (as shown on p. 582). Hence the throttle experiment with ideal gases proceeds along an isotherm. But for real gases, αT may be larger or indeed smaller than 1. (For low temperatures the attractive forces between the molecules are the stronger ones, so cooling by decompression is possible, while at high temperatures the repulsive forces are the stronger ones, so the gas heats up under decompression. However, under normal conditions, only hydrogen and the noble gases have $\alpha T < 1$.) In the (T, p) plane the two regions are separated by the *inversion curve*. We shall also investigate all this more precisely for a van der Waals gas (Sect. 6.6.2).

It is not the enthalpy, but the *free energy* F that is important for isothermal, reversible processes, e.g., if the system is coupled to a heat bath. With $dT = 0$, we have $dF = -p\,dV$. Thus the free energy F changes here by performing work. The free energy is the part of the internal energy which, for an isothermal, reversible process, can be extracted, while the rest $U - F = TS$ is the *energy bound* in the irregular motion. In contrast, for an adiabatic isolated system, $dS = 0$ holds, and thus $-p\,dV = dU$.

A very important example is the energy density of electromagnetic fields. According to electrostatics, a potential energy $\frac{1}{2} \int dV\,\rho\,\Phi = \frac{1}{2} \int dV\,\mathbf{E} \cdot \mathbf{D}$ is associated with a charge density ρ and a potential Φ (see Sect. 3.1.8), while the magnetic field is associated with the energy $\frac{1}{2} \int dV\,\mathbf{j} \cdot \mathbf{A} = \frac{1}{2} \int dV\,\mathbf{H} \cdot \mathbf{B}$ (see Sect. 3.3.5). Here it is assumed that temperature and volume remain unchanged by (quasi-statically) bringing the charges and currents from infinity to their respective positions—only afterwards can the charge and current density change. Therefore, with $\frac{1}{2}\,(\mathbf{E} \cdot \mathbf{D} + \mathbf{H} \cdot \mathbf{B})$, we have identified the density of the free energy.

We can also arrive at the free energy if we derive the state variables from the canonical partition function Z_C. Sections 6.3.6 and 6.3.8 give in particular $S = k\,(\ln Z_C + \lambda_E U)$, with $\lambda_E = (kT)^{-1}$, and thus $-kT \ln Z_C = U - TS = F$:

$$F = -kT \ln Z_C , \quad \text{or} \quad Z_C = \exp\frac{-F}{kT} .$$

To compute this, T, V, and N are normally given. The conjugate variables follow using $dF = -S\,dT - p\,dV + \mu\,dN$:

$$S = -\left(\frac{\partial F}{\partial T}\right)_{V,N} , \qquad p = -\left(\frac{\partial F}{\partial V}\right)_{T,N} , \qquad \mu = +\left(\frac{\partial F}{\partial N}\right)_{T,V} .$$

The other thermodynamic potentials then result from

$$U = F + TS , \qquad G = F + pV , \qquad H = U + pV ,$$

but the internal energy U, according to pp. 554 and 559, thus comes directly from

$$U = -\frac{\partial \ln Z_C}{\partial \lambda_E} = kT^2 \left(\frac{\partial \ln Z_C}{\partial T}\right)_{V,N} .$$

We can thus derive the *thermal equation of state* for p, V and T, and likewise the *canonical equation of state* for U, F, H and G, from the canonical partition function.

6.4.9 Irreversible Alterations

In this section, we have considered only reversible changes of state, even though at the beginning, in Sects. 6.4.1 and 6.4.2, we also allowed for irreversible ones. If we fix $dt > 0$ as there, then we generally have

$$
\begin{aligned}
dU &\leq +T\,dS - p\,dV + \mu\,dN , \\
dH &\leq +T\,dS + V\,dp + \mu\,dN , \\
dF &\leq -S\,dT - p\,dV + \mu\,dN , \\
dG &\leq -S\,dT + V\,dp + \mu\,dN .
\end{aligned}
$$

The first inequality was already proven in Sect. 6.4.1. The second follows from there with $H = U + pV$, the third with $F = U - TS$, and the fourth from the third with $G = F + pV$.

The last two inequalities are particularly important, because it is not the entropy changes dS that are of interest, but the temperature differences dT. If we keep, e.g., T, p, and N fixed for an irreversible process, then the free enthalpy nevertheless decreases, i.e., $dG < 0$, because the system was not yet in equilibrium. *Stable equilibrium states are the minima of the thermodynamic potentials.* This means the free energy for fixed T, V, and N, and the free enthalpy for fixed T, p, and N. Of course, in each case, the entropy is also then as large as possible. We have already made use of this for the phase transition (Sect. 6.4.7): only the phase with the smaller free enthalpy is stable for given T and p.

6.4.10 Summary: General Theorems of Thermodynamics

We have derived relations between the macroscopic state variables T, S, p, V, μ, N, U, H, F, and G, including equations for equilibrium states and reversible processes and inequalities for non-equilibrium states and irreversible processes. This all follows from the main theorems of thermodynamics, which can be justified microscopically or required axiomatically, but which in either case must be tested by experience. Basic for the first and second main theorems is the relation

$$dU \leq T\,dS - p\,dV + \mu\,dN\,, \quad \text{for} \quad dt > 0\,,$$

where U has the natural variables S, V, and N. This implies, for example, $T = (\partial U/\partial S)_{V,N}$ and $p = -(\partial U/\partial V)_{S,N}$ as well as Maxwell's integrability condition $(\partial T/\partial V)_{S,N} = -(\partial p/\partial S)_{V,N}$. Other thermodynamic potentials like $F = U - T\,S$, $H = U + pV$, and $G = H - T\,S$ follow from Legendre transformations (with other natural variables) and deliver further similar constraints.

6.5 Results for the Single-Particle Model

6.5.1 Identical Particles and Symmetry Conditions

In the last section, we presented macroscopic thermodynamics and derived general relations between observable quantities. Now we want to restrict ourselves to equilibrium states and special cases with known partition functions. Then according to p. 576, we may derive all thermal and canonical equations of states.

Identical particles without correlations are particularly simple. Then the same one-particle potential acts on all particles, and the probability distribution of the many-particle problem splits into a product of one-particle distributions. These depend on the one-particle states or on the cells in phase space of each individual particle (μ-space). We order them with respect to their energy e_i, and degenerate ones in some arbitrary way.

Now it is suggestive to assign to every particle its state, and thus fix the many-body state. This leads to *Maxwell–Boltzmann statistics*, although it contains an internal contradiction. In particular, we have assumed the ability to distinguish between the individual particles, otherwise we cannot decide how a given particle behaves in the course of time. Then distinguishing features are necessary, and therefore the particles cannot be completely identical.

This contradiction does not occur in quantum theory, because there we have to account for the exchange symmetry. Consider two particles in the states $|\alpha\rangle$ and $|\beta\rangle$. For *bosons*, only the symmetric state

$$|\alpha,\,\beta\rangle_{\mathrm{s}} = +|\beta,\,\alpha\rangle_{\mathrm{s}} \quad \propto \quad |\alpha\rangle\,|\beta\rangle + |\beta\rangle\,|\alpha\rangle$$

is permitted, and for *fermions*, only the antisymmetric state

$$|\alpha,\,\beta\rangle_{\mathrm{a}} = -|\beta,\,\alpha\rangle_{\mathrm{a}} \quad \propto \quad |\alpha\rangle\,|\beta\rangle - |\beta\rangle\,|\alpha\rangle\,.$$

In both cases, the first particle occurs with the same probability in the state $|\alpha\rangle$ as in the state $|\beta\rangle$, and the second, of course, likewise.

Two bosons may occupy the same one-particle state, but not fermions, because this contradicts the antisymmetry (Pauli principle). If n_i is the occupation number of the ith one-particle state, then we have the *occupation-number representation* (see Sect. 5.3.5):

$$\begin{array}{lll} \text{bosons} & |z\rangle_s \mathrel{\hat{=}} |n_1, n_2, \ldots\rangle_s & \text{with } n_i \in \{0, 1, \ldots\}, \\ \text{fermions} & |z\rangle_a \mathrel{\hat{=}} |n_1, n_2, \ldots\rangle_a & \text{with } n_i \in \{0, 1\}. \end{array}$$

Correspondingly, for bosons, we have *Bose–Einstein statistics*, and for fermions, *Fermi–Dirac statistics*.

In the classical Maxwell–Boltzmann statistics, several particles may occupy the same one-particle state. However, there the many-body state does not have to be symmetric under particle exchange. There are classically more states (by the factor $N!/n_1!\ldots$) than in Bose–Einstein-statistics, because classically each permutation counts as a new state. If all states are occupied just a little bit (all $n_i = 0$ or 1), then according to Stirling's formula, this produces an additional term $k \ln N! \approx Nk \ln N$ in the entropy $S = k \ln Z_{\mathrm{MC}}$. This addition does not increase in proportion to N, even though it has to be an extensive variable. This contradiction, occasionally called *Gibbs' paradox*, can only be removed by replacing $Z \to Z/N!$ in classical statistics. This leads to the *corrected Boltzmann statistics*.

6.5.2   Partition Functions in Quantum Statistics

This is best evaluated for the grand canonical ensemble, for which the energy and particle number are given only on average. For a sharp particle number, the calculation is rather involved (see the textbook by Reif in the reading list on p. 620), and soluble only with an approximation, which is in effect the transition from the canonical to the grand canonical ensemble. Note that the volume should also be given, because the one-particle energies depend on it.

If the ith one-particle state contains n_i particles of energy e_i, then according to the single-particle model, we have

$$N = \sum_i n_i \quad \text{and} \quad E = \sum_i n_i \, e_i \, ,$$

with $n_i \in \{0, 1, 2, \ldots\}$ for bosons and $n_i = 0$ or 1 for fermions. Note that N and E do not stand for the mean values here. For the grand canonical partition function $Z_{\mathrm{GC}} = \mathrm{tr}[\exp\{-(E - \mu N)/kT\}]$, with $z \mathrel{\hat{=}} \{n_1, n_2, \ldots\}$, we obtain

$$Z_{\mathrm{GC}} = \sum_{\{n_1, n_2, \ldots\}} \exp \frac{-\sum_i n_i \, (e_i - \mu)}{kT} \, .$$

The exponential function of a sum is equal to the product of the exponential functions:

$$Z_{GC} = \sum_{\{n_1, n_2, \ldots\}} \prod_i \exp \frac{-n_i (e_i - \mu)}{kT} .$$

In each term, the first the factor is $\exp\{-n_1(e_1 - \mu)/kT\}$, then we have the factor with $i = 2$, and then the remaining ones, whence we may write:

$$Z_{GC} = \prod_i \sum_{n_i} \exp \frac{-n_i (e_i - \mu)}{kT} .$$

For example, with $a = \exp\{-(e_1 - \mu)/kT\}$ and $b = \exp\{-(e_2 - \mu)/kT\}$, we have initially $Z_{GC} = a^0 b^0 + a^0 b^1 + \cdots + a^1 b^0 + a^1 b^1 + \cdots + \cdots$, but this sum of products may be written as product of simple sums $Z_{GC} = (a^0 + a^1 + \cdots)(b^0 + b^1 + \cdots)$.

For bosons, we thus obtain the geometric series of $\{1 - \exp(-(e_i - \mu)/kT)\}^{-1}$, where the chemical potential μ keeps the average particle number finite, and thus the geometric series converges. For fermions, on the other hand, we arrive at the sum $1 + \exp(-(e_i - \mu)/kT)$. Therefore, the result may be reformulated as

$$Z_{GC} = \prod_i \left(1 \mp \exp \frac{-(e_i - \mu)}{kT}\right)^{\mp 1} ,$$

or again,

$$\ln Z_{GC} = \mp \sum_i \ln\left(1 \mp \exp \frac{-(e_i - \mu)}{kT}\right) ,$$

where the upper sign holds for bosons and the lower one for fermions. We will also keep to this notation in the following.

According to p. 556, the natural variables of the grand canonical partition function are λ_E, λ_N, and V, or according to Sect. 6.3.8, T, μ, and V. Here, according to Sect. 6.3.6, the entropy S is given by $k \ln Z_{GC} + (U - \mu N)/T$. Consequently, $-kT \ln Z_{GC} = F - \mu N$ holds, and by the discussion on p. 567, this is the *grand canonical potential* J:

$$J \equiv -kT \ \ln Z_{GC} = F - \mu N = G - pV - \mu N ,$$

with

$$dJ = -S \, dT - p \, dV - N \, d\mu .$$

Using $Z_{GC}(T, V, \mu)$, the quantities S, p, and N may be derived immediately, and then also the other potentials U, H, F, and G may be determined. According to the Gibbs–Duhem relation, homogeneous systems have $G = \mu N$ and thus $J = -pV$.

6.5.3   Occupation of One-Particle States

So far we have viewed the grand canonical partition function as a function of T, V, and μ, but in the single-particle model, the energies $\{e_i\}$ replace the volume. These depend not only on V, but also on the average one-particle potential. Therefore, from

$$N = -\left(\frac{\partial J}{\partial \mu}\right)_{T,\{e_i\}} = kT\left(\frac{\partial \ln Z_{\mathrm{GC}}}{\partial \mu}\right)_{T,\{e_i\}} = -kT\sum_i \left(\frac{\partial \ln Z_{\mathrm{GC}}}{\partial e_i}\right)_{T,\{e_{k\neq i}\},\mu}$$
$$= \sum_i \langle n_i \rangle \,,$$

we deduce the *average occupation number* of the ith one-particle state as

$$\langle n_i \rangle = \left(\frac{\partial J}{\partial e_i}\right)_{T,\{e_{k\neq i}\},\mu} = \left(\exp\frac{e_i - \mu}{kT} \mp 1\right)^{-1} .$$

One-particle states of high energy ($e_i \gg \mu + kT$) are thus barely occupied. In addition, as required by the Pauli principle,

$$0 \le \langle n_i \rangle \le 1 \,, \quad \text{for fermions}\,,$$

while for bosons $\langle n_i \rangle$ may be greater than 1. But for the latter, due to the constraint $N \ge \langle n_i \rangle \ge 0$, the chemical potential μ is restricted to $\mu < \min e_i$, and so is never positive for $e_0 = 0$. In a grand canonical ensemble and for $e_i < e_j$, for both sorts of particles, we have $\langle n_i \rangle > \langle n_j \rangle$.

Since $\exp\{(e_i - \mu)/kT\} = \langle n_i \rangle^{-1} \pm 1$ and with the average occupation numbers $\langle n_i \rangle$, the partition function Z_{GC} is given by

$$\ln Z_{\mathrm{GC}} = \mp \sum_i \ln\left(1 \mp \frac{1}{\langle n_i \rangle^{-1} \pm 1}\right) = \mp \sum_i \ln \frac{\langle n_i \rangle^{-1}}{\langle n_i \rangle^{-1} \pm 1} = \pm \sum_i \ln\left(1 \pm \langle n_i \rangle\right) .$$

Using this for the ith one-particle state, we may also give the probability for its occupation by n particles. Here we write n instead of n_i. The partition function is clearly equal to $(1 \pm \langle n \rangle)^{\pm 1}$ and from $\rho = Z^{-1}\exp\{-(E - \mu N)/kT\}$ (see p. 555 and Sect. 6.3.8), it follows that

$$\rho_n = \frac{\exp\{-n(e_i - \mu)/kT\}}{(1 \pm \langle n \rangle)^{\pm 1}} = \frac{\langle n \rangle^n}{(1 \pm \langle n \rangle)^{n\pm 1}} .$$

For bosons, since $\rho_{n+1}/\rho_n = \langle n \rangle/(1 + \langle n \rangle) < 1$, the state without particles always has the highest probability, and that with $n \approx \langle n \rangle$ is not special at all. The situation is quite different for fermions: for them, $0 \le \langle n \rangle \le 1$ and in addition $\rho_0 = 1 - \langle n \rangle$ and $\rho_1 = \langle n \rangle$.

The relation $U - \mu N = \sum_i \langle n_i \rangle (e_i - \mu) = kT \sum_i \langle n_i \rangle \ln (\langle n_i \rangle^{-1} \pm 1)$ implies

$$S = k \ln Z_{GC} + (U - \mu N)/T ,$$

whence

$$S = \pm k \sum_i \left(\ln (1 \pm \langle n_i \rangle) \pm \langle n_i \rangle \ln \frac{1 \pm \langle n_i \rangle}{\langle n_i \rangle} \right)$$

$$= -k \sum_i \left(\langle n_i \rangle \ln \langle n_i \rangle \mp (1 \pm \langle n_i \rangle) \ln (1 \pm \langle n_i \rangle) \right) .$$

Since $x \ln x = 0$ for $x = 0$ and $x = 1$, the unoccupied states do not contribute to the entropy, and likewise for fermion states with $\langle n_i \rangle = 1$. This can also be justified by considering the uncertainty of the occupation number because, for the squared fluctuation of the particle number in the ith one-particle state, using $\lambda_N = -\mu/kT$ and

$$(\Delta n_i)^2 = -\frac{\partial \langle n_i \rangle}{\partial \lambda_N} = kT \frac{\partial \langle n_i \rangle}{\partial \mu} = \frac{\exp\{(e_i - \mu)/kT\}}{[\exp\{(e_i - \mu)/kT\} \mp 1]^2} ,$$

we obtain the noteworthy result (see Fig. 6.18)

$$(\Delta n_i)^2 = \langle n_i \rangle \left(1 \pm \langle n_i \rangle \right) .$$

This vanishes when $\langle n_i \rangle = 0$ and also for fermions when $\langle n_i \rangle = 1$, while for bosons, when $\langle n_i \rangle \gg 1$, the error width is $\Delta n_i \approx \langle n_i \rangle$, not $\sqrt{\langle n_i \rangle}$, as would be expected classically. Note also that, for fermions with $\langle n_i \rangle = \frac{1}{2}$, the error width is $\frac{1}{2}$.

With decreasing temperature the states of higher energy become ever more depopulated. In the limit $T \approx 0$, fermions only occupy one-particle states with $e_i \leq \mu$, while the states above stay empty. Then we have a degenerate Fermi gas with

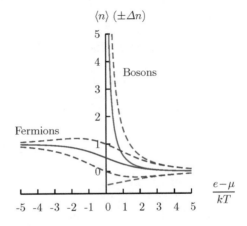

Fig. 6.18 Occupation number of the one-particle states as a function of $(e-\mu)/kT$ for bosons (*red curve*) and fermions (*blue curve*). We also show $\langle n \rangle \pm \Delta n$ (*dashed curves*) for bosons and for fermions. With $(\Delta n)^2 = -\partial \langle n \rangle / \partial \lambda_N$ and $\lambda_N = -\mu/kT$, the uncertainty is greater, the more rapidly $\langle n(x) \rangle$ decreases. Note that the base line here appears shifted to negative values!

$\mu(T = 0)$ as the *Fermi energy* e_F. We shall return to this in Sect. 6.5.6. With decreasing temperature, bosons crowd into the one-particle state of lowest energy e_0. Their chemical potential for $T \approx 0$ is thus determined by the constraint $\langle N \rangle \approx \langle n_0 \rangle$, which yields $\mu \approx e_0 - kT \ln(1 + \langle N \rangle^{-1}) \approx e_0 - kT/N$. More on that in Sect. 6.6.6.

6.5.4 Ideal Gases

For high temperatures, a great many states are occupied with nearly equal probability. For $\langle N \rangle$ to remain finite, we must then have $\exp\{(e_i - \mu)/kT\} \gg 1$ for all i, and hence $-\mu \gg kT$. But then Bose–Einstein and Fermi–Dirac statistics no longer differ because the exchange symmetry is no longer respected if all one-particle states are barely occupied. According to the above remarks, we then have

$$-\frac{J}{kT} = \ln Z_{GC} \approx \sum_i \exp\left(-\frac{e_i - \mu}{kT}\right) \approx \sum_i \langle n_i \rangle = \langle N \rangle \ .$$

If we make use here of the Gibbs–Duhem relation for homogeneous systems, thus $J = -pV$, we obtain the *Gay-Lussac law*, which is just the *thermal equation of state for ideal gases*, viz.,

$$pV = NkT \ .$$

Then using the results $\alpha \equiv V^{-1}(\partial V/\partial T)_{p,N}$, $\kappa_T \equiv -V^{-1}(\partial V/\partial p)_{T,N}$, $\beta = \alpha/\kappa_T$, and $C_p - C_V = \alpha\beta TV$, we obtain

$$\alpha = \frac{1}{T} \ , \qquad \kappa_T = \frac{1}{p} \ , \qquad \beta = \frac{p}{T} = \frac{Nk}{V} \ , \qquad C_p - C_V = Nk \ .$$

Hence for ideal gases with $(\partial U/\partial V)_T = -p + \beta T$ and $(\partial H/\partial p)_T = (1 - \alpha T)V$ (see p. 570), both $(\partial U/\partial V)_T$ and $(\partial H/\partial p)_T$ are zero. For (reversible) isothermal processes in ideal gases, when the volume changes, the internal energy is conserved, and when the pressure changes, the enthalpy is conserved. Consequently, for ideal gases, there is no Joule–Thomson effect, something we commented on already on p. 575.

Clearly, the canonical partition function for a particle may be extracted from the above-mentioned equation $N \approx \sum_i \exp\{-(e_i - \mu)/kT\}$, and we denote this by $Z_C(1)$, whence $Z_C(1)/N$ is an intensive variable:

$$N = Z_C(1) \, \exp\frac{\mu}{kT} \ .$$

The factor $\exp(\mu/kT)$ is called the *fugacity*, and in physical chemistry, the *absolute activity* of the material. We shall soon determine $Z_C(1)$ for important examples, and hence also μ via the Gibbs–Duhem relation G:

$$\mu = -kT \ln \frac{Z_C(1)}{N} \quad \text{and} \quad G = -NkT \ln \frac{Z_C(1)}{N} .$$

Hence we obtain the free energy $F = G - pV = G - NkT$ if we also use the Gay-Lussac law. The internal energy

$$U = F + TS = F - T(\partial F/\partial T)_{V,N} = -T^2(\partial(F/T)/\partial T)_{V,N}$$

yields

$$U = NkT^2 \left(\frac{\partial \ln Z_C(1)}{\partial T}\right)_{V,N},$$

and the enthalpy $H = U + NkT$. For the entropy, we obtain

$$S = -\left(\frac{\partial F}{\partial T}\right)_{V,N} = Nk \left\{\ln \frac{Z_C(1)}{N} + 1 + T\left(\frac{\partial \ln Z_C(1)}{\partial T}\right)_{V,N}\right\}.$$

Here we have required $-\mu \gg kT$, and hence $\ln Z_C(1)/N \gg 1$, but it is not necessary that it should be very much greater than 3, as can be seen from Fig. 6.19. The canonical partition function $Z_C(1)$ is determined according to the internal degrees of freedom of the given gas.

For the *ideal monatomic gases*, up to rather high temperatures (1 eV $\hat{=} 11\,600$ K), there is no internal excitation of the atoms (the electronic degrees of freedom are frozen), so what is important for e_i is only the kinetic energy $p_i^2/2m$ of their centers of mass. Here, according to p. 525, a particle confined to a cube of volume $V = L^3$ has the momentum eigenvalues $\mathbf{p}_i = \mathbf{n}_i \hbar \pi/L$, where $\mathbf{n}_i$ may have only natural numbers as Cartesian components, and not even negative integers. If we insert this into the canonical partition function and replace the sum by an integral, we obtain

$$Z_C(1) = \int \frac{d^3 n}{8} \exp \frac{-(n\hbar\pi/L)^2}{2m\,kT} = \frac{4\pi}{8} \int_0^\infty dn \; n^2 \exp \frac{-\pi^2\hbar^2 n^2}{2m\,kT\,L^2},$$

and therefore, since $\int_0^\infty dx\, x^2 \exp(-ax^2) = \frac{1}{4}\sqrt{\pi/a^3}$,

$$Z_C(1) = \left(\frac{kT}{4\pi\hbar^2/2m}\right)^{3/2} V \equiv \frac{V}{\lambda^3},$$

where the *thermal de Broglie wavelength* is defined by

$$\lambda \equiv \frac{h}{\sqrt{2\pi mkT}} .$$

However, the Maxwell distribution for $\langle h/mv \rangle$ delivers twice this value (see Problem 6.11), so the name is not quite satisfying. The result holds for high temperatures, and not only for a cube. For $V \gg \lambda^3$, other restrictions deliver the same value for the partition function $Z_C(1)$.

Fig. 6.19 The single-particle model for ideal monatomic gases yields the equations mentioned in the text, if $T \gg T_0$ with $kT_0 \equiv 4\pi\,(N/V)^{2/3}\,\hbar^2/2m$. But for $T \approx T_0$, the exchange symmetry contributes. The *upper curve* is for bosons and the *lower curve* for fermions. We return to this in Fig. 6.22

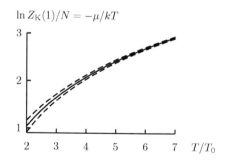

$\ln Z_K(1)/N = -\mu/kT$

Consequently, for ideal monatomic gases, we find $(\partial \ln Z_C(1)/\partial T)_V = \frac{3}{2}T^{-1}$ and, as expected by the equidistribution law (p. 559),

$$U = \tfrac{3}{2}\,NkT\,, \qquad H = \tfrac{5}{2}\,NkT\,, \qquad S = Nk\left(\ln\frac{Z_C(1)}{N} + \frac{5}{2}\right).$$

Hence with $C_V = (\partial U/\partial T)_{V,N} = C_p - Nk$ and $\kappa_S = \kappa_T\,C_V/C_p$, we have

$$C_V = \tfrac{3}{2}\,Nk\,, \qquad C_p = \tfrac{5}{2}\,Nk\,, \qquad \text{and} \quad \kappa_S = \tfrac{3}{5}\,\kappa_T = \tfrac{3}{5}\,p^{-1}\,.$$

If we relate to a mole, then according to p. 572, we have to take the gas constant R instead of Nk.

For *ideal diatomic gases*, the molecules rotate and oscillate. As long as its moment of inertia does not change notably despite the oscillations, the canonical partition function of a molecule may be written as the product of the canonical partition functions for the the center-of-mass motion, the rotations, and the oscillations, disregarding electronic degrees of freedom, which do not contribute anything (as established above).

At room temperature, in addition to the electronic excitations, the oscillations are also frozen. The rotations of diatomic molecules for constant moment of inertia Θ have the energy $j(j+1)\,\hbar^2/2\Theta$, and each level is $(2j+1)$-fold degenerate due to the isotropy. Therefore, we have

$$Z_C\,\mathrm{rot}(1) = \sum_j (2j+1)\,\exp\!\left(-\frac{\hbar^2}{2\Theta}\,\frac{j(j+1)}{kT}\right).$$

We evaluate this sum again via an integral, and use the continuous variable

$$x = (j+\tfrac{1}{2})\,\hbar/\sqrt{2\Theta\,kT}\,.$$

For molecules containing two identical atoms, however, the states with odd angular momentum do not occur, and this halves the partition function. Without this factor of

$\frac{1}{2}$ (thus in the case of non-identical atoms), with $\int_0^\infty dx\, 2x \exp(-x^2) = 1$, we obtain

$$Z_C \text{ rot}(1) = \frac{kT}{\hbar^2/2\Theta} \exp\frac{\hbar^2/2\Theta}{4kT} \approx \frac{kT}{\hbar^2/2\Theta} + \frac{1}{4}, \quad \text{for } kT \gg \frac{\hbar^2}{2\Theta}.$$

For sufficiently high temperatures, the product of the partition functions is

$$Z_C(1) = \frac{kT}{\hbar^2/2\Theta} \left(\frac{kT}{4\pi\hbar^2/2m}\right)^{3/2} V,$$

and thus now $(\partial \ln Z_C(1)/\partial T)_V = \frac{5}{2} T^{-1}$. For all diatomic molecules (of identical or non-identical atoms) and for sufficiently high temperatures, it thus follows that

$$U = \frac{5}{2} NkT, \quad H = \frac{7}{2} NkT, \quad S = Nk\left(\ln\frac{Z_C(1)}{N} + \frac{7}{2}\right).$$

This result does not contradict the equidistribution law, because for a diatomic molecule, the moment of inertia about the symmetry axis is then small compared to the other two, so this rotation is frozen. Therefore, for the symmetric top (see p. 145), we only have $H_{\text{tot}} = (p_\beta^2 + p_\alpha^2/\sin^2\beta)/2\Theta$. Each of the N molecules thus contributes to the internal energy $\frac{3}{2} kT$ from the translational motion and also $\frac{2}{2} kT$ from the rotation. Note that the factor of p_α^2 is not fixed but depends on β, but this does not affect the equidistribution law, as was shown by its proof on p. 559. We thus obtain

$$C_V = \frac{5}{2} Nk, \quad C_p = \frac{7}{2} Nk, \quad \text{and} \quad \kappa_S = \frac{5}{7}\kappa_T,$$

with $\kappa_T = p^{-1}$, as for all ideal gases.

These expressions are of course only valid for ideal diatomic gases as long as the oscillations are frozen. Otherwise, we must consider

$$Z_K \text{ vib}(1) = \sum_n \exp\left(-\frac{\hbar\omega}{kT}\left(n + \frac{1}{2}\right)\right) = \frac{\exp(-\frac{1}{2}\hbar\omega/kT)}{1 - \exp(-\hbar\omega/kT)} = \frac{1}{2\sinh(\hbar\omega/2kT)}.$$

If this degree of freedom is fully thawed, i.e., $kT \gg \hbar\omega$, then this results in $kT/\hbar\omega$, whence

$$Z_C(1) = \frac{kT}{\hbar\omega} \frac{kT}{\hbar^2/2\Theta} \left(\frac{kT}{4\pi\hbar^2/2m}\right)^{3/2} V.$$

Then,

$$U = \frac{7}{2} NkT, \quad H = \frac{9}{2} NkT, \quad S = Nk\left(\ln\frac{Z_C(1)}{N} + \frac{9}{2}\right),$$

and

$$C_V = \frac{7}{2} Nk, \quad C_p = \frac{9}{2} Nk, \quad \kappa_S = \frac{7}{9}\kappa_T.$$

If the molecules consist of two identical atoms, then in fact the above-mentioned factor $\frac{1}{2}$ changes the expression for the state sum $Z_C(1)$ by a factor of 2, which modifies μ only by $\Delta\mu = kT \ln 2$ and S by $\Delta S = -Nk \ln 2$. Also unimportant according to the equidistribution law is whether the molecules consist of identical or non-identical atoms.

6.5.5 Mixing Entropy and the Law of Mass Action

Mixtures of several materials may be evaluated rather simply as long as no correlations have to be accounted for. To begin with, we consider a segregated equilibrium state, with the same temperature and pressure everywhere. Each part has its particle number N_i corresponding to its volume V_i and entropy $S_i(T,\ p,\ N_i)$. The total volume is $V = \sum_i V_i$, the energy $U = \sum_i U_i$, and the entropy $S = \sum_i S_i$. If we now allow for a complete mixture with fixed U and V, then the entropy increases, because the number of accessible states increases with the volume. We restrict ourselves here to ideal gases. Then the chemical potential changes with $\mu = -kT \ln (Z_C(1)/N)$ and $Z_C(1) \propto V$ by $-kT \ln (V/V_i) = -kT \ln (N/N_i)$, and the entropy S_i by $N_i k \ln (N/N_i)$. Consequently, the *mixing entropy* amounts to

$$S_M = -k \sum_i N_i \ln \frac{N_i}{N} > 0 .$$

The mixing is an irreversible process, because the entropy increases. Since N_i/N is the probability ρ_i for the component i, we find $S_M/N \equiv s_M = -k \sum_i \rho_i \ln \rho_i$ for the mixing entropy per particle. This fits very well with the notion of information entropy (Sect. 6.1.6).

The mixing entropy depends only on the different particle numbers, not on the consistency. This leads to *Gibbs' paradox*. According to classical conceptions the difference between the particle types would have to vanish continuously. Even though a mixture would then no longer be conceivable, the last equation would still be valid. According to quantum theory, the transition is not continuous, however.

We found the Gibbs–Duhem relation $G = \mu N$ for pure homogeneous systems in Sect. 6.4.6 and now want to generalize it to systems of different materials (as long as they do not react chemically). For homogeneous mixtures of different particles (e.g., solutions), the equilibrium condition for given T and p is

$$\mu_i = \left(\frac{\partial G}{\partial N_i} \right)_{T,p,\{N_{k\neq i}\}} .$$

G is a homogeneous function of first order in the particle numbers N_i, since thermodynamic potentials of homogeneous systems are extensive variables. For arbitrary $x > 0$, we have $x\, G(T,\ p,\ N_1, N_2, \ldots) = G(T,\ p,\ x N_1,\ x N_2, \ldots)$. If we differentiate this with respect to x at the position 1 and make use of *Euler's theorem for homogeneous functions*, we may deduce the important *generalized Gibbs–Duhem relation*

$$G = \sum_i \mu_i \, N_i \; .$$

Here the mixing entropy also affects the free enthalpy, and in particular, $g_i = G_i/N_i$ denotes the free enthalpy per particle for pure systems ($G_{\text{pure}} = \sum_i G_i$). Then for mixtures (of ideal gases), we have $G = G_{\text{pure}} - T \, S_{\text{M}}$, and hence,

$$G = \sum_i N_i \left(g_i + kT \, \ln \frac{N_i}{N} \right) .$$

From the comparison with the generalized Gibbs–Duhem relation, we conclude that

$$\mu_i = g_i + kT \, \ln \frac{N_i}{N} < g_i \; .$$

The mixing entropy thus lowers the chemical potential, which is now different from the free enthalpy.

We can exemplify the above by considering the thawing of ice with salt, assuming that the salt is dissolved only in the water, but not also in the ice. At the transition point, both phases have to have the same chemical potential. If, in a similar way to p. 563, we denote the solid phase by [ ] and the liquid by ( ), then at the freezing temperature of pure water, we have $g_{[\,]}(T, p) = g_{(\,)}(T, p)$, in contrast to the freezing temperature of salt water:

$$g_{[\,]}(T + \Delta T, p) = g_{(\,)}(T + \Delta T, p) + k \, (T + \Delta T) \, \ln \frac{N_{\text{W}}}{N_{\text{W}} + N_{\text{S}}} \; .$$

Therefore, to the first approximation,

$$\Delta T \left\{ \left(\frac{\partial (g_{[\,]} - g_{(\,)})}{\partial T} \right)_p + k \, \ln \left(1 + \frac{N_{\text{S}}}{N_{\text{W}}} \right) \right\} = -kT \, \ln \left(1 + \frac{N_{\text{S}}}{N_{\text{W}}} \right) ,$$

where, since $dG = -S \, dT + V \, dp$ and $dH = T \, dS + V \, dp$, we may use

$$\left(\frac{\partial (g_{[\,]} - g_{(\,)})}{\partial T} \right)_p = s_{(\,)} - s_{[\,]} = \frac{\Delta h}{T} \quad (> 0) \; .$$

The *reduction in the freezing temperature* is thus

$$\Delta T = - \frac{kT^2 \, \ln(1 + N_{\text{S}}/N_{\text{W}})}{\Delta h + kT \, \ln(1 + N_{\text{S}}/N_{\text{W}})} \; .$$

For small salt concentrations and for one mole, it follows that

$$\Delta T \approx -\frac{N_S}{N_W} \frac{RT^2}{\Delta H} ,$$

where ΔH is the melting heat of water per mole (6 kJ). Every percent of salt lowers the freezing temperature by one degree centigrade.

If we now also allow for *chemical reactions*, then the equilibrium condition $\sum_i \nu_i \mu_i = 0$ on p. 561 initially delivers the equation $\sum_i \nu_i g_i = -kT \sum_i \ln (N_i/N)^{\nu_i}$. Hence, we have the *law of mass action*, viz.,

$$\prod_i \left(\frac{N_i}{N}\right)^{\nu_i} = \exp \frac{-\sum_i \nu_i g_i}{kT} \equiv K(T, p) ,$$

with given fixed temperature and pressure. Of interest is then the difference between the free enthalpies before and after the reaction, in contrast to the difference between the free energies for isochoric instead of isobaric processes. The *equilibrium constant* K depends on the chemical consistency of the materials, but not on the concentration (which is of course the important aspect of the law of mass action).

The temperature dependence of the chemical reaction follows from

$$\left(\frac{\partial \ln K}{\partial T}\right)_p = -\sum_i \nu_i \left(\frac{\partial(g_i/kT)}{\partial T}\right)_p .$$

Hence, with $(\partial g/\partial T)_p = -s$ and $g + Ts = h$, we obtain

$$\left(\frac{\partial \ln K}{\partial T}\right)_p = \frac{\sum_i \nu_i h_i}{kT^2} .$$

For constant pressure, heating thus shifts the reaction equilibrium in favor of the enthalpy-rich side (*endothermic reaction*).

6.5.6   Degenerate Fermi Gas and Conduction Electrons in Metals

For typical temperatures, the conduction electrons in metals form a degenerate Fermi gas. According to the considerations on p. 582, their chemical potential μ for the temperature $T = 0$ is equal to the *Fermi energy* $e_F = p_F^2/2m$. On p. 525, we determined the number of motional states whose energies e_i are smaller than the Fermi energy:

$$\Omega = \frac{V}{h^3} \cdot \frac{4\pi}{3} \left(2m \, e_F\right)^{3/2} = \frac{V}{6\pi^2} \left(\frac{2m}{\hbar^2} e_F\right)^{3/2} .$$

Furthermore, two spin states are associated with each of these states, so for N electrons in the volume V, we obtain the Fermi energy

Fig. 6.20 Fermi distributions for $T/T_0 = \frac{1}{2}$ (*red curve*), 1 (*blue curve*), and 2 (*green curve*). Note that, in Fig. 6.18, there is only a single curve, because for each temperature a different energy unit was taken. Here, the one-particle ground state energy lies very far to the left!

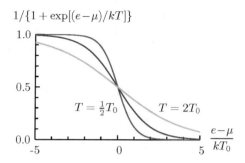

$$e_{\mathrm{F}} = \frac{\hbar^2}{2m} \left(3\pi^2 \frac{N}{V}\right)^{2/3} .$$

In metals, this energy is very much higher than kT (even at 1000 K) and the electron gas is therefore degenerate (see the Fermi distribution function in Fig. 6.20).

When computing mean values for Fermi gases, we always encounter expressions like

$$\langle A \rangle = \sum_i a_i \langle n_i \rangle = \sum_i a_i \left(\exp \frac{e_i - \mu}{kT} + 1\right)^{-1},$$

for which we shall now give a useful computational method for low temperatures. For high temperatures, we would have an ideal gas. If the values a_i depend only weakly on the index i and if sufficiently many states contribute, the sum may be replaced by an integral:

$$\langle A \rangle = \int_0^\infty \frac{a(e)\, g(e)\, de}{\exp\{(e - \mu)/kT\} + 1},$$

where $g(e)$ is the density of states for a particle. Note that we have to add an argument e in order to avoid confusions with the free enthalpy per particle. For $T = 0$, and therefore $\mu = e_{\mathrm{F}}$, only the integral from 0 to e_{F} is important—the denominator there is equal to one. However, with increasing temperature, the states for $e \approx e_{\mathrm{F}}$ are reshuffled (see the last figure).

For the expansion in terms of powers of T, we consider the expression

$$F = \int_0^\infty \frac{f(x)\, dx}{\exp\{\beta(x - x_0)\} + 1}, \quad \text{with } \beta > 0 \text{ and } \beta x_0 \gg 1,$$

i.e., actually for $\mu \gg kT$, which applies to a degenerate Fermi gas. With $F(x)$ as "anti-derivative" to $f(x)$ passing through zero, thus $f(x) = dF/dx$ and $F(0) = 0$, after integration by parts, we obtain

$$F = \frac{F(x)}{\exp\{\beta(x - x_0)\} + 1}\bigg|_0^\infty - \int_0^\infty F(x) \frac{d}{dx} \frac{1}{\exp\{\beta(x - x_0)\} + 1}\, dx .$$

The first term on the right vanishes because $F(0) = 0$ and the denominator for $x \to \infty$ is too large, while it is clear that only the integrand near $x \approx x_0$ contributes to the second. Therefore, we expand $F(x)$ in a Taylor expansion about this position to obtain

$$F = \sum_{n=0}^{\infty} \frac{1}{n!} \frac{d^n F}{dx^n} \bigg|_{x=x_0} \int_0^{\infty} (x-x_0)^n \frac{d}{dx} \frac{-1}{\exp\{\beta(x-x_0)\}+1} \, dx \ .$$

With $z = \beta(x-x_0)$ and $d(e^z+1)^{-1}/dz = -(e^z+1)^{-2} e^z$, it follows that

$$\int_0^{\infty} (x-x_0)^n \frac{d}{dx} \frac{-1}{\exp\{\beta(x-x_0)\}+1} \, dx = \beta^{-n} \int_{-\beta x_0}^{\infty} \frac{z^n \, dz}{(e^z+1)(e^{-z}+1)} \ .$$

Because of the denominator, the important contributions to the integrand come only from $z \approx 0$, since we assumed $\beta x_0 \gg 1$. Therefore, the lower integration limit may be taken as $-\infty$. Then terms with n odd do not contribute, and for n even,

$$\int_{-\infty}^{\infty} \frac{z^n \, dz}{(e^z+1)(e^{-z}+1)} = -2 \int_0^{\infty} z^n \frac{d}{dz} \frac{1}{e^z+1} \, dz \ ,$$

which gives 1 for $n = 0$. For $n > 0$, we integrate by parts and use $z^n/(e^z+1)|_0^{\infty} = 0$:

$$\int_{-\infty}^{\infty} \frac{z^n \, dz}{(e^z+1)(e^{-z}+1)} = 2n \int_0^{\infty} \frac{z^{n-1} \, dz}{e^z+1} \ .$$

In the next section (on bosons), we shall arrive at nearly the same integral, except that there, -1 occurs in the denominator instead of $+1$. Therefore, for $n \in \{1, 2, \ldots\}$, we consider here the two denominators simultaneously and expand $e^{-z}/(1 \mp e^{-z})$ in a geometric series:

$$\int_0^{\infty} \frac{z^{n-1} \, dz}{e^z \mp 1} = \sum_{k=0}^{\infty} (\pm)^k \int_0^{\infty} z^{n-1} \, e^{-(1+k)z} \, dz = (n-1)! \sum_{k=0}^{\infty} \frac{(\pm)^k}{(1+k)^n} \ .$$

Both sums lead to *Riemann's zeta function* (see Fig. 6.21):

$$\zeta(z) = \sum_{k=0}^{\infty} \frac{1}{(1+k)^z} \ , \quad \text{for } \mathrm{Re} z > 1 \ ,$$

because the alternating sum (for fermions) is equal to $(1 - 2(\frac{1}{2})^n) \, \zeta(n)$, given that $1+k$ is even for all negative terms and their sum leads to $(\frac{1}{2})^n \zeta(n)$. We need $\zeta(2) = \pi^2/6$ and in the next section $\zeta(4) = \pi^4/90$, but later also $\zeta(3)$, $\zeta(\frac{3}{2})$, and $\zeta(\frac{5}{2})$. The two values for $\zeta(2)$ and $\zeta(4)$ result from a Fourier expansion of the meander curve [9].

Table 6.2 Riemann zeta function for $1 \leq x \leq 4$. See also Fig. 6.21

x	$\zeta(x)$
1.0	∞
1.5	2.612375
2.0	1.644934
2.5	1.341487
3.0	1.202057
3.5	1.126734
4.0	1.082323

Fig. 6.21 Riemann zeta function for $1 \leq x \leq 4$. See also Table 6.2

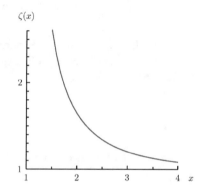

We thus obtain the expression up to order $n = 2$ (and 3):

$$F = F(x_0) + \frac{1}{6} \frac{\pi^2}{\beta^2} \frac{df}{dx}\bigg|_{x=x_0},$$

or for the Fermi distribution, as the weight function[2]

$$\frac{1}{\exp\{\beta(x - x_0)\} + 1} \approx \varepsilon(x_0 - x) - \frac{\pi^2}{6\beta^2} \delta'(x - x_0) + \cdots$$

in an integral, with the step function $\varepsilon(x)$ mentioned on p. 18 and the derivative $\delta'(x)$ of the Delta function.

Putting all this together, we thus have for the degenerate Fermi gas,

$$\langle A \rangle \approx A(\mu) + \frac{\pi^2}{6} (kT)^2 \frac{\partial}{\partial e}\big(a(e)g(e)\big)\bigg|_{e=\mu},$$

with $A(\mu) = \int_0^\mu a(e') g(e') \, de'$. Here, since $dA/de = ag(e)$, $A(\mu)$ differs from $A(e_F) = \langle A \rangle_{T=0}$ by approximately $(\mu - e_F) a(e_F) g(e_F)$. In order to evaluate the chemical potential $\mu(T)$, we consider the particle number, which does not depend on

---

[2]In nuclear physics, the radial distribution of nuclear matter is similar to a Fermi distribution [10].

the temperature, and hence take $a(e) = 1$. Then $(\mu - e_F) g(e_F) + \frac{1}{6}\pi^2 (kT)^2 g'(e_F) \approx 0$. If we use this in

$$\langle A \rangle - \langle A \rangle_{T=0} \approx (\mu - e_F) a(e_F) g(e_F) + \frac{1}{6}\pi^2 (kT)^2 \{a'(e_F)g(e_F) + a(e_F)g'(e_F)\} ,$$

then all terms on the right cancel out except for the term $\frac{1}{6}\pi^2 (kT)^2 a'(e_F) g(e_F)$. The only thing missing is the density of states $g(e_F)$. Here, $\Omega(e) \propto e^{3/2}$, and the further factor is equal to $N\, e_F^{-3/2}$, so $g(e_F) = \frac{3}{2}N/e_F$. From this, for $T \approx 0$, we find the important result

$$\langle A \rangle \approx \langle A \rangle_{T=0} + \frac{\pi^2}{4}\frac{N}{e_F} a'(e_F) (kT)^2 .$$

If we take this expression for the internal energy, then $a(e) = e$ and thus $a' = 1$. Near the origin, the internal energy of a degenerate Fermi gas increases with the square of the temperature. Hence,

$$C_V \equiv \left(\frac{\partial U}{\partial T}\right)_{VN} \approx \frac{\pi^2}{2} Nk \frac{kT}{e_F} .$$

For the chemical potential $\mu(T) \approx e_F - \frac{1}{6}\pi^2(kT)^2 g'(e_F)/g(e_F)$, using

$$g(e) \approx \frac{3}{2} N/e_F \sqrt{e/e_F} ,$$

and thus $g'(e)/g(e) \approx \frac{1}{2}e^{-1}$, we find (see Fig. 6.22)

$$\mu(T) \approx e_F \left\{1 - \frac{\pi^2}{12}\left(\frac{kT}{e_F}\right)^2\right\} .$$

Thus, it varies as T^2 for a Fermi gas near the zero temperature, whereas it varies linearly with T for a Bose gas because according to p. 582, we then have $\mu(T) \approx e_0 - kT/N$. As expected according to p. 572, the chemical potential decreases with increasing temperature in both cases.

The "high-temperature expansion" in Fig. 6.22 relies on $Z_C(1) = V/\lambda^3$ (see p. 583), but uses a more precise expression for the chemical potential, and in particular, one which differentiates between bosons and fermions. For sufficiently high temperatures in

$$N = \sum_i \left(\exp\frac{e_i - \mu}{kT} \mp 1\right)^{-1} ,$$

we have $\mu < 0$ and hence $e_i - \mu > 0$. After multiplying by $\exp\{-(e_i - \mu)/kT\}$, each term can be expanded in a geometric series. After reordering the series, it follows that

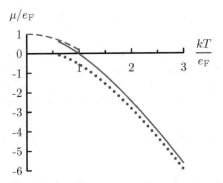

Fig. 6.22 The chemical potential of an ideal monatomic Fermi gas as a function of temperature, relative to the Fermi energy $e_F = (9\pi/16)^{1/3} kT_0$ with kT_0 from Fig. 6.19. The *continuous red curve* corresponds to the high-temperature expansion, the *dashed magenta curve* to the low-temperature expansion, and the *dotted blue curve* to a Bose gas (see Fig. 6.29 for Bose–Einstein condensation)

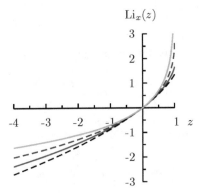

Fig. 6.23 The logarithm $\text{Li}_x(z) = \sum_{n=1}^{\infty} \frac{z^n}{n^x}$ for $|z| < 1$, *continuous for $x = 1$ (green)* and 2 (*red*), *dashed for $x = \frac{3}{2}$ (blue)* and $\frac{5}{2}$ (*black*). The name stems from $\text{Li}_1(z) = -\ln(1 - z)$. Then $\text{Li}_2(z)$ is also called the *dilogarithm*. Furthermore, $\text{Li}_x(1) = \zeta(x)$ and $\frac{d\,\text{Li}_x}{dz} = \frac{\text{Li}_{x-1}(z)}{z}$ (also for $|z| \geq 1$)

$$N = \sum_{n=0}^{\infty} (\pm)^n \exp\left((n+1)\frac{\mu}{kT}\right) \sum_i \exp\left(-(n+1)\frac{e_i}{kT}\right).$$

We may write the last sum as $V\lambda^{-3}(T/(n + 1))$. With $\lambda(T) \propto T^{-1/2}$ and the abbreviation $\sigma = \exp(\mu/kT)$ for the fugacity, we obtain an implicit equation for the determination of the chemical potential, which contains the *polylogarithm* $\text{Li}_x(z)$ (see Fig. 6.23):

$$N = \pm \frac{V}{\lambda^3(T)} \sum_{n=1}^{\infty} \frac{(\pm\sigma)^n}{n^{3/2}} = \pm \frac{V}{\lambda^3(T)} \text{Li}_{3/2}\left(\pm\exp\frac{\mu}{kT}\right).$$

6.5.7 *Electromagnetic Radiation in a Cavity*

An interesting and important system consists of photons in a cavity of volume V. They may be absorbed or emitted by the walls so the particle number is not fixed, not even on average. Therefore, there is no chemical potential ($\mu = 0$), and the canonical ensemble suffices with the free energy as thermodynamic potential:

$$F = -kT \ln Z_C = kT \sum_i \ln\left(1 - \exp\frac{-e_i}{kT}\right) .$$

The second equation holds, because we are dealing with bosons. They move with the speed of light. Therefore, we have $e_i = \hbar\omega_i = \hbar c k_i$ with $\mathbf{k}_i = \mathbf{n}_i\,\pi/L$, as on p. 525, so $\omega_i = n_i\,\pi c/L$. Since there are two polarization possibilities (helicities), the number of states follows from

$$2 \cdot \frac{4\pi}{8}\, n^2\, dn = \pi\left(\frac{L}{\pi c}\right)^3 \omega^2\, d\omega = \frac{V}{\pi^2 c^3}\, \omega^2\, d\omega .$$

If we replace the partition function by an integral, we obtain

$$\frac{F}{V} = \frac{kT}{\pi^2 c^3} \int_0^\infty \ln\left(1 - \exp\frac{-\hbar\omega}{kT}\right) \omega^2\, d\omega = \frac{kT}{\pi^2}\left(\frac{kT}{\hbar c}\right)^3 \int_0^\infty \ln\left(1 - e^{-x}\right) x^2\, dx .$$

According to the last section, integration by parts with $x^3 \ln(1 - e^{-x})|_0^\infty = 0$ yields

$$\int_0^\infty \ln\left(1 - e^{-x}\right) x^2\, dx = -\frac{1}{3} \int_0^\infty \frac{x^3\, dx}{e^x - 1} = -2\,\zeta(4) = -\frac{\pi^4}{45} .$$

With the *Stefan–Boltzmann constant* (see p. 623)

$$\sigma \equiv \frac{\pi^2}{60}\frac{k^4}{c^2\hbar^3} ,$$

the result reads

$$F = -\frac{4\sigma}{3c}\, VT^4 .$$

For the *radiation pressure* $p = -(\partial F/\partial V)_T$ and the entropy $S = -(\partial F/\partial T)_V$, this gives

$$p = -\frac{F}{V} = \frac{4\sigma}{3c}\, T^4 \quad\text{and}\quad S = -\frac{4F}{T} = \frac{16\sigma}{3c}\, VT^3 .$$

The pressure does not depend on the volume. For the free enthalpy $G = F + pV$, we obtain the value 0, as expected from the Gibbs–Duhem relation with $\mu = 0$. Clearly, $TS = -4F = 4pV$ and thus

$$U = -3F = 4 \frac{\sigma}{c} V T^4 \quad \text{and} \quad p = \frac{1}{3} \frac{U}{V} \,.$$

For ideal gases, we also have $p \propto U/V$, but with the factor $\frac{2}{3}$ for the monatomic gas—for $v \ll c$ the pressure is twice as large as for $v \approx c$. The frequency of collisions of the molecules is proportional to their speed, and the recoil proportional to their momentum. The product of velocity times momentum is important for the pressure. In the relativistic regime, it is equal to the energy (see p. 245), but twice as large in the non-relativistic regime.

If the wall has a hole of area A, then the energy per unit time that flows from the cavity is the area times the light intensity, viz.,

$$A \cdot I = \frac{AcU}{V} \cdot \frac{1}{4\pi} \int_{2\pi} \cos\theta \; d\Omega = A \, 4\sigma T^4 \cdot \frac{1}{2} \int_0^1 \cos\theta \; d\cos\theta \,,$$

where θ is the angle between the current direction and the normal to the area. This then leads to the *Stefan–Boltzmann equation*

$$I = \sigma \, T^4 \,,$$

where the Stefan–Boltzmann constant σ was already introduced above.

According to p. 580, the average number of (polarized) photons in the ith one-particle state is given by the *Planck distribution*:

$$\langle n_i \rangle = \frac{1}{\exp(\hbar\omega_i/kT) - 1} \,.$$

For the frequency interval $d\omega$, the energy density is therefore (see Fig. 6.24)

$$\frac{dU}{V} = \frac{\hbar\omega}{\exp(\hbar\omega/kT) - 1} \frac{\omega^2 \, d\omega}{\pi^2 c^3} \,.$$

This *Planck radiation formula* freezes high frequencies, while for low frequencies it goes over to the *Rayleigh–Jeans law*

$$\frac{dU}{V} \approx kT \frac{\omega^2 \, d\omega}{\pi^2 c^3} \,,$$

which was originally derived for classical oscillators. According to the equidistribution law, each one contributes kT to the internal energy. But this led to the ultraviolet catastrophe: U/V was not finite.

The maximum of the energy density as a function of the wavelength $\lambda = 2\pi c/\omega$ follows with $|\omega^3 d\omega| = (2\pi c)^4 \lambda^{-5} d\lambda$ from $\widehat{x} \equiv hc/(kT\widehat{\lambda}) = 5\{1 - \exp(-\widehat{x})\}$ as $\widehat{x} = 4.965114231745$. Together with the *second radiation constant* $c_2 \equiv hc/k$ (see Fig. 6.24), this leads to

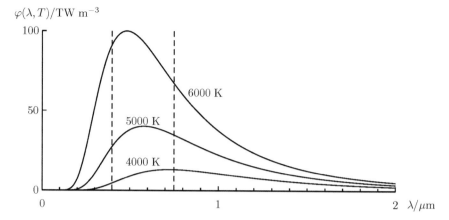

Fig. 6.24 Planck's radiation distribution $\varphi(\lambda, T) = c_1\lambda^{-5}/\{\exp(c_2/(\lambda T)) - 1\}$ with the first radiation constant $c_1 \equiv 2\pi hc^2$ and the second radiation constant $c_2 \equiv hc/k$. Here φ is the radiation flux density emitted into a half space, viz., $\varphi = \frac{1}{4}c\ du/d\lambda$. The factor $\frac{1}{4}c$ was derived for the Stefan–Boltzmann equation. Three isotherms are shown. The visible light range (400 nm $\leq \lambda \leq$ 750 nm) is indicated by *dashed lines*. The temperature of the surface of the Sun is such that a lot of visible light is emitted (adaption of the eye)

$$\widehat{\lambda} = \frac{1}{\widehat{x}}\frac{hc}{kT} = \frac{c_2}{4.965114231745\ T}\ .$$

This is *Wien's displacement law*—the higher the temperature, the shorter the most intense wavelength. As a function of the angular frequency ω (or the energy $\hbar\omega$), the maximum follows from $\widehat{x} \equiv \hbar\widehat{\omega}/(kT)$ as $\widehat{x} = 3\{1 - \exp(-\widehat{x})\} = 2.821439372122$.

Incidentally, according to the above equation for $\langle n_i\rangle$, the total number of photons in the volume V may be evaluated from $N/V = 2\zeta(3)\pi^{-2}(kT/\hbar c)^3$ with $\zeta(3) = 1.202$. This depends strongly on the temperature. With this value, we find $U = \pi^4/(30\zeta(3))\ NkT \approx 2.7\ NkT$ and hence the average energy per photon.

6.5.8 Lattice Vibrations

In a solid, each of the N atoms may oscillate about its equilibrium site. Here we may restrict ourselves to harmonic oscillations with small displacements and introduce $3N$ normal coordinates (see Sect. 2.3.9). We can then describe the motion of the atoms as $3N$ decoupled oscillations—sound waves, corresponding to phonons as quanta, without fixing their number. They obey Bose–Einstein statistics. In contrast to the photons in the last section, we have only a finite number ($3N$) of eigen frequencies, in particular a limiting frequency ω_{max}.

The excitation energy of the states $|n_1, n_2, \ldots\rangle_{\mathrm{s}}$ is $\sum_{i=1}^{3N} n_i\ \hbar\omega_i$. Since the number of phonons is not limited, we consider—as for photons—the canonical partition function

$$Z_{\mathrm{C}} = \sum_{\{...n_i...\}} \exp \frac{-\sum_i n_i \, \hbar\omega_i}{kT} = \prod_{i=1}^{3N} \frac{1}{1 - \exp(-\hbar\omega_i/kT)} \;,$$

or $\ln Z_{\mathrm{C}} = -\sum_{i=1}^{3N} \ln\big(1 - \exp(-\hbar\omega_i/kT)\big)$. The energy is therefore

$$U = -\frac{\partial \ln Z_{\mathrm{C}}}{\partial \lambda_E} = \sum_{i=1}^{3N} \frac{\hbar\omega_i}{\exp(\hbar\omega_i/kT) - 1} \;,$$

and the heat capacity at constant volume (fixed frequencies) is

$$C_V = \left(\frac{\partial U}{\partial T}\right)_V = \frac{1}{kT^2} \sum_{i=1}^{3N} \left(\frac{\hbar\omega_i}{\exp(\hbar\omega_i/kT) - 1}\right)^2 \exp\frac{\hbar\omega_i}{kT} \;.$$

For $kT \gg \hbar\omega_{\max}$, we have the *Dulong–Petit law* $C_V \approx 3\,Nk$, which follows from the equidistribution law for all temperatures.

With decreasing temperature, ever more degrees of freedom freeze, and for low temperatures, only the low frequency eigen oscillations are important, i.e., the normal oscillations with longer wavelength. These wavelengths are essentially longer than the interatomic distances, and we may make an ansatz for the density of states $\propto \omega^2$ (according to Debye) like the one for the electromagnetic radiation in a cavity. However, we have to account for the fact that there is an upper bound $\omega_{\max}$ for the eigen frequencies:

$$g_{\mathrm{D}}(\omega) = \begin{cases} 9N\,\omega_{\mathrm{D}}^{-3}\,\omega^2 & \text{for } \omega \le \omega_{\mathrm{D}} \equiv \omega_{\max} \;, \\ 0 & \text{otherwise} \;. \end{cases}$$

The factor $9N\omega_{\mathrm{D}}^{-3}$ follows from the constraint $3N = \int_0^\infty g_{\mathrm{D}}(\omega)\,d\omega$. This yields

$$U = \int_0^\infty \hbar\omega \, \{\exp(\hbar\omega/kT) - 1\}^{-1} \, g_{\mathrm{D}}(\omega)\,d\omega \;,$$

for the energy, or

$$U = 9\,NkT \; f_{\mathrm{D}}(\hbar\omega_{\mathrm{D}}/kT) \;,$$

with the *Debye function* $f_{\mathrm{D}}(x)$, which is displayed in Fig. 6.25.

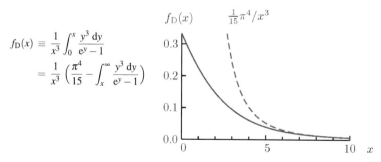

$$f_D(x) \equiv \frac{1}{x^3} \int_0^x \frac{y^3\, dy}{e^y - 1}$$

$$= \frac{1}{x^3} \left(\frac{\pi^4}{15} - \int_x^\infty \frac{y^3\, dy}{e^y - 1} \right)$$

Fig. 6.25 Debye function (*continuous red curve*) and its approximation $\frac{1}{15}\pi^4/x^3$ (*dashed blue curve*)

Fig. 6.26 Temperature dependence of the lattice energy. For $T \ll T_D$, we have $U \approx -3F \approx \frac{3}{5}\pi^4 N k T_D\, (T/T_D)^4$

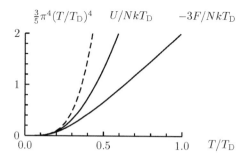

It is also common to introduce a *Debye temperature* $T_D \equiv \hbar\omega_D/k$ (200–300 K). For $T \ll T_D$, the last integral is not important. Then, for the heat capacity,

$$C_V \approx \frac{12\pi^4}{5}\, Nk \left(\frac{T}{T_D} \right)^3 .$$

In fact, for low temperatures, $C_V \propto T^3$ is observed, except for metals at very low temperature. (There the conduction electrons contribute, and their heat capacity is proportional to T according to p. 592.) Integrating by parts, the free energy is obtained from

$$F = -kT \ln Z_C = kT \int_0^\infty \ln \left(1 - \exp \frac{-\hbar\omega}{kT} \right) g_D(\omega)\, d\omega$$

$$= 3NkT \left\{ \ln \left(1 - \exp \frac{-T_D}{T} \right) - f_D \left(\frac{T_D}{T} \right) \right\} .$$

For low temperatures, $F = -\frac{1}{3}U$ and $S = \frac{1}{3}C_V \propto T^3$ (see Fig. 6.26), like for electromagnetic radiation in a cavity for all temperatures. Note that, for the harmonic oscillations about fixed positions we are concerned with here, F does not depend on the volume, so a pressure cannot be derived for phonons.

6.5.9 Summary: Results for the Single-Particle Model

In this section, we calculated partition functions for several examples and thereby
derived the equation of states, thus verifiable statements, which were not always
obvious for the original many-particle problem, where quantum theory was always
necessary. Classical physics leads to internal contradictions, e.g., to Gibbs' paradox
(the entropy has to be an extensive variable) and to the ultraviolet catastrophe. Here
we have restricted ourselves to examples which can all be described in the single-
particle model of independent quanta: gases, conduction electrons, electromagnetic
radiation, and lattice oscillations. Here the first two examples were treated as grand
canonical ensembles, because the particle number is an important parameter for them,
and the last two as canonical ensembles, because the number of oscillation quanta
(photons, phonons) cannot be given as a fixed variable in those cases.

6.6 Phase Transitions

6.6.1 Van der Waals Equation

The equation of state of ideal gases assumes sufficiently high temperatures, because
real gases behave differently at lower temperatures, when interactions between the
molecules may no longer be neglected. These interactions are strongly repulsive for
small distances and weakly attractive for large distances. If the electronic shells of
two molecules overlap, they repel each other strongly, so we assign a volume b to each
molecule which is inaccessible to the others. Then the volume in the gas equation
must be replaced by $V - Nb = N(v - b)$. At greater distances, on the other hand,
the molecules attract each other weakly like electric dipoles. It is not necessary for
permanent dipole moments to exist here. Before the quantum mechanical average, all
molecules have dipole moments, whose coupling does not vanish under the averaging
process. This attraction reduces the pressure on the outer walls and is proportional to
the product of the molecular densities in the interior of the volume and at the surface,
hence proportional to v^{-2}. Therefore, in the gas equation, we have to replace the
pressure by $p + av^{-2}$. We thus generalize the equation $pv = kT$ for ideal gases to
the *van der Waals equation*

$$\left(p + \frac{a}{v^2}\right)(v - b) = kT \ .$$

These additional terms contribute only for comparably small $v = V/N$.

Of course, the equation only makes sense for $v \geq b$. But it does not generally hold
even then, because it is an equation of third order in $v(p, T)$, viz.,

$$pv^3 - (bp + kT)\,v^2 + av - ab = 0 \ ,$$

and therefore allows for three different densities N/V. In fact, the van der Waals equation describes not only real gases rather well, but to some extent also liquids. It only gets things wrong for the phase transition. This is not so surprising, because so far we have assumed homogeneous systems rather than a spatially separated gas and liquid with their different densities.

How should the van der Waals solution be modified in order to describe the phase transition without contradictions? Here we argue that, of three real solutions $v(p, T)$, the one with the highest density (lowest v) should hold for the liquid and the one with the lowest density (highest v) for the gas. For given p and T, the two phases exist simultaneously between these densities. For the phase transition, despite a change in v, we nevertheless expect p and T to remain constant. If we take, e.g., isotherms as functions $p(v)$, then the van der Waals solution in this ambiguous regime should be replaced by a horizontal straight line segment.

In order to determine the pressure at which this straight line segment is to be taken, we have to respect the free enthalpy and the equilibrium condition $\mu_1 = \mu_2$ for the phase transition. We have $dN_1 = -dN_2$ and $dT = 0$ and therefore $dG = V\,dp$. The area $\int V\,dp$ between the van der Waals isotherm and the straight line segment has to be (*Maxwell construction*) chosen such that $\int dG$ vanishes, because G is a state variable.

The van der Waals equation does not therefore always deliver $(\partial p/\partial v)_T < 0$, as it actually should according to p. 560 with $(\Delta V)^2 > 0$. Given that

$$(\partial p/\partial v)_T = -kT/(v-b)^2 + 2a/v^3 ,$$

in particular, the stability condition requires $2a\,(v-b)^2/v^3 \le kT$. This is not always satisfied for low temperatures. The stable phase becomes unstable if we have equality here and in addition $(\partial^2 p/\partial v^2)_T$ vanishes, which leads to $kT = 3a\,(v-b)^3/v^4$. At the *critical point* for the stability, it is clear that $kT_c = 2a\,(v_c-b)^2/v_c{}^3 = 3a\,(v_c - b)^3/v_c{}^4$, whence

$$v_c = 3b , \qquad kT_c = \frac{8a}{27b} , \quad \text{and} \quad p_c = \frac{a}{27b^2} ,$$

and thereby $p_c v_c = \tfrac{3}{8}kT_c$, in contrast to an ideal gas. Note that the van der Waals equation holds only approximately here. Instead of $\tfrac{3}{8} = 0.375$, we observe 0.31 for O_2, 0.29 for N_2, and 0.23 for H_2O. With the reduced quantities $v_r = v/v_c$, $T_r = T/T_c$, and $p_r = p/p_c$, the van der Waals equation reads (see Fig. 6.27)

$$\left(p_r + \frac{3}{v_r^2}\right)(3v_r - 1) = 8T_r .$$

The parameters a and b are then hidden.

Fig. 6.27 Van der Waals isotherms with $T_r = 1.2, 1.0$, and 0.8. The *middle red curve* is the critical one, while the *lower curve* corresponds to a phase transition. Also shown here is the unstable solution of the van der Waals equation (*dashed curve*)

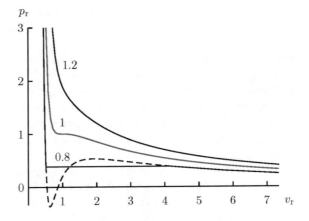

6.6.2 Conclusions Regarding the van der Waals Equation

For the stress coefficients $\beta = (\partial p/\partial T)_v$, the van der Waals equation implies

$$\beta = \frac{k}{v-b} = \frac{1}{T}\left(p + \frac{a}{v^2}\right).$$

According to p. 570, $(\partial U/\partial V)_T = -p + \beta T$. This is now equal to a/v^2. Thus the potential energy of the cohesive forces between the molecules contributes to the internal energy. This addition depends in fact on the volume per particle, but not on the temperature. Therefore, we also find

$$\frac{\partial C_V}{\partial V} = \frac{\partial^2 U}{\partial V\,\partial T} = 0\,,$$

as for an ideal gas.

On the other hand, according to the equation for $(\partial p/\partial v)_T$ mentioned in the last section, the isothermal compressibility is

$$\kappa_T = -\left\{v\left(\frac{\partial p}{\partial v}\right)_T\right\}^{-1} = f\frac{(v-b)^2}{vkT - 2a\,(1 - b/v)^2}\,,$$

so for the expansion coefficient, we have

$$\alpha = \beta\,\kappa_T = \frac{1}{T}\frac{v-b}{v - (2a/kT)(1 - b/v)^2}\,.$$

According to p. 575, $1 - \alpha T$ is important for the Joule–Thomson experiment, because $(\partial T/\partial p)_H$ contains only the extra factor $-V/C_p$:

$$1 - \alpha T = \frac{b - (2a/kT)(1 - b/v)^2}{v - (2a/kT)(1 - b/v)^2} \;.$$

If we keep only terms of first order in a and b, then this is equal to $(b-2a/kT)/v$. It is negative for low temperatures and delivers $(\partial T/\partial p)_H > 0$. All real gases may be cooled to low temperatures by decompression $(dp < 0)$. But for normal temperatures, this does not hold for hydrogen and the noble gases. Their cohesive forces are then weak (a is small), so for normal temperatures these gases heat up under decompression. Indeed, highly compressed hydrogen ignites upon streaming out of leaks.

We can only differentiate the remaining thermal coefficients if we know the entropy or one of the thermodynamic potentials. As for the ideal gases, the internal degrees of freedom of the molecules are important, and here we proceed as for the ideal gases. For the change, we account only for the center-of-mass motion.

Here we disregard the feedback of a given molecule on the others and describe the coupling by an effective one-particle potential $V(\mathbf{r})$. Note that, in order to avoid confusion we shall always indicate the position with the volume V. Then the classical canonical partition function due to the center-of-mass motion of a molecule is

$$Z_C(1) = \frac{1}{h^3} \int \exp\left\{-\frac{1}{kT}\left(\frac{p^2}{2m} + V(\mathbf{r})\right)\right\} d^3r \, d^3p \;,$$

and according to p. 583,

$$Z_C(1) = \lambda^{-3} \int \exp\frac{-V(\mathbf{r})}{kT} d^3r \;, \quad \text{with} \quad \lambda = \frac{h}{\sqrt{2\pi \, mkT}} \;.$$

If at first we neglect the attractive forces and account only for the strong repulsion, then the integral yields $N(v - b)$. The weak attraction is approximated by the mean value $\overline{V}(\mathbf{r}) \approx -a/v$:

$$Z_C(1) = \lambda^{-3} \, N(v - b) \, \exp\frac{a/v}{kT} \;.$$

In addition, for independent particles, according to the corrected Boltzmann statistics (see p. 578), we have

$$\ln Z_C(N) = N \, \ln\frac{Z_C(1)}{N} \;.$$

With this we obtain the free energy

$$F = -kT \, \ln Z_C = N\left(kT \, \ln\frac{\lambda^3}{v - b} - \frac{a}{v}\right),$$

and $p = -(\partial F/\partial V)_{T,N} = -N^{-1}(\partial F/\partial v)_{T,N} = kT/(v - b) - a/v^2$ for the pressure. Thus we have derived the van der Waals equation in a different way. (For molecules containing more atoms, F also contains additions, and according to Sect. 6.5.4, these in fact depend upon T, but not on V, whence we obtain the same pressure.) But the entropy $S = -(\partial F/\partial T)_{V,N}$ for a real gas is lower than for an ideal one:

$$S_{real} - S_{ideal} = Nk \ln \frac{v-b}{v} = Nk \ln\left(1 - \frac{b}{v}\right).$$

In addition, the chemical potential $\mu = (\partial F/\partial N)_{T,V}$ is different:

$$\mu_{real} - \mu_{ideal} = -kT \ln \frac{v-b}{v} + kT \frac{b}{v-b} - \frac{2a}{v}.$$

6.6.3 Critical Behavior

The free enthalpy depends on the aggregation state and determines whether a probe exists in the form of gas or liquid (or solid)—only the phase with the lowest free enthalpy is stable, as we already stressed in Fig. 6.17. For fixed pressure $p < p_c$, the (monotonically decreasing) function $G(T)$ has a kink at the transition temperature, and likewise, for fixed temperature $T < T_c$, the function $G(p)$ has a kink at the transition pressure. The first derivatives $(\partial G/\partial T)_p$ and $(\partial G/\partial p)_T$ have a discontinuity for this *discontinuous phase transition*, and likewise the entropy and the volume:

$$S_+ - S_- = -\left(\frac{\partial G_+}{\partial T}\right)_{p,N} + \left(\frac{\partial G_-}{\partial T}\right)_{p,N},$$

$$V_+ - V_- = \left(\frac{\partial G_+}{\partial p}\right)_{T,N} - \left(\frac{\partial G_-}{\partial p}\right)_{T,N}.$$

Here we also speak of a *first order phase transition*, because the first derivatives of G are discontinuous. Such phase transitions have a transition enthalpy (the pressure remains constant) $H_+ - H_- = T(S_+ - S_-) \neq 0$ and obey the Clausius–Clapeyron equation

$$\frac{dp}{dT} = \frac{S_+ - S_-}{V_+ - V_-} = \frac{1}{T} \frac{H_+ - H_-}{V_+ - V_-},$$

discussed on p. 573.

According to Sect. 6.6.1, the isotherm $p(V)$ has a horizontal tangent at the phase transition, i.e., $(\partial p/\partial V)_T = 0$. Therefore, the volume (and density) uncertainty is infinitely large there. Otherwise, it is negligibly small for macroscopic bodies, e.g., for an ideal gas, we have $(\Delta V/V)^2 = 1/N$ (since $(\partial V/\partial p)_T = -V/p = -V^2/NkT$):

$$(\Delta V)^2 = -\left(\frac{\partial V}{\partial \lambda_V}\right)_T = -kT \left(\frac{\partial V}{\partial p}\right)_T = -kT \bigg/ \left(\frac{\partial p}{\partial V}\right)_T.$$

The density therefore fluctuates enormously at the phase transition. Hence, the isothermal compressibility $\kappa_T = -V^{-1}(\partial V/\partial p)_T$ is infinite there, too, and likewise (if a transition enthalpy is involved) the isobaric heat capacity $C_p = T (\partial S/\partial T)_p$ and the expansion coefficient $\alpha = V^{-1} (\partial V/\partial T)_p = -V^{-1}(\partial S/\partial p)_T$.

At the critical point, S_+ and S_- agree with each other, as do V_+ and V_-. A transition heat is unnecessary, and the first derivatives of G are continuous. But with $(\partial V/\partial p)_T = (\partial^2 G/\partial p^2)_T$, the second derivative of the free enthalpy is infinite. Then we have a *second order phase transition* (a *continuous phase transition*). At the critical point, the volume is very unsharp, as for a phase transition of first order— the density fluctuates strongly. At the critical point, an otherwise transparent body scatters light very strongly and appears opaque (*critical opalescence*).

We shall now investigate the behavior near the critical point. According to Cardani's formula, the *cubic equation* $v^3 + 3Av^2 + Bv + C = 0$ has the three solutions $v_i = x_i - A$ with

$$x_0 = R_+ + R_- \quad \text{and} \quad x_{\pm 1} = -\frac{R_+ + R_-}{2} \pm i\sqrt{3}\,\frac{R_+ - R_-}{2}\,,$$

and the abbreviations

$$R_\pm = \sqrt[3]{-Q \pm \sqrt{Q^2 + P^3}}\,, \quad \text{with} \quad Q = A^3 + \frac{C - AB}{2}\,, \quad P = \frac{B}{3} - A^2\,,$$

where the third root is taken such that $R_+ R_- = -P$. For real coefficients, there are three real solutions with $Q^2 + P^3 < 0$, and hence $R_- = R_+{}^*$. For the reduced van der Waals equation, we have $A = -\frac{8}{9}T_r/p_r - \frac{1}{9}$, $B = 3/p_r$, and $C = -1/p_r$, and hence, $Q = A^3 - \frac{1}{2}(3A + 1)/p_r$ and $P = 1/p_r - A^2$. Therefore, near the critical point with $\Delta T = T_r - 1$ and $\Delta p = p_r - 1$, we have

$$A \approx -1 + \tfrac{8}{9}\Delta p - \tfrac{8}{9}\Delta T\,, \qquad Q \approx \tfrac{1}{3}\Delta p - \tfrac{4}{3}\Delta T\,, \qquad P \approx \tfrac{7}{9}\Delta p - \tfrac{16}{9}\Delta T\,.$$

We reach the critical point along $Q = 0$, i.e., $\Delta p = 4\Delta T$. This delivers $R_\pm \approx \pm 2\sqrt{\Delta T/3}$, and hence for $\Delta T < 0$, i.e., $T < T_c$, the two solutions $v_r - 1 \approx \pm 2\sqrt{1 - T_r}$ at the phase boundary. For the density $\rho_r \propto v_r^{-1}$, it follows that

$$|\rho - \rho_c| \propto (T_c - T)^{1/2}\,.$$

The density ρ is called an *order parameter* for the considered system since it has a discontinuity at the phase transition, and from the last relation, the *critical exponent* $\frac{1}{2}$ for this order parameter is extracted from the van der Waals equation.

For the isothermal compressibility, $p_r = 8T_r/(3v_r - 1) - 3v_r^{-2}$ implies

$$\left(\frac{\partial p_r}{\partial v_r}\right)_T = -\frac{24T_r}{(3v_r - 1)^2} + \frac{6}{v_r^3} \approx -6T_r\left(1 - 3\Delta v + \tfrac{27}{4}(\Delta v)^2\right) + 6\left(1 - 3\Delta v + 6(\Delta v)^2\right).$$

For $T \geq T_c$ and $\Delta v \approx 0$, this leads to $\kappa_T^{-1} = 6p_c\,(T_r - 1)$, but for $T \leq T_c$ and $(\Delta v)^2 \approx 4(1 - T_r)$, to $\kappa_T^{-1} = 12p_c\,(1 - T_r)$. In total, this gives

$$\kappa_T \propto |T - T_c|^{-1} \; ,$$

where the proportionality factor for $T > T_c$ is equal to $\frac{1}{6}T_c/p_c$ and for $T < T_c$ half as large. We usually set $\kappa_T \propto |T - T_c|^{-\gamma}$. According to the van der Waals equation, the critical exponent here is $\gamma = 1$.

6.6.4 Paramagnetism

Magnetism also provides an example of a phase transition. As for gases, we begin by neglecting the interaction between the atoms (*paramagnetism*), and include them in the next section in the molecular field approximation due to Weiss.

We thus start from the magnetic moment $mg\mu_B$ of an atom with μ_B the Bohr magneton (see p. 327), g the Landé factor, which is equal to $(2j+1)/(2l+1)$ for the angular momentum $j = l \pm \frac{1}{2}$, according to p. 373, and m the directional (magnetic) quantum number along the magnetic-field direction. The potential energy is then

$$W_{\text{pot}} = -mg\mu_B\mu_0 H = -m\,\eta\,kT \; , \quad \text{with} \quad \eta \equiv g\,\frac{\mu_B\mu_0 H}{kT} \; .$$

In vacuum, we have $\mathbf{B} = \mu_0\mathbf{H}$ and the energy $-\boldsymbol{\mu} \cdot \mathbf{B}$ due to the coupling of the magnetic moment to the magnetic field. Nevertheless, here we investigate the magnetization induced by the magnetic field, and use now $\mu_0\mathbf{H}$ instead of $\mathbf{B}$ (see Sect. 3.2.6).

For a given magnetic field, the eigenstates of the energy are evenly spaced at distances ηkT from each other. However, there are only $2j + 1$ of them and not infinitely many as for a harmonic oscillator. Hence the directional quantum number m in the canonical partition function $\sum_m \exp(m\eta)$ takes the values from $-j$ to $+j$ in even-numbered steps. Now

$$x^{-j}\,(1 + x + \cdots + x^{2j}) = x^{-j}\,\frac{1 - x^{2j+1}}{1 - x} = \frac{x^{j+1/2} - x^{-j-1/2}}{x^{1/2} - x^{-1/2}} \; .$$

Hence, for the canonical partition function, we find

$$Z_C = \sum_{m=-j}^{j} e^{m\eta} = \frac{\sinh((j + \frac{1}{2})\eta)}{\sinh(\frac{1}{2}\eta)} \; ,$$

and clearly, $\rho_m = Z_C^{-1}\,\exp(m\eta)$ for the occupation probability of the states with the directional quantum number m.

Fig. 6.28 Brillouin function
$B_j(\eta)$ for $j = \frac{1}{2}, \frac{3}{2}$, and $\frac{5}{2}$.
For $\eta \approx 0$, it depends
linearly on j, viz.,
$B_j(\eta) \approx \frac{1}{3}(j+1)\eta$, and for
$\eta \gg 1$, $B_j(\eta) \approx 1$
(saturation)

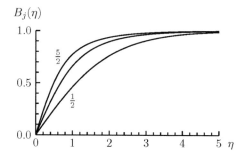

For the average magnetic moment, we obtain

$$\overline{m} = \frac{\sum_m m \exp(m\eta)}{\sum_m \exp(m\eta)} = \frac{d}{d\eta} \ln \frac{\sinh((j + \frac{1}{2})\eta)}{\sinh(\frac{1}{2}\eta)} .$$

The polarization $\overline{m}/j$ is therefore given by the *Brillouin function* (see Fig. 6.28)

$$B_j(\eta) \equiv \frac{1}{j} \frac{d}{d\eta} \ln \frac{\sinh((j + \frac{1}{2})\eta)}{\sinh(\frac{1}{2}\eta)} = \frac{(j + \frac{1}{2})\coth((j + \frac{1}{2})\eta) - \frac{1}{2}\coth(\frac{1}{2}\eta)}{j} .$$

For $j = \frac{1}{2}$, in particular, $B_{1/2}(\eta) = \tanh(\frac{1}{2}\eta)$ holds. Generally, $B_j(\eta)$ is a monotonically increasing function—the stronger the magnetic field H and the lower the
temperature T, the better the orientation.

For the magnetization from mutually independent moments, we obtain N/V times
the mean value of $mg\mu_B$:

$$M = \frac{N}{V} \overline{m} \, g\mu_B = \frac{Njg\mu_B}{V} B_j\left(\frac{g\mu_B\mu_0 H}{kT}\right) .$$

So for paramagnetism at low temperatures ($\eta \gg 1$),

$$M \approx \frac{N}{V} j \, g\mu_B , \quad \text{for} \ \ kT \ll g\mu_B\mu_0 H .$$

Then it depends neither on the temperature nor on the magnetic field, and the system has reached saturation: all moments are oriented and the magnetization cannot
increase any further. In contrast, at high temperatures, we obtain $M \propto H$ and hence
for the *magnetic susceptibility*

$$\chi \equiv \frac{M}{H} \approx \frac{N}{V} \frac{j(j+1)(g\mu_B)^2\mu_0}{3kT} \quad \text{for} \ \ kT \gg g\mu_B\mu_0 H .$$

It is thus proportional to the reciprocal of the temperature, which is *Curie's law*.

6.6.5 Ferromagnetism

The correlation between the atoms neglected so far (for paramagnetism) is decisive for ferromagnetism. Here what is important is not so much the magnetic coupling between the dipole moments, as the exchange symmetry of the fermion states, where position and spin states are important, because their product has to be antisymmetric under particle exchange. For this reason even the electric coupling of two electrons depends on the spin states. This leads to the *Ising model*

$$W_{ik} = -2J \, m_i \, m_k \, ,$$

where only nearest neighbors i and k interact, although actually the parameter J depends on the distances. It is adjusted, and even the sign is not the same for all materials.

We follow P. Weiss with the *molecular field approximation* and assume an average one-particle potential. The coupling to the n nearest neighbors is then given simply by $-2n J \overline{m} m$, and for the average directional quantum number $\overline{m}$, we found $j B_j(\eta)$ in the last section. The field at the position of the test particle is now composed of the external field and the remaining part. Thus we obtain

$$W_{\text{pot}} = -m \, \{g\mu_B\mu_0 H + 2nj \, B_j(\eta) \, J\} \, .$$

As we have already done for paramagnetism, we may therefore set

$$W_{\text{pot}} = -m \, \eta \, kT \quad \text{and} \quad M = \frac{N}{V} \, g\mu_B \, j B_j(\eta) \, ,$$

but where η now follows from a new equation:

$$\eta = \frac{g\mu_B\mu_0 H + 2nj \, B_j(\eta) \, J}{kT} \quad \Longleftrightarrow \quad B_j(\eta) = \frac{kT\eta - g\mu_B\mu_0 H}{2nj \, J} \, .$$

We have thus to find the points where the Brillouin curve crosses a straight line. Here the solution with the largest $\eta > 0$ is stable, because it has the smallest free energy, given that the partition function $Z_C = \sinh((j + \tfrac{1}{2})\eta)/\sinh(\tfrac{1}{2}\eta)$ increases monotonically with η, and therefore $F = -kT \ln Z_C$ decreases.

The case $J > 0$ is particularly instructive, so we shall now restrict ourselves to this. For $H = 0$, in addition to the crossing point for $\eta = 0$, there is another for $\eta > 0$ if

$$\left. \frac{dB_j(\eta)}{d\eta} \right|_0 = \frac{j+1}{3} > \frac{kT}{2nj \, J} = \frac{j+1}{3} \frac{T}{T_C} \, , \quad \text{with} \quad kT_C \equiv \tfrac{2}{3}nj(j+1) \, J \, .$$

Below the *Curie temperature* T_C, we also find spontaneous magnetization for $H = 0$, because for $J > 0$ the parallel orientation is convenient for the magnetic moments.

The slope of the above-mentioned straight line is proportional to the temperature, and therefore its crossing point with the Brillouin curve moves from $T \to 0$ to ever higher values of η. But then we may set $B_j(\eta) \approx 1$ and find again the *saturation magnetization*. In contrast, for $T \to T_C$, the crossing point moves towards the origin. The magnetization vanishes for $T = T_C$. In this case, we have to evaluate $B_j(\eta)$ to a higher accuracy than we have done so far, because now also the curvature of the Brillouin curve is important:

$$ B_j(\eta) \approx \frac{j+1}{3}\,\eta - \frac{(j+\frac{1}{2})^4 - (\frac{1}{2})^4}{45\,j}\,\eta^3 \,. $$

The crossing point with the straight line $\frac{1}{3}(j+1)\,(T/T_C)\,\eta$ then leads to

$$ \eta^2 \propto 1 - T/T_C \,, $$

and therefore to

$$ M \propto \sqrt{T_C - T} \,. $$

For $T > T_C$ and $H = 0$, there is no solution $\eta \ne 0$.

For $H \ne 0$ this changes, because then the straight line is shifted downwards and therefore always cuts the Brillouin curve with $\eta > 0$, thus for $T > T_C$. At least for these temperatures and for $H \approx 0$, we also find $\eta \approx 0$, and therefore we may set $B_j(\eta) \approx \frac{1}{3}(j+1)\,\eta$. This delivers $\eta = g\mu_B\mu_0 H/(k(T - T_C))$, and hence for the magnetic susceptibility,

$$ \chi = \frac{N}{V}\,\frac{j(j+1)\,(g\mu_B)^2\mu_0}{3k\,(T - T_C)} \,, \quad \text{for} \quad T > T_C \,. $$

This *Curie–Weiss law* reproduces the observation for $T \gg T_C$ very well, but not close to the Curie temperature, where the molecular field approximation is too coarse. This means that the phase transition occurs not exactly at T_C, if we have determined this parameter using the Curie–Weiss law for higher temperatures. For $T < T_C$, η is larger than for $H = 0$ and the same temperature. Furthermore, the magnetization and the susceptibility are larger, but the saturation values remain the same.

6.6.6 Bose–Einstein Condensation

We have in fact already considered a photon gas and lattice vibrations, both examples of Bose gases, but in both cases the (average) particle number was not given. Now we shall go back to that case, but start with the grand canonical ensemble and take

$$ J = -kT\,\ln Z_{GC} = kT \sum_{i=0} \ln\!\left(1 - \exp\frac{-(e_i - \mu)}{kT}\right) . $$

We choose e_0 as the zero energy and once again write σ for the fugacity $\exp(\mu/kT)$. The term $i = 0$ then contributes $kT \ln(1-\sigma)$, with $0 \le \sigma < 1$. So far we have not accounted for this in the high-temperature expansions in Sects. 6.5.4 and 6.5.6, because replacing the partition function by an integral with the density of states, a state with the zero energy has no weight:

$$g(e) = \frac{d}{de} \frac{V}{6\pi^2} \left(\frac{2me}{\hbar^2}\right)^{3/2} = \frac{V}{(2\pi)^2} \left(\frac{2m}{\hbar^2}\right)^{3/2} \sqrt{e} = \frac{V}{\lambda^3} \frac{2}{\sqrt{\pi}} \frac{\sqrt{e/kT}}{kT} \ .$$

The internal degrees of freedom are in fact frozen at low temperatures and do not need to be considered here, but a potential energy would have an effect. In this sense, we are greatly simplifying here. We now obtain

$$\ln Z_{GC} = -\ln(1-\sigma) - \frac{V}{\lambda^3} \frac{2}{\sqrt{\pi}} \int_0^\infty \sqrt{x} \ \ln(1 - \sigma e^{-x}) \, dx \ ,$$

where $x \equiv e/kT$. Here, integrating by parts, we find

$$\int_0^\infty \sqrt{x} \ \ln(1 - \sigma e^{-x}) \, dx = -\frac{2\sigma}{3} \int_0^\infty \frac{x^{3/2} \, dx}{e^x - \sigma} = -\frac{\sqrt{\pi}}{2} \sum_{n=1}^\infty \frac{\sigma^n}{n^{5/2}} \ .$$

Thus with the polylogarithm $\mathrm{Li}_{5/2}(\sigma)$ (see Fig. 6.23), we obtain

$$J = kT \ \ln(1-\sigma) - kT \ \frac{V}{\lambda^3} \ \mathrm{Li}_{5/2}(\sigma) \ .$$

Hence it follows that

$$\langle N \rangle = -\left(\frac{\partial J}{\partial \mu}\right)_{T,V} = -\left(\frac{\partial J}{\partial \sigma}\right)_{T,V} \left(\frac{\partial \sigma}{\partial \mu}\right)_{T,V} = \frac{\sigma}{1-\sigma} + \frac{V}{\lambda^3} \ \mathrm{Li}_{3/2}(\sigma) \ .$$

The first term on the right gives the particle number $\langle n_0 \rangle$ in the ground state and the rest then the number of particles in excited states (N^*). We divide this equation by N and introduce a critical temperature

$$T_c \equiv \left(\frac{\lambda^3}{V} \frac{N}{\mathrm{Li}_{3/2}(1)}\right)^{2/3} = \frac{h^2}{2\pi mk} \left(\frac{N/V}{\zeta(\frac{3}{2})}\right)^{2/3} \ .$$

This increases with increasing density N/V. Hence,

$$1 - \frac{\sigma}{N\,(1-\sigma)} = \frac{\mathrm{Li}_{3/2}(\sigma)}{\zeta(\frac{3}{2})} \left(\frac{T}{T_c}\right)^{3/2} \ .$$

This equation fixes $\sigma(T)$ for given T_c. In particular, $\sigma(0) = N/(N+1) \approx 1$. For $N \gg 1$, this barely changes up to $T = T_c$. In particular, on the left-hand side,

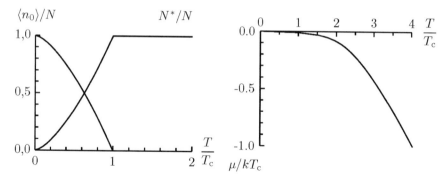

Fig. 6.29 Bose–Einstein condensation and its dependence on the temperature T relative to the critical temperature T_c. *Left*: The number of particles in the ground state $\langle n_0 \rangle$ or in excited states N^* relative to the total number N. *Right*: The chemical potential μ, represented for $N = 100$

$\sigma = 1 - 1/\sqrt{N}$ delivers approximately $1 - 1/\sqrt{N} \approx 1$, and the right-hand side with $T = T_c$ and $\sigma = 1$ thus yields 1. Here with $\langle n_0 \rangle = \sigma/(1 - \sigma)$, the whole expression is equal to $1 - \langle n_0 \rangle/N = N^*/N$. Thus for $T \geq T_c$, it always stays equal to one, and compared with the number N of particles in the ground state, i.e., $\langle n_0 \rangle$, this is clearly negligible (see Fig. 6.29):

$$\frac{N^*}{N} = \begin{cases} (T/T_c)^{3/2} & \text{for} \quad T \leq T_c\,, \\ 1 & \text{for} \quad T \geq T_c\,. \end{cases}$$

Here, of course, there are always more bosons in the ground state than in any other one-particle state—only the sum of numbers over the many excited states may be greater than the number in the ground state for higher temperatures.

These considerations thus lead to $\sigma \approx 1$ for $T \leq T_c$ and to $\mathrm{Li}_{3/2}(\sigma) = \lambda^3 N^*/V$ for $T \geq T_c$, so $\mathrm{Li}_{3/2}(\sigma) = \zeta(\frac{3}{2})\,(T_c/T)^{3/2}$. If we differentiate this with respect to T, then on the left, we have $\sigma^{-1}\mathrm{Li}_{1/2}(\sigma) \cdot d\sigma/dT$ according to the chain rule, and the polylogarithm diverges for $\sigma = 1$ (more strongly than $-\ln x$ at the origin). On the right, for $T = T_c$, we obtain the finite value $-\frac{3}{2}\zeta(\frac{3}{2})/T_c$. The derivative of σ with respect to T thus vanishes at T_c, and is continuous (as is the chemical potential μ).

From the generalized grand canonical potential, the pressure and entropy may also be derived:

$$p = -\left(\frac{\partial J}{\partial V}\right)_{T,\mu} = \frac{kT}{\lambda^3}\,\mathrm{Li}_{5/2}(\sigma)\,,$$

$$S = -\left(\frac{\partial J}{\partial T}\right)_{V,\mu} = -k\,\ln(1-\sigma) + \frac{\frac{5}{2}pV - \mu N}{T}\,.$$

The bosons in the ground state do not contribute to the pressure, and for fixed T and μ, σ is also constant. For $T \leq T_c$, it depends only on the temperature (and the mass of the bosons) (proportional to $T^{5/2}$), but not on the density. With decreasing volume,

Fig. 6.30 Influence of the
Bose–Einstein condensation
on the pressure coefficients β
(and the isochoric heat
capacity $C_V = \frac{3}{2} V \beta$). At
$T = T_c$, we have
$\beta = \frac{5}{2}\zeta(\frac{5}{2})/\zeta(\frac{3}{2}) \, Nk/V$.
The *dashed line* is for an
ideal gas

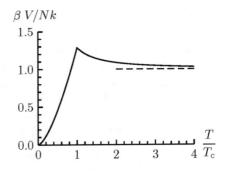

T_c increases and hence also $\langle n_0 \rangle$. In other words, the particles condense. This also
holds for the internal energy. From $U = J + TS + \mu N$, we obtain $U = \frac{3}{2} pV$.

Clearly, the second derivatives of p and U with respect to T are discontinuous at
T_c, and so also are the first derivative of the pressure coefficients β and the isochoric
heat capacity C_V, as well as the isothermal compressibility κ_T. Then, for the pressure
coefficients, we obtain $\beta = (\partial p / \partial T)_{VN}$ (see Fig. 6.30)

$$
\beta = \frac{Nk}{V}
\begin{cases}
\dfrac{5}{2} \dfrac{\zeta(\frac{5}{2})}{\zeta(\frac{3}{2})} \left(\dfrac{T}{T_c}\right)^{3/2} & \text{for } T \le T_c \,, \\[2ex]
\dfrac{5}{2} \dfrac{\mathrm{Li}_{5/2}(\sigma)}{\zeta(\frac{3}{2})} \left(\dfrac{T}{T_c}\right)^{3/2} - \dfrac{3}{2} \dfrac{\mathrm{Li}_{3/2}(\sigma)}{\mathrm{Li}_{1/2}(\sigma)} & \text{for } T \ge T_c \,.
\end{cases}
$$

From this, we also have the isochoric heat capacity C_V, because with $U = \frac{3}{2} pV$, this
is equal to $\frac{3}{2} V\beta$ here.

6.6.7 Summary: Phase Transitions

As examples of phase transitions and critical behavior, we have investigated in some
detail the van der Waals gas, magnetism in Weiss's molecular field approximation,
and Bose–Einstein condensation. Here the van der Waals equation had to be amended
by the Maxwell construction, to make the volume a unique function of pressure and
temperature.

A phase transition of nth order has a discontinuity in the nth derivative of the free
enthalpy. The Clausius–Clapeyron equation holds for phase transitions of first order.
At the critical point, there is a phase transition of second order. Here the density ρ or
the magnetization M are taken as the order parameter. Below the critical temperature,
their value jumps at the phase transition, but it is continuous above. At the critical
temperature, the isothermal compressibility κ_T and the susceptibility χ are infinite.

Problems

Problem 6.1 Legend tells us that the inventor of chess asked for $S = \sum_{z=0}^{63} 2^z$ grains of rice as a wage: one grain on the first square, two on the second, and twice as many on each subsequent square. Compare the sum S for all the squares with the Loschmidt number $N_L \approx 6 \times 10^{23}$. How often can the surface of the Earth be covered with S grains, if 10 of them are equivalent to 1 square centimeter? By the way, 29% of the surface of the Earth is covered by land.
(3 P)

Problem 6.2 Justify Stirling's formula $n! = (n/e)^n \sqrt{2\pi n}$ with the help of the equation $n! = \int_0^\infty x^n \exp(-x)\, dx$, using a power series expansion of $n \ln x - x$ about the maximum and also by comparing with $\ln(n!)$, $n \ln(n/e)$, and $n \ln(n/e) + \frac{1}{2}\ln(2\pi n)$ for $n = 5$, 10, and 50.
(9 P)

Problem 6.3 Draw the binomial distribution $\rho_z = \binom{Z}{z} p^z (1-p)^{Z-z}$ when $Z = 10$ for $p = 0.5$ and $p = 0.1$. Compare this with the associated Gauss distribution (equal to $\langle z \rangle$ and Δz) and for $p = 0.1$ with the associated Poisson distribution. Note that the Gauss and Poisson distributions also assign values for $z > 10$, but which we do not want to consider. For comparisons, set up tables with three digits after the decimal point, no drawings.
(8 P)

Problem 6.4 From the binomial distribution for $Z \gg 1$, derive the Gauss distribution if the probabilities p and $q = 1 - p$ are not too small compared to one.

Hint: Here it is useful to investigate the properties of the binomial distribution near its maximum and let ρ depend continuously on z.
(8 P)

Problem 6.5 How high is the probability for z decays in 10 seconds in a radioactive source with an activity of 0.4 Bq? Give in particular the values $\rho(z)$ for $z = 0$ to 10 with two digits after the decimal point.
(6 P)

Problem 6.6 Which probability distribution $\{\rho_z\}$ delivers the highest information measure $I = -\sum_{z=1}^{Z} \rho_z \operatorname{lb} \rho_z$?

Hint: Note the constraint $\sum_{z=1}^{Z} \rho_z = 1$.

How does I change if initially Z_1 states are occupied with equal probability and then $Z_2 < Z_1$? *Freezing of degrees of freedom*: Determine ΔI for $Z_1 = 10$ and $Z_2 = 2$. For two possibilities, I may be written as a function of just $p = \rho(z_1)$. Set up a table of values with the step width 0.05.
(6 P)

Problem 6.7 In phase space, every linear harmonic oscillation proceeds along an ellipse. How does the area of this ellipse depend on the energy and oscillation period? By how much do the areas of the ellipses of two oscillators differ when their energies differ by $\hbar\omega$? Determine the probability density $\rho(x)$ for a given oscillation amplitude x_0 and equally distributed phases φ.

Hint: Thus we may set $x = x_0 \sin(\omega t + \varphi)$. Actually, the probability density should be taken at time t. Why is this unnecessary here? (7 P)

Problem 6.8 A molecule in a gas travels equal distances l between collisions with other molecules. We assume that the molecules are of the same kind, but always at rest, a useful simplification which does not falsify the result. Here all directions occur with equal probability. Determine the average square of the distance from the initial point after n elastic collisions, and express the result as a function of time. (4 P)

Problem 6.9 Does $\rho(t, \mathbf{r}) = \sqrt{4\pi Dt}^{-3} \exp(-r^2/4Dt)$ solve the diffusion equation $\partial\rho/\partial t = D\Delta\rho$, and does it obey the initial condition $\rho(0, \mathbf{r}) = \delta(\mathbf{r})$? What is the time dependence of $\langle r^2 \rangle$? Compare with Problem 6.8. How do the solutions $\rho(t, \mathbf{r})$ read in one and two dimensions? (9 P)

Problem 6.10 Consider N interaction-free molecules each of which is equally probable in any of two equal sections of a container. What is the probability for all N molecules to be in just one of the sections? If each of the possibilities since the existence of the world (2×10^{10}) has occurred corresponding to its probability, how long have 100 molecules (very, very few for macroscopic processes!) been in one section? (2 P)

Problem 6.11 Given the Maxwell distribution

$$\rho(v) = 4\pi \, v^2 (2\pi kT/m)^{-3/2} \, \exp(-mv^2/2kT) \, ,$$

determine the most frequent and the average velocities ($\widehat{v}$, $\langle v \rangle$), kinetic energies ($\widehat{E}$, $\langle E \rangle$), and de Broglie wavelengths ($\widehat{\lambda}$, $\langle \lambda \rangle$).

Hint:

$$\int_0^\infty \exp(-\alpha x^2) \, dx = \frac{1}{2}\sqrt{\frac{\pi}{\alpha}} \, ,$$

$$\int_0^\infty x^{2n} \exp(-\alpha x^2) \, dx = (-)^n \frac{\partial^n}{\partial\alpha^n} \int_0^\infty \exp(-\alpha x^2) \, dx = \frac{(2n-1)!!}{2^{n+1}\alpha^n}\sqrt{\frac{\pi}{\alpha}} \, ,$$

$$\int_0^\infty x^{2n+1} \exp(-\alpha x^2) \, dx = \frac{1}{2}\int_0^\infty y^n \exp(-\alpha y) \, dy = \frac{n!}{2\alpha^{n+1}} \, .$$

The first integral is half as large as $\int_{-\infty}^\infty$ and the latter equal to the square root of the surface integral

$$\iint_{-\infty}^\infty \exp\{-\alpha(x^2 + y^2)\} \, dx \, dy = 2\pi \int_0^\infty \exp(-\alpha r^2) r \, dr = \pi \int_0^\infty e^{-\alpha x} dx \, .$$

(8 P)

Problem 6.12 Consider the 1D diffusion equation $\partial y / \partial t = D \, \partial^2 y / \partial x^2$ with the boundary condition $y(t, 0) = c(0) \exp(-i\omega t)$. Which differential equation follows for $c(x)$, and what are its physical solutions for $x > 0$? (Example: seasonal ground temperature.)   (3 P)

Problem 6.13 Under what circumstances do the Maxwell equations yield a diffusion equation for the electric field strength? How large is the diffusion constant under such circumstances?   (3 P)

Problem 6.14 For a molecular beam, all velocities $\mathbf{v}$ outside of a small solid angle $d\Omega$ around the beam direction are suppressed. How large is the number of suppressed molecules with velocities between v and $v + dv$ per unit time and unit area? Determine the most frequent and the average velocity in the beam.   (4 P)

Problem 6.15 According to quantum theory, the phase space cells cover the area h. Therefore, according to Problem 6.7, the number of states of *one* linear oscillator up to the highest excitation energy E is equal to $\Omega(E, 1) = E/\hbar\omega + 1 = n + 1$, with the *oscillator quantum number* n. Determine $\Omega(E, 2)$ for distinguishable oscillators and then $\Omega(E, N)$ by counting. Simplify the result for the case $n \gg N$. Is the density of states for this system equal to $\frac{1}{N!} E^{N-1} (\hbar\omega)^{-N}$?

Hint: The *binomial coefficients* for natural m and arbitrary x are given by

$$\binom{x}{m} = \frac{x \cdot (x-1) \cdots (x-m+1)}{m!} = \frac{x-m+1}{m} \binom{x}{m-1}.$$

Consequently,

$$\binom{x}{1} = x = x \binom{x}{0}, \qquad \binom{m}{m} = 1 = \frac{1}{m} \binom{m}{m-1}, \qquad \text{and for } n < m, \quad \binom{n}{m} = 0.$$

In addition,

$$\binom{x+1}{m} = \binom{x}{m} + \binom{x}{m-1}, \qquad \text{and hence} \qquad \binom{n+1}{m+1} = \sum_{k=0}^{n-m} \binom{n-k}{m}.$$

(6 P)

Problem 6.16 From the expression found for $\Omega(E, N)$ in Problem 6.15, determine the canonical partition function and hence the average energy $\langle E \rangle$ and the squared relative fluctuation $(\Delta E/\langle E \rangle)^2$.   (4 P)

Problem 6.17 The energy of N non-interacting spin-$\frac{1}{2}$ particles with magnetic moments μ in the magnetic field is $E = (n_{\downarrow\uparrow} - n_{\uparrow\uparrow}) \, \mu B$. What is the microcanonical partition function of this system?   (4 P)

Problem 6.18 Take the result of the last problem as a binomial distribution (with the energy as state variable), and approximate it by the Gauss distribution for $\mu B \ll dE \ll E$. Thereby determine the entropy. How does the entropy differ from the one found in Problem 6.17, obtained with the Stirling formula for $E \ll N\mu B$?   (6 P)

Problem 6.19 Determine, as for the equidistribution law, $\langle p_n \dot{x}^m \rangle$ and $\langle x^m \dot{p}_n \rangle$ for canonical ensembles of particles which are enclosed between impenetrable walls. Why are these considerations not also valid for unbound particles?   (4 P)

Problem 6.20 For an N-particle system, the expression $\sum_{i=1}^{N} \mathbf{r}_i \cdot \mathbf{F}_i$ is called the *virial* of the force. What follows for its expectation value? Compare the result with $\langle E_{\text{kin}} \rangle = N \frac{m}{2} \langle \dot{\mathbf{x}} \cdot \dot{\mathbf{x}} \rangle$ and with the virial theorem of classical mechanics. Note that this holds for the mean value over the time (!), and in fact for "quasi-periodic" systems, i.e., x and p always have to stay finite.   (5 P)

Problem 6.21 Consider the 1D diffusion equation $\partial y / \partial t = D \, \partial^2 y / \partial x^2$. How do its solutions read with the *initial condition* $y(0, x) = f(x)$ instead of the *boundary condition* of Problem 6.12?   (2 P)

Problem 6.22 The gas pressure p on the walls can be determined from the momentum change due to the elastic collision of the molecules. Determine the pressure as a function of the average energy of the individual molecules. Here the same assumptions are made as for the derivation of the Boltzmann equation. Do we need the Maxwell distribution? What follows for $\langle E \rangle$ if the ideal gas equation $pV = NkT$ holds?   (6 P)

Problem 6.23 In a galvanometer, a quartz fiber with the torque $\delta = 10^{-13}$ J supports a plane mirror. How large is the directional uncertainty at $20\,°C$ from the Brownian motion of the air molecules? How much does a reflected light beam fluctuate on a target scale at 1 m distance?   (3 P)

Problem 6.24 For an ideal monatomic gas, $pV^{5/3}$ is a constant for isentropic processes. How much does the internal energy U change if the volume increases from V_0 to V? Does U increase or decrease?   (3 P)

Problem 6.25 Consider a *cycle* in an (S, T) diagram. What area corresponds to the usable work and what area to the heat energy input? Consider a heat engine with the heat input $Q_+ = T_+ \Delta S_1$ at the temperature T_+ and $Q_0 = T_0 \Delta S_2$ at $T_0 < T_+$, as well as heat output $Q_- = T_- \, (\Delta S_1 + \Delta S_2)$ at $T_- < T_0$. Determine the efficiency $\eta(Q_+, Q_0, Q_-)$ and compare it with the efficiency of an ideal Carnot process (η_C with $Q_0 = 0$). Express the result as a function of η_C, Q_0/Q_+, and T_0/T_+. Determine a least upper bound for the efficiency of a cycle process with heat reservoirs at several input and output temperatures.   (9 P)

Problem 6.26 Why do we have to do work to pump heat from a cold to a hot medium? Investigate this with an ideal cycle. Under ideal constraints, let the work A be necessary in order to keep a house at the temperature T_+ inside, while the temperature outside is T_-. How are these three quantities connected with the heat loss Q_+? How is the input heat Q'_+ in an ideal power plant related to the heat loss Q_+ considered above if it works between the temperatures T'_+ and T'_-? Neglect the losses in the power plant that delivers the electric energy. Take as an example $T'_+ = 800\,°C$, $T_+ = 20\,°C$, and $T_- = T'_- = 0\,°C$.   (8 P)

Problem 6.27 Determine the functional determinant

$$\frac{\partial(S, T)}{\partial(V, p)} = \left(\frac{\partial S}{\partial V}\right)_p \left(\frac{\partial T}{\partial p}\right)_V - \left(\frac{\partial S}{\partial p}\right)_V \left(\frac{\partial T}{\partial V}\right)_p .$$

(2 P)

Problem 6.28 Express the derivatives of S with respect to T, V, and p, with the other parameters kept fixed, in terms of the thermal coefficients and V and T. Express the derivatives of T with respect to S, V, and p in terms of the quantities above. Express $(\partial F/\partial T)_p$ and $(\partial G/\partial T)_V$ in terms of these quantities.   (6 P)

Problem 6.29 Are $(\partial^2 U/\partial S^2)_V$, $(\partial^2 U/\partial V^2)_S$, $(\partial^2 G/\partial T^2)_p$, and $(\partial^2 G/\partial p^2)_T$ always positive?   (4 P)

Problem 6.30 If a charge dq is inserted isothermally and isochorically into a reversibly working galvanic element at the open circuit voltage Φ, the work $\delta A = \Phi\, dq$ is done. How does its internal energy change for given $\Phi(T)$?

Hint: Note the integrability condition for the free energy F. In addition, we should have $\delta A = \varphi dQ$, if upper-case letters always stand for extensive quantities and lower-case letters for intensive quantities.   (4 P)

Problem 6.31 What vapor pressure $p(T)$ is obtained from the Clausius–Clapeyron equation if we assume a constant transition heat Q, neglecting the volume of the liquid compared to the volume of the gas, and using the equation $pV = NkT$ for an ideal gas?   (4 P)

Problem 6.32 One liter of water at $20\,°C$ and normal pressure ($1013\,hPa$) is subject to a pressure twenty times the normal pressure. Here the compressibility is $0.5/GPa$ on average and the expansion coefficient $2 \times 10^{-4}/K$. Determine V/V_0 as a function of p and p_0 (give values in numbers as well). How much work is necessary for the change of state? By how much does the internal energy change?   (6 P)

Problem 6.33 At the freezing temperature, ice has the density 0.918 g/cm^3 and water the density 0.99984 g/cm^3. An energy of 6.007 kJ/mole is needed to melt ice. How large are the discontinuities in the four thermodynamic potentials for this phase transition (relative to one mole)? (4 P)

Problem 6.34 What is the connection between $(\partial U/\partial V)_T$ and $(\partial \frac{p}{T}/\partial T)_V$? Can $(\partial C_V/\partial V)_T$ be uniquely determined for a given thermal equation of state? Transfer the results to the enthalpy and C_p. (6 P)

Problem 6.35 For a given heat capacity $C_V(T, V)$ and thermal equation of state, is the entropy uniquely defined? Can we then also determine the thermodynamic potentials? (4 P)

Problem 6.36 From thermal coefficients for ideal gases, derive the relation

$$pV^{\kappa_T/\kappa_S} = \text{const.} ,$$

for isentropic processes. Determine $V(T)$ and $p(T)$ for adiabatic changes in ideal gases. How does the sound velocity c in an ideal gas depend on T, and what is obtained for nitrogen at 290 K? (6 P)

Problem 6.37 For a mole of ^{4}He at 1 bar and 290 K, determine the thermal de Broglie wavelength λ, the fugacity $\exp(\mu/kT)$, the free enthalpy (in J), and the entropy (in J/K). Here, helium may be taken as an ideal gas. (4 P)

Problem 6.38 How is the thermal equation of state for ideal monatomic gases to be modified in order to account to first order for the difference in $\ln Z_{GC}$ between bosons and fermions?

Hint: We may expand pV/kT in powers of the fugacity and express this in terms of N, V, and λ.

Compare the pressures of the Bose and Fermi gases with that of a classical gas. (8 P)

Problem 6.39 How do the pressure and temperature of the air depend on the height for constant gravitational acceleration if heat conduction is negligible compared to convection and therefore each mass element keeps its entropy? This is more realistic than the assumption of constant temperature. (2 P)

Problem 6.40 Consider the heating of a house as an isobaric–isochoric situation: the air expands with increasing temperature and escapes through leakages. Assuming an ideal gas, how does the number of molecules in the house change, and how does the internal energy change, assuming that there are no internal excitations of the molecules? Does the entropy increase or decrease. Or is this clear anyway from the entropy law? (Heating is not an energy problem, but an entropy problem!) (5 P)

Fig. 6.31 Diesel cycle.
Idealized cycle from 1 to 2
and from 3 to 4 along
isentropic (adiabatic) curves
of an ideal gas, between
either isobaric (2 → 3) or
isochoric (4 → 1) curves.
Contrast with twice isochoric
for the *Otto cycle* and twice
isobaric for the *Joule cycle*
(gas turbine)

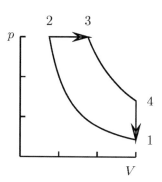

Problem 6.41 To extend a surface by dA, work $\delta W = \sigma\, dA$ has to be done against the attraction between the molecules, where σ is the *surface tension*. What sign does $(\partial \sigma / \partial T)_A$ have? How does the free energy change for an isothermal surface (without volume change) and how does the internal energy change? How much heat is involved in an isothermal surface extension assuming that $\sigma(T, A)$ is given? (6 P)

Problem 6.42 For four-stroke engines (suction, compression, combustion, ejection), only two cycles are assumed to be idealized. For example, Fig. 6.31 shows the *diesel cycle*. Note that diesel engines are "compression–ignition engines": the fuel burns at approximately constant pressure. Which two cycles are related to the diesel cycle (why?), and which path is taken by the one and the other in Fig. 6.31? What is the efficiency of the idealized diesel engine as a function of the compression $K = V_1/V_2$ and expansion $E = V_4/V_3$, assuming a single ideal diatomic gas, i.e., assuming the air to be pure nitrogen? Note that, clearly, $K > E > 1$. Begin by expressing $Q_\pm$ in terms of the relevant temperatures. The compression depends on the construction, but the expansion does not. It is determined by the "heat of combustion" (combustion enthalpy). Determine the ratio K/E of the enthalpies.   (9 P)

List of Symbols

We stick closely to the recommendations of the *International Union of Pure and Applied Physics* (IUPAP) and the *Deutsches Institut für Normung* (DIN). These are listed in *Symbole, Einheiten und Nomenklatur in der Physik* (Physik-Verlag, Weinheim 1980) and are marked here with an asterisk. However, one and the same symbol may represent different quantities in different branches of physics. Therefore, we have to divide the list of symbols into different parts (Table 6.3).

Table 6.3 Symbols used in thermodynamics and statistics

	Symbol	Name	Page number
*	Q	Amount of heat	513
*	A	Work	513
*	V	Volume	9
*	p	Pressure	560
*	N	Particle number	552
*	μ	Chemical potential	560
*	S	Entropy	523
*	T	Temperature	558
*	U	Internal energy	556
*	$F = U - TS$	(Helmholtz) Free energy	567
*	$H = U + pV$	Enthalpy	567
*	$G = H - TS$	(Gibbs) Free enthalpy	567
	$J = F - \mu N$	Grand canonical potential	567
*	$\alpha = \frac{1}{V}\left(\frac{\partial V}{\partial T}\right)_p$	(Volume-) Expansion coefficient	569
*	$\beta = \left(\frac{\partial p}{\partial T}\right)_V$	Pressure coefficient	569
*	$C_p = T\left(\frac{\partial S}{\partial T}\right)_p$	Isobaric heat capacity	569
*	$C_V = T\left(\frac{\partial S}{\partial T}\right)_V$	Isochoric heat capacity	569
*[a]	$\kappa_T = -\frac{1}{V}\left(\frac{\partial V}{\partial p}\right)_T$	Isothermal compressibility	569
	$\kappa_S = -\frac{1}{V}\left(\frac{\partial V}{\partial p}\right)_S$	Adiabatic compressibility	569
*	c	Sound velocity	570
	ρ_z	Probability for the state z	515
	Ω	Partition function up to limiting energy	525, 550
	Z	Partition function	549, 556
*[b]	Z_C	Canonical partition function	554
*[b]	Z_{MC}	Micro-canonical partition function	549
*[b]	Z_{GC}	Macro-canonical partition function	555

(continued)

Table 6.3 (continued)

	Symbol	Name	Page number
*	τ	Relaxation time	527
*	k	Boltzmann constant	623
*	N_A	Avogadro constant	623
*	R	Gas constant	572
*	ν	Stoichiometric coefficient	561

[a]For this compressibility, the abbreviation κ is recommended. However, we also use it for the isentropic exponent $-(V/p)\,(\partial p/\partial V)_S = 1/(p\kappa_S)$. For an ideal gas it is equal to the ratio $\kappa_T/\kappa_S = C_p/C_V$

[b]The abbreviations "C", "MCC", and "GC" stand for *canonical*, *micro-canonical*, and *grand canonical*, and we also use them for the probabilities ρ_C, ρ_{MC}, and ρ_{GC}

References

1. C. Caratheodory, Sitzungsber. Preu. Akad. **33** (3 July 1919)
2. A. Sommerfeld, *Lectures on Theoretical Physics 5–Thermodynamics and Statistical Mechanics* (Academic, London-Elsevier, Amsterdam, 1964)
3. Ch. Kittel, H. Krämer, *Thermal Physics*, 2nd edn. (W.H. Freeman, San Francisco, 1980)
4. F. Reif, *Fundamentals of Statistical and Thermal Physics* (McGraw-Hill, New York NY, 1965—Waveland Press, Long Grove, 2010)
5. R. Zurmühl, *Matrizen* (Springer, Berlin, 1964). in German
6. C. Syros, The linear Boltzmann equation properties and solutions. Phys. Rep. **45**, 211–300 (1978)
7. H. Risken, *The Fokker-Planck Equation* (Springer, Berlin, 1989)
8. J.R. Rumble (Ed.), *CRC Handbook of Chemistry and Physics*, 98th edn. (CRC Press, Taylor & Francis, London, 2017)
9. A. Sommerfeld, *Lectures on Theoretical Physics 6–Partial Differential Equations in Physics* (Academic, London-Elsevier, Amsterdam, 1964)
10. A. Bohr, B.R. Mottelson, *Nuclear structure*, Vol. 1 (Benjamin 1969—World Scientific 1998)

Suggestions for Textbooks and Further Reading

11. R. Baierlein, *Thermal Physics* (Cambridge University Press, Cambridge, 1999)
12. S.J. Blundell, K.M. Blundell, *Concepts in Thermal Physics*, 2nd edn. (Oxford University Press, Oxford, 2010)
13. N.N. Bogolubov, N.N. Bogolubov Jr., *Introduction to Quantum Statistical Mechanics* (World Scientific, Singapore, 1982)
14. W. Greiner, L. Heise, H. Stöcker, *Thermodynamics and Statistical Mechanics* (Springer, New York, 1995)
15. L.P. Kadanov, G. Baym, *Quantum Statistical Mechanics* (Benjamin, New York, 1982)
16. D. Kondepudi, *Introduction to Modern Thermodynamics* (Wiley, Chichester, 2008)
17. L.D. Landau, E.M. Lifshitz, *Course of Theoretical Physics Vol. 5—Statistical Physics* 3rd edn., (Butterworth–Heinemann, Oxford, 1980)
18. E.M. Lifshitz, L.P. Pitaevskii, *Course of Theoretical Physics Vol. 9—Statistical Physics Part 2—Theory of the Condensed State* (Butterworth–Heinemann, Oxford, 1980)
19. W. Nolting, *Theoretical Physics 5–Thermodynamics* (Springer, Berlin, 2017)

20. B.N. Roy, *Fundamentals of Classical and Statistical Thermodynamics* (Wiley, Chichester, 2002)
21. F. Scheck, *Statistical Theory of Heat* (Springer, Berlin, 2016)
22. D.V. Schroeder, *An Introduction to Thermal Physics* (Addison-Wesley, San Francisco, 2000)

Appendix A
Important Constants

This appendix contains four tables. Table A.1 gives the names for different powers of 10, Tables A.2 and A.3 give some important constants, and Table A.4 gives some derived quantities. The generally accepted CODATA values are taken from http://www.physics.nist.gov/cuu/Constants/Table/allascii.txt
Energy conversion units: $J = kg\,m^2/s^2 = N\,m = W\,s = V\,A\,s = V\,C = A\,Wb = Pa\,m^3$.

Table A.1 Terminology for powers of 10

Factor	Prefix	Abbreviation	Factor	Prefix	Abbreviation
10^{-1}	deci	d	10^{+1}	deca	da
10^{-2}	centi	c	10^{+2}	hecto	h
10^{-3}	milli	m	10^{+3}	kilo	k
10^{-6}	micro	μ	10^{+6}	mega	M
10^{-9}	nano	n	10^{+9}	giga	G
10^{-12}	pico	p	10^{+12}	tera	T
10^{-15}	femto	f	10^{+15}	peta	P
10^{-18}	atto	a	10^{+18}	exa	E

© Springer Nature Switzerland AG 2018
A. Lindner and D. Strauch, *A Complete Course
on Theoretical Physics*, Undergraduate Lecture Notes in Physics,
https://doi.org/10.1007/978-3-030-04360-5

Table A.2 Important constants in vacuum by choice of the units (m, A). The mass unit (like the units of meter, second, and ampere) is expected to be a quantity defined by independent elementary quantities in the near future as from May 20, 2019 on

Quantity	Symbol	Value	Unit
Light velocity	c_0	299,792,458	m/s
Magnetic field constant	μ_0	$4\pi \times 10^{-7}$	N/A^2
		$12.566370614359 \times 10^{-7}$	$N/A^2 = H/m$
Electric field constant	$\varepsilon_0 = 1/\mu_0 c_0{}^2$	$8.854187817622 \times 10^{-12}$	F/m
Elementary charge	e	$1.602176634 \times 10^{-19}$	C
Planck constant	h	$6.62607015 \times 10^{-34}$	J s
Action quantum	$\hbar = h/2\pi$	$1.054571818\ldots \times 10^{-34}$	J s
Boltzmann constant	k	1.380649×10^{-23}	J/K
Avogadro constant	N_A	$6.02214076 \times 10^{23}$	1/mol
Atomic mass constant	u	$1.66053922\ldots \times 10^{-27}$	kg

Table A.3 Further constants

Quantity	Symbol	Value	Unit
Gravitational constant	G	$6.67408(31) \times 10^{-11}$	$m^3/kg\ s^2$
Electron mass	m_e	$9.10938356(16) \times 10^{-31}$	kg
		$5.48579909070(16) \times 10^{-4}$	u
Proton mass	m_p	$1.672621898(21) \times 10^{-27}$	kg
		$1.007276466789(91)$	u
Neutron mass	m_n	$1.674927471(21) \times 10^{-27}$	kg
		$1.00866491588(49)$	u

Table A.4 Derived quantities

Quantity	Symbol	Value	Unit
Fine structure constant	$\alpha = \mu_0 c_0 e^2/2h$	$7.2973525664(17) \times 10^{-3}$ $= 1/137.0359991\ldots 4$	
Bohr magneton	$\mu_B = e\hbar/2m_e$	$9.2740089994(57) \times 10^{-24}$	J/T
Stefan–Boltzmann constant	$\sigma = \pi^2 k^4/60\hbar^3 c_0^2$	$5.670367(13) \times 10^{-8}$	$W/m^2\ K^4$

Index

© Springer Nature Switzerland AG 2018
A. Lindner and D. Strauch, *A Complete Course
on Theoretical Physics*, Undergraduate Lecture Notes in Physics,
https://doi.org/10.1007/978-3-030-04360-5

Printed in the United States
By Bookmasters